Signal Transduction in Mast Cells and Basophils

Springer
New York
Berlin
Heidelberg
Barcelona
Hong Kong
London
Milan
Paris
Singapore
Tokyo

Ehud Razin
Juan Rivera

Editors

Signal Transduction in Mast Cells and Basophils

With 65 Illustrations

 Springer

Ehud Razin
Department of Biochemistry
Haddasah Medical School
Hebrew University
Jerusalem, 91120 Israel

Juan Rivera
Arthritis and Rheumatism Branch
National Institute of Arthritis and
    Musculoskeletal and Skin Diseases
National Institutes of Health
Bethesda, MD 20892-1820, USA

Library of Congress Cataloging-in-Publication Data
Signal transduction in mast cells and basophils / [edited by] Ehud
    Razin, Juan Rivera.
        p. cm.
    Includes bibliographical references and index.
    ISBN 0-387-98625-1 (hbk. : alk. paper)
    1. Mast cells.   2. Basophils.   3. Cellular signal transduction.
    I. Razin, Ehud, 1947– . II. Rivera, Juan, 1956–
    QR185.8.M35S56   1998
    616.07′9—dc21   98-30561

Printed on acid-free paper.

Production coordinated by Chernow Editorial Services, Inc., and managed by Tim
Taylor; manufacturing supervised by Thomas King.
Typeset by Best-set Typesetter Ltd., Hong Kong.
Printed and bound by Maple-Vail Book Manufacturing Group, York, PA.
Printed in the United States of America.

9 8 7 6 5 4 3 2 1

ISBN 0-387-98625-1 Springer-Verlag New York Berlin Heidelberg   SPIN 10692257

Preface

The recent advances in the field of receptor-mediated molecular signaling were a major impetus in the compilation of this book. In particular, we believed the time was appropriate to compile the current knowledge in signal transduction in mast cells and basophils, not necessarily because of a major therapeutic breakthrough in the treatment of allergies, but mostly because a summary of recent work may provide new perspectives that may serve to advance the field and delineate new areas of research. Hope for the development of new areas with therapeutic potential has driven much of the ongoing effort, but to date success based on the targeting of particular molecular events has been limited. However, a clearer picture is evolving from the study of signal transduction in various systems. The redundancy in many signaling pathways makes the effort to find a cell-specific target molecule akin to the search for a "needle in a haystack." Furthermore, the daunting task of cell-specific delivery of therapeutic agents is an issue that pervades the fields of molecular and genetic therapies.

Regardless of the limited success in the area of therapeutics, the study of molecular signals has allowed major advances in our understanding of the biochemical events that govern cell responses and has driven the rapid growth of technology for study of protein structure and function. Our knowledge of the fundamental mechanisms that govern protein–protein interactions, enzyme–substrate reactions, cellular ontogeny and differentiation, and other areas has been greatly enhanced by the study of molecular signaling events. It is in this context that we have set out to compile the current information in the area of molecular signals in mast cells and basophils. We believe this monograph reflects the advances made in this field and may serve as a resource tool for the expert and novice alike.

We have arbitrarily divided the book into four sections to provide the reader with a framework of the areas we wish to cover. The first section discusses the molecular events that govern the ontogeny, growth, differentiation, and physiological function of mast cells. The second section focuses on the early events mediated primarily through the activation of the high-affinity receptor for IgE on mast cells and basophils. As a major route of mast cell and basophil activation, this topic has been the most extensively studied and is therefore emphasized. The third section reviews what is known about events that link the receptor activation to the effector function of mast cells and basophils. Finally, the last section considers the molecular events that govern the secretion of preformed and newly formed allergic mediators. Each topic area has been introduced by a perspective chapter from one of the leaders in the particular area of research. We hope the perspective chapters serve to provide the reader with a useful framework for evaluating the following chapters and stimulating thought and discussion. It will be evident to the reader that the four topic areas have much in common as well as distinct identities. We wish to leave the reader with the task of merging the useful information in each of these areas that may benefit the reader's purpose in perusing this book.

This book represents the wide-ranging effort of many individual scientists who pursue studies in various aspects of mast cell and basophil biology with a concentration on molecular signaling events. We are extremely grateful to the contributing authors without whom this comprehensive effort would not be possible. We hope you, the reader, will gain an up-to-date perspective on advances relevant to allergic disease and inflammation and that you will find your time has been well spent.

Ehud Razin
Jerusalem, Israel

Juan Rivera
Bethesda, Maryland

Contents

Section III

Contributors

Jonathan P. Arm, MD
Division of Rheumatology, Immunology, and Allergy, Brigham and Women's Hospital, Harvard Medical School, Boston, MA 02115, USA

K. Frank Austen, MD
Division of Rheumatology, Immunology, and Allergy, Brigham and Women's Hospital, Harvard Medical School, Boston, MA 02115, USA

Barbara Baird, PhD
Department of Chemistry and Chemical Biology, Baker Laboratory, Cornell University, Ithaca, NY 14853, USA

Sheryll A. Barker, PhD
Department of Pathology, Asthma Research Center, Cancer Research and Treatment Center, University of New Mexico, Albuquerque, NM 87131, USA

Thomas Baumruker, PhD
Novartis Research Institute, A-1235 Vienna, Austria

Michael A. Beaven, PhD
Laboratory of Molecular Immunology, National Heart, Lung, and Blood Institute, National Institutes of Health, Bethesda, MD 20892, USA

Clifton O. Bingham III, MD
Division of Rheumatology, Immunology, and Allergy, Brigham and Women's Hospital, Harvard Medical School, Boston, MA 02115, USA

Melissa A. Brown, PhD
Department of Experimental Pathology, Emory University School of Medicine, Atlanta, GA 30322, USA

David S. Cissel, PhD
Laboratory of Molecular Immunology, National Heart, Lung, and Blood Institute, National Institutes of Health, Bethesda, MD 20892, USA

Kenneth A. Field, PhD
Department of Chemistry and Chemical Biology, Baker Laboratory, Cornell University, Ithaca, NY 14853, USA

Andrea Fleig, PhD
Queens Medical Center Neuroscience Institute, University of Hawaii, Honolulu, HI 96822, USA

Paul F. Fraundorfer, PhD
Laboratory of Molecular Immunology, National Heart, Lung, and Blood Institute, National Institutes of Health, Bethesda, MD 20892, USA

Daniel S. Friend, MD
Division of Rheumatology, Immunology, and Allergy, Brigham and Women's Hospital, Harvard Medical School, Boston, MA 02115, USA

Stephen J. Galli, MD
Department of Pathology, Beth Israel Deaconess Medical Center and Harvard Medical School, Boston, MA 02215, USA

Timothy E. Graham, MD
Department of Pathology, Asthma Research Center, Cancer Research and Treatment Center, University of New Mexico, Albuquerque, NM 87131, USA

Karin Hartmann, MD
Laboratory of Allergic Diseases, National Institute of Allergy and Infectious Diseases, National Institutes of Health, Bethesda, MD 20892, USA

David Holowka, PhD
Department of Chemistry and Chemical Biology, Baker Laboratory, Cornell University, Ithaca, NY 14853, USA

Tomoko Jippo, PhD
Department of Pathology, Osaka University Medical School, Suita, Osaka 565-0871, Japan

Howard R. Katz, PhD
Division of Rheumatology, Immunology, and Allergy, Brigham and Women's Hospital, Harvard Medical School, Boston, MA 02115, USA

Toshiaki Kawakami, MD, PhD
Division of Allergy, La Jolla Institute for Allergy and Immunology,
San Diego, CA 92121, USA

Yuko Kawakami, MD, PhD
Division of Allergy, La Jolla Institute for Allergy and Immunology,
San Diego, CA 92121, USA

Teruaki Kimura, MD, PhD
Receptors and Signal Transduction Section, National Institute of
Dental Research, National Institutes of Health, Bethesda, MD
20892, USA

Yukihiko Kitamura, MD
Department of Pathology, Osaka University Medical School, Suita,
Osaka 565-0871, Japan

Anna Koffer, PhD
Physiology Department, University College London, London
WC1E 6JJ, UK

Steven A. Krilis, MB, BS, PhD, FRACP
Department of Immunology, Allergy and Infectious Disease,
University of New South Wales School of Medicine, St. George
Hospital, Kogarah, New South Wales, 2217, Australia

Chris S. Lantz, PhD
Department of Pathology, Beth Israel Deaconess Medical Center,
Harvard Medical School, Boston, MA 02215, USA

Lixin Li, MD
Department of Immunology, Allergy and Infectious Disease,
University of New South Wales School of Medicine, St. George
Hospital, Kogarah, New South Wales, 2217, Australia

Susan M. MacDonald, MD
Johns Hopkins Asthma and Allergy Center, Johns Hopkins University School of Medicine, Baltimore, MD 21224, USA

Donald MacGlashan, Jr., MD, PhD
Johns Hopkins Asthma and Allergy Center, Johns Hopkins University School of Medicine, Baltimore, MD 21224, USA

Chris Mathes, PhD
Queens Medical Center Neuroscience Institute, University of
Hawaii, Honolulu, HI 96822, USA

Michael A. McCloskey, PhD
Department of Zoology and Genetics, Iowa State University,
Ames, IA 50011, USA

Yoseph A. Mekori, MD
Sackler School of Medicine, Tel-Aviv University, Department of
Medicine, Meir General Hospital, Kfar-Saba, 44281 Israel

Francisco Mendez, PhD
Queens Medical Center Neuroscience Institute, University of
Hawaii, Honolulu, HI 96822, USA

Dean D. Metcalfe, MD
Laboratory of Allergic Diseases, National Institute of Allergy and
Infectious Diseases, National Institutes of Health, Bethesda, MD
20892, USA

Henry Metzger, MD
Arthritis and Rheumatism Branch, National Institute of Arthritis
and Musculoskeletal and Skin Diseases, National Institutes of
Health, Bethesda, MD 20892, USA

Eiichi Morii, MD, PhD
Department of Pathology, Osaka University Medical School, Suita,
Osaka 565-0871, Japan

Hovav Nechushtan, MD, PhD
Department of Biochemistry, Hadassah Medical School, Hebrew
University, Jerusalem, 91120 Israel

Sandra Odom, MS
Arthritis and Rheumatism Branch, National Institute of Arthritis
and Musculoskeletal and Skin Diseases, National Institutes of
Health, Bethesda, MD 20892, USA

Janet M. Oliver, PhD
Department of Pathology, Asthma Research Center, Cancer
Research and Treatment Center, University of New Mexico, Albu-
querque, NM 87131, USA

Matthew J. Peirce, PhD
Arthritis and Rheumatism Branch, National Institute of Arthritis
and Musculoskeletal and Skin Diseases, National Institutes of
Health, Bethesda, MD 20892, USA

Reinhold Penner, MD, PhD
Queens Medical Center Neuroscience Institute, University of
Hawaii, Honolulu, HI 96822, USA

Janet R. Pfeiffer, MS
Department of Pathology, Asthma Research Center, Cancer
Research and Treatment Center, University of New Mexico, Albu-
querque, NM 87131, USA

Eva E. Prieschl, PhD
Novartis Research Institute, A-1235 Vienna, Austria

Ehud Razin, PhD
Department of Biochemistry, Hadassah Medical School, Hebrew University, Jerusalem, 91120 Israel

Juan Rivera, PhD
Arthritis and Rheumatism Branch, National Institute of Arthritis and Musculoskeletal and Skin Diseases, National Institutes of Health, Bethesda, MD 20892, USA

Ronit Sagi-Eisenberg, PhD
Department of Cell Biology and Histology, Sackler School of Medicine, Tel Aviv University, Tel Aviv, 69978 Israel

John W. Schrader, PhD
Biomedical Research Centre, University of British Columbia, Vancouver, BC V6T 1Z3, Canada

Melanie A. Sherman, PhD
Department of Experimental Pathology, Emory University School of Medicine, Atlanta, GA 30322, USA

Reuben P. Siraganian, MD, PhD
Receptors and Signal Transduction Section, National Institute of Dental Research, National Institutes of Health, Bethesda, MD 20892, USA

Richard L. Stevens, PhD
Division of Rheumatology, Immunology and Allergy, Brigham and Women's Hospital, Harvard Medical School, Boston, MA 02115, USA

Richard Sullivan, MD
Physiology Department, University College London, London WC1E 6JJ, UK

Patrick G. Swann, PhD
Arthritis and Rheumatism Branch, National Institute of Arthritis and Musculoskeletal and Skin Diseases, National Institutes of Health, Bethesda, MD 20892, USA

Mindy Tsai, DMSc
Department of Pathology, Beth Israel Deaconess Medical Center, Harvard Medical School, Boston, MA 02215, USA

Bridget S. Wilson, PhD
Department of Pathology, Asthma Research Center, Cancer Research and Treatment Center, University of New Mexico, Albuquerque, NM 87131, USA

Guang W. Wong, BS
Division of Rheumatology, Immunology and Allergy Brigham
and Women's Hospital, Harvard Medical School, Boston, MA
02115, USA

Libo Yao, MD, PhD
Biochemistry Department, Xi'an Molecular Biology Institute,
Xi'an, China

Cheng Zhang, PhD
Laboratory of Molecular Immunology, National Heart, Lung, and
Blood Institute, National Institutes of Health, Bethesda, MD
20892, USA

Juan Zhang, MD, PhD
Receptors and Signal Transduction Section, National Institute of
Dental Research, National Institutes of Health, Bethesda, MD
20892, USA

Xiao-Tong Zhang, MD
Department of Immunology, Allergy and Infectious Disease,
University of New South Wales School of Medicine, St. George
Hospital, Kogarah, New South Wales, 2217, Australia

Section I

1
Signals in the Regulation of Mast Cell Growth and Development: A Perspective

Stephen J. Galli

Biological systems can be wonderfully and, at times, frustratingly, complex. This is true not only in general terms, but also with respect to highly specialized areas of biology. The regulation of mast cell growth and development is one such area. This short "overview" chapter is meant to introduce this topic in two ways: (1) by providing a personal perspective on certain aspects of the field, and (2) by providing a preview of the chapters that discuss in detail particular areas of this subject. The first objective, however, is to further define the context of the discussion by commenting on the importance of research in this area.

Mast Cells: Versatile Effector Cells in Health and Disease

The mast cells of humans and other mammals express the high-affinity Fc receptor for IgE (FcεRI) on their surface and can be activated to secrete a variety of potent, biologically active mediators in response to challenge with multivalent antigens that are recognized by their FcεRI-bound IgE.[1-3] It is now widely believed that the release of such mediators by mast cells represents a critical component of many, and perhaps essentially all, clinically important IgE-dependent acute allergic reactions, including those which are responsible for acute respiratory responses to allergen in patients with atopic asthma, or for systemic anaphylactic reactions in subjects with allergies to foods,

pharmaceutical agents, or components of insect venoms.[4-6] It is very likely that mast cells also contribute to the pathogenesis of clinically significant late-phase reactions and at least some aspects of chronic allergic inflammation, in such settings as allergic rhinitis, asthma, and, perhaps, atopic dermatitis.[6,7]

Given the enormous human and economic impact of allergic diseases,[8] it seems clear that a better understanding of the signals that can regulate mast cell growth and development may be helpful to efforts to devise new strategies to manage or treat mast cell-related diseases. In addition to IgE-associated allergic disorders, such diseases also include a relatively uncommon but clinically troublesome group of disorders (known generically as "mastocytosis") that are characterized by the local or systemic development of excessive numbers of mast cells.[9]

Moreover, it is now clear that, even in a single species, mast cells in different anatomical microenvironments can vary significantly in multiple aspects of their phenotype.[10-13] Notably, these include the cells' level of expression of FcεRI and other cell-surface structures through which mediator release can be elicited (e.g., receptors for complement components or the c-*kit* receptor, which binds the major mast cell developmental/survival factor, stem cell factor (SCF; also known as kit ligand [KL], mast cell growth factor [MGF], steel locus factor [SLF], and steel factor),[14] resulting in variation in the sensitivity of these cells to activation by IgE and antigen or other agonists of mediator release.[15-17] Different populations of mast cells can also vary in the amounts and types of preformed mediators, lipid products, and cytokines that they can secrete, as well as in their sensitivity to various drugs which can influence their function.[10-17] Because factors which can influence mast cell development or proliferation can also directly or indirectly modulate the cells' mediator content, surface expression of FcεRI and other receptors that can activate mediator secretion, and other important aspects of their phenotype, a better understanding of the regulation of mast cell growth and development will also enhance understanding of the regulation and expression of mast cell function.

While the role of mast cells as important effector cells in allergic diseases appears certain, there is increasing support for the notion that mast cells have a "good side" as well as a "bad side." The immune responses that are elicited by many different parasites are associated with mast cell hyperplasia and activation at sites of infection, as well as high levels of parasite-specific and nonspecific IgE production.[18,19] Even though it can be difficult to prove that a particular effector cell, such as the mast cell, is critical for the expression of complex immune responses to particular parasites,[20] it has been shown that IgE and mast cells are essential for normal immunologically mediated resistance to the cutaneous feeding of certain larval ticks in mice.[21] Mast cells probably also represent an important component of effective host immunity to certain enteric nematodes, at least in mice.[22] However, recent work (also performed in mice) has shown that mast cells, and specifically, the mast cell-dependent orchestration of a local inflammatory response that includes the recruitment of neutrophils, can represent critical aspects of innate or "natural" immune responses to bacterial infection.[23-26]

There is as yet no direct evidence that mast cells importantly contribute to innate immunity to infection with bacteria or other pathogens in humans. But, to me, this represents a real possibility. Furthermore, it has been reported that patients with genetically determined or human immunodeficiency virus- (HIV-) induced immunodeficiency diseases can exhibit significant

reductions in the numbers of mast cells in enteric mucosal tissues.[27] These observations raise the intriguing question of whether, in certain clinical contexts, it would be advantageous to *enhance* mast cell numbers or function, at least in particular anatomical sites. This question cannot answered definitively without much additional work. However, at the time of writing this chapter, it would certainly seem that a better understanding of the regulation of mast cell development and survival is of interest not only because it may be clinically desirable, in the management of some diseases, to reduce mast cell numbers or suppress mast cell function, but also because it may be advantageous to increase their numbers or function in other settings.

Finally, studies of the mechanisms that influence mast cell growth and development may continue to provide general insights into the function of growth factor receptors and the molecules that can modulate their activity, as well as the regulation of signaling pathways which can influence cell survival, differentiation/maturation, or function[14] (see also Chapter 6 in this volume).

Mast Cell Development: A Brief Overview

It is remarkable that definitive proof of the hematopoietic origin of mouse mast cells was not published until 1977,[28] and that methods to grow essentially homogeneous populations of lineage-committed normal, if "immature," mast cells in vitro were not reported until 1981.[29-33] Since then, much has been learned about the growth factors that can regulate the development of mast cells from hematopoietic precursors or influence the phenotypic and functional characteristics of these cells (Chapters 2, 3, 4, 5, 6 and 7). However, I would like to emphasize here a few particularly important conclusions derived from work in this area. Although mast cells are derived from hematopoietic progenitors, unlike basophils and other granulocytes, mast cells do not ordinarily circulate in the blood in fully mature form but complete a substantial amount of their maturation in vascularized connective tissues and, in rodents, within serosal cavities.[10,11,14] During mouse fetal development, a circulating pro-mastocyte can be identified that is apparently committed to the mast cell lineage but which lacks certain important characteristics of mature mast cells, such as surface expression of FcεRI and large numbers of cytoplasmic granules.[34] However, these pro-mastocytes can mature and differentiate further, either in vitro in the presence of IL-3 and SCF, or in vivo, in the permissive microenvironment of the genetically mast cell-deficient Kit^W/Kit^{W-v} mouse.[34] As discussed elsewhere in this volume (see Chapters 2, 3, 4, 5, 6 and 7), this process of "further maturation/differentiation" appears to be subject to exquisitely fine regulation, resulting in the generation of mast cells that can vary in patterns of expression of serine proteases and other mediators as well as in many other aspects of their phenotype. Moreover, combined in vitro and in vivo studies in mice have indicated that, in clonally derived mast cell populations, some aspects of phenotype can be modulated reversibly.[35] It seems plausible, although as yet essentially unproven, that such differences in phenotype permit various mast cell "subpopulations" to express distinct roles in health or disease. In addition, the ability of mast cell populations to express reversible changes in phenotype may permit them to express different functions at different stages of a protective host response or disease process.

In Vivo Veritas: In Vitro Versus In Vivo Models of Mast Cell Development

The basic observation that aspects of mast cell phenotype can vary according to anatomical site is at least 93 years old.[36] It is also now widely accepted, as the result of the work of many different groups, that particular phenotypic characteristics of mast cells can be differentially or coordinately regulated by signals, such as cytokines and growth factors, which vary in their expression in different microenvironments.[10,11,13] Clearly, the "microenvironment" of mast cells that develop, or are maintained, in vitro differs considerably from the microenvironments in which these cells develop and ultimately reside in vivo. Therefore, it is to be expected that in vitro-versus in vivo-derived mast cells may differ in many aspects of their phenotype, not all of which have yet been identified or fully characterized.

Once these points are accepted, certain implications naturally follow. Perhaps the most important of these is that conclusions about the physiological, or disease, relevance of observations about mast cell development or function that are based on observations made in vitro must be qualified, pending the results of experiments which can test these conclusions or hypotheses critically in vivo. An example that illustrates this point is presented in the next chapter of this volume (see Chapter 2). Briefly, several lines of evidence, derived from both in vitro and in vivo studies, indicated that IL-3 can represent an important mast cell growth factor in the mouse[22] (see also Chapter 2). Moreover, the addition of IL-3 can greatly enhance mast cell development in response to the c-*kit* ligand, SCF, in certain settings in vitro.[22] However, we recently found that IL-3 –/– mice have few or no abnormalities in levels of mast cells in multiple tissues in vivo.[22] Perhaps more remarkably, the increased numbers of mast cells that developed in various tissues of IL-3 –/– mice that had been injected with recombinant SCF daily for 21 days were roughly the same as or, in some tissues, significantly *exceeded*, those in the identically treated IL-3 +/+ mice.[22]

Neither of these two in vivo findings, that is, normal levels of tissue mast cells at baseline in IL-3 –/– mice, and normal or enhanced levels of tissue mast cell hyperplasia in response to subcutaneous (s.c.) injection of recombinant SCF in vivo, would have been predicted based solely on the in vitro finding of impaired SCF-driven mast cell development from IL-3 –/– mouse bone marrow cells. On the other hand, in vitro studies have identified one possible mechanism by which endogenous IL-3 might suppress SCF-induced mast cell hyperplasia: IL-3-dependent downregulation of mast cell c-*kit* surface expression.[37] Whether this occurs in vivo, and the extent to which this mechanism explains the findings in SCF-injected IL-3 −/− mice, remain to be determined.

Mice or Humans? Species Differences in Mast Cell Development and Phenotype

The pioneering work of Yukihiko Kitamura and his colleagues in *Kit* and *Mgf* (*Scf*) mutant mice in vivo,[10] (see also Chapter 3), and the development of methods to generate essentially pure populations of lineage-committed mouse mast cell populations from hematopoietic precursor cells in vitro,[29–33] clearly illustrate the power of using mouse models to investigate mast cell development. Also, work with human mast cells indicates that many "major themes" in mast cell development are expressed in both mice and humans (Table 1.1). However, it cannot be denied that *Mus musculus* and *Homo sapiens* are distinct (each with their own tissue microenvironments), and one must therefore expect that some of the "details" of mast cell development may differ in these rodent and primate species.

Two major examples of such differences are the role of SCF versus IL-3 in mast cell development and the pattern of mast cell serine protease development. Based on both in vitro and in vivo data, SCF appears to be a critical mast cell developmental and survival factor in both mice and humans, and recombinant SCF can promote the generation of populations of mast cells from the hematopoietic progenitor cells of each species[14] (see also Chapters 2–5). By contrast, while IL-3 can be used as the only exogenous growth factor to generate immature mast cells from mouse hematopoietic precursor cells in vitro,[38] this has generally not been true for human cells, at least in vitro (see Chapters 2, 5). While studies of rhesus macaques that have received recombinant human IL-3 suggest that IL-3 may contribute to the development of mast cell hyperplasia in primates under certain circumstances in vivo, whether this is by direct or indirect mechanisms is not yet clear.[39]

TABLE 1.1. Major similarities between mast cell development in mice and humans.

1. Mast cells are derived from hematopoietic progenitor cells, but do not ordinarily circulate in mature form.
2. Major steps in the differentiation and maturation of the mast cell lineage occur in peripheral (nonhematopoietic) tissues, or, for some mast cells, in the bone marrow itself.
3. Interactions between c-*kit* and its ligand (SCF, stem cell factor; MGF, mast cell growth factor; KL, kit ligand) are necessary for mast cell differentiation/ maturation/survival.
4. Mast cell development/maturation/survival can be modulated by cytokines/growth factors in addition to SCF.
5. Apparently "mature" mast cells residing in the tissues can be induced to proliferate, at least under certain circumstances.
6. Mast cells of different "subpopulations," e.g., in different anatomical microenvironments, or at different stages of differentiation/maturation, can vary in multiple important aspects of phenotype, including mediator content, susceptibility to activation for mediator release or proliferation, and sensitivity to drugs that affect mast cell function.
7. The numbers, phenotypic characteristics, and anatomical distribution of mast cells may change, sometimes dramatically, at sites of immune responses or disease processes.

Note: Mice and humans also express differences in certain aspects of their development and phenotype (see text).

As discussed in Chapter 4 by Wong et al., mouse mast cells can express various combinations of at least 9 serine proteases, of which 2 are tryptases and 7 are chymases.[40] Lutzelschwab et al.[41] recently reported that rat mast cells can express mRNA for at least 10 different serine proteases, of which 2 are tryptases and 8 are chymases, and one group of closely related rat mast cell proteases, so far characterized at the level of cDNA (RMCP-8, -9, and -10), may be encoded by three to five genes. By contrast, human mast cells express genes for two closely related tryptases (α- and "β-like") and, apparently, only a single chymase.[42,43] The levels of human mast cell tryptase and chymase expression in individual cells vary, with some cells containing both proteases, some containing predominantly tryptase, and some apparently containing chymase with little or no tryptase[42] (see also

Chapter 5). While there appear to be some minor polymorphisms in the human "β-like" tryptase gene, these are of uncertain functional significance.[42,43]

Why are we, as a species, so outdistanced by murine rodents in the diversity of our mast cell proteases? Does this fact necessarily indicate that the functions of mast cell proteases in health or disease are different in rats and mice as opposed to humans? If so, what are the implications with respect to the preclinical development of potential therapeutic agents that target mast cell proteases? While the answers to such questions are not yet in hand, they do raise the possibility that differences in the "details" of mast cell development and phenotype in humans and other species may be of significant clinical, as well as "academic," interest.

Conclusions

Many important points about the regulation of mast cell growth and development have already been revealed, but many others remain unresolved. For example, studies in murine rodents, experimental primates, and humans leave little doubt that SCF–c-*kit* interactions are critical for mast cell development and survival and can also promote mast cell proliferation, but SCF can also elicit the secretion of mast cell mediators[14] (see Chapter 2). Indeed, in terms of activation of mitogen-activated protein (MAP) kinases and of pp90[rsk] and pp70-S6 kinases, SCF–c-*kit* interactions can result in the initiation of patterns of intracellular signaling that overlap greatly with those originating by aggregation of FcεRI.[44] Yet the cues by which the interaction between c-*kit* and its ligand prompts the mast cell to express these disparate cellular responses are not fully understood. Is it simply a matter of c-*kit* "signal strength," as determined by the concentration or configuration (soluble versus membrane-associated) of ligand? How important, in the selection of an appropriate cellular response to SCF–c-*kit* interactions, is the occurrence of simultaneous signaling via other receptors, and which are the critical ones?

Although it is important to define further how mast cell growth, development, and phenotype are regulated by SCF, IL-3, and the many other factors that are already known to influence these processes, it seems unlikely that all the players which can affect mast cell development, survival, or phenotype have been recognized. Indeed, the work of Krilis and his colleagues indicates that a soluble factor may exist in humans that can preferentially promote the development of chymase[+] mast cells which express little or no tryptase[45] (see also Chapter 5). These are fascinating observations that make the identification and (if it is a previously undescribed protein) the cloning and characterization of this factor (and its receptor) high priorities for future work.

As with other cell types, the size of mast cell populations can be regulated by factors that decrease their survival, as well as those which enhance their development or proliferation. Clinically important examples of each type of mechanism have already been reported. Gain-of-function mutations of c-*kit*[46–48] (see also Chapter 3), and perhaps dysregulated production or metabolism of SCF,[49] may contribute to excessive mast cell production in individual subjects with various forms of mastocytosis. The administration of recombinant SCF can also enhance mast cell development, and can result in a striking mast cell hyperplasia that is largely reversible upon cessation of SCF dosing.[50–52] By contrast, Finotto et al. have provided in vivo and in vitro evidence that the ability of glucocorticosteroids to reduce populations of mast cells in vivo may reflect their ability to diminish SCF production[53] (also see Chapter 7).

Taken together, these observations identify the SCF–c-*kit* interaction, and other processes that regulate mast cell survival or apoptosis, as potential "targets" for manipulating mast cell numbers for therapeutic ends. The extent to which this represents a feasible, safe, and effective therapeutic approach, and, if so, in which circumstances, remains to be seen however. One obvious problem in such strategies is to achieve "specificity" for the mast cell, as opposed to other hematopoietic or c-*kit*[+] cell types.

Substantial progress is being made in defining the transcription factors, such as that encoded by the *mi* locus of mice (MITF) (see Chapter 3), which help to regulate mast cell gene expression, in working out the complex interplay of cytokines that may regulate the differential expression of specific mast cell secretory products, such as the cytoplasmic granule-associated serine proteases (see Chapter 4), and in defining the factors which influence the development of mast cell hyperplasia during immune responses in vivo (see Chapter 2). There is no doubt that such work, now performed primarily in the mouse, will reveal important new insights into the regulation of mast cell development, phenotype, and function. In addition, the generation of knockout mice that lack specific mast cell proteases, and the development of novel bioengineering approaches to obtain large amounts of such proteases for structural and functional analyses, should greatly facilitate efforts to clarify the function of these products in vivo (see Chapter 4). However, exploring the extent to which findings derived from such murine systems apply to mast cell biology in humans will represent a difficult challenge.

The following chapters in this section provide a detailed discussion of several important aspects of mast cell growth and development. They also point to a number of interesting areas for future research. The results of such studies will be awaited with great anticipation, as they are sure to unravel further the enduring "riddle of the mast cell".[54]

References

1. Ishizaka T, Ishizaka K. Activation of mast cells for mediator release through IgE receptors. Prog Allergy 1984;34:188–235.
2. Beaven MA, Metzger H. Signal transduction by Fc receptors: the Fc epsilon RI case. Immunol Today 1993;14:222–226.
3. Ravetch JV, Kinet J-P. Fc receptors. Annu Rev Immunol 1991;9:457–492.
4. Bochner BS, Lichtenstein LM. Anaphylaxis. N Engl J Med 1991;324:1785–1790.
5. Schwartz LB, Huff TF. Biology of mast cells and basophils. In: Middleton E, Reed CE, Ellis EF, Adkinson NF, Yunginger JW, Busse WW, eds. Allergy: Principles and Practice. Vol. I. St. Louis: Mosby-Year Book, 1993:135–168.
6. Galli SJ. New concepts about the mast cell. N Engl J Med 1993;328:257–265.
7. Galli SJ. Complexity and redundancy in the pathogenesis of asthma: reassessing the roles of mast cells and T cells. J Exp Med 1997;186:343–347.
8. Goldstein RA, Paul WE, Metcalfe DD, Busse WW, Reece ER. Asthma. Ann Intern Med 1994;121:698–708.
9. Metcalfe DD. Classification and diagnosis of mastocytosis: current status. J Invest Dermatol 1991;96:2S–4S.
10. Kitamura Y. Heterogeneity of mast cells and phenotypic change between subpopulations. Annu Rev Immunol 1989;7:59–76.
11. Galli SJ. New insights into "the riddle of the mast cell": microenvironmental regulation of mast cell development and phenotypic heterogeneity. Lab Invest 1990;62:5–33.
12. Bienenstock J, Befus AD, Denburg JA. Mast cell heterogeneity: basic questions and clinical implications. In: Befus AD, Bienenstock J, Denburg JA, eds. Mast Cell Differentiation and Heterogeneity. New York: Raven Press, 1986:391–402.
13. Stevens RL, Austen KF. Recent advances in the cellular and molecular biology of mast cells. Immunol Today 1989;10:381–386.
14. Galli SJ, Zsebo KM, Geissler EN. The kit ligand, stem cell factor. Adv Immunol 1994;55:1–96.
15. Yamaguchi M, Lantz CS, Oettgen HC, et al. IgE enhances mouse mast cell FcεRI expression in vitro and in vivo. Evidence for a novel amplification mechanism in IgE-dependent reactions. J Exp Med 1997;185:663–672.
16. Yano K, Yamaguchi M, de Mora F, et al. Production of macrophage inflammatory protein-1α by human mast cells: increased anti-IgE-dependent secretion after IgE-dependent enhancement of mast cell IgE binding ability. Lab Invest 1997;77:185–193.
17. Gagari E, Tsai M, Lantz CS, Fox LG, Galli SJ. Differential release of mast cell interleukin-6 via c-*kit*. Blood 1997;89:2654–2663.
18. Jarrett EEE, Miller HRP. Production and activities of IgE in helminth infection. Prog Allergy 1982;31:178–233.
19. Miller HRP, Huntley JF, Newlands GFJ, et al. Mast cell granule proteases in mouse and rat: a guide to mast cell heterogeneity and activation

in the gastrointestinal tract. In: Galli SJ, Austen KF, eds. Mast Cell and Basophil Differentiation and Function in Health and Disease. New York: Raven Press, 1989:81–91.

20. Reed ND. Function and regulation of mast cells in parasite infections. In: Galli SJ, Austen KF, eds. Mast Cell and Basophil Differentiation and Function in Health and Disease. New York: Raven Press, 1989:205–215.

21. Matsuda H, Watanabe N, Kiso Y, et al. Necessity of IgE antibodies and mast cells for manifestation of resistance against larval *Haemaphysalis longicornis* ticks in mice. J Immunol 1990;144:259–262.

22. Lantz CS, Boesiger J, Song CH, et al. Role for interleukin-3 in mast cell and basophil development and parasite immunity. Nature (Lond) (in press). 1998;392:90–93.

23. Echtenacher B, Männel DN, Hültner L. Critical protective role of mast cells in a model of acute septic peritonitis. Nature (Lond) 1996;381:75–77.

24. Malaviya R, Ikeda T, Ross E, Abraham SN. Mast cell modulation of neutrophil influx and bacterial clearance at sites of infection through TNF-α. Nature (Lond) 1996;381:77–80.

25. Galli SJ, Wershil BK. The two faces of the mast cell. Nature (Lond) 1996;381:21–22.

26. Prodeus AP, Zhou X, Maurer M, Galli SJ, Carroll MC. Impaired mast cell-dependent natural immunity in complement C3-deficient mice. Nature (Lond) 1997;390:172–175.

27. Irani AM, Craig SS, DeBlois G, Elson CO, Schechter NM, Schwartz LB. Deficiency of the tryptase-positive, chymase-negative mast cell type in gastrointestinal mucosa of patients with defective T lymphocyte function. J Immunol 1987;138:4381–4386.

28. Kitamura Y, Shimada M, Hatanaka K, Miyano Y. Development of mast cells from grafted bone marrow cells in irradiated mice. Nature (Lond) 1977;268:442–443.

29. Nabel GJ, Galli SJ, Dvorak AM, Dvorak HF, Cantor H. Inducer T lymphocytes synthesize a factor that stimulates proliferation of cloned mast cells. Nature (Lond) 1981;291:332–334.

30. Nagao K, Yokoro K, Aaronson SA. Continuous lines of basophil/mast cells derived from normal mouse bone marrow. Science 1981;212:333–335.

31. Razin E, Cordon-Cardo C, Good RA. Growth of a pure population of mast cells in vitro with conditioned medium derived from concanavalin A-stimulated splenocytes. Proc Natl Acad Sci USA 1981;78:2559–2561.

32. Schrader JW, Lewis SJ, Clark-Lewis I, Culvenor JG. The persisting (P) cell: histamine content, regulation by a T cell-derived factor, orgin from a bone marrow precursor, and relationship to mast cells. Proc Natl Acad Sci USA 1981;78:323–327.

33. Tertian G, Yung YP, Guy-Grand D, Moore MA. Long-term in vitro culture of murine mast cells. I. Description of a growth factor-dependent culture technique. J Immunol 1981;127:788–794.

34. Rodewald H-R, Dessing M, Dvorak AM, Galli SJ. Identification of a committed precursor for the mast cell lineage. Science 1996;271:818–822.

35. Kanakura Y, Thompson H, Nakano T, et al. Multiple bidirectional alterations of phenotype and changes in proliferative potential during the in vitro and in vivo passage of clonal mast cell populations derived from mouse peritoneal mast cells. Blood 1988;72:877–885.

36. Maximow A. Uber die Zellformen des lockeren Bindegewebes. Arch Mikrosk Anat Entw Mech 1905;67:680–757.

37. Welham MJ, Schrader JW. Modulation of c-*kit* mRNA and protein by hemopoietic growth factors. Mol Cell Biol 1991;11:2901–2904.

38. Ihle JN, Keller J, Orsozlan S, et al. Biological properties of homogeneous interleukin 3. I. Demonstration of WEHI-3 growth factor activity, mast cell growth factor activity, P cell-stimulating factor activity, and histamine producing factor activity. J Immunol 1983;131:282–287.

39. Volc-Platzer B, Valent P, Radaszkiewicz T, Mayer P, Bettelheim P, Wolff K. Recombinant human interleukin 3 induces proliferation of inflammatory cells and keratinocytes in vivo. Lab Invest 1991;64:557–566.

40. Hunt JE, Friend DS, Gurish MF, et al. Mouse mast cell protease 9, a novel member of the chromosome 14 family of serine proteases that is selectively expressed in uterine mast cells. J Biol Chem 1997;272:29158–29166.

41. Lutzelschwab C, Pejler G, Aveskogh M, Hellman L. Secretory granule proteases in rat mast cells. Cloning of 10 different serine proteases and a carboxypeptidase A from various rat mast cell populations. J Exp Med 1997;185:13–29.

42. Schwartz LB. Mast cell tryptase: properties and roles in human allergic responses. In: Caughey GH, ed. Mast Cell Proteases in Immunology and Biology. Vol. 6. New York: Marcel Dekker, 1995:9–23.

43. Caughey GH. Mast cell chymases and tryptases: Phylogeny, family relations, and biogenesis. In: Caughey GH, ed. Mast Cell Proteases in Immunology and Biology. Vol. 6. New York: Marcel Dekker, 1995:305–329.

44. Tsai M, Chen R-H, Tam S-Y, Blenis J, Galli SJ. Activation of MAP kinases, pp90[rsk] and pp70-S6 kinases in mouse mast cells by signaling through the c-*kit* receptor tyrosine kinase or FcεRI: rapamycin inhibits activation of pp70-S6 kinase and proliferation in mouse mast cells. Eur J Immunol 1993;23:3286–3291.

45. Li L, Meng X, Krilis SA. Mast cells expressing chymase but not tryptase can be derived by culturing human progenitors in conditioned medium obtained from a human mastocytosis cell strain with c-*kit* ligand. J Immunol 1996;156:4839–4844.

46. Furitsu T, Tsujimura T, Tono T, et al. Identification of mutations in the coding sequence of the proto-oncogene c-*kit* in a human mast cell leukemia cell line causing ligand-independent activation of c-*kit* product. J Clin Invest 1993;92:1736–1744.

47. Nagata H, Worobec AS, Oh CK, et al. Identification of a point mutation in the catalytic domain of the protooncogene c-*kit* in peripheral blood mononuclear cells of patients who have mastocytosis with an associated hematologic disorder. Proc Natl Acad Sci USA 1995;92:10560–10564.

48. Longley BJ, Tyrrell L, Lu SZ, et al. Somatic c-*kit* activating mutation in urticaria pigmentosa and aggressive mastocytosis: establishment of clonality in a human mast cell neoplasm. Nat Genet 1996;12:312–314.

49. Longley BJ Jr, Morganroth GS, Tyrrell L, et al. Altered metabolism of mast-cell growth factor (c-*kit* ligand) in cutaneous mastocytosis. N Engl J Med 1993;328:1302–1307.

50. Galli SJ, Iemura A, Garlick DS, Gamba-Vitalo C, Zsebo KM, Andrews RG. Reversible expansion of primate mast cell populations in vivo by stem cell factor. J Clin Invest 1993;91:148–152.

51. Ando A, Martin TR, Galli SJ. Effects of chronic treatment with the c-*kit* ligand, stem cell factor, on immunoglobulin E-dependent anaphylaxis in mice: genetically mast cell-deficient *Sl/Sl^d* mice acquire anaphylactic responsiveness, but the congenic normal mice do not exhibit augmented responses. J Clin Invest 1993;92:1639–1649.

52. Costa JJ, Demetri GD, Harrist TJ, et al. Recombinant human stem cell factor (kit ligand) promotes human mast cell and melanocyte hyperplasia and functional activation in vivo. J Exp Med 1996;183:2681–2686.

53. Finotto S, Mekori YA, Metcalfe DD. Glucocorticoids decrease tissue mast cell number by reducing the production of the c-kit ligand, stem cell factor, by resident cells: in vitro and in vivo evidence in murine systems. J Clin Invest 1997;99:1721–1728.

54. Riley JF. The riddle of the mast cells—a tribute to Paul Ehrlich. Lancet 1954;i:841.

2

The Regulation of Mast Cell and Basophil Development by the Kit Ligand, SCF, and IL-3

Stephen J. Galli, Mindy Tsai, and Chris S. Lantz

As noted in the introduction to Chapter 1, the regulation of mast cell and basophil development, and the related area of "mast cell heterogeneity," are large topics to which many groups have made important contributions. In this chapter, we review current understanding of the effects of two cytokines, the kit ligand, stem cell factor (SCF), and interleukin-3 (IL-3), on mast cell and basophil development. Several other chapters in this book also address aspects of this subject (see Chapters 3, 4, 5, 6, and 7), and many of the topics that are considered in detail in these other reviews are omitted, or discussed only briefly, herein.

We focus here particularly on findings obtained in analyses of the effects of SCF or IL-3 on mast cell and basophil development in vivo. In part, this emphasis reflects the fact that recombinant human SCF (rhSCF) is already in clinical testing in humans. rhSCF is being developed as an agent to enhance hematopoiesis and to facilitate the harvesting of hematopoietic progenitor cells, not as a factor to promote mast cell hyperplasia.[1,2] Nevertheless, one of the effects of rhSCF is to induce hyperplasia of human mast cells in vivo,[2] which represents the first clear example of the fruits of analyses of the cytokine-dependent regulation of mast cell development making the long trip "from the bench to the bedside."

However, our focus on in vivo findings is also prompted by three other considerations. First, studies of mice with spontaneous or targeted mutations that affect SCF or IL-3 production, or the receptors for these cytokines, have provided information about the actual importance of endogenous SCF or IL-3 in mast cell and basophil development in vivo. Second, a number of studies have analyzed the effects of recombinant forms of SCF or IL-3 on mast cell and basophil development in mice and other mammalian species in vivo. Finally, as is shown here, in some cases it has not been possible to use even extensive amounts of in vitro data to predict the results obtained when genetic or other approaches are used to analyze the effects of SCF or IL-3 on mast cell or basophil development in vivo.

Discovery of SCF, a Ligand for the Receptor Encoded at c-kit

It has been known for decades that mice with a double dose of mutations at either *W* or *Sl* exhibit a hypoplastic, macrocytic anemia, sterility, and a lack of cutaneous melanocytes.[3,4] Transplantation and embryo fusion studies employing *W* or *Sl* mutant and congenic normal mice, or in vitro analyses employing cells or tissues derived from these animals, indicated that the deficits in the *W* mutant mice are expressed by the cells in the affected lineages, whereas those in the *Sl* mutant animals are expressed by microenvironmental cells necessary for the normal development of the affected lineages.[3,4] The complementary nature of the phenotypic abnormalities expressed by *W* or *Sl* mutant mice suggested that the *W* locus might encode a receptor expressed by hematopoietic cells, melanocytes, and germ cells, whereas the *Sl* locus might encode the corresponding ligand.[3]

Kitamura et al.[5] and Kitamura and Go[6] made the important observation that mutations at *W* or *Sl* also profoundly affect mast cell development. They demonstrated that the virtual absence of mast cells in *W/W^v* mice, like the anemia of these animals, reflected an abnormality intrinsic to the affected lineage[5] whereas the mast cell deficiency of *Sl/Sl^d* mice, which could not be corrected by bone marrow transplantation from the congenic normal (+/+) mice, reflected an abnormality in the microenvironments necessary for normal mast cell development.[6] Kitamura's finding that transplantation of bone marrow cells from the congenic +/+ mice or from beige (C57BL/6-*bg/bg*) mice, whose mast cells can be identified unequivocally because of their giant cytoplasmic granules, repaired the mast cell deficiency of the *W/W^v* mice provided clear evidence that mast cells were derived from precursors that reside in the bone marrow. This work also showed that mutations at *W* had a more profound effect on the mast cell than on any other hematopoietic lineage.

In light of Yukihiko Kitamura's many contributions to our understanding of the biology of the *W* gene product, it seems fitting (albeit only when one considers the English spelling of his name) that *W* would ultimately be shown to encode the kit tyrosine kinase receptor.[7,8] Shortly after this discovery, in confirmation of Elizabeth Russell's prediction,[3] three groups simultaneously reported that *Sl* encodes the corresponding ligand, which were named (in alphabetical order) kit ligand (KL),[9] mast cell growth factor (MGF),[10-12] steel factor (SLF or SF),[13,14] and stem cell factor (SCF).[15-17] In this chapter, the terms SCF for the ligand and kit for the receptor are used.

The gene for SCF encodes two transmembrane proteins of 220 or 248 amino acids, which are generated by alternative splicing; both forms may be proteolytically cleaved to produce soluble forms of the molecule that retain biological activity and which spontaneously form noncovalently linked dimers in solution.[18-20] While native SCF is glycosylated, the nonglycosylated, *E. coli*-derived soluble recombinant forms of the extracellular ligand domain of the molecule (rSCF[164]), which were used for many of the studies that we review herein, have significant biological activity.[19,20]

By transducing extracellular signals transmitted by their cognate ligands, receptor tyrosine kinases can regulate cell survival, proliferation, and differentiation.[21] As predicted based on the phenotypic abnormalities expressed by *W* or *Sl* mutant mice, SCF has been shown to promote hematopoiesis and mast cell development, as well as melanocyte survival and proliferation, and to influence the survival and proliferation of primordial germ cells.[19,20] Other findings, such as the expression of high levels of kit or SCF in the central nervous system, or the expression of kit on lymphocytes, had not been expected, because *W* or *Sl* mutant mice were not known to exhibit central nervous system abnormalities and, in general, these mutants have normal numbers of mature lymphocytes and normal B-cell and T-cell function.[19,20] Similarly, it has only recently become apparent that SCF–kit interactions have a critical role in the development of the interstitial cells of Cajal, which generate intestinal electrical pacemaker activity.[22-24]

Effects of SCF in Mast Cell Biology

SCF can promote the in vitro survival of early hematopoietic progenitor cells and can act synergistically with other hematopoietic growth factors to promote the further differentiation of multiple hematopoietic lineages.[18-20,25] However, unlike most other hematopoietic lineages, mast cells retain significant expression of the SCF receptor (kit) into maturity, and thus exhibit responsiveness to SCF not only during their development but also, in all likelihood, throughout their mature lifespan.[19]

The results of in vitro analyses, which in many instances have been confirmed by in vivo studies, indicate that SCF can have many effects in mast cell development and function: it can maintain mast cell survival, promote chemotaxis or haptotaxis of mast cells and their precursors, promote the proliferation of immature or mature mast cells, promote the maturation of mast cell precursors or immature mast cells and alter the phenotype and mediator content of these cells, directly promote the degranulation and secretion of mediators by mast cells, enhance mast cell ability to secrete mediators in response to other signals, including IgE and specific antigen, and alter the expression of other receptors, including those for extracellular matrix components and neuropeptides.[19,20]

However, the specific effects of SCF on mast cell biology that are expressed under individual circumstances can be influenced significantly by other factors. For example, long-term dosing with recombinant rat SCF (rrSCF) can induce mast cell hyperplasia in multiple organs in normal rats.[26] The pattern of expression of the mast cell-associated proteases (RMCP). RMCP-I and RCMP-II, by these mast cells varied according to the specific anatomical sites analyzed.[26] Thus, rrSCF promoted the development of mast cells that expressed predominantly RMCP-I in the skin and peritoneal cavity, whereas those in the small intestinal mucosa expressed predominantly RMCP-II.[26] Subsequently, work in mice,[27] and later in rats,[28] showed that IL-3 represents one of the additional cytokines that can influence the proliferation, and the serine protease phenotype, of mast cells which have been exposed to SCF.

Recombinant Human SCF Promotes Human Mast Cell Hyperplasia In Vivo

As reviewed elsewhere[19] (and see Chapters 1, 5), the regulation of mast cell development in mice and humans may differ in important details. Moreover, rhSCF is now in clinical testing to determine the extent to which this cytokine, when used together with granulocyte-colony-stimulating factor (G-CSF), can enhance the production and harvesting of hematopoietic progenitor cells.[1] Thus, it was (and still is) of particular interest to evaluate the effects of rhSCF on human mast cell development and function.

In vitro studies demonstrated that recombinant human SCF (rhSCF) can promote the development of mast cells from various sources of human hematopoietic progenitor cells.[29–32] Subsequently, a phase I study of E. coli-derived rhSCF showed that the administration of rhSCF (at 5–50 μg kg^{-1} day^{-1}, subcutaneously, for 14 days) to patients with advanced breast carcinoma resulted in a significant increase, by about 70%, in the numbers of cutaneous mast cells at sites that had not been directly injected with the agent.[2,33,34] In addition, the patients exhibited increased urinary levels of the major histamine metabolite, methyl-histamine,[2,33,34] and markedly increased (by 100%–1220%) serum levels of mast cell α-tryptase, as detected by an assay that can measure both the α and β forms of this protease.[2] The latter finding, when taken together with the observation that, in cynomolgus monkeys, rhSCF dosing induced much higher levels of mast cell development in the liver, spleen, and lymph nodes than in the skin,[35] suggested that the effect of rhSCF dosing on numbers of cutaneous mast cells may have greatly underestimated the effects of the agent on mast cell populations at other anatomical sites. In any event, this work identified rhSCF as the first cytokine that can induce human mast cell hyperplasia in vivo, and also showed that humans may be more sensitive to this action of rhSCF than are cynomolgus monkeys.[2,33–35]

SCF/Kit Signaling and the Regulation of the Survival and Size of Mast Cell Populations In Vivo

While all the biological effects of SCF on mast cells are of interest, none of them can be expressed unless the survival of the lineage is maintained. Work in both genetically mast cell-deficient SCF/MGF mutant Sl/Sl^d (Mgf^{Sl}/ Mgf^{Sl-d}) mice[15,26] and in cynomolgus monkeys[35] demonstrated that rSCF can promote the survival of the mast cell lineage in vivo. Thus, cessation of rhSCF dosing in cynomolgus monkeys was followed by a rapid decline of tissue mast cell numbers, in some cases to nearly baseline levels.[35]

Subsequently, three studies established that SCF can promote mast cell survival by suppressing apoptosis, either in vitro[36–38] or in vivo.[38] Indeed, the study by Iemura et al.[38] indicated that apoptosis represents a mechanism that can account for striking (up to 50 fold) and rapid reductions in the sizes of mast cell populations in vivo, apparently without significant associated inflammation.

By contrast, the *increased* numbers of mast cells present in mast cell neoplasms or examples of naturally occurring mastocytosis may in part reflect enhanced mast cell survival. Two SCF/kit-dependent mechanisms that may account for increased mast cell survival in such settings in vivo have been described: (1)"gain-of-function" mutations affecting c-*kit* itself,[39–41] and (2) altered production and/or biodistribution of endogenous SCF.[42] A third mechanism that can enhance kit-dependent mast cell development has been defined in recent studies of mice with various combinations of mutations at c-*kit* and *me*, which encodes the SH2-containing nontransmembrane protein tyrosine phosphatase, SHP1.[43,44] This work showed that the decrease in dermal mast cell numbers in W^v/W^v (Kit^{W-v}/Kit^{W-v}) mice was significantly improved by superimposition of the *me/me* genotype, which results in diminished negative regulation of kit signaling.[43,44]

In principle, these findings suggest that agents that can interfere with kit-dependent signaling in mast cells might be effective in diminishing the size of mast cell populations in vivo. However, kit is expressed on hematopoietic progenitor cells, melanocytes, germ cells, and many other cell types, including certain neurons.[18–20] Accordingly, the development of effective and safe approaches for manipulating the SCF/kit receptor–ligand interaction to reduce mast cell numbers in vivo will require achieving either adequate target cell selectivity or clinically acceptable control of the agent's bioavailability. For example, in vitro studies and analyses performed in vivo in mice indicate

that the ability of glucocorticoids to reduce mast cell populations may reflect the agent's ability to suppress local SCF production[45] (also see Chapter 6). However, in humans, glucocorticoid treatment protocols that result in diminished numbers of dermal mast cells also can produce significant cutaneous atrophy.[46]

SCF Can Regulate Mast Cell Secretory Function

Because the cell lineages that are most profoundly affected by *W* or *Sl* mutations (affecting kit or SCF, respectively,[18,19] ordinarily are essentially missing in the mutant animals, it was not generally suspected that SCF might regulate the secretory function of cells which express kit. However, Wershil et al.[47] showed that SCF can induce mouse skin mast cell degranulation in vivo in doses as low as 140 fmol/site and that this response is kit dependent, in that it occurs when dermal mast cells express the wild-type kit but not in phorbol myristate acetate- (PMA-) induced dermal mast cells in W/W^v (Kit^W/Kit^{W-v}) mice which express the Kit^{W-v} mutant receptor. The receptor encoded by Kit^{W-v} has a normal extracellular ligand binding domain, but a point mutation in the kinase domain results in markedly reduced tyrosine kinase activity upon ligand engagement.[48]

Subsequently, it was shown that SCF can also induce mediator release in vitro from rat[49] or mouse[50] peritoneal mast cells and from human skin mast cells.[51] At even lower concentrations in vitro, SCF can augment IgE-dependent activation of mouse peritoneal mast cell[50] or human lung[52] or skin[51] mast cells. SCF treatment can also enhance the responsiveness of mouse mast cells to the neuropeptides substance P[53] and pituitary adenylate cyclase activating polypeptide (PACAP)[54] in vitro. These findings suggest that SCF may be able to influence neuroimmune interactions by regulating the expression of neuropeptide receptors on mast cells.

The ability of SCF to promote mast cell secretion directly, and to enhance mast cell activation via FcεRI, prompted experiments to compare the signaling pathways that were activated in mast cells which had been stimulated via kit as opposed to FcεRI.[55] This work showed that patterns of activation of MAP kinases and pp90[rsk] and pp70-S6 kinases were very similar in mouse mast cells that were activated through these structurally distinct receptors,[55] and suggested that kit- or FcεRI-dependent signaling pathways may exhibit more overlap than had previously been suspected.[55]

It should be emphasized, however, that the effects of SCF on mast cell secretory function are potentially complex and may vary not only according to species and type of mast cell population[19] but also according to duration of exposure to SCF and class of mast cell mediators. For example, in purified mouse peritoneal mast cells, short-term exposure to rrSCF can both induce serotonin release directly and enhance IgE-dependent serotonin release.[50] However, in immature mouse mast cells generated in vitro, short-term incubation with r-mouse SCF induces little or no mediator release.[56] By contrast, longer-term incubation of such cells with r-mouse SCF *enhances* IgE-dependent prostaglandin D_2 (PGD_2) generation, at least in part through effects on hematopoietic PGD_2 synthase, but simultaneously *diminishes* the cell's ability to release the granule-associated mediator, β-hexosaminidase.[56] Also, in studies employing rrSCF, immature mouse mast cells that had been generated in IL-3-containing medium in vitro secreted IL-6 and, to a lesser extent, tumor necrosis factor-α (TNF-α), in response to challenge with soluble rrSCF, whereas concentrations of rrSCF that were effective in inducing the release of IL-6 resulted in little or no specific release of serotonin or histamine.[57]

Although the molecular basis for the differences in responsiveness of various mast cell populations to the effects of SCF on mediator secretion are not yet well understood, the diversity of the secretory responses induced in different mast cell populations by challenge with SCF in vitro made it difficult to predict whether administering this cytokine to human subjects in vivo would provoke mast cell degranulation. Nevertheless, in a phase I study of rhSCF,[2,33,34] we found that subcutaneous injections of rhSCF at 5–50 μg/kg induced a wheal and flare response in each of the 10 subjects tested and at each rhSCF injection site, and that these reactions, when examined by transmission electron microscopy, exhibited evidence of extensive, anaphylactic-type, degranulation of dermal mast cells.[2,58] Moreover, a few subjects developed adverse events after rhSCF dosing that were consistent with the induction of systemic activation of mast cell populations.[2,33,34] These findings strongly suggest that rhSCF can directly induce human mast cell degranulation in vivo, as it can in vitro.

It has recently been shown that chymase, a major cytoplasmic granule-associated protease of human cutaneous mast cells, can cleave human SCF at a novel site that results in the release of a soluble, but biologically active, fragment of SCF which is 7 amino acids shorter at the C-terminus than previously characterized soluble SCF.[59] Thus, in addition to initiating a local, mast cell-dependent inflammatory response, injection of SCF might also initiate a mast cell chymase-dependent mechanism that results in local changes in the proportion of cell membrane-associated versus soluble SCF.

The adverse effects of the mast cell activation that is induced at rhSCF injection sites in vivo can be largely ameliorated by pretreatment of subjects with H_1 and H_2 antihistamines.[1,2] However, a recent study reports that it may be possible to modify SCF in a way that enhances its ability to promote hematopoiesis but does not increase its ability to enhance mast cell mediator secretion.[60] Specifically, a soluble disulfide-linked dimer of mouse SCF (murine KL covalent dimer, or KL-CD), in comparison to the native, noncovalently linked KL dimer, exhibited significantly enhanced growth-promoting activity in colony-forming assays of mouse hematopoietic progenitor cells (CFU-GM) and in assays of [³H]thymidine incorporation by immature, IL-3-derived mouse mast cells in vitro, and increased mobilization of CFU-GM in the blood and spleen of mice in vivo, without exhibiting a significant change in its ability to enhance either FcεRI-dependent release of hexosaminidase from IL-3-derived immature mouse mast cells in vitro or induce degranulation of dermal mast cells in the mouse ear in vivo.[60] Whether the same (or other) chemical modification of rhSCF would result in an agent with an enhanced therapeutic profile (i.e., more hematopoietic cell growth-promoting activity, but unchanged or diminished ability to promote mast cell secretion) remains to be seen.

Interleukin-3

Interleukin-3 is a 28-kDa glycoprotein that was first characterized as a factor which can induce the expression of 20-α-hydroxysteroid dehydrogenase in the splenocytes of nude mice in vitro.[61] It was later shown that this cytokine can promote the in vitro differentiation and proliferation of hematopoietic progenitor cells, leading to the generation of multipotential blast cells, mast cells, basophils, neutrophils, macrophages, eosinophils, erythrocytes, megakarocytes, and dendritic cells.[62–64] Indeed, Ihle et al.[65] demonstrated that IL-3 represented the critical factor present in the various "conditioned media" that were used by different groups to generate populations of immature mast cells from mouse hematopoietic cells in vitro.[66–70]

This work, and many other in vitro studies, indicated that IL-3 might represent a major mast cell developmental/growth factor in the mouse, as well as a T-cell-derived factor that can contribute to the enhanced development of other hematopoietic effector cells during immune responses to pathogens.[71] However, studies with human hematopoietic cells indicated that IL-3 promoted the development of basophils in vitro, but had little if any ability, under most circumstances, to induce mast cell development.[72–74]

In vivo studies showed that the administration of IL-3 can enhance hematopoiesis in mice[75,76] and experimental primates[73,77], and can markedly enhance levels of circulating basophils in rhesus macaques.[73] In addition, widespread cutaneous inflammation developed at sites of recombinant human IL-3 injection in rhesus macaques, and this was associated with a modest increase in numbers of dermal mast cells at such sites.[78] However, in part because of apparently conflicting data on the extent to which human mast cells can express receptors for IL-3 (see Chapter 5), it was not clear whether the ability of IL-3 to influence human mast cell development in vitro[74] or in rhesus macaques in vivo[78] represented direct or indirect effects of the cytokine. On the other hand, both the demonstration that IL-3 can markedly enhance mast cell development in the intestines of nude mice infected with the nematode *Strongyloides ratti*[79] and the demonstration that neutralizing antibodies to IL-3 can partially suppress (by ~50%) the mast cell hyperplasia that develops in the intestines of mice infected with the nematode *Nippostrongylus brasiliensis*[80,81] supported the hypothesis that IL-3 can contribute to the hyperplasia of mucosal mast cells which occurs in murine rodents during T-cell-dependent immune responses to certain parasites.

In addition to its effects on the development of mast cells, basophils, and other hematopoietic cells, many studies have shown that IL-3 can also enhance antigen presentation for T-cell-dependent responses, augment macrophage cytotoxicity and adhesion, and promote the secretory function of eosinophils, basophils, and mast cells.[80,82–89] Taken together, these findings supported the hypothesis that IL-3 derived from T cells (and perhaps other sources) represents a critical link between the immune and hematopoietic systems, and may be particularly important for promoting the development, survival, and effector function of tissue mast cells and blood basophils. However, the actual importance of such potential functions of IL-3 in vivo remained unclear. For example, mice carrying an inactivating mutation in the α-chain of the heterodimeric IL-3 receptor are apparently normal, and hematopoiesis can occur in vitro in the absence of IL-3.[90,91] And even though mouse T lymphocytes and mast cells[92–94] can produce IL-3 in vitro, the conditions in which IL-3 is expressed in vivo, and the sources of this cytokine in these settings, are not fully understood.[95]

Assessing the Role of IL-3 in Mouse Mast Cell and Basophil Development Using IL-3 –/– Mice

Mice that lack IL-3 were produced using gene targeting in embryonic stem cells.[96] IL-3 –/– mice are healthy and, unlike W/W^v (Kit^W/Kit^{W-v}) mice, are fertile. Moreover, like mice that carry an inactivating mutation in the α-chain of the heterodimeric IL-3 receptor[91] or which lack both IL-3 and the common β-subunit of the receptors for IL-3, IL-5, and GM-CSF,[97] IL-3 –/– mice exhibit no detectable abnormalities in multiple aspects of hematopoiesis in vitro or in vivo.[96] Thus, in comparison to mice with mutations resulting in impaired kit or SCF expression or function, the phenotype of IL-3 –/– mice was remarkably normal.

However, we found that IL-3 –/– mice did exhibit abnormalities in mast cell development in vitro and in vivo. In accord with previous work indicating that exogenous IL-3 can augment SCF-dependent mast cell development in vitro,[19,27,28,98–101] we found that SCF induced fewer mast cells to develop in vitro in suspension cultures of bone marrow cells derived from IL-3 –/– mice as opposed to IL-3 +/+ mice.[102] By contrast, substantially higher, and essentially equivalent, numbers of mast cells developed when bone marrow cells from either IL-3 –/– or IL-3 +/+ mice were maintained in vitro in exogenous SCF plus IL-3.[102]

Our cell culture studies thus showed that endogenous IL-3 can enhance, but is not required for, mast cell development from bone marrow progenitors in the presence of exogenous SCF in vitro. To assess the role of IL-3 in mast cell development in vivo, we quantified mast cells in the tissues of IL-3 –/– versus wild-type mice at baseline or after 21 daily s.c. injections of recombinant rat SCF (rrSCF, at $100\,\mu g\,kg^{-1}\,day^{-1}$) or vehicle alone.[102] The results of these experiments showed that endogenous IL-3 is not essential for the development of mast cells under physiological conditions in vivo. Indeed, in all sites examined, levels of tissue mast cells at baseline in adult IL-3 –/– mice were very similar to those in the corresponding sites in IL-3

+/+ mice. Moreover, in contrast to our observations in the in vitro system, we found that endogenous IL-3 was not required for rrSCF-induced mast cell hyperplasia in vivo. In fact, in certain tissues, mast cell levels after rrSCF-treatment were significantly *greater* (by up to 140%) in IL-3 –/– mice than in the corresponding wild-type mice.[102]

To assess the role of IL-3 in the augmented development of basophils and mast cells that is observed during parasite infections, we quantified tissue mast cells and bone marrow basophils, and assessed the expression of parasite immunity, in IL-3 –/– and corresponding wild-type mice which had been infected with the intestinal nematode *Strongyloides venezuelensis* (*S.v.*). *S.v.* is a naturally occurring parasite of murine rodents that is rejected by a T-cell-dependent immune response which is associated with extensive mast cell hyperplasia in the intestinal mucosa.[103,104] In three separate experiments, one of which is shown in Fig. 2.1a, we found that IL-3 –/– mice that had been inoculated with 2000 *S.v.* L3, in comparison to the corresponding wild-type mice, exhibited both significantly delayed expulsion of the adult worms (data not shown) and significantly prolonged production of parasite eggs (Fig. 2.1a).

In addition, the *S.v.*-infected IL-3 –/– mice exhibited striking abnormalities in their basophil and mast cell responses to the parasite. First, while baseline percentages of bone marrow basophils were essentially identical in IL-3 –/– versus wild-type mice, *S.v.* infection induced a significant increase in bone marrow basophil levels in the wild-type mice but not in the IL-3 –/– mice (Fig. 2.1b). These findings confirm the hypothesis, which had been based largely on analyses of the effects of recombinant IL-3,[72,73,77] that endogenous IL-3 can function to expand basophil populations in vivo. However, they also show that IL-3 is not required for baseline levels of bone marrow basophil production in mice.

Second, we found that endogenous IL-3 was required for a substantial proportion (~76%), but not all, of the mast cell hyperplasia that occurred in C57BL/6 mice near the major site of *S.v.* infection, the jejunum (Fig. 2.1c). IL-3 ap-

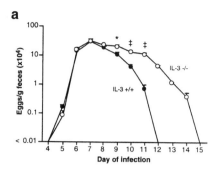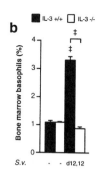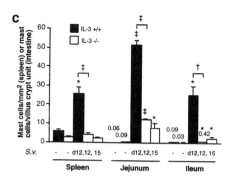

FIGURE 2.1. Defective parasite immunity (a) and parasite-enhanced bone marrow basophil (b) and tissue mast cell (c) development in C57BL/6 (4th backcross generation) IL-3 –/– vs. IL-3 +/+ mice. (a) Kinetics of *Strongyloides venezuelensis* (*S.v.*) egg production in male C57BL/6 IL-3 –/– vs. IL-3 +/+ mice that had been inoculated with 2000 *S.v.* L3.[102] All data are mean ± SEM (*n* = 4–11 mice per point); *, *p* < 0.05, ‡, *p* < 0.0001 vs. corresponding values for IL-3 +/+ mice. All IL-3 –/– or IL-3 +/+ mice cleared the infection on the same day. Similar results were obtained in two other experiments with C57BL/6 or BALB/c IL-3 –/– vs. IL-3 +/+ mice.[102] (b) Percentage of bone marrow basophils at baseline (–) or at day 12 after *S.v.* infection in the IL-3 +/+ and IL-3

–/– mice shown in (a). All data are mean ± SEM (*n* = 3–6 per group). ‡, *p* < 0.0001 vs. corresponding baseline values or (as indicated by brackets) vs. corresponding values for mice of the other genotype. (c) Numbers of mast cells in the spleen, proximal jejunum, and ileum at baseline (–) or at day 12 or 15 after *S.v.* infection in the IL-3 +/+ and IL-3 –/– mice shown in (a). All data are mean ± SEM, except that only mean values are given for very low values (*n* =3–6 per group). *, *p* < 0.05; †, *p* < 0.001; ‡, *p* < 0.0001 vs. corresponding baseline values or (as indicated by brackets) vs. corresponding values for mice of the other genotype. (From Lantz et al.,[102] with permission.)

peared to make a lesser contribution to jejunal mast cell hyperplasia (~50%) during *S.v.* infection in mice on the BALB/c background,[102] perhaps because of strain-dependent differences in levels of other cytokines that can influence mast cell development in mice. However, IL-3 was required for essentially all the increases in mast cells that developed in the ileum or spleen of *S.v.*-infected C57BL/6 mice (Fig. 2.1c) or BALB/c mice (data not shown).

Kit^W/Kit^{W-v}, IL-3 –/– Mice Exhibit a Profound Impairment of Mucosal Mast Cell Development and Immunity During Infection with *Strongyloides venezuelensis*

Previous work showed that host immunity to *S.v.* is even more impaired in Kit^W/Kit^{W-v} mice[104] than in IL-3 –/– mice. Because their c-kit

mutations result in markedly reduced SCF/kit signaling,[48,105] Kit^W/Kit^{W-v} mice are ordinarily profoundly mast cell deficient.[5] On the other hand, mature human basophils[51] or mouse bone marrow basophils[106,107] exhibit low or undetectable levels of kit surface expression, and Kit^W/Kit^{W-v} mice have been reported to have normal levels of blood basophils[108] as well as apparently adequate T-cell function.[109] Moreover, IL-3 can induce mast cell development in Kit^W/Kit^{W-v} mice,[110] which has been thought to account, at least in part, for the modest numbers of mast cells that develop in the intestines of these animals during infections with some parasites, including *Trichinella spiralis*[111] and *S.v.*[104] Finally, studies with neutralizing antibodies to SCF indicate that adequate SCF/kit signaling is required for the intestinal mast cell hyperplasia induced by *T. spiralis* infection in mice, as well as for the expression of normal immunity to this helminth.[112]

Taken together, these and other findings suggested that interactions between SCF- and

IL-3-dependent signaling mechanisms might importantly influence mast cell development, and immunity to certain parasites, at least in mice. However, parasite egg production during a primary infection with *Nippostrongylus brasiliensis* was significantly *less* in c-*kit* mutant *Ws/Ws* mast cell-deficient rats than in the corresponding wild-type rats[113] and normal rats that were treated with an anti-SCF antibody, as compared to rats which had been treated with a control antibody preparation, exhibited both significantly diminished intestinal mucosal mast cell hyperplasia and significantly *diminished* parasite egg production during primary infection with *N. brasiliensis*.[114] These findings raised the possibility that, under some circumstances, SCF- and/or IL-3-dependent mucosal mast cell hyperplasia may have consequences which are more advantageous to the parasite than to the host.

To examine *S.v.* infection in mice that both have markedly impaired c-*kit* function and cannot make IL-3, we produced Kit^W/Kit^{W-v}, IL-3 –/– mice.[102] Adult Kit^W/Kit^{W-v}, IL-3 –/– mice were clinically healthy and resembled Kit^W/Kit^{W-v}, IL-3 +/+ mice in hematocrit and percentage of bone marrow basophils at baseline (Fig. 2.2d). However, Kit^W/Kit^{W-v}, IL-3 –/– mice exhibited a much more profound defect in their ability to reject *S.v.* than did either Kit^W/Kit^{W-v}, IL-3 +/+ mice (Fig. 2.2a,b) or *Kit* +/+, IL-3 –/– mice (cf. Fig. 2.2b with Fig. 2.1a).

Kit^W/Kit^{W-v}, IL-3 –/– mice, unlike Kit^W/Kit^{W-v}, IL-3 +/+ or wild-type mice, also exhibited little or no enhancement of bone marrow basophil production during *S.v.* infection (Fig. 2.2d). By contrast, Kit^W/Kit^{W-v}, IL-3 +/+ mice developed a significant enhancement of bone marrow basophil levels during the course of *S.v.* infection, findings that are in accord with previous work indicating that c-*kit* mutant *Ws/Ws* mast cell-deficient rats exhibit little or no impairment of the blood basophilia which is induced in rats by infection with the parasite *N. brasiliensis*.[115] On the other hand, we found that levels of bone marrow basophils were about 50% lower at baseline, and about 35% lower at the time of parasite expulsion, in Kit^W/Kit^{W-v}, IL-3 +/+ mice than in wild-type mice (Fig. 2.2d).

Strongyloides venezuelensis-infected Kit^W/Kit^{W-v}, IL-3 –/– mice exhibited levels of histologically detectable mast cells in the jejunum, ileum, and spleen that were even more profoundly reduced (vs. the corresponding levels in wild-type mice) than those in the tissues of *S.v.*-infected Kit^W/Kit^{W-v}, IL-3 +/+ mice (Fig. 2.2e,f). Indeed, while levels of mast cells in the spleen, jejunum, or ileum of *S.v.*-infected Kit^W/Kit^{W-v}, IL-3 +/+ mice were about 23%, 3.5%, or 3.2% those in the corresponding tissues of *S.v.*-infected wild-type mice, mast cells were virtually absent from the corresponding tissues of *S.v.*-infected Kit^W/Kit^{W-v}, IL-3 –/– mice (Fig. 2.2e,f).

These findings demonstrate that IL-3 importantly contributes to the mast cell hyperplasia and enhanced bone marrow basophil development observed in mice during *S.v.* infection, as well as to the expression of normal host immunity to this nematode. Our data also indicate that IL-3 and SCF may express overlapping and/or synergistic roles in maintaining an adequate immune response to this parasite. However, it should be emphasized that our findings do not prove that the abnormalities in mast cell and basophil development in IL-3 –/– or Kit^W/Kit^{W-v}, IL-3 –/– mice are the sole basis for (or, necessarily, even significantly contribute to) the impaired immunity to *S.v.* expressed by these mice. The expression of contact hypersensitivity reactions (but not T-cell-dependent immunity to tumor cells) is moderately reduced in IL-3 –/– mice,[96] indicating that IL-3 –/– mice may express defects in some T-cell-dependent responses that are not caused solely by problems with mast cell or basophil mobilization or function. In addition, Kit^W/Kit^{W-v} mice virtually lack interstitial cells of Cajal, which generate gut electrical pacemaker activity,[22,24] and can exhibit reduced numbers of $\gamma\delta$ T cells in the gastrointestinal tract.[116] Thus, it is possible that abnormalties in addition to those affecting their mast cell and basophil responses contributed to the delayed resolution of *S.v.* infections in IL-3 –/– or Kit^W/Kit^{W-v}, IL-3 –/– mice.

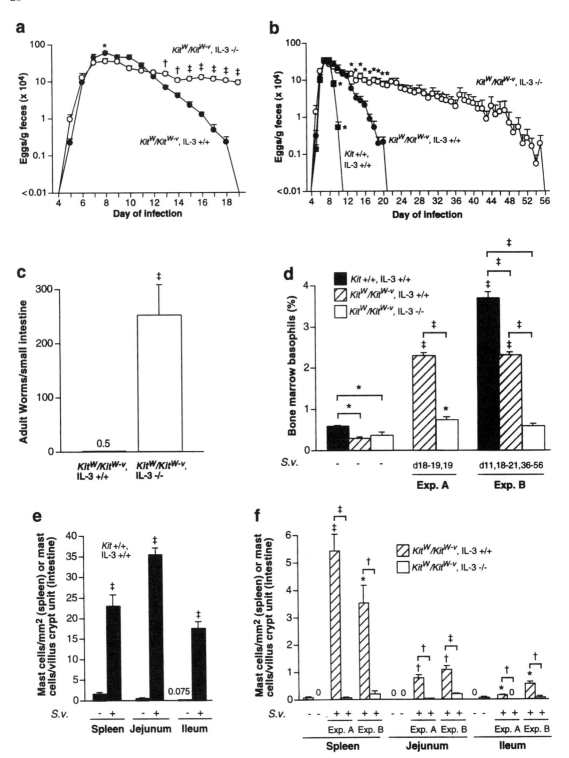

FIGURE 2.2. Markedly defective responses to *S.v.* infection in Kit^W/Kit^{W-v}, IL-3 –/– mice. Kinetics of *S.v.* egg production in (a) 9 Kit^W/Kit^{W-v}, IL-3 +/+ mast cell-deficient mice (5 male, 4 female) vs. 6 Kit^W/Kit^{W-v}, IL-3 –/– mice (3 male, 3 female) (all killed on day 18 or 19 of infection) and (b) in groups of 6 male WBB6F$_1$-

SCF and IL-3 Have Distinct Roles in the Regulation of Physiological Mast Cell and Basophil Development and Immunologically Induced Mast Cell and Basophil Hyperplasia

Our current model of the roles of SCF and IL-3 in mouse mast cell and basophil development is outlined in Fig. 2.3, and some explanatory notes are provided in the legend. The model specifically refers to the mouse system because the availability of mice that are defective in SCF/kit signaling and/or which lack IL-3 permit a direct assessment of the importance of SCF and IL-3 in mast cell and basophil development in this species in vivo. However, in vitro studies, and analyses of the effects of rhSCF in vivo, indicate that many of the major themes of mast cell and basophil development (as shown in Fig. 2.3) are very similar in mice and humans[19] (also see Chapters 1, 5).

On the other hand, our understanding of these processes still has significant gaps. For example, the "pro-mastocyte," which represents the earliest mast cell-committed precursor to be identified during ontogeny, has so far been identified only in the mouse.[100] The mouse pro-mastocyte is defined by the phenotype Thy-1^{lo} c-Kit^{hi}, contains small numbers of cytoplasmic granules that are very similar (by ultrastructure) to those which had previously been

identified in immature mouse mast cells generated in IL-3-containing medium in vitro, and expresses mRNAs encoding mouse mast cell-associated proteases (MC-CPA, MMCP-4, and MMCP-2).[100] However, this cell lacks expression, at the mRNA level, of FcεRI. Purified pro-mastocytes can generate functionally competent mast cells at high frequencies in vitro but do not exhibit developmental potential for other hematopoietic lineages.

Notably, the development of mast cells from pro-mastocytes in vitro occurred in cultures that had been supplemented with SCF and IL-3, but not in cultures that had been supplemented with only one of these cytokines. However, given our finding of normal levels of tissue mast cells in IL-3 −/− mice,[102] it is likely that cytokines other than IL-3 can function together with SCF to regulate the maturation of pro-mastocytes in vivo. When transferred intraperitoneally, pro-mastocytes can reconstitute the peritoneal mast cell compartment of Kit^W/Kit^{W-v} mice to wild-type levels; moreover, these pro-mastocyte-derived peritoneal mast cells exhibit certain phenotypic characteristics of "mature" peritoneal mast cells. The fetal blood pro-mastocyte population was first detected on day 14.5 of mouse gestation and, on day 15.5 of gestation, pro-mastocytes represented approximately 1/40 of the CD45$^+$ leukocyte fraction in the peripheral blood of these animals. However, the numbers of pro-mastocytes in the fetal blood declined from day 15.5 until birth. The origin

Kit +/+, IL-3 +/+ wild-type mice, WBB6F$_1$-Kit^W/Kit^{W-v}, IL-3 +/+ mice or Kit^W/Kit^{W-v}, IL-3 −/− mice inoculated with 2000 S.v. L3 (all the mice were killed on the day of clearance of infection). All data are mean ± SEM. *, $p < 0.05$; †, $p < 0.001$; ‡, $p < 0.0001$ vs. corresponding values for WBB6F$_1$-Kit^W/Kit^{W-v}, IL-3 +/+ mice. In (b), the day parasite egg production fell to zero was day 11 for all the wild-type mice, day 18 (2 mice), and day 21 (4 mice) for Kit^W/Kit^{W-v}, by IL-3 +/+ mice, and day 36 (1 mouse), day 38 (2 mice), day 48 (1 mouse), and day 56 (2 mice) for Kit^W/Kit^{W-v}, IL-3 −/− mice. Similar results were obtained in another experiment. (c) Numbers of adult S.v. per small intestine on day 18 or 19 of infection in the mice shown in experiment (a). ‡, $p < 0.0001$ vs. values for Kit^W/

Kit^{W-v}, IL-3 +/+ mice. (d) Percentage of bone marrow basophils at baseline ($n = 3$–7 mice) or at various days after infection in the mice shown in (a) (Exp. A) and (b) (Exp. B). *, $p < 0.05$; ‡, $p < 0.0001$ vs. corresponding baseline values for mice of the same genotype or vs. values indicated by the bracket. (e,f) Numbers of mast cells in the spleen, proximal jejunum, and ileum at baseline ($n = 3$ or 4 mice) or at various days after S.v. infection in the mice shown in (a,b). *, $p < 0.05$; †, $p < 0.001$; ‡, $p < 0.0001$ vs. corresponding values for uninfected mice of the same genotype or vs. values indicated by the bracket. (a–f). All data are mean ± SEM, except that only mean values are shown for very low values. (From Lantz et al.[102] with permission.)

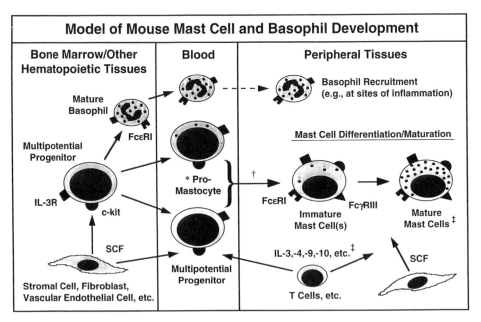

FIGURE 2.3. A highly simplified model of mouse mast cell and basophil development. Basophils arise from c-kit+ multipotential hematopoietic progenitor cells, but typically mature in the bone marrow before entering the peripheral circulation. Basophils then can be recruited from the blood to sites of inflammatory or immune responses. Both human and mouse basophils generally express relatively low levels of kit on their surface. Mast cells also arise from multipotential hematopoietic progenitor cells, but complete major parts of their differentiation/maturation in the peripheral tissues. Unlike basophils, mast cells express high levels of kit on their surface throughout their development; such kit receptors can interact with either membrane-associated or soluble forms of SCF. *, A committed precursor of tissue mast cells, the pro-mastocyte, has been identified in mouse fetal blood. Fetal and adult mouse blood also contains multipotential hematopoietic progenitor cells which, under appropriate circumstances, can give rise to mast cells as well as other cell types. Mouse blood, hematopoietic tissues, and certain other tissues may also contain unipotential mast cell progenitors that are distinct from the pro-mastocyte (see text). †, It is unclear to what extent, either in fetal or adult mouse tissues, immature mast cells that reside in the tissues are derived from pro-mastocytes as opposed to circulating multipotential hematopoietic progenitor cells. ‡, Note that the phenotype of mature mast cells can vary considerably in different anatomical sites, based in part on local levels of stem cell factor (SCF) and other cytokines (e.g., IL-3, -4, -9, -10), and that the phenotypic characteristics of mast cells may vary (in some cases, reversibly) during the course of immune responses or inflammatory processes.

of pro-mastocytes is uncertain; specifically, these cells have not yet been identified in mouse fetal liver in either mid- or late gestation.

Although a mast cell-committed precursor cell that was functionally or morphologically distinct from multipotent hematopoietic stem cells had not previously been purified from mouse bone marrow or blood,[117,118] prior work had established that mast cell precursor activity could be identified in day 9.5 mouse embryonic yolk sac,[119] in adult mouse bone marrow,[120] and

in adult mouse or rat peripheral blood,[117,121–123] as well as in the mesenteric lymph nodes of *Nippostrongylus brasiliensis*-infected mice.[99,118] In humans, in vitro analyses indicate that mast cells can arise from a circulating c-kit+, CD34+, Ly−, CD14−, CD17− hematopoietic progenitor cell[124,125] that lacks detectable expression of FcεRI,[124] but not from basophils or other differentiated hematopoietic lineages.[125] However, it is not known whether an equivalent to the mouse pro-mastocyte exists in humans or, indeed, to what extent pro-mastocytes are present in the blood or other tissues of adult mice.

In addition, as indicated in Fig. 2.3, it is not clear (even in the mouse) whether pro-mastocytes or Thy-1⁻ c-Kit⁺ multipotential hematopoietic progenitor cells (or some other population of mast cell progenitors) represents the most important source of the immature mast cells that are found in the peripheral tissues. For example, assays of mast cell colony formation in vitro indicate that unipotential precursors of mast cells (i.e., colony formation unit-mast or CFU-Mast) may occur in the blood and other tissues of mice and rats,[117,118] but the exact relationship of these cells to the pro-mastocyte (or to the Thy1⁻ Kit⁺ multipotential hematopoietic progenitor cell) is not yet clear. It is possible that tissue mast cells can be derived from either pro-mastocytes or less-differentiated precursors, and that the proportion of tissue mast cells which are derived from these various potential precursor populations varies at different stages of development, or in the context of different inflammatory or immune responses that are associated with changes in numbers of mast cells.

Nevertheless, it is clear, both in mice and in humans, that mature mast cells do not ordinarily circulate (mast cells may appear in the circulation after long-term dosing with rSCF[35], or in subjects with mastocytosis or mast cell leukemia[126]). Accordingly, in both mice and humans, much of the mast cell differentiation/maturation process occurs in the peripheral tissues; these processes are regulated by SCF and many other cytokines in the mouse and, probably, also in humans. By contrast, the weight of current evidence indicates that basophil maturation is completed (or nearly completed) before the cells are released into the peripheral circulation.[127–129] And, unlike mast cells, basophils do not ordinarily reside in large numbers in peripheral tissues, but can be recruited to sites of inflammatory or immune responses.[19,129]

development in mice, as mast cells are ordinarily essentially absent in the tissues of Kit^W/Kit^{W-v} mice. Nevertheless, modest numbers of mast cells (generally, only about 3%–25% of the numbers in similarly treated congenic +/+ mice) can appear in the gastrointestinal mucosal tissues or spleen of Kit^W/Kit^{W-v} mice that have been infected with various nematodes, and studies in IL-3 −/− and Kit^W/Kit^{W-v}, IL-3 −/− mice show that IL-3 importantly contributes to this example of "SCF-independent" mast cell hyperplasia.[102] Indeed, this may represent the "in vivo equivalent" of the mast cell development that occurs when bone marrow cells derived from Kit^W/Kit^{W-v} mice are placed in IL-3-containing medium in vitro.[130–132]

On the other hand, limited development of "mucosal mast cells" occurred in *Strongyloides venezuelensis*-infected mice that were totally devoid of IL-3, indicating that other cytokines can also promote the development of such mast cells.[102] Based primarily on the results of in vitro analyses, as well as a limited number of in vivo experiments, cytokines that may promote the development of certain mast cell populations in mice include IL-4,[133,134] IL-6,[135] IL-9,[136,137] IL-10,[138] TNF-α,[135] and nerve growth factor (NGF).[139–141] Thus, while SCF may be the most important of the mast cell developmental/growth factors in mice, IL-3 (and perhaps many other cytokines) can also contribute to mast cell development in this species.

The finding of essentially normal levels of bone marrow basophils in IL-3 −/− mice indicates that IL-3 is not required for the production of this granulocyte in mice.[102] However, virtually all the basophil hyperplasia that occurred in the bone marrow of mice that were infected with *Strongyloides venezuelensis* was IL-3 dependent.[102] This finding confirms a large body of evidence, derived from studies in mice, humans, and other species, which indicates that IL-3 probably represents a major (if not the major) cytokine responsible for basophil hyperplasia in vivo.

Conclusions

SCF and IL-3 have distinct roles in mouse mast cell and basophil development. SCF/kit interactions are required for physiological mast cell

Acknowledgments. Some of the work reviewed herein was supported by U.S. Public Health Service grants (AI-23990, CA-72074, AI-

33372), the Beth Israel Hospital Pathology Foundation, Inc., and AMGEN Inc. S.J.G. performs research funded by, and consults for, AMGEN Inc., under terms that are in accord with Beth Israel Deaconess Medical Center and Harvard Medical School conflict of interest policies.

References

1. Moskowitz CH, Stiff P, Gordon MS, et al. Recombinant methionyl human stem cell factor and filgrastim for peripheral blood progenitor cell mobilization and transplantation in non-Hodgkin's lymphoma patients—results of a phase I/II trial. Blood 1997;89:3136–3147.
2. Costa JJ, Demetri GD, Harrist TJ, et al. Recombinant human stem cell factor (kit ligand) promotes human mast cell and melanocyte hyperplasia and functional activation in vivo. J Exp Med 1996;183:2681–2686.
3. Russell ES. Hereditary anemias of the mouse: a review for geneticists. In: Caspari EW, ed. Advances in Genetics. Vol. 20. New York: Academic Press, 1979.
4. Silvers WK. Dominant spotting, patch and rump-white; steel, flexed tail, splotch and varitint-waddler. In: The Coat Colors of Mice: A Model for Gene Action and Interaction. New York: Springer-Verlag, 1979:206–241.
5. Kitamura Y, Go S, Hatanaka K. Decrease of mast cells in W/W^v mice and their increase by bone marrow transplantation. Blood 1978; 52:447–452.
6. Kitamura Y, Go S. Decreased production of mast cells in Sl/Sl^d anemic mice. Blood 1979; 53:492–497.
7. Chabot B, Stephenson DA, Chapman VM, Besmer P, Bernstein A. The proto-oncogene c-*kit* encoding a transmembrane tyrosine kinase receptor maps to the mouse *W* locus. Nature (Lond) 1988;335:88–89.
8. Geissler EN, Ryan MA, Housman DE. The dominant-white spotting (*W*) locus of the mouse encodes the c-*kit* proto-oncogene. Cell 1988;55:185–192.
9. Huang E, Nocka K, Beier DR, et al. The hematopoietic growth factor KL is encoded by the *Sl* locus and is the ligand of the c-*kit* receptor, the gene product of the *W* locus. Cell 1990; 63:225–233.
10. Williams DE, Eisenman J, Baird A, et al. Identification of a ligand for the c-*kit* proto-oncogene. Cell 1990;63:167–74.
11. Copeland NG, Gilbert DJ, Cho BC, et al. Mast cell growth factor maps near the steel locus on mouse chromosome 10 and is deleted in a number of steel alleles. Cell 1990;63:175–183.
12. Anderson DM, Lyman SD, Baird A, et al. Molecular cloning of mast cell growth factor, a hematopoietin that is active in both membrane bound and soluble forms. Cell 1990;63:235–243.
13. Witte ON. Steel locus defines new multipotent growth factor. Cell 1990;63:5–6.
14. Williams DE, de Vries P, Namen AE, Widmer MB, Lyman SD. The steel factor. Dev Biol 1992;151:368–376.
15. Zsebo KM, Williams DA, Geissler EN, et al. Stem cell factor (SCF) is encoded at the *Sl* locus of the mouse and is the ligand for the c-*kit* tyrosine kinase receptor. Cell 1990;63:213–224.
16. Zsebo KM, Wypych J, McNiece IK, et al. Identification, purification, and biological characterization of hematopoietic stem cell factor from buffalo rat liver-conditioned medium. Cell 1990;63:195–210.
17. Martin TR, Suggs SV, Langley KE, et al. Primary structure and functional expression of rat and human stem cell factor DNAs. Cell 1990; 63:203–211.
18. Besmer P. The *kit* ligand encoded at the murine *Steel* locus: a pleiotropic growth and differentiation factor. Curr Opin Cell Biol 1991;3:939–946.
19. Galli SJ, Zsebo KM, Geissler EN. The kit ligand, stem cell factor. Adv Immunol 1994; 55:1–96.
20. Broudy VC. Stem cell factor and hematopoiesis. Blood 1997;90:1345–1364.
21. Ullrich A, Schlessinger J. Signal transduction by receptors with tyrosine kinase activity. Cell 1990;61:203–212.
22. Maeda H, Yamagata A, Nishikawa S, et al. Requirement of c-*kit* for development of intestinal pacemaker system. Development (Camb) 1992;116:369–375.
23. Huizinga JD, Thuneberg L, Kluppel M, Malysz J, Mikkelsen HB, Bernstein A. *W/Kit* gene required for interstitial cells of Cajal and for intestinal pacemaker activity. Nature (Lond) 1995;373:347–349.
24. Ward S, Burns A, Torihashi S, Sanders K. Mutation in the proto-oncogene c-*kit* blocks devel-

opment of interstitial cells and electrical rhythm in murine intestine. J Physiol 1994; 480:91–97.

25. Morstyn G, Brown S, Gordon M, et al. Stem cell factor is a potent synergistic factor in hematopoiesis. Oncology (Basel) 1994;51:205–214.

26. Tsai M, Shih L, Newlands GFJ, et al. The rat c-*kit* ligand, stem cell factor, induces the development of connective tissue-type and mucosal mast cells in vivo. Analysis by anatomical distribution, histochemistry, and protease phenotype. J Exp Med 1991;174:125–131.

27. Gurish MF, Ghildyal N, McNeil HP, Austen KF, Gillis S, Stevens RL. Differential expression of secretory granule proteases in mouse mast cells exposed to interleukin 3 and c-*kit* ligand. J Exp Med 1992;175:1003–1012.

28. Haig DM, Huntley JF, Mackellar A, et al. Effects of stem cell factor (kit-ligand) and interleukin-3 on the growth and serine proteinase expression of rat bone marrow-derived or serosal mast cells. Blood 1994;83:72–83.

29. Valent P, Spanblochl E, Sperr WR, et al. Induction of differentiation of human mast cells from bone marrow and peripheral blood mononuclear cells by recombinant human stem cell factor/kit-ligand in long-term culture. Blood 1992;80:2237–2245.

30. Kirshenbaum AS, Goff JP, Kessler SW, Mican JM, Zsebo KM, Metcalfe DD. Effect of IL-3 and stem cell factor on the appearance of human basophils and mast cells from CD34+ pluripotent progenitor cells. J Immunol 1992; 148:772–777.

31. Irani A-MA, Nilsson G, Miettinen U, et al. Recombinant human stem cell factor stimulates differentiation of mast cells from dispersed human fetal liver cells. Blood 1992;80: 3009–3021.

32. Mitsui H, Furitsu T, Dvorak AM, et al. Development of human mast cells from umbilical cord blood cells by recombinant human and murine c-*kit* ligand. Proc Natl Acad Sci USA 1993;90:735–739.

33. Costa JJ, Demetri GD, Hayes DF, Merica EA, Menchaca DM, Galli SJ. Increased skin mast cells and urine methyl histamine in patients receiving recombinant methionyl human stem cell factor. Proc Am Assoc Cancer Res 1993; 34:211 (abstract).

34. Costa JJ, Demetri GD, Harrist TJ, et al. Recombinant human stem cell factor (rhSCF) induces cutaneous mast cell activation and hyperplasia, and hyperpigmentation in humans in vivo. J Allergy Clin Immunol 1994;93:225 (abstract).

35. Galli SJ, Iemura A, Garlick DS, Gamba-Vitalo C, Zsebo KM, Andrews RG. Reversible expansion of primate mast cell populations *in vivo* by stem cell factor. J Clin Invest 1993;91:148–152.

36. Mekori YA, Oh CK, Metcalfe DD. IL-3-dependent murine mast cells undergo apoptosis on removal of IL-3: Prevention of apoptosis by c-*kit* ligand. J Immunol 1993; 151:3775–3784.

37. Yee NS, Paek I, Besmer P. Role of kit-ligand in proliferation and suppression of apoptosis in mast cells: Basis for radiosensitivity of *White Spotting* and *Steel* mutant mice. J Exp Med 1994;179:1777–1787.

38. Iemura A, Tsai M, Ando A, Wershil BK, Galli SJ. The c-*kit* ligand, stem cell factor, promotes mast cell survival by suppressing apoptosis. Am J Pathol 1994;144:321–328.

39. Furitsu T, Tsujimura T, Tono T, et al. Identification of mutations in the coding sequence of the proto-oncogene c-*kit* in a human mast cell leukemia cell line causing ligand-independent activation of c-*kit* product. J Clin Invest 1993;92:1736–1744.

40. Nagata H, Worobec AS, Oh CK, et al. Identification of a point mutation in the catalytic domain of the protooncogene c-*kit* in peripheral blood mononuclear cells of patients who have mastocytosis with an associated hematologic disorder. Proc Natl Acad Sci USA 1995; 92:10560–10564.

41. Longley BJ, Tyrrell L, Lu SZ, et al. Somatic c-*kit* activating mutation in urticaria pigmentosa and aggressive mastocytosis: establishment of clonality in a human mast cell neoplasm. Nat Genet 1996;12:312–314.

42. Longley BJ Jr, Morganroth GS, Tyrrell L, et al. Altered metabolism of mast-cell growth factor (c-*kit* ligand) in cutaneous mastocytosis. N Engl J Med 1993;328:1302–1307.

43. Lorenz U, Bergemann AD, Steinberg HN, et al. Genetic analysis reveals cell type-specific regulation of receptor tyrosine kinase c-Kit by the protein tyrosine phosphatase SHP1. J Exp Med 1996;184:1111–1126.

44. Paulson RF, Vesely S, Siminovitch KA, Bernstein A. Signalling by the *W/Kit* receptor tyrosine kinase is negatively regulated in vivo

by the protein tyrosine phosphatase SHP1. Nat Genet 1996;13:309–315.

45. Finotto S, Mekori YA, Metcalfe DD. Glucocorticoids decrease tissue mast cell number by reducing the production of the c-*kit* ligand, stem cell factor, by resident cells: in vitro and in vivo evidence in murine systems. J Clin Invest 1997;99:1721–1728.

46. Lavker RM, Schechter NM. Cutaneous mast cell depletion results from topical corticosteroid usage. J Immunol 1985;135:2368–2373.

47. Wershil BK, Tsai M, Geissler EN, Zsebo KM, Galli SJ. The rat c-*kit* ligand, stem cell factor, induces c-*kit* receptor-dependent mouse mast cell activation in vivo. Evidence that signaling through the c-*kit* receptor can induce expression of cellular function. J Exp Med 1992;175:245–255.

48. Nocka K, Tan JC, Chiu E, et al. Molecular bases of dominant negative and loss of function mutations at the murine c-*kit*/white spotting locus: W^{37}, W^v, W^{41}, and W. EMBO J 1990;9:1805–1813.

49. Nakajima K, Hirai K, Yamaguchi M, et al. Stem cell factor has histamine releasing activity in rat connective tissue-type mast cells. Biochem Biophys Res Commun 1992;183:1076–1083.

50. Coleman JW, Holliday MR, Kimber I, Zsebo KM, Galli SJ. Regulation of mouse peritoneal mast cell secretory function by stem cell factor, IL-3 or IL-4. J Immunol 1993;150:556–562.

51. Columbo M, Horowitz EM, Botana LM, et al. The human recombinant c-*kit* receptor ligand, rhSCF, induces mediator release from human cutaneous mast cells and enhances IgE-dependent mediator release from both skin mast cells and peripheral blood basophils. J Immunol 1992;149:599–608.

52. Bischoff SC, Dahinden CA. c-*kit* ligand: a unique potentiator of mediator release by human lung mast cells. J Exp Med 1992;175:237–244.

53. Wershil BK, Lavigne JA, Galli SJ. Stem cell factor (SCF) can influence neuroimmune interactions: bone marrow-derived mast cells (BMCMC) maintained in SCF acquire the ability to release histamine and tumor necrosis factor-alpha (TNF-α) in response to substance P (SP). FASEB J 1994;8:A742 (abstract).

54. Schmidt-Choudhury A, Goetzl EJ, Xia M, et al. Mouse mast cells grown in stem cell factor (SCF) express type II pituitary adenylate cyclase activating polypeptide (PACAP) receptors. J Invest Dermatol 1995;10:A1268.

55. Tsai M, Chen R-H, Tam S-Y, Blenis J, Galli SJ. Activation of MAP kinases, pp90[rsk] and pp70-S6 kinases in mouse mast cells by signaling through the c-*kit* receptor tyrosine kinase or FcεRI: rapamycin inhibits activation of pp70-S6 kinase and proliferation in mouse mast cells. Eur J Immunol 1993;23:3286–3291.

56. Murakami M, Matsumoto R, Urade Y, Austen KF, Arm JP. c-Kit ligand mediates increased expression of cytosolic phospholipase A$_2$, prostaglandin endoperoxide synthase-1, and hematopoietic prostaglandin D$_2$ synthase and increased IgE-dependent prostaglandin D$_2$ generation in immature mouse mast cells. J Biol Chem 1995;270:3239–3246.

57. Gagari E, Tsai M, Lantz CS, Fox LG, Galli SJ. Differential release of mast cell interleukin-6 via c-*kit*. Blood 1997;89:2654–2663.

58. Dvorak AM, Costa JJ, Morgan ES, Monahan-Earley RA, Galli SJ. Diamine oxidase-gold ultrastructural localization of histamine in human skin biopsies containing mast cells stimulated in degranulate in vivo by exposure to recombinant human stem cell factor. Blood 1997;90:2893–2900.

59. Longley BJ, Tyrrell L, Ma Y, et al. Chymase cleavage of stem cell factor yields a bioactive, soluble product. Proc Natl Acad Sci USA 1997;94:9017–9021.

60. Nocka KH, Levine BA, Ko J-L, et al. Increased growth promoting but not mast cell degranulation potential of a covalent dimer of c-*kit* ligand. Blood 1997;90:3874–3883.

61. Ihle JN, Weinstein Y. Immunological regulation of hematopoietic/lymphoid stem cell differentiation by interleukin 3. Adv Immunol 1986;39:1–50.

62. Rennick DM, Lee FD, Yokota T, Arai KI, Cantor H, Nabel GJ. A cloned MCGF cDNA encodes a multilineage hematopoietic growth factor: multiple activities of interleukin 3. J Immunol 1985;134:910–914.

63. Suda T, Suda J, Ogawa M, Ihle JN. Permissive role of interleukin 3 (IL-3) in proliferation and differentiation of multipotential hemopoietic progenitors in culture. J Cell Physiol 1985;124:182–190.

64. Caux C, Vanbervliet B, Massacrier C, Durand I, Banchereau J. Interleukin-3 cooperates with tumor necrosis factor-alpha for the development of human dendritic Langerhans cells from cord blood CD34(+) hematopoietic progenitor cells. Blood 1996;87:2376–2385.

65. Ihle JN, Keller J, Orsozlan S, et al. Biological properties of homogeneous interleukin 3. I. Demonstration of WEHI-3 growth factor activity, mast cell growth factor activity, P cell stimulating factor activity and histamine producing factor activity. J Immunol 1983; 131:282–287.

66. Nabel GJ, Galli SJ, Dvorak AM, Dvorak HF, Cantor H. Inducer T lymphocytes synthesize a factor that stimulates proliferation of cloned mast cells. Nature (Lond) 1981;291:332–334.

67. Nagao K, Yokoro K, Aaronson SA. Continuous lines of basophil/mast cells derived from normal mouse bone marrow. Science 1981; 212:333–335.

68. Razin E, Cordon-Cardo C, Good RA. Growth of a pure population of mast cells in vitro with conditioned medium derived from concanavalin A-stimulated splenocytes. Proc Natl Acad Sci USA 1981;78:2559–2561.

69. Schrader JW, Lewis SJ, Clark-Lewis I, Culvenor JG. The persisting (P) cell: histamine content, regulation by a T cell-derived factor, orgin from a bone marrow precursor, and relationship to mast cells. Proc Natl Acad Sci USA 1981;78:323–327.

70. Tertian G, Yung YP, Guy-Grand D, Moore MA. Long-term in vitro culture of murine mast cells. I. Description of a growth factor-dependent culture technique. J Immunol 1981; 127:788–794.

71. Ihle JN. Interleukin-3 and hematopoiesis. Chem Immunol 1992;51:65–106.

72. Saito H, Hatake K, Dvorak AM, et al. Selective differentiation and proliferation of hematopoietic cells induced by recombinant human interleukins. Proc Natl Acad Sci USA 1988; 85:2288–2292.

73. Mayer P, Valent P, Schmidt G, Liehl E, Bettelheim P. The in vivo effects of recombinant human interleukin-3: demonstration of basophil differentiation factor, histamine-producing activity, and priming of GM-CSF-responsive progenitors in nonhuman primates. Blood 1989;74:613–621.

74. Kirshenbaum AS, Goff JP, Dreskin SC, Irani AM, Schwartz LB, Metcalfe DD. IL-3-dependent growth of basophil-like cells and mast-like cells from human bone marrow. J Immunol 1989;142:2424–2429.

75. Metcalf D, Begley CG, Johnson NA, Nicola NA, Lopez AF, Williamson DJ. Effects of purified bacterially synthesized murine multi-CSF

76. Kindler V, Thorens B, de Kossodo S, et al. Stimulation of hematopoiesis in vivo by recombinant bacterial murine interleukin 3. Proc Natl Acad Sci USA 1986;83:1001–1005.

77. Donahue RE, Seehra J, Metzger M, et al. Human IL-3 and GM-CSF act synergistically in stimulating hematopoiesis in primates. Science 1988;241:1820–1823.

78. Volc-Platzer B, Valent P, Radaszkiewicz T, Mayer P, Bettelheim P, Wolff K. Recombinant human interleukin 3 induces proliferation of inflammatory cells and keratinocytes in vivo. Lab Invest 1991;64:557–566.

79. Abe T, Nawa Y. Worm expulsion and mucosal mast cell response induced by repetitive IL-3 adminstration in Strongyloides ratti-infected nude mice. Immunology 1988;63:181–185.

80. Madden KB, Urban JF, Ziltener HJ, Schrader JW, Finkelman FD, Katona IM. Antibodies to IL-3 and IL-4 supress helminth-induced intestinal mastocytosis. J Immunol 1991;147:1387–1391.

81. Finkelman FD, Madden KB, Morris SC, et al. Anti-cytokine antibodies as carrier proteins. Prolongation of in vivo effects of exogenous cytokines by injection of cytokine-anti-cytokine antibody complexes. J Immunol 1993;151:1235–1244.

82. Hirai K, Morita Y, Misaki Y, et al. Modulation of human basophil histamine release by hemopoietic growth factors. J Immunol 1988; 141:3958–3964.

83. Rothenberg M, Owen W, Silberstein D, et al. Human eosinophils have prolonged survival, enhanced functional properties, and become hypodense when exposed to human interleukin-3. J Clin Invest 1988; 81:1986–1992.

84. Kimoto M, Kindler V, Higaki M, Ody C, Izui S, Vassalli P. Recombinant murine IL-3 fails to stimulate T or B lymphopoiesis in vivo, but enhances immune responses to T cell-dependent antigens. J Immunol 1988;140:1889–1894.

85. Cannistra SA, Vellenga E, Groshek P, Rambaldi A, Griffin JD. Human granulocyte-macrophage colony-stimulating factor and interleukin-3 stimulate monocyte cytotoxicity through a tumor necrosis factor-dependent mechanism. Blood 1988;71:672–676.

86. Schleimer RP, Derse CP, Friedman B, et al. Regulation of human basophil mediator release by cytokines. I. Interaction with

antiinflammatory steroids. J Immunol 1989; 143:1310–1317.

87. Brunner T, Heusser CH, Dahinden CA. Human peripheral blood basophils primed by interleukin 3 (IL-3) produce IL-4 in response to immunoglobulin E receptor stimulation. J Exp Med 1993;177:605–611.

88. MacGlashan DW Jr, Schroeder JT. Basophils as the target and source of cytokines. In: Kitamura Y, Yamamoto S, Galli SJ, Greaves MW, eds. Biological and Molecular Aspects of Mast Cell and Basophil Differentiation and Function. New York: Raven Press, 1995:51–63.

89. Pulaski BA, Yeh KY, Shastri N, et al. Interleukin 3 enhances cytotoxic T lymphocyte development and class I major histocompatibility complex representation of exogenous antigen by tumor-infiltrating antigen-presenting cells. Proc Natl Acad Sci USA 1996;93:3669–3674.

90. Li CL, Johnson GR. Stimulation of multipotential, erythroid and other murine haematopoietic progenitor cells by adherent cell lines in the absence of detectable multi-CSF (IL-3). Nature (Lond) 1985;316:633–636.

91. Ichihara M, Hara T, Takagi M, Cho LC, Gorman DM, Miyajima A. Impaired interleukin-3 (IL-3) response of the A/J mouse is caused by a branch point deletion in the IL-3 receptor α subunit gene. EMBO J 1995;14:939–950.

92. Plaut M, Pierce JH, Watson CJ, Hanley-Hyde J, Nordan RP, Paul WE. Mast cell lines produce lymphokines in response to cross-linkage of FcεRI or to calcium ionophores. Nature (Lond) 1989;339:64–67.

93. Wodnar-Filipowicz A, Heusser CH, Moroni C. Production of the haemopoietic growth factors GM-CSF and interleukin-3 by mast cells in response to IgE receptor-mediated activation. Nature (Lond) 1989;339:150–152.

94. Burd PR, Rogers HW, Gordon JR, et al. Interleukin 3-dependent and -independent mast cells stimulated with IgE and antigen express multiple cytokines. J Exp Med 1989; 170:245–257.

95. Svetic A, Madden KB, Zhou X, et al. A primary intestinal helminthic infection rapidly induces a gut-associated elevation of Th2-associated cytokines and IL-3. J Immunol 1993;150:3434–3441.

96. Mach N, Lantz CS, Galli SJ, et al. Involvement of interleukin-3 in delayed-type hypersensitivity. Blood 1998;91:778–783.

97. Nishinakamura R, Miyajima A, Mee PJ, Tybulewicz VLJ, Murray R. Hematopoiesis in mice lacking the entire granulocyte-macrophage colony-stimulating factor/interleukin-3/interleukin-5 functions. Blood 1996;88: 2458–2464.

98. Rennick D, Hunte B, Holland G, Thompson-Snipes L. Cofactors are essential for stem cell factor-dependent growth and maturation of mast cell progenitors: comparative effects of interleukin-3 (IL-3), IL-4, IL-10, and fibroblasts. Blood 1995;85:57–65.

99. Lantz CS, Huff TF. Differential responsiveness of purified mouse c-kit+ mast cells and their progenitors to IL-3 and stem cell factor. J Immunol 1995;155:4024–4029.

100. Rodewald H-R, Dessing M, Dvorak AM, Galli SJ. Identification of a committed precursor for the mast cell lineage. Science 1996;271:818–822.

101. Tsuji K, Zsebo KM, Ogawa M. Murine mast cell colony formation supported by IL-3, IL-4, and recombinant rat stem cell factor, ligand for c-kit. J Cell Physiol 1991;148:362–369.

102. Lantz CS, Boesiger J, Song CH, et al. Role for interleukin-3 in mast cell and basophil development and parasite immunity. Nature (Lond) 1998;392:90–93.

103. Nawa Y, Ishikawa N, Tsuchiya K, et al. Selective effector mechanisms for the expulsion of intestinal helminths. Parasite Immunol (Oxf) 1994;16:333–338.

104. Khan AI, Horii Y, Tiuria R, Sato Y, Nawa Y. Mucosal mast cells and the expulsive mechanisms of mice against Strongyloides venezuelensis. Int J Parasitol 1993;23:551–555.

105. Hayashi S, Kunisada T, Ogawa M, Yamaguchi K, Nishikawa S. Exon skipping by mutation of an authentic splice site of c-kit gene in W/W mouse. Nucleic Acids Res 1991;19:1267–1271.

106. Dvorak AM, Seder RA, Paul WE, Morgan ES, Galli SJ. Effects of interleukin-3 with or without the c-kit ligand, stem cell factor, on the survival and cytoplasmic granule formation of mouse basophils and mast cells in vitro. Am J Pathol 1994;144:160–170.

107. Lantz CS, Yamaguchi M, Oettgen HC, et al. IgE regulates mouse basophil FcεRI expression in vivo. J Immunol 1997;158:2517–2521.

108. Jacoby W, Cammarata PV, Findlay S, Pincus SH. Anaphylaxis in mast cell-deficient mice. J Invest Dermatol 1984;83:302–304.

109. Galli SJ, Kitamura Y. Animal models of human disease. Genetically mast-cell-deficient W/W^v and Sl/Sl^d mice: their value for the analysis of the roles of mast cells in biological responses in vivo. Am J Pathol 1987;127:191–198.

110. Ody C, Kindler V, Vassalli P. Interleukin 3 perfusion in W/W^v mice allows the development of macroscopic hematopoietic spleen colonies and restores cutaneous mast cell numbers. J Exp Med 1990;172:403–406.

111. Alizadeh H, Murrell KD. The intestinal mast cell response to *Trichinella spiralis* infection in mast cell-deficient W/W^v mice. J Parasitol 1984;70:767–773.

112. Donaldson LE, Schmitt E, Huntley JF, Newlands GFJ, Grencis RK. A critical role of stem cell factor and c-*kit* in host protective immunity to an intestinal helminth. Int Immunol 1996;8:559–567.

113. Arizono N, Kasugai T, Yamada M, et al. Infection of *Nippostrongylus brasiliensis* induces development of mucosal-type but not connective tissue-type mast cells in genetically mast cell-deficient Ws/Ws rats. Blood 1993;81:2572–2578.

114. Newlands GFJ, Miller HRP, MacKellar A, Galli SJ. Stem cell factor contributes to intestinal mucosal mast cell hyperplasia in rats infected with *Nippostrongylus brasiliensis* or *Trichinella spiralis*, but anti-stem cell factor treatment decreases parasite egg production during *N. brasiliensis* infection. Blood 1995;86:1968–1976.

115. Kasugai T, Okada M, Morimoto M, et al. Infection of *Nippostrongylus brasiliensis* induces normal increases of basophils in mast cell-deficient Ws/Ws rats with a small deletion at the kinase domain of c-*kit*. Blood 1993;81:2521–2529.

116. Puddington L, Olson S, Lefrancois L. Interactions between stem cell factor and c-*kit* are required for intestinal immune system homeostasis. Immunity 1994;1:733–739.

117. Kasugai T, Tei H, Okada M, et al. Infection with *Nippostrongylus brasiliensis* induces invasion of mast cell precursors from peripheral blood to small intestine. Blood 1995;85:1334–1340.

118. Huff TF, Lantz CS, Ryan JJ, Leftwich JA. Mast cell-committed progenitors. In: Kitamura Y, Yamamoto S, Galli SJ, Greaves MW, eds. Biological and Molecular Aspects of Mast Cell Differentiation and Function. New York: Raven Press, 1995:105–117.

119. Sonoda T, Hayashi C, Kitamura Y. Presence of mast cell precursors in the yolk sac of mice. Dev Biol 1983;97:89–94.

120. Kitamura Y, Shimada M, Hatanaka K, Miyano Y. Development of mast cells from grafted bone marrow cells in irradiated mice. Nature (Lond) 1977;268:442–443.

121. Kitamura Y, Hatanaka K, Murakami M, Shibata H. Presence of mast cell precursors in peripheral blood of mice demonstrated by parabiosis. Blood 1979;53:1085–1088.

122. Zucker-Franklin D, Grusky G, Hirayama N, Schnipper E. The presence of mast cell precursors in rat peripheral blood. Blood 1981;58:544–551.

123. Sonoda T, Ohno T, Kitamura Y. Concentration of mast-cell progenitors in bone marrow, spleen, and blood of mice determined by limiting dilution analysis. J Cell Physiol 1982;112:136–140.

124. Rottem M, Okada T, Goff JP, Metcalfe DD. Mast cells cultured from the peripheral blood of normal donors and patients with mastocytosis originate from CD34+/FcεRI- cell populations. Blood 1994;84:2489–2496.

125. Agis H, Willheim M, Sperr WR, et al. Monocytes do not make mast cells when cultured in the presence of SCF. Characterization of the circulating mast cell progenitor as a c-*kit*+, CD34+, Ly-, CD14-, CD17-, colony-forming cell. J Immunol 1995;151:4221–4227.

126. Metcalfe DD. Classification and diagnosis of mastocytosis: current status. J Invest Dermatol 1991;96:2S–4S.

127. Dvorak AM, Seder RA, Paul WE, Kissell-Rainville S, Plaut M, Galli SJ. Ultrastructural characteristics of FcεR-positive basophils in the spleen and bone marrow of mice immunized with goat anti-mouse IgD antibody. Lab Invest 1993;68:708–715.

128. Seder RA, Paul WE, Dvorak AM, et al. Mouse splenic and bone marrow cell populations that express high-affinity Fc epsilon receptors and produce interleukin 4 are highly enriched in basophils. Proc Natl Acad Sci USA 1991;88:2835–2839.

129. Costa JJ, Galli SJ. Mast cells and basophils. In: Rich R, Fleisher TA, Schwartz BD, Shearer WT, Strober W, eds. Clinical Immunology: Principles and Practice. St. Louis: Mosby-Year Book, 1996:408–430.

130. Yung Y-P, Moore MAS. Long-term in vitro culture of murine mast cells. III. Discrimination of mast cell growth-factor and granulocyte CSF. J Immunol 1982;129:1256–1261.

131. Suda T, Suda J, Spicer SS, Ogawa M. Proliferation and differentiation in culture of mast cell progenitors derived from mast cell-deficient mice of genotype *W/W^v*. J Cell Physiol 1985;122:187–192.

132. Nakano T, Sonoda T, Hayashi C, et al. Fate of bone marrow-derived cultured mast cells after intracutaneous, intraperitoneal, and intravenous transfer into genetically mast cell-deficient *W/W^v* mice. Evidence that cultured mast cells can give rise to both connective tissue type and mucosal mast cells. J Exp Med 1985;162:1025–1043.

133. Mosmann TR, Bond MW, Coffman RL, Ohara J, Paul WE. T-cell and mast cell lines respond to B-cell stimulatory factor 1. Proc Natl Acad Sci USA 1986;83:5654–5658.

134. Smith CA, Rennick DM. Characterization of a murine lymphokine distinct from interleukin 2 and interleukin 3 (IL-3) possessing a T-cell growth factor activity and a mast-cell growth factor activity that synergizes with IL-3. Proc Natl Acad Sci USA 1986;83:1857–1861.

135. Hu Z-Q, Kobayashi K, Zenda N, Shimamura T. Tumor necrosis factor-α- and interleukin-6-triggered mast cell development from mouse spleen cells. Blood 1997;89:526–533.

136. Hültner L, Druez C, Moeller J, et al. Mast cell growth-enhancing activity (MEA) is structurally related and functionally identical to the novel mouse T cell growth factor P40/TCGFIII (interleukin 9). Eur J Immunol 1990;20:1413–1416.

137. Khalil RM, Luz A, Mailhammer R, et al. *Schistosoma mansoni* infection in mice augments the capacity for interleukin 3 (IL-3) and IL-9 production and concurrently enlarges progenitor pools for mast cells and granulocytes-macrophages. Infect Immun 1996;64:4960–4966.

138. Thompson-Snipes L, Dhar V, Bond MW, Mosmann TR, Moore KW, Rennick D. Interleukin-10: a novel stimulatory factor for mast cells and their progenitors. J Exp Med 1991;173:507–510.

139. Aloe L, Levi-Montalcini R. Mast cells increase in tissues of neonatal rats injected with the nerve growth factor. Brain Res 1977;133:358–366.

140. Aloe L. The effect of nerve growth factor and its antibody on mast cells in vivo. J Neuroimmunol 1988;18:1–12.

141. Matsuda H, Kannan Y, Ushio H, et al. Nerve growth factor induces development of connective tissue-type mast cells in vitro from murine bone marrow cells. J Exp Med 1991;174:7–14.

3
The c-*kit* Receptor and the *mi* Transcription Factor: Two Important Molecules for Mast Cell Development

Yukihiko Kitamura, Eiichi Morii, and Tomoko Jippo

The development of hematopoietic cells is controlled through the cooperative effects of growth factors that permit cellular proliferation and nuclear transcription factors that integrate the signals from growth factors and activate lineage-specific genes.[1] For the development of mast cells, the signals generated by binding of stem cell factor (SCF) to the c-kit receptor tyrosine kinase (KIT) is essential. Loss-of-function mutations at either the SCF (*Sl*) or the c-*kit* (*W*) genes result in a deficiency of mast cells.[2,3] In contrast, gain-of-function mutations of c-*kit* result in the neoplastic transformation of mast cells.[4] Many transcription factors probably are involved in the development of mast cells, but only a few of them have been reported to activate mast cell-specific genes. We found that the transcription factor encoded by the *mi* locus of mice (hereafter called MITF) is involved in the transactivation of various genes in mast cells including the c-*kit* gene.[5,6] Because kit and MITF appear to be essential molecules for the development of mast cells, we review here our recent studies on kit and MITF.

Kit

Loss-of-Function Mutation of c-*kit*

Early in our investigation of mast cells, we identified mast cell-deficient mice. Mice of W/W^v and Sl/Sl^d genotypes were known to have three abnormalities, that is, lack of melanocytes,

depletion of germ cells, and anemia[7]; our work highlighted the fourth abnormality, the deficiency of mast cells.[2,3] More recently, the fifth abnormality of W/W^v and Sl/Sl^d mice was found, the deficiency of the interstitial cells of Cajal that regulate the autonomic movement of the gastrointestinal tract.[8-10]

The *W* locus and the *Sl* locus are completely different, yet the phenotypes of W/W^v and Sl/Sl^d mice are indistinguishable. Our study looked at the mechanism of mast cell deficiency of W/W^v and Sl/Sl^d mice and found that mast cells were defective in W/W^v mice whereas stromal cells supporting the development of mast cells were defective in Sl/Sl^d mice.[2,3] The *W* and *Sl* genes were cloned, and molecular characterization of the *W* and *Sl* gene products provided the answer for the similarity in the phenotypes of W/W^v and Sl/Sl^d mice, and identified the mechanism underlying the expression of the defect of W/W^v mice in mast cells and that of Sl/Sl^d mice in stromal cells.

The *W* locus encodes kit,[11,12] and the *Sl* locus encodes the ligand of kit.[13-16] Although many cell types express kit, only five cell types are deficient in W/W^v and Sl/Sl^d mice. Signals from kit appear to be essential for development of melanocytes, erythrocytes, germ cells, mast cells, and interstitial cells of Cajal, in contrast to some of the neurons within the cerebellum and the dorsal root ganglion that express kit but whose development can be supported by other ligand–receptor systems.[17,18]

Previous studies on mast cells have employed rats more often than mice. Our studies primarily focused on mast cell-deficient mice, but we were also able to identify mast cell-deficient rats. Dr. Kondo reported a rat mutant whose appearance was very similar to that of W/W^v and Sl/Sl^d mice. Histologically, the mutant rat lacked mast cells, and a deletion of 12 bases was found at the c-*kit* locus of this mutant.[19,20]

Survival of Mast Cells

The signals generated by the binding of SCF to KIT are essential not only for the development of mast cells but also for their survival. W^{sh} is a mutant allele of the W (c-*kit*) locus, although no abnormalities were found in the coding region of the W^{sh} allele. The W^{sh} allele was considered a mutation at the regulatory region necessary for normal transcription of the c-*kit* gene in mast cells.[21,22]

Homozygous W^{sh}/W^{sh} mice have an appreciable number of mast cells at birth, but the number drops exponentially after birth. The exponential decrease in mast cells of W^{sh}/W^{sh} mice was attributed to the abrogation of the c-*kit* transcription.[23] On day 150 after birth, the number of mast cells in the skin of W^{sh}/W^{sh} mice was only 1% that of +/+ mice, indicating the important role of the SCF-kit system for mast cell survival. Although mast cells, melanocytes, and interstitial cells of Cajal decreased remarkably in tissues of W^{sh}/W^{sh} mice, nearly normal numbers of erythrocytes and germ cells were found in W^{sh}/W^{sh} mice. This result is consistent with the fact that c-*kit* mRNA was detectable in the erythropoietic tissue (spleen) and the testis of W^{sh}/W^{sh} mice.[21] The promoter of c-*kit* used in mast cells appears to be common with the promoter used in melanocytes and interstitial cells of Cajal but not with the promoter used in erythrocytes, germ cells, and neurons.

Neoplastic Transformation of Mast Cells

Binding of SCF activates kit and leads to autophosphorylation of kit on tyrosine and to association of kit with substrates such as phosphatidylinositol 3-kinase (PI3K). In the human mast cell leukemia cell line HMC-1, kit was constitutively phosphorylated on tyrosine, activated, and associated with PI3K without the addition of SCF. Furitsu et al.[4] found that the c-*kit* gene of HMC-1 cells was composed of a normal, wild-type allele and a mutant allele with point mutations resulting in amino acid substitutions of Gly-560 for Val and Val-816 for Asp. Amino acid sequences in the regions of the two mutations are completely conserved in all of mouse, rat, and human kit.

To determine the causal role of these mutations in the constitutive activation, murine mutant c-*kit* genes encoding Gly-559 or Val-814, corresponding to human Gly-560 or Val-816, were constructed and expressed in a human embryonic kidney cell line, 293T cells. In the transfected cells, kit with the Val-814 mutation was abundantly phosphorylated on tyrosine and activated in immune complex kinase reaction in the absence of SCF, whereas tyrosine phosphorylation and activation of kit with the Gly-559 mutation or wild-type c-*kit* kinase was modest or little, respectively. The conversion of Asp-816 to Val in human kit appears to be an activation mutation and responsible for the constitutive activation of kit in HMC-1 cells.[4]

Tsujimura et al.[24,25] found the corresponding mutation in the P-815 mouse mastocytoma cell line (Asp-814 to Tyr) and the RBL-2H3 rat mast cell leukemia cell line (Asp-817 to Tyr). Both P-815 and RBL-2H3 cells show constitutive activation of kit without the addition of SCF. The substitution of the same amino acid has not been found among various loss-of-function mutants of mice, rats, and humans, but was found in the tumor mast cell lines of all mice, rats, and humans. There is a possibility that this mutation has induced mast cell neoplasms in these three species. In fact, the Asp-816 to Val mutation has been found in various types of mast cell neoplasms of human patients.[26,27]

To examine the transforming and differentiation-inducing potential of the activation mutation of the c-*kit* gene, we used the murine IL-3-dependent IC-2 mast cell line as a transfectant.[28] The IC-2 cells contained few basophilic granules and did not express kit on the surface. The Val-814 or Gly-559 mutant c-*kit* gene was introduced into IC-2 cells using a retroviral vector. The mutant kit proteins expressed in IC-2 cells were constitutively phosphorylated on tyrosine and demonstrated kinase activity in the absence of SCF. IC-2 cells expressing either the Val-814 or G-559 mutation showed factor-independent growth in suspension culture and produced tumors in nude athymic mice. In addition, IC-2 cells with the Val-814 or Gly-559 mutation showed a more mature phenotype compared with the phenotype of the original IC-2 cells, especially after transplantation into nude mice. The number of basophilic granules and the content of histamine increased remarkably. The Val-814 and Gly-559 mutations also influenced the transcriptional phenotype of mouse mast cell protease (MMCP) in IC-2 cells. The expression of MMCP-2, MMCP-4, and MMCP-6 was much greater in the IC-2 cells that expressed c-*kit* with the Val-814 or Gly-559 mutation than in the original IC-2 cells, indicating that constitutively activated kit had not only oncogenic activity but also differentiation-inducing activity in mast cells.[28]

Introduction of the c-*kit* with Val-814 or Gly-559 mutation also resulted in the malignant transformation of the Ba/F3 murine Pro-B cell line and the FDC-P1 murine myeloid cell line.[29] However, the direct effects of the c-*kit*-activating mutation on the neoplastic transformation and aberrant differentiation of normal hematopoietic cells were not clear because the IL-3-dependent hematopoietic cell lines had already acquired the ability to self-renew by unknown transformation events. Moreover, the differentiation of the experimental cell lines was not complete and did not reflect normal differentiation. Kitayama et al.[30] introduced wild-type and mutant forms of c-*kit* genes into normal bone marrow cells to investigate how an aberrantly expressed and constitutively activated kit can stimulate the in vitro growth of hematopoietic stem cells in the presence or absence of IL-3 or SCF. Moreover, the retrovirally infected bone marrow cells were injected to *W/W^v* mice to examine the development of hematopoietic malignancies. The retroviral infection of c-*kit* with the Gly-559 mutation induced development of granulocyte-

macrophage and mast cell colonies in vitro without the addition of exogenous growth factors. The c-*kit* Val-814 mutation induced the factor-independent growth of various types of hematopoietic progenitor cells, resulting in the development of mixed colonies containing erythroid cells, myeloid cells, and mast cells in addition to granulocyte-macrophage and mast cell colonies. Moreover, the transplantation of bone marrow cells with the c-*kit* Val-814 or Gly-559 mutation led to development of acute leukemia in 6 of 10 and 1 of 10 transplanted mice, respectively. No mice developed hematopoietic malignancies after transplantation of wild-type c-*kit*-infected cells. Moreover, transgenic mice expressing the c-*kit* Val-814 mutation developed acute leukemia or malignant lymphoma.[30] These results demonstrated a direct role of the activating c-*kit* mutants, particularly the Val-814 mutation, in tumorigenesis of hematopoietic cells.

Val-814 is a point mutation at the tyrosine kinase domain of the c-*kit* gene whereas Gly-559 is at the juxtamembrane domain. Chemical cross-linking analysis showed that a substantial fraction of the phosphorylated kit with the Gly-559 mutation underwent dimerization even in the absence of SCF, whereas the phosphorylated c-*kit* with the Val-814 mutation did not, suggesting that distinct mechanisms resulted in the constitutive activation of c-*kit* by the Gly-559 and Val-814 mutations. Tsujimura et al.[31] found another gain-of-function mutation at the juxtamembrane domain of FMA3 murine mastocytoma cells. The c-*kit* cDNA of FMA3 cells carried an in-frame deletion of 21 base pairs (bp). The FMA3-type c-*kit* cDNA with the 21-bp deletion was introduced into the IC-2 cell line. The FMA3-type kit was constitutively phosphorylated on tyrosine and activated. Moreover, the FMA3-type kit was dimerized without stimulation by SCF. The FMA3-type kit, which spontaneously dimerized without SCF binding, was not internalized even though it was activated, as is also the case with the Gly-559 kit. IC-2 cells expressing the FMA3-type kit grew in suspension culture without IL-3 and SCF and became leukemic in nude athymic mice. Although the Gly-559 mutation and the FMA3-type mutation were different in nature, a point mutation and a 21-bp deletion, respectively, their biological effects appeared comparable. It is likely that the normal juxtamembrane domain may inhibit the dimerization of kit, and some mutations at the juxtamembrane domain may induce SCF-independent, constitutive dimerization of kit.

MITF

Abnormalities in Mast Cells of *mi/mi* Mice

A double gene dose of mutant alleles at the *mi* locus produces the pleiotropic effects of microphthalmia, depletion of pigment in both hair and eyes, and osteopetrosis. Moreover, Stevens and Loutit[32] and Stechschulte et al.[33] reported depletion of mast cells in *mi/mi* mice.

In addition to the decrease in mast cell number, the phenotype of skin mast cells is abnormal in *mi/mi* mice. The staining property of skin mast cells changed from Alcian blue$^+$/berberine sulfate$^-$ to Alcian blue$^+$/berberine sulfate$^+$ in the skin of +/+ and W^v/W^v mice. In contrast, this change did not occur in the skin of *mi/mi* mice, in the skin of which the number of mast cells was comparable to that of W^v/W^v mice (Table 3.1).[34,35] The heparin content and histamine content per *mi/mi* skin mast cell were estimated to be 34% and 18% of those of a +/+ skin mast cell, respectively.[35] The low heparin content of *mi/mi* skin mast cells was consistent with the Alcian blue$^+$/berberine sulfate$^-$ staining property of these cells. The expression of genes encoding mast cell-specific enzymes was examined by northern blotting and in situ hybridization. Messenger (m) RNA of mast cell carboxypeptidase A was expressed by most +/+, W^v/W^v, or *mi/mi* skin mast cells, whereas mRNA of mouse mast cell protease-6 (MMCP-6) was expressed by approximately one-half of +/+ or W^v/W^v skin mast cells but by only 3% of *mi/mi* skin mast cells (Table 3.1).[35]

Osteopetrosis of *mi/mi* mice can be cured by bone marrow transplantation from +/+ donors, and its cause is attributed to a defect of osteoclasts. Both osteoclasts and mast cells

TABLE 3.1. Mean number of mast cells in the skin of various mutant mice[a].

Genotype	Mean number per 1-cm skin section[b]	
	Alcian blue$^+$	MMCP-6 mRNA$^+$
+/+	319	160
W/W	0	0
W^v/W^v	76	34
mi/mi	91	3

[a] Serial sections were made; one section was stained with Alcian blue and the other section was used for in situ hybridization to detect mRNA of mouse mast cell protease (MMCP-6).[34,35]
[b] Mean number of 6 mice.

are the progeny of multipotential hematopoietic stem cells. We started our studies of the mechanism of mast cell deficiency in *mi/mi* mice by coculturing spleen-derived cultured mast cells (CMCs) with fibroblasts. Spleen cells of *mi/mi* mice were cultured in the presence of IL-3. In spite of the depletion of mast cells in the tissues of *mi/mi* mice, CMCs did develop as described in the case of W/W^v and Sl/Sl^d mice. The response of *mi/mi* CMCs to T-cell-derived growth factors was normal, but the proliferation of *mi/mi* CMCs was not induced by coculture with fibroblasts derived from +/+ embryos.[34] Thus the mast cell deficiency of *mi/mi* mice appeared to result from the inability of *mi/mi* mast cells to respond to the proliferative stimulus presented by fibroblasts.

Fibroblasts produce SCF, which is the most important growth factor for the development of mast cells. Recombinant SCF induced a dose-dependent proliferation of +/+ CMCs. Also, SCF induced +/+ CMCs to acquire a phenotype like that of connective tissue-type mast cells (CTMCs) in medium containing IL-3. In contrast, SCF neither stimulated the proliferation of *mi/mi* CMCs nor induced a phenotypic change from immature mast cells to mature CTMC-like mast cells.[36] Immunoblotting with antiphosphotyrosine antibody showed that SCF induced considerable tyrosine phosphorylation of kit in +/+ CMCs, whereas tyrosine phosphorylation of kit was barely detectable in *mi/mi* CMCs.[36] Northern blot and flow cytometry analysis showed that *mi/mi* CMCs expressed significantly less c-*kit* at both the protein and mRNA levels than +/+ CMCs. Moreover, *mi/mi* CMCs were found to differ from +/+ CMCs in the expression of MMCP-6 mRNA, as in the case of *mi/mi* skin mast cells. These results suggest that the gene product of the *mi* locus may be important in regulating the expression of gene products such as kit and MMCP-6, and that the mast cell deficiency of *mi/mi* mice appears to be caused by impaired signaling through kit.

Jippo-Kanemoto et al.[37] described another abnormality of *mi/mi* CMCs. Conditioned medium (CM) of BALB/3T3 fibroblasts induced the migration of +/+ CMCs but not of *mi/mi* CMCs. However, *W/W* CMCs, which did not express KIT on the surface, did migrate normally toward BALB/3T3 CM. An antibody to the extracellular domain of kit did not inhibit the migration of +/+ CMCs toward BALB/3T3 CM. Although BALB/3T3 CM contained SCF, SCF itself did not appear to represent the major chemoattractive activity. The mechanism of the deficient migration of *mi/mi* CMCs toward BALB/3T3 CM remains to be clarified.

Transactivation of Mast Cell-Related Genes by MITF

Hodgkinson et al.[38] and Hughes et al.[39] demonstrated that the *mi* locus of mice encodes a novel member of the basic helix-loop-helix-leucine zipper (bHLH-Zip) protein family of transcription factors. Although c-*kit* gene expression by CMCs from *mi/mi* mice is deficient at both the mRNA and protein levels, it remained to be examined whether c-*kit* gene expression was also deficient in mast cells in the tissues of *mi/mi* mice. We examined c-*kit* gene expression by *mi/mi* skin mast cells using in situ hybridization and immunohistochemistry. Moreover, we also examined c-*kit* expression by various cells other than mast cells in the tissues of *mi/mi* mice. We found that c-*kit* gene expression was deficient in mast cells but not in erythroid precursors, testicular germ cells, and neurons of *mi/mi* mice.[40] This result suggested that the regulation of the c-*kit* gene transcription by MITF was cell type dependent.

We investigated the effect of MITF on the transcription of the c-*kit* gene. First, we intro-duced cDNA encoding normal (+) MITF or mutant (*mi*) MITF into *mi/mi* CMCs using a retroviral vector. Overexpression of +-MITF but not *mi*-MITF normalized the expression of the c-*kit* gene and the poor response of *mi/mi* CMCs to SCF, indicating the involvement of +-MITF in c-*kit* gene transactivation.[5] Second, we analyzed the promoter of the c-*kit* gene. Three CANNTG motifs recognized by bHLH-Zip-type transcription factors were conserved between the mouse and human c-*kit* promoters. Among these three CANNTG motifs, only a CACCTG motif was specifically bound by +-MITF. When the luciferase gene under the control of c-*kit* promoter was contransfected into NIH/3T3 fibroblasts with cDNA encoding +-MITF or *mi*-MITF, the luciferase activity significantly increased only when +-MITF cDNA was cotransfected. The deletion of the promoter region containing the CACCTG motif or the mutation of the CACCTG to CTCCAG abolished the transactivation effect of +-MITF, indicating that +-MITF transactivated the c-*kit* gene through the CACCTG motif.[5] When the luciferase gene under the control of the c-*kit* promoter was introduced into FMA3 mastocytoma and FDC-P1 myeloid cell lines, remarkable luciferase activity was observed only in FMA3 cells.[5] The involvement of +-MITF in the c-*kit* transactivation appeared to be specific to the mast cell lineage.

Because the expression of the MMCP-6 gene is remarkably reduced in both skin mast cells and CMCs of *mi/mi* mice, we investigated the effect of MITF on the transcription of the MMCP-6 gene.[6] First, we introduced the cDNA encoding +-MITF or *mi*-MITF into *mi/mi* CMCs using a retroviral vector. Overexpression of +-MITF but not *mi*-MITF normalized the expression of the MMCP-6 gene, indicating the involvement of +-MITF in the MMCP-6 gene transactivation. Second, we analyzed the promoter of the MMCP-6 gene in a transient cotransfection assay. The luciferase construct under the control of the MMCP-6 promoter and the cDNA encoding +- or *mi*-MITF were cotransfected into NIH/3T3 fibroblasts. The coexpression of +-MITF but not of *mi*-MITF increased the luciferase activity 10 fold. We found both CACATG and

CATCTG motifs in the MMCP-6 promoter, both of which are generally recognized by bHLH-Zip-type transcription factors. We also found a GACCTG motif that was strongly bound by +-MITF. These three motifs were necessary for the 10-fold transcription activity of the MMCP-6 promoter by +-MITF. Mutations of each motif significantly reduced the transactivation, suggesting that +-MITF directly transactivated the MMCP-6 gene through these three motifs.[6]

The *mi* mutant allele represents a deletion of an arginine in the basic domain of MITF. With this basic domain, bHLH-Zip type transcription factors bind DNA, and the *mi*-MITF did not bind DNA because of the deletion.[41] Takebayashi et al.[42] produced anti-MITF antibody and found another abnormality of *mi*-MITF. Immunocytochemistry and immunoblotting revealed that more than 99% of +-MITF located in the nuclei, whereas *mi*-MITF was predominantly located in the cytoplasm. When immunoglobulin G (IgG)-conjugated peptides representing a part of the DNA-binding domain of +-MITF or *mi*-MITF were microinjected into the cytoplasm of fibroblasts, wild-type peptide–IgG conjugate localized in nuclei, but *mi*-type peptide–IgG conjugate was detectable only in the cytoplasm. Thus, in addition to having deficient DNA-binding activity, *mi*-MITF is also deficient in nuclear localization potential.

Poor Response of *mi/mi* Mast Cells to Nerve Growth Factor

The addition of nerve growth factor (NGF) to a suboptimal dose of IL-3 increased the plating efficiency of +/+ CMCs,[43] but not of *mi/mi* CMCs.[44] Although +/+ CMCs were Alcian blue[+]/ berberine sulfate[-] when cultured with IL-3 alone, +/+ CMCs became Alcian blue[+]/ berberine sulfate[+] when cultured in the presence of both IL-3 and NGF, which suggested the increased heparin content. In contrast, NGF did not influence the phenotype of *mi/mi* CMCs. +/+ CMCs significantly bound ^{125}I-NGF, but *mi/mi* CMCs did not, suggesting a defect of NGF receptors in *mi/mi* CMCs.[44] Both p75 and p140 molecules are known to be involved in the formation of the NGF receptor. Although the expression of p140 mRNA was comparable between +/+ and *mi/mi* CMCs, the expression of p75 mRNA was significantly lower in *mi/mi* CMCs than in +/+ CMCs.[44] Therefore, the poor response of *mi/mi* CMCs to NGF appeared to be attributable to the impaired transcription of the p75 gene. In fact, introduction of cDNA encoding +-MITF into *mi/mi* CMCs normalized both the expression of p75 mRNA and the poor responses of *mi/mi* CMCs to NGF. In the 5′-upstream region of the p75 gene, there is a CACTTG motif that was bound by +-MITF but not by *mi*-MITF. The binding of the +-MITF transactivated the luciferase gene under the control of the p75 promoter (Jippo et al., unpublished data).

Conclusions

Kit and MITF appear to represent key molecules for development of mast cells. There is a functional correlation between kit and MITF; MITF is necessary for transcription of kit in mast cells. Although activation of kit by binding of SCF may activate some transcription factors for transduction of signals to the nucleus, such transcription factors have not yet been identified. On the other hand, there is a possibility that some growth factors may control the production and function of MITF, but such growth factors have not yet been identified. To advance our understanding of mast cell development, it will be important to clarify the cascade of molecules that contain kit and MITF as critical components.

References

1. Orkin SH. GATA-binding transcription factors in hematopoietic cells. Blood 1992;80:575–581.
2. Kitamura Y, Go S. Decreased production of mast cells in *Sl/Sl^d* anemic mice. Blood 1979;53:492–497.
3. Kitamura Y, Go S, Hatanaka K. Decrease of mast cells in *W/W^v* mice and their increase by bone marrow transplantation. Blood 1987; 52:447–452.
4. Furitsu T, Tsujimura T, Tono T, et al. Identification of mutations in the coding region of the

proto-oncogene c-*kit* in a human mast cell leukemia cell line causing ligand-independent activation of c-*kit* product. J Clin Invest 1993;92:1736–1744.

5. Tsujimura T, Morii M, Nozaki E, et al. Involvement of transcription factor encoded by the *mi* locus in the expression of c-*kit* receptor tyrosine kinase in cultured mast cells of mice. Blood 1996;88:1225–1233.

6. Morii E, Tsujimura T, Jippo T, et al. Regulation of mouse mast cell protease 6 gene expression by transcription factor encoded by the *mi* locus. Blood 1996;88:2488–2494.

7. Russel ES. Hereditary anemias of the mouse: a review for geneticists. Adv Genet 1979;20:357–459.

8. Maeda H, Yamagata A, Nishikawa S, et al. Requirement of c-*kit* for development of intestinal pacemaker system. Development (Camb) 1992;116:369–375.

9. Huizinga JD, Thuneberg L, Klüppel M, et al. *W/kit* gene required for interstitial cells of Cajal and for intestinal pacemaker activity. Nature (Lond) 1995;373:347–349.

10. Isozaki K, Hirota S, Nakama A, et al. Disturbed intestinal movement, bile reflex to the stomach, and deficiency of c-*kit*-expressing cells in *Ws/Ws* mutant rats. Gastroenterology 1995;109:456–464.

11. Chabot B, Stephenson DA, Chapman VM, et al. The proto-oncogene c-*kit* encoding a transmembrane tyrosine kinase receptor maps to the mouse *W* locus. Nature (Lond) 1988;335:88–89.

12. Geissler EN, Ryan MA, Houseman DE. The dominant white spotting (*W*) locus of the mouse encodes the c-*kit* proto-oncogene. Cell 1988;55:185–192.

13. Williams DE, Eisenman J, Barid A, et al. Identification of a ligand for the c-*kit* proto-oncogene. Cell 1990;63:167–174.

14. Flanagan JG, Leder P. The *kit* ligand: a cell surface molecule altered in steel mutant fibroblasts. Cell 1990;63:185–194.

15. Zsebo KM, Williams DA, Geissler EN, et al. Stem cell factor is encoded at the *Sl* locus of the mouse and is the ligand for the c-*kit* tyrosine kinase receptor. Cell 1990;63:213–224.

16. Huang E, Nocka K, Beier DR, et al. The hematopoietic growth factor KL is encoded by the *Sl* locus and is the ligand of the c-*kit* receptor, the gene product of the *W* locus. Cell 1990;63:225–233.

17. Morii E, Hirota S, Kim HM, et al. Spacial expression of genes encoding c-*kit* receptors and their ligands in mouse cerebellum as revealed by in situ hybridization. Dev Brain Res 1992; 65:123–126.

18. Hirata T, Morii E, Morimoto M, et al. Stem cell factor induces outgrowth of c-*kit*-positive neurites and supports the survival of c-*kit*-positive neurons in dorsal root ganglia of mouse embryos. Development (Camb) 1993; 119:49–56.

19. Niwa Y, Kasugai T, Ohno K, et al. Anemia and mast cell depletion in mutant rats that are homozygous at "White spotting (*Ws*)" locus. Blood 1991;78:1936–1941.

20. Tsujimura T, Hirota S, Nomura S, et al. Characterization of *Ws* mutant allele of rats: A 12 base deletion in tyrosine kinase domain of c-*kit* gene. Blood 1991;78:1942–1946.

21. Tono T, Tsujimura T, Koshimizu U, et al. Deficient transcription of c-*kit* gene in cultured mast cells of W^{sh}/W^{sh} mice that have nearly normal number of erythrocytes and normal c-*kit* coding region. Blood 1992;80:1448–1453.

22. Duttlinger RS, Manova K, Berrozpe G, et al. The W^{sh} and *Ph* mutations affect the c-*kit* expression profile: c-*kit* misexpression in embryogenesis impairs melanogenesis in W^{sh} and *Ph* mutant mice. Proc Natl Acad Sci USA 1995;92:3754–3758.

23. Yamazaki M, Tsujmura T, Isozaki K, et al. c-*kit* gene is expressed by skin mast cells in embryos but not in puppies of W^{sh}/W^{sh} mice: age-dependent abolishment of c-*kit* gene expression. Blood 1994;83:3509–3516.

24. Tsujimura T, Furitsu T, Morimoto M, et al. Ligand-independent activation of c-*kit* receptor tyrosine kinase in a murine mastocytoma cell line P-815 generated by a point mutation. Blood 1994;83:2619–2626.

25. Tsujimura T, Furitsu T, Morimoto M, et al. Substitution of an aspartic acid results in constitutive activation of c-*kit* receptor tyrosine kinase in a rat tumor mast cell line RBL-2H3. Int Arch Allergy Immunol 1995;106:377–385.

26. Nagata H, Worobec AS, Oh CK, et al. Identification of a point mutation in the catalytic domain of the protooncogene c-*kit* in peripheral blood mononuclear cells of patients who have mastocytosis with an associated hematologic disorder. Proc Natl Acad Sci USA 1995;92:10560–10564.

27. Longley BJ, Tyrrell L, Lu SZ, et al. Somatic c-*kit* activating mutation in urticaria pigmentosa and aggressive mastocytosis: establishment of clonality in a human mast cell neoplasm. Nat Genet 1996;12:312–314.

28. Hashimoto K, Tsujimura T, Moriyama Y, et al. Transforming and differentiation-inducing potential of constitutively activated c-*kit* mutant genes in the IC-2 murine interleukin-3-dependent mast cell line. Am J Pathol 1996; 148:189–200.

29. Kitayama H, Kanakura Y, Furitsu T, et al. Constitutively activating mutations of c-*kit* receptor tyrosine kinase confer factor-independent hematopoietic cell lines. Blood 1995;85:790–798.

30. Kitayama H, Tsujimura T, Matsumura I, et al. Neoplastic transformation of normal hematopoietic cells by constitutively activating mutations of c-*kit* receptor tyrosine kinase. Blood 1996;88:995–1004.

31. Tsujimura T, Morimoto M, Hashimoto K, et al. Constitutive activation of c-*kit* in FMA3 murine mastocytoma cells caused by deletion of seven amino acids at the juxtamembrane domain. Blood 1996;87:273–283.

32. Stevens J, Loutit JF. Mast cells in spotted mutant mice (*W, Ph, mi*). Proc R Soc Lond 1982; 215:405–409.

33. Stechschulte DJR, Sharma KN, Dileepan KM, et al. Effect of the *mi* allele on mast cells, basophils, natural killer cells, and osteoclasts in C57BL/6J mice. J Cell Physiol 1987;132:565–570.

34. Ebi Y, Kasugai T, Seino Y, et al. Mechanism of mast cell deficiency in mutant mice of *mi/mi* genotype: an analysis by co-culture of mast cells and fibroblasts. Blood 1990;75:1247–1251.

35. Kasugai T, Oguri K, Jippo-Kanemoto T, et al. Deficient differentiation of mast cells in the skin of *mi/mi* mice: usefulness of in situ hybridization for evaluation of mast cell phenotype. Am J Pathol 1993;143:1337–1347.

36. Ebi Y, Kanakura Y, Jippo-Kanemoto T, et al. Low c-*kit* expression of cultured mast cells of *mi/mi* genotype may be involved in their defective responses to fibroblasts that express the ligand for c-*kit*. Blood 1992;80:1454–1462.

37. Jippo-Kanemoto T, Adachi S, Ebi Y, et al. BALB/3T3 fibroblast conditioned medium attracts cultured mast cells derived from *W/W* but not from *mi/mi* mutant mice, both of which are deficient in mast cells. Blood 1992;80:1933–1936.

38. Hodgkinson CA, Moore KJ, Nakayama A, et al. Mutations at the mouse microphthalmia locus are associated with defects in a gene encoding a novel basic helix-loop-helix-zipper protein. Cell 1993;74:395–404.

39. Hughes JJ, Lingrel JB, Krakowsky JM, et al. A helix-loop-helix transcription factor-like gene is located at the *mi* locus. J Biol Chem 1993;268:20687–20690.

40. Isozaki K, Tsujimura T, Nomura S, et al. Cell type-specific deficiency of c-*kit* gene expression in mutant mice of *mi/mi* genotype. Am J Pathol 1994;145:827–836.

41. Morii E, Takebayashi K, Motohashi H, et al. Loss of DNA binding ability of the transcription factor encoded by the mutant *mi* locus. Biochem Biophys Res Commun 1994;205:1299–1304.

42. Takebayashi K, Chida K, Tsukamoto I, et al. The recessive phenotype displayed by a dominant negative microphthalmia-associated transcription factor mutant is a result of impaired nuclear localization potential. Mol Cell Biol 1996;16:1203–1211.

43. Matsuda H, Kannan Y, Ushio H, et al. Nerve growth factor induces development of connective tissue-type mast cells in vitro from murine bone marrow cells. J Exp Med 1991;174:7–14.

44. Jippo T, Ushio H, Hirota S, et al. Poor response of cultured mast cells derived from *mi/mi* mutant mice to nerve growth factor. Blood 1994; 84:2977–2983.

4
Mouse and Rat Models of Mast Cell Development

Guang W. Wong, Daniel S. Friend, and Richard L. Stevens

Based on histochemical and morphological properties, Enerbäck concluded in 1966 that the mast cells (MC) residing in the jejunal mucosa of rats at the height of helminth infection are phenotypically different from the MC residing in the skin of uninfected animals.[1] The biochemical basis for these variations was subsequently shown to be the different neutral proteases[2-4] and proteoglycans[5,6] stored in the secretory granules of the MC. Because hybrid MC with different stainable granules are not routinely seen in vivo, it was originally thought that mucosal and cutaneous MC are developmentally distinct.

The first indication that the granule differentiation of a mature MC was not fixed occurred in 1982 with the demonstration that p-nitrophenyl-β-D-xyloside could induce peritoneal MC to rapidly switch from the biosynthesis of heparin to the biosynthesis of chondroitin sulfate E.[7] Adoptive transfer experiments of unfractionated bone marrow cells,[8] isolated peritoneal MC,[9] and cultured mouse bone marrow-derived MC (mBMMC)[10,11] into MC-deficient WBB6F$_1$-W/W^v mice also suggested that certain populations of MC are developmentally related and originate from a FcεRI^{lo}/Thy-1^{lo}/c-kit^{hi} (FcεRI, high-affinity Fc receptor for IgE) progenitor[12] that has few granules.[13]

The purification of granule proteins,[14-16] the cloning of cDNAs that encode 10 different mouse MC proteases,[17-27] and the generation of protease-specific antibodies[27-32] provided investigators with additional reagents to uncover the complexities of MC heterogeneity and even the fate of a MC that has outlived its usefulness. In vitro and in vivo studies carried out during the past decade have revealed that the eventual phenotype of a mouse MC is dependent on both the current and previous microenvironments in which the mature cell and its progenitor reside. The ability of the MC to acutely alter its phenotype has important functional implications and indicates that a nomenclature based on a fixed phenotype is no longer appropriate for this versatile cell type. Some of the rat and mouse models that have been used to assess MC development are reviewed here.

Identification of the Major Granule Constituents of Mouse MC

As assessed by N-terminal amino acid analysis of granule proteins resolved by sodium dodecyl sulfate-polyacrylamide gel electrophoresis, 26- to 36-kDa proteases that are enzymatically active at neutral pH represent the major protein constituents of the secretory granules of all mouse MC.[14-16] Nine serine proteases [designated mouse MC protease (mMCP) -1 to mMCP-9] that are derived from distinct genes have been identified in the mouse.[17-27] Based on their relative similarities to pancreatic chymotrypsin and trypsin, mMCP-6[21,24] and mMCP-7[22,24] are tryptases; the other mMCPs are chymases. The genes that encode all but mMCP-3[16] have been cloned and sequenced.

Each chymase gene is approximately 2.5 kb in size, consists of 5 exons, and resides on chromosome 14[26,27] at the complex that also contains the genes that encode cathepsin G and granzymes B–F.[33–35] Some of these genes are close together on the chromosome. For example, the mMCP-1 and mMCP-9 genes are separated only by about 7 kb of flanking DNA.[27]

All chymases are synthesized as zymogens containing 226 to 247 amino acids with an 18- or 19-residue signal peptide and a Glu/Gly-Glu activation peptide. mMCP-5[20] exhibits the greatest dissimilarity within the mouse chymase family; it is the mMCP that is most homologous to human mast cell chymase. Crystallographic analysis of purified rat MCP-II[36] and comparative protein modeling analysis of the varied mouse MC chymases[37] revealed that each member of the chromosome 14 family of serine proteases possesses structural features near the active site that restricts its substrate specificity.

The genes that encode the two tryptases, mMCP-6 and mMCP-7, reside on chromosome 17.[26,38] Although both tryptase genes are less than 2.5 kb in size, the mMCP-7 gene contains 1 less exon than the mMCP-6 gene. A comparison of the 5′-end of the transcript with the genomic sequence indicated that the region corresponding to the first intron in the mMCP-6 gene is not spliced during transcription of the mMCP-7 gene because of a point mutation at the intron 1 acceptor splice site. The putative propeptides of mMCP-6 and mMCP-7 consist of 10 amino acids each and are 60% identical. They begin with an Ala-Pro sequence and end with a Gly residue. Mature mMCP-6 and mMCP-7 share 71% sequence identity but possess very different substrate-binding pockets.[24,31] Using a novel bioengineering strategy, both mMCP-6 and mMCP-7 have been expressed in insect cells to help deduce their physiological substrates. Analysis of a tryptase-specific, phage-display peptide library with these recombinant tryptases revealed that the preferred amino acid sequences cleaved by mMCP-6 are very different from those cleaved by mMCP-7.[133] In vitro studies carried out with recombinant mMCP-7 and in vivo studies performed on the V3 mastocytosis mouse revealed that fibrinogen is a physiological substrate of mMCP-7.[39] The physiological substrate(s) of mMCP-6 remains to be determined, but recombinant mMCP-6 can induce neutrophil extravasation and accumulation in tissues by inducing endothelial cells to selectively increase their expression of the interleukin (IL) -8 family of chemokines.[133]

The mouse MC carboxypeptidase A (mMC-CPA) gene resides on chromosome 3, is greater than 30 kb in size, and consists of 11 exons.[17,26] mMC-CPA is translated as a zymogen containing a 15-residue signal peptide and a 94-residue activation peptide. Although its propeptide is quite large, the amino acid sequences that are cleaved to activate mMC-CPA and the chymases are nearly identical. Thus, the same processing mechanism probably is used in the posttranslational modification of these different granule proteases. After removal of the signal and activation peptides, mature mMC-CPA has a molecular weight of about 36 kDa. Because mMCP-5 and mMC-CPA are cordinately expressed in varied populations of MC[16,17,20] and because the disruption of the mMCP-5 gene affects the granule expression of mMC-CPA,[40] it has been proposed that these two proteases form a binary complex inside the granule so that they can sequentially degrade common protein substrates after their exocytosis.

Serglycin proteoglycans represent the other major constituent of the MC secretory granule.[5,6,41–46] Because of their high content of sulfate and carboxylic acid, serglycin proteoglycans are the most negatively charged molecules in the body. The serglycin gene encodes the common peptide core of this family of proteoglycans. Depending on the population of MC, a specific type of glycosaminoglycan (e.g., heparin, chondroitin sulfate E, or other highly sulfated chondroitin) is chosen for covalent attachment to the Ser-Gly repeat region of the peptide core. At the granule pH of about 5.5,[47] mMC-CPA and all mMCPs appear to be ionically bound to the glycosaminoglycan side chains of serglycin proteoglycans. Comparative protein modeling and electrostatic calculations disclosed that each mMCP contains a prominent Lys/Arg/His-rich domain on its surface away from the active site that enables these proteases to ionically bind to serglycin proteoglycans.[31,37,48] When MC degranulate, most of the exocytosed proteoglycan/protease macromolecular complexes stay intact outside of the MC for hours. mMCP-7 is an exception because its proteoglycan-binding domain consists primarily of His residues rather than Lys and Arg residues.[48] mMCP-7 loses most of its positive charge after its release into a pH 7.0 microenvironment, thereby enabling it to dissociate from the exocytosed protease/proteoglycan complexes. It quickly makes its way into the blood, where it circulates for more than 1 h in its mature, enzymatically active state.[31]

Granule Characterization of Varied Populations of MC in Normal Mice

Because mouse MC express different combinations of at least 10 different neutral proteases, protease expression has become the primary means of assessing MC heterogeneity in this species. The MC that reside in the spleen of the healthy BALB/c mouse express all known granule proteases except mMCP-9.[27,49] In contrast, the MC that reside in the uterus preferentially express mMCP-4, mMCP-5, and mMCP-9,[27] whereas those that reside in the peritoneal cavity preferentially express mMCP-4, mMCP-5, mMCP-6, and mMC-CPA.[16–21] Although histochemically identical to serosal MC, the MC that reside in the skin and ears also express mMCP-7.[30,50] This diversity in protease expression argues against distinct subclasses of MC that are irreversibly programmed to express a specific granule phenotype. Rather, the local tissue microenvironment appears to play a major role in dictating the eventual phenotype of this cell.

Strain-dependent differences compound the issue of MC heterogeneity. For example, the WBB6F$_1$-+/+ mouse differs from the BALB/c mouse in that its serosal, ear, and skin MC additionally express mMCP-2.[50] In contrast, the

C57BL/6 mouse differs from most other strains in that its ear and skin MC do not express mMCP-7.[30] The genetic basis for the strain-dependent expression of mMCP-2 remains to be determined, but the C57BL/6 mouse cannot express mMCP-7 because its gene possesses a point mutation in its exon 2/intron 2 splice site that causes rapid nonsense-mediated degradation of the defective transcript.[51] The functional significance of strain-dependent expression of proteases in tissues has not been determined. However, the BALB/c mouse differs considerably from the C57BL/6 mouse in its susceptibility to infection and in its immunological response to a variety of inflammatory agents.[52,53]

Helminth-Induced Mastocytosis in the Jejunum

Because mice and rats that have been infected with a helminth such as *Trichinella spiralis* or *Nippostrongylus brasiliensis* experience a transient, but pronounced, T-lymphocyte-dependent mastocytosis in their jejunal mucosa,[32,54–58] helminth infection models have been particularly useful for investigating MC development in vivo during inflammatory reactions. As assessed by histochemistry and enzyme cytochemistry, MC first appear in the submucosa of the helminth-infected animal. They remain confined to the crypt area of the mucosa for the first 7 days, but at the height of the infection (generally at week 2) the number of MC in the mucosa increases more than 25 fold, and many MC are found between the epithelial cells in the tips of the villi. During the 2-week recovery phase of the disease, MC slowly disappear from the mucosal epithelium and lamina propria. By 4 weeks, the remaining MC are located primarily in the crypts and the lower halves of the villi. One to 2 weeks after BALB/c mice have been infected with *T. spiralis* or *N. brasiliensis*, the jejunum contains high levels of mMCP-1 and mMCP-2 but not mMCP-4 or mMCP-5 proteins and transcripts.[15,18,20,29,32,57–59] These data indicate that most, if not all, of the MC that reside in the jejunal mucosa of the helminth-infected mouse are phenotypically different

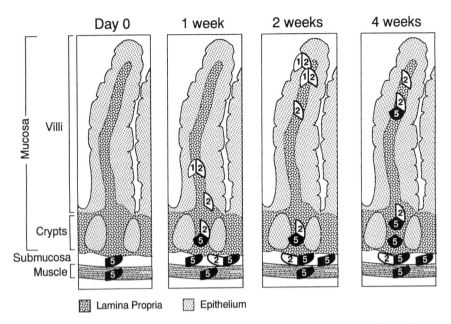

FIGURE 4.1. Chymase expression in mouse jejunal mast cells (MC) during helminth infection. The numbers *1*, *2*, and *5* in this figure, which was modified from that in Friend et al.,[32] refer to the chymases (mMCP, mouse MC protease) mMCP-1, mMCP-2, and mMCP-5, respectively. Although the figure does not portray the quantitative relationships of the different types of MC at a particular location, it shows schematically which of the three chymases these MC express during the varied stages of the helminth infection. mMCP-9 is not expressed in the jejunal MC of noninfected mice (data not shown). However, during the recovery phase of the helminth infection, most MC that reside in the submucosa and lower portion of the lamina propria express this chymase.

from the MC that reside in the spleen, peritoneal cavity, and skin of normal, noninfected mice.

Subsequent morphological, histochemical, immunohistochemical, and ultrastructural analyses conducted on the MC present in various regions of the jejunum during the mastocytosis that develops and then subsides in *T. spiralis*-infected BALB/c mice revealed that jejunal MC sequentially and reversibly alter their granule morphology and protease expression in response to the onset and subsidence of the helminth-induced inflammation.[32,60] By examining the stellate-shaped granules of intraepithelial and lamina propria MC during the height and resolution phases of the infection, Friend and coworkers concluded that at least some mature MC migrate through the varied strata of the jejunum.

Immunohistochemical analyses of serial sections disclosed that jejunal MC acquire distinct patterns of protease expression as they migrate in this tissue. In terms of their expression of three chymases (Fig. 4.1), most of the MC located in the muscle express mMCP-5 without mMCP-1 or mMCP-2, whereas most of the MC located in the epithelium express mMCP-1 and mMCP-2 without mMCP-5.[32] However, accompanying these two populations are transitional forms in the submucosa that express mMCP-5 and mMCP-2 without mMCP-1, and in the lamina propria that express mMCP-2 alone. These findings suggest that jejunal MC express mMCP-2, cease expressing mMCP-5, and then express mMCP-1 as they appear progressively in the submucosa, lamina propria, and epithelium, respectively. In the recovery phase of *T. spiralis* infection, MC in the BALB/c mouse cease expressing mMCP-1, express mMCP-5, and finally cease expressing mMCP-2 as they present in the epithelium at the tips of the villi, the lamina propria at the base of the villi, and the submucosa, respectively. Recent studies have revealed that the majority of the MC that reside in the lower villus and submucosa of the mouse during the recovery phase of the helminth infection also express the highly restricted chymase mMCP-9.[60]

Reversible changes in tryptase expression also occur in jejunal MC during the various stages of the mastocytosis.[60] Early in the inflammation, the MC in the submucosa express both mMCP-6 and mMCP-7. However, as these cells move into the tips of the villi, they sequentially cease expressing mMCP-7 and then mMCP-6. In the recovery phase of the disease, many of the MC in the mucosa reacquire their expression of mMCP-6 or mMCP-7. The cumulation of these findings highlights the complexity of MC heterogeneity and indicates that a mature MC is capable of undergoing reversible changes in its granule phenotype as it migrates through different compartments of a tissue during an inflammatory response. Immunohistochemical analyses of human tissues have revealed that occasionally MC that differ in their chymase expression can be found near each other in the same tissue microenvironment.[61] Although this finding and others led to the concept that protease expression in the human MC is predetermined before the progenitor cell leaves the bone marrow, data from a recent characterization study of the transformed but clonal MC present in varied tissue sites of a human mastocytosis patient[62] indicate that the granule protease phenotype of a human MC probably also is not fixed in vivo. The recent studies of the phenotypic changes in jejunal MC in the mouse explain how two MC in the same environment can be phenotypically different. Because it takes time for MC to catabolize their older granules, the former location of the MC as well as its current location must be considered.

The mMCP-specific antibodies enabled Friend and coworkers to address what is the ultimate fate of a MC during an inflammatory reaction.[60] It had been known for some time that MC slowly disappear from the jejunum during the recovery phase of the helminth infection. While it had been assumed that these MC underwent apoptosis locally, substantial numbers of apoptotic MC were not observed in the jejunum. Rather, the number of MC increased substantially in the spleen precisely when the

MC began to disappear from the jejunum. Not only were these splenic MC mature, they were not dividing. Although the MC in the jejunum and spleen of noninfected mice do not express mMCP-9,[27] it was discovered that more than 50% of the MC in the jejunal lamina propria and almost all the MC in the jejunal submucosa and spleen during the recovery phase of the mastocytosis expressed this highly restricted chymase.[60] The observation that mMCP-9+ MC can be found occasionally in splenic blood vessels now indicates that the spleen is a repository for some tissue MC that have outlived their usefulness. Because a jejunal MC can travel through the blood to the spleen, the presence of a mature metachromatic cell in the blood of a mouse during an inflammatory reaction gives no insight as to whether this cell is a basophil or a MC.

mBMMC

Cytokine-Developed mBMMC

Probably the most significant advance in the study of MC has been the development of in vitro techniques to obtain pure populations of nontransformed and transformed rodent MC. In 1963, Ginsburg and Sachs[63] derived a mixed population of mouse MC by culturing dispersed lymph node cells in T-lymphocyte-derived conditioned medium in the presence of a primary monolayer of mouse embryonic cells that morphologically resembled fibroblasts. Then in 1981, several laboratories simultaneously derived more convenient and reproducible methods of obtaining pure populations of nontransformed mouse MC by culturing progenitor cells from bone marrow, fetal liver, or lymph node tissue for about 3 weeks in the presence of T-lymphocyte-derived conditioned medium or WEHI-3 cell conditioned medium.[64–68] Haig and coworkers[69] generated similar populations of T-lymphocyte-dependent MC from rat progenitors. The essential factor in these conditioned media that promotes the growth and differentiation of committed progenitors into immature MC is IL-3.[70,71]

In vitro-differentiated mBMMC resemble those MC found in the jejunal epithelium in terms of histochemistry, IL-3 dependence, and arachidonic acid metabolism. Nevertheless, mBMMC possess many vacuolated granules, low levels of histamine, and low levels of serine protease enzymatic activity. In addition, they contain low to undectable levels of the transcripts that encode mMCP-1 and mMCP-2,[18,29,59] but high levels of the transcripts that encode mMCP-5, mMCP-6, mMCP-7, and mMC-CPA.[16,17,20,22] Although mBMMC in IL-3-enriched conditioned medium can maintain their high levels of the mMCP-5, mMCP-6, and mMC-CPA transcripts for many weeks of culture, the steady-state level of the mMCP-7 transcript progressively decreases the longer these cells are maintained during the cultured period.[22] Even though IL-3-developed mBMMC continue to be thought of by some investigators as the in vitro analog of mucosal MC, they clearly are not identical to the MC that reside in the jejunal mucosa at the height of helminth infection. This realization emphasizes the fact that MC cannot be categorized on the basis of histochemistry or the expression of a single phenotypic marker.

IL-3-developed mBMMC are not phenotypically analogous to any population of mature MC so far identified in the BALB/c mouse, but the in vitro culture techniques developed by Razin[64,71] and others[29,59,65–68] have permitted the generation of large numbers of nontransformed mouse MC for study. When mBMMC developed from WBB6F$_1$-+/+ mice are adoptively transferred into MC-deficient WBB6F$_1$-W/W^v mice, the transferred MC eventually resemble histochemically the endogenous MC of the WBB6F$_1$-+/+ mouse at specific sites.[10,11] Although treated WBB6F$_1$-W/W^v mice only partially reconstitute their cutaneous MC,[72] the adoptive transfer experiments of Kitamura and coworkers[10] demonstrated that mBMMC generated from WBB6F$_1$-+/+ mice can change their histochemical phenotype and give rise to different populations of MC in vivo.

Because in vitro-differentiated, nontransformed, and transformed MC have substantially more mRNA per cell than in vivo-differentiated MC[73] and because they can be obtained free of other cell types, culture-derived MC are more suitable than mature,

in vivo-differentiated cells for the identification and characterization of MC-specific genes. mBMMC, for example, have been used to clone the cDNAs and genes that encode plasma membrane glycoproteins,[74] serglycin,[45] and mMCP-5.[20]

The study of the microenvironmental regulation of MC heterogeneity can be approached by determining the accessory factors needed for the induction of MC-specific genes in cultured cells. When cultured in the presence of c-kit ligand (KL; also known as stem cell factor), IL-3-developed BALB/c mBMMC continue to express mMCP-5, mMCP-6, mMCP-7, and mMC-CPA but also express mMCP-4 and preferentially synthesize heparin-containing serglycin proteoglycans.[75,76] In contrast, they additionally express mMCP-1 and mMCP-2 when cultured in the presence of either IL-9[77] or IL-10.[29,59,78] The expression of mMCP-1 and mMCP-2 by the MC that reside in the jejunal epithelium at the height of infection is consistent with the production of IL-10 by Th2 lymphocytes during helminth infections, and the expression of mMCP-4 by cutaneous MC is consistent with the production of KL by fibroblasts.[79]

The observations that IL-3, IL-4, and glucocorticoids can suppress the differentiation-promoting effects of IL-9, IL-10, and KL enabled investigators to determine whether the granule phenotype of a nontransformed MC can be reversibly altered in vitro.[29,59,75,77,78,80] mBMMC contain high levels of mMCP-2 mRNA 24 h after their exposure to IL-10 but substantial amounts of mMCP-2 protein are not detected in these cells unless they are cultured in the presence of IL-10 for about 7 days.[29] The level of mMCP-2 mRNA in IL-10-treated mBMMC decreases dramatically 24 h after the IL-10 is removed from the culture medium, but approximately 5 days are required before the level of mMCP-2 protein declines. Thus, although mBMMC can acutely change their protease expression at the mRNA level, the fact that it takes a number of days for a mBMMC to completely metabolize its older granules emphasizes that the phenotype of a MC at any moment in its life span is also a consequence of its previous microenvironment.

With transient transfection approaches, *cis*-acting elements have been identified in the 5'-flanking regions of the rat MCP-II,[81] mMC-CPA,[82] mMCP-5,[83] and mouse serglycin[84] genes that control their expression in MC. For the mMC-CPA gene, the GATA family of transcription factors appears to be essential for the regulated expression of this granule protease. It is possible that differential rates of transcription contribute somewhat to the overall steady-state levels of cytokine-inducible chymase transcripts. However, recent studies have revealed that chymase expression in MC is also controlled by a posttranscriptional mechanism.[80,85] Although mBMMC developed in IL-3 do not contain high levels of mMCP-1, mMCP-2, and mMCP-4 transcripts in their cytoplasm, nuclear run-on analysis disclosed that the three chymase genes are transcribed at high rates in these cultured cells. Pulse-chase experiments indicated that mMCP-2 protein is not detected because its transcript is rapidly degraded. IL-3, IL-4, and glucocorticoids induce mBMMC to rapidly catabolize their mMCP-1 and mMCP-2 transcripts, whereas IL-10 appears to induce expression of *trans*-acting factors that slow down the rate of degradation of these chymase transcripts. A repetitive "AUUUA" motif in the 3'-untranslated regions of numerous cytokine transcripts regulates their stability in MC and other cell types.[86] Because the "AUUUA" motif is not present in any mMCP transcript, a different *cis*-acting element must control the stability of chymase transcripts in mBMMC. The fact that the mMCP-1, mMCP-2, mMCP-4, and mMCP-9 transcripts all contain a repetitive "UGXCCCC" sequence in their 3'-untranslated regions implies that this motif participates in the posttranscriptional regulation of chymases.

mBMMC and various IL-3-dependent mouse MC lines were the first populations of MC found to express multiple types of cytokines and chemokines when their high-affinity Fc receptors for IgE (FcεRI) are cross-linked with antigen.[87–89] Lu-Kuo and coworkers[90] discovered that the FcεRI-mediated expression of IL-6 in mBMMC is controlled, in part, by a cytokine-regulated posttranscriptional mechanism, as Wodnar-Filipowicz and coworkers also noted for IL-3 expression in certain mouse MC lines.[87] Although KL and IL-10 are able to induce an approximate twofold increase in the level of the IL-6 transcript in these MC, the IL-6 transcript still has a relatively short half-life. When IL-1 is added to the culture medium, the steady-state levels of IL-6 transcript and protein increase dramatically. As assessed by nuclear run-on analysis, IL-1 does not increase transcription. Rather, this cytokine inhibits the catabolism of IL-6 mRNA. Because the inductive effects of IL-1 can be blocked by treatment of the cells with cycloheximide, IL-1 appears to induce the expression of a protein that stabilizes IL-6 mRNA in mBMMC, possibly through the "AUUUA" motif in its 3'-untranslated region.

With respect to arachidonic acid pathways, FcεRI-activated rat peritoneal MC preferentially produce and release prostaglandin D_2.[91] mBMMC developed in IL-3 were the first MC found to metabolize arachidonic acid primarily to leukotriene C_4.[92] As discussed elsewhere in this volume (see Chapter 26), IL-1, IL-3, IL-9, IL-10, and KL regulate the biosynthesis of prostaglandin D_2 and leukotriene C_4 in MC at a number of levels.[93–97] Glucocorticoids, IL-1, IL-3, IL-4, IL-9, IL-10, and KL are not the only factors that regulate the growth and differentiation of mouse and rat MC. Transforming growth factor-β_1 promotes MC chemotaxis[98] and inhibits the IL-3- and IL-4-dependent proliferation of cultured mouse and rat MC.[99,100] Peritoneal MC express the high-affinity nerve growth factor receptor *trk*A,[101] and nerve growth factor promotes the in vitro survival of peritoneal MC[102] and induces a histochemical change in IL-3-developed mBMMC.[103] Substance P induces mouse MC to selectively express transforming growth factor-β.[104]

Coculture of mBMMC with Fibroblasts

The potential influence of mesenchymal cells on the differentiation and maturation of the MC has been studied extensively in different in vitro systems. The viability and phenotype of purified rat and mouse peritoneal MC were found to be maintained for at least 30 days when these mature cells were cocultured ex vivo on a monolayer of Swiss albino mouse skin-derived 3T3 fibroblasts[105] or Swarm rat chondrosarcoma chondrocytes[106] in the absence of growth factors other than those in fetal calf serum. Peritoneal MC cocultured with fibroblasts continue to synthesize heparin rather than chondroitin sulfate E.[105] When activated through FcεRI, cocultured peritoneal MC release a higher percentage of their preformed mediators than do freshly isolated cells, but the activated cells continue to metabolize arachidonic acid to prostaglandin D_2 rather than to leukotriene C_4.

IL-3-developed mBMMC cocultured with mouse 3T3 fibroblasts for 2 weeks slowly become safranin positive, express globopentaosylceramide, and preferentially synthesize serglycin proteoglycans that contain heparin chains instead of chondroitin sulfate E chains.[107,108] Cocultured mBMMC also undergo pronounced granule maturation events, storing more than 50-fold more histamine and mMC-CPA in their granules than noncocultured mBMMC.[107,109-111] The fact that cocultured mBMMC do not increase their aminopeptidase activity indicates that fibroblasts selectively regulate which proteases are stored in the MC granule.[112] When cocultured, the plasma membranes of the MC and fibroblast do not form recognizable cell junctions, but the two cell types do come within 20 nm of each other.[109] The importance of integrins[113-116] in the interaction of mBMMC and fibroblasts remains to be determined, but c-kit on the surface of the MC must bind to KL on the surface of the fibroblast.[117] Although numerous cytokines induce mBMMC to change their granule phenotype, mBMMC do not undergo pronounced granule maturation unless they are cocultured with fibroblasts. During this coculture, the MC also influence the fibroblasts, inducing them to lose contact inhibition[118] and to alter their glycolipid expression.[108]

Transformed MC Lines

Eccleston and coworkers[119] generated a leukemia composed of basophil-like cells in rats by exposing them to the chemical carcinogen β-chlorethylamine. Like the mouse mastocytoma cell line P815, rat basophilic leukemia cell lines (e.g., RBL-1 and RBL-2H3) are able to grow in culture in the absence of MC-regulatory cytokines because of a point mutation in the kinase domain of c-kit that renders the receptor constitutively active.[120] RBL cells are similar to rat jejunal mucosal MC in terms of their histochemistry, proteoglycan expression, and chymase expression.[42,121] RBL cells were invaluable for characterizing arachidonic acid metabolites,[122] granule proteases,[73,121,123] serglycin proteoglycans,[42,46] and FcεRI.[124-126] The recent cloning of cDNAs that encode 10 different granule proteases from RBL cell cDNA and rat peritoneal MC cDNA libraries indicates that the complexity of protease expression observed first in the mouse also occurs in the rat.[123] Whereas mBMMC are difficult to transfect, RBL cells can be transfected easily with plasmid DNA. Therefore, RBL cells have been extremely useful for elucidating at the molecular level the steps involved in FcεRI expression[127] and FcεRI-mediated signal transduction,[128] as well as the *trans*-acting factors and *cis*-acting elements that regulate transcription of genes which encode varied constituents of the secretory granule.[81-84]

Fibroblast coculture helped to address the role of mesenchymal cells in MC development, but 2 to 4 weeks of coculture are required to induce an mBMMC to change its granule phenotype and to mature. Because enzymatic treatment and density-gradient centrifugation also are required to separate the two cell types, attempts were made to generate immortalized cutaneous-like MC lines that can proliferate in the absence of other cell types. During the 1980s, immortalized mouse MC lines were derived with Kirsten sarcoma virus, Harvey sarcoma virus, or Abelson murine leukemia virus.[129-131] Because the Kirsten sarcoma virus-immortalized cell lines were derived with fibroblast coculture, many of the resulting lines that initially developed contained relatively mature granules.[131] Although these transformed cell lines allowed investigators to isolate, clone, and characterize a number of MC-specific genes and proteins, the chronic production and release of infectious retrovirus particles hindered their use in vivo. Thus, an immortalized, lineage-restricted MC line (V3-MC) was derived from a mouse that had developed systemic mastocytosis after transplantation of bone marrow cells transduced with the v-*abl* oncogene.[49] Because the v-*abl*-immortalized V3-MC line does not release virus, it can be used to study MC development in vivo. Cultured V3-MC contain low levels of histamine and undectable levels of the transcripts that encode mMCP-1, mMCP-2, mMCP-4, and mMCP-7. However, V3-MC contain high levels of the transcripts that encode mMCP-5, mMCP-6, and mMC-CPA, and therefore resemble mBMMC that have been cultured for more than 6 weeks in the presence of IL-3.

When the V3-MC line is adoptively transferred into BALB/c mice, a reproducible mastocytosis develops quickly in the liver and spleen, and then more slowly in the stomach, lung, small intestine, and other tissues of the recipient mice.[49] V3-MC do not localize to the skin when they are administered intravenously. Similar results are obtained when mBMMC or unfractionated bone marrow cells from +/+ mice are adoptively transferred into MC-deficient *W/W^v* mice.[72] The underlying reasons for these observations remain to be determined, but the skin appears to have a mechanism that dominantly controls the number of MC-committed progenitors it can support. The V3-MC that populate the liver and spleen of 2- to 3-week-old mastocytosis mice differentiate and mature. Analogous to the indigenous MC in the spleen of normal BALB/c mice, the V3-MC that differentiate in the spleen express all examined mMCPs except mMCP-9.[27,49] The V3-MC that populate the jejunum express the same panel of mMCPs as the indigenous MC found in the jejunal lamina propria of helminth-infected mice. These jejunal-localized V3-MC cease expressing mMCP-5, mMCP-6, and mMC-CPA but express mMCP-1 and mMCP-2 as they progress into the jejunal mucosa and epithelium. The fact that the granule phenotype of the V3-MC that differentiate in the jejunal mucosa is distinct from that of the starting population of cultured cells and from that of the cells that differentiate in the spleen and liver supports the in vitro and in vivo findings that the tissue microenvironment regulates the growth, differentiation, and maturation of MC.

Future Directions

Most future studies on rodent MC will be carried out at the molecular level. The transfectable V3-MC line should be invaluable for elucidating the mechanisms that control MC development in mouse tissues. It has been proposed that human basophils and MC are developmentally distinct cell types. Nevertheless, the recent discovery that mouse MC can reside in the blood for a short period of time,[60] coupled with the recent discovery that mouse MC can possess segmented/lobular nuclei,[132] now indicates that the developmental relationship of all metachromatic/FcεRI⁺ cells in the body must be reexamined.

The biological significance of MC heterogeneity in the mouse and rat needs to be understood. With various transgenic approaches, MC-specific genes will be disrupted and the consequences will be addressed. How the recently identified transcriptional, posttranscriptional, and posttranslational mechanisms work together to control protease and cytokine expression and MC heterogeneity will be deduced. Because many of the genes that encode the varied protease constituents of the MC granule have been cloned and sequenced, these proteases can be expressed in insect cells and other cell types to study their substrate specificities and to design inhibitors of their activity in vivo. During the last decade, most investigators have focused their attention on understanding the growth, differentiation, and maturation of committed progenitors into MC. Because MC are such important effector cells of the immune response, it is now clear that greater attention must be placed on understanding the mechanisms at the molecular level by which unwanted MC are eventually disposed of in the body.

References

1. Enerbäck L. Mast cells in rat gastrointestinal mucosa. 2. Dye-binding and metachromatic properties. Acta Pathol Microbiol Scand 1966;66:303–312.

2. Woodbury RG, Katunuma N, Kobayashi K, et al. Covalent structure of a group-specific protease from rat small intestine. Biochemistry 1978;17:811–819.

3. Le Trong H, Parmelee DC, Walsh KA, et al. Amino acid sequence of rat mast cell protease I (chymase). Biochemistry 1987;26:6988–6994.

4. Cole KR, Kumar S, Le Trong H, et al. Rat mast cell carboxypeptidase: amino acid sequence and evidence of enzyme activity within mast cell granules. Biochemistry 1991;30:648–655.

5. Yurt RW, Leid RW Jr, Austen KF, et al. Native heparin from rat peritoneal mast cells. J Biol Chem 1977;252:518–521.

6. Stevens RL, Lee TDG, Seldin DC, et al. Intestinal mucosal mast cells from rats infected with *Nippostrongylus brasiliensis* contain protease-resistant chondroitin sulfate di-B proteoglycans. J Immunol 1986;137:291–295.

7. Stevens RL, Austen KF. Effect of *p*-nitrophenyl-β-D-xyloside on proteoglycan and glycosaminoglycan biosynthesis in rat serosal mast cell cultures. J Biol Chem 1982;257:253–259.

8. Kitamura Y, Go S, Hatanaka K. Decrease of mast cells in *W/W^v* mice and their increase by bone marrow transplantation. Blood 1978;52:447–452.

9. Sonoda S, Sonoda T, Nakano T, et al. Development of mucosal mast cells after injection of a single connective tissue-type mast cell in the stomach mucosa of genetically mast cell-deficient *W/W^v* mice. J Immunol 1986;137:1319–1322.

10. Nakano T, Sonoda T, Hayashi C, et al. Fate of bone marrow-derived cultured mast cells after intracutaneous, intraperitoneal, and intravenous transfer into genetically mast cell deficient *W/W^v* mice. J Exp Med 1985;162:1025–1043.

11. Otsu K, Nakano T, Kanakura Y, et al. Phenotypic changes of bone marrow-derived mast cells after intraperitoneal transfer into *W/W^v* mice that are genetically deficient in mast cells. J Exp Med 1987;165:615–627.

12. Rodewald HR, Dessing M, Dvorak AM, et al. Identification of a committed precursor for the mast cell lineage. Science 1996;271:818–822.

13. Yung Y-P, Wang S-Y, Moore MAS. Characterization of mast cell precursors by physical means: dissociation from T cell and T cell precursors. J Immunol 1983;130:2843–2848.

14. Newlands GF, Gibson S, Knox DP, et al. Characterization and mast cell origin of a chymotrypsin-like proteinase isolated from intestines of mice infected with *Trichinella spiralis.* Immunology 1987;62:629–634.

15. Le Trong H, Newlands GFJ, Miller HRP, et al. Amino acid sequence of a mouse mucosal mast cell protease. Biochemistry 1989;28:391–395.

16. Reynolds DS, Stevens RL, Lane WS, et al. Different mouse mast cell populations express various combinations of at least six distinct mast cell serine proteases. Proc Natl Acad Sci USA 1990;87:3230–3234.

17. Reynolds DS, Stevens RL, Gurley DS, et al. Isolation and molecular cloning of mast cell carboxypeptidase A: a novel member of the carboxypeptidase gene family. J Biol Chem 1990;264:20094–20099.

18. Serafin WE, Reynolds DS, Rogelj S, et al. Identification and molecular cloning of a novel mouse mucosal mast cell serine protease. J Biol Chem 1990;265:423–429.

19. Serafin WE, Sullivan TP, Conder GA, et al. Cloning of the cDNA and gene for mouse mast cell protease-4. Demonstration of its late transcription in mast cell subclasses and analysis of its homology to subclass-specific neutral proteases of the mouse and rat. J Biol Chem 1991;266:1934–1941.

20. McNeil HP, Austen KF, Somerville LL, et al. Molecular cloning of the mouse mast cell protease-5 gene. A novel secretory granule protease expressed early in the differentiation of serosal mast cells. J Biol Chem 1991;266:20316–20322.

21. Reynolds DS, Gurley DS, Austen KF, et al. Cloning of the cDNA and gene of mouse mast cell protease-6. Transcription by progenitor mast cells and mast cells of the connective tissue subclass. J Biol Chem 1991;266:3847–3853.

22. McNeil HP, Reynolds DS, Schiller V, et al. Isolation, characterization, and transcription of the gene encoding mouse mast cell protease 7. Proc Natl Acad Sci USA 1992;89:11174–11178.

23. Chu W, Johnson DA, Musich PR. Molecular cloning and characterization of mouse mast cell chymases. Biochim Biophys Acta 1992;1121:83–87.

24. Johnson DA, Barton GJ. Mast cell tryptases: examination of unusual characteristics by multiple sequence alignment and molecular modeling. Protein Sci 1992;1:370–377.

25. Huang R, Blom T, Hellman L. Cloning and structural analysis of mMCP-1, mMCP-4 and

mMCP-5, three mouse mast cell-specific serine proteases. Eur J Immunol 1991;21:1611–1621.

26. Gurish MF, Nadeau JH, Johnson KR, et al. A closely linked complex of mouse mast cell-specific chymase genes on chromosome 14. J Biol Chem 1993;268:11372–11379.

27. Hunt J, Friend DS, Gurish MF, et al. Mast cell protease 9, a novel member of the chromosome 14 family of serine proteases that is selectively expressed in uterine mast cells. J Biol Chem 1997;272:29158–29166.

28. McNeil HP, Frenkel DP, Austen KF, et al. Translation and granule localization of mouse mast cell protease-5: immunodetection with specific antipeptide Ig. J Immunol 1992;149:2466–2472.

29. Ghildyal N, Friend DS, Nicodemus CF, et al. Reversible expression of mouse mast cell protease-2 mRNA and protein in cultured mast cells exposed to IL-10. J Immunol 1993;151:3206–3214.

30. Ghildyal N, Friend DS, Freelund R, et al. Lack of expression of the tryptase mouse mast cell protease 7 in mast cells of the C57BL/6J mouse. J Immunol 1994;153:2624–2630.

31. Ghildyal N, Friend DS, Stevens RL, et al. Fate of two mast cell tryptases in V3 mastocytosis and normal BALB/c mice undergoing passive systemic anaphylaxis. Prolonged retention of exocytosed mMCP-6 in connective tissues and rapid accumulation of enzymatically active mMCP-7 in the blood. J Exp Med 1996;184:1061–1073.

32. Friend DS, Ghildyal N, Austen KF, et al. Mast cells that reside at different locations in the jejunum of mice infected with *Trichinella spiralis* exhibit sequential changes in their granule ultrastructure and chymase phenotype. J Cell Biol 1996;135:279–290.

33. Brunet JF, Dosseto M, Denizot F, et al. The inducible cytotoxic T-lymphocyte-associated gene transcript CTLA-1 sequence and gene localization to mouse chromosome 14. Nature (Lond) 1986;322:268–271.

34. Crosby JL, Bleackley RC, Nadeau JH. A complex of serine protease genes expressed preferentially in cytotoxic T-lymphocytes is closely linked to the T-cell receptor-α and δ-chain genes on mouse chromosome 14. Genomics 1990;6:252–259.

35. Heusel JW, Scarpati EM, Jenkins NA, et al. Molecular cloning, chromosome location, and tissue-specific expression of the mouse cathepsin G gene. Blood 1993;81:1614–1623.

36. Remington SJ, Woodbury RG, Reynolds RA, et al. The structure of rat mast cell proease II at 1.9-Å resolution. Biochemistry 1988;27:8097–8105.

37. Sali A, Matsumoto R, McNeil HP, et al. Three-dimensional models of four mouse mast cell chymases. Identification of proteoglycan-binding regions and protease-specific antigenic epitopes. J Biol Chem 1993;268:9023–9034.

38. Gurish MF, Johnson KR, Webster MJ, et al. Location of the mouse mast cell protease 7 gene (*Mcpt7*) to chromosome 17. Mamm Genome 1994;5:656–657.

39. Huang C, Wong GW, Ghildyal N, et al. The tryptase, mouse mast cell protease 7, exhibits anticoagulant activity *in vivo* and *in vitro* due to its ability to degrade fibrinogen in the presence of the diverse array of protease inhibitors in plasma. J Biol Chem 1997;272:31885–31893.

40. Stevens RL, Qui D, McNeil HP, et al. Transgenic mice that possess a disrupted mast cell protease 5 (mMCP-5) gene cannot store carboxypeptidase A in their granules. FASEB J 1996;10:1307.

41. Razin E, Stevens RL, Akiyama F, et al. Culture from mouse bone marrow of a subclass of mast cells possessing a distinct chondroitin sulfate proteoglycan with glycosaminoglycans rich in *N*-acetylgalactosamine-4,6-disulfate. J Biol Chem 1982;257:7229–7236.

42. Seldin DC, Austen KF, Stevens RL. Purification and characterization of protease-resistant secretory granule proteoglycans containing chondroitin sulfate di-B and heparin-like glycosaminoglycans from rat basophilic leukemia cells. J Biol Chem 1985;260:11131–11139.

43. Bourdon MA, Oldberg A, Pierschbacher M, et al. Molecular cloning and sequence analysis of a chondroitin sulfate proteoglycan cDNA. Proc Natl Acad Sci USA 1985;82:1321–1325.

44. Avraham S, Austen KF, Nicodemus CF, et al. Cloning and characterization of the mouse gene that encodes the peptide core of secretory granule proteoglycans and expression of this gene in transfected rat-1 fibroblasts. J Biol Chem 1989;264:16719–16726.

45. Avraham S, Stevens RL, Nicodemus CF, et al. Molecular cloning of a cDNA that encodes the peptide core of a mouse mast cell secretory granule proteoglycan and comparison with the analogous rat and human cDNA. Proc Natl Acad Sci USA 1989;86:3763–3767.

46. Avraham S, Stevens RL, Gartner MC, et al. Isolation of a cDNA that encodes the peptide core of the secretory granule proteoglycan of rat basophilic leukemia-1 cells and assessment of its homology to the human analogue. J Biol Chem 1988;263:7292–7296.

47. De Young MB, Nemeth EF, Scarpa A. Measurement of the internal pH of mast cell granules using microvolumetric fluorescence and isotopic techniques. Arch Biochem Biophys 1987;254:222–233.

48. Matsumoto R, Sali A, Ghildyal N, et al. Packaging of proteases and proteoglycans in the granules of mast cells and other hematopoietic cells. A cluster of histidines on mouse mast cell protease 7 regulates its binding to heparin serglycin proteoglycans. J Biol Chem 1995;270:19524–19531.

49. Gurish MF, Pear WS, Stevens RL, et al. Tissue-regulated differentiation and maturation of a v-abl-immortalized mast cell-committed progenitor. Immunity 1995;3:175–186.

50. Stevens RL, Friend DS, McNeil HP, et al. Strain-specific and tissue-specific expression of mouse mast cell secretory granule proteases. Proc Natl Acad Sci USA 1994;91:128-132.

51. Hunt JE, Stevens RL, Austen KF, et al. Natural disruption of the mouse mast cell protease 7 gene in the C57BL/6 mouse. J Biol Chem 1996;271:2851–2855.

52. Levitt RC, Mitzner W. Autosomal recessive inheritance of airway hyperreactivity to 5-hydroxytryptamine. J Appl Physiol 1989; 67: 1125–1132.

53. Mitchell GF, Hogarth-Scott RS, Edwards RD, et al. Studies on immune responses to parasite antigens in mice. Int Arch Allergy Appl Immunol 1976;52:64–78.

54. Wells PD. Mast cell, eosinophil, and histamine levels in Nippostrongylus brasiliensis-infected rats. Exp Parasitol 1962;12:82–101.

55. Murray M, Miller HRP, Jarrett WFH, et al. The globule leukocyte and its derivation from the subepithelial mast cell. Lab Invest 1968;19:222–234.

56. Miller HRP, Jarrett WFH. Immune reactions in mucous membranes. Intestinal mast cell response during helminth expulsion in the rat. Immunology 1971;20:277–288.

57. Miller HRP, Huntley JF, Newlands GFJ, et al. Granule proteinases define mast cell heterogeneity in the serosa and the gastrointestinal mucosa of the mouse. Immunology 1988;65:559–566.

58. Huntley JF, Gooden C, Newlands GFJ, et al. Distribution of intestinal mast cell proteinase in blood and tissues of normal and Trichinella-infected mice. Parasite Immunol 1990;12:85–95.

59. Ghildyal N, McNeil HP, Stechschulte S, et al. IL-10 induces transcription of the gene for mast cell protease-1, a serine protease preferentially expressed in the mucosal mast cells of Trichinella spiralis-infected mice. J Immunol 1992;149:2123–2129.

60. Friend DS, Ghildyal N, Austen KF, et al. Tryptase expression in jejunal mouse mast cells (MC) during T. spiralis infection. Mol Biol Cell 1996;7:143a.

61. Irani AMA, Schechter NM, Craig SS, et al. Two types of human mast cell that have distinct neutral protease compositions. Proc Natl Acad Sci USA 1986;83:4464–4468.

62. Longley J, Tyrrell L, Lu S, et al. Chronically kit-stimulated clonally-derived human mast cells show heterogeneity in different tissue microenvironments. J Invest Dermatol 1997;108:792–795.

63. Ginsburg H, Sachs L. Formation of pure suspensions of mast cells in tissue culture by differentiation of lymphoid cells from the mouse thymus. J Natl Cancer Inst 1963;31:1–40.

64. Razin E, Cordon-Cardo C, Good RA. Growth of a pure population of mouse mast cells in vitro with conditioned medium derived from concanavalin-A stimulated splenocytes. Proc Natl Acad Sci USA 1981;78:2559–2561.

65. Schrader JW, Lewis SJ, Clark-Lewis I, et al. The persisting (P) cell: Histamine content, regulation by a T cell-derived factor, origin from a bone marrow precursor, and relationship to mast cells. Proc Natl Acad Sci USA 1981;78:323–327.

66. Nabel G, Galli SJ, Dvorak AM, et al. Inducer T lymphocytes synthesize a factor that stimulates proliferation of cloned mast cells. Nature (Lond) 1981;291:332–334.

67. Nagao K, Yokoro K, Aaronson SA. Continuous lines of basophil/mast cells derived from normal mouse bone marrow. Science 1981;212:333–335.

68. Tertian G, Yung Y-P, Guy-Grand D, et al. Long-term in vitro culture of mouse mast cells. I. Description of a growth factor-dependent culture technique. J Immunol 1981;127:788–794.

69. Haig DM, McKee TA, Jarrett EEE, et al. Generation of mucosal mast cells is stimulated in vitro by factors derived from T cells of

helminth-infected rats. Nature (Lond) 1982; 300:188-190.

70. Ihle JN, Keller J, Oroszlan S, et al. Biologic properties of homogeneous interleukin-3. I. Demonstration of WEHI-3 growth factor activity, mast cell growth factor activity, P cell-stimulating factor activity, colony-stimulating factor activity, and histamine-producing cell-stimulating factor activity. J Immunol 1983;131:282–287.

71. Razin E, Ihle JN, Seldin D, et al. Interleukin-3: a differentiation and growth factor for the mouse mast cell that contains chondroitin sulfate E proteoglycan. J Immunol 1984;132:1479–1486.

72. Du T, Friend DS, Austen KF, et al. Time-dependent differences in the asynchronous appearance of mast cells in normal mice and in congenic mast cell-deficient mice after infusion of normal bone marrow cells. Clin Exp Immunol 1996;103:316–321.

73. Benfey PN, Yin FH, Leder P. Cloning of mast cell protease, RMCP-II. Evidence for cell-specific expression and a multi-gene family. J Biol Chem 1987;262:5377–5384.

74. Arm JP, Gurish MF, Reynolds DS, et al. Molecular cloning of gp49, a cell-surface antigen that is preferentially expressed by mouse mast cell progenitors and is a new member of the immunoglobulin superfamily. J Biol Chem 1991;266:15966–15973.

75. Gurish MF, Ghildyal N, McNeil HP, et al. Differential expression of secretory granule proteases in mouse mast cells exposed to interleukin 3 and c-kit ligand. J Exp Med 1992;175:1003–1012.

76. Tsai M, Takeishi T, Thompson H, et al. Induction of mast cell proliferation, maturation, and heparin synthesis by rat c-kit ligand, stem cell factor. Proc Natl Acad Sci USA 1991;88:6382–6386.

77. Eklund KK, Ghildyal N, Austen KF, et al. Induction by IL-9 and suppression by IL-3 and IL-4 of the levels of chromosome 14-derived transcripts that encode late-expressed mouse mast cell proteases. J Immunol 1993;151:4266–4273.

78. Ghildyal N, McNeil HP, Gurish MF, et al. Transcriptional regulation of the mucosal mast cell-specific protease gene, mMCP-2, by interleukin 10 and interleukin 3. J Biol Chem 1992;267:8473–8477.

79. Flanagan JG, Leder P. The kit ligand: a cell surface molecule altered in steel mutant fibroblasts. Cell 1990;63:185–194.

80. Eklund KK, Humphries DE, Xia Z, et al. Glucocorticoids inhibit the cytokine-induced proliferation of mast cells, the high-affinity IgE receptor-mediated expression of TNF-α, and the IL-10-induced expression of chymases. J Immunol 1997;158:4373–4380.

81. Sarid J, Benfey PN, Leder P. The mast cell-specific expression of a protease gene, RMCP-II, is regulated by an enhancer element that binds specifically to mast cells trans-acting factors. J Biol Chem 1989;264:1022–1026.

82. Zon LI, Gurish MF, Stevens RL, et al. GATA-binding transcription factors in mast cells regulate the promoter of the mast cell carboxypeptidase A gene. J Biol Chem 1991;266:22948–22953.

83. Gurish MF, Fernandez T, Austen KF, et al. Location of a cis-acting element that promotes the transcription of the gene that encodes mouse mast cell protease 5 (mMCP-5). J Allergy Clin Immunol 1997;99:S370.

84. Avraham S, Avraham H, Austen KF, et al. Negative and positive cis-acting regulatory elements in the 5′ flanking region of the mouse gene that encodes the serine/glycine-rich peptide core of secretory granule proteoglycan. J Biol Chem 1992;267:610–617.

85. Xia Z, Ghildyal N, Austen KF, et al. Post-transcriptional regulation of chymase expression in mast cells. A cytokine-dependent mechanism for controlling the expression of granule neutral proteases of hematopoietic cells. J Biol Chem 1996;271:8747–8753.

86. Shaw G, Kamen R. A conserved AU sequence from the 3′ untranslated region of GM-CSF mRNA mediates selective mRNA degradation. Cell 1986;46:659–667.

87. Wodnar-Filipowicz A, Heusser CH, Moroni C. Production of the haemopoietic growth factor GM-CSF and interleukin-3 by mast cells in response to IgE receptor-mediated activation. Nature (Lond) 1989;339:150–152.

88. Burd PR, Rogers HW, Gordon JR, et al. Interleukin 3-dependent and -independent mast cells stimulated with IgE and antigen express multiple cytokines. J Exp Med 1989;170:245–257.

89. Plaut M, Pierce JH, Watson CJ, et al. Mast cell lines produce lymphokines in response to cross-linkage of FcεRI or to calcium ionophores. Nature (Lond) 1989;339:64–67.

90. Lu-Kuo JM, Austen KF, Katz HR. Post-transcriptional stabilization by interleukin-1 of interleukin-6 mRNA induced by c-kit ligand

and interleukin-10 in mouse bone marrow-derived mast cells. J Biol Chem 1996;271: 22169–22174.

91. Lewis RA, Soter NA, Diamond PT, et al. Prostaglandin D_2 generation after activation of rat and human mast cells with anti-IgE. J Immunol 1982;129:1627–1631.

92. Razin E, Mencia-Huerta JM, Stevens RL, et al. IgE-mediated release of leukotriene C_4, chondroitin sulfate E proteoglycan, β-hexosaminidase, and histamine from cultured bone marrow-derived mouse mast cells. J Exp Med 1983;157:189–201.

93. Murakami M, Austen KF, Arm JP. The immediate phase of c-*kit* ligand stimulation of mouse bone marrow-derived mast cells elicits rapid leukotriene C_4 generation through post-translational activation of cytosolic phospholipase A_2 and 5-lipoxygenase. J Exp Med 1995;182:197–206.

94. Murakami M, Austen, KF, Bingham CO III, et al. Interleukin-3 regulates development of the 5-lipoxygenase/leukotriene C_4 synthase pathway in mouse mast cells. J Biol Chem 1995;270:22653–22656.

95. Murakami M, Bingham CO III, Matsumoto R, et al. IgE-dependent activation of cytokine-primed mouse cultured mast cells induces a delayed phase of PGD_2 generation via prostaglandin endoperoxide synthase-2. J Immunol 1995;155:4445–4453.

96. Murakami M, Matsumoto R, Austen KF, et al. Prostaglandin endoperoxide synthase-1 and -2 couple to different transmembrane stimulito generate prostaglandin D_2 in mouse bone marrow-derived mast cells. J Biol Chem 1994;269:22269–22275.

97. Murakami M, Matsumoto R, Urade Y, et al. c-*kit* ligand mediates increased expression of cytosolic phospholipase A_2, prostaglandin endoperoxide synthase-1, and hematopoietic prostaglandin D_2 synthase and increased IgE-dependent prostaglandin D_2 generation in immature mouse mast cells. J Biol Chem 1995;270:3239–3246.

98. Gruber BL, Marchese MJ, Kew RR. Transforming growth factor-β1 mediates mast cell chemotaxis. J Immunol 1994;152:5860–5867.

99. Broide DH, Wasserman SI, Alvaro-Gracia J, et al. 1989. Transforming growth factor 1 selectively inhibits IL-3-dependent mast cell proliferation without affecting mast cell function or differentiation. J Immunol 1989;143:1591–1597.

100. Toyota N, Hashimoto Y, Matsuo S, et al. Transforming growth factor β, 1 inhibits IL-3- and IL-4-dependent mouse connective tissue-type mast cell proliferation. Arch Dermatol Res 1995;287:198–201.

101. Horigome K, Pryor JC, Bullock ED, et al. Mediator release from mast cells by nerve growth factor. Neurotrophin specificity and receptor mediation. J Biol Chem 1993;268:14881–14887.

102. Horigome K, Bullock ED, Johnson EM. Effects of nerve growth factor on rat peritoneal mast cells. Survival promotion and immediate-early gene induction. J Biol Chem 1994;269:2695–2702.

103. Matsuda H, Kannan Y, Ushio H, et al. Nerve growth factor induces development of connective tissue-type mast cells *in vitro* from murine bone marrow cells. J Exp Med 1991;174:7–14.

104. Ansel JC, Brown JR, Pyan DG, et al. Substance P selectively activates TNF-α gene expression in mouse mast cells. J Immunol 1993;l50:4478–4485.

105. Levi-Schaffer F, Austen KF, Caulfield JP, et al. Fibroblasts maintain the phenotype and viability of the rat heparin-containing mast cell in vitro. J Immunol 1985;135:3454–3462.

106. Stevens RL, Somerville LL, Sewell D, et al. Serosal mast cells maintain their viability and promote the metabolism of cartilage proteoglycans when cocultured with chondrocytes. Arthritis Rheum 1992;35:325–335.

107. Levi-Schaffer F, Austen KF, Gravallese PM, et al. Coculture of interleukin-3-dependent mouse mast cells with fibroblasts results in a phenotypic change of the mast cells. Proc Natl Acad Sci USA 1986;83:6485–6488.

108. Katz HR, Dayton ET, Levi-Schaffer F, et al. Coculture of mouse IL-3-dependent mast cells with 3T3 fibroblasts stimulates synthesis of globopentaosylceramide (Forssman glycolipid) by fibroblasts and surface expression on both populations. J Immunol 1988;140:3090–3097.

109. Levi-Schaffer F, Dayton ET, Austen KF, et al. Mouse bone marrow-derived mast cells cocultured with fibroblasts: morphology and stimulation-induced release of histamine, leukotriene B_4, leukotriene C_4, and prostaglandin D_2. J Immunol 1987;139:3431–3441.

110. Dayton ET, Pharr P, Ogawa M, et al. 3T3 fibroblasts induce cloned interleukin 3-dependent

mouse mast cells to resemble connective tissue mast cells in granular constituency. Proc Natl Acad Sci USA 1988;85:569–572.

111. Serafin WE, Dayton ET, Gravallese PM, et al. Carboxypeptidase A in mouse mast cells: Identification, characterization, and use as a differentiation marker. J Immunol 1987;139:3771–3776.

112. Serafin WE, Guidry UA, Dayton ET, et al. Identification of aminopeptidase activity in the secretory granules of mouse mast cells. Proc Natl Acad Sci USA 1991;88:5984–5988.

113. Dastych J, Costa JJ, Thompson HL, et al. Mast cell adhesion to fibronectin. Immunology 1991;73:478–484.

114. Ducharme LA, Weis JH. Modulation of integrin expression during mast cell differentiation. Eur J Immunol 1992;22:2603–2607

115. Gurish MF, Bell AF, Smith TJ, et al. Expression of murine β_7, α_4, and β_1 integrin genes by rodent mast cells. J Immunol 1992;149:1964–1972.

116. Metcalfe DD. Interaction of mast cells with extracellular matrix proteins. Int Arch Allergy Immunol 1995;107:60–62.

117. Adachi S, Ebi Y, Nishikawa S, et al. Necessity of extracellular domain of W (c-*kit*) receptors for attachment of mouse cultured mast cells to fibroblasts. Blood 1992;79:650–656.

118. Dayton ET, Caulfield JP, Hein A, et al. Regulaion of the growth rate of mouse fibroblasts by IL-3-activated mouse bone marrow-derived mast cells. J Immunol 1989; 142:4307–4313.

119. Eccleston E, Leonard BJ, Lowe JS, et al. Basophilic leukaemia in the albino rat and a demonstration of the basopoietin. Nat New Biol 1973;244:73–76.

120. Tsujimura T, Furutsu T, Morimoto M, et al. Ligand-independent activation of c-*kit* receptor tyrosine kinase in a mouse mastocytoma cell line P-815 generated by a point mutation. Blood 1994;83:2619–2626.

121. Seldin DC, Adelman S, Austen KF, et al. Homology of the rat basophilic leukemia cell and the rat mucosal mast cell. Proc Natl Acad Sci USA 1985;82:3871–3875.

122. Örning L, Hammarström S, Samuelsson B. Leukotriene D: a slow reacting substance from rat basophilic leukemia cells. Proc Natl Acad Sci USA 1980;77:2014–2017.

123. Lützelschwab C, Pejler G, Aveskogh M, et al. Secretory granule proteases in rat mast cells. Cloning of 10 different serine proteases and a carboxypeptidase A from various rat mast cell populations. J Exp Med 1996;185:13–29.

124. Isersky C, Rivera J, Triche TJ, et al. Characterization of the receptors for IgE on membranes isolated from rat basophilic leukemia cells. Mol Immunol 1982;19:925–941.

125. Kinet JP, Metzger H, Hakimi J, et al. A cDNA presumptively coding for the subunit of the receptor with high affinity for immunoglobulin E. Biochemistry 1987;26:4605–4610.

126. Shimizu A, Tepler I, Benfey PN, et al. Human and rat mast cell high-affinity immunoglobulin E receptors: characterization of putative α-chain gene products. Proc Natl Acad Sci USA 1988;85:1907–1911.

127. Ra C, Jouvin MH, Kinet JP. Complete structure of the mouse mast cell receptor for IgE (FcϵRI) and surface expression of chimeric receptors (rat-mouse-human) on transfected cells. J Biol Chem 1989;264:15323–15327.

128. Jouvin MH, Adamczewski M, Numerof R, et al. Differential control of the tyrosine kinases Lyn and Syk by the two signaling chains of the high affinity immunoglobulin E receptor. J Biol Chem 1994;269:5918–5925.

129. Pierce JH, Di Fiore PP, Aaronson SA, et al. Neoplastic transformation of mast cells by Abelson-MuLV: abrogation of IL-3 dependence by a nonautocrine mechanism. Cell 1985;41:685–693.

120. Rein A, Keller J, Schultz AM, et al. Infection of immune mast cells by Harvey sarcoma virus: immortalization without loss of requirement for interleukin-3. Mol Cell Biol 1985;5:2257–2264.

131. Reynolds DS, Serafin WE, Faller DV, et al. Immortalization of mouse connective tissue-type mast cells at multiple stages of their differentiation by coculture of splenocytes with fibroblasts that produce Kirsten sarcoma virus. J Biol Chem 1988;263:12783–12791.

132. Gurish MF, Friend DS, Webster M, et al. Mouse mast cells that possess segmented/multi-lobular nuclei. Blood 1997;90:382–390.

133. Huang C, Friend DS, Qiu W-T, et al. Induction of a selective and persistent extravasation of neutrophils into the peritoneal cavity by the tryptase mouse mast cell protease 6. J Immunol 1998;160:1910–1919.

5
Factors That Affect Human Mast Cell and Basophil Growth

Lixin Li, Xiao-Tong Zhang, and Steven A. Krilis

Mast cells and basophils are specialized effector cells of the immune system and both cells express FcεRI, stain metachromatically, synthesize a number of cytokines, and have comparable amounts of histamine. It has been assumed that the two cells share a common progenitor, or that one cell originates from the other. This notion is unlikely to be correct because basophils appear to represent a terminally differentiated cell rather than a peripheral blood mast cell precursor. In addition, mast cells and basophils have distinguishing features with respect to morphology, mediator content, response to distinct cytokines, and specific cell-surface molecules that identify these two cells as distinct.

Like that of other hemopoietic progenitor cells, mast cell and basophil growth and differentiation are regulated by a complex network of cytokines and microenvironmental factors. This review examines the development of basophils and mast cells and the specific growth factors that have been implicated in this process.

Ontogeny of Mast Cells and Basophils

Mast cells and basophils originate from hemopoietic stem cells. Basophils complete their differentiation in the bone marrow, circulate in the peripheral blood, and are not primarily found in tissues. In contrast, mast cells arise from mast cell-committed precursors in the bone marrow, which are distinct from the monocyte/macrophage precursors and would appear to form a unique cell lineage within the hemopoietic system.[1,2] They then traverse the vascular space and enter the tissues where they complete their differentiation and maturation process, which is most likely regulated by local microenvironmental factors (Fig. 5.1). The mast cell-committed precursor in the peripheral blood has recently been reported as a nongranulated CD34+ mononuclear cell.[3] Because mature mast cells are not normally found in the circulation, their tissue specific localization following the release of progenitors from the bone marrow must be an extremely efficient process. Although the factors that regulate transendothelial migration of mast cell precursors to the tissues have not been clearly identified, it would appear that mast cells (at least in rodents) share many of the mucosal homing molecules that are expressed by T cells.[4] Mature human mast cells derived from dispered lung and uterus have been demonstrated to express integrin molecules CD29, CD49d, CD49e, CD51, and CD61.[5] Cultured mast cells derived from fetal liver express functional CD51/CD61 integrin receptors at high density.[6] Mast cells reside in tissues and basophils in the circulation; however, basophils may emigrate into tissues at sites of inflammation,[7] and mast cells have been identified in peripheral blood in some patients with acute myeloid leukemia.[8]

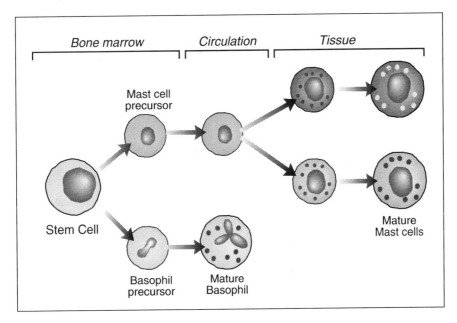

FIGURE 5.1. Differentiation pathway for human mast cells and basophils.

General Characteristics of Mast Cells and Basophils

Mast cells have been distinguished from basophils by morphological criteria, their neutral protease content, and the expression of certain cell-surface molecules. Under electron microscopy, mature human mast cells are much larger than basophils, presenting elongated plasma membrane processes with narrow folds and distinct patterns of electron-dense granules. In contrast, basophils do not exhibit such narrow surface folds and typical granule patterns.[9] In addition to cell morphology, the expression of neutral proteases termed tryptase and chymase are the most selective markers for human mast cells,[10] whereas human basophils do not express neutral proteases.[11,12] Furthermore, the cell-surface membrane antigen profile has demonstrated that both mast cells and basophils are positive for CD43 (leukosialin), CD44 (Pgp-1), CD45, CD54 (intracellular adhesion molecule 1, ICAM-1), and FcεRI.[13–15] Human blood basophils express the myeloid markers CD11b/CD18, CD13, CD17, CD35, CD25 (IL-2 α-chain), and CD38.[16] In contrast, human mast cells do not express CD17, CD25, or CD38 but bind to anti-c-kit antibody (CD117).[17] Al-though the Bsp-1 antigen is expressed on human basophils, it is not expressed on human mast cells.[14,17–19] Although the studies quoted imply that human mast cells and basophils are derived from specific cell lineages, there has been no comprehensive characterization of human mast cell and basophil progenitors to clearly establish that they arise from different lineages.

Mast Cell Phenotypes In Vivo

Like T cells, mast cells exhibit substantial structural and functional heterogeneity in both humans and mice. Mast cells can be divided into three subtypes by immunohistochemical detection of secretory granule protease expression. Mast cells that express tryptase, along with chymase, carboxypeptidase A, and a cathepsin G-like protease, are termed MC_{TC} and are found predominantly in connective tissue of skin, intestinal submucosa, breast parenchyma, lymph nodes, and synovium.[20,21] Mast cells with tryptase only are designated MC_T and are found predominantly in intestinal mucosa and lung.[10,22] Recently, a third type of mast cell, which stains with antibodies to chymase but not

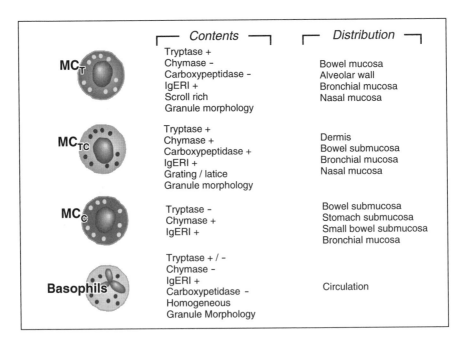

FIGURE 5.2. Characteristics of mast cells and basophils and their distribution in vivo.

tryptase and is designated MC_C, has been identified in human tissue sections (Fig. 5.2).[23] Although mast cells originate from a common precursor, some authors have suggested that the MC_T and MC_{TC} subtypes originate along distinct developmental pathways and that commitment occurs early in the bone marrow where progenitors for MC_T and MC_{TC} can be identified by morphological criteria. The exact relationship between the MC_{TC}, MC_T, and the recently identified MC_C subtypes is not clear. For example, it is not clear whether one subtype can switch into another. Human basophils do not contain any detectable chymase, carboxypeptidase, or cathespin G but have been demonstrated to have negligible amounts of tryptase (Fig. 5.2).[12]

Development In Vitro of Human Basophils and Mast Cells

Effect of IL-3

Both mast cells and basophils, like other hematopoietic cells, are derived from bone marrow stem cells, and their growth and differentiation is regulated by cytokines. In particular, interleukin-3 (IL-3) has been demonstrated as the principal cytokine for human basophil growth and differentiation.[24] IL-3 selectively induces a significant formation of metachromatic cells from human bone marrow, and peripheral blood. Under certain circumstances other growth factors such as IL-4, IL-5, granulocyte macrophage-colony stimulating factor (GM-CSF), and transforming growth factor-β (TGF-β) function as regulators of human basophil differentiation.[25–29] Nerve growth factor (NGF) synergizes with GM-CSF or IL-5 and TGF-β synergizes with IL-3 to regulate basophil growth and differentiation.[26–29] Differentiation of human basophils from bone marrow precursors grown in IL-3 can be inhibited with very low concentrations of interferon (IFN-α and IFN-γ).

Both IL-4 and NGF can function as cofactors with IL-3 to support basophil production from human umbilical cord blood cells.[30] IL-4 has not been demonstrated to be a growth factor for human basophils either alone or in combination with IL-3 under culture conditions using serum. However, the ability of IL-4 to directly stimulate basophil growth in serum-free culture conditions is in accordance with the demonstration that immature basophils appear to express four times as many IL-4 receptors as mature basophils.[31] Although IL-5 has been considered to be an exclusive eosinophilopoieten by some workers,[32,33] it is clear that IL-5 can induce basophil differentiation from cord blood cultures and peripheral blood cells.[34] On the other hand, IL-6 does not exert any stimulation of histamine production by cord blood cells either by itself or in combination with IL-3 in culture systems with or without serum.

In contrast, it has been difficult to induce human progenitor cells to preferentially differentiate in vitro into mature mast cells. IL-3, IL-4, IL-5, IL-9, IL-10, GM-CSF, and NGF can each regulate the growth, differentiation, and granule maturation of mouse mast cells[35,45] but not mast cells from human progenitors.[46–51] Attempts to obtain, in liquid culture, significant amounts of mast cells from human precursors with human IL-3 as the only exogenous growth factor have been uniformly unsuccessful.

Human mast cell growth in vitro has been reported in response to IL-3 only with bone marrow cells in a liquid–solid interface culture system.[48] The major difference in the biological response to IL-3 between human and murine mast cells is reported to be a loss during evolution of mast cell high-affinity IL-3-binding sites.[52] However, recent studies have demonstrated receptors for IL-3 on in vitro-derived human mast cells.[53]

Effect of Kit Ligand and Stromal Cells

Furitsu and coworkers[54] demonstrated that development of mature mast cells can be achieved by coculture of umbilical cord blood mononuclear cells with a confluent layer of murine 3T3 fibroblast cell line. After 8 weeks of culture, the mast cells expressed both tryptase and chymase (MC_{TC}) and the high-affinity receptor for IgE. There was no synergy observed with IL-1, IL-2, IL-3, IL-4, IL-5, IL-6, or GM-CSF.[54] However, in subsequent studies these investigators found that the fibroblast-conditioned medium induced stem cells from human cord blood to differentiate into a population of immature mast cells that expressed tryptase but not chymase.[55] The fibroblast-conditioned media contained soluble kit ligand (KL). Because neither recombinant nor purified KL alone induced human progenitor cells to fully differentiate into mature mast cells, the fibroblast-induced differentiation and maturation process of cord blood progenitor cells into mast cells is dependent on either membrane-bound KL or another fibroblast-derived factor. In contrast, the mast cells derived from human fetal liver cells cocultured with murine 3T3 fibroblasts were predominantly of the MC_T phenotype, did not express FcεRI on their cell surface, and appeared immature by electron microscopy.[56] This implies either that the progenitors from the two sources were different or that there were other cells in the cord blood culture providing accessory mast cell regulatory cytokines which induced chymase expression.

Human mast cells have been induced in culture from a variety of sources such as bone marrow, umbilical cord blood, fetal liver cells, and peripheral blood with recombinant human (rh)KL.[55–60] KL is the ligand for the receptor that is encoded by the c-kit protooncogene. KL is produced either as a mature protein expressed on stromal cell surfaces or as a soluble form. The membrane form of the molecule is determined by tissue-specific alternative splicing and may function as a homing receptor for cells expressing receptors for KL. KL is a member of the tyrosine kinase receptor family.[61] rhKL maintains mast cell survival, promoting the proliferation of immature mast cells. The fact that mast cells generated with rhKL are immature suggests that a factor(s) other than soluble KL is needed to achieve full mast cell maturation.

Although consistent results have shown that rhKL alone promotes selective differentiation to immature mast cells, Li et al. have demonstrated that progenitor cells can be induced to differentiate into morphologically mature mast cells that expressed FcεRI and mast cell-specific proteases tryptase, chymase, and carboxypeptidase from bone marrow and umbilical cord blood cells cultured in conditioned medium from a human adherent cell strain, HBM-M.[60] The HBM-M cell strain was originally derived form a patient with mastocytosis. The experiments showed that HBM-M-conditioned medium (HBM-M-CM) could induce human progenitor cells to differentiate into a mast cell lineage after only 1 week of culture, in the absence of exogenous KL. However, rhKL was required to maintain the long-term survival of the mast cells.[60] The mast cells derived from human bone marrow using rhKL alone did not bind human IgE in contrast to the cells that were grown in HBM-M-CM. Interestingly, the addition of rhKL to HBM-M-CM resulted in downregulation of IgE binding. In both culture systems, reverse transcriptase-polymerase cham reaction (RT-PCR) analysis revealed the presence of the transcripts that encode the α-chain of FcϵRI. Mast cells derived by culturing human fetal liver cells in the presence of rhKL alone also do not bind human IgE but occasionally in some cultures have the FcϵRI transcript.

In contrast to these studies, Valent and co-workers induced differentiation of human mast cells from bone marrow using rhKL in long-term cultures and reported that these cells bound human IgE.[59] In subsequent studies, Li et al.[62] performed double immunoenzymatic analysis to determine the sequential development of expression of tryptase and chymase in umbilical cord blood cells cultured with HBM-M-CM and rhKL. It has been suggested by Schwartz and coworkers that, in the human, MC_T and MC_{TC} cells develop along distinct lineage pathways. The evidence they present is, first, that in T-cell immunodeficiency states such as acquired immunodeficiency syndrome there is a selective decrease in MC_T and not in the MC_{TC} subtype,[63] and second, that immature human mast cells have specific ultrastructural characteristics consistent with distinct progenitors for MC_T and MC_{TC}. The MC_T have scroll patterns in their granules whereas the MC_{TC} exhibit a lattice substructure while only regions lacking this protease exhibit complete scroll patterns. However, Weidner and Austen[64] have presented data to show that individual mast cell granules possess mixed morphology with fea-

tures of complete scrolls and lattice patterns in the same granule. The model proposed by Schwartz and colleagues is at odds with that proposed by other workers for mast cell differentiation in rodents, in which one can reversibly alter the expression of serine proteases during in vitro differentiation of rodent mast cell progenitors.

The exact relationship of MC_{TC} to MC_T has not been clearly elucidated. However the study by Li et al.[62] using HBM-M-CM and rhKL provides evidence that different mast cell subtypes develop depending on the cytokine content of the culture. KL induced mast cells predominantly of two immunophenotypes, MC_T and MC_{TC}. In contrast, the HBM-M-CM induced mast cell subtype MC_{TC} and a third subtype MC_C, positive for chymase but negative for tryptase from both bone marrow and umbilical cord blood progenitors. In umbilical cord blood cultures supplemented with both HBM-M-CM and rhKL, the number of MC_C cells was upregulated from 2.7% to 34.0% of the total cells in the culture. In contrast, the number of MC_T cells declined from 54.3% to 21.5% at day 49 of culture. In bone marrow cultures supplemented with HBM-M-CM alone, the MC_C subtype represented 100% of the mast cells at day 10 of culture.[62] This study clearly demonstrated that factors other than KL are critical for inducing the differentiation and maturation of mast cells. Also, a third type of mast cell MC_C expressing chymase without concomitant expression of tryptase can be induced in vitro from human progenitors. In addition tryptase and chymase, two mast cell-specific proteases, can be differentially expressed in, in vitro-derived human mast cells by changing the cytokine combination of the culture. In the studies by Li et al.,[60,62] the metachromatic cells staining positive for tryptase and chymase with high-affinity receptor for IgE, FcϵRI, possessed multilobed nuclei resembling basophils. Nuclear morphology has been used as one of the criteria to assess the lineage of metachromatic cells of hemopoietic origin in different species, including humans. Thus, one should be cautious about categorizing in vitro-derived cell populations as mast cells or basophils based on nuclear morphological criteria.

Effect of Kit Ligand, IL-6 and IL-3, IL-4, IL-13, and PGE$_2$

Kit ligand has been identified as a critical cytokine for the development of human mast cells in culture. However, the cells derived with KL as the only exogenous cytokine are immature using a number of criteria.

Mature human mast cells have been derived from CD34+ or mononuclear cell populations from umbilical cord blood in the presence of KL and IL-6.[53] After 8 weeks of culture, all the cells were reported to be immunoreactive for tryptase but only 20 to 30% were immunoreactive for chymase. IL-6 enhanced KL-dependent mast cell growth from the purified CD34+ population but not when cord blood mononuclear cells were cultured instead of CD34+ cells. Prostaglandin E (PGE$_2$) was necessary for enhancing KL- and IL-6-dependent development of mast cells from cord blood mononuclear cells, probably by blocking GM-CSF secretion from accessory cells.[65]

The cells had IgE receptors that could be sensitized to release mediators on challenge with anti-IgE antibody.[53] IL-3, IL-4, IL-5, or IL-6 promoted the in vitro survival of the mast cells that were generated from cord blood cultures in the presence of rhKL and IL-6. Although IL-3, IL-4, IL-5, and IL-6 synergized with rhKL in inducing mast cells, these cytokines by themselves failed to induce mast cell growth. The mast cells were shown to express m-RNA for IL-3 receptor (R) as well as IL-4R, IL-5R, and IL-6R α-chains. In contrast IL-2, IL-9, IL-10, IL-11, TNF-α and TGF-β$_1$ and NGF did not have any mast cell survival-promoting activity. Radioligand binding studies also revealed that these mast cells possessed receptors for IL-3. These results are in contrast to those reported by Valent and coworkers.[52] Durand and coworkers[66] have also developed long-term cultures of immature human mast cells in serum-free medium in the presence of rhKL and IL-3. Removal of either rhKL or IL-3 from the culture induced apoptosis, indicating that the mast cells were dependent on both rhKL and IL-3 for their survival. However, human lung mast cells have been reported not to have detectable levels of IL-3R.[52] It is possible that IL-3R may be present on mast cell precursors and that during mast cell development the density of IL-3R is decreased and that mature mast cells do not express detectable amounts of IL-3R. This is further supported by the finding that IL-3 can induce KL receptors in human bone marrow precursors.[66] Razin and colleagues have reported that human bone marrow cells grown in a conditioned medium derived from lectin-activated human peripheral white blood cells differentiated into cells that were characterized as mast cells.[67,68] Interestingly, these cells also responded by proliferating when IL-3 was added to the conditioned medium, implying that human mast cells have functional IL-3 receptors.[68]

Nilsson et al. reported that addition of IL-4 with rhKL resulted in a downregulation of c-kit expression and inhibition of the development of tryptase-positive cells whereas a combination of IL-3 and KL did not facilitate human mast cell differentiation.[58] In contrast, IL-13, which has a number of activities similar to IL-4, did not downregulate kit expression or tryptase during differentiation of cord blood-derived mast cells in the presence of KL.[69] Li et al. however derived mature mast cells from human umbilical cord blood cells in the absence of exogenous KL.[60]

Although KL is an essential growth factor for human mast cells, the studies outlined here indicate that it requires the presence of additional cytokines such as IL-6, IL-3, and other putative mast cell growth factors for maturation to occur. Because IL-3, IL-4, IL-5, and IL-6 are mainly produced by T-helper 2 lymphocytes, it suggests that in allergic or inflammatory conditions the T-helper 2 phenotype may regulate the function of human mast cells in vivo via specific receptors.

In Vitro Studies on Mast Cells and Basophils Utilizing Cell Lines

A number of in vitro studies have utilized the putative human mast cell line HMC-1, which was originally considered to be typical of human mast cells.[70] However, caution should be taken in interpreting data obtained with the HMC-1 cells as these cells lack a number of features characteristic of human mast cells.[71] We have recently demonstrated (unpublished observations) that these cells also bind the putative basophil-specific antigen Bsp-1. It is noteworthy that the HMC-1 cell line has been demonstrated to have two point mutations within the protein kinase domain of the receptor for KL (see Chapter 3). Interestingly, this same mutation has been identified in some patients with mastocytosis.

Development of Human Mast Cells and Basophils In Vivo

The majority of studies quoted here giving data on specific cytokines inducing mast cell proliferation have been obtained from in vitro work. However, this does not constitute conclusive evidence that in vitro activity can be translated to in vivo function, as is illustrated by the in vivo results obtained with IL-3 administered for primates. Valent and coworkers administered IL-3 to rhesus monkeys who developed a marked peripheral basophilia and increased production of basophilic/eosinophil progenitors from bone marrow. Interestingly, when these investigators examined the local inflammatory reaction it was noted that there were increased numbers of mast cells,[72] lending further support for a role of IL-3 in primate mast cell growth.

Both in humans and in monkeys, recombinant human KL administered subcutaneously resulted in mast cell proliferation. In monkeys administered KL, mast cell populations increased dramatically in a variety of tissues and organs; this was maintained until treatment with KL was discontinued, when a rapid decline in the number of mast cells was demonstrated.[73] In humans, 7 to 14 days of subcutaneous administration of KL resulted in a significant increase in dermal mast cells at sites distal to the injection site.[74]

Conclusions

Although rodent mast cells have been differentiated in vitro from a variety of sources using IL-3 and KL, human mast cells have been extremely difficult to differentiate using in vitro culture systems. However, in recent years, a number of groups have been able to differentiate human mast cells in vitro using a number of culture systems (Fig. 5.3). In particular, KL, as the only exogenous cytokine apart from those found in serum, promotes selective differentiation of human progenitors to mast cells. However, the mast cells derived from human progenitors when KL alone is used are immature. The mast cells derived by KL are mainly of the MC_T type, with a few MC_{TC}. Mature human mast cells have been derived from CD34+ or mononuclear cell populations from umbilical cord blood in the presence of both KL and IL-6. These cells were functionally mature and could be activated through their IgE receptor. These cells were also shown to possess functional IL-3 receptors. In contrast, Li et al.[60,62] have derived mature mast cells expressing both tryptase and chymase from both umbilical cord blood and bone marrow precursors in the absence of KL using a conditioned medium from an adherent cell strain. These cells appeared mature and, by double immunoenzymatic analysis, were mainly of the MC_{TC} and MC_C types. Interestingly, the mast cells derived with KL alone did not express receptors for IgE and did not have message for the high-affinity α-chain of the IgE receptor. Furitsu and coworkers[54] have derived mature mast cells using umbilical cord mononuclear cells with 3T3 fibroblasts in a coculture system. Although human IL-3 is a growth factor for basophils, there is some experimental evidence to suggest that IL-3 may influence mast cell differentiation both in vitro and in vivo, possibly by indirect effects. The demonstration of IL-3 receptors on in vitro-derived human mast cells further provides evidence that IL-3 may affect human mast cell differentiation. Current evidence would suggest that KL is an important growth factor for human mast cells, as demonstrated in vitro and in vivo; however, other factors must be involved for mast cells to reach full maturity (Fig. 5.4).

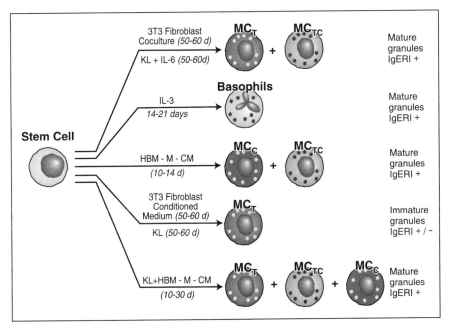

FIGURE 5.3. Human mast cell phenotypes defined by tryptase and chymase expression in mast cells derived in vitro from human stem cells derived from a variety of sources.

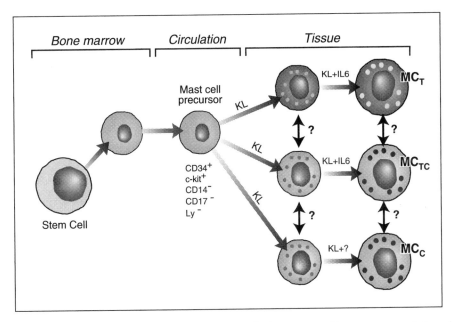

FIGURE 5.4. Proposed differentiation pathway of stem cells into mast cell precursors and the effect of kit ligand and hypothetical tissue differentiation of mast cell subtypes in vivo.

References

1. Fodinger M, Fritsch G, Winkler K, et al. Origin of human mast cell: development from transplanted haematopoietic stem cells after allogeneic bone marrow transplantation. Blood 1994;84:2954–2959.

2. Agis H, Willheim M, Sperr WR, et al. Monocytes do not make mast cells when cultured in the presence of SCF. Characterization of the circulating mast cell progenitor as a c-kit$^+$, CD34$^+$, Ly$^-$, CD14$^-$, CD17$^-$, colony-forming cell. J Immunol 1995;151:4221–4227.

3. Rottem M, Okada T, Goff JP, et al. Mast cells cultured from the peripheral blood of normal donors and patients with mastocytosis originate from a CD34$^+$/FcεRI$^-$ cell population. Blood 1994;84:2489–2496.

4. Smith TJ, Weis JH. Mucosal T cells and mast cells share common adhesion receptors. Immunol Today 1996;17:60–64.

5. Guo CB, Kagey-Sobotka A, Lichtenstein LM, et al. Immunophenotyping and functional analysis of purified human uterine mast cells. Blood 1992;79:708–712.

6. Shimizu Y, Irani AM, Brown EJ, et al. Human mast cells derived from fetal liver cells cultured with stem cell factor express a functional CD51/CD61 (αvβ3) integrin. Blood 1995; 86:930–939.

7. Guo CB, Liu MC, Galli SJ, et al. The histamine containing cells in the late phase response in the lung are basophils. J Allergy Clin Immunol 1990;85:172.

8. Rothenburg ME, Caulfield JP, Austen KF, et al. Biochemical and morphological characterization of basophilic leucocytes from two patients with myelogenous leukemia. J Immunol 1987;138: 2616–2625.

9. Dvorak AM. Basophil and mast cell degranulation and recovery. In: Harris RJ, ed. Blood Cell Biochemistry. New York: Plenum Press, 1991; 4:1.

10. Irani AA, Schechter NM, Craig SS, et al. Two types of human mast cells that have distinct neutral protease compositions. Proc Natl Acad Sci USA 1986;83:4464–4468.

11. Irani AM, Schwartz LB. Mast cell heterogeneity. Clin Exp Allergy 1989;19:143–155.

12. Castells MC, Irani AM, Schwartz LB. Evaluation of human peripheral blood leukocytes for mast cell tryptase. J Immunol 1987;138:2184–2189.

13. Valent P, Besemer J, Kishi K, et al. IL-3 promotes basophilic differentiation of KU812 cells through high affinity binding sites. J Immunol 1990;145:1885–1889.

14. Valent P, Majdic O, Maurer D, et al. Further characterization of surface membrane structures expressed on human basophils and mast cells. Int Arch Allergy Appl Immunol 1990;91:198–203.

15. Bochner BS, McKelvey AA, Sterbinsky SA, et al. IL-3 augments adhesiveness for endothelium and CD11b expression in human basophils but not neutrophils. J Immunol 1990;145:1832–1837.

16. Stain C, Stockinger H, Scharf M, et al. Human blood basophils display a unique phenotype including activation linked membrane structures. Blood 1987;70:1872–1879.

17. Lerner NB, Nocka KH, Cole SR, et al. Monoclonal antibody YB5.B8 identifies the human c-kit protein product. Blood 1991;77:1876–1883.

18. Bodger MP, Newton LA. The purification of human basophils: their immunophenotype and cytochemistry. Br J Haematol 1987;67:281–284.

19. Bodger MP, Mounsey GL, Nelson J, et al. A monoclonal antibody reacting with human basophils. Blood 1987;69:1414–1418.

20. Irani AM, Goldstein SM, Wintroub BU, et al. Human mast cell carboxypeptidase: selective localization to MC$_{TC}$ cells. J Immunol 1991; 147:247–253.

21. Schechter NM, Irain AA, Sprows JL, et al. Identification of a cathepsin G-like proteinase in the MC$_{TC}$ type of human mast cell. J Immunol 1990;145:2652–2661.

22. Schwartz LB. Monoclonal antibodies against human mast cell tryptase demonstrate shared antigenic sites on subunits of tryptase and selective localization of the enzyme to mast cells. J Immunol 1985;134:526–531.

23. Weidner N, Austen KF. Heterogeneity of mast cells at multiple body sites: fluorescent determination of avidin binding and immunofluorescent determination of chymase, tryptase, and carboxypeptidaase content. Pathol Res Pract 1993;189:156–162.

24. Valent P, Schmidt G, Besemer J, et al. Interleukin-3 is a differentiation factor for human basophils. Blood 1989;73:1763–1769.

25. Denburg JA, Silver JE, Abrams JS. Interleukin-5 is a human basophilopoietin: induction of histamine content and basophilic differentiation of HL-60 cells and of peripheral blood basophil-eosinophil progenitors. Blood 1991;77:1462–1468.

26. Tsuda T, Wong D, Dolovich J, et al. Synergistic effects of nerve growth factor and granulocyte-macrophage colony-stimulating factor on human

basophilic cell differentiation. Blood 1991;77: 971–979.

27. Tsuda T, Switzer J, Bienenstock J, et al. Interactions of hemopoietic cytokines on differentiation of HL-60 cells. Nerve growth factor is a basophilic lineage-specific co-factor. Int Arch Allergy Appl Immunol 1990;91:15–21.

28. Matsuda H, Coughlin MD, Bienenstock J, et al. Nerve growth factor promotes human hemopoietic colony growth and differentiation. Proc Natl Acad Sci USA 1988;85:6508–6512.

29. Sillaber C, Geissler K, Eher R, et al. Type β transforming growth factor promotes IL-3 dependent differentiation of human basophils but inhibits IL-3 dependent differentiation of human eosinophils. Blood 1992;80:634–641.

30. Richard A, McColl SR, Pelletier G. Interleukin-4 and nerve growth factor can act as cofactors for interleukin-3-induced histamine production in human umbilical cord blood cells in serum-free culture. Br J Haematol 1992;81:6–11.

31. Valent P, Besemer J, Kishi K, et al. Human basophils express interleukin-4 receptors. Blood 1990;76:1734–1738.

32. Campbell HD, Tucker WQ, Hort Y, et al. Molecular cloning, nucleotide sequence, and expression of the gene encoding human eosinophil differentiation factor (interleukin 5). Proc Natl Acad Sci USA 1987;84:6629–6633.

33. Yokota T, Coffman RL, Hagiwara H, et al. Isolation and characterization of lymphokine cDNA clones encoding mouse and human IgA-enhancing factor and eosinophil colony-stimulating factor activities: relationship to interleukin 5. Proc Natl Acad Sci USA 1987; 84:7388–7392.

34. Denburg JA, Silver JE, Abrams JS. Interleukin-5 is a human basophilopoietin: induction of histamine content and basophilic differentiation of HL-60 cells and of peripheral blood basophil-eosinophil progenitors. Blood 1991;77:1462–1468.

35. Ihle JN, Keller J, Oroszlan S, et al. Biological properties of homogeneous interleukin 3. I. Demonstration of WEHI-3 growth-factor activity, mast cell growth-factor activity. P cell-stimulating factor activity and histamine-producing factor activity, colony-stimulating factor activity, and histamine-producing cell-stimulating factor activity. J Immunol 1983;131:282–287.

36. Razin E, Ihle JN, Seldin D, et al. Interleukin 3: a differentiation and growth factor for the mouse mast cell that contains chondroitin sulfate E proteoglycan. J Immunol 1984;132: 1479–1486.

37. Lee F, Yokota T, Otsuka T, et al. Isolation and characterization of a mouse interleukin cDNA clone that expresses B-cell stimulatory factor 1 activities and T-cell- and mast-cell-stimulating activities. Proc Natl Acad Sci USA 1986;83: 2061–2065.

38. Hamaguchi Y, Kanakura Y, Fujita J, et al. Interleukin-4 as an essential factor for in vitro clonal growth of murine connective tissue-type mast cells. J Exp Med 1987;165:268–273.

39. Hultner L, Druez C, Moeller J, et al. Mast cell growth-enhancing activity (MEA) is structurally related and functionally identical to the novel mouse T cell growth factor P40/TCGFIII (interleukin 9). Eur J Immunol 1990;20:1413–1416.

40. Eklund KK, Ghildyal N, Austen KF, et al. Induction by IL-9 and suppression by IL-3 and IL-4 of the levels of chromosome 14-derived transcripts that encode late-expressed mouse mast cell proteases. J Immunol 1993;151:4266–4273.

41. Thompson-Snipes L, Dhar V, Bond MW, et al. Interleukin 10: a novel stimulatory factor for mast cells and their progenitors. J Exp Med 1991;173:507–510.

42. Ghildyal N, McNeil HP, Gurish MF, et al. Transcriptional regulation of the mucosal mast cell-specific protease gene, MMCP-2, by interleukin 10 and interleukin 3. J Biol Chem 1992; 267:8473–8477.

43. Ghildyal N, McNeil HP, Stechschulte S, et al. IL-10 induces transcription of the gene for mouse mast cell protease-1, a serine protease preferentially expressed in mucosal mast cells of Trichinella spiralis-infected mice. J Immunol 1992;149:2123–2129.

44. Matsuda H, Kaman Y, Ushio H, et al. Nerve growth factor induces development of connective tissue-type mast cells in vitro from murine bone marrow cells. J Exp Med 1989;174: 7–14.

45. Bressler RB, Thompson HL, Keffer JM, et al. Inhibition of the growth of IL-3-dependent mast cells from murine bone marrow by recombinant granulocyte macrophage-colony-stimulating factor. J Immunol 1989;143:135–139.

46. Ishizaka T, Dvorak AM, Conrad DH, et al. Morphologic and immunologic characterization of human basophils developed in cultures of cord blood mononuclear cells. J Immunol 1985;134:532–540.

47. Ishizaka T, Conrad DH, Huff TF, et al. Unique features of human basophilic granulocytes de-

veloped in in vitro culture. Int Arch Allergy Appl Immunol 1985;77:137–143.

48. Kirshenbaum AS, Goff JP, Dreskin SC, et al. IL-3-dependent growth of basophil-like cells and mastlike cells from human bone marrow. J Immunol 1989;142:2424–2429.

49. Kirshenbaum AS, Goff JP, Kessler SW, et al. Effect of IL-3 and stem cell factor on the appearance of human basophils and mast cells from CD34+ pluripotent progenitor cells. J Immunol 1992;148:772–777.

50. Saito H, Hatake K, Dvorak AM, et al. Selective differentiation and proliferation of hematopoietic cells induced by recombinant human interleukins. Proc Natl Acad Sci USA 1988;85: 2288–2292.

51. Mayer P, Valent P, Schmidt G, et al. The in vivo effects of recombinant human interleukin-3: demonstration of basophil differentiation factor, histamine-producing activity, and priming of GM-CSF-responsive progenitors in nonhuman primates. Blood 1989;74:613–621.

52. Valent P, Besemer J, Sillaber C, et al. Failure to detect IL-3-binding sites on human mast cells. J Immunol 1990;145:3432–3437.

53. Yanagida M, Fukamachi H, Ohgami K, et al. Effects of T-helper 2-type cytokines, interleukin-3 (IL-3), IL-4, IL-5, and IL-6 on the survival of cultured human mast cells. Blood 1995;86:3705–3714.

54. Furitsu T, Saito H, Dvorak A, et al. Development of human mast cells in vitro. Proc Natl Acad Sci USA 1989;86:10039–10043.

55. Mitsui H, Furitsu T, Dvorak AM, et al. Development of human mast cells from umbilical cord blood cells by recombinant human and murine c-kit ligand. Proc Natl Acad Sci USA 1993;90: 735–739.

56. Irani AA, Craig SS, Nilsson G, et al. Characterization of human mast cells developed in vitro from fetal liver cells cocultured with murine 3T3 fibroblasts. Immunology 1992;77:136–143.

57. Irani AM, Nilsson G, Miettinen U, et al. Recombinant human stem cell factor stimulates differentiation of mast cells from dispersed human fetal liver cells Blood 1992;80:3009–3021.

58. Nilsson G, Miettinen U, Ishizaka T, et al. Interleukin-4 inhibits the expression of Kit and tryptase during stem cell factor-dependent development of human mast cells from fetal liver cells. Blood 1994;84:1519–1527.

59. Valent P, Spanblochl E, Sperr WR, et al. Induction of differentiation of human mast cells from bone marrow and peripheral blood mono-

nuclear cells by recombinant human stem cell factor/kit-ligand in long-term culture. Blood 1992;80:2237–2245.

60. Li L, Macpherson JJ, Adelstein S, et al. Conditioned media from a cell strain derived from a patient with mastocytosis induces preferential development of cells that possess high affinity IgE receptors and the granule protease phenotype of mature cutaneous mast cells. J Biol Chem 1995;270:2258–2263.

61. Galli SJ, Zsebo KM, Geissler EN. The kit ligand, stem cell factor. Adv Immunol 1994;55:1–96.

62. Li L, Meng X, Krilis SA. Mast cells expressing chymase but not tryptase can be derived by culturing human progenitors in conditioned medium obtained from a human mastocytosis cell strain with c-kit ligand. J Immunol 1996;156: 4839–4844.

63. Irani AA, Craig SS, DeBlois G, et al. Deficiency of the tryptase-positive, chymase-negative mast cell type in gastrointestinal mucosa of patients with defective T lymphocyte function. J Immunol 1987;138:4381–4386.

64. Weidner N, Austen KF. Ultrastructural and immunohistochemical characterization of normal mast cells at multiple body sites. J Invest Dermatol 1991;96:26s–30s.

65. Saito H, Ebisawa M, Tachimoto H, et al. Selective growth of human mast cells induced by steel factor, IL-6, and prostaglandin E_2 from cord blood mononuclear cells. J Immunol 1996l;1557: 343–350.

66. Durand B, Migliaccio G, Yee NS, et al. Long-term generation of norman mast cells in serum-free cultures of CD34+ cord blood cells stimulated with stem cell factor and interleukin-3. Blood 1994;84:3667–3674.

67. Ratajczak MZ, Luger SM, DeRiel K, et al. Role of the KIT protooncogene in normal and malignant human hematopoiesis. Proc Natl Acad Sci USA 1992;89:1710–1714.

68. Gilead L, Bibi O, Razin E. Fibroblasts induce heparin synthesis in chondroitin sulfate E containing human bone marrow-derived mast cells. Blood 1990;76:1188–1195.

69. Nilsson G, Nilsson K. Effects of interleukin (IL)-13 on immediate-early response gene expression, phenotype and differentiation of human mast cells. Comparison with IL-4. Eur J Immunol 1995;25:870–873.

70. Butterfield JH, Weiler D, Dewald G, et al. Establishment of an immature mast cell line from a patient with mast cell leukemia. Leuk Res 1988;12:345–355.

71. Nilsson G, Blom T, Kusche-Gullberg M, et al. Phenotypic characteristic of the human mast cell line HMC-1. Scand J Immunol 1994;39:489–498.

72. Volk-Platzer B, Valent P, Radaszkiewicz T, et al. Recombinant human interleukin 3 induces proliferation of inflammatory cells and keratinocytes in vivo. Lab Invest 1991;64:557–566.

73. Galli SJ, Iemura A, Garlick DS, et al. Reversible expansion of primate mast cell populations in vivo by stem cell factor. J Clin Invest 1993;91:148–152.

74. Costa JJ, Demetri GD, Harrist TJ, et al. Recombinant human stem cell factor (Kit ligand) promotes human mast cell and melanocyte hyperplasia and functional activation in vivo. J Exp Med 1996;183:2681–2686.

6
Signal Transduction by Cytokines

John W. Schrader

Mast cells, like other cells in the body, receive multiple signals from their environment that regulate their generation, differentiation, and survival. These signals determine the number and type of mast cells in particular tissues in health and disease, as well as regulating their state of activity. Most of these extracellular stimuli are sensed by cell-surface receptors that activate the complex networks of intracellular signals which ultimately regulate expression of genes or control processes such as endocytosis, adhesion, motility, or secretion.

The cell-surface receptors on mast cells and their precursors interact with a wide range of molecules, some of which are soluble and others of which are integral components of the membranes of other cells or associated with extracellular matrices. These receptors themselves belong to diverse structural families; they include members of the families of tyrosine kinase receptors, hemopoietin receptors, G protein-linked receptors, integrins, and the immunoreceptor tyrosine-based activating motif (ITAM) immunoreceptors including lymphocyte antigen receptors and the receptors for the Fc fragments of immunoglobins.

In general, few if any of the intracellular signaling paths that are activated by a particular receptor are unique to that receptor. The activation of common intracellular signals downstream of different types of receptors in part reflects the fact that most ligand–receptor interactions regulate general cellular processes or "public functions" such as growth or apoptosis, as well as more ligand-specific, "private,"

aspects of cellular differentiation or function. It is only in a minority of instances that these private effects of a particular cytokine can be clearly attributed to its ability to trigger a distinctive intracellular signal. The best example of an intracellular signal that is specifically activated by two closely related cytokines and accounts for some of their distinctive functions is the activation of the transcriptional activator, STAT-6.[1-3] This occurs only in response to the interleukins (IL) IL-4 and IL-13 and results directly in up- or downregulation of genes with STAT-6-binding sites in their promoters. In most cases, however, the characteristic effects of a particular cytokine or combination of cytokines appear to reflect a distinctive quantitative and qualitative summation of signals, none of which alone are unique to activation of that receptor. Given the combinatorial nature of many of the molecular signals that regulate processes such as growth, apoptosis, and gene expression, it is not surprising that combinations of cytokines exhibit striking synergies and in some cases antagonisms.

Mast Cells and Cytokines

The murine mast cell can be readily grown as primary cultures in vitro from bone marrow precursors using IL-3.[4-8] Moreover, these mast cells can be induced to differentiate into a variety of phenotypes in vitro by exposure to different cytokines or mixtures of cytokines.[9-15] These primary cells have provided a useful model system to study signal transduction and in particular the similarities and differences between the actions of a range of cytokines. Mast cells played important roles in the early history of many of the cytokines discussed in this chapter, for example, IL-3, IL-4, IL-9, steel locus factor (SLF), IL-13, IL-15, and IL-10. Here I review general features of cytokines and their receptors and then discuss in detail IL-3-stimulated signal transduction. This is a useful paradigm for signal transduction by other cytokines, reflecting the fact that we understand far more about the similarities between signaling by different cytokines than we do about the differences.

The Four-Helix Bundle Cytokine Family

Cytokines such as IL-3, IL-4, IL-9, IL-13, and SLF are now all recognized as members of a large evolutionary "family" that includes not only cytokines, but also conventional hormones such as growth hormones (GH) and prolactin (PRL). The family includes many cytokines active in the lymphohemopoietic system such as IL-2, IL-3, IL-4, IL-5, IL-6, IL-7, IL-9, IL-11, IL-12, IL-13, IL-15, granulocyte-colony stimulating factor (G-CSF), granulocyte macrophage-CSF (GM-CSF), SLF, colony-stimulating factor-1 (CSF-1), Flt-3 ligand (Flt-3L), erythropoietin (EPO), thrombopoietin (TPO), oncostatin-M, and cytokines affecting other tissues such as leukemia inhibitory factor (LIF), ciliary neurophic factor, cardiotrophin, and leptin. Members of the family thus serve diverse functions, regulating not only inflammation and immunity, for example, IL-3, GM-CSF, IL-2, IL-6, IL-5, IL-12, IL-13, and IL-4, but also development, for example, IL-7, SLF, CSF-1, and Flt-3L, and homeostasis, such as GH, PRL, EPO, TPO, and leptin.

These molecules differ widely in their primary amino acid sequences, but are easily recognizable as members of a widely diverged evolutionary family by virtue of a common three-dimensional structure. This consists of a characteristic arrangement of four α-helices, first recognized in growth hormone. Cytokines of this family can be divided into two subgroups, each of which acts through a distinct class of cell-surface receptors. Both receptors ultimately act via activation of protein tyrosine kinase activity; however, in the one group this enzymatic activity is intrinsic to the cytoplasmic domain of the receptor, while in the other group the receptors belong to a novel class that acts through associations with cytoplasmic tyrosine kinases.

The Cytokine–Receptor Superfamily

Most of the "four-helix bundle cytokines" bind to this novel class of cell-surface receptors, which collectively have been termed the "cytokine-" or "hemopoietin-receptor superfamily".[16,17] Members of this family share characteristic structural motifs, both in their extracellular, ligand-binding domains and in their intracellular domains, which mediate signal transduction. The intracellular domains have no intrinsic enzymatic activity, but as discussed, ligand binding results in activation of a novel class of cytoplasmic tyrosine kinases, the Janus or JAK family kinases.[18,19] Members of this subgroup of cytokines are monomers, with the exception of IL-5, which has a unique structure of two interlinked four-helix bundles involving interdigitation of the α-helixes of the two polypeptide chains.[20]

Members of the smaller, second group of four-helix bundle cytokines bind to receptors with intrinsic tyrosine kinase activity and include SLF (also called stem cell factor), colony-stimulating factor-1 (CSF-1), and Flt-3 ligand (Flt3L). These receptors resemble those for platelet-derived growth factors and fibroblast growth factors in having extracellular regions made up of a series of immunoglobulin domains and cytoplasmic regions with a protein tyrosine kinase activity. Members of this subgroup of four-α-helix bundle cytokines also differ in that they are homodimers, not monomers.

The evolutionary history of this split in the four-helix bundle family of cytokines into the dimeric, cytokines interacting with tyrosine kinase receptors, and the monomers interacting with cytokine superfamily receptors, is obscure.

Interestingly, the four-helix cytokines that interact with protein tyrosine kinase receptors all appear to have a role in embryonic development; for example, SLF is critical for aspects of development of the nervous system, skin, and lymphohemopoietic system, including the development of mast cells.[21] Both SLF and CSF-1 occur in both soluble and membrane-bound forms.[22]

Ligand-Induced Oligomerization of Cytokine Receptors

Binding of a homodimeric four-helix bundle cytokine such as SLF to its receptor is followed by binding of this ligand–receptor complex to a second receptor. The resultant homodimerization of the two receptors results in approximation of their intracytoplasmic catalytic domains, which become tyrosine phosphorylated and activated.

The larger group of monomeric cytokines that bind to receptors of the cytokine receptor superfamily also activate their receptors by inducing oligomerization. The simplest receptors of the family are those for growth hormone or erythropoietin, and are made up of single subunits that dimerize following ligand binding. The best-studied example is that of growth hormone.[23] After GH binds to its receptor subunit, the low-affinity GH-GHR complex then interacts with a second identical receptor subunit. This occurs through a second binding site on GH and a second site on the GHR, resulting in a high-affinity ligand–receptor complex with a stoichiometry of 1:2. Formation of this stable oligomeric complex leads to the approximation of the intracellular domains of the receptor subunits, which are weakly associated with members of the JAK kinase family. Ligand binding stabilizes association of JAK kinases with the receptor complex and leads to tyrosine phosphorylation of the JAK kinase and consequent activation of their tyrosine kinase activity.

In most cases, however, the functional high-affinity receptors for four-helix bundle cytokines are more complex hetero-oligomers, composed of two or more subunits of the cytokine receptor family. Typically, one of these subunits has detectable affinity for the ligand and functions as a low-affinity ligand-specific receptor. This ligand-binding subunit is usually termed the α-chain of the receptor. The complex of ligand and its low-affinity receptor then binds to a second chain, resulting in increased affinity of binding. This second chain usually has no measurable affinity for ligand and is termed a β-chain. This trimeric complex of ligand and α- and β-receptor subunits may in some cases constitute the functional receptor. However, in some instances the functional receptor is formed by interaction of the ligand–α-β complex with a third subunit, or, as is the case with IL-6, by interaction of two ligand–α-β complexes to form a hexameric complex.[24,25]

Typically, individual β-chains can interact with multiple ligand–α-chain complexes and thus participate in the formation of high-affinity receptors for multiple cytokines. In this way, a single β-chain can be involved in the function of receptors for a family of cytokines, and, because the β-chain in general plays a dominant role in signal transduction, this will result in strong similarities in the intracellular signals generated in response to cytokines of this family. For example, β_c is a component of the receptors for IL-3, IL-5, and GM-CSF, and likewise the IL-2Rβ is a component of the receptors for IL-2 and IL-15. There are many variations on these themes. The α-chain of the IL-4 receptor binds IL-4 but not IL-13, but nevertheless serves as a β-chain for the high-affinity IL-13 receptor, accounting for the many similarities in signals triggered by IL-4 and IL-13 (see following). The receptors for IL-2 and IL-15 are slightly different in that they each contain a third ligand-binding subunit that is not a member of the cytokine receptor superfamily.

Interferons and IL-10

The interferons share broad structural similarities with the four-helix bundle cytokines and receptors of the cytokine superfamily. Interleukin-10 (IL-10), which has effects on mast cell growth and differentiation, is a member of this interferon superfamily. The three-dimensional structure of IL-10 resembles that of the four-helix bundle cytokines, although, as in the case of interferon-α, there are additional helices.[26] The receptors for members of the interferon family are likewise structurally related to the receptors of the cytokine superfamily. Moreover, the functional high-affinity receptors are once again heterodimers of structurally related subunits that are associated with kinases of the JAK family. Binding of ligand to these receptors also results in dimerization of receptor subunits and activation of the JAK kinases and members of the STAT family.[18,19]

IL-3

IL-3 is a typical four-helix-bundle cytokine[27] and is a key mediator of the T-cell-dependent expansion of mast cell numbers that is characteristic of many parasitic infections and allergic responses. Some of these mast cells are generated by the expansion and differentiation of morphologically undifferentiated committed mast cell precursors present in tissues.[28,29] However, IL-3 also acts on pluripotential hemopoietic stem cells and most of their progeny.[30] Thus, IL-3 stimulates, alone or in combination with other factors, the growth and differentiation of progenitors of mast cells, erythrocytes, megakaryocytes, neutrophils, macrophages, and B lymphocytes. IL-3 also regulates the survival and function of mature cells such as mast cells, eosinophils, macrophages, and megakaryocytes. Like SLF, IL-3 potentiates the release of mediators from mast cells or basophils.[10,31–34] Depending on its target IL-3 stimulates growth, promotes survival, and regulates the state of cellular differentiation or effector function.

IL-3-Induced Receptor Oligomerization

The IL-3 receptor is made up of two subunits.[35–39] IL-3 binds with relatively low affinity to a specific α-chain termed the IL-3 receptor-alpha (IL-3Rα). A high-affinity complex is formed by interaction of the IL-3/IL-3Rα complex with β_c,

another member of the hemopoietin receptor superfamily.[37] β_c also interacts with complexes of IL-5 and GM-CSF with their respective specific α-chains, and is thus a component of the high-affinity receptors for IL-5 and GM-CSF as well as that for IL-3.[39,40] The situation in the mouse is a little more complex; the β_c gene has been duplicated giving rise to β_{IL-3}, which is closely related to β_c but interacts exclusively with IL-3Rα and itself has a low affinity for IL-3.[41,42] The cytoplasmic domain of the IL-3Rα is relatively short and contains a Pro Pro-X-Pro motif characteristic of the cytoplasmic domains of receptors of the hemopoietin super family, termed Box-1. The cytoplasmic domain of IL-3Rα is essential for function of the receptor (Orban and Schrader, unpublished data), although its precise function is unclear. It is often assumed that it binds JAK-2 kinase, which is known to associate with β_c.[43,44] IL-3-induced dimerization of IL-3Rα and β_c could thus lead to approximation of two molecules of JAK kinase and their cross-phosphorylation and activation.

An alternative view is that JAK-2 kinase is only associated with β_c or β_{IL-3} and that activation of intracellular signaling involves dimers of the β-chain. In favor of this view are experiments showing that in certain cell lines dimerization of the cytoplasmic domain of β_c can generate a growth signal.[45–47] There is however also some experimental evidence in favor of the notion that a simple approximation of the cytoplasmic domains of IL-3Rα and β_c is sufficient to initiate intracellular signaling. Thus, expression in IL-3-dependent cells of artificially generated heterodimers of the cytoplasmic domains of IL-3Rα and β_c, generated signals that maintained cellular growth and viability in the absence of IL-3 (Orban et al., unpublished data). These experiments involved chimeric proteins made up of the extracellular region of human CD8 and the cytoplasmic domain of either IL-3Rα or β_{IL-3}. The extracellular domains of CD8 spontaneously form disulfide-linked dimers, thus generating simple homo- or heterodimers of the cytoplasmic domains of IL-3Rα and β_c. The choice of the extracellular domain of CD8 as a dimerization agent is important; experiments involving chimeric proteins in which the cytoplasmic domain of β_c or β_{IL-3} is fused to the extracellular domain of another receptor of the hemopoietin receptor family are difficult to interpret because it is hard to exclude the possibility that the extracellular domain is recruiting another chain. Cells expressing either CD8/IL-3Rα or CD8/β_{IL-3} chimeric proteins alone, failed to live in the absence of IL-3, indicating that simple homodimers of the cytoplasmic domains of either IL-3Rα or β_{IL-3} were unable to initiate signals for survival and growth. In contrast, cells coexpressing both chimeric proteins grew continuously in the absence of IL-3, suggesting that the presence of heterodimers of the cytoplasmic domains of IL-3Rα and β_{IL-3} was sufficient for the initiation of signaling. While these results do not exclude the possibility that the cytoplasmic domains of IL-3Rα and β_c formed higher-order aggregates, they show that homodimers of β_c are not sufficient for signaling. This point is significant because in the case of the IL-6 receptor there is physical evidence that the functional receptor is a hexamer made up of two copies of each of the ligand and the α- and β-chain of the receptor.[24,25]

Tyrosine Phosphorylation Events

The precise sequence of events following receptor oligomerization is uncertain but activation of tyrosine kinases occurs rapidly.[34,35,48] It is clear that tyrosine phosphorylation and activation of JAK-2 kinase is a very early event.[49] Tyrosine kinases of other families have also been implicated in signaling by IL-3 or its close relatives, including Fyn kinase,[50] Lyn kinase,[51] and Btk kinase.[52] While it is not certain which kinases phosphorylate which substrates, IL-3 binding is rapidly followed by tyrosyl phosphorylation of many different proteins. It should be noted that the β_c and β_{IL-3} of the IL-3 receptor are also rapidly phosphorylated on Ser/Thr following IL-3 binding, and this Ser/Thr phosphorylation results in a major reduction in its electrophoretic mobility.[37,53]

Tyrosine phosphorylation of β_c occurs early and creates docking sites for a series of cytoplasmic proteins with domains that recognize specific phosphotyrosines, either Src homology-2 (SH-2) domains or phosphotyrosine-binding (PTB) domains.[54] Many of these proteins in turn become phosphorylated on tyrosine. Identification of these tyrosine-phosphorylated proteins has been instrumental in implicating a range of the signal transduction proteins in IL-3-stimulated signaling. Here I discuss four such paths, all initiated by and dependent upon, tyrosine phosphorylation events; these are activation of Ras proteins, activation of MAP family kinases, activation of the PI-3 kinase pathway, and activation of STAT-5. I also mention two recently identified components of IL-3 signaling: a lipid phosphatase, SHIP, that may have an important role in downregulating signaling, and OP18/Stathmin, a protein which appears to directly regulate mitosis.

Activation of Ras Proteins

Binding of IL-3 to cells results in rapid activation of Ras proteins.[55,56] The Ras proteins are members of a large family of small GTPases that are encoded by three closely related, highly conserved genes: Ha, N, and Ki-Ras. They bind guanosine diphosphate (GDP) or guanosine triphosphate (GTP) and act as molecular switches that are triggered when Ras-bound GDP is exchanged for GTP.[57,58] The key to their function as biological switches lies in the fact that, when GDP is exchanged for GTP, two portions of the Ras protein, "switch I" and "switch II," undergo allosteric changes. These changes allow Ras molecules to bind to a series of downstream effector molecules normally present in the cytoplasm. This binding results in recruitment of these effectors to the membrane, as the Ras proteins are associated with the inner leaflet of the plasma membrane through a carboxy-terminal prenyl group (Fig. 6.1). The effectors that bind to GTP-bound Ras are diverse, both in structure and function. Some are enzymes like the p110 catalytic unit of PI-3 kinase or the Ser/Thr protein kinases Raf-1, A-Raf, B-Raf, or MEKK1. Recruitment of these enzymes by Ras results in their

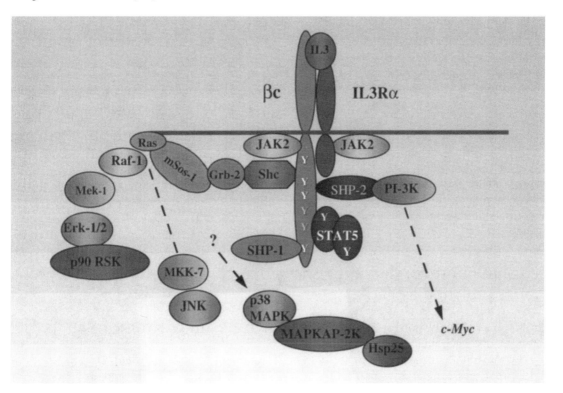

FIGURE 6.1. Signal Transduction Paths downstream of the IL-3 Receptor The two components of the receptor for IL-3, the β-common chain (βc) and the α-chain (IL-3Rα), are shown in the functional complex generated by interaction of IL-3 with their extracellular domain. Within the cell membrane (horizontal black line) the intracellular domains are shown interacting through their membrane proximal regions with JAK-2 kinases. Activation of the JAK-2 kinases is followed by phosphorylation of tyrosine residues in βc and these phosphotyrosines and surrounding amino acids are recognized by SH-2 domains of SHP-1, SHP-2 or STAT-5, or the PTB domain of SHC. SHC functions as an adaptor to recruit Grb-2-m-Sos-1 complexes to the membrane where they activate Ras and the downstream kinase cascade. The PI-3 kinase (PI-3K) pathway is activated by translocation of this enzyme to the receptor complex and thus to the membrane and its substrates. SHP-2 is one candidate for the adaptor that mediates this translocation. The so-called "stress-activated" kinases, JNK and p38 MAPK are probably downstream of Ras but the precise components of these paths are unclear.

activation by mechanisms that are not well understood. These enzymes in turn trigger signal transduction cascades, discussed next, that regulate cell division, the suppression of apoptosis, and the expression of genes.

Another set of proteins that bind activated Ras are termed GTP-ase activating proteins (GAPs). These GAPs bind specifically to the GTP-bound form of Ras and catalyze hydrolysis of the GTP to GDP, thus returning Ras to its inactive GDP-bound state. However, GAPs themselves also have effector activity. For example, p120 GAP interacts in turn with a protein termed p190, which is a GAP for the Rho family of GTPases that regulate cell shape and other processes. Activated Ras also binds to Ral GDS, a protein which promotes the exchange of guanine nucleotides on Ral, a GTPase that in turn regulates the Rho family of GTPases, which regulate cell shape.

The activation of Ras-GDP is mediated by interaction with a series of cytoplasmic proteins termed guanine-nucleotide exchange factors (GEF). This interaction promotes the exchange of Ras-bound GDP for GTP. The best-characterized GEF is termed mSos-1. GEFs, which are normally cytoplasmic proteins, can only activate membrane-bound Ras when they are translocated from the cytoplasm to the membrane. Translocation of GEF to the membrane occurs following activation of many types of cell-surface receptor and is mediated by so-called adaptor proteins.[59] Adaptor proteins have specialized domains that mediate interactions, including src homology (SH-2) or phosphotyrosine-binding (PTB) domains,[54] which bind to phosphotyrosines on receptor-associated molecules, and in some cases SH-3 domains, which bind to proline-rich motifs, for example, on GEF.

Tyrosine Phosphorylation of Shc and Its Function as an Adaptor

The role in the activation of Ras by growth factors of one such adaptor protein, Shc, was first demonstrated in mast cells. The 55-kDa protein that was prominently phosphorylated on tyrosine in response to stimulation of mast cells by IL-3,[53,60] proved to be Shc.[61] Shc can bind to specific phosphotyrosines through either a PTB or an SH-2 domain. Binding of Shc to β_c occurs via the PTB and is dependent on the presence in β_c of a particular tyrosine, Y577.[62] After binding to phosphorylated Y577, Shc is then itself tyrosine phosphorylated by tyrosine kinases associated with the activated IL-3R. The IL-3-stimulated tyrosine phosphorylation of Shc generates a docking site for another adaptor protein named Grb-2. Grb-2 has an SH-2 domain that recognizes phosphotyrosines on activated Shc and also has two SH-3 domains through which it associates with the GEF mSos-1. Thus, the activated IL-3 receptor associates indirectly, through Shc and Grb-2, with mSos-1, thereby recruiting mSos-1 from the cytoplasm to the membrane and into proximity with Ras (Fig. 6.1).

The notion that the IL-3-activated receptor is the nidus of a stable complex of associated proteins at the cell membrane is probably an oversimplification. Thus, after stimulation of mast cells with IL-3, most of the tyrosine-phosphorylated Shc with its associated Grb-2 is still found in this cytoplasm.[58] Likewise, most of the mSos-1 remains in the cytoplasm, although much of it becomes modified by Ser/Thr phosphorylation.[61]

Investigation of other cytokines including GM-CSF, IL-5, and SLF showed that they too stimulated tyrosine phosphorylation of Shc and activation of Ras.[56,63] However, there were two informative exceptions, IL-4 and its close relative IL-13. The same studies showed that IL-4 failed to induce the reduction in the electrophoretic mobility of mSos-1 observed following stimulation with IL-3 or SLF.[63] The fact that IL-4 also failed to activate Ras[55,56] or the Erk-1/2 MAP kinases downstream of Ras[63,64] correlated growth factor-induced tyrosine phosphorylation of Shc and its association with Grb-2 with activation of Ras.

MAP Family Kinases

Treatment of mast cells with IL-3 or SLF, but not with IL-4 or IL-13, results in activation of all the three major families of MAP kinases, the Erk-1 and -2 MAP kinases,[64] the JNK/SAP kinases,[65] and the p38 MAP kinases.[66] The MAP kinases are an ancient evolutionary family and are also found in yeast and plants. The JNK/SAP kinases and p38 MAP kinases are known as "stress-activated" kinases because they are activated by stresses such as heat, UV light, or hyperosmolarity, whereas the Erk or "extracellular regulated kinases" are activated by activation of multiple families of cell-surface receptors. However, as discussed next, it is now clear that members of all three families of MAP kinases are activated in mast cells by growth factors and ligation of a variety of cell-surface molecules. Kinases of all three families are activated by phosphorylation of the threonine and tyrosine residues in a TXY motif in the activating loop of the catalytic site of the enzyme. This motif is a TEY in the case of Erk kinases, a TPY in the case of JNK/SAP kinases, and TGY in the case of the p38 MAP kinases.

Erk MAP Kinases

The kinase cascade that begins downstream of Ras with activation of Raf-1, A-Raf, or B-Raf[67] culminates in activation of Erk-1 and Erk-2 MAP kinases. Binding of B-Raf to activated Ras results directly in stimulation of kinase activity; activation of Raf-1 and A-Raf seems to require phosphorylation of regulatory tyrosine residues. The Raf kinases phosphorylate and activate downstream MAP kinase kinases (MKK), named MEK-1 and MEK-2. MEK-1 and MEK-2 are dual-specificity kinases and phosphorylate specific threonine and tyrosine residues in the characteristic TEY motif present in the activation loop of the catalyst domain of Erk-1 and -2 MAP kinases. Constitutively activated forms of Raf-1 act as oncogenes, promoting growth and suppressing apoptosis in the absence of exogenous growth and survival factors. In addition to upregulating Erk-1 and Erk-2 activity, Raf-1 also may phosphorylate proteins such as Bad to directly regulate apoptosis. Erk kinases phosphorylate and activate the S6-kinase p90[rsk] (see Fig. 6.1), as well as transcription factors.

p38 MAP Kinases

Two-dimensional electrophoresis of proteins that were tyrosine phosphorylated in mast cells stimulated with either IL-3 or SLF revealed an unidentified 38-kDa protein.[57] The identity of this 38-kDa protein was established only recently as a result of work in a seemingly unrelated field. This research involved identification of the molecular target of a series of compounds had been empirically identified through their ability to block the release of tumor necrosis factor-α (TNF-α) and IL-1 from lipopolysaccharide-stimulated monocytes.[68] The target was shown to be a novel MAP family Ser/Thr kinase that was a mammalian homolog of a protein in *Saccharomyces cerevisiae*, HOG-1 MAP kinase, which was required for the survival of these yeast in hyperosmotic conditions. The mammalian protein is now termed p38 MAP kinase[69] and is a member of a family of four genes termed α, β, γ, and δ. Like HOG-1, all members of the family have a characteristic TGY motif in the activation loop of their catalytic domains. As with other members of the MAP family, enzymatic activity is increased following phosphorylation of the threonine and

tyrosine residues in this motif by a dual-specificity MAP kinase kinase (MKKs). Two such MAP kinase kinases termed MKK3 and MKK6 have been shown to phosphorylate and activate p38 MAP kinase in vivo. A third, termed MKK4, phosphorylates and activates p38 MAP kinase under artificial conditions, but under physiological conditions is probably only an activator of the JNK/SAP kinases. IL-3, GM-CSF, and SLF, but not IL-4, induce tyrosine phosphorylation and enzymatic activation of p38 MAP kinase. IL-3, however, has been reported to be unable to activate either MKK3 or MKK6; if verified, this result raises the possibility of the existence of an additional upstream activator of p38 MAP kinases.

JNK/SAP Kinases

The *jun* N-terminal kinases (JNK) or stress-activated protein kinases (SAPK) are encoded by three genes, which, through splicing variations, generate 10 isoforms. Like other Ser/Thr kinases of the MAP tyrosine family, they are activated by phosphorylation of specific threonine and tyrosine residues in the activation loop of the kinase domain, in this instance the characteristic motif being TPY. Two of the substrates of JNKs are Ser-63 and Ser-73 in the N-terminus of c-Jun, phosphorylation of which activates its transcriptional activity. However, JNKs also phosphorylate other transcription factors including ATF-2. IL-3 and SLF are potent activators of JNK/SAPK activity in normal bone marrow-derived mast cells and the mast cell line MC/9[66] (also, Foltz et al., unpublished data).

JNK/SAP kinase are activated by the dual-specificity kinase MKK4, both in vitro and in transient transfection systems. Treatment of MC/9 mast cells with SLF resulted in activation of MKK4.[66] Endogenous MKK4 is also reported to be activated in macrophages stimulated with TNF-α.[70] Recently we and others have identified a novel MKK termed MKK7, and shown that it phosphorylates and activates JNK/SAP kinase but not p38 kinase.[71,72,118] We have shown that endogenous MKK7 and MKK4 are both activated by IL-3. Ligation of the FcR on the mast cell line MC/9 also activates both MKK4 and MKK7.[118]

The activation of JNK by hemopoietins such as IL-3 appears to involve activation of Ras plus another unknown path. In an epithelial cell line, coexpression of activated Ras and MKK7 was sufficient to activate MKK7, but this was not the case in an IL-3-dependent cell line, Baf-3.[118] Coexpression of activated Cdc-42 or activated Rac-1, however, did result in activation of MKK7 in both epithelial and hemopoietic cell lines. The failure of IL-4 to activate either JNK/SAPK[66] or MKK7[118] is consistent with the postulated involvement of Ras.

Functions of MAP Family Kinases in Mast Cells

Many extracellular stimuli result in activation of not only the Erk-1 and Erk-2 MAP kinases but also the JNK/SAP kinases and p38 MAP kinases. There is evidence that Erk-1 and Erk-2 kinases are essential for growth. For example, an inhibitor of MEK-1 and MEK-2 kinases that blocks activation of Erk-1 and Erk-2 activation inhibits IL-3-stimulated proliferation. Likewise, expression of a dominant inhibitory active mutant of MEK-1 in IL-3-dependent cells results in decreased sensitivity to IL-3.[73]

As discussed, extracellular stimuli such as growth factors like SLF or IL-3, or ligation of FcR, also activate members of the two families of so-called stress-activated MAP family kinases, that is, the JNK and p38 MAP kinases. Their activation in response to growth factors such as IL-3 and SLF raises the question of whether these "stress-activated" kinases also have a role in growth. Likewise, their activation by cross-linking of FcR raises the possibility of their playing a role in processes such as exocytosis or cytokine secretion.

There is considerable evidence suggesting that c-Jun activity may be needed for growth; for example, the constitutively active v-*Jun* is an oncogene and promotes growth. The existence of three genes encoding multiple splice variants of related JNK family members raises the possibility of redundant functions, and it may take some time to clarify the functional roles of members of this family in particular cell types.

The same considerations apply to assessment of the role of p38-MAP kinases in growth. They too are activated in mast cells by IL-3 and SLF. The fact that SB203580, a specific inhibitor of the activity of both the α- and β-isoforms of p38-MAP kinase, blocks cellular growth and DNA synthesis in response to IL-3 (Lee and Schrader, unpublished data), suggests that the activity of one or both of these enzymes has a critical role in mitogenesis. Further work using genetic approaches to modifying p38 activity and JNK activity is required to clarify this issue.

The products of the four p38 MAP kinase genes appear to differ in their tissue distribution, activation, and substrate preferences,[74] although much remains to be learned. In particular the pattern of expression in mast cells is unknown, although it is clear from our data[60,66] that one or both of the α- and β-forms must be present. Thus, in hemopoietic cells both IL-3 and SLF activate MAP-KAP kinase 2 (see Fig. 6.1) via an isoform of p38-MAP kinase that is inhibited by SB203580, an analog of the antiinflammatory compounds that were originally shown to target p38-MAP kinase.[66] The existence of highly specific inhibitors of the α- and β-forms of p38-MAP-kinase such as SB203580 has provided some clues to the function of these enzymes. Inhibition of p38-MAP kinase activity in human monocytes dramatically inhibits the production of TNF-α and IL-1 by endotoxin-stimulated monocytes.[68] Likewise, IL-6 production by the murine endothelial cell line L929 in response to stimulation with TNF-α is blocked by inhibition of p38-MAP kinase activity by SB203580.[75]

In that cross-linking of FcR activates p38-MAP family kinases in macrophages[76] and mast cells (Foltz, unpublished data), it might have been predicted that the production of cytokines such as TNF-α by mast cells following cross-linking of FcR would also be sensitive to inhibition of p38-MAP kinase activity. There is evidence, however, that inhibition of p38-MAP kinase activity does not inhibit production of TNF-α by mast cells in response to ligation of FcεRI.[77] There is evidence that JNK kinase activity is required for TNF-α production in mast cells[70] and in macrophages in response to lipopolysaccharide.[78] Erk kinase activity is required for TNF-α modulation in response to ligation of the FcR on macrophages.[76,79]

The Tyrosine Phosphatase SHP-2

One of the major proteins that is tyrosine phosphorylated in response to IL-3 in mast cells is a 68-kDa protein,[57] subsequently identified as the tyrosine phosphatase SHPTP2, Syp, PTP-1D, or SHP-2.[80] IL-3 stimulation resulted in association of SHP-2 with the β_c of the IL-3R (Fig. 6.1) through Y577 and other sites[81,82] and an increase in its tyrosine phosphatase activity.[80] These observations suggest that SHP-2 might have a negative role in IL-3 signaling by dephosphorylating tyrosines in the activated IL-3R complex. However, there is evidence that SHP-2 has a positive role in growth factor action. Homozygous SHP-2-null mouse embryos die in utero, and the *Drosophila* ortholog of SHP-2, *Corkscrew*, appears to positively regulate signals downstream of tyrosine kinase receptors. In keeping with this notion, stimulation of cells with IL-3 results in association of tyrosine phosphatase SHP-2 with the SH-2 domain of Grb-2. Grb-2 is itself associated with the exchange factor mSos-1. Thus, SHP-2 may be acting as an adaptor protein, which by docking to β_c and becoming itself phosphorylated on tyrosine, provides an additional docking site for Grb-2–mSos-1 complexes, and thus an additional mechanism for bringing mSos-1 to the membrane and into proximity with Ras. SHP-2 is not phosphorylated on tyrosine in response to IL-4,[80] consistent with the failure of IL-4 to activate the Ras pathway and to induce tyrosine phosphorylation of Shc, another of the adaptor molecules involved in recruiting of Grb-2–Sos-1 complexes to activated receptors. IL-3 also induces associaton of SHP-2 with the p85 subunit of PI-3 kinase[80] (see Fig. 6.1), suggesting that SHP-2 may also function in activation of this pathway.

JAK-STAT Pathway

Stimulation of cells with IL-3 results in rapid tyrosine phosphorylation of JAK-2 kinase and of the two isoforms of STAT-5, STAT-5a and STAT-5b.[83,84] These STATs (signal transducers and activation of transcription) are members of a family of transcriptional activators that are characterized by SH-2 domains, which allow STATs to dock onto specific phosphotyrosines present in activated receptors of the interferon- or cytokine receptor superfamilies.[18,19] STATs are activated by tyrosine phosphorylation, which results in homo- or heterodimerization and subsequent translocation to the nucleus. Phosphorylation of STAT-5 is assumed to occur when it docks via its SH-2 domain onto phosphotyrosines present in the activated IL-3 receptor complex (Fig. 6.1). A number of the tyrosine residues present in β_c appear to provide suitable docking sites for the SH-2 domain of STAT-5. After docking to the receptor, STAT-5 is itself phosphorylated on a tyrosine residue that generates a docking site for the SH-2 domain of another STAT-5 monomer. This leads to the formation of a homodimer of STAT-5, which rapidly translocates to the nucleus.

A number of genes are known to be directly regulated by STAT-5, including the cytokine oncostatin-M and Cis.[85] The latter is a member of a novel family of proteins that contain an SH-2 domain and function as inhibitors of cytokine action. Other members of this family have recently been identified and have been termed SOCS,[86] JAB,[87] or SS-1.[88] These proteins appear to damp down responses to cytokines by blocking docking sites on JAK kinase or the activated receptor complexes.

PI-3 Kinase Pathway

PI-3 kinases are heterodimeric molecules, made up of a 110-kDa catalytic unit (p110) and an 85-kDa regulatory subunit, p85. Stimulation of cells with IL-3, IL-4, or SLF leads to the production of inositol 3,4-phosphate and inositol 3,4,5-phosphate.[89] The p110 catalytic subunit of PI-3 kinase can be activated directly by binding to Ras-GTP.[90] In addition, the p85 subunit of PI-3 kinase has two SH-2 domains, and binding of these to their target phosphotyrosine-containing motifs leads to conformational changes that activate the enzyme. Following stimulation of cells with IL-3, p85 associates with components of the activated receptor complex including SHP-2.

The action of PI-3 kinase leads to activation of the Ser/Thr protein kinases p70 S6-kinase and Akt/PKB. IL-3 stimulation is known to result in activation of Akt,[91] which may protect against apoptosis by phosphorylation of Bad and release of Bcl-X_L from complexes with Bad. Increases in PI-3 kinase action induced by IL-3 result in upregulation of expression of c-*myc* (Weiler and Schrader, unpublished observations). There is also evidence that products of PI-3 kinase activity lead to activation of the small GTPase Rac-1, which is involved in membrane ruffling. Mast cells expressing mutants of the Kit receptor that fail to activate PI-3 kinase fail to respond to SLF with membrane ruffling.[92]

SHIP

A lipid phosphatase has recently been shown to be a component of IL-3 signalling. The phosphatidylinositol 3,4,5-trisphosphate 5-phosphatase SH-2-containing inositol phosphatase (SHIP) becomes associated with the PTB of Shc after stimulation of cells with IL-3.[93,94] Overexpression of SHIP interfered with growth stimulated by IL-3 or CSF-1[95] and reduced the viability of IL-3-dependent cells under certain conditions.[94] It is possible that this reflects the enzymatic action of SHIP in hydrolyzing the product of PI-3 kinase activity, PI-3,4,5-P_3.

OP 18/Stathmin

A novel approach to identifying molecules modified during signal transduction has identified a novel component of the IL-3 and SLF signal transduction paths (Quadroni and Schrader, unpublished data). In these experiments, a relatively small number of IL-3-dependent cells (2×10^7) were stimulated with IL-3 for 10 min and then were lysed and the proteins digested with trypsin. The resultant complex mixture of peptides was applied to an immobilized metal-affinity column loaded with ferric ions, to enrich for phosphopeptides.

Electrospray mass spectrometry was then used to analyze the pattern of masses of the phosphopeptides that were present in the IL-3-treated cells but not in the untreated cells. One peptide with a mass:charge ratio of 735 was a prominent feature of the spectrum obtained from the IL-3-treated cells. The fragmentation pattern of this peptide generated after it was collided with gas atoms was used, in conjunction with the mass data, to search the protein database for proteins that, when treated with trypsin, should generate such a phosphopeptide. This search identified the source of the phosphopeptide as a protein known to be elevated in leukemia cells termed OP 18/Stathmin. Antibodies to OP 18/Stathmin were then used to verify that IL-3 induced phosphorylation of this protein. Mass spectrometric analysis of immunoprecipitated OP18/Stathmin showed that IL-3 induced phosphorylation on two serines, Ser 25 and Ser 16. Stimulation of a mast cell line, MC-9, with SLF also induced phosphorylation of OP18/Stathmin, whereas stimulation with IL-4 did not. The latter observation was consistent with data that IL-3-induced phosphorylation of Ser 25 of OP18/Stathmin was dependent on activation of the Erk MAP kinases, which are not activated by IL-4. Phosphorylation of OP18/Stathmin appears to be required to allow cells to pass the G_2/M checkpoint, probably because it inactivates the destabilizing effect of OP18/Stathmin on microtubules, thus allowing formation of the mitotic spindle.[96]

The IL-4 and IL-13 Receptors

IL-4 and IL-13 share a set of unique biological properties.[97] Both promote the generation of IgE antibodies and the differentiation of the Th2 subclass of helper T cells, and both can downregulate the production of proinflammatory cytokines by monocytes and macrophages. IL-4 and IL-13 both promote the growth of mast cells, acting in synergy with other factors such as IL-3.[98-100] In vitro studies have shown that IL-4 synergizes strongly with SLF in promoting the growth and differentiation of mast cell progenitors.[101] Likewise, IL-4 was critical in permitting the in vitro growth of connective tissue mast cells through a synergistic interaction with IL-3.[102,103] IL-4 also exerts powerful effects on the differentiation state of mast cells, upregulating expression of cell-surface molecules such as LFA-12/ICAM[104] and FcεRI,[14] regulating expression of mast cell proteases,[9,15] and upregulating responsiveness to endothelin-1 and secretory function.[10]

Investigations of signal transduction by IL-4 and IL-13 revealed similarities. First, neither IL-4 nor IL-13 activated the Ras-Erk MAP kinase pathway, at least in mast cells and other cells of hemopoietic origin,[63] nor did IL-4[65,66] or IL-13 (Foltz, unpublished data) stimulate activation of p38-MAP kinases or JNK stress-activated MAP-family kinases. IL-4 also failed to activate either the known upstream activator of JNKs, MKK4,[66] or the novel JNK kinase MKK7.[118] Both IL-4 and IL-13 induced tyrosine phosphorylation and activation of STAT-6.[1] Studies in mice with homozygous deletions of the STAT-6 gene indicate that it has roles in upregulation of IgE production, in generation of Th2 cells, and in regulation of specific genes such as that encoding the IL-4 receptor.[2,3]

Both IL-4 and IL-13 also induced prominent tyrosine phosphorylation of a 170-kDa protein that is a homolog of insulin receptor substrate-1 (IRS-1) and is now known as IRS-2.[105–107] IRS-2 associates via phosphotyrosine residues with the SH-2 domains of the p85 subunit of PI-3 kinase, with which it coprecipitates from IL-4- or IL-13-stimulated cells.[107] There is some evidence that IRS-2 is involved in the growth-promoting activity of IL-4.[105] However, IL-4 can also promote growth in cell lines lacking both IRS-2 and IRS-1.[107] Moreover, normal murine mast cells grown from bone marrow using IL-3 and IL-4 lack detectable levels of IRS-1 or IRS-2,[107] yet still respond to IL-4 with enhanced growth.[98,99]

Some cells, such as T cells, bind to and respond to IL-4 but not IL-13, whereas other cells, such as macrophages and mast cells, bind to and respond to both.[97] IL-4 binds to a receptor termed IL-4Rα that belongs to the hemopoietin receptor superfamily and is expressed on many lymphohemopoietic cells, as well as on endothelial cells and fibroblasts.[97] The γc-subunit of the IL-2 receptor was shown to bind the complex of IL-4Rα and IL-4, forming a high-affinity, functional IL-4 receptor. Binding of IL-4 to this receptor resulted in activation of JAK-1 kinase and JAK-3 kinase, the latter known to be associated with γc.[108] Studies have identified tyrosine residues in IL-4Rα that are critical for activation of STAT-6 and others for tyrosine phosphorylation of IRS-2.[109]

IL-13 fails to bind to IL-4Rα, yet in cells able to bind both IL-13 and IL-4, IL-13 can compete with IL-4 for binding sites, suggesting that a component of the IL-4 and IL-13 receptors was shared. One hypothesis was that there was an IL-13-binding chain, analogous to IL-4Rα, which also interacted with γc. However, this is not the case. When cloned, the IL-13-binding receptor (IL-13R) proved to resemble more closely the α-chains of the IL-3/IL-5/GM-CSF receptors than the IL-4Rα, having only a short cytoplasmic domain.[110] Moreover, biochemical studies showed that unlike IL-4, IL-13 failed to activate JAK-3 kinase, instead activating Tyk-2 kinase.[106] In addition IL-13 stimulated tyrosine phosphorylation of the IL-4Rα.[106,111] These results suggested that the complex of IL-13 and the IL-13 receptor interacted not with γc but with IL-4Rα. This was supported by evidence, some from cells lacking γc, suggesting that IL-4 could also interact with a receptor that did not contain γc but instead the IL-13R.[106]

Thus there appear to be two types of functional IL-4 receptor, one made up of IL-4Rα and γc and exclusive to IL-4, and the other, composed of IL-4Rα and IL-13R, able to interact with either IL-4 or IL13. Many cells, such as B cells and mast cells, can express both types of IL-4 receptor. The small cytoplasmic domain of IL-13R is essential for its role in IL-13- or IL-4-mediated signaling.[112] Overexpression of an IL-13R lacking the cytoplasmic domain in a cell expressing IL-4 receptors and γc resulted in a dominant inhibitory effect on the response to IL-4.[112] This demonstrated that there was competition for complexes of IL-4 and IL-4Rα between γc and the truncated, nonfunctional IL-13R. This effect was exacerbated by the addition of IL-13, demonstrating that complexes of IL-13 and the nonfunctional IL-13R competed with IL-4 for binding to IL-4Rα.[112] In that mast cells express both γ_c and the IL-13R, they can generate two species of functional IL-4 receptors depending on whether the complex of IL-4 and the IL-4Rα is associated with γ_c or the IL-13R. Currently, there is no evidence of functional differences between the signals triggered by IL-4 in mast cells by these two receptors, only one of which would be shared with IL-13.

Similarities and Differences in Signal Transduction by Different Cytokines

More is known about signal transduction by IL-3, SLF, and IL-4 than is the case for other cytokines affecting mast cells such as IL-9,[113] IL-15,[114] and IL-10.[115] IL-3 and SLF have similar effects on mast cell growth and survival in vitro, reflecting the fact that both activate important intracellular signaling molecules such as Ras, members of all three families of MAP kinases, and PI-3 kinase. There is evidence that, at least for some cells, the ability of SLF to stimulate growth may be absolutely dependent on the presence of cofactors like IL-4.[101] One clear difference between SLF and IL-3 lies in the fact that only IL-3 activates STAT-5, and thus can directly regulate the expression of a specific subset of STAT-5-responsive genes. The implications of other differences are less obvious. For example, whereas IL-3 induces marked tyrosine phosphorylation of SHP-2 in mast cells, we could not detect tyrosine phosphorylation of SHP-2 in normal mast cells in response to SLF.[60,80]

IL-4 and IL-13 clearly differ from other cytokines in two important respects. First, they are unusual in failing to activate Ras or members of any of the three families of MAP kinases, that is, of the Erk, p38, or JNK families. The significance of this for their effects on cell differentiation is unclear. It probably does explain the failure of IL-4 or IL-13 to act alone as growth factors. Second, IL-4 and IL-13 are unique among cytokines in activating STAT-6, thus providing an obvious mechanism for regulating genes with promoters binding STAT-6. However, like IL-3 and SLF, IL-4 does activate PI-3 kinase.[89] This may account for the ability of IL-4 to promote cell survival through the Bcl-2 family, and for its ability to synergize with other factors in promoting cell growth by upregulating expression of c-*myc*.

IL-10 probably fails to activate Ras and any of the MAP family kinases, although the data are incomplete. IL-10 does activate JAK family kinases and STAT-3, although so do other cytokines such as IL-15. In T cells and NK cells, IL-15 utilizes the IL-2Rβ chain and the γ_c and is thus likely to engage many of the same signal transduction paths as IL-2, including activation of Ras and kinases of the MAP-families, PI-3 kinase and STAT-5. However in mast cells, IL-15 is reported to use a distinct receptor that does not utilize γ_c.[114]

The signal transduction paths activated via cell-surface receptors for other ligands, such as the Fc fragment of immunoglobins discussed elsewhere in this volume, share many similarities with those activated by cytokines; for example, activation of Ras, the three families of MAP kinases and of PI-3 kinase. These similarities may account for the fact that pretreatment of mast cells with cytokines such as SLF or IL-3 can prime cells for subsequent responses triggered by ligation of FcR.[10,31,32] Similar considerations may apply to interactions between signals generated by adhesion molecules and cytokines. Thus activation of PI-3 kinase by SLF has been proposed to enhance the adhesion of mast cells to fibronectin[116] that is enhanced by SLF.[117]

Conclusions

The rapid growth in the number of cytokines and receptors known to regulate the development and differentiation of mast cells has fortunately been accompanied by new insights into their evolutionary history and structural and functional relationships. It is evident that the broad structural similarities in the receptors correlate with many similarities in the signal transduction paths they activate. Some differences, for example, in the STATs that are used by different receptors, promise to illuminate the differential effects of cytokines such as IL-3, IL-4, and IL-10 on mast cell differentiation. It would not be surprising if more new families of signal transduction molecules emerged, but the likelihood that there will be one-to-one correspondence between a particular family member and a particular cytokine receptor like that of STAT-6 and phosphorylation of the IL-4Rα seems low. Answers to the question of how a cytokine or combination of cytokines results in a particular phenotype may in many cases involve understanding of the complex question of how multiple paths are integrated.

References

1. Hou J, Schindler U, Henzel WJ, et al. An interleukin-4-induced transcription factor: IL-4 stat. Science 1994;265:1701–1706.

2. Shimoda K, van Deursen J, Sangster MY, et al. Lack of IL-4 induced Th2 response and IgE class switching in mice with disrupted Stat6 gene. Nature (Lond) 1996;380(6575):630–633.

3. Kaplan MH, Schindler U, Smiley ST, et al. Stat6 is required for mediating responses to IL-4 and for development of Th2 cells. Immunity 1996;4(3):313–319.

4. Nabel G, Galli SJ, Dvorak AM, Dvorak HF, Cantor H. Inducer T lymphocytes synthesize a factor that stimulates proliferation of cloned mast cells. Nature (Lond) 1981;291:332–334.

5. Razin E, Cordon-Cardo C, Good RA. Growth of a pure population of mast cells in vitro with conditioned medium derived from concanavalin-A stimulated splenocytes. Proc Natl Acad Sci USA 1981;78:2559–2561.

6. Schrader JW, Nossal GJV. Strategies for the analysis of accessory-cell function: the in vitro cloning and characterization of the P cell. Immunol Rev 1980;53:61–85.

7. Schrader JW. In in vitro production and cloning of the P cell, a bone marrow-derived null cell that expresses H-2 and Ia-antigens, has mast cell-like granules, and is regulated by a factor released by activated T cells. J Immunol 1981;126:452–458.

8. Tertian G, Yung T-P, Guy-Grand D, et al. Long-term in vitro culture of murine-mast cells. I. Description of a growth factor dependent culture technique. J Immunol 1981;127:788–794.

9. Eklund KK, Ghildyal N, Austen KF, et al. Induction by IL-9 and suppression by IL-3 and IL-4 of the levels of chromosome 14-derived transcripts that encode late-expressed mouse mast cell proteases. J Immunol 1993;151:4266–4273.

10. Coleman JW, Holliday MR, Kimber I, et al. Regulation of mouse peritoneal mast cell secretory function by stem cell factor, IL-3 or IL-4[1]. J Immunol 1993;150:556–562.

11. Ghildyal N, McNeil HP, Gurish MF, et al. Transcriptional regulation of the mucosal mast cell-specific protease gene, MMCP-2, by interleukin 10 and interleukin 3. J Biol Chem 1991;267:8473–8477.

12. Ghildyal N, McNeil HP, Stechschulte S, et al. Interleukin-10 induces transcription of the gene for mouse mast cell protease-1, a serine protease preferentially expressed in mucosal mast cells of *Trichinella spiralis*-infected mice. J Immunol, 1992;149: 2123–2129.

13. Ghildyal N, Friend DS, Nicodemus CF, et al. Reversible expression of mouse mast cell protease 2 mRNA and protein in cultured mast cells exposed to IL-10. J Immunol 1993;151:3206–3214.

14. Toru H, Ra C, Nonoyama S, et al. Induction of the high-affinity IgE receptor (Fc epsilon RI) on human mast cells by IL-4. Int Immunol 1996;8:1367–1373.

15. Xia HZ, Du Z, Craig S, et al. Effect of recombinant human IL-4 on tryptase, chymase, and Fc epsilon receptor type I expression in recominant human stem cell factor-dependent fetal liver-derived human mast cells. J Immunol 1997;159:2911–2921.

16. Bazan JF. Haemopoietic receptors and helical cytokines. Immunol Today 1990;11:350–355.

17. Bazan JF. Structural design and molecular evolution of a cytokine receptor superfamily. Proc Natl Acad Sci USA 1990;87:6934–6938.

18. Ihle JN, Kerr IM. Jaks and Stats in signaling by the cytokine receptor superfamily. Trends Genet 1995;11(2):69–74.

19. Schindler C, Darnell JE Jr, Transcriptional responses to polypeptide ligands: the JAK-STAT pathway. Annu Rev Biochem 1995;64:621–651.

20. Milburn MV, Hassell AM, Lambert MH, et al. A novel dimer configuration revealed by the crystal structure at 2.4 Å resolution of human interleukin-5. Nature (Lond) 1993;363:172–176.

21. Silvers WK. White spotting, patch and rump-white. In: The Coat Colours of Mice: A Model

for Gene action and Interaction. New York: Springer-Verlag, 1979;206–241.

22. Anderson DM, Lyman SD, Baird A, et al. Molecular cloning of mast cell growth factor, a hemopoietin that is active in both membrane bound and soluble forms. Cell 1990;63:235–243.

23. deVos AM, Ultsch M, and Kossiakoff AA. Human growth hormone and extracellular domain of its receptor: Crystal structure of the complex. Science 1992;255:306–312.

24. Paonessa G, Graziani R, De Serio A, et al. Two distinct and independent sites on IL-6 trigger gp130 dimer formation and signalling. EMBO J 1995;14:1942.

25. Ward LD, Howlett GJ, Discolo G, et al. High affinity interleukin-6 receptor is a hexameric complex consisting of two molecules each of interleukin-6, interleukin-6 receptor and gp-130. J Biol Chem 1994;269:23286–23289.

26. Zdanov A, Schalk-Hihi C, Menon S, et al. Crystal structure of Epstein-Barr virus protein BCRF1, a homolog of cellular interleukin-10. J Mol Biol 1997;268:460–467.

27. Feng Y, Klein BK, Vu L, et al. 1H, 13C and 15N NMR resonance assignments, secondary structure, and backbone topology of a variant of human interleukin-3. Biochemistry 1995;34:6540–6551.

28. Crapper RM, Schrader JW. Frequency of mast-cell precursors in normal tissues determined by an in vitro assay: antigen induces parallel increases in the frequency of P-cell precursors and mast cells. J Immunol 1983;131:923–928.

29. Crapper RM, Schrader JW. Evidence for the in vivo production and release into the serum of a T-cell lymphokine, persisting-cell stimulating factor (PSF), during graft-versus-host reactions. Immunology 1986;57:553–558.

30. Schrader JW, Clark-Lewis I, Crapper RM, et al. The physiology and pathology of pan-specific hemopoietin (IL-3) in interleukin 3. The panspecific hemopoietin. In: Schrader JW (ed) Lymphokines 15. San Diego: Academic Press, 1988:281–311.

31. Bischoff, SC, Dahinden CA. c-kit ligand: A unique potentiator of mediator release by human lung mast cells. J Exp Med 1992;175:237–244.

32. Columbo M, Horowitz EM, Botana LM, et al. The human recombinant c-kit receptor ligand, rhSCF, induces mediator release from human cutaneous mast cells and enhances IgE-dependent mediator release from both skin mast cells and peripheral blood basophils. J Immunol 1992;149:599–608.

33. Kurimoto Y, de Weck AL, et al. Interleukin-3 dependent mediator release in basophils triggered by C5a. J Exp Med 1989;170:467–479.

34. Hirai K, Morita Y, Misaki Y, et al. Modulation of human basophil histamine release by hemopoietic growth factors. J Immunol 1988;141:3958–3964.

35. Isfort RJ, Stevens D, May WS, et al. Interleukin 3 binds to a 140-kDa phosphotyrosine-containing cell surface protein. Proc Natl Acad Sci USA 1988;85:7982–7986.

36. Koyasu S, Tojo A, Miyajima A, et al. Interleukin 3-specific tyrosine phosphorylation of a membrane glycoprotein of Mr 150,000 in multi-factor-dependent myeloid cell lines. EMBO J 1987;6:3979–3984.

37. Duronio V, Clark-Lewis I, Schrader JW. Two polypeptides identified by interleukin-3 crosslinking represent distinct components of the interleukin-3 receptor. Exp Hematol 1992;20:505–511.

38. Kitamura T, Miyajima A. Functional reconstitution of the human interleukin-3 receptor. Blood 1992;80:84–90.

39. Kitamura T, Sato N, Arai K, et al. Expression cloning of the human IL-3 receptor cDNA reveals a shared beta subunit for the human IL-3 and GM-CSF receptors. Cell 1991;66:1165–1174.

40. Miyajima A. Molecular structure of the IL-3, GM-CSF and IL-5 receptors, Int J Cell Cloning 1992;10:126–134.

41. Gorman DM, Itoh N, Kitamura T, et al. Cloning and expression of a gene encoding an interleukin 3 receptor-like protein: identification of another member of the cytokine receptor gene family. Proc Natl Acad Sci USA 1990;87:5459–5463.

42. Gorman DM, Itoh N, Jenkins NA, et al. Chromosomal localization and organization of the murine genes encoding the β subunits (AIC2A and AIC2B) of the interleukin 3, granulocyte/macrophage colony-stimulating factor and interleukin 5 receptors. J Biol Chem 1992;267:15842–15848.

43. Quelle FW, Sato N, Witthuhn BA, et al. JAK2 associates with the βc chain of the receptor for granulocyte-macrophage colony-stimulating factor, and its activation requires the membrane-proximal region. Mol Cell Biol 1994;14:4335–4341.

44. Rao P, Mufson RA. A membrane proximal domain of the human interleukin-3 receptor βc subunit that signals DNA synthesis in NIH 3T3 cells specifically binds a complex of Src and Janus family tyrosine kinases and phosphatidylinositol 3-kinase. J Biol Chem 1995;270: 6886–6893.

45. Patel N, Herrman JM, Timans JC, et al. Functional replacement of cytokine receptor extracellular domains by leucine zippers. J Biol Chem 1996;271:30386–30391.

46. Muto A, Watanabe S, Miyajima A, et al. High affinity chimeric human granulocyte-macrophage colony-stimulating factor receptor carrying the cytoplasmic domain of the beta subunit but not the alpha subunit transduces growth promoting signals in Ba/F3 cells. Biochem Biophys Res Commun 1995;208:368–375.

47. Takai S, Yamada K, Hirayama NN, et al. Mapping of the human gene encoding the mutual signal-transducing subunit (beta-chain) of granulocyte-macrophage colony-stimulating factor (GM-CSF), interleukin-3 (IL-3), and interleukin-5 (IL-5) receptor complexes to chromosome 22q13.1. Hum Genet 1994;93:198–200.

48. Morla AO, Schreurs J, Miyajima A, et al. Hematopoietic growth factors activate the tyrosine phosphorylation of distinct sets of proteins in interleukin-3-dependent murine cell lines. Mol Cell Biol 1988;8:2214–2218.

49. Silvennoinen O, Witthuhn B, Quella FW, et al. Structure of the murine Jak2 protein-tyrosine kinase and its role in interleukin 3 signal transduction. Proc Natl Acad Sci USA 1993;90:8429–8433.

50. Appleby MW, Kerner JD, Chien S, et al. Involvement of p59fynT in IL-5 receptor signaling. J Exp Med 1995;182:811–820.

51. Torigoe T, O'Connor R, Santoli D, et al. Interleukin-3 regulates the activity of the Lyn protein-tyrosine kinase in myeloid-committeed leukemic cell lines. Blood 1992;80:617–624.

52. Sato S, Katagiri T, Takaki S, et al. IL-5 receptor-mediated tyrosine phosphorylation of SH2/SH3-containing proteins and activation of Bruton's tyrosine and Janus 2 kinases. J Exp Med 1994;180:2101–2111.

53. Duronio V, Clark-Lewis I, Federsppiel B, et al. Tyrosine phosphorylation of receptor beta subunits and common substrates in response to interleukin-3 and granulocyte-macrophage colony-stimulating factor. J Biol Chem 1992;267:21856–21863.

54. Pawson T. Protein modules and signalling networks. Nature (Lond) 1995;373(6561):573–580.

55. Satoh T, Nakafuku M, Miyajima A, et al. Involvement of ras p21 protein in signal-transduction pathways from interleukin 2, interleukin 3, and granulocyte/macrophage colony-stimulating factor, but not from interleukin 4. Proc Natl Acad Sci USA 1991;88:3314–3318.

56. Duronio V, Welham M, Abraham S, et al. p21 ras activation via haempoietin receptors and c-kit requires tyrosine kinase activity but not tyrosine phosphorylation of GAP. Proc Natl Acad Sci USA 1992;89:1587–1591.

57. Katz ME, McCormick F. Signal transduction from multiple Ras effectors. Curr Opin Genet Dev 1997;7(1):75–79.

58. Downward J. Control of ras activation. Cancer Surv 1996;27:87–100.

59. Buday L, Downard J, Epidermal growth factor regulates p21ras through the formation of a complex of receptor, Grb2 adapter protein, and Sos nucleotide exchange factor. Cell 1993;73:611–620.

60. Welham MJ, Schrader JW. Steel factor-induced tyrosine phosphorylation in murine mast cells. Common elements with IL-3-induced signal transduction pathways. J Immunol 1992;149:2772–2783.

61. Welham MJ, Duronio V, Leslie KB, et al. Multiple hemopoietins, with the exception of interleukin-4, induce modification of shc and mSos1, but not their translocation. J Biol Chem 1994;269:21165–21176.

62. Pratt JC, Weiss M, Sieff CA, et al. Evidence for a physical association between the Shc-PTB domain and the βc chain of the granulocyte-macrophage colony-stimulating factor receptor. J Biol Chem 1996;271:12137–12140.

63. Welham MJ, Duronio V, Schrader JW. Interleukin-4-dependent proliferation dissociates $p44^{erk-1}$, $p42^{erk-2}$ and $p21^{ras}$ activation from cell growth. J Biol Chem 1994;269:5865–5873.

64. Welham MJ, Duronio V, Sanghera J, et al. Multiple hemopoietic growth factors stimulate activation of mitogen-activated protein kinase family members. J Immunol 1992;149:1683–1693.

65. Foltz IN, Schrader JW. Activation of the stress-activated protein kinases by multiple hematopoietic growth factors with the exception of interleukin-4. Blood 1997; 89:3092–3096.

66. Foltz IN, Lee JC, Young PR, et al. Hemopoietic growth factors with the exception of interleukin-4 activate the p38 mitogen-activated protein kinase pathway. J Biol Chem 1997;272:3296–3301.

67. Morrison DK, Cutler RE. The complexity of Raf-1 regulation. Curent Opin Cell Biol 1997;9(2):174–179.

68. Lee JC, Laydon JT, McDonnell PC, et al. A protein kinase involved in the regulation of inflammatory cytokine biosynthesis. Nature (Lond) 1994;372:739–746.

69. Han J, Lee J-D, Bibbs L, et al. A MAP kinase targeted by endotoxin and hyperosmolarity in mammalian cells. Science 1994;265:808–810.

70. Winston BW, Chan ED, Johnson GL, et al. Activation of p38mapk, MKK3, and MKK4 by TNF-alpha in mouse bone marrow-derived macrophages. J Immunol 1997;159:4491–4497.

71. Tournier C, Whitmarsh AJ, Cavanagh J, et al. Mitogen-activated protein kinase kinase 7 is an activator of the c-Jun NH2-terminal kinase. Proc Natl Acad Sci USA 1997;94(14):7337–7342.

72. Holland PM, Magali S, Campbell JS, et al. MKK7 is a stress-activated mitogen-activated protein kinase functionally related to hemipterous. J Biol Chem 1997;272(40):24994–24998.

73. Perkins GR, Marshall CJ, Collins MK. The role of MAP kinase kinase in interleukin-3 stimulation of proliferation. Blood 1996;87(9):3669–3675.

74. Jaing Y, Chen C, Li Z, et al. Characterization and function of a new mitogen-activated protein kinase (p38β) J Biol Chem 1996;271:17920–17926.

75. Beyaert R, Cuenda A, Vanden Berghe W, et al. The p38/RK mitogen-activated protein kinase pathway regulates interleukin-6 synthesis response to tumor necrosis factor. EMBO J 1996;15:1914–1923.

76. Rose DM, Winston BW, Chan ED, et al. Fc gamma receptor cross-linking activates p42, p38, and JNK/SAPK mitogen-activated protein kinases in murine macrophages: role for p42MAPK in Fc gamma receptor-stimulated TNF-alpha synthesis. J Immunol 1997;158:3433–3438.

77. Ishizuka T, Terada N, Gerwins P, et al. Mast cell tumor necrosis factor α production is regulated by MEK kinases. Proc Natl Acad Sci USA 1997;94:6358–6363.

78. Swantek JL, Cobb MH, Geppert T. Jun N-terminal kinase/stress-activated protein kinase (JNK/SAPK) is required for lipopolysaccharide stimulation of tumor necrosis factor alpha (TNFα) translation: glucocorticoids inhibit TNFα translation by blocking JNK/SAPK. Mol Cell Biol 1997;17(11):6274–6282.

79. Trotta R, Kanakaraj P, Perussia B. FcγR-dependent mitogen-activated protein kinase activation in leukocytes: a common signal transduction event necessary for expresion of TNF-α and early activation genes. J Exp Med 1996;184:1027–1035.

80. Welham MJ, Dechert U, Leslie KB, et al. IL-3 and GM-CSF, but not IL-4, induce tyrosine phosphorylation, activation and association of SH-PTP2 with grb2 and PI3 kinase. J Biol Chem 1994;269:23764–23768.

81. Itoh T, Muto A, Watanabe S, et al. Granulocyte-macrophage colony-stimulating factor provokes RAS activation and transcription of c-fos through different modes of signalling. J Biol Chem 1996;271:7587–7592.

82. Bone H, Dechert U, Jirik F, et al. SHP1 and SHP2 protein tyrosine phosphatases associate with the IL-3 receptor β subunit after IL-3 induced receptor tyrosine phosphorylation: identification of potential binding sites and substrates. J Biol Chem 1997;272:14470–14476.

83. Azam M, Erdjument-Bromage H, Kreider BL, et al. Interleukin-3 signals through multiple isoforms of Stat5. EMBO J 1995;14:1402–1411.

84. Mui AL-F, Wakao H, O'Farrell A, et al. Interleukin-3, granulocyte-macrophage colony stimulating factor and interleukin-5 signals through two STAT5 homologs. EMBO J 1995;14:1166–1175.

85. Mui AL-F, Wakao H, Kinoshita T, et al. Suppression of interleukin-3-induced gene expression of a C-terminal truncated Stat5: role of Stat5 in proliferation. EMBO J 1996;15:2425–2433.

86. Starr R, Willson TA, Viney EM, et al. A family of cytokine-inducible inhibitors of signalling. Nature (Lond) 1997;387(6636):917–921.

87. Endo TA, Masuhara M, Yokouchi M, et al. A new protein containing an SH2 domain that inhibits JAK kinases. Nature (Lond) 1997;387:921–924.

88. Naka T, Narazaki M, Hirata M, et al. Structure and function of a new STAT-induced STAT inhibitor. Nature (Lond) 1997;387:924–929.

89. Gold M, Duronio V, Saxena S, et al. Multiple cytokines activate phosphatidylinositol 3-kinase in hemopoietic cells: association of the

enzyme with various tyrosine-phosphorylated proteins. J Biol Chem 1994;269:5403–5412.

90. Davenport SE, Mergey M, Cherqui G, et al. Deregulated expression and function of CFTR and Cl- secretion after activation of the Ras and Src/PyMT pathways in Caco-2 cells. Biochem Biophys Res Commun 1996;229:663–672.

91. Songyang Z, Baltimore D, Cantley LC, et al. Interleukin 3-dependent survival by the Akt protein kinase. Proc Natl Acad Sci USA 1997;94:11345–11350.

92. Vosseller K, Stella G, Yee NS, et al. c-*kit* receptor signaling through its phosphatidylinositide-3′-kinase-binding site and protein kinase C: role in mast cell enhancement of degranulation, adhesion and membrane rufffling. Mol Biol Cell 1997;8(5):909–922.

93. Liu L, Jefferson AB, Zhang X, et al. A novel phosphatidylinositol-3,4,5-triphosphate 5-phosphatase associates with the interleukin-3 receptor. J Biol Chem 1996;271(47):29729–29733.

94. Liu L, Damen JE, Hughes MR, et al. The Src homology 2 (SH2) domain of Sh2 containing inositol phosphatase (SHIP) is essential for tyrosine phosphorylation of SHIP, its association with Shc, and its induction of apoptosis. J Biol Chem 1997;272(14):8983–8988.

95. Lioubin MN, Algate PA, Tsai S, et al. p150 Ship, a signal transduction molecule with inositol polyphosphate-5-phosphatase activity. Genes Dev 1996;10(9):1084–1095.

96. Belmont L, Mitchison T, Deacon HW. Catastrophic revelations about Op18/stathmin. Trends Biochem Sci 1996;21(6):197–198.

97. Zurawski G, de Vries JE. Interleukin 13, an interleukin 4-like cytokine that acts on monocytes and B cells, but not on T cells. Immunol Today 1994;15(1):19–26.

98. Mosmann TR, Bond MW, Coffman RL, et al. T cell and mast cell lines respond to B cell stimulatory factor-1. Proc Natl Acad Sci USA 1986;83:5654–5658.

99. Smith CA, Rennick DM. Characterization of a murine lymphokine distinct from interleukin-2 and interleukin-3 (IL-3) possessing a T cell growth factor activity and a mast cell growth factor activity that synergizes with IL-3. Proc Natl Acad Sci USA 1986;83:1857–1861.

100. Schmitt E, Fassbender B, Beyreuther K, et al. Characterization of a T cell-derived lymphokine that acts synergistically with IL-3 on the growth of murine mast cells and is identical with IL-4. Immunobiology 1987;174:406–419.

101. Rennick D, Hunte B, Holland G, et al. Cofactors are essential for stem cell factor-dependent growth and maturation of mast cell progenitors: comparative effects of interleukin-3 (IL-3), IL-4, IL-10, and fibroblasts. Blood 1995;85:57–65.

102. Hamaguchi Y, Kanakura Y, Fujita J, et al. Interleukin 4 as an essential factor for in vitro clonal growth of murine connective tissue-type mast cells. J Exp Med 1987;165:268–273.

103. Tsuji K, Nakahata T, Takagi M, et al. Effects of interleukin-3 and interleukin-4 on the development of "connective tissue-type" mast cells: interleukin-3 supports their survival and interleukin-4 triggers and supports their proliferation synergistically with interleukin-3. Blood 1990;75: 421–427.

104. Toru H, Kinashi T, Ra C, et al. Interleukin-4 induces homotypic aggregation of human mast cells by promoting LFA-1/ICAM-1 adhesion molecules. Blood 1997;89:3296–3302.

105. Wang LM, Myers MG Jr, Sun XJ, et al. IRS-1:essential for insulin- and IL-4-stimulated mitogenesis in hematopietic cells. Science 1993;261(5128):1591–1594.

106. Welham MJ, Learmonth L, Bone H, et al. IL-13 signal transduction in lymphohemopoietic cells: Similarities and differences in signal transduction with IL-4 and insulin. J Biol Chem 1995;269:23764–23768.

107. Welham MJ, Bone H, Levings M, et al. Insulin receptor substrate-2 is the major 170-kDa protein phosphorylated on tyrosine in response to cytokines in murine lymphohemopoietic cells. J Biol Chem 1997;272(2):1377–1381.

108. Keegan AD, Johnston JA, Tortolani PJ, et al. Similarities and differences in signal transduction by interleukin 4 and interleukin 13: analysis of Janus kinase activation. Proc Nat Acad Sci USA 1995;92(17):7681–7685.

109. Wang HY, Paul, WE, Keegan AD. IL-4 function can be transferred to the IL-2 receptor by tyrosine containing sequences found in the IL-4 receptor alpha chain. Immunity 1996;4(2):113–121.

110. Hilton DJ, Zhang JG, Metcalf D, et al. Cloning and characterization of a binding subunit of the interleukin 13 receptor that is also a component of the interleukin 4 receptor. Proc Natl Acad Sci USA 1996;93(1):497–501.

111. Smerz-Bertling C, Duschl A. Both interleukin 4 and interleukin 13 induce tyrosine phosphorylation of the 140-kDa subunit of the interleukin 4 receptor. J Biol Chem 1995;270(2):966–970.

112. Orchansky PL, Ayres SD, Hilton DJ, et al. An interleukin (IL)-13 receptor lacking the cytoplasmic domain fails to transduce IL-13-induced signals and inhibits responses to IL-4. J Biol Chem 1997;272(36):22940–22947.

113. Hültner L, Druez C, Moeller J, et al. Mast cell growth-enhancing activity (MEA) is structurally related and functionally identical to the novel mouse T cell growth factor P 40/TCGF III (interleukin 9). Eur J Immunol 1990;20:1413–1416.

114. Tagaya Y, Burton JD, Miyamoto Y, et al. Identification of a novel receptor/signal transduction pathway for IL-15/T in mast cells. EMBO J 1996;15(8):4928–4939.

115. Thompson-Snipes L, Dhar V, Bond MW, et al. Interleukin 10: a novel stimulatory factor for mast cells and their progenitors. J Exp Med 1991;173:507–510.

116. Serve H, Yee NS, Stella G, et al. Differential roles of PI3-kinase and Kit tyrosine 821 in Kit receptor-mediated proliferation, survival and cell adhesion in mast cells. EMBO J 1995;14(3):473–483.

117. Kinashi T, Springer TA. Steel factor and c-*kit* regulate cell-matrix adhesion. Blood 1994;83:1033–1038.

118. Foltz IN, Gerl RE, Wieler JS, Luckach M, Salmon RA, Schrader JW. Human mitogen-activated protein kinase kinase 7 (MKK7) is a highly conserved c-Jun N-terminal kinase/stress-activated protein kinase (JNK/SAPK) activated by environmental stresses and physiological stimuli. J Biol Chem 1998;273:9344–9351.

7
Mast Cell Apoptosis and Its Regulation

Yoseph A. Mekori, Karin Hartmann, and Dean D. Metcalfe

Mast cells are found almost exclusively in tissues in which they reach phenotypic maturation under certain microenvironmental influences.[1,2] The number of mast cells in these tissues under normal conditions is relatively constant.[3] Mast cell hyperplasia occurs, however, in several conditions including host responses to parasites and neoplasia; in tissue repair and fibrosis; and in chronic inflammatory conditions associated with the pathology of diseases such as Crohn's disease and rheumatoid arthritis.[4,5] The growth factors required for mast cell proliferation or maturation are well studied, and include IL-3 and the c-*kit* ligand stem cell factor (SCF) as the principal growth-promoting cytokines. However, less is known about the mechanisms that regulate the viability of mature mast cells and permit mast cell apoptosis.

Maintenance of homeostasis of cell systems is achieved by a balance between the rate of cell proliferation and cell loss. Cell death under physiological conditions most often occurs by apoptosis, an active, inherently programmed process.[6–9] Cells undergoing apoptosis display characteristic morphological changes manifested by blebbing of the plasma membrane and loss of microvilli, condensation of the cytoplasm, cytoplasmic vacuolization, dense chromatin, and segmentation of the nucleus (Fig. 7.1). These changes are often accompanied by fragmentation of the DNA into regular subunits of 180 to 200 bp, the apparent result of double-stranded DNA breaks in the linker regions between nucleosomal cores.[6,7] It is widely assumed that the key event in apoptosis

is activation of a Ca^{2+}- and Mg^{2+}-dependent endogenous endonuclease that is inhibited by Zn, but its full biochemical characterization has yet to be determined.[7] The "ladder" pattern, seen in DNA electrophoresis, is not a feature of necrotic cell death, which results from a gross insult to the cell (i.e., hypoxia, toxins, lytic viruses, complement attack, etc.). In such instances, there may be digestion of chromatin into a continuous spectrum of sizes before disintegration, indicating destruction of histones by proteases and exposure of the entire length of DNA to endonucleases.

External stimuli that initiate apoptosis are diverse and include growth factor deprivation,[6,10,11] certain cytokines, drugs such as glucocorticoids (GC) and some cytotoxic agents,[12] exposure to physical influences such as ionizing irradiation,[11,12] and hyperthermia.[13] Also, cytotoxic T lymphocytes and natural killer cells kill the target cells by inducing apoptosis.[14] In some of the examples just described, as for steroid-induced apoptosis of thymocytes, new gene expression after exposure to the apoptotic stimulus is required and apoptosis is blocked if mRNA or protein synthesis is inhibited. The apoptotic mechanisms that involve gene expression are collectively referred to as "induction." This mechanism seems to be the most important in true physiological cell death.[9,15] In other situations, such as with human leukemic HL-60 cells, apoptosis is, in contrast, triggered by the inhibition of mRNA or protein synthesis.[16] Because these cells behave as though the suicide program is constitutively present, but

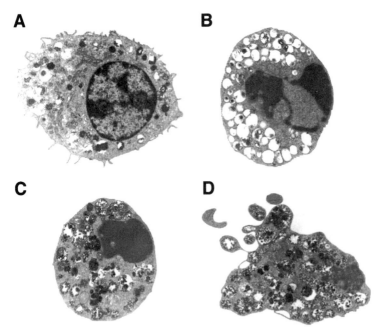

A **B**

C **D**

FIGURE 7.1. Ultrastructure of apoptotic BMCMC undergoing apoptosis after treatment with anti-Fas mAb and actinomycin D. (a) Cells incubated with an isotype control Ab; normal morphology with a corrugated cell surface and regularly distributed nuclear chromatin. (b) After anti-Fas and actinomycin D treatment; early apoptotic stages are characterized by disappearance of surface folds, vacuolization and the beginning of chromatin condensation. (c) Later stages of apoptosis include almost complete chromatin condensation, (d) blebbing of the cell membrane, as well as formation of apoptotic bodies.

cannot be expressed until inhibitory factors are removed, this mechanism has been referred to as "release".[15] Finally, there are systems in which inhibitors of protein synthesis have no effect, such as with the activation of apoptosis in target cells by cytotoxic T cells. This pathway has been called "transduction." Regardless of the triggering mechanism, all forms of apoptosis are morphologically similar, indicating that different stimuli may lead into a "final common pathway" of apoptosis.[15]

Intracellular inducers of apoptosis include several genes that are either activated in cells undergoing apoptosis or their modulation affects the process. Among the latter, the c-*myc* proto-oncogene may play a part in regulating the choice between proliferation and apoptosis. For example, fibroblasts that express c-*myc* do not undergo growth arrest in low serum concentration as do wild-type fibroblasts, but rather undergo apoptosis.[17] Also, in T-cell hybridoma cells that are induced to die by T-cell receptor (TCR) cross-linking, a process which depends on RNA or protein synthesis, c-*myc* antisense oligonucleotides have been reported to interfere with activation-induced apoptosis without affecting lymphokine production in these cells.[18] Other genes inducing apoptosis have been identified in the nematode

Caenorhabditis elegans. During development of the nematode, cell death is controlled by the genes *ced-3* and *ced-4*; expression of *ced-9* blocks death.[19] *ced-3* codes for a cysteine protease, with homology to a growing family of eukaryotic proteases related to ICE (IL-1β-converting enzyme). The ICE homolog CPP32 (now called apopain), and another cysteine protease, calpain, which is not a member of the ICE family, have recently been shown to be required for induction of apoptosis in several models.[8,15]

The tumor suppresser gene p53 has also been associated with apoptosis. The product of this gene arrests cell proliferation and may switch the cell to a differentiation mode. In many cell lineages, induction of differentiation will eventually lead to apoptotic death. p53 will thus cause apoptosis in association with differentiation when expressed in myeloid or epithelial cell lines.[20] A requirement for p53 in radiation-induced apoptosis of thymocytes has also been demonstrated.[12,21]

Fas is a gene whose product is a cell-surface receptor protein of the tumor necrosis factor/nerve growth factor (TNF/NGF) receptor family. Fas is identical to the human cell surface molecule identified as APO-1 (CD95). Apoptosis mediated by the Fas protein may be triggered by agonistic antibodies or by aggregation with the natural ligand (FasL).[14] Mutation in the Fas gene in mice (*lpr*) leads to features of autoimmune disease. "Lymphoaccumulation" is observed, as the lymph nodes fill with cells that do not die as they would eventually do in a normal animal. Fas is also a target molecule for cytotoxic T cells that bear the Fas ligand.[14]

Several mechanisms of active suppression of apoptosis are also known, including trophic growth factors,[10,11,22,23] intercellular contacts,[24] contact with extracellular matrix,[25] and protooncogenes such as bcl-2 (homologous to *ced-9*) and its relatives.[26,27] For example, bcl-2 deregulation appears to spare IL-3-dependent hemopoietic cell lines from death induced by growth factor withdrawal.[28] There is a growing family of gene products that are related to or interact with bcl-2. Bax is a pro-apoptotic relative that dimerizes with bcl-2 and thereby may play a role in regulating apoptosis.[29]

The ability of growth factors to promote cell survival by suppressing apoptosis relates to their function in promoting cell proliferation and differentiation. It should be noted, however, that although a cytokine is able to rescue one cell from apoptosis, the same cytokine may enhance apoptosis in a different cell type.[30,31] Because mast cells reside mainly in extravascular tissues, it is reasonable to assume that the regulation of mast cell number may largely be related to the local control of mast cell proliferation or apoptosis.

Physical Triggers of Mast Cell Apoptosis

Mild hyperthermia is known to enhance apoptosis in both normal and neoplastic cell populations. In the murine mastocytoma cell line P-815, heating (43°C, 30 min) results in apoptotic cell death as defined by both electron microscopic and DNA electophoresis analyses.[13] The heat-induced apoptosis was prevented by the presence of zinc sulfate, a putative inhibitor of endonucleases, but was unaffected by the protein synthesis inhibitor cycloheximide. These data indicate that active protein synthesis is not involved in executing

apoptosis by hyperthermia in such cells. Cells subjected to hyperthermia are also known to exhibit an increase in intracellular Ca^{2+}, a release from internal stores being implicated in this process. A rapid increase in inositol triphosphate (IP_3) also occurs, and precedes the rise in Ca^{2+} mobilization in heated cells.[32] An increase in cytosolic calcium has been suggested as critical for initiation of apoptosis.[33] It thus is reasonable to suggest that an IP_3-mediated increase in cytosolic Ca^{2+} may be a critical event in the activation of apoptosis by mild hyperthermia.[13]

γ-Irradiation (2500 rad) of murine bone marrow-derived cultured mast cells (BMCMC) has been found to induce apoptosis with relatively rapid kinetics, such that greater than 90% of the irradiated cells became apoptotic within 12 h.[11] Unlike hyperthermia, γ-irradiation-induced apoptosis of mast cells was suppressed by cycloheximide. This result indicates that continuous protein synthesis is at least in part required for inducing apoptosis by this physical trigger.

Cytokines Regulating Mast Cell Apoptosis

Mast cell precursors migrate into tissues where they acquire their mature phenotypes. At present, several cytokines are known to influence such changes in mast cells. Mast cells cultured in the presence of IL-3 acquire some features of the "mucosal" phenotype, whereas those cultured in the presence of SCF or on fibroblasts with or without exogenous IL-3 resemble "connective tissue mast cells" (CTMC).[2,34] SCF is aided by other growth factors and perhaps interactions with connective matrix components, through integrins, in promoting the proliferation and differentiation of specific cell lineages.[35] In addition, SCF promotes mast cell adhesion to fibroblasts and extracellular matrix components, thus influencing mast cell migration and distribution.[36]

The regulation of tissue mast cell numbers may largely relate to the local control of mast cell apoptosis. Mast cells are known to undergo apoptosis upon withdrawal of IL-3, an effect that is suppressed by cycloheximide.[10,11] The apoptotic changes following IL-3 elimination are prevented by the addition of SCF to the cell cultures (Fig. 7.2). Whereas apoptosis induced by γ-irradiation is an inductive process and follows a high order of kinetics, apoptosis induced by growth factor deprivation in BMCMC was found to be a stochastic process and follows zero-order kinetics. The addition of SCF to growth factor-deprived mast cells could be delayed up to 1h after removal of growth factors, after which cells became irreversibly committed to apoptosis.[11] It was therefore proposed that SCF promotes mast cell survival by suppressing apoptosis. The microenvironment would thus be capable of regulating mast cell numbers by modulating SCF production, providing a mechanism to allow mast cells to survive and differentiate in the relative absence of IL-3. This finding of SCF-dependent regulation of mouse mast cell apoptosis has been con-

firmed in vivo.[37] Regulation of suppression of apoptosis (and proliferation) by SCF is dose dependent, such that at suboptimal concentrations, SCF becomes a limiting factor in determining the responses elicited.[11]

Similarly, the number of functional cell-surface kit receptors may be important in controlling the responsiveness of mast cells. Indeed, using anti-kit mAb or IL-3-dependent BMCMC obtained from mice with mutations at the W locus that encodes kit (i.e., W,W^v and W^{42}), it was found that SCF exerts its antiapoptotic effect through kit.[10,11] The action of SCF was at least in part mediated by tyrosine kinases, in that the tyrosine kinase inhibitors herbimycin or tyrphostin B42 inhibited the ability of SCF to prevent apoptosis in IL-3-deprived cells.[10,38] SCF was also found to induce a rapid and sustained alkalinization of the intracellular pH by activating the Na^+/H^+ antiporter that is required for its biological

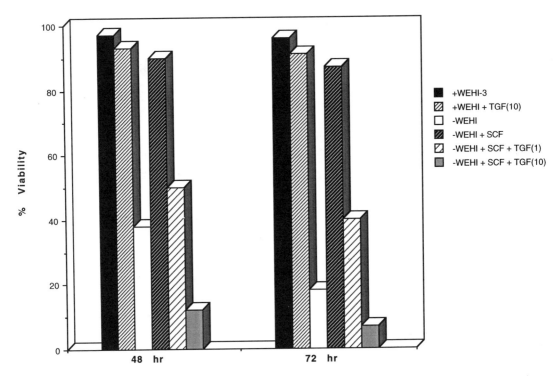

FIGURE 7.2. The effect of SCF and TGF-β on mast cell apoptosis induced by IL-3 deprivation. BMCMC were incubated in the presence or absence of WEHI-3 supernatant as a source of IL-3, SCF (40 ng/ml) either with or without TGF-β₁ (1 or 10 ng/ml, as indicated in parentheses) was added to some of the cultures at the initiation of incubation. Cell viability was assessed by flow cytometric analysis of propidium iodide uptake.

effects.[38] Neither dexamethasone nor cyclosporin A, previously shown to inhibit mast cell-derived cytokine poroduction,[39] suppressed the "rescue" effect of SCF.[10,47] Thus, the "rescue" appeared to be mediated directly by SCF and not through the induction of other cytokines known to support mast cell growth. Interestingly, SCF was also found to suppress apoptosis of BMCMC in vitro induced by either cytotoxic effects of nitric oxide or γ-irradiation.[11,40] The latter observation is in agreement with a role for SCF as a radioprotective agent in vivo, where *W* and *Sl* mutant mice display an increased sensitivity toward the lethal effects of γ-irradiation.[41] Insulin-like growth factor (IGF), an essential component in the control of somatic cell growth, has also been shown to prevent BMCMC apoptosis following IL-3 removal.[42] Because IL-3, SCF, and IGF stimulate tyrosine phosphorylation of cellular substrates, it is possible that the pathway leading to activation of the endonuclease could be regulated by a set of kinase/phosphatase activities.

The effect of SCF on mast cell apoptosis appears to be modulated by other cytokines that are known to be synthesized in tissues. Thus, transforming growth factor-β_1 (TGF-β_1) (1 and 10 ng/ml; see Fig. 7.2), known to be an important regulator of cell growth and function, was found to inhibit SCF-mediated "rescue" from apoptosis of IL-3-deprived murine bone marrow-derived mast cells.[43] In contrast, TGF-β_1 had no substantial effect on viability of mast cells cultured in the presence of IL-3. TGF-β_1 also had no noticeable effect on viability and proliferation of a growth factor-independent mast cell line. This inhibitory effect of TGF-β could be neutralized by a specific anti-TGF-β mAb.[43] The specific effect of TGF-β_1 on the SCF-mediated survival pathway may relate to its ability to downregulate the expression of cell-surface kit.[44] Thus, TGF-β and SCF may act in concert to regulate mast cell numbers under physiological or pathological conditions.

Rodent CTMC, such as rat peritoneal mast cells, maintain viability in the presence of SCF or when cocultured with fibroblasts as a source of SCF. In the absence of SCF, CTMC manifest the typical morphological and biochemical characteristics of apoptosis. Addition of NGF or SCF decreases the number of apoptotic cells in a dose-dependent manner.[45-47] The ability of NGF to promote peritoneal mast cell survival was found to be cell-density dependent and mediated by interaction with the NGF high-affinity p140[trk] tyrosine kinase receptor. Thus, addition of tyrosine kinase inhibitors resulted both in inhibition of the NGF-induced phosphorylation of this receptor and in suppression of the antiapoptotic effect.[45,46] Recently, it has been reported that NGF causes an increase in the mRNA expression for cytokines such as IL-3, IL-4, IL-10, TNF-α, and GM-CSF in rat peritoneal mast cells.[47] Cyclosporin A blocked both NGF-activated survival promotion and cytokine induction but not survival dependent on IL-3 or SCF, suggesting that NGF enhanced peritoneal mast cell survival by increasing cytokine production.[47] Thus, NGF may affect mast cell viability indirectly by inducing cytokine release with autocrine survival activity. Unlike SCF, NGF did not induce proliferative activity.[45,47] Indeed, NGF is known to synergize with IL-3 in promoting colony formation of murine IL-3-dependent BMCMC, without a significant effect on survival after IL-3 withdrawal.[48] NGF also induces a phenotypic change from BMCMC to CTMC.[48] Treatment of newborn rats with NGF causes a dramatic increase in the number of mast cells in various tissues.[49] These data in total suggest that the effects of NGF on mast cell numbers in vivo may in part be the result of enhanced survival.

Human mast cell cultures have been successfully developed from cord blood-derived mononuclear cells in the presence of SCF and IL-6. These cells were found to be immunohistologically positive for tryptase (MC$_T$) and functionally active, whereas only 20% to 30% of the cells were immunoreactive for chymase. As expected, upon withdrawal of SCF or IL-6 from the culture medium, the cells exhibited characteristic apoptotic changes, and viability declined rapidly over 6 days of culture.[50] In the presence of SCF alone, almost 100% of cells remained viable, while the viability of cells in the presence of IL-6 alone gradually declined but was higher than that cultured in the absence of any cytokine.[51] The addition of IL-3, IL-4, IL-5, IL-6, or IFN-γ to the cultures maintained in medium alone prolonged their survival in a dose-dependent manner, whereas the same concentrations of other cytokines such as IL-2, IL-9, IL-10, IL-11, TNF-α, TGF-β[1], and NGF had no significant survival-promoting effect.[50,51] As some of the former cytokines (e.g., IL-4, IL-5, IL-6, or IFN-γ) enhanced IgE-mediated degranulation of these cells, it is possible that their viability-promoting effect might be attributed at least in part to autocrine effects of secreted growth factors. This information thus indicates that various mast cell populations, derived from different origins (and species), may respond differently to survival-promoting cytokines.

Genetic Control of Mast Cell Apoptosis

SCF, NGF, and IGF prevent apoptosis of mast cells by interacting with their specific tyrosine kinase receptors. IL-3 also stimulates tyrosine phosphorylation of cellular substrates.[52] In this respect, it has been shown that transformation of IL-3-dependent cells with tyrosine kinase oncogenes abrogated IL-3 dependence through signaling pathways involving c-myc.[53] In contrast, overexpression of protein kinase C did not play a role in IL-3-mediated growth control, and was insufficient in inducing factor independence in IL-3-dependent myeloid cells.[54] In hemopoietic cells and MC-9 mast cells, phosphatidylinositol 3-OH kinase (PI-3 kinase) was found to provide a signal necessary for some cytokines to prevent apoptosis. Thus, two inhibitors of PI-3 kinase, wortmannin and LY294002, rapidly induced apoptosis in cells incubated in the presence of IL-3, IL-4, or SCF, while cells incubated with GM-CSF and IL-5 could bypass the effect of these two PI-3 kinase inhibitors.[55] However, the MAP kinase p70 S6, a downstream effector of PI-3 kinase and a protein critical for progression through the cell cycle, is not required for the prevention of apoptosis in that MC-9 mast cells grown with IL-3 or IL-4 could survive in the presence of rapamycin at concentrations that block p70 S6 kinase activity.[56] In contrast, addition of PI-3 kinase inhibitors resulted in cell death,

indicating that the downstream signals generated by PI-3 kinase required for survival are divergent from signals downstream of p70 S6. Thus, other targets of PI-3 kinase may have a role in the regulation of mast cell apoptosis.

In examining the regulatory mechanisms of mast cell apoptosis, we found that BMCMC undergo apoptosis on withdrawal of IL-3 coincident with a decrease in endogenous bcl-2 mRNA.[57] We have also found that SCF does not induce expression of bcl-2 when added to IL-3-deprived mast cells. These results are in agreement with a recent report that apoptosis induced by IL-3 removal was only partially dependent on p53 and that the effect of SCF in preventing apoptosis was not associated with the induction of bcl-2 mRNA.[11] However, SCF induced the expression of bax, while bcl-2 expression (and to a lesser extent that of bax), was induced by IL-3.[11] Therefore, although SCF and IL-3 share components of signal transduction pathways,[52] they appear to facilitate cell survival by different mechanisms. Bcl-2 may not be the key regulator mediating the "rescue" effect of SCF, but when overexpressed, it prolongs survival of bcl-2-transfected mast cells on withdrawal of IL-3.[57] The precise mechanism by which SCF suppresses apoptosis remains to be determined.

In rat peritoneal mast cells, SCF, IL-3, and NGF increased the level of bcl-2 message. While NGF was able to markedly increase bcl-2 expression, it did not affect the expression of other apoptosis-regulating genes such as bcl-x, ICE, or cpp-32.[47] The induction of bcl-2 mRNA by NGF was not blocked by cyclosporine A, while cyclosporine A was found to inhibit NGF-induced cytokine expression. This observation might therefore indicate that induced cytokine gene expression, but not increased expression of bcl-2, mediates NGF survival promotion.[47]

In variance with growth factor deprivation-induced apoptosis in mast cells, γ-irradiation-induced apoptosis was shown to be dependent on p53 in that the kinetics of apoptosis in γ-irradiated p53-deficient (p53[-/-]) BMCMC resembled that in nonirradiated cells.[11]

Primary BMCMC, peritoneal mast cells, and several mast cell lines constitutively express Fas antigen as detected by flow cytometry.[58–60] No Fas expression was detected when mast cells were cultured from Fas-deficient *lpr* mice. Aggregation of the Fas antigen with anti-Fas mAb resulted in the characteristic changes of apoptosis in C57 and P815 mast cell lines, while BMCMC were resistant to stimulation with this antibody.[58–60] Although SCF, TGF-β, or FcεRI aggregation enhanced Fas expression on mast cells, Fas-mediated apoptosis was not augmented by these cytokines or the aggregation of the FcεRI. However, addition of the transcription inhibitor actinomycin D enhanced apoptosis in C57 cells and to a lesser degree in BMCMC.[60] This effect suggests that inhibition of RNA synthesis may disrupt the synthesis or action of proteins ordinarily directed to protecting the cell against Fas-mediated apoptosis. On the other hand, overexpression of bcl-2 protein in the murine mastocytoma P815 cells suppressed Fas-mediated cell death, indicating that Fas-mediated apoptosis as well as other apoptotic pathways in mast cells are modulated by *bcl-2*.[59]

Mast Cell Apoptosis In Vivo: Clinical Correlates

Administration of SCF to both rodents and primates leads to both local and systemic mast cell hyperplasia.[61,62] Subcutaneous administration of recombinant human SCF to patients with advanced breast carcinoma also resulted in a significant increase of dermal mast cell density including sites distant to those injected with the cytokine, accompanied by an increase in histamine metabolites in the urine and in the tryptase level in the blood. Hyperplasia was found to be reversible on cessation of cytokine dosing.[63] This observation supports the view that the increased production or bioavailability of endogenous SCF may contribute to certain diseases associated with mast cell hyperplasia. Indeed, it has been reported that there is an increase in SCF in both clinically normal and lesional skin of patients with urticaria pigmentosa (UP), a form of cutaneous mastocytosis. While SCF was found to be cell bound (to keratinocytes and some dermal cells) in the skin of normal subjects, in patients with UP it

was also found free in the dermis and in extracellular spaces between keratinocytes, suggesting the (abnormal) presence of a soluble form of this cytokine.[64] As SCF maintains mast cell survival by suppressing apoptosis,[10,11] its presence may contribute to the characteristic accumulation of mast cells in the skin found in UP. Thus, some forms of mastocytosis may represent reactive hyperplasia rather than mast cell neoplasia.[64] However, more recently a mutation that results in constitutive activation and expression of c-*kit* has been identified in patients with aggressive mastocytosis.[65,66] This suggests that an activation c-*kit* mutation may permit the clonal proliferation of mast cells observed in mastocytosis.

Glucocorticoids (GC) delivered at potential or ongoing sites of tissue inflammation decrease the number of resident tissue mast cells, an effect that contributes to the antiinflammatory properties of these agents.[67,68] It has recently been demonstrated that one GC effect on mast cell number in tissues is through the regulation of SCF production by cells in the microenvironment.[69] Local application of GC to mouse skin resulted in a decrease in dermal mast cell number, which was associated with an increase in mast cell apoptosis. This did not appear to be caused by a direct effect of the GC on mast cells, as the addition of dexamethasone to mast cell cultures that were grown in the presence of either IL-3 or SCF did not enhance mast cell death.[10,69] However, addition of dexamethasone to cultured fibroblasts in vitro or fluocinonide to mouse skin in vivo did result in a downregulation of SCF production.[69] Administration of SCF at sites of fluocinonide treatment abolished the mast cell-depleting effect of this GC. Thus, GC decrease mast cell number in tissues at least in part by inducing apoptosis that results from downregulation of SCF production required for the survival of local mast cells.

Conclusions

The microenvironment, by producing cytokines with effect on mast cell survival and apoptosis such as SCF, TGF-β, and NGF, may thus play an important role in the regulation of mast cell viability and number in both homeostatic and pathological conditions, even independently of any effect on the number of mast cell precursors arriving from the bone marrow via the blood and lymphatics. Understanding the mechanisms by which mast cells undergo programmed cell death has obvious implications in understanding the etiology of mast cell hyperplasia in various pathological processes and in developing strategies to reverse such abnormal increases in mast cell number.

References

1. Galli SJ. New concepts about the mast cell. N Engl J Med 1993;328:257–265.
2. Metcalfe DD, Mekori YA, Rottem M. Mast cell ontogeny and apoptosis. Exp Dermatol 1995;4:227–230.
3. Garriga MM, Friedman M, Metcalfe DD. A survey of the number and distribution of mast cells in the skin of patients with mast cell disorders. J Allergy Clin Immunol 1988;82:425–430.
4. Mekori YA, Zeidan Z. Mast cells in non-allergic inflammatory processes. Isr J Med Sci 1990;26:337–340.
5. Nechushtan H, Razin E. Regulation of mast cell growth and proliferation. Crit Rev Oncol Hematol 1996;23:131–150.
6. Cohen JJ, Duke RC, Fadok VA. Apoptosis and programmed cell death in immunity. Annu Rev Immunol 1992;10:267–293.
7. Arends MJ, Morris RG, Wyllie AH. Apoptosis. The role of the endonuclease. Am J Pathol 1990;136:593–608.
8. Stewart BW. Mechanisms of apoptosis: integration of genetic, biochemical and cellular indicators. J Natl Cancer Inst 1994;86:1286–1295.
9. Cohen JJ. Apoptosis. Immunol Today 1993;14:126–130.
10. Mekori YA, Oh CK, Metcalfe DD. IL-3-dependent mast cells undergo apoptosis on removal of IL-3. Prevention of apoptosis by c-kit ligand. J Immunol 1993;151:3775–3784.
11. Yee NS, Paek I, Besmer P. Role of c-kit ligand in proliferation and suppression of apoptosis in mast cells. Basis for radiosensitivity of white spotting and steel mutants. J Exp Med 1994;179:1777–1787.
12. Clarke AR, Purdie CA, Harrison DJ, et al. Thymocyte apoptosis induced by p53-dependent and independent pathways. Nature (Lond) 1993;362:849–852.
13. Takano YS, Harmon BV, Kerr JFR. Apoptosis induced by mild hyperthermia in human and murine tumor cell lines: a study using electron microscopy and DNA gel electrophoresis. J Pathol 1991;163:329–336.
14. Nagata S. Fas-mediated apoptosis. Adv Exp Med Biol 1996;406:119–124.
15. Cohen JJ. Apoptosis and its regulation. Adv Exp Med Biol 1996;406:11–20.
16. Martin SJ, Lennon SV, Bonham AM, et al. Induction of apoptosis (programmed cell death) in human leukemic HL-60 cells by inhibition of RNA or protein synthesis. J Immunol 1990;145:1859–1857.
17. Evan GI, Wyllie AH, Gilbert CS, et al. Induction of apoptosis in fibroblasts by c-myc protein. Cell 1992;69:119–128.
18. Shi Y, Glynn JM, Guilbert LJ, et al. Role for c-myc in activation-induced apoptosis cell death in T cell hybridomas. Science 1992;257:212–214.
19. Armstrong RC, Aja T, Xiang J, et al. Fas-induced activation of the cell death-related protease CPP32 is inhibitied by Bcl-2 and by ICE family protease inhibitors. J Biol Chem 1996;271:16850–16855.
20. Yonish E, Resnitzky D, Lottem J, et al. Wild type p53 induces apoptosis of myeloid leukemic cells that is inhibited by interleukin 6. Nature (Lond) 1991;352:345–348.
21. Lowe SW, Schmitt EM, Smith SW, et al. p53 is required for radiation-induced apoptosis in mouse thymocytes. Nature (Lond) 1993;362:847–849.
22. Brach MA, de Vos S, Gruss H-J, et al. Prolongation of survival of human polymorphonucelar neutrophils by GM-CSF is caused by inhibition of programmed cell death. Blood 1992;80:2920–2924.
23. Dancescu M, Rubio-Trujillo M, Biron G, et al. Interleukin 4 protects chronic lymphocytic leukemic B cells from death by apoptosis and upregulates Bcl-2 expression. J Exp Med 1992;176:1319–1326.
24. Bates RC, Buret A, van Helden DF, et al. Apoptosis induced by inhibition of intercellular contacts. J Cell Biol 1994;125:403–415.
25. Meredith JE, Fazeli B, Schwartz MA. The extracellular matrix as a cell survival factor. Mol Biol Cell 1993;4:953–961.
26. Korsmeyer SJ. Bcl-2 repressor of lymphocyte death. Immunol Today 1992;13:285–288.

27. Boise LH, Gonzalez Garcia M, Postema CE, et al. *bcl-x*, a *bcl-2*-related gene that functions as a dominant regulator of apoptotic cell death. Cell 1993;74:597–608.

28. Nunez G, London L, Hockenbery D, et al. Deregulated Bcl-2 gene expression selectively prolongs survival of growth factor deprived hemopoietic cell lines. J Immunol 1990; 144:3602–3607.

29. Oltvai ZN, Milliman CL, Korsmeyer SJ. Bcl-2 heterodimerizes in vivo with conserved homolog, Bax, that accelerates programed cell death. Cell 1993;74:609–619.

30. Zubiaga AM, Munoz E, Huber BT. IL-4 and IL-2 selectively rescue Th cell subsets from glucocorticoid-induced apoptosis. J Immunol 1992;149:107–112.

31. Mangan DF, Robertson B, Wahl SM. IL-4 enhances programmed cell death (apoptosis) in stimulated human monocytes. J Immunol 1992;148:1812–1816.

32. Calderwood SK, Stevenson MA, Hahn GM. Effects of heat on cell calcium and inositol metabolism. Radiat Res 1988;113:414–425.

33. Smith CA, Williams GT, Kingston R, et al. Antibodies to CD3/T cell receptor complex induce death by apoptosis in immature T cells. Nature (Lond) 1989;337:181–184.

34. Tsai M, Takashi T, Thompson H, et al. Induction of mast cell proliferation maturation and herptin synthesis by the rat c-*kit* ligand, stem cell factor. Proc Natl Acad Sci USA 1991;88:6382–6386.

35. Bianchine PJ, Burd PR, Metcalfe DD. IL-3-dependent mast cells attach to plate-bound vitronectin: demonstration of augmented proliferation in response to signals transduced via cell surface vitronectin receptors. J Immunol 1992;149:3665–3671.

36. Dastych J, Metcalfe DD. Stem cell factor induces mast cell adhesion to fibronectin. J Immunol 1994;152:213–219.

37. Iemura A, Tsai M, Ando A, et al. The c-*kit* ligand, stem cell factor, promotes mast cell survival by suppressing apoptosis. Am J Pathol 1994;144:321–328.

38. Caceres-Cortes J, Rajotte D, Dumouchel J, et al. Product of the steel locus suppresses apoptosis in hemopoietic cells. J Biol Chem 1994; 269:12084–12091.

39. Hatfield SM, Roehm NW. Cyclosporine and FK506 inhibition of mast cell cytokine production. J Pharmacol Exp Ther 1992;260:680–688.

40. Park SJ, Jun CD, Choi BM, et al. Stem cell factor protects bone marrow-derived cultured mast cells (BMCMC) from cytocidal effect of nitric oxide secreted by fibrobasts in murine BMCMC-fibroblasts coculture. Biochem Mol Biol Int 1996;40:721–729.

41. Zsebo KM, Smith KA, Hartley CA, et al. Radioprotection of mice by rSCF. Proc Natl Acad Sci USA 1992;90:9464–9468.

42. Rodriguez-Traduchy G, Collins MKL, Garcia I, et al. Insulin-like growth factor-1 inhibits apoptosis in IL-3-dependent hemopoietic cells. J Immunol 1992;149:535–540.

43. Mekori YA, Metcalfe DD. Transforming growth factor-β prevents stem cell factor-mediated rescue of mast cells from apoptosis after IL-3 deprivaton. J Immunol 1994;153:2194–2203.

44. Dubois CM, Ruscetti FW, Stankova J, et al. TGF-β regulates c-*kit* message stability and cell-surface protein expression in hemopoietic progenitors. Blood 1994;83:3138–3145.

45. Horigome K, Bullock ED, Johnson EM. Effects of nerve growth factor on rat peritoneal mast cells. J Biol Chem 1994;269:2695–2702.

46. Kawamoto K, Okada T, Kannan Y, et al. Nerve growth factor prevents apoptsis of rat peritoneal mast cells through the *trk* proto-oncogene receptor. Blood 1995;86:4638–4644.

47. Bullock ED, Johnson EM. Nerve growth factor induces the expression of certain cytokine genes and bcl-2 in mast cells. Potential role in survival promotion. J. Biol Chem 1996;271:27500–27508.

48. Matsuda H, Kannan Y, Ushio H, et al. Nerve growth factor induces development of connective tissue type mast cells in vitro from murine bone marrow cells. J Exp Med 1991;174:7–14.

49. Aloe L, Levi-Montalcini R. Mast cells increase in tissues of neonatal rats injected with the nerve growth factor. Brain Res 1977;133:358–366.

50. Yanagida M, Fukamachi H, Ohgami K, et al. Effects of T-helper 2-type cytokines, interleukin-3 (IL-3), IL-4, IL-5, and IL-6 on the survival of cultured human mast cells. Blood 1995;86:3705–3714.

51. Yanagida M, Fukamachi H, Takei M, et al. Interferon-γ promotes the survival and FcεRI-mediated histamine release in cultured human mast cells. Immunology 1996;89:547–552.

52. Welham MJ, Schrader JW. Steel factor-induced tyrosine phosphorylation in murine mast cells: common elements with IL-3 induced signal transduction pathways. J Immunol 1992;149:2772–2778.

53. Cleveland JL, Dean M, Rosenberg N, et al. Tyrosine kinase oncogenes abrogate interleukin-3

dependence of murine myeloid cells through signaling pathways involving c-*myc*: conditional regulation of c-*myc* transcription by temperature-sensitive v-*abl*. Mol Cell Biol 1989;9:5685–5695.

54. Kraft AS, Wagner F, Housey GM. Overexpression of protein kinase C beta 1 is not sufficient to induce factor independence in the interleukin-3-dependent myeloid cell line FDC-P1. Oncogene 1990;5:1243–1246.

55. Scheid MP, Lauener RW, Duronio V. Role of phosphatidylinositol 3-OH-kinase activity in the inhibition of apoptosis in haemopoietic cells: phosphatydylinositol 3-OH kinase inhibitors reveal a difference in signalling between interleukin-3 and granulocyte-macrophage colony stimulating factor. Biochem J 1995;312:159–162.

56. Scheid MP, Charlton L, Pelech SL, et al. Role of p70 S6 kinase in cytokine-regulated hemopoietic cell survival. Biochem Cell Biol 1996;74:595–600.

57. Mekori YA, Oh CK, Dastych J, et al. Characterization of a mast cell line which lacks the extracellular domain of membrane c-*kit*. Immunology 1997;90:518–525.

58. Hughes DPM, Crispe IN. A naturally occurring soluble isoform of murine Fas generated by alternative splicing. J Exp Med 1995;182:1395–1401.

59. Schroter M, Lowin B, Borner C, Tschopp J. Regulation of Fas (Apo-1/CD95)- and perforin-mediated lytic pathways of primary cytotoxic T lymphocytes by the protooncogene bcl-2. Eur J Immunol 1995;25:3509–3513.

60. Hartmann K, Wagelie-Steffen A, von-Stebut E, et al. Fas (CD95,APO 1) antigen expression and function in murine mast cells. J Immunol 1997;159:4006–4014.

61. Tsai M, Shih LS, Newlands GFJ, et al. The rat c-*kit* ligand, stem cell factor induces the development of connective tissue-type and mucosal mast cell in vivo. J Exp Med 1992;174:125–131.

62. Galli SJ, Iemura A, Garlick DS, et al. Reversible expansion of primate mast cell population in vivo by stem cell factor. J Clin Invest 1993;91:148–152.

63. Costa JJ, Demetri GD, Harrist TJ, et al. Recombinant human stem cell factor (kit ligand) promotes human mast cell and melanocytes hyperplasia and functional activation in vivo. J Exp Med 1996;183:2681–2686.

64. Longley BJ Jr, Morganroth GS, Tyrrell L, et al. Altered metabolism of mast cell growth factor (c-*kit* ligand) in cutaneous mastocytosis. New Engl J Med 1993;328:1302–1307.

65. Nagata H, Worobec AS, Oh CK, et al. Identification of a point mutation in the catalytic domain of the protooncogene c-*kit* in peripheral blood mononuclear cells of patients who have mastocytosis with an associated hematologic disorder. Proc Natl Acad Sci USA 1995;92:10560–10564.

66. Longley BJ, Tyrrell L, Lu SZ, et al. Somatic c-*kit* activating mutation in urticaria pigmentosa and aggressive mastocytosis: establishment of clonality in a human mast cell neoplasm. Nat Genet 1996;12:312–314.

67. Lavker RM, Schechter NM. Cutaneous mast cell depletion results from topical corticosteroids usage. J Immunol 1985;135:2368–2373.

68. Goldsmith P, McGarity B, Walls AF, et al. Corticosteroid treatment reduces mast cell numbers in inflammatory bowel disease. Dig Dis Sci 1990;35:1409–1413.

69. Finotto S, Mekori YA, Metcalfe DD. Glucocorticoids decrease tissue mast cell number by reducing the production of the c-*kit* ligand, stem cell factor, by resident cells. J Clin Invest 1997;99:1721–1728.

Section II

8
Early Signals in Mast Cell Activation: A Perspective

Henry Metzger

In this brief overview, I consider some selected aspects of the activation of mast cells, limiting my discussion to activation triggered through the receptors with a distinctively high affinity for IgE (FcεRI). Like many other aspects of signal transduction in mast cells and basophils, the receptor itself and the mechanism by which it initiates a cellular response have many features in common with the structure and mechanisms of action of other receptors peculiar to the immune system. When I first entered this field 25 years ago, it was more faith than facts that caused me to believe that such parallels would exist, and it was principally because the mast cell–IgE system appeared to be particularly susceptible to biochemical analysis that I was attracted to it. The numerous similarities of this system with the other central immune receptors were certainly not anticipated by any of us, and this provides a rare and pleasant opportunity for scientists who do not often collaborate . . . those interested in the afferent and efferent pathways of the immune response . . . to learn from each other. It was a review by Achsah Keegan and William Paul that most succinctly itemized the extraordinary similarities among the clonotypic receptors on B and T lymphocytes, FcεRI, and other Fc receptors, and in the 5 years since they suggested it, the usefulness of considering these proteins as part of a family of "multichain immune recognition receptors (MIRRs)"[1] has been validated.

I have had a long-standing interest in clarifying the mechanism by which ligands binding to immunoglobulin-like molecules initiate a variety of effector functions, and through reading, writing, and research have explored the molecular details for more than 25 years. Evidence for the importance of aggregation (clustering) began to be gathered over 35 years ago, and although the possibility that subtler conformational changes also participate continues to attract some investigators, the hypothetical changes have so far eluded any persuasive direct experimental demonstration. On the other hand, not only is there a wealth of evidence that aggregation is necessary,[2] but now at least two plausible alternative mechanisms have been proposed by which aggregation could play its initiating role.

Role of Aggregation: Two Models

One of these models is described at length in Chapter 9. That model, which I shall refer to as the "co-localization model," posits that aggregation shifts the spectrum of lipids that surround the receptors from predominantly one similar to undifferentiated regions of the plasma membrane to one characteristic of certain specialized microdomains. These domains are typically resistant to solubilization by detergents such as Triton X-100 because they are enriched in sphingolipids, gangliosides, and cholesterol. They preferentially accumulate glycosyl phosphoinositide-anchored proteins and proteins, such as the Src family kinase(s), that are anchored in the inner leaflet of the membrane because of their modification with myristic and palmityl fatty acids, and certain distinctive proteins such as caveolin.[3] Whether the aggregation drives the receptor into such regions or causes such domains to form around the clustered receptors is unclear and could represent only a semantic difference. The important consideration is that the mechanism by which this is postulated to occur is through a clustering of some site on the receptor that selectively acts with a component preferentially associated with the specialized domains. Such a site might be in the transmembrane domains and more or less specific for a particular lipid, or a site in the extracellular domain of the α-chain capable of interacting with the extracellular portions of a particular glycolipids,[4] or even sites in the intracellular domains or some combination. The model further proposes that it is simply this shift that now brings the receptors into contact with a heightened concentration of Src family kinase and thereby promotes the phosphorylation of the receptors.

Our own group has presented evidence for an alternative model ("transphosphorylation model"). We propose that a small fraction of the receptors are constitutively associated with Lyn kinase, which is anchored in the inner leaflet of the plasma membrane bilayer. Direct experiments using chemical cross-linking and indirect experiments involving competition protocols suggest that about 5% of the unaggregated receptors are associated with the kinase at any one time.[5,6] Even the receptors constitutively associated with Lyn kinase remain unphosphorylated so long as they are unaggregated, but upon aggregation, rapid phosphorylation of tyrosines on the β- and γ-chains of the receptors occurs[7] principally and perhaps exclusively by a process of *transphosphorylation* when the receptors are brought into proximity by aggregation.[5,8] Our experiments provided evidence that the phosphorylated receptors now attract additional molecules of Lyn kinase[5] possibly because the SH2 regions on the latter interact with phosphorylated tyrosines on the receptors. Such recruited enzyme appears to be more firmly bound to the receptor, but there is no evidence

for a major change in its intrinsic enzymatic activity.

Some progress has been made in defining the putative sites on the receptor and Lyn that interact constitutively. Using transfection techniques, our evidence suggests that the so-called SH4 region of the "unique" domain mediates the binding.[9] Comparable experiments to define the sites on the receptor have not yet been reported, but other experimental approaches implicate the C-terminal cytoplasmic domain of the β-chain.[9–12] Related studies using affinity matrices in vitro have revealed comparable interactions between a Src family kinase or kinase domain and one or more MIRR subunits.[13] Admittedly, whether such interactions are quantitatively significant in vivo is not yet clear, but if so, then the enhanced phosphorylation of the receptor by the kinase may not simply result from their nonspecific association in special lipid domains as predicated by the colocalization model.

Our experiments in which we employed transfection of wild-type and mutant Lyn kinase provided additional support for the proposal that, under certain conditions, the relative shortage of the kinase available to the receptor can limit the intensity of the signals generated.[6,14] Although it is not impossible to accommodate the observations described in our competition[14] and inhibition[9] studies by the colocalization model, it takes a bit of stretching. An attractive notion is that the molecule to which the aggregated receptors binds in the special microdomains is Lyn itself, that aggregation induces the multivalency of the constitutive binding site for Lyn on the receptor and that this drives the receptor into the Lyn-enriched domains. However, we failed to observe such a (phosphorylation-independent) phenomenon when we looked for it explicitly.[5] Clearly more incisive experiments must be devised to resolve these questions.

Ligand Affinity and Signaling

It is important to consider the colocalization and the transphosphorylation models in relation to one of the most fundamental aspects of all responses of the immune system, namely the influence of the affinity of the ligand for the receptor. In the case of the clonotypic receptors it is the affinity of the ligand (free, or bound in the groove of a major histocompatibility antigen) for the receptor itself on the B or T lymphocyte; in the IgE–mast cell system, it is the affinity of the ligand (free, or bound to one or another cell surface) for the IgE, which is firmly bound to the FcεRI.

It has been discussed in several publications[15,16] that in any mechanism in which the binding of a ligand induces a change in a receptor that is followed by additional "downstream" changes, the residence time of the individual receptor in its altered state may be the critical factor that determines the efficacy of signal transduction. More recently, McKeithan has explicitly and very lucidly applied the "kinetic proofreading" scheme[17] to systems stimulated by MIRR.[18] The essentials of this formulation are shown in Fig. 8.1, which is slightly modified from McKeithan's Fig. 1. The essential features are as follows. The initial event is when a second molecule of receptor-bound IgE binds to a multivalent antigen (or a

second molecule of antigen-bound IgE binds to a cell-bound receptor). This interaction by definition leads to a dimer of receptor ($n = 2$ in nR). This polymerization step has never been measured directly but is estimated to be very fast,[19] and so k_{ass} and k_{dis} are virtually identical to the on and off intrinsic rate constants for the binding of the antibody-combining site on the IgE to the epitope on the antigen.

The next step we can consider is the phosphorylation of nR. The experimental evidence indicates that the increase in the number of phosphorylated receptors results from an enhanced rate of phosphorylation rather than a decreased rate of dephosphorylation. In the transphosphorylation model we postulate that the aggregation promotes the formation of the Michaelis–Menton complex and have shown that the phosphorylated aggregates ($nR-P_1$) are no less rapidly dephosphorylated than disaggregated receptors ($R-P_1$).[20] Similarly, when the receptors are aggregated under conditions known to drive the receptors into the special microdomains, the phosphorylated receptors are just as rapidly dephosphorylated as disaggregated receptors if the action of the kinase(s)

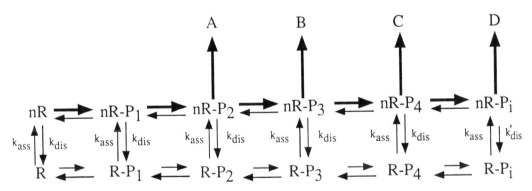

FIGURE 8.1. Kinetic proofreading scheme. This figures is a modified version of Fig. 1 drawn by McKeithan.[18] Here R refers to the receptor and P either to a modified receptor or a modified complex of receptor plus some associated molecule. The initial step shown at the *left* by the reaction generating nR from R is caused by the binding of a second molecule of IgE to a multivalent antigen. The *arrows* between R and nR represent the values of the intrinsic rate constants for association and dissociation between individual binding sites on the IgE and the epitopes on the antigen. They are drawn equal in size

only for convenience. The one exception is at the *far right*. In the reaction shown there, the *arrow* representing dissociation is shown smaller to indicate the possibility that a modified species such as $nR-P_i$ may be stabilized because of the modification i. The differences in the arrows that generate $R-P_1$ from R and $nR-P_1$ from nR are meant to show that the driving force is the difference in phosphorylation and not dephosphorylation.[20] A, B, C, and D represent different biochemical sequelae as a result of the reaction having proceeded to the step indicated by the subscripts on P.

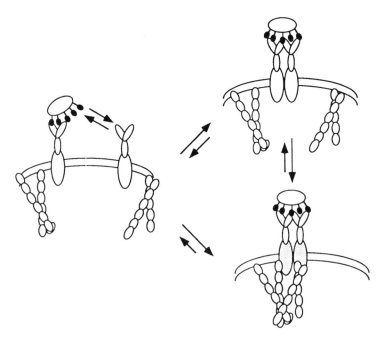

FIGURE 8.2. Moleuclar model of stabilization of modified receptors. The FcεRI are shown as ellipses with their bound IgE (*Y-shaped figures*). Multivalent antigen is shown to aggregate the receptors leading to one or more modifications (*shading*). Interaction of the modified receptors with, for example, a polymeric cytoskeletal constituent can slow the dissociation of individual receptors from the clusters.

is halted.[20] This argues against a model in which phosphorylation would be promoted by sequestering the aggregates in an environment protected from the action of phosphatases.

The species $nR-P_1$ serves as a nidus for further changes (see Chapter 10), and the question we want to consider is the likelihood that subsequent species such as $nR-P_2$, $nR-P_3$, $nR-P_4$, and $nR-P_i$ will form. The principal insight provided by the kinetic proofreading formulation is that as k_{dis} becomes relatively large compared to the rate constants for progression from $nR-P_2$ to $nR-P_3$ and so on, the probability of achieving $nR-P_i$ or even $nR-P_4$ markedly decreases. McKeithan has provided some plausible numerical examples.[18]

Figure 8.1 incorporates a number of simplifying assumptions. In particular, it suggests that the progressively distal intermediates retain the same affinity for the ligand (i.e., that the ratio k_{ass}/k_{dis} is the same). But suppose this were not the case; suppose that at some intermediate

stage, $nR-P_i$, dissociation to $R-P_i$ is inhibited. As illustrated by McKeithan, even a modest reduction can markedly enhance the likelihood that signal "D" will be generated. At first it may seem hard to envision how a modification of some protein attached to the cytoplasmic domain of one of the subunits of FcεRI could affect the intrinsic affinity of the IgE combining site which is tens of nanometers away. It cannot, but what it can do is prevent the diffusing away of the dissociated pair and effectively stabilize the complex. Figure 8.2 illustrates a perfectly plausible molecular mechanism.

Conclusions

Even the initial events related to the cellular perturbations initiated by the binding of antigen to surface-bound IgE involve an enormous number of discrete events. It is readily calculable that for a single dimer of receptor there are 4096 discrete states of phosphorylation involving only the canonical tyrosines in the receptor immunoreceptor tyrosine-based activation motifs.[6] The events that these phosphorylated oligomers initiate must involve tens of thousands of additional reactions. Defining these reactions and discovering which of them serve as the critical control points will initially involve defining the molecular species and their specific sites of interaction. Quantitative analysis not only will provide us with the specific rate constants and concentrations[6] but is the most rigorous test of the molecular schemes we invent to describe what we cannot yet see. The biggest challenge for the coming decades will be to develop techniques by which some of this complex molecular machinery can be directly visualized.

References

1. Keegan AD, Paul WE. Multichain immune recognition receptors: Similarities in structure and signaling pathways. Immunol Today 1992;13:63–68.
2. Metzger H. Transmembrane signaling: The joy of aggregation. J Immunol 1992;149:1477–1487.
3. Lisanti MP, Scherer PE, Tang Z, et al. Caveolae, caveolin and caveolin-rich membrane domains: a signalling hypothesis. Trends Cell Biol 1997; 4:231–235.
4. Repetto B, Bandara G, Kado-Fong H, et al. Functional contributions of the FcεRI α and FcεRI γ subunit domains in Fc epsilon RI-mediated signaling in mast cells. J Immunol 1996;156: 4876–4883.
5. Yamashita T, Mao S-Y, Metzger H. Aggregation of the high-affinity IgE receptor and enhanced activity of p53/56^{lyn} protein-tyrosine kinase. Proc Natl Acad Sci USA 1994;91:11251–11255.
6. Wofsy C, Torigoe C, Kent UM, et al. Exploiting the difference between extrinsic and intrinsic kinases: Implications for regulation of signaling by immunoreceptors. J Immunol 1997;159:5984–5992.
7. Paolini R, Jouvin M-H, Kinet J-P. Phosphorylation and dephosphorylation of the high-affinity receptor for immunoglobulin E immediately after receptor engagement and disengagement. Nature (Lond) 1991;353:855–858.
8. Pribluda VS, Pribluda C, Metzger H. Transphosphorylation as the mechanism by which the high-affinity receptor for IgE is phosphorylated upon aggregation. Proc Natl Acad Sci USA 1994;91:11246–11250.
9. Vonakis BM, Chen H, Haleem-Smith H, et al. The unique domain as the site on lyn kinase for its constitutive association with the high affinity receptor for IgE. J Biol Chem 1997;272:24072–24080.
10. Jouvin M-HE, Adamczewski M, Numerof R, et al. Differential control of the tyrosine kinases lyn and syk by the two signaling chains of the high affinity immunoglobulin E receptor. J Biol Chem 1994;269:5918–5925.
11. Kihara H, Siraganian RP. Src homology 2 domains of Syk and Lyn bind to tyrosine-phosphorylated subunits of the high affinity IgE receptor. J Biol Chem 1994;269:22427–22432.
12. Wilson BS, Kapp N, Lee RJ, et al. Distinct functions of the FcεRI γ and β subunits in the control of FcεR1-mediated tyrosine kinase activation and signaling responses in RBL-2H3 mast cells. J Biol Chem 1995;270:4013–4022.
13. Pleiman CM, Abrams C, Gauen LT, et al. Distinct p53/56^{lyn} and p59^{fyn} domains associate with nonphosphorylated and phosphorylated Ig-α. Proc Natl Acad Sci USA 1994;91:4268–4272.
14. Torigoe C, Goldstein B, Wofsy C, et al. Shuttling of initiating kinase between discrete aggregates of the high affinity receptor for IgE regulates the cellular response. Proc Natl Acad Sci USA 1997;94:1372–1377.
15. DeLisi C. The biophysics of ligand-receptor interactions. Q Rev Biophys 1980;13:201–230.
16. Metzger H. A comment on the "speculation" of Jarvis and Voss. Mol Immunol 1982;19:1071.
17. Burgess SM, Guthrie C. Beat the clock: paradigms for NTPases in the maintenance of biological fidelity. Trends Biochem Sci 1993;18: 381–384.
18. McKeithan TW. Kinetic proofreading in T-cell receptor signal transduction. Proc Natl Acad Sci USA 1995;92:5042–5046.
19. Wofsy C, Kent UM, Mao S-Y, et al. Kinetics of tyrosine phosphorylation when IgE dimers bind to Fc receptors on rat basophilic leukemia cells. J Biol Chem 1995;270:20264–20272.
20. Mao S-Y, Metzger H. Characterization of protein-tyrosine phosphatases that dephosphorylate the high affinity IgE receptor. J Biol Chem 1997;272:14067–14073.

9
FcεRI Signaling in Specialized Membrane Domains

Kenneth A. Field, David Holowka, and Barbara Baird

Background

FcεRI Activation

Aggregation of FcεRI, the receptor with high affinity for IgE found on mast cells and basophils, initiates a signaling cascade culminating in the secretion of granules containing inflammatory mediators from these cells, as well as the production of other inflammatory agents. The earliest detectable events following receptor aggregation involve the tyrosine phosphorylation of FcεRI within immunoreceptor tyrosine-based activation motifs (ITAMs). This initial phosphorylation step, referred to here as *receptor activation*, is mediated by the protein tyrosine kinase (PTK) Lyn, and it is regulated by unidentified tyrosine phosphatases. The phosphorylated ITAMs recruit SH-2 domain-containing signaling proteins, including Syk, a ZAP-70-related PTK that binds the phosphorylated receptor γ-chains. The recruitment and activation of Syk then lead to further downstream events, as discussed in detail elsewhere in this volume.

Although FcεRI activation has been intensively studied in the rat mucosal mast cell line RBL-2H3, the molecular mechanism of the initial ITAM phosphorylation has been difficult to define. Syk-negative RBL cells phosphorylate β and γ on FcεRI aggregation,[1] suggesting that Lyn is responsible for this initial step. By reconstituting FcεRI in a nonhematopoietic cell line, Kinet and colleagues have shown that Lyn, but not Syk, can mediate this ITAM phosphorylation.[2] The aggregation-dependent phosphoryla-tion of FcεRI-β was severalfold lower in the absence of Lyn coexpression, demonstrating that the endogenous Src present in fibroblasts cannot efficiently substitute for Lyn. Interestingly, Fyn or Lck can substitute for Lyn in phosphorylating FcεRI transfected into T-cell lines.[3] A possible explanation for this difference, as we discuss here, is that Src, because it lacks a site for palmitoylation, cannot associate with membrane domains in the manner of Lyn, Fyn, and Lck.[4]

Other studies investigating FcεRI activation in RBL cells have attempted to detect Lyn association with unaggregated receptors. This preassociated Lyn was hypothesized to mediate the *trans*-phosphorylation of adjacent receptors upon aggregation.[5] PTK activity associated with monomeric FcεRI was first reported by Eiseman and Bolen, who used sensitive in vitro kinase assays to detect Lyn coimmunoprecipitating with the receptor.[6] Jouvin et al. showed similar results using chimeric β subunits for coimmunoprecipitation,[7] implicating the ITAM-containing cytoplasmic tail of β in this constitutive association. Other studies have had difficulty quantitating the amount of Lyn bound to monomeric FcεRI at the cell surface. Two studies from Metzger's lab were able to demonstrate this interaction using special conditions to stabilize the association. Yamashita et al. used a chemical cross-linker on permeabilized RBL cells to quantify Lyn association with FcεRI.[8] Before receptor aggregation, they found that 3% to 4% of FcεRI appears to associate with Lyn. It is unclear, however, whether Lyn is directly associated with FcεRI, because this cross-linking stabilizes large receptor-associated complexes.[9] Pribluda et al. stabilized the association of tyrosine kinase activity with monomeric FcεRI by carefully controlling the ratio of detergent to cell lipid used to solubilize the receptor.[5] Interestingly, this associated kinase only phosphorylated FcεRI if the receptor was aggregated. This kinase activity presumably results from preassociated Lyn, but again, it is not certain that the PTK directly associates with the receptor under their conditions of low detergent to cell lipid ratios. We have subsequently found these conditions stabilize the interaction of FcεRI with membrane domains.[10]

These studies are consistent with a role for the preassociation of Lyn and FcεRI, but they do not provide a structural basis for this interaction. Weak interactions between Lyn and FcεRI β-subunits have been detected in vitro that do not depend on the tyrosine phosphorylation of the β-ITAM.[7] Such direct interaction between Lyn and β may play a role in receptor activation but it does not appear to be the only, or perhaps even the primary, means by which FcεRI couples to Lyn. Several observations suggest that an additional mechanism exists for this activation step. When FcεRI is expressed without the β-subunit ($\alpha\gamma_2$), Lyn maintains the capacity to phosphorylate aggregated receptors.[11,12] Naturally occurring IgE receptors lacking β-subunits have been found on Langerhans cells where they can stimulate cellular signaling.[13] In addition, chimeric receptors expressing only the intracellular domains of the γ-subunit can elicit at least a partial response on aggregation.[14–16] Thus, the β-subunit of FcεRI is clearly dispensable for receptor activation and appears to function as a signal amplifier.[15] A previously undetected interaction of Lyn with the γ-subunit could be involved or, alternatively, a lipid-mediated interaction of FcεRI with Lyn-containing membrane domains could initiate receptor activation.

Plasma Membrane Domains

The plasma membrane consists of a variety of components, including phospholipids, glycolipids, cholesterol, transmembrane proteins, and lipid-anchored proteins. These molecules are not uniformly distributed within the bilayer but cluster according to their physical properties. Protein–protein interactions often dictate this arrangement, as seen for focal contacts, clathrin-coated pits, or tight junctions.

Recent studies have demonstrated that lipid-based interactions can also lead to the organization of distinct regions of the plasma membrane. Brown and Rose originally characterized one type of membrane domain, showing that it could be isolated based on its insolubility in nonionic detergents such as Triton X-100.[17] These detergent-resistant plasma membrane domains include flasklike structures identified as caveolae by electron microscopy 40 years earlier.[18] It has subsequently been shown that caveolae are organized by both lipid- and protein-based interactions provided by the oligomerization of the structural protein caveolin.[19,20] Cells that do not express caveolin, including hematopoietic cell lines,[21–23] do not display caveolae on their surface.[21,23] However, these cells do contain detergent-resistant membrane domains enriched in some of the same molecules as caveolae. For the purposes of this discussion, the term membrane domain refers to noncaveolar detergent-resistant membrane domains.

These membrane domains likely form as a result of favorable packing of the predominantly saturated acyl chains of certain lipids.[24] Domains isolated from lysed cells are enriched in sphingomyelin, gangliosides, and other glycolipids.[17] Cholesterol, an important structural component,[24] is also enriched in these structures.[17] Reconstitution of membrane domains in model membranes reveals that they form in the absence of any structural proteins and exist before the addition of nonionic detergents.[24] Further characterization of these domains from lysed cells revealed that they are enriched in proteins linked to the plasma membrane either by glycosylphosphatidylinositol (GPI) or by myristate and palmitate.[17,25–27] Interestingly, unaggregated GPI-linked proteins appear to associate with membrane domains but are not concentrated in caveolae.[28,29] The aggregation of certain GPI-linked proteins on the outer leaflet of the membrane bilayer is known to activate some Src family PTKs anchored to the inner leaflet. This enigmatic signaling probably occurs through these membrane domains, which are enriched in both of these components.[30,31] The association of Src family members with membrane domains is determined by the presence of two tandem, saturated acyl chains. Myristoylation, a constitutive modification of all Src PTKs, together with palmitoylation, a reversible modification, are both necessary and sufficient for this interaction.[25,32] Other proteins, including some heterotrimeric G protein subunits, may use similar acyl modifications to mediate their association with membrane domains.

The presence of these doubly acylated signaling proteins within membrane domains and caveolae led to the suggestion that they might serve as concentrated signaling centers.[33] A variety of observations support this hypothesis, including the association of additional signaling proteins,[22,27,34,35] lipid metabolites,[36] and activated receptors[37–39] with the domains. Most of these studies were performed on preparations that probably contained both caveolae and membrane domains. Distinct roles for each of these structures can now be studied because it is possible to separate them.[29] Some signaling proteins have been shown to associate with membrane domains isolated from cells lacking caveolae, and these include Fyn, Yes, Lyn, Lck, certain G proteins, and GPI-linked proteins.[21,22,40,41] These results are consistent with the well-characterized association of GPI-linked proteins and Src family PTKs in large, detergent-resistant complexes in lymphocytes.[42–47] In fact, these complexes containing Src family kinases and either GPI-linked proteins[44] or gangliosides[47] were some of the earliest detergent-resistant membrane domains isolated. Their relationship to the membrane domains isolated in sucrose gradients is supported not only by the presence of these components but by the similarity of the PTK substrates seen following in vitro kinase assays.[10,40,42,44,47] These observations make a compelling argument for an important role of membrane domains in regulating cellular signaling. They have the capacity to serve either as sites for sequestering inactive signaling proteins until they are needed, or as activation sites rich in downstream effectors for cell-surface receptors.

FcεRI Activation in Membrane Domains

The phosphorylation of FcεRI ITAMs is a finely controlled step in receptor activation. Clusters of FcεRI as small as dimers can, in some cases, initiate cellular signaling,[48] and dissociation of aggregated receptors results in their rapid dephosphorylation.[49] The related T-cell receptor utilizes a coreceptor (either CD4 or CD8) with a constitutively associated Src family PTK to phosphorylate MHC-engaged receptors selectively.[50] However, FcεRI lacks a similar coreceptor and appears to use a different strategy for the initial engagement of Src family PTKs.

A possible mechanism for FcεRI activation is depicted in Fig. 9.1. This model of domain-mediated receptor activation is especially attractive because of several recent observations from our laboratory that are summarized here. The model shows that FcεRI monomers are not stably associated with membrane domains (top panel). The unaggregated receptors may weakly associate with certain components of the domains, such as gangliosides or Lyn. Indeed, such interactions with GD_{1b} ganglioside derivatives[51] and Lyn[5] have been described. However, these are not sufficiently stable in detergent lysates to allow efficient coimmunoprecipitation.[47] The domains themselves are probably dynamic in composition in the resting plasma membrane and are not sufficiently large or stable to visualize with fluorescence microscopy.[52] Aggregation of FcεRI (Fig. 9.1, lower panel) rapidly stabilizes the receptor–domain associations, perhaps because the increased valency of FcεRI aggregates leads to the cooperative binding to small clusters of domain components. This causes the formation of larger, more stable membrane domains around the receptor clusters. These domains are enriched in Lyn, and the locally high concentration favors FcεRI phosphorylation. Once phosphorylated, the β-ITAM binds directly to the SH2 domain of Lyn, and this interaction appears to amplify the signal.[12] Lyn

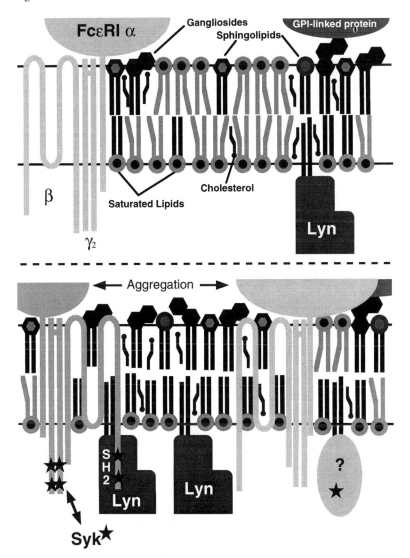

FIGURE 9.1. Model of FcεRI signaling in membrane domains. Before receptor stimulation (*top panel*), FcεRI is only weakly or transiently associated with membrane domain components (shown as *darker* lipids with *straight acyl chains*). In this situation, the domains are small and dynamic in composition and contain a significant amount of cholesterol, Lyn, and GPI-linked proteins. Aggregation of FcεRI (*lower panel*) stabilizes the association of receptors with the domain components, and the domains coalesce between and around the aggregated receptors. The locally high concentration of active Lyn allows the phosphorylation of the domain-associated FcεRI (shown as ★). Following phosphorylation of the receptor ITAMs, additional Lyn, Syk, and other signaling proteins with appropriate SH2 domains are recruited, perhaps coupling to additional domain-associated proteins.

phosphorylation of the γ-ITAM leads to the recruitment, phosphorylation, and activation of Syk.[53]

The formation of membrane domains around aggregated FcεRI may facilitate additional signaling steps. Other proteins involved in these processes, particularly tyrosine phosphatases and serine/threonine kinases and phosphatases, may be enriched or excluded from the domains. Enzymes associated with the domains may have restricted-access to other membrane-associated substrates. For example, lipid-metabolizing enzymes that are recruited to these receptor complexes would have altered substrate access as the result of the local enrichment of domain-associated lipids. In addition, recent evidence suggests that these membrane domains interact directly with the microfilament cytoskeleton[54] and that this may serve to regulate FcεRI signaling.[55]

Evidence Supporting Domain-Mediated FcεRI Activation

The most compelling evidence for the model of receptor activation in Fig. 9.1 is the observation that tyrosine-phosphorylated FcεRI are selectively associated with membrane domains.[10] In this study, FcεRI were aggregated for 2 min at 37°C, followed by cell lysis and sucrose gradient analysis of tyrosine-phosphorylated substrates under conditions that preserve the phosphorylation state present when the cells were lysed. We found that only the β- and γ-subunits of aggregated receptors that were associated with isolated domains were phosphorylated; the receptors from the same cells that were not domain associated were not phosphorylated. The aggregation-dependent association of FcεRI with membrane domains occurs rapidly, faster than receptor phosphorylation, and association occurs to the same extent when stimulated tyrosine phosphorylation is prevented.[10] A large fraction of FcεRI aggregated with an optimal dose of multivalent ligand associate with the domains (50%–60% within 2 min),[10] and correspondingly less associate when suboptimal doses of ligand are used (unpublished results). Taken together, these observations provide strong support for the model in Fig. 9.1 that tyrosine phosphorylation of FcεRI is a consequence of its association with membrane domains.

Thus far, it has been difficult to prove that FcεRI phosphorylation is the *direct result* of domain association. However, we have been able to reproduce this step in vitro with isolated membrane domains.[10] In these experiments, the tyrosine phosphorylation of the β- and γ-receptor subunits was found to depend critically on the absence of micellar Triton X-100 in the kinase assay. We found that micellar Triton

extracts FcεRI from the domains, thereby decoupling Lyn from the receptor even though active Lyn is still present in the domains.[10] This sequestration of Lyn from the receptor ITAMs is analogous to what may occur at the cell surface before receptor aggregation and is an important aspect of our model.

Several Src family PTKs appear to be regulated by their association with membrane domains. However, reports differ about whether domain association has a positive[56,57] or negative[58] effect on kinase activity. We find that at least 30% of Lyn is stably associated with membrane domains isolated from unstimulated RBL cells.[40] Using in vitro kinase assays, we observe that the domain-associated Lyn is highly active, and this is only modestly influenced by FcεRI stimulation.[10,40] However, these experiments on Lyn confined to isolated membrane domains may overestimate the basal activity of Lyn in the intact cells by promoting kinase activation via *trans*-autophosphorylation of adjacent kinases in these isolated domains. Thus, on the cell surface, Lyn within small membrane domains may have low activity until the domains are coalesced by FcεRI aggregation. The amount of Lyn and, proportionally, the PTK activity isolated in domains increases upon receptor stimulation.[40] This Lyn recruitment does not depend on the continued presence of FcεRI in the domains, suggesting that FcεRI signaling can modulate the composition or stability of the domains, possibly by influencing the palmitoylation levels of associated proteins.

Our model predicts the recruitment of Syk to membrane domain-associated FcεRI following stimulation, but we have not yet detected this association in the sucrose gradients (unpublished results). Because the majority of phosphorylated, activated Syk is found unassociated with FcεRI in cell lysates,[59] perhaps it is not surprising that this interaction is not preserved during the overnight isolation procedure. In fact, there are very few tyrosine kinase substrates present in the isolated membrane domains.[10,40] Even following in vitro tyrosine kinase assays, the most prominent substrates are Lyn, a 40-kDa protein, and FcεRI,[10] demonstrating that these domains are highly enriched in a very limited subset of membrane associated proteins. The presence of most of the receptor-stimulated tyrosine kinase substrates in the solubilized (40% sucrose) fractions of the gradient[10] shows that these proteins, like Syk, are not stably associated with the domains during the isolation. Of course, this does not preclude their association with membrane domains in intact cells before detergent lysis.

The model in Fig. 9.1 can be considered an extension of the *trans*-phosphorylation model proposed by Metzger and others.[5] The major difference is the mechanism of FcεRI–Lyn coupling before the ITAM phosphorylation. Our model shows a lipid-mediated association (described further next) that is entirely consistent with the observations used to support the earlier model. In fact, the detergent sensitivity of the FcεRI-associated kinase activity described by Pribluda et al. first suggested to us lysis conditions that could stabilize FcεRI association with membrane domains.[40] Our model is also consistent with the observed role of the β subunit as a nonessential signal amplifier of receptor activation.[12] Protein–protein interactions between Lyn and FcεRI subunits may enhance receptor activation, but they are not a critical mechanism for signal initiation in our model.

The domain formation process shown in Fig. 9.1 has been indirectly observed by fluorescence microscopy techniques. Membrane domains can be fluorescently labeled with lipid analogs or with antibodies to GPI-linked proteins or gangliosides. We have found that DiI-C_{16}, a fluorescent lipid analog with saturated acyl chains, coaggregates at the cell surface with the ganglioside derivatives recognized by the AA4 antibody[52] or the GPI-linked protein Thy-1 (unpublished results). As expected, both Thy-1 and the AA4 antigen associate with membrane domains isolated from RBL cells.[40] Interestingly, all three of these membrane domain markers coredistribute from a uniform distribution in unstimulated cells into patches around aggregated FcεRI[52,60] (and unpublished data). The observation that the DiI-C_{16} located in patches around aggregated receptors is immobilized relative to that located elsewhere on

the plasma membrane[52] suggests that receptor aggregation can cause the coalescence of less fluid membrane domains, a property predicted for detergent-resistant membrane domains from model studies.[24]

Important support for this model of FcεRI activation could come from a method for disrupting a critical step in this process, such as Lyn association with the domains or domain formation itself. By modulating the composition of the plasma membrane, we can attempt to interfere or enhance the coupling of FcεRI or Lyn with the domains. When plasma membrane cholesterol is lowered in RBL cells by sphingomyelin liposome treatment, degranulation is enhanced.[61] Subsequent microscopic analysis has shown that this treatment promotes the formation of large membrane domains even in the absence of receptor aggregation,[54] possibly facilitating receptor–domain coupling. Further experiments are in progress to understand the physical basis of these results.

Structural Basis of the FcεRI–Domain Interaction

A central question for understanding the mechanism of FcεRI–domain interactions and the possible role of membrane domains in the function of other receptors is the structural basis for this association. A significant clue is the detergent sensitivity of the interaction. As first defined for the association of the β- and γ-subunits with the α-subunit,[62] the ratio of detergent to cell lipid can be critical for maintaining functional interactions involving FcεRI after cell lysis. The description of the detergent-sensitive association of kinase activity with FcεRI[5] implied an additional membrane-associated interaction that modulates Lyn–FcεRI coupling. As we subsequently demonstrated, the addition of too much detergent disrupts the weak interaction of receptor aggregates with the domains and hence with Lyn.[10]

There are several possible explanations for the aggregation-dependent association of FcεRI with domains. We have looked at three other receptors to determine if this association is a nonspecific consequence of receptor aggregation.[88] Neither the type 1 interleukin-1 (IL-1) receptor nor α_4 integrins associate with domains before or after aggregation, but IL-2 α receptors (p55) are found to associate with domains upon aggregation. Interestingly, IL-2 receptors are known to interact with Src family PTKs during signal transduction.[63] These results indicate that there are receptor-specific requirements for this interaction. We have established that domain association of FcεRI does not depend on receptor phosphorylation or other signaling steps by showing that it occurs on permeabilized, EDTA-treated cells,[10] in the presence of several kinase inhibitors (unpublished results), and also with mutant receptors lacking the β- or γ-ITAM regions.[88] Indeed, the dissimilarity of FcεRI and the IL-2α receptor indicates that this association is not likely the result of a specific sequence in these receptors. It is notable that this IL-2 receptor is often used to make chimeric receptors containing ITAM sequences. The capacity of these chimeras to mediate signaling on aggregation[14,15] may be related to the capacity of the IL-2 receptor to interact with membrane domains.

Cell-type-specific components also do not appear to be critical for FcεRI association with membrane domains that we have seen on both hematopoietic (RBL and P815) and nonhematopoietic (CHO) cell lines (Field et al., in manuscript). Possible candidates for mediating this interaction are the GD_{1b} ganglioside derivatives that are recognized by the AA4 monoclonal antibody.[64] Both this antibody and its monovalent Fab fragment can block IgE binding to FcεRI,[51] suggesting that these gangliosides are closely associated with unaggregated receptors at the cell surface and could be involved in mediating receptor–domain interactions. However, these particular gangliosides are found only on mast cells,[64] and on CHO cells other gangliosides might substitute for the AA4 antigen. The association of receptors with membrane domains could also be facilitated by palmitoylation of FcεRI, which was previously demonstrated,[65] and this could help the transmembrane receptor interact with the tightly packed acyl chains of the domains.

The working model for this interaction involves the weak association of FcεRI with gangliosides or similar domain components. This association could involve either the transmembrane or extracellular portions of the receptor but does not depend on the β- and γ-ITAMs. The aggregation of the receptor, by coclustering small patches of domain components, nucleates the formation of more stable membrane domains. These features, also incorporated into Fig. 9.1, are consistent with the observed functional results, fluorescence microscopy, and biophysical data on the structure of membrane domains described in more detail in the following section.

Plasma Membrane Organization

Several lines of evidence have indicated that the plasma membrane is not adequately described by a "fluid mosaic",[66] but rather contains patches and diffusion barriers to limit the mobility of associated molecules.[67] FcεRI is a particularly well-studied membrane protein, and a great deal is known about its mobility in the plasma membrane.[68] Fluorescence photobleaching recovery experiments on monomeric FcεRI first showed that the diffusion coefficient of this receptor, with seven membrane-spanning domains, was at least an order of magnitude lower than predicted by theory.[69,70] Similar measurements made of the rotational motion of FcεRI within the membrane[71–73] also showed slower rotational diffusion for the intact receptor than was predicted. Dimerization of FcεRI results in the unexpected loss of rotational mobility,[71] indicating the interaction of receptor aggregates with less mobile structures, possibly membrane domains. Interestingly, mutant receptors lacking the C-terminal β or γ cytoplasmic tail exhibited faster rotational motion,[74] suggesting that protein–protein interactions contribute a portion of the retarding effects. One possible explanation for these results comes from the demonstration that isolated FcεRI is associated with both covalently and noncovalently attached lipids,[65] suggesting that the receptor moves through the membrane with an associated lipid "skirt" that slows its mobility. For this to slow the receptor sufficiently, however, this skirt would have to be very large or else interact with other membrane components. A lipid-anchored protein such as Lyn could contribute to this effect, because it might enlarge the skirt of receptor-associated lipids and provide additional "drag" by interacting with other proteins.

Another outcome of the fluorescence photobleaching experiments was the identification of an apparently immobile fraction of FcɛRI, constituting up to 40% of monomeric receptors.[69,70] This fraction increases dramatically on receptor aggregation,[75,76] even for small aggregates that would not be expected to affect lateral diffusion so profoundly. Larger aggregates also exhibit association with the cellular cytoskeleton, as shown by their well-characterized association with Triton-insoluble nuclear/cytoskeletal residues.[77–80] From these various results, immobile receptors have been postulated to be associated with a membrane skeleton,[77] the actin cytoskeleton,[79] or an extracellular matrix or glycocalyx.[81] Aggregation-dependent association with membrane domains can also account for these observations as these domains appear to be capable of coupling to the cytoskeleton (see following).

Similar biophysical studies have been performed on gangliosides and GPI-linked proteins. These studies also show diffusion coefficients and immobile fractions larger than predicted for a simple lipid attachment.[67,82] This could also be explained by membrane domain interactions that slow or immobilize these molecules. Glycolipids have been demonstrated to form small clusters in model membranes,[83] perhaps representing the membrane domains on resting cells. Single-particle tracking experiments on GPI-linked proteins support this conclusion, demonstrating several types of anomalous diffusion apparently created by "corrals" in the plasma membrane. The molecular nature of these diffusion barriers has been uncertain, and several models have been proposed, including association with membrane domains. The size of the corrals, suggested to be about 300 nm in diameter[84] is consistent with this explanation, but equally likely are protein barriers, either inside or outside the plasma membrane. Together, these

studies indicate that, although these membrane domain components generally appear to be uniformly distributed at the resolution level of fluorescence microscopy, they are restricted by molecular interactions of a dynamic nature. These studies, together with the fluorescence studies of Thy-1, AA4, and DiI-C_{16} described earlier, support the hypothesis that FcɛRI aggregation, or other stimuli, can shift the balance toward more stable membrane domain formation.

In recent confocal microscopy experiments, we have observed coredistribution of polymerized actin with aggregated membrane domain components.[54] This interaction is lost in sphingomyelin-treated cells, and we also find that aggregated FcɛRI do not coredistribute actin filaments on these cells as seen in untreated cells. This suggests that these domains can couple to the actin cytoskeleton and possibly provide a mechanism for the cytoskeletal association of FcɛRI. Upon sufficient aggregation, FcɛRI becomes associated with a detergent-insoluble complex.[77–80] Isolated membrane domains alone do not pellet under these conditions of low-speed microfuge centrifugations (unpublished results), so it is likely that a cytoskeletal interaction is maintained following cell lysis under conditions of excess cross-linking. This detergent insolubility has been correlated with the downregulation or desensitization of FcɛRI signaling,[80] and it is possible that the cytoskeleton could regulate the coupling of membrane domains with FcɛRI aggregates or interfere with more downstream signaling steps.[55]

Together, these studies on the organization of plasma membrane domains point to a dynamic and highly regulated structure. Membrane domains could provide an explanation for both the function of FcɛRI and the structural associations revealed by physical studies on this receptor.

Future Directions

Clearly, one of the most important current issues in FcɛRI signaling is the mechanism of signal initiation immediately following receptor aggregation. It should be possible to clarify the participation of membrane domains in FcεRI activation using several, complementary approaches. Manipulation of the lipid composition of the cells is one tool that may help to reveal components which are critical in these interactions. Mutational approaches are also valuable, and it may be possible to identify specific Lyn or receptor mutants that do not interact with membrane domains but still localize to the plasma membrane. The capacity of these mutant proteins to function in FcεRI activation could indicate whether membrane domains are required for signal initiation or function in an accessory role. The outcome of these studies should help define the precise role of membrane domains in early FcεRI signaling events.

Another aspect of FcεRI activation that remains unresolved is the identity and role of protein tyrosine phosphatases which regulate this process. CD45-like phosphatases are presumably involved, because CD45 is necessary to reconstitute FcεRI signaling in T-cell transfectants.[3] However, an analogous phosphatase on RBL cells has not yet been identified. When the identity of this critical component is revealed, it will be interesting to see if it is selectively excluded or enriched in the membrane domains. Such studies on CD45 have been contradictory, with one report showing the exclusion of CD45 from membrane domains in a T-cell line[58] and another showing its domain association in peripheral lymphocytes and a T-cell line.[22] Careful study of this in RBL cells may help clarify the possibility that the exclusion of phosphatases from certain plasma membrane compartments is used to modulate signaling. One attractive possibility is that FcεRI association with membrane domains recruits a Lyn-activating phosphatase to initiate signaling.

In addition to kinases and phosphatases, many other signaling proteins have been reported to associate with isolated membrane domains or caveolae. It will be valuable to identify which of these are associated with domains in RBL cells and to determine their role in downstream signaling events. Attractive candidates include lipid signaling enzymes,[36,39] Ca^{2+}

signaling machinery,[85] serine/threonine kinases,[27,39] heterotrimeric G proteins,[27,34] and vesicle fusion proteins.[27,86] The localization of some of these components could be involved in generating a local gradient of second messengers that allows the directed delivery of secretory granule contents that has been observed in mast cells.[87]

Many questions also remain about the composition and structure of these membrane domains at the cell surface. We are using several biophysical approaches to study this problem, including electron spin resonance (ESR) and additional fluorescence studies. Preliminary ESR experiments have shown that isolated membrane domains can incorporate spin-labeled gangliosides (M. Ge and J.H. Freed, unpublished results), and we are now using these probes to assess the local environment within the membrane domains compared with established model membranes. Fluorescence resonance energy transfer experiments have identified local associations of certain membrane domain components (unpublished results), and we will continue to use this technique to study the dynamics of domain structure.

One final question raised by these studies that remains to be answered is the extent to which the observations summarized here extend to other cell-surface receptors. This issue is particularly relevant for other receptors that are known to utilize Src family PTKs in an early signaling step, such as the IL-2 receptor. Because seven of the nine Src family PTKs contain the consensus site for palmitoylation,[4] it seems likely that domain association will be important in modulating intracellular signaling pathways involving these kinases. Recent observations that receptor tyrosine kinases, such as the EGF[38] and PDGF[37,39] receptors, utilize caveolae to generate local activation signals point to the possibility that compartmentalized receptor activation is a more general phenomena for mitogenic receptors. As our appreciation of the complexity of cellular signaling networks grows, we find that localizing signals becomes increasingly attractive as a mechanism for the cell to interpret the many stimuli it receives simultaneously.

Conclusions

Recent studies have led to the hypothesis that signal transduction immediately following the aggregation of FcεRI involves the association of this receptor with specialized subdomains of the plasma membrane. These membrane domains appear to mediate the phosphorylation of the receptor by providing a high concentration of active Lyn in the vicinity of the aggregated receptor. Domain-mediated FcεRI activation provides a model for signal initiation that invokes selective protein–lipid interactions rather than direct protein–protein interactions between the unphosphorylated receptor and Lyn kinase. This aggregation-dependent IgE receptor compartmentalization represents a mechanism for signal regulation that is increasingly appreciated as having relevance to signaling by other receptors.

Acknowledgments. This chapter was published as portions of a doctoral dissertation at Cornell University by K.A. Field. This work was supported by grants AI22449, AI18306, and GM07273 (to K.A.F.) from the National Institutes of Health.

References

1. Zhang J, Berenstein EH, Evans RL, et al. Transfection of Syk protein tyrosine kinase reconstitutes high affinity IgE receptor-mediated degranulation in a Syk-negative variant of rat basophilic leukemia RBL-2H3 cells. J Exp Med 1996;184:71–79.
2. Scharenberg AM, Lin S, Cuenod B, et al. Reconstitution of interactions between tyrosine kinases and the high affinity IgE receptor which are controlled by receptor clustering. EMBO J 1995;14:3385–3394.
3. Adamczewski M, Numerof RP, Koretzky GA, et al. Regulation by CD45 of the tyrosine phosphorylation of high affinity IgE receptor beta- and gamma-chains. J Immunol 1995;154:3047–3055.
4. Resh MD. Myristylation and palmitylation of Src family members: the fats of the matter. Cell 1994;76:411–413.
5. Pribluda VS, Pribluda C, Metzger H. Transphosphorylation as the mechanism by which the high-affinity receptor for IgE is phosphorylated upon aggregation. Proc Natl Acad Sci USA 1994;91:11246–11250.
6. Eiseman E, Bolen JB. Engagement of the high-affinity IgE receptor activates src protein-related tyrosine kinases. Nature (Lond) 1992;355:78–80.
7. Jouvin MH, Adamczewski M, Numerof R, et al. Differential control of the tyrosine kinases Lyn and Syk by the two signaling chains of the high affinity immunoglobulin E receptor. J Biol Chem 1994;269:5918–5925.
8. Yamashita T, Mao S-Y, Metzger H. Aggregation of the high affinity IgE receptor and enhanced activity of p53/56lyn protein-tyrosine kinase. Proc Natl Acad Sci USA 1994;91:11251–11255.
9. Mao SY, Yamashita T, Metzger H. Chemical cross-linking of IgE-receptor complexes in RBL-2H3 cells. Biochemistry 1995;34:1968–1977.
10. Field KA, Holowka D, Baird B. Compartmentalized activation of the high affinity immunoglobulin E receptor within membrane domains. J Biol Chem 1997;272:4276–4280.
11. Alber G, Miller L, Jelsema CL, et al. Structure-function relationships in the mast cell high affinity receptor for IgE. Role of the cytoplasmic domains and of the beta subunit. J Biol Chem 1991;266:22613–22620.
12. Lin S, Cicala C, Scharenberg AM, et al. The FcεRIβ subunit functions as an amplifier of FcεRIγ-mediated cell activation signals. Cell 1996;85:985–995.
13. Juergens M, Wollenberg A, Hanau D, et al. Activation of human epidermal langerhans cells by engagement of the high affinity receptor for IgE, Fc-epsilon-RI. J Immunol 1995;155:5184–5189.
14. Eiseman E, Bolen JB. Signal transduction by the cytoplasmic domains of Fc-epsilon-RI-gamma and TCR-zeta in rat basophilic leukemia cells. J Biol Chem 1992;267:21027–21032.
15. Wilson BS, Kapp N, Lee RJ, et al. Distinct functions of the Fc-epsilon-R1 gamma and beta subunits in the control of Fc-epsilon-RI-mediated tyrosine kinase activation and signaling responses in RBL-2H3 mast cells. J Biol Chem 1995;270:4013–4022.
16. Repetto B, Bandara G, Kado-Fong H, et al. Functional contributions of the FcεRIα and FcεRIγ subunit domains in FcεRI-mediated signaling in mast cells. J Immunol 1996;156:4876–4883.
17. Brown DA, Rose JK. Sorting of GPI-anchored proteins to glycolipid-enriched membrane

subdomains during transport to the apical cell surface. Cell 1992;68:533–544.

18. Yamada E. The fine structure of the gall bladder epithelium of the mouse. J Biophys Biochem Cytol 1955;1:445–458.

19. Rothberg KG, Heuser JE, Donzell WC, et al. Caveolin, a protein component of caveolae membrane coats. Cell 1992;68:673–682.

20. Lisanti MP, Tang ZL, Sargiacomo M. Caveolin forms a hetero-oligomeric protein complex that interacts with an apical GPI-linked protein: implications for the biogenesis of caveolae. J Cell Biol 1993;123:595–604.

21. Fra AM, Williamson E, Simons K, et al. Detergent-insoluble glycolipid microdomains in lymphocytes in the absence of caveolae. J Biol Chem 1994;269:30745–30748.

22. Parolini I, Sargiacomo M, Lisanti MP, et al. Signal transduction and glycophosphatidylinositol-linked proteins (lyn, lck, CD4, CD45, G proteins, and CD55) selectively localize in Triton-insoluble plasma membrane domains of human leukemic cell lines and normal granulocytes. Blood 1996;87:3783–3794.

23. Fra AM, Williamson E, Simons K, et al. De novo formation of caveolae in lymphocytes by expression of VIP21-caveolin. Proc Natl Acad Sci USA 1995;92:8655–8659.

24. Schroeder R, London E, Brown D. Interactions between saturated acyl chains confer detergent resistance on lipids and glycosyl-phosphatidylinositol (GPI)-anchored proteins: GPI-anchored proteins in liposomes and cells show similar behavior. Proc Natl Acad Sci USA 1994;91:12130–12134.

25. Shenoy-Scaria AM, Dietzen DJ, Kwong J, et al. Cysteine 3 of Src family protein tyrosine kinase determines palmitoylation and localization in caveolae. J Cell Biol 1994;126: 353–363.

26. Arreaza G, Melkonian KA, Lafevre-Bernt M, et al. Triton X-100-resistant membrane complexes from cultured kidney epithelial cells contain the src family protein tyrosine kinase p62-yes. J Biol Chem 1994;269:19123–19127.

27. Sargiacomo M, Sudol M, Tang Z, et al. Signal transducing molecules and glycosyl-phosphatidylinositol-linked proteins form a caveolin-rich insoluble complex in MDCK cells. J Cell Biol 1993;122:789–807.

28. Mayor S, Rothberg KG, Maxfield FR. Sequestration of GPI-anchored proteins in caveolae triggered by cross-linking. Science 1994;264: 1948–1951.

29. Schnitzer JE, McIntosh DP, Dvorak AM, et al. Separation of caveolae from associated microdomains of GPI-anchored proteins. Science 1995;269:1435–1439.

30. Shenoy-Scaria AM, Gauen LK, Kwong J, et al. Palmitylation of an amino-terminal cysteine motif of protein tyrosine kinases p56lck and p59fyn mediates interaction with glycosyl-phosphatidylinositol-anchored proteins. Mol Cell Biol 1993;13:6385–6392.

31. Brown D. The tyrosine kinase connection: how GPI-anchored proteins activate T cells. Curr Opin Immunol 1993;5:349–354.

32. van't Hof W, Resh MD. Rapid plasma membrane anchoring of newly synthesized p59fyn: selective requirement for NH_2-terminal myristoylation and palmitoylation at cysteine-3. J Cell Biol 1997;136:1023–1035.

33. Lisanti MP, Scherer PE, Tang Z, et al. Caveolae, caveolin and caveolin-rich membrane domains: a signalling hypothesis. Trends Cell Biol 1994;4:231–235.

34. Chang WJ, Ying YS, Rothberg KG, et al. Purification and characterization of smooth muscle cell caveolae. J Cell Biol 1994;126:127–138.

35. Lisanti MP, Scherer PE, Vidugiriene J, et al. Characterization of caveolin-rich membrane domains isolated from an endothelial-rich source: implications for human disease. J Cell Biol 1994;126:111–126.

36. Liu P, Anderson RGW. Compartmentalized production of ceramide at the cell surface. J Biol Chem 1995;270:27179–27185.

37. Liu P, Ying Y, Ko YG, et al. Localization of platelet-derived growth factor-stimulated phosphorylation cascade to caveolae. J Biol Chem 1996;271:10299–10303.

38. Mineo C, James GL, Smart EJ, et al. Localization of epidermal growth factor-stimulated ras/raf-1 interaction to caveolae membrane. J Biol Chem 1996;271:11930–11935.

39. Liu J, Oh P, Horner T, et al. Organized endothelial cell surface signal transduction in caveolae distinct from glycosylphosphatidylinositol-anchored protein microdomains. J Biol Chem 1997;272:7211–7222.

40. Field KA, Holowka D, Baird B. FcεRI-mediated recruitment of p53/56lyn to detergent resistant membrane domains accompanies cellular signaling. Proc Natl Acad Sci USA 1995;92:9201–9205.

41. Gorodinsky A, Harris DA. Glycolipid-anchored proteins in neuroblastoma cells form detergent-

resistant complexes without caveolin. J Cell Biol 1995;129:619–627.

42. Draberova L, Draber P. Thy-1 glycoprotein and src-like protein-tyrosine kinase p53/p56lyn are associated in large detergent-resistant complexes in rat basophilic leukemia cells. Proc Natl Acad Sci U S A 1993;90:3611–3615.

43. Draberova L, Amoui M, Draber P. Thy-1-mediated activation of rat mast cells: the role of thy-1 membrane microdomains. Immunology 1996;87:141–148.

44. Stefanova I, Horejsi V, Ansotegui IJ, et al. GPI-anchored cell-surface molecules complexed to protein tyrosine kinases. Science 1991;254:1016–1019.

45. Cinek T, Horejsi V. The nature of large noncovalent complexes containing glycosyl-phosphatidylinositol-anchored membrane glycoproteins and protein tyrosine kinases. J Immunol 1992;149:2262–2270.

46. Shenoy-Scaria AM, Kwong J, Fujita T, et al. Signal transduction through decay-accelerating factor. Interaction of glycosyl-phosphatidylinositol anchor and protein tyrosine kinases p56lck and p59fyn. J Immunol 1992;149:3535–3541.

47. Minoguchi K, Swaim WD, Berenstein EH, et al. Src family tyrosine kinase p53/56lyn, a serine kinase and Fc epsilon RI associate with alpha-galactosyl derivatives of ganglioside GD1b in rat basophilic leukemia RBL-2H3 cells. J Biol Chem 1994;269:5249–5254.

48. Wofsy C, Kent UM, Mao SY, et al. Kinetics of tyrosine phosphorylation when IgE dimers bind to Fc-epsilon receptors on rat basophilic leukemia cells. J Biol Chem 1995;270:20264–20272.

49. Paolini R, Numerof R, Kinet JP. Phosphorylation-dephosphorylation of high-affinity IgE receptors: a mechanism for coupling-uncoupling a large signaling complex. Proc Natl Acad Sci USA 1992;89:10733–10737.

50. Chan AC, Desai DM, Weiss A. The role of protein tyrosine kinases and protein tyrosine phosphatases in T cell antigen receptor signal transduction. Annu Rev Immunol 1994;12:555–592.

51. Basciano LK, Berenstein EH, Kmak L, et al. Monoclonal antibodies that inhibit IgE binding. J Biol Chem 1986;261:11823–11831.

52. Thomas JL, Holowka D, Baird B, et al. Large-scale co-aggregation of fluorescent lipid probes with cell surface proteins. J Cell Biol 1994;125:795–802.

53. Beaven MA, Baumgartner RA. Downstream signals initiated in mast cells by FcεRI and other receptors. Curr Opin Immunol 1996;8:766–772.

54. Holowka D, Hine C, Baird B. Alterations in cellular lipid composition affect the interactions of aggregated Fc-epsilon-RI with p53-56-lyn and the microfilament cytoskeleton. FASEB J 1996;10:A1214.

55. Pierini L, Harris NT, Holowka D, et al. Evidence supporting a role for microfilaments in regulating the coupling between poorly-dissociable IgE-FcεRI aggregates and downstream signaling pathways. Biochemistry 1997;36:7447–7456.

56. Schuh SM, Lublin DM. The Triton-insoluble fraction of p56-lck has increased protein tyrosine kinase activity. FASEB J 1995;9:A1302.

57. Yan SR, Fumagalli L, Berton G. Activation of src family kinases in human neutrophils. Evidence that p58c-fgr and p53/56lyn redistributed to a Triton X-100-insoluble cytoskeletal fraction, also enriched in the caveolar protein caveolin, display an enhanced kinase activity. FEBS Lett 1996;380:198–203.

58. Rodgers W, Rose JK. Exclusion of CD45 inhibits activity of p56-lck associated with glycolipid-enriched membrane domains. J Cell Biol 1996;135:1515–1523.

59. Benhamou M, Ryba NJ, Kihara H, et al. Protein-tyrosine kinase p72syk in high affinity IgE receptor signaling. Identification as a component of pp72 and association with the receptor gamma chain after receptor aggregation. J Biol Chem 1993;268:23318–23324.

60. Pierini L, Holowka D, Baird B. Fc-epsilon-RI-mediated association of 6-μm beads with RBL-2H3 mast cells results in exclusion of signaling proteins from the forming phagosome and abrogation of normal downstream signaling. J Cell Biol 1996;134:1427–1439.

61. Chang EY, Zheng Y, Holowka D, et al. Alteration of lipid composition modulates Fc-epsilon-RI signaling in RBL-2H3 cells. Biochemistry 1995;34:4376–4384.

62. Kinet JP, Alcaraz G, Leonard A, et al. Dissociation of the receptor for immunoglobulin E in mild detergents. Biochemistry 1985;24:4117–4124.

63. Minami Y, Taniguchi T. IL-2 signaling: recruitment and activation of multiple protein tyrosine kinases by the components of the IL-2 receptor. Curr Opin Cell Biol 1995;7:156–162.

64. Guo NH, Her GR, Reinhold VN, et al. Monoclonal antibody AA4, which inhibits binding of IgE to high affinity receptors on rat basophilic leukemia cells, binds to novel alpha-galactosyl

derivatives of ganglioside GD1b. J Biol Chem 1989;264:13267–13272.

65. Kinet JP, Quarto R, Perez MR, et al. Non-covalently and covalently bound lipid on the receptor for immunoglobulin E. Biochemistry 1985;24:7342–7348.

66. Singer SJ, Nicolson GL. The fluid mosaic model of the structure of cell membranes. Science 1972;175:720–731.

67. Edidin M, Kuo SC, Sheetz MP. Lateral movements of membrane glycoproteins restricted by dynamic cytoplasmic barriers. Science 1991;254:1379–1382.

68. Holowka D, Baird B. Antigen-mediated IgE receptor aggregation and signaling: a window on cell surface structure and dynamics. Annu Rev Biophys Biomol Struct 1996;25:79–112.

69. Schlessinger J, Webb WW, Elson EL, et al. Lateral motion and valence of Fc receptors on rat peritoneal mast cells. Nature (Lond) 1976;264:550–552.

70. Mao SY, Varin BN, Edidin M, et al. Immobilization and internalization of mutated IgE receptors in transfected cells. J Immunol 1991;146:958–966.

71. Myers JN, Holowka D, Baird B. Rotational motion of monomeric and dimeric immunoglobulin E-receptor complexes. Biochemistry 1992;31: 567–575.

72. Pecht I, Ortega E, Jovin TM. Rotational dynamics of the Fc-epsilon receptor on mast cells monitored by specific monoclonal antibodies and IgE. Biochemistry 1991;30:3450–3458.

73. Zidovetzki R, Bartholdi M, Arndt JD, et al. Rotational dynamics of the Fc receptor for immunoglobulin E on histamine-releasing rat basophilic leukemia cells. Biochemistry 1986; 25:4397–4401.

74. Chang EY, Mao SY, Metzger H, et al. Effects of subunit mutation on the rotational dynamics of Fc-epsilon-RI, the high affinity receptor for IgE, in transfected cells. Biochemistry 1995;34:6093–6099.

75. Menon AK, Holowka D, Webb WW, et al. Clustering mobility and triggering activity of small oligomers of immunoglobulin E on rat basophilic leukemia cells. J Cell Biol 1986;102:534–540.

76. Menon AK, Holowka D, Webb WW, et al. Cross-linking of receptor-bound immunoglobulin E to aggregates larger than dimers leads to rapid immobilization. J Cell Biol 1986;102:541–550.

77. Apgar JR. Antigen-induced cross-linking of the IgE receptor leads to an association with the detergent-insoluble membrane skeleton of rat basophilic leukemia RBL-2H3 cells. J Immunol 1990;145:3814–3822.

78. Apgar JR. Association of the cross-linked IgE receptor with the membrane skeleton is independent of the known signaling mechanisms in rat basophilic leukemia cells. Cell Regul 1991;2:181–192.

79. Robertson D, Holowka D, Baird B. Cross-linking of immunoglobulin E-receptor complexes induces their interaction with the cytoskeleton of rat basophilic leukemia cells. J Immunol 1986;136:4565–4572.

80. Seagrave J, Oliver JM. Antigen-dependent transition of IgE to a detergent-insoluble form is associated with reduced IgE receptor-dependent secretion from RBL-2H3 mast cells. J Cell Physiol 1990;144:128–136.

81. Zhang F, Crise B, Su B, et al. Lateral diffusion of membrane-spanning and glycosylphosphatidylinositol-linked proteins: toward establishing rules governing the lateral mobility of membrane proteins. J Cell Biol 1991;115: 75–84.

82. Hannan LA, Lisanti MP, Rodriguez-Boulan E, et al. Correctly sorted molecules of a GPI-anchored protein are clustered and immobile when they arrive at the apical surface of MDCK cells. J Cell Biol 1993;120:353–358.

83. Rock P, Allietta M, Young WWJ, et al. Organization of glycosphingolipids in phosphatidylcholine bilayers: use of antibody molecules and Fab fragments as morphologic markers. Biochemistry 1990;29:8484–8490.

84. Sheets ED, Lee GM, Simson R, et al. Transient confinement of a glycosylphosphatidylinositol-anchored protein in the plasma membrane. Biochemistry 1997;36:12449–12458.

85. Schnitzer JE, Oh P, Jacobson BS, et al. Caveolae from luminal plasmalemma of rat lung endothelium: microdomains enriched in caveolin, Ca^{2+}-ATPase, and inositol trisphosphate receptor. Proc Natl Acad Sci USA 1995;92:1759–1763.

86. Schnitzer JE, Liu J, Oh P. Endothelial caveolae have the molecular transport machinery for vesicle budding, docking, and fusion including VAMP, NSF, SNAP, annexins, and GTPases. J Biol Chem 1995;270:14399–14404.

87. Lawson D, Fewtrell C, Raff MC. Localized mast cell degranulation induced by concanavalin A-sepharose beads. J Cell Biol 1978;79:394–400.

88. Field KA, Holowka D, Baird B. Structural aspects of the association of FcεRI with detergent resistant membranes. J Biol Chem 1999, in press.

10
Regulation and Function of Protein Tyrosine Kinase Syk in FcεRI-Mediated Signaling

Reuben P. Siraganian, Juan Zhang, and Teruaki Kimura

The FcεRI is an oligomeric structure that lacks any known enzymatic activity, and therefore this receptor depends on associated molecules for transducing intracellular signals. The COOH-terminal cytoplasmic domains of both the β and the γ subunits contain a motif with the amino acid sequence $(D/E)x_2Yx_2Lx_{6-8}Yx_2(L/I)$ that is critical for cell activation.[1–3] This immunoreceptor tyrosine-based activation motif (ITAM) is also present in the ζ, γ, ε, and δ subunits of the T-cell receptor complex and in Igα and Igβ of the B-cell receptor.[4–8] Following stimulation of cells, phosphorylation of the tyrosine residues within the ITAM then creates binding sites for other proteins.

There are at least nine distinct families of nonreceptor protein tyrosine kinases. Among these cytoplasmic tyrosine kinases, members of the following families are involved in FcεRI-mediated reactions: Src, Csk, FAK, BTK(Tec/Itk), and Syk. As is discussed here, Syk is essential for the receptor-initiated signaling events.

Protein Tyrosine Phosphorylation

In studies in the 1970s and 1980s, it was observed that stimulation of rat mast cells for secretion results in the phosphorylation of several proteins.[9–11] In early studies with rat basophilic leukemia RBL-2H3 cells, histamine release was accompanied by the rapid phosphorylation of several proteins.[12–15] As the identity of the phosphorylated amino acids was not determined, we cannot know the class of the involved kinases.

The first observations on protein tyrosine phosphorylation in mast cells was that aggregation of the FcεRI results in rapid protein tyrosine phosphorylations, the most prominent being a 72-kDa protein.[16] The phosphorylation of this 72-kDa protein (pp72) is correlated by both time course and antigen dose with histamine release. Protein kinase C activation by phorbol ester or an increase of intracellular Ca^{2+} by ionophore A23187 did not induce tyrosine phosphorylation of pp72. Furthermore, in the absence of extracellular Ca^{2+}, there was FcεRI-mediated phosphorylation of pp72 but no degranulation. The tyrosine kinase inhibitor genistein efficiently blocked both tyrosine phosphorylation of pp72 and secretion, suggesting that this was an early event induced specifically by aggregation of the FcεRI.[17] After Syk was identified, it was found that FcεRI activation of RBL-2H3 cells results in rapid protein tyrosine phosphorylation and activation of this kinase and that Syk is a component of pp72.[18–20] The very rapid time course of this tyrosine phosphorylation, the coupling to FcεRI aggregation, and the fact that tyrosine kinase inhibitors block secretion indicated the importance of protein tyrosine phosphorylation in FcεRI-mediated secretion.[17,21–23] Other studies that defined the role of Syk in signal transduction in mast cells are described here.

The tyrosine phosphorylation of proteins can result in conformational changes and, if the molecule is an enzyme, have effects on its enzymatic activity. Tyrosine phosphorylation may also change the interaction of a protein with other proteins. Such interactions are usually mediated by the Src homology-2 (SH-2) domain present on many protein tyrosine kinases and other molecules.[24–26]

This chapter discusses the structural characteristics of the protein tyrosine kinase Syk, the mechanism of its activation and regulation, with emphasis on studies in basophils and mast cells.

Importance of Syk in Cellular Signaling

The important role of Syk in cell signaling was first demonstrated in studies with chimeric proteins in which the extracellular domain of CD16 was attached to Syk.[27] Clustering of these chimeras induced protein tyrosine phosphorylations, Ca^{2+} mobilization, and effector functions of the cells. In contrast, clustering similar chimeras that had Src family kinases as the intracellular domain were ineffective. Further evidence for a critical role of Syk in signal transduction were studies in B-cell-receptor signaling.[28,29] In a Syk negative avian B-cell line, receptor aggregation did not induce tyrosine phosphorylation of phospholipase C (PLC-γ2), generation of inositol 1,4,5-trisphosphate, or Ca^{2+} mobilization. In contrast, in Lyn-negative B cells activation of Syk still occurred, although the increase in intracellular Ca^{2+} was delayed. Similarly, in mast cells from mice in which Lyn has been inactivated by homologous recombination, FcεRI-induced histamine release still occurs, although the level of tyrosine phosphorylation of Syk is dramatically decreased.[30] These data suggest that Syk, more than Lyn, is critical for signal transduction. Syk has also been introduced into RBL-2H3 cells as a chimeric transmembrane protein with the extracellular domain of CD16 and the transmembrane domain of CD7.[31] Antibodies to CD16 clustered Syk, stimulated tyrosine phosphorylation of PLC-γ1, and activated downstream signals and degranulation. However, several inhibitors at concentrations that block the function of Syk did not inhibit ty-

rosine phosphorylation of the receptor subunits but blocked downstream signaling events, including Ca^{2+} mobilization.[32,33] These results demonstrated that Syk could activate downstream signaling events.

The importance of Syk for stimulation by different hematopoetic receptors was also demonstrated by experiments in which the *syk* gene was inactivated by homologous recombination.[34,35] Mice that were homozygous for *syk* mutation (Syk$^{-/-}$) developed severe systemic hemorrhage and died shortly before or after birth. These experiments suggest a critical role for Syk in platelet function. There were also severe immunological abnormalities with blocked differentiation of B cells. Therefore, Syk is important for B-cell development and function as well as the function of other hematopoetic cells such as platelets and mast cells.

A variant of the rat basophilic RBL-2H3 mast cell line that has no detectable Syk was identified.[36] Aggregation of FcεRI in these cells induced tyrosine phosphorylation of the β- and γ-subunits of the receptor but did not result in Ca^{2+} mobilization or secretion. Most of the receptor-induced tyrosine phosphorylations were also absent in these cells. Transfection and expression of Syk in these cells reconstituted the FcεRI-mediated signaling events and histamine release. Such cells also demonstrate that Syk activates a large number of downstream pathways. Mast cells have also been isolated from the Syk$^{-/-}$ mice by culture of fetal liver cells.[37] As with the Syk-negative RBL-2H3 cells, there was no FcεRI-induced Ca^{2+} mobilization or secretion, although tyrosine phosphorylation of the β- and γ-subunits of FcεRI and Lyn was normal. These results conclusively demonstrate that Syk is essential for FcεRI-induced signal transduction.

Present Model for Early Events in FcεRI Signaling

In the present model for mast cell signaling, the aggregation of FcεRI results in the tyrosine phosphorylation of the ITAMs of the β- and γ-subunits of the receptor.[38,39] This phosphorylation is probably by Lyn or another Src family kinase that associates with the receptor. Although Lyn is associated with the receptor in quiescent cells, the mechanism for tyrosine phosphorylation of the receptor is not fully established.[40–42] The tyrosine-phosphorylated ITAMs in the COOH-terminal cytoplasmic domain of FcεRIβ and the cytoplasmic domain of FcεRIγ function as a scaffold for binding of additional signaling molecules. These include adapter molecules that have no catalytic activity but have multiple binding sites and can recruit additional molecules to the receptor and a second group of molecules with catalytic domains which regulate more distal signaling events. There is preferential binding of different downstream signaling molecules to the β and γ phosphorylated ITAMs.[43–46] For example, a synthetic tyrosine-diphosphorylated peptide based on the ITAM sequence of the β-subunit of FcεRI precipitates Syk, SHIP, SHP-2, Shc, PLC-γ1, and Lyn whereas the similar peptide based on the ITAM of γ precipitates only Syk.[44,46–48] The binding of Syk to the tyrosine-phosphorylated ITAM, mediated by its two SH-2 domains, results in a conformational change in Syk, with an increase in its enzymatic activity and the downstream propagation of signals such as the tyrosine phosphorylation of PLC-γ1, PLC-γ2, and the influx of Ca^{2+}.[49] Therefore, tyrosine phosphorylation of the ITAM of the γ-subunit recruits Syk, which is crucial in downstream activating signals. In contrast, the ITAM based on the β-subunit, once phosphorylated, recruits other molecules that are critical for activating and regulating signaling events. The γ-subunit is critical for activation whereas the β functions as an amplifier.[50] The clustering of different molecules on the two receptor subunits would also allow the kinases associated with one subunit to tyrosine phosphorylate signaling molecules on the other subunit.

Structure of Syk

General Characteristics

A 40-kDa protein tyrosine kinase was purified from pig spleen in the laboratory of Prof. Yamamura. Based on the amino acid sequence, the cDNA of this molecule was isolated and called Syk, for spleen tyrosine kinase. The 40-kDa unit is a proteolytic fragment of the 72-kDa mature protein. At about the same time the laboratory of Dr. Geahlen isolated a similar tyrosine kinase that they named PTK72.[52,53] After the isolation of Syk it was shown that PTK72 and Syk were identical.

The isolation of Syk defined a new family of nonreceptor tyrosine kinases that, unlike the Src kinases, do not have a N-terminal myristylation site and therefore are present in the cytoplasm. Shortly afterward another member of this family was found in T cells and called ZAP-70 (for zeta chain-associated 70-kDa protein).[54] The primary structure of ZAP-70 is very similar to Syk (57% sequence identity, 73% amino acid homology), and these two molecules are the only members of this family of protein tyrosine kinases. The striking features of this family are the presence of two tandem SH-2 domains at the N-terminal region, the lack of an SH-3 region, and no clear regulatory carboxy domain that may regulate the function of the molecule (Fig. 10.1). The two SH-2 do-

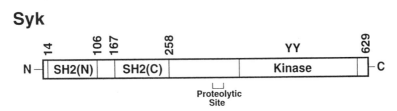

FIGURE 10.1. Schematic diagram of protein tyrosine kinase Syk. The *numbers* refer to the rat amino acid sequence of the different domains; *YY* refers to the autophosphorylation tyrosines in the activation loop. The proteolytic cleavage site is also marked.

mains are connected by a 65-residue segment called the inter-SH-2 region, followed by a linker sequence between the second SH-2 domain and the catalytic region of the molecule. Since the original isolation of pig Syk, the cDNA for human, rat, and mouse Syk have been isolated.[19,35,55] The sequences are very similar in all these species, with most of the variation being in the linker region between the SH-2 and the kinase domains. This conservation in sequence suggests the importance of structure in the functioning of this molecule. The *syk* locus has been mapped to chromosome 9 at band q22 and the mouse to the B-C2 region on chromosome 13.[55,56]

There are two alternatively spliced forms of Syk cDNA that result in two distinct isoforms differing by the presence of a 23-amino-acid insert located within the linker region between the second SH-2 and the catalytic domain.[57,58] Both mRNAs arising from this splicing encode functional protein tyrosine kinases and are expressed, although it is not clear how much of each of these proteins is actually present in cells. However, when expressed by transfection both the alternatively spliced forms are enzymatically active.[59]

By both northern and western blot analysis, Syk is expressed to varying amounts in almost all hematopoetic cells. It is expressed at high levels in spleen and at lower levels in the thymus. It is present in B cells, basophils, mast cells, monocytes, polymorphonuclear cells, platelets, and red blood cells but in more variable amounts in T cells. In cells, it is localized not only to the cytosol but also to particulate fractions.

In Vitro Enzymatic Properties and Substrates

In vitro, Syk can phosphorylate several substrates including tubulin, myelin basic protein, H2B histone, and [Val5]angiotensin II.[60] The best in vitro substrates for Syk contain tyrosines surrounded by an extensive number of acidic amino acids. For example, using a phage display system, although Lyn preferred substrates with the structure I/L-Y-D/E-x-L, a

sequence present in ITAM, Syk preferred substrates that had aspartic acid at position −1 and glutamic acid in position +1 relative to phosphorylated tyrosine.[61] One such Syk substrate that has been identified is the cytoplasmic domain of human erythrocyte anion transport channel, band 3, which may be a natural substrate in red blood cells.[62] This is phosphorylated near the N-terminus at two positions (EELQDDpYEDMM and LEQEEpYEDPD) where a tyrosine is contained within many acidic residues. Syk also phosphorylates HS1 (at PEGDpYEEVL) and Vav (at DEIpYEDLM) proteins at sites that have acidic residues at −1, +1, and +2 positions.[63] Interestingly, the Syk autophosphorylation sequence in the activation loop (as the DENYYKA peptide) was a poor substrate. In contrast, Lyn and Src family kinases are at least 1000 fold less efficient in phosphorylating substrates that have an acidic residue at −1. Other proteins in which this -D-Y-E- motif appear and could be potential substrates of Syk include ezrin and CD22, a molecule known to associate with Syk.[64,65]

The in vitro enzymatic activity of intact Syk is much lower than that of the 40-kDa proteolytic fragment,[60] suggesting that there are structural constraints on the activity of the molecule. This is discussed further in the section on the regulation of Syk.

Comparison of Syk and ZAP-70 Activities

Syk and ZAP-70 are structurally similar molecules and may be able to substitute at the functional level for each other. For example, in a Syk-negative B-cell line, the expression of ZAP-70 reconstitutes early signaling events including the rise in intracellular Ca^{2+}.[66] However, there are differences in the activities of ZAP-70 and Syk.[27] Expression of Syk or ZAP-70 as a chimeric molecule attached to a transmembrane protein suggests that Syk is more active than ZAP-70. Aggregation of these chimeric proteins induces Ca^{2+} mobilization; however, although cross-linking of the chimeric Syk induces these downstream effects, ZAP-70 requires Fyn or Lck, which could be the result of the differences in intrinsic activity of these two kinases. T-cell receptor (TCR) aggregation in different CD45-negative T-cell lines results in signal transduction if they contain Syk but not ZAP-70.[67] Similarly, in a Lck-negative but CD45-positive cell line, Syk but not ZAP-70 results in signal transduction, suggesting that there are differences in activation requirements between Syk and ZAP-70. Finally in a transient transfection system, the enzymatic activity of Syk is at least 100 fold greater than that of ZAP-70. Chimeras between ZAP-70 and Syk suggest that this is an intrinsic activity of the kinase domain.[59]

Tyrosine Phosphorylation Sites in Syk

The substrates of protein tyrosine kinases are other molecules that also could be tyrosine kinases. Some protein tyrosine kinases also autophosphorylate, which may be by inter- or intramolecular reactions. Autophosphorylation then regulates not only the activity of the protein tyrosine kinase but also the interaction of the kinase with other molecules (see following on regulation of Syk).

The tyrosine residues that are conserved between Syk and ZAP-70 would be expected to be those that are important for the function of these molecules. There are 18 tyrosines conserved between ZAP-70 and Syk, another 9 tyrosines that are unique to ZAP-70, and 9 which are only in Syk. However, only 10 of these tyrosines are phosphorylated in Syk after an in vitro kinase reaction.[68] The position of these tyrosines are as follows: 1 tyrosine in the inter-SH-2 region, five sites in the linker region between the second SH-2 domain and the catalytic region, 2 adjacent tyrosines in the activation loop (autophosphorylation site), and 2 tyrosines at the extreme carboxy-terminal region. The tyrosine between the two SH-2 domains may regulate the activation of Syk and its dissociation from pITAM.[69] All except one of these tyrosines that are phosphorylated are also conserved in ZAP-70, suggesting that they are functionally important. While the activation loop tyrosines in ZAP-70 require a Src family kinase for phosphorylation, they are

autophosphorylated in Syk.[70] This could explain the fundamental difference between Syk and ZAP-70 in the requirement of a Src family kinase for activation.[71]

There are some data on the in vivo function of these tyrosine residues. Syk expressed as a chimeric protein with both the activation loop tyrosines (autophosphorylation site) mutated had decreased kinase activity and did not induce the tyrosine phosphorylation of PLC-γ1, whereas substitution of two tyrosines in the linker region of Syk did not affect the kinase activity of the chimera. In a Syk-deficient B-cell line, the expression of Syk in which both of the tyrosines in the activation loop were replaced with phenylalanine was also defective in signal transduction, but this mutation did not affect recruitment of Syk to the phosphorylated ITAM.[73] However, there was still some tyrosine phosphorylation of this mutated Syk after receptor aggregation. In studies with transient transfection and overexpression of mutated Syk in RBL-2H3 cells, it was concluded that phosphorylation by a Src kinase of the activation loop tyrosines of Syk was critical in activating Syk.[74] However, the phosphorylation of these residues was also caused by transphosphorylation by Syk. The role of activation loop tyrosine phosphorylation in regulating the function of Syk is discussed in the section on regulation of Syk function. The different phosphorylated tyrosines in Syk function as docking sites for the binding of SH-2-containing molecules (see section on Association of Syk with Other Proteins).

Interaction of Syk with ITAM

The approximately 100-amino-acid residue SH-2 domain is present on many cytosolic signaling molecules and binds to specific phosphorylated tyrosine residues in other proteins. The three-dimensional structure of SH-2 domains in different proteins is very similar and has a ligand-binding site that is split into two distinct pockets, one for the phosphotyrosine (pTyr) and the other to accommodate the amino acids that are C-terminal to the pTyr. Screening of random phosphotyrosine libraries also demonstrate that the three residues C-terminal to the pTyr are the principal determinants of the peptide selectivity for the binding of different SH-2 domains. The preferred binding with the COOH-terminal SH-2 domain of Syk expressed as a fusion protein is Y + 1 Q/T/E; Y + 2 E/Q/T and Y + 3 L.[75,76] However, there is much weaker binding with the N-terminal SH-2 domain, which is Y + 1 T; Y + 2 T; and Y + 3 I/L/M. This weaker binding can be explained on the basis of the data on the crystal structure of ZAP-70 (see following).

After the identification of Syk, it was demonstrated that it coprecipitates with FcεRI, especially after cell activation.[18,19] Fusion proteins containing the SH-2 domains of Syk show pronounced selectivity in binding of tyrosine-phosphorylated proteins, precipitating only three phosphoproteins, two of which were the β- and γ-subunits of FcεRI from lysates of stimulated RBL-2H3 cells.[43,44] In contrast, Lyn SH-2, in addition to precipitating the same FcεRI components, binds to many other phosphoproteins. The tandem Syk SH-2 domains are much more active than either of the two SH-2 domains alone, indicating the cooperative nature of the binding. There is direct interaction of the Syk SH-2 domains with the γ-tyrosine-phosphorylated subunit. The importance of this interaction is demonstrated by in vivo experiments in which the association of Syk with FcεRI and degranulation was inhibited by the addition of a protein containing the SH-2 domains of Syk in strepolysin-O-permeablized cells.[77] Synthetic nonphosphorylated and phosphorylated peptides based on the ITAM of different signaling molecules have also been used to study these interactions. From lysate of RBL-2H3 cells, Syk is precipitated predominantly by the tyrosine-diphosphorylated FcεRIγ ITAM peptide, but much less by γ-monophosphorylated peptides or diphosphorylated β-peptide, and there is no binding to nonphosphorylated ITAM peptides.[44,46] This binding requires both SH-2 domains interacting with a diphosphorylated ITAM with much less binding with the individual SH-2 domains of Syk or ZAP-70. The binding of Syk or ZAP-70 to tyrosine-diphosphorylated ITAM is with high affinity, but binding to monophosphorylated ITAM-

based peptides is with affinities that are 100 to 1000 times lower.[78–81] Both SH-2 domains from one Syk or ZAP-70 molecule bind to both the phosphorylated tyrosines in an ITAM. The high-affinity binding then results from the dramatic advantage from this bidentate interaction.

Although the ITAM sequences of the different immune receptors are similar, there are differences in their interaction with Syk or ZAP-70. For example, there is as much as 30-fold difference in the affinity of ZAP-70 binding to the different phospho-ITAMs of ζ.[82] The distance between the two pTyr in the ITAM may be important for the interaction with the two SH-2 domains. Although the ITAM of the β-subunit has one less amino acid between the two pTyr, it can still bind the SH-2 domains of Syk. The difference in the amino acid sequence of the ITAMs in the β- and γ-subunits of FcεRI could explain the binding of different downstream signaling molecules. The binding of Syk is preferentially with the γ-subunit and with much lower affinity with the β, whereas the β-subunit binds also Shc, PLC-γ1, SHIP, and SHP-2[46–48]

Crystal Structure

Although there is some structural information on Syk SH-2 domains complexed with a phosphotyrosine peptide,[83] our present structural information is derived from determination of the crystal structure of the tandem SH-2 domains of ZAP-70 bound to the diphosphorylated ITAM peptide based on the ζ-chain of the TCR.[84] The structure suggests that the two SH-2 domains do not function independently and explains the selectivity for diphosphorylated ITAM sequences that was observed in binding studies. The carboxy-terminal SH-2 domain binds to the amino-terminal YxxL sequence of the ITAM by a typical SH-2 domain to phosphotyrosine interaction, with one pocket for the pTyr and another pocket for the leucine that is three amino acids distal. However, the amino-terminal SH-2 domain is unusual and has an incomplete pTyr-binding pocket, which is completed by residues provided by the adjacent carboxy-terminal SH-

2 domain. Therefore, the N-terminal SH-2 domain if expressed alone is incomplete and cannot bind phosphotyrosine residues well. To bind both ITAM phosphotyrosines simultaneously, the two SH-2 domains must interact to form this unusal pTyr-binding pocket. The distance between the two pYXXL in the ITAM provides properly spaced partners for the pair of SH-2 domains that are tethered by the inter-SH-2 coiled-coil loop. This coiled-coil loop formed by the 65 amino acid residues that are between the two SH-2 domains may be important in the stability of the SH-2–SH-2 interactions with the phosphorylated ITAM sequence. It is also possible that binding to the pITAM could influence other domains with the coiled inter-SH-2 region interacting with the kinase domain of the molecule. Such effects may explain the conformational changes that occur after Syk binding to pITAM (see following). The crystal structure of the full ZAP-70 or Syk molecule may help answer these questions.

Regulation of the Function of Syk

The activation of Syk appears to result from conformational changes that are induced either by tyrosine phosphorylation or by the binding to the tyrosine-phosphorylated ITAM (see Fig. 10.2). Tyrosine-phosphorylated Syk is in a conformation different from that of nonphosphorylated Syk, as demonstrated by changes in immunoreactivity.[49] The binding of diphosphorylated ITAM also results in a

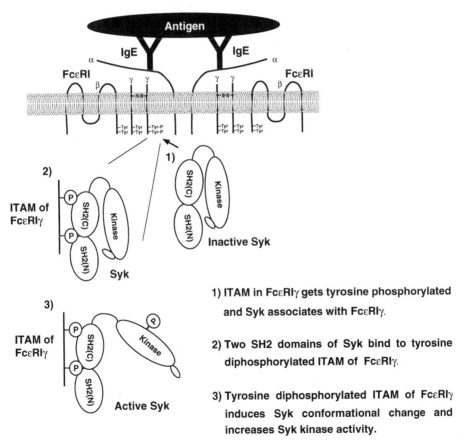

1) ITAM in FcεRIγ gets tyrosine phosphorylated and Syk associates with FcεRIγ.

2) Two SH2 domains of Syk bind to tyrosine diphosphorylated ITAM of FcεRIγ.

3) Tyrosine diphosphorylated ITAM of FcεRIγ induces Syk conformational change and increases Syk kinase activity.

FIGURE 10.2. Model of activation of Syk by binding to FcεRI. ITAM, immunoreceptor tyrosine-based activation motif.

conformational change in nonphosphorylated Syk and stimulates Syk kinase activity at least 10 fold.[85,86] This results in increased autophosphorylation of Syk in the absence of a Src family kinase and requires that both Syk SH-2 domains be functional. The diphosphorylated γ-ITAM is more active than the β-ITAM in inducing these changes. These observations indicate that after activation there is a conformational change so that the carboxy-terminal part of Syk becomes accessible to antibodies.[49] Therefore, the catalytic domain of Syk is probably regulated by intramolecular interactions with adjacent domains, and the binding of Syk to the tyrosine-phosphorylated ITAM stimulates Syk kinase activity. Further these results support the concept that the activation of Syk does not require a Src family kinase.

The increased autophosphorylation of Syk induced by binding to the ITAM could then regulate Syk function. There are at least two mechanisms for the regulation of tyrosine kinase activity by tyrosine phosphorylation. The Src family kinases are regulated through the tyrosine phosphorylation of a carboxy-terminal site.[87–89] This site interacts with the SH-2 domain to induce an inactive conformation of the kinase with the SH-3 domain binding to the linker region between the SH-2 and kinase domains. Dephosphorylation of this carboxy-terminal site results in activation of the kinase. Although Syk has a number of tyrosines in the carboxy-terminal region that can become tyrosine phosphorylated, there is no evidence that they are involved in regulating the function of this molecule. A second mechanism that may activate protein tyrosine kinases is by phosphorylation of tyrosine residues within the activation loop of the kinase domain. The crystal structure of two kinase domains of receptor tyrosine kinases has been determined.[90,91] In the case of the insulin receptor, there are two tyrosines (Y1162 and Y1163) in the activation loop; Y1162 but not Y1163 lies within the catalytic center such that it prevents ATP or substrate binding. The phosphorylation of this Y1162 induces a conformational change that would move Y1162 out of the catalytic center

and allow kinase activity. The structural studies suggest that activation could occur optimally only by transphosphorylation. The structure of the fibroblast growth factor receptor kinase domain suggests an alternative model in which Y653 in the activation loop interferes with substrate binding but not with ATP binding. Phosphorylation of Y653 is thought to induce a conformational change that would allow substrate access to the catalytic center. Syk has a similar activation loop with two tyrosines that may be important in regulating the kinase activity, which when mutated have an effect on the capacity of the molecule to signal. However, the results are not clear on whether mutation of these tyrosines has an effect on the in vitro kinase activity of Syk. These activation loop tyrosines can be either autophosphorylated or phosphorylated by other protein tyrosine kinases. Interestingly, transient cotransfection of Syk and Lyn results in tyrosine phosphorylation of Syk and increase in enzymatic activity of Syk.[92,93]

This model for activation of Syk by binding to pITAM is probably different than that for ZAP-70 activation. ZAP-70 binds to pITAM without activation, suggesting a requirement for a Src family kinase. In transient transfection experiments, a chimera of a transmembrane protein and the cytoplasmic domain of the γ-subunit of FcεRI activated Syk, whereas ZAP-70 required the cotransfection with Lyn.[93] When Syk is recruited by the tyrosine-phosphorylated ITAM of the γ-subunit of FcεRI, it would be close to Lyn that is associated with the β-subunit. This location of Syk in the membrane together with the change in its conformation may also allow it to be further phosphorylated by Lyn or other kinases, resulting in further activation of Syk. For example, such stepwise phosphorylation by several different protein tyrosine kinases has been demonstrated in HS-1.[94] Although Syk is activated by the γ-subunit, this activation is enhanced by the presence of the β-subunit.[50,95] Therefore, Syk may be activated by several mechanisms that could include conformational change from ITAM binding, autophosphorylation, and phosphorylation by other kinases.

The function of Syk in cells can also be regulated by other molecules such as p120[c–cbl] (Cbl), which is expressed in many hematopoetic cells.[96,97] Cbl is a 120-kDa protein that has a putative nuclear localization sequence, a "ring finger" motif typical of DNA-binding proteins, and several proline-rich sequences in the C-terminal half that may serve as SH-3-binding sites. Cbl associates with a variety of molecules including Syk, PI-3-kinase, Lyn, Fyn, Crk, Grb2, and with Syk. Although some of these interactions are SH-2 mediated, the association with Syk is independent of receptor aggregation and the tyrosine phosphorylation of Syk. The interaction of Syk and Cbl is between the proline-rich region of Cbl and a region of Syk composed of the two SH-2 domains and intradomain linker, but does not require the tyrosine phosphorylation of Cbl. Whether this is caused by indirect interaction is not clear as there is also a complex of Cbl, Syk, and Lyn.[98] There is also FcεRI-induced tyrosine phosphorylation of Cbl, which is predominantly downstream of Syk.[96] This phosphorylation of Cbl is mainly at sites distal to the proline-rich region within the COOH-terminal 250 amino acids. The overexpression of Cbl in RBL-2H3 cells inhibits the receptor-induced tyrosine phosphorylation and activation of Syk, resulting in a decrease in downstream phosphorylation and secretion.[97] These results suggest that Cbl can regulate the kinase activity of Syk by regulating the level of its tyrosine phosphorylation.

The function of Syk is also probably regulated by protein tyrosine phosphatases and other molecules with which Syk interacts. In cells, the deaggregation of FcεRI results in rapid dephosphorylation of the receptor subunits and of Syk.[16,38,99] The tyrosine phosphatases associated with FcεRI, which include SHP-1 and SHP-2, could be responsible for this dephosphorylation of Syk that is bound to the ITAM.[48,100] SHP-2 can also be imunoprecipitated with hyperphosphorylated Syk (unpublished observations). Therefore, this or other tyrosine phosphatases could dephosphorylate Syk and control the extent of signaling in cells.

Association of Syk with Other Proteins

Syk can associate with Src family protein tyrosine kinases.[43,101–103] For example, Lyn and Syk can be coprecipitated, although in B-cell lines this may depend on the stage of cell differentiation.[101] Some of this coprecipitation may be indirect because of complexes of multiple proteins such as Lyn, Syk, and Cbl.[104] However, direct interaction between Syk and Src family kinases is suggested by experiments in which the SH-2 domain of many proteins, including Src family kinases, precipitates tyrosine-phosphorylated Syk.[43,103,105] The binding of Lyn SH-2 to Syk is inhibited by a phosphorylated peptide (GpYESPL), based on the sequence from the linker region of Syk.[105] However, the relationship between Syk and Src family tyrosine kinases may be complex. The SH-2 domain of Lck binds to the two tandem Tyr in the activation loop of Syk when both these tyrosines are phosphorylated.[106] This association results in increased tyrosine phosphorylation of Lck and of cellular proteins in transient transfection experiments.[107,108] The transient coexpression of Lck and Syk does not alter the catalytic activity of Syk but phosphorylates Lck at Tyr-192 and increases the activity of Lck.[107] TCR-induced activation of Syk in Jurkat T cells that are defective in Lck suggests that Syk does not require the presence of a Src family kinase.[108] However, the increased tyrosine phosphorylation of cellular proteins when both Lck and Syk are present suggests synergistic effects in cells. Therefore, Lyn or another Src family kinase associated with Syk may phosphorylate Syk and, as discussed in the previous section, regulate its activity.

Vav is another molecule that both associates with and is a substrate of Syk. Vav is a hematopoietic cell-specific proto-oncogene that contains a number of structural motifs, including leucine-rich, Dbl-homology, pleckstrin-homology, zinc-finger, SH-2, and SH-3 domains. These domains are important for its interaction with molecules such as Grb2, Lck, ZAP-70, SLP-76, and tubulin. The Dbl-homology domain serves as a GDP-GTP exchange factor for the Rac/Rho family of small GTP-binding proteins.[109,110] Rho regulates the assembly of focal adhesion and actin stress fibers whereas Rac regulates membrane ruffling[111] and both Rac and Rho regulate mast cell secretion.[112] Vav is tyrosine phosphorylated after activation of receptors such as the BCR, TCR, FcγR, and FcεRI.[113-116] Tyrosine-phosphorylated Vav, but not the nonphosphorylated protein, catalyzes GDP–GTP exchange on Rac-1.[110] Vav associates with ZAP-70 and with Syk.[117,118] The interaction with Syk was shown in the yeast two-hybrid system and is by the SH2 domain of Vav binding to two tyrosines (Tyr-341 and or Tyr 345) in the linker region of Syk. Coprecipitation of Vav with FcεRI may therefore be indirectly mediated by Syk binding to the γ-subunit of FcεRI.[116]

Syk also associates with a 120-kDa protein and PLC-γ1.[101] The 120-kDa protein was initially thought to function as a linker between Syk and PLC-γ1, although more recent studies indicate that it is probably protein kinase Cμ.[119] Cross-linking of B-cell antigen receptor (BCR) results in the activation of protein kinase Cμ, which associates with Syk, PLC-γ1, and PLC-γ2. Protein kinase Cμ can phosphorylate Syk in vitro and decrease the enzymatic activity of Syk. Activation of protein kinase Cμ in these cells is downstream of Syk. Association of Syk with PLC-γ1 could be the interaction of the kinase with its substrate. Syk and PLC-γ1 coprecipitate and fusion proteins containing the SH-2 domains of PLC-γ1 bind to tyrosine-phosphorylated Syk.[68,72,101,120] The interaction with PLC-γ1 requires two tyrosines in the linker region of Syk. In vitro, Syk isolated from activated but not from nonactivated B cells phosphorylates PLC-γ1 on Tyr-771 and Tyr-783, which are important regulatory residues. In contrast, Lyn phosphorylates only Tyr-771.

A number of other molecules coprecipitate with Syk and could also be substrates of Syk. These include the 80- to 85-kDa protein cortactin that is also a substrate of Src, the 54-kDa protein α-tubulin, and the anion transporter, band 3.[62,121,122] Syk also associates with other proteins. For example, in the yeast two-hybrid system Syk interacts not only with Vav

and Lck but also with the p85 regulatory subunit of phosphatidylinositol 3-kinase.[118] The SH-2 domains of Syk bind directly to CD22, and by coprecipitation the complex of CD22 includes SHP-1, PLC-γ1, and Syk.[64,121,123]

Downstream Molecules: Substrates

The clustering of Syk to FcεRI at the membrane may be important for the subsequent downstream signaling events. As discussed in previous sections, the binding of Syk to the tyrosine-phosphorylated ITAM, the conformational change, autophosphorylation, and phosphorylation by Lyn could all contribute to the increase in the enzymatic activity of Syk. Syk bound to FcεRI may then phosphorylate membrane proteins in the vicinity of the receptor. Some of these proteins could have also been recruited to the receptor after it is aggregated or tyrosine phosphorylated. For example, the tyrosine-phosphorylated ITAM of FcεRIβ binds not only Syk but also other proteins that have SH-2 domains including Shc, PLC-γ1, SHIP, and SHP-2. The tyrosine phosphorylation of Syk could also recruit other proteins that bind to it through SH-2-mediated interactions.

In the Syk-negative variant of the RBL-2H3 cell line, FcεRI aggregation induced minimal increase in protein tyrosine phosphorylations of total cellular proteins.[36] Therefore, most of the FcεRI-induced cellular protein tyrosine phosphorylations require Syk. However, these results do not imply that Syk is directly phosphorylating these proteins. This section describes molecules that are tyrosine phosphorylated upstream or downstream of Syk. The details of these pathways are described in other chapters in this volume.

The FcεRI-induced tyrosine phosphorylation of a number of proteins is independent of the presence of Syk in cells. These include the following: the β- and γ- subunits of FcεRI, the SH-2 domain containing inositol phosphate polyphosphate 5-phosphatase, SHIP, and the SH-2-containing protein tyrosine phosphatases SHP-1 and SHP-2.[47,48] The β- and γ-subunits of

FcεRI are tyrosine phosphorylated after receptor aggregation in the Syk-deficient cells, and the overexpression of a dominant negative Syk in RBL-2H3 cells does not decrease this phosphorylation.[36,92] SHIP associates with FcγRIIB, and coclustering of the immune receptor with FcγRIIB negatively regulates both B-cell and mast cell functions.[124–126] The ITAM of the β-subunit of FcεRI can recruit SHIP to the receptor.[47,127] SHIP is tyrosine phosphorylated after FcεRI aggregation even in Syk-negative cells, indicating that this is an early event and is upstream of the activation of Syk.[47] Aggregation of FcεRI induces an increase in the tyrosine phosphorylation of both SHP-1 and SHP-2 that is independent of Syk.[48] Interestingly, these molecules that are tyrosine phosphorylated independently of Syk may regulate the downstream signals generated by the activation of Syk.

In contrast, the FcεRI-induced tyrosine phosphorylation of a large number of proteins is downstream of Syk. These include PLC-γ1, PLC-γ2, Vav, Shc, MAP kinase, JNK kinase, the focal adhesion kinase pp125^{FAK} (FAK), paxillin, the adhesion molecule PECAM-1(CD31), Cbl, rasGAP, SLP-76, and HS-1.

PLC-γ1 and PLC-γ2 are SH2-, SH-3-containing proteins that are rapidly tyrosine phosphorylated after FcεRI aggregation and catalyze the hydrolysis of phosphatidylinostiol 4,5-bisphosphate, resulting in the generation of inositol 1,4,5-trisphosphate and 1,2-diacylglycerol. These second messengers release Ca^{2+} from internal stores and activate PKC, respectively. Both events are essential for FcεRI-mediated secretion and are dependent on Syk.[36] Tyrosine phosphorylation of PLC-γ1 results in increased catalytic activity.[128] Experiments summarized in the previous section suggest that there is association of PLC-γ1 with Syk, and a 44-kDa protein tyrosine kinase associates with PLC-γ1 in mouse bone marrow culture-derived mast cells.[129] In vitro, this 44-kDa kinase phosphorylates PLC-γ1 on tyrosine and is possibly a proteolytic fragment of Syk.

The tyrosine phosphorylation of Vav is also downstream of Syk. (Zhang et al., in manuscript). In RBL-2H3 cells, transient overexpression of wild-type Syk enhanced

FcεRI-induced tyrosine phosphorylation of Vav, whereas overexpression of a dominant negative form of Syk blocked this phosphorylation.[130] In Syk-negative cells, Vav lost both its constitutive and its receptor-mediated tyrosine phosphorylation, and Syk transfection reconstituted both of these tyrosine phosphorylations[37] (and Zhang et al., in manuscript). The tyrosine phosphorylation of Vav may play an important role in regulating the function of Rac-1 and thereby affect cell morphology, membrane ruffles, and secretion. Therefore Vav may link tyrosine phosphorylation with subsequent signal transduction events. The major tyrosine-phosphorylated protein that associates with Vav is SLP-76, a protein that also associates with Grb2.[131] SLP-76 is tyrosine phosphorylated after FcεRI aggregation downstream of Syk, but in the absence of extracellular Ca^{2+}.[132] SLP-76 has an amino-terminal region with several potential tyrosine phosphorylation sites, a central proline-rich region that interacts with Grb2, and a single carboxy-terminal SH-2 domain. The SH-2 domain of Vav binds to tyrosine-phosphorylated SLP-76 and may enhance its downstream effects. SLP-76 associates with ZAP-70, but it does not appear to interact directly with Syk.

Shc has an SH-2 domain that binds to the tyrosine-phosphorylated β-ITAM, and its increased tyrosine phosphorylation and association with Grb2 and other proteins are also dependent on Syk.[133] GRB-2 associates with the tyrosine-phosphorylated Shc via its own SH-2 domain, and GRB-2 constitutively associates by its SH-3 domain with the GDP–GTP exchange protein Sos. The Syk-dependent tyrosine phosphorylation of Shc and its association with Grb2 may provide a pathway through Sos for activation of Ras by FcεRI, leading to the MAPK signaling cascade. Although the activation of MAPK and c-Jun NH$_2$-terminal kinase (JNK) are both downstream of Syk, only the JNK pathway also requires Vav.[134] Tyrosine-phosphorylated Vav is a guanine nucleotide exchange factor for Rac-1 that leads to the activation of JNK.[110] This pathway may be important for the induction of cytokine genes. The tyrosine phosphorylation of HS1 is also Syk dependent. This protein, which has an

SH-3 domain close to its C-terminus, a putative nuclear localization signal, and repetitive helix-turn-helix motifs found in many DNA-binding proteins, could function in nuclear signaling.[20,135,136]

FcεRI-induced activation of the MAP kinase pathway requires Syk, as demonstrated by the following studies. Aggregation of Syk expressed as a chimeric protein in RBL-2H3 cells induced activation of MAP kinase.[31] Similarly, activation of this pathway is enhanced by transient overexpression of a wild-type form of Syk and blocked by overexpression of a dominant negative form of Syk.[130] Definitive proof was the finding that FcεRI-induced activation of the MAP kinase pathway is defective in the Syk-negative cells and is restored after transfection of Syk[37] (and Zhang et al., in manuscript). The Ras-GTPase activating protein (rasGAP) is probably a downregulator of Ras by stimulating its weak intrinsic GTPase activity and may also function to regulate cell shape.[137] The FcεRI-induced tyrosine phosphorylation of rasGAP also requires Syk (Zhang et al., in manuscript). Thus, Syk functions in activating and regulating the function of the MAP kinase pathway.

Aggregation of FcεRI on adherent RBL-2H3 cells also increases the tyrosine phosphorylation of the focal adhesion kinase pp125^{FAK} (FAK), a cytosolic nonreceptor protein tyrosine kinase.[138] Although FAK does not have any SH-2 or SH-3 domains, it interacts with many molecules and has binding sites for the SH-2 domains of Src family kinases such as Src, Fyn, and Lyn, the p85 subunit of phosphatidylinositol 3'-kinase, Grb2, Crk, and Csk. From platelet lysates, Syk is coprecipitated with FAK, which could result from the interaction of both FAK and Syk with the cytoskeleton. In Syk-negative adherent cells there is still constitutive tyrosine phosphorylation of FAK, but the FcεRI-induced phosphorylation requires Syk (Zhang et al., in manuscript). The FcεRI-induced tyrosine phosphorylation of FAK may be important for further propagation of signaling in the late steps of degranulation. For example, in vitro FAK can phosphorylate paxillin, a cytoskeletal protein that accumulates at focal adhesion sites.

Paxillin has multiple binding sites for SH-2 and SH-3 domains, and forms complexes with many proteins including protein tyrosine kinases and cytoskeletal proteins such as FAK, Lyn, vinculin, α-actinin, and talin.[139] In RBL-2H3 cells, FAK and paxillin become coordinately phosphorylated on tyrosine in response to FcεRI aggregation,[140] and like FAK, the FcεRI-induced tyrosine phosphorylation of paxillin is also dependent on the expression of Syk. The coordinate tyrosine phosphorylation of paxillin and FAK may be important in the development of actin stress fibers and the reorganization of F-actin[141,142] and thus may play a role in signal transduction from receptors to the cytoskeleton.

By scanning electron microscopy, aggregation of FcεRI results in cell spreading, transformation of the cell surface to a lamellar topography with deep folds and ruffles that may be secondary to the reorganization of F-actin and its appearance in ruffles. These morphological changes require Syk (Zhang et al., in manuscript) and could be caused by the tyrosine phosphorylation of Vav that regulates Rac-1, which is important for membrane ruffling.[111] Therefore, the biochemical events resulting in morphological changes induced by receptor aggregation are downstream of Syk.

Platelet/endothelial cell adhesion molecule 1 (PECAM-1 or CD31) is an adhesion molecule that is required for the transmigration of cells across the endothelium into sites of inflammation. The FcεRI-induced tyrosine phosphorylation of PECAM-1 is an early event, independent of Ca^{2+} influx or of the activation of protein kinase C. Tyrosine phosphorylation of PECAM-1 is partly independent of Syk, although it is much stronger in Syk-positive than in Syk-negative cells.[143]

Conclusions

The data reviewed here demonstrate that Syk is essential for propagating intracellular signals following FcεRI aggregation. The pathway downstream of Syk leads to the activation of phospholipases, the generation of secondary mediators that result in the increase in intracellular Ca^{2+} and activation of protein kinase C. The changes in intracellular Ca^{2+} result in the sustained influx of Ca^{2+} from the extracellular medium and the regulation of Ca^{2+}/calmodulin-dependent events. The protein kinase C family of molecules phosphorylates proteins on serine or threonine residues and is an essential transducer of signals for secretion. Other pathways are also downstream of Syk. Tyrosine phosphorylation of the FAK may play an important role in the late steps of degranulation, while the activation of the MAP kinase and JNK pathways is important for the release of arachidonic acid and nuclear events such as the induction of cytokine genes. These changes in the cell results in cytoskeletal reorganization, the movement of granules, and degranulation.

This review has concentrated on the role of Syk in mast cells. However, Syk is also essential for the function of other cells and is tyrosine phosphorylated and activated in platelets and B cells. Thus, Syk is essential for the function of several immune receptors.

References

1. Alber G, Miller L, Jelsema CL, et al. Structure-function relationships in the mast cell high affinity receptor for IgE. Role of the cytoplasmic domains and of the β subunit. J Biol Chem 1991;266:22613–22620.

2. Reth M. Antigen receptor tail clue. Nature (Lond) 1989;338:383–384.

3. Chan AC, Desai DM, Weiss A. The role of protein tyrosine kinases and protein tyrosine phosphatases in T cell antigen receptor signal transduction. Annu Rev Immunol 1994;12:555–592.

4. Letourneur F, Klausner RD. T-cell and basophil activation through the cytoplasmic tail of T-cell-receptor ζ family proteins. Proc Natl Acad Sci USA 1991;88:8905–8909.

5. Romeo C, Amiot M, Seed B. Sequence requirements for induction of cytolysis by the T cell antigen/Fc receptor ζ chain. Cell 1992;68:889–897.

6. Irving BA, Chan AC, Weiss A. Functional characterization of a signal transducing motif present in the T cell antigen receptor ζ chain. J Exp Med 1993;177:1093–1103.

7. Cambier JC, Pleiman CM, Clark MR. Signal transduction by the B cell antigen receptor and its coreceptors. Annu Rev Immunol 1994;12:457–486.

8. Gauen LKT, Zhu Y, Letourneur F, et al. Interactions of p59^{fyn} and ZAP-70 with T-cell receptor activation motifs: defining the nature of a signalling motif. Mol Cell Biol 1994;14:3729–3741.

9. Sieghart W, Theoharides TC, Alper SL, et al. Calcium-dependent protein phosphorylation during secretion by exocytosis in the mast cell. Nature (Lond) 1978;275:329–331.

10. Theoharides TC, Sieghart W, Greengard P, et al. Antiallergic drug cromolyn may inhibit histamine secretion by regulating phosphorylation of a mast cell protein. Science 1980;207:80–82.

11. Sieghart W, Theoharides TC, Douglas WW, et al. Phosphorylation of a single mast cell protein in response to drugs that inhibit secretion. Biochem Pharmacol 1981;30:2737–2738.

12. Hattori Y, Siraganian RP. Rapid phosphorylation of a 92,000 MW protein on activation of rat basophilic leukemia cells for histamine release. Immunology 1987;60:573–578.

13. Teshima R, Ikebuchi H, Terao T. Ca^{2+}-dependent and phorbol ester activating phosphorylation of a 36K-dalton protein of rat basophilic leukemia cell membranes and immunoprecipitation of the phosphorylated protein with IgE-anti IgE system. Biochem Biophys Res Commun 1984;125:867–874.

14. Sagi-Eisenberg R, Mazurek N, Pecht I. Ca^{2+} fluxes and protein phosphorylation in stimulus-secretion coupling of basophils. Mol Immunol 1984;21:1175–1181.

15. Essani N, Essani K, Siraganian RP. Protein phosphorylation in rat basophilic leukemia cells following stimulation for histamine release. Fed Proc 1986;45:243.

16. Benhamou M, Gutkind JS, Robbins KC, et al. Tyrosine phosphorylation coupled to IgE receptor-mediated signal transduction and histamine release. Proc Natl Acad Sci USA 1990;87:5327–5330.

17. Stephan V, Benhamou M, Gutkind JS, et al. FcεRI-induced protein tyrosine phosphorylation of pp72 in rat basophilic leukemia cells (RBL-2H3). Evidence for a novel signal transduction pathway unrelated to G protein activation and phosphatidylinositol hydrolysis. J Biol Chem 1992;267:5434–5441.

18. Hutchcroft JE, Geahlen RL, Deanin GG, et al. FcεRI-mediated tyrosine phosphorylation and

activation of the 72-kDa protein-tyrosine kinase, PTK72, in RBL-2H3 rat tumor mast cells. Proc Natl Acad Sci USA 1992;89:9107–9111.

19. Benhamou M, Ryba NJP, Kihara H, et al. Protein-tyrosine kinase p72^{syk} in high affinity IgE receptor signaling. Identification as a component of pp72 and association with the receptor γ chain. J Biol Chem 1993;268:23318–23324.

20. Minoguchi K, Benhamou M, Swaim WD, et al. Activation of protein tyrosine kinase p72^{syk} by FcεRI aggregation in rat basophilic leukemia cells: p72^{syk} is a minor component but the major protein tyrosine kinase of pp72. J Biol Chem 1994;269:16902–16908.

21. Lavens SE, Peachell PT, Warner JA. Role of tyrosine kinases in IgE-mediated signal transduction in human lung mast cells and basophils. Am J Respir Cell Mol Biol 1992;7:637–644.

22. Kawakami T, Inagaki N, Takei M, et al. Tyrosine phosphorylation is required for mast cell activation by FcεRI cross-linking. J Immunol 1992;148:3513–3519.

23. Benhamou M, Siraganian RP. Protein-tyrosine phosphorylation: an essential component of FcεRI signaling. Immunol Today 1992;13:195–197.

24. Koch CA, Anderson D, Moran MF, et al. SH2 and SH3 domains: elements that control interactions of cytoplasmic signaling proteins. Science 1991;252:668–674.

25. Mayer BJ, Baltimore D. Signalling through SH2 and SH3 domains. Trends Cell Biol 1993;3:8–13.

26. Pawson T. Protein modules and signalling networks. Nature (Lond) 1995;373:573–580.

27. Kolanus W, Romeo C, Seed B. T cell activation by clustered tyrosine kinases. Cell 1993;74:171–183.

28. Takata M, Sabe H, Hata A, et al. Tyrosine kinases Lyn and Syk regulate B cell receptor-coupled Ca^{2+} mobilization through distinct pathways. EMBO J 1994;13:1341–1349.

29. Kurosaki T, Takata M, Yamanashi Y, et al. Syk activation by the Src-family tyrosine kinase in the B cell receptor signaling. J Exp Med 1994;179:1725–1729.

30. Nishizumi H, Yamamoto T. Impaired tyrosine phosphorylation and Ca^{2+} mobilization, but not degranulation, in Lyn-deficient bone marrow-derived mast cells. J Immunol 1997;158:2350–2355.

31. Rivera VM, Brugge JS. Clustering of Syk is sufficient to induce tyrosine phosphorylation and release of allergic mediators from rat basophilic leukemia cells. Mol Cell Biol 1995; 15:1582–1590.

32. Oliver JM, Burg DL, Wilson BS, et al. Inhibition of mast cell FcεRI-mediated signaling and effector function by the Syk-selective inhibitor, piceatannol. J Biol Chem 1994;269:29697–29703.

33. Valle A, Kinet J. N-Acetyl-L-cysteine inhibits antigen-mediated Syk, but not Lyn tyrosine kinase activation in mast cells. FEBS Lett 1995;357:41–44.

34. Cheng AM, Rowley B, Pao W, et al. Syk tyrosine kinase required for mouse viability and B-cell development. Nature (Lond) 1995;378:303–306.

35. Turner M, Mee PJ, Costello PS, et al. Perinatal lethality and blocked B-cell development in mice lacking the tyrosine kinase Syk. Nature (Lond) 1995;378:298–302.

36. Zhang J, Berenstein EH, Evans RL, et al. Transfection of Syk protein tyrosine kinase reconstitutes high affinity IgE receptor mediated degranulation in a Syk negative variant of rat basophilic leukemia RBL-2H3 cells. J Exp Med 1996;184:71–79.

37. Costello PS, Turner M, Walters AE, et al. Critical role for the tyrosine kinase Syk in signalling through the high affinity IgE receptor of mast cells. Oncogene 1996;13:2595–2605.

38. Paolini R, Jouvin MH, Kinet JP. Phosphorylation and dephosphorylation of the high-affinity receptor for immunoglobulin E immediately after receptor engagement and disengagement. Nature (Lond) 1991;353:855–858.

39. Pribluda VS, Pribluda C, Metzger H. Biochemical evidence that the phosphorylated tyrosines, serines, and threonines on the aggregated high affinity receptor for IgE are in the immunoreceptor tyrosine-based activation motifs. J Biol Chem 1997;272:11185–11192.

40. Jouvin MH, Adamczewski M, Numerof R, et al. Differential control of the tyrosine kinases Lyn and Syk by the two signaling chains of the high affinity immunoglobulin E receptor. J Biol Chem 1994;269:5918–5925.

41. Yamashita T, Mao S, Metzger H. Aggregation of the high-affinity IgE receptor and enhanced activity of p53/56lyn protein-tyrosine kinase. Proc Natl Acad Sci USA 1994;91:11251–11255.

42. Pribluda VS, Pribluda C, Metzger H. Transphosphorylation as the mechanism by which the high-affinity receptor for IgE is phosphorylated upon aggregation. Proc Natl Acad Sci USA 1994;91:11246–11250.

43. Kihara H, Siraganian RP. Src homology 2 domains of Syk and Lyn bind to tyrosine-phosphorylated subunits of the high affinity IgE receptor. J Biol Chem 1994;269:22427–22432.

44. Shiue L, Green J, Green OM, et al. Interaction of p72^{syk} with the γ and β subunits of the high-affinity receptor for immunoglobulin E, FcɛRI. Mol Cell Biol 1995;15:272–281.

45. Wilson BS, Kapp N, Lee RJ, et al. Distinct functions of the FcɛRI γ and β subunits in the control of FcɛRI-mediated tyrosine kinase activation and signaling responses in RBL-2H3 mast cells. J Biol Chem 1995;270:4013–4022.

46. Kimura T, Kihara H, Bhattacharyya S, et al. Downstream signaling molecules bind to different phosphorylated immunoreceptor tyrosine-based activation motif (ITAM) peptides of the high affinity IgE receptor. J Biol Chem 1996; 271:27962–27968.

47. Kimura T, Sakamoto H, Appella E, et al. The negative signaling molecule inositol polyphosphate 5-phosphatase, SHIP, binds to tyrosine phosphorylated β subunit of the high affinity IgE receptor. J Biol Chem 1997;272: 13991–13996.

48. Kimura T, Zhang J, Sagawa K, et al. Syk independent tyrosine phosphorylation and association of the protein tyrosine phosphatase SHP-1 and SHP-2 with the high affinity IgE receptor. J Immunol 1997;159:4426–4434.

49. Kimura T, Sakamoto H, Appella E, et al. Conformational changes induced in the protein tyrosine kinase p72^{syk} by tyrosine phosphorylation or by binding of phosphorylated immunoreceptor tyrosine-based activation motif peptides. Mol Cell Biol 1996;16:1471–1478.

50. Lin SQ, Cicala C, Scharenberg AM, et al. The FcɛRIβ subunit functions as an amplifier of FcɛRIγ-mediated cell activation signals. Cell 1996;85:985–995.

51. Taniguchi T, Kobayashi T, Kondo J, et al. Molecular cloning of a porcine gene *syk* that encodes a 72-kDa protein-tyrosine kinase showing high susceptibility to proteolysis. J Biol Chem 1991;266:15790–15796.

52. Zioncheck TF, Harrison ML, Isaacson CC, et al. Generation of an active protein-tyrosine kinase from lymphocytes by proteolysis. J Biol Chem 1988;263:19195–19202.

53. Hutchcroft JE, Harrison ML, Geahlen RL. B lymphocyte activation is accompanied by phosphorylation of a 72-kDa protein-tyrosine kinase. J Biol Chem 1991;266:14846–14849.

54. Chan AC, Iwashima M, Turck CW, et al. ZAP-70: a 70 kd protein-tyrosine kinase that associates with the TCR ζ chain. Cell 1992;71: 649–662.

55. Law CL, Sidorenko SP, Chandran KA, et al. Molecular cloning of human Syk. A B cell protein-tyrosine kinase associated with the surface immunoglobulin M-B cell receptor complex. J Biol Chem 1994;269:12310–12319.

56. Ku G, Malissen B, Mattei MG. Chromosomal location of the Syk and ZAP-70 tyrosine kinase genes in mice and humans. Immunogenetics 1994;40:300–302.

57. Yagi S, Suzuki K, Hasegawa A, et al. Cloning of the cDNA for the deleted Syk kinase homologous to ZAP-70 from human basophilic leukemia cell line (KU812). Biochem Biophys Res Commun 1994;200:28–34.

58. Rowley RB, Bolen JB, Fargnoli J. Molecular cloning of rodent p72^{syk}. Evidence of alternative mRNA splicing. J Biol Chem 1995;270: 12659–12664.

59. Latour S, Chow LML, Veillette A. Differential intrinsic enzymatic activity of Syk and Zap-70 protein-tyrosine kinases. J Biol Chem 1996;271: 22782–22790.

60. Shimomura R, Sakai K, Tanaka Y, et al. Phosphorylation sites of myelin basic protein by a catalytic fragment of non-receptor type protein-tyrosine kinase p72^{syk} and comparison with those by insulin receptor kinase. Biochem Biophys Res Commun 1993;192:252–260.

61. Schmitz R, Baumann G, Gram H. Catalytic specificity of phosphotyrosine kinases Blk, Lyn, c-Src and Syk as assessed by phage display. J Mol Biol 1996;260:664–677.

62. Harrison ML, Isaacson CC, Burg DL, et al. Phosphorylation of human erythrocyte band 3 by endogenous p72^{syk}. J Biol Chem 1994;269: 955–959.

63. Brunati AM, Donella-Deana A, Ruzzene M, et al. Site specificity of p72^{syk} protein tyrosine kinase: efficient phosphorylation of motifs recognized by Src homology 2 domains of the Src family. FEBS Lett 1995;367:149–152.

64. Wienands J, Freuler F, Baumann G. Tyrosine-phosphorylated forms of Igβ, CD22, TCRζ and HOSS are major ligands for tandem SH2 domains of Syk. Int Immunol 1995;7:1701–1708.

65. Leprince C, Draves KE, Geahlen RL, et al. CD22 associates with the human surface IgM-B-cell antigen receptor complex. Proc Natl Acad Sci USA 1993;90:3236–3240.

66. Kong GH, Bu JY, Kurosaki T, et al. Reconstitution of Syk function by the ZAP-70 protein tyrosine kinase. Immunity 1995;2:485–492.

67. Chu DH, Spits H, Peyron JF, et al. The Syk protein tyrosine kinase can function independently of CD45 or Lck in T cell antigen receptor signaling. EMBO J 1996;15:6251–6261.

68. Furlong MT, Mahrenholz AM, Kim KH, et al. Identification of the major sites of autophosphorylation of the murine protein-tyrosine kinase Syk. Biochim Biophys Acta Mol Cell Res 1997;1355:177–190.

69. Keshvara LM, Isaacson C, Harrison ML, et al. Syk activation and dissociation from the B-cell antigen receptor is mediated by phosphorylation of tyrosine 130. J Biol Chem 1997;272: 10377–10381.

70. Watts JD, Affolter M, Krebs DL, et al. Identification by electrospray ionization mass spectrometry of the sites of tyrosine phosphorylation induced in activated Jurkat T cells on the protein tyrosine kinase ZAP-70. J Biol Chem 1994;269:29520–29529.

71. Iwashima M, Irving BA, Van Oers NSC, et al. Sequential interactions of the TCR with two distinct cytoplasmic tyrosine kinases. Science 1994;263:1136–1139.

72. Law CL, Chandran KA, Sidorenko SP, et al. Phospholipase C-γ1 interacts with conserved phosphotyrosyl residues in the linker region of Syk and is a substrate for Syk. Mol Cell Biol 1996;16:1305–1315.

73. Kurosaki T, Johnson SA, Pao L, et al. Role of the Syk autophosphorylation site and SH2 domains in B cell antigen receptor signaling. J Exp Med 1995;182:1815–1823.

74. El-Hillal O, Kurosaki T, Yamamura H, et al. Syk kinase activation by a src kinase-initiated activation loop phosphorylation chain reaction. Proc Natl Acad Sci USA 1997;94:1919–1924.

75. Songyang Z, Shoelson SE, McGlade J, et al. Specific motifs recognized by the SH2 domains of Csk, 3BP2, fps/fes, GRB-2, HCP, SHC, Syk, and Vav. Mol Cell Biol 1994;14:2777–2785.

76. Zhou S, Cantley LC. Recognition and specificity in protein tyrosine kinase-mediated signalling. Trends Biochem Sci 1995;20:470–475.

77. Taylor JA, Karas JL, Ram MK, et al. Activation of the high-affinity immunoglobulin E receptor FcεRI in RBL-2H3 cells is inhibited by Syk SH2 domains. Mol Cell Biol 1995;15:4149–4157.

78. Koyasu S, Tse AGD, Moingeon P, et al. Delineation of a T-cell activation motif required for

binding of protein tyrosine kinases containing tandem SH2 domains. Proc Natl Acad Sci USA 1994;91:6693–6697.

79. Bu JY, Shaw AS, Chan AC. Analysis of the interaction of ZAP-70 and syk protein-tyrosine kinases with the T-cell antigen receptor by plasmon resonance. Proc Natl Acad Sci USA 1995;92:5106–5110.

80. Chen T, Repetto B, Chizzonite R, et al. Interaction of phosphorylated FcεRIγ immunoglobulin receptor tyrosine activation motif-based peptides with dual and single SH2 domains of p72^{syk}. Assessment of binding parameters and real-time binding kinetics. J Biol Chem 1996;271:25308–25315.

81. Osman N, Lucas SC, Turner H, et al. A comparison of the interaction of Shc and the tyrosine kinase ZAP-70 with the T cell antigen receptor ζ chain tyrosine-based activation motif. J Biol Chem 1995;270:13981–13986.

82. Isakov N, Wange RL, Burgess WH, et al. ZAP-70 binding specificity to T cell receptor tyrosine-based activation motifs: the tandem SH2 domains of ZAP-70 bind distinct tyrosine-based activation motifs with varying affinity. J Exp Med 1995;181:375–380.

83. Narula SS, Yuan RW, Adams SE, et al. Solution structure of the C-terminal SH2 domain of the human tyrosine kinase Syk complexed with a phosphotyrosine pentapeptide. Structure (Lond) 1995;3:1061–1073.

84. Hatada MH, Lu X, Laird ER, et al. Molecular basis for interaction of the protein tyrosine kinase ZAP-70 with the T-cell receptor. Nature (Lond) 1995;377:32–38.

85. Rowley RB, Burkhardt AL, Chao HG, et al. Syk protein-tyrosine kinase is regulated by tyrosine-phosphorylated Igα/Igβ immunoreceptor tyrosine activation motif binding and autophosphorylation. J Biol Chem 1995;270: 11590–11594.

86. Shiue L, Zoller MJ, Brugge JS. Syk is activated by phosphotyrosine-containing peptides representing the tyrosine-based activation motifs of the high affinity receptor for IgE. J Biol Chem 1995;270:10498–10502.

87. Xu WQ, Harrison SC, Eck MJ. Three-dimensional structure of the tyrosine kinase c-Src. Nature (Lond) 1997;385:595–602.

88. Sicheri F, Moarefi I, Kuriyan J. Crystal structure of the Src family tyrosine kinase Hck. Nature (Lond) 1997;385:602–609.

89. Moarefi I, LaFevre-Bernt M, Sicheri F, et al. Activation of the Src-family tyrosine kinase

Hck by SH3 domain displacement. Nature (Lond) 1997;385:650–653.

90. Hubbard SR, Wei L, Ellis L, et al. Crystal structure of the tyrosine kinase domain of the human insulin receptor. Nature (Lond) 1994;372:746–754.

91. Mohammadi M, Schlessinger J, Hubbard SR. Structure of the FGF receptor tyrosine kinase domain reveals a novel autoinhibitory mechanism. Cell 1996;86:577–587.

92. Scharenberg AM, Lin S, Cuenod B, et al. Reconstitution of interactions between tyrosine kinases and the high affinity IgE receptor which are controlled by receptor clustering. EMBO J 1995;14:3385–3394.

93. Zoller KE, MacNeil IA, Brugge JS. Protein tyrosine kinases Syk and ZAP-70 display distinct requirements for Src family kinases in immune response receptor signal transduction. J Immunol 1997;158:1650–1659.

94. Ruzzene M, Brunati AM, Marin O, et al. SH2 domains mediate the sequential phosphorylation of HS1 protein by p72^{syk} and Src-related protein tyrosine kinases. Biochemistry 1996;35: 5327–5332.

95. Jouvin MH, Adamczewski M, Numerof R, et al. Differential control of the tyrosine kinases Lyn and Syk by the two signaling chains of the high affinity immunoglobulin E receptor. J Biol Chem 1994;269:5918–5925.

96. Ota Y, Beitz LO, Scharenberg AM, et al. Characterization of Cbl tyrosine phosphorylation and a Cbl-Syk complex in RBL-2H3 cells. J Exp Med 1996;184:1713–1723.

97. Ota Y, Samelson LE. The product of the proto-oncogene c-cbl: a negative regulator of the Syk tyrosine kinase. Science 1997;276:418–420.

98. Marcilla A, Rivero-Lezcano OM, Agarwal A, et al. Identification of the major tyrosine kinase substrate in signaling complexes formed after engagement of Fcγ receptors. J Biol Chem 1995;270:9115–9120.

99. Paolini R, Serra A, Kinet JP. Persistence of tyrosine-phosphorylated FcεRI in deactivated cells. J Biol Chem 1996;271:15987–15992.

100. Swieter M, Berenstein EH, Siraganian RP. Protein tyrosine phosphatase activity associates with the high affinity IgE receptor and dephosphorylates the receptor subunits, but not Lyn or Syk. J Immunol 1995;155:5330–5336.

101. Sidorenko SP, Law CL, Chandran KA, et al. Human spleen tyrosine kinase p72^{syk} associates with the Src-family kinase p53/56Lyn and a 120-kDa phosphoprotein. Proc Natl Acad Sci USA 1995;92:359–363.

102. Thome M, Duplay P, Guttinger M, et al. Syk and ZAP-70 mediate recruitment of p56^{lck}/CD4 to the activated T cell receptor/CD3/ζ complex. J Exp Med 1995;181:1997–2006.

103. Aoki Y, Kim YT, Stillwell R, et al. The SH2 domains of Src family kinases associate with Syk. J Biol Chem 1995;270:15658–15663.

104. Cantley LC, Songyang Z. Specificity in recognition of phosphopeptides by src-homology 2 domains. J Cell Sci 1994;107(suppl 18):121–126.

105. Amoui M, Draberova L, Tolar P, et al. Direct interaction of Syk and Lyn protein tyrosine kinases in rat basophilic leukemia cells activated via type I Fcε receptors. Eur J Immunol 1997; 27:321–328.

106. Couture C, Deckert M, Williams S, et al. Identification of the site in the Syk protein tyrosine kinase that binds the SH2 domain of Lck. J Biol Chem 1996;271:24294–24299.

107. Couture C, Baier G, Oetken C, et al. Activation of p56^{lck} by p72^{syk} through physical association and N-terminal tyrosine phosphorylation. Mol Cell Biol 1994;14:5249–5258.

108. Couture C, Baier G, Altman A, et al. p56^{lck}-independent activation and tyrosine phosphorylation of p72^{syk} by T-cell antigen receptor/CD3 stimulation. Proc Natl Acad Sci USA 1994;91:5301–5305.

109. Khosravi-Far R, Chrzanowska-Wodnicka M, Solski PA, et al. Dbl and Vav mediate transformation via mitogen-activated protein kinase pathways that are distinct from those activated by oncogenic Ras. Mol Cell Biol 1994;14:6848–6857.

110. Crespo P, Schuebel KE, Ostrom AA, et al. Phosphotyrosine-dependent activation of Rac-1 GDP/GTP exchange by the vav proto-oncogene product. Nature (Lond) 1997;385: 169–172.

111. Hall A. Small GTP-binding proteins and the regulation of the actin cytoskeleton. Annu Rev Cell Biol 1994;10:31–54.

112. Price LS, Norman JC, Ridley AJ, et al. The small GTPases Rac and Rho as regulators of secretion in mast cells. Curr Biol 1995;5:68–73.

113. Gulbins E, Coggeshall KM, Baier G, et al. Tyrosine kinase-stimulated guanine nucleotide exchange activity of Vav in T cell activation. Science 1993;260:822–825.

114. Bustelo XR, Ledbetter JA, Barbacid M. Product of vav proto-oncogene defines a new class

of tyrosine protein kinase substrates. Nature (Lond) 1992;356:68–71.

115. Margolis B, Hu P, Katzav S, et al. Tyrosine phosphorylation of *vav* proto-oncogene product containing SH2 domain and transcription factor motifs. Nature (Lond) 1992;356:71–74.

116. Song JS, Gomez J, Stancato LF, et al. Association of a p95 Vav-containing signaling complex with the FcεRIγ chain in the RBL-2H3 mast cell line. Evidence for a constitutive in vivo association of Vav with Grb2, Raf-1, and ERK2 in an active complex. J Biol Chem 1996;271: 26962–26970.

117. Katzav S, Sutherland M, Packham G, et al. The protein tyrosine kinase ZAP-70 can associate with the SH2 domain of proto-Vav. J Biol Chem 1994;269:32579–32585.

118. Deckert M, Tartare-Deckert S, Couture C, et al. Functional and physical interactions of Syk family kinases with the Vav proto-oncogene product. Immunity 1996;5:591–604.

119. Sidorenko SP, Law CL, Klaus SJ, et al. Protein kinase Cμ (PKCμ) associates with the B cell antigen receptor complex and regulates lymphocyte signaling. Immunity 1996;5:353–363.

120. Sillman AL, Monroe JG. Association of p72^{syk} with the src homology-2 (SH2) domains of PLCγ1 in B lymphocytes. J Biol Chem 1995; 270:11806–11811.

121. Maruyama S, Kurosaki T, Sada K, et al. Physical and functional association of cortactin with Syk in human leukemic cell line K562. J Biol Chem 1996;271:6631–6635.

122. Peters JD, Furlong MT, Asai DJ, et al. Syk, activated by cross-linking the B-cell antigen receptor, localizes to the cytosol where it interacts with and phosphorylates α-tubulin on tyrosine. J Biol Chem 1996;271:4755–4762.

123. Law CL, Sidorenko SP, Chandran KA, et al. CD22 associates with protein tyrosine phosphatase 1C, Syk, and phospholipase C-γ1 upon B cell activation. J Exp Med 1996;183:547–560.

124. Chacko GW, Tridandapani S, Damen JE, et al. Negative signaling in B lymphocytes induces tyrosine phosphorylation of the 145-kDa inositol polyphosphate 5-phosphatase, SHIP. J Immunol 1996;157:2234–2238.

125. Ono M, Bolland S, Tempst P, et al. Role of the inositol phosphatase SHIP in negative regulation of the immune system by the receptor FcγRIIB. Nature (Lond) 1996;383:263–266.

126. Daeron M, Latour S, Malbec O, et al. The same tyrosine-based inhibition motif, in the intracytoplasmic domain of FcγRIIB, regulates

negatively BCR-, TCR-, and FcγR-dependent cell activation. Immunity 1995;3:635–646.

127. Osborne MA, Zenner G, Lubinus M, et al. The inositol 5′-phosphatase SHIP binds to immunoreceptor signaling motifs and responds to high affinity IgE receptor aggregation. J Biol Chem 1996;271:29271–29278.

128. Nishibe S, Wahl MI, Hernandez-Sotomayor SM, et al. Increase of the catalytic activity of phospholipase C-γ1 by tyrosine phosphorylation. Science 1990;250:1253–1256.

129. Fukamachi H, Kawakami Y, Takei M, et al. Association of protein-tyrosine kinase with phospholipase C-γ1 in bone marrow-derived mouse mast cells. Proc Natl Acad Sci USA 1992;89:9524–9528.

130. Hirasawa N, Scharenberg A, Yamamura H, et al. A requirement for Syk in the activation of the microtubule-associated protein kinase/ phospholipase A2 pathway by FcεRI is not shared by a G protein-coupled receptor. J Biol Chem 1995;270:10960–10967.

131. Wu J, Motto DG, Koretzky GA, et al. Vav and SLP-76 interact and functionally cooperate in IL-2 gene activation. Immunity 1996;4:593–602.

132. Hendricks-Taylor LR, Motto DG, Zhang J, et al. SLP-76 is a substrate of the high affinity IgE receptor-stimulated protein tyrosine kinases in rat basophilic leukemia cells. J Biol Chem 1997;272:1363–1367.

133. Jabril-Cuenod B, Zhang C, Scharenberg AM, et al. Syk-dependent phosphorylation of Shc— a potential link between FcεRI and the Ras/ mitogen-activated protein kinase signaling pathway through Sos and Grb2. J Biol Chem 1996;271:16268–16272.

134. Teramoto H, Salem P, Robbins KC, et al. Tyrosine phosphorylation of the *vav* proto-oncogene product links FcεRI to the Rac1-JNK pathway. J Biol Chem 1997;272:10751–10755.

135. Fukamachi H, Yamada N, Miura T, et al. Identification of a protein, SPY75, with repetitive helix-turn-helix motifs and an SH3 domain as a major substrate for protein tyrosine kinase(s) activated by FcεRI cross-linking. J Immunol 1994;152:642–652.

136. Yamanashi Y, Fukuda T, Nishizumi H, et al. Role of tyrosine phosphorylation of HS1 in B cell antigen receptor-mediated apoptosis. J Exp Med 1997;185:1387–1392.

137. McGlade J, Brunkhorst B, Anderson D, et al. The N-terminal region of GAP regulates cytoskeletal structure and cell adhesion. EMBO J 1993;12:3073–3081.

138. Hamawy MM, Mergenhagen S, Siraganian RP. Tyrosine phosphorylation of pp125^{FAK} by the aggregation of high affinity immunoglobulin E receptors requires cell adherence. J Biol Chem 1993;268:6851–6854.

139. Minoguchi K, Kihara H, Nishikata H, et al. Src family tyrosine kinase Lyn binds several proteins including paxillin in rat basophilic leukemia cells. Mol Immunol 1994;31:519–529.

140. Hamawy MM, Swaim WD, Minoguchi K, et al. The aggregation of the high affinity IgE receptor induces tyrosine phosphorylation of paxillin, a focal adhesion protein. J Immunol 1994;153:4655–4662.

141. Burridge K, Turner CE, Romer LH. Tyrosine phosphorylation of paxillin and pp125^{FAK} accompanies cell adhesion to extracellular matrix: a role in cytoskeletal assembly. J Cell Biol 1992;119:893–903.

142. Pfeiffer JR, Oliver JM. Tyrosine kinase-dependent assembly of actin plaques linking FcεRI cross-linking to increased cell substrate adhesion in RBL-2H3 tumor mast cells. J Immunol 1994;152:270–279.

143. Sagawa K, Swain SL, Zhang J, et al. Aggregation of the high affinity IgE receptor results in the tyrosine phosphorylation of the surface adhesion protein PECAM-1 (CD31). J Biol Chem 1997;272:13412–13418.

11

The Role of Protein Phosphatases in Cell Signaling by the High-Affinity Receptor for Immunoglobulin E

Matthew J. Peirce

The addition or removal of phosphate at particular amino acids may profoundly alter the catalytic activity or localization of a diverse array of proteins. The protein kinases that catalyze protein phosphorylations and the protein phosphatases which catalyze dephosphorylations therefore represent crucial points of control in the coordination of cell activity.

The antigen-mediated aggregation of the high-affinity receptor for IgE (FcɛRI) promotes the accumulation of phosphate on tyrosine, serine and threonine residues of the β- and γ-chains of the receptor itself, as well as of a diverse array of signaling proteins.[1-3] The tyrosine phosphorylations appear to be initiated by p56[Lyn], a member of the Src tyrosine kinase family, and propagated by p72[Syk], a member of the Syk/Zap70 kinase family. An equally striking observation, however, is that disruption of receptor aggregates by the addition of monovalent hapten results in the rapid dephosphorylation of the receptor subunits and the downstream targets of Lyn and Syk.[1,4] Furthermore, in permeabilized cells antigen-induced receptor phosphorylation is rapidly reversed if receptor-associated kinase activity is blocked with EDTA.[4]

These data highlight the role played by protein phosphatases in FcɛRI signaling. They demonstrate that protein tyrosine phosphatases (PTPs) are involved not only in resetting the system to baseline following disaggregation of FcɛRI but also regulate the state of phosphorylation of kinase substrates during an ongoing response to antigen. Thus, in the mast cell, as in many other systems, there is a growing appreciation that the importance of protein phosphatases in the regulation of cell function may equal that of protein kinases.

In this chapter I discuss the current understanding of the role of protein phosphatases in the regulation of mast cell function and FcɛRI-mediated signaling. I first consider the role of the transmembrane protein phosphatase, CD45, in regulating mast cell activation. I continue by assessing the importance of the SH-2 domain-containing PTPs, SHP-1 and SHP-2, as well as the SH2 domain-containing inositol phosphatase, SHIP. I will then address the role of protein phosphatases in the downregulation of FcɛRI signaling through interactions with inhibitory receptors such as gp49b1 and FcγRIIB1. Finally, I discuss the importance of protein serine/threonine phosphatases in the regulation of mast cell function and consider some more general issues, such as targeting and specificity of phosphatase activity.

Protein Tyrosine Phosphatases

CD45

The common leukocyte antigen, CD45, is a transmembrane protein tyrosine phosphatase (PTP) containing two adjacent PTP catalytic domains in the N-terminal, cytoplasmic half of the molecule.[5] Alternative use of exon six of the primary CD45 sequence yields a number of splice variants differing in the C-terminal extracellular domain,[6,7] indicating the possibility that different isoforms of CD45 interact with distinct ligands. Several data indicate that expression of different CD45 isoforms is developmentally regulated and may be modified following cell activation, suggesting functional differences may exist between CD45 isoforms.[5,8] Despite the intriguing potential of these findings, physiological ligands for the distinct extracellular domains of CD45 remain to be identified.

Many of the data regarding the role of CD45 in cellular signaling have come from studies of B and T lymphocytes, and these support the idea that CD45 has a positive regulatory role in lymphocyte signaling and development.[5,9,10] For example, CD45-deficient mice have dramatically reduced numbers of mature CD4+ and CD8+ single positive T-cells,[11] suggesting an important role for CD45 in the regulation of T-cell development. While the major deficit in CD45-deficient mice is in the T-cell compartment, B cells are also affected.[12–14] B cells from CD45-negative mice are resistant to the autoantigen-induced clonal deletion observed in wild-type B cells.[15] CD45-negative B cells also exhibit markedly attenuated B-cell-receptor-derived signals compared to wild-type B cells or those transfected with a CD45 construct.[15] These data led to the hypothesis that CD45 deficiency, by attenuating B-cell receptor signals generated upon exposure to autoantigen, prevented the negative selection of these cells.

Both B-cell- and T-cell-receptor-mediated cellular responses are deficient or absent in CD45 knockout mice.[13,16,17] These deficits can be ameliorated by transfection of the deficient B or T cells with the catalytic domains of CD45.[13,16] Cell signals derived from the B- and T-cell receptors depend critically upon members of the Src family of tyrosine kinases including p56[Lck], p59[Fyn], and p53/56[Lyn].[18,19] Phosphorylation of a C-terminal tyrosine residue, found in all Src family kinases, enables it to interact with the N-terminal SH2 domain, thereby blocking access to the catalytic site of the kinase.[20–22] Several data indicate that CD45, by dephosphorylating the C-terminal tyrosine

residue, disrupts this inhibitory interaction, thereby activating Src family kinases.[23,24] For example, in vitro CD45 interacts with and dephosphorylates the C-terminal tyrosine residues of Lck and Fyn, activating kinase activity.[24–26] Furthermore, the Src family kinases Lck and Fyn, isolated from CD45-deficient T-cells, are hyperphosphorylated on the negative regulatory tyrosine and exhibit markedly reduced catalytic activity.[26–28]

Although a persuasive body of evidence supports a role for CD45 in regulating the activity of Src family kinases,[5] recent data indicate that other PTPs may subserve the same function. An SH2 domain-mediated interaction between P60[Src] and the SH2 domain-containing PTP, SHP-1, has recently been demonstrated.[29] In vitro, it was noted that the PTP acted preferentially to dephosphorylate the negative regulatory tyrosine residue of Src. The same authors showed that Src isolated from SHP-1-negative thymocytes was hyperphosphorylated at this site and exhibited abrogated catalytic activity, suggesting a physiological role for SHP-1 in the regulation of Src. The functional significance of this finding is undermined by the demonstration that expression of a membrane-targeted SHP-1/HLA-2 chimeric protein was unable to rescue T-cell (TCR) receptor mediated signals in a CD45–negative variant of the Jurkat T cell line.[30] A similar construct containing the N-terminal catalytic domain of CD45 successfully restored TCR signaling in these cells. In addition, other Src family members, Lck and Fyn, are hyperactivated not inhibited in SHP-1 null thymocytes. These findings suggest that the activity of SHP-1 toward the negative regulatory tyrosine is restricted to Src. Similarly, the catalytic activities of the Src family kinases Lck, Fyn, and Src are differentially affected by the loss of CD45 in CD45-deficient T-cell and B-cell lines.[31,32] Thus the importance of CD45 relative to other PTPs in regulating signals mediated by immune receptors may depend crucially upon the profile of Src kinases utilized by a particular receptor pathway in a particular cell type.

In the mast cell system there are conflicting data regarding the role of CD45 in signaling from FcεRI. One study demonstrated a defective FcεRI-mediated secretory response in mast cells isolated from CD45-deficient mice, whereas the same cells responded normally to receptor-independent stimuli.[33] These data suggested an essential and selective role for CD45 in receptor-mediated secretory responses. Furthermore, an anti-CD45 antibody similarly blocked FcεRI-mediated secretory responses but only when the antibody was biotinylated, and antibody–CD45 complexes were further aggregated using streptavidin.[33] These data are in partial agreement with those of Hook et al.,[34] who demonstrated attenuated FcεRI-dependent mediator release from human basophils following extensive preincubation with anti-CD45 antibodies. Although the effects of CD45 aggregation on its PTP activity were not assessed in either of these studies, their findings may be explained by the recent solving of the X-ray crystal structure of a related receptor tyrosine phosphatase (RPTP), PTP-α.[35] This RPTP crystalized as a dimer in which the catalytic domains of each member of the dimer was predicted to engage in a reciprocal interaction with a helix-turn-helix motif of the other monomer. In this conformation, the enzyme is predicted to be catalytically inactive. By analogy one predicts that by promoting the aggregation of CD45 into dimers or oligomers, anti-CD45 antibodies would inhibit CD45 PTP activity. Data in support of this possibility come from the T-cell system. A CD45-deficient variant of the Jurkat cell line was transfected with a chimeric construct consisting of the extracellular and transmembrane portion of the epidermal growth factor (EGF) receptor coupled to the catalytic domain of CD45. While the monomeric construct rescued TCR signaling, dimerizing the CD45 chimera by adding EGF prevented the restoration of TCR signaling.[36]

Irrespective of mechanistic details, the cumulative data from these mast cell studies indicate that CD45 may regulate FcεRI signaling. However, in the rodent mucosal mast cell line RBL-2H3, FcεRI signals in CD45-deficient variants of the cell line are found to be normal while responses to a calcium ionophore are attenuated.[37]

The apparent differences in the role of CD45 in the RBL-2H3 cell line as compared with human basophils or mouse peritoneal mast cells highlights the differences between mast cells isolated from different sources. The recent finding just discussed, that the SH2 domain-containing PTP, SHP-1, may dephosphorylate the negative regulatory tyrosine residue in the Src tyrosine kinase, suggests that in some circumstances PTPs other than CD45 are responsible for regulating signaling via Src family kinase-dependent pathways. The requirement for CD45 in a particular system may depend upon the availability of other PTPs capable of fulfilling the same role. In addition, given that different members of the Src kinase family appear to be differentially regulated by CD45 and SHP-1, the requirement for CD45 will also depend upon the member of the Src family utilized in a particular cell type. In this regard, while Lyn kinase appears to have a pivotal and well-documented role in FcεRI signaling in RBL-2H3,[2,3] the kinase(s) utilized in other mast cells have yet to be unequivocally demonstrated. These considerations also impinge upon the findings of Adamczewski et al.[38] who reconstituted FcεRI in wild-type or CD45-deficient variants of the Jurkat T-cell line. These authors demonstrated that in Jurkat T cells the presence of CD45 was required for FcεRI-mediated calcium responses and furthermore that CD45 expression enhanced the phosphorylation of FcεRI chains following receptor aggregation. They also demonstrated that loss of CD45 expression led to enhanced basal levels of autophosphorylation of the Src family kinase, Lck. This study clearly demonstrates that CD45 may impinge upon FcεRI signaling. However, given the heterogeneity in the effect of CD45 on different members of the Src kinase family, it remains to be seen whether similar effects can be demonstrated in mast cells wherein Src kinases other than Lck may be important.

SH2 Domain-Containing PTP

Numerous data from mast cell studies, as well as from the related B-cell and T-cell systems, suggest that the SH2 domain-containing PTPs

SHP-1 and SHP-2 may have important and contrasting roles in the regulation of signaling through the family of multichain immune recognition receptors (MIRRs).

SHP-1

SHP-1 is expressed primarily in hematopoietic cells and has previously been called hematopoietic cell phosphatase (HCP) as well as PTP1C.[39,40] Mutations in the SHP-1 gene give rise to the motheaten (*me*) or motheaten viable (*me^v*) phenotype.[41-43] The *me* (SHP-1 null) and *me^v* (catalytically deficient SHP-1) phenotypes are characterized by a profound dysregulation in lymphocyte development, primarily in the B-cell compartment, and by death 2 to 3 weeks after birth.[44,45] B cells from *me* mice exhibit exaggerated cellular responses to stimuli insufficient in normal littermates to provoke cell activation.[46,47] While disruption of T-cell development is less apparent, proliferation and secretion of IL-2 in response to TCR stimulation are similarly enhanced in *me* relative to normal mice whereas responses to IL-2 are unchanged.[48,49]

It is postulated that the particular lack of mature B cells associated with the motheaten phenotype is attributable to the inappropriately vigorous responses of developing B cells exposed to normally subthreshold autoantigen stimuli leading to large-scale deletion of maturing B cells.[46] These in vivo data, suggesting a negative role for SHP-1 in the regulation of B- and T-cell-receptor signaling, are supported by data at the cellular and molecular level. Calcium responses to B-cell-receptor-mediated stimuli are larger, more rapid, and more prolonged in SHP-1-deficient cells than in normal B cells.[46] Similarly, activation of the Src family kinases Fyn and Lck following engagement of the TCR is enhanced in T cells derived from motheaten mice.[48] In addition, SHP-1 has been shown in vitro to dephosphorylate and inhibit the Src family kinase Lck[50] and the Syk family kinase Zap-70,[51] as well as the ζ-chain of CD3.[50]

SHP-1 is expressed in RBL-2H3 cells, and a role for SHP-1 in FcεRI signaling may be inferred from studies demonstrating its tyrosine phosphorylation on receptor aggregation.[52] In addition, a constitutive, although apparently indirect, association between SHP-1 and the β-chain of the IgE receptor has recently been demonstrated.[53]

Set against this is the finding that FcεRI-mediated signaling appears normal in mast cells isolated from motheaten (SHP-1-null) mice.[54] However, this may be attributable, at least in part, to the very powerful aggregating stimulus employed to activate the cells in this study. Given that SHP-1 is predicted to modulate the threshold of stimulation required to generate a response,[46] a more accurate picture of the role of SHP-1 in FcεRI signaling might be achieved by comparing responses to weak stimuli, such as small oligomers of IgE, in SHP-1-deficient and wild-type mast cells.

As previously discussed, SHP-1 may have opposing effects on different members of the Src kinase family. Thus, as for CD45, the role of SHP-1 in mast cells may depend upon the identity of the Src family kinase utilized in a particular cell type. While persuasive circumstantial evidence points to the involvement of SHP-1 in the regulation of FcεRI signaling, its precise role and important substrates remain to be clarified.

SHP-2

In contrast to SHP-1, SHP-2, a second SH2 domain-containing PTP (previously called PTP1D) is ubiquitously expressed and has been demonstrated in several systems to subserve a positive regulatory role in cytokine[55] and growth factor receptor[56,57] signaling.

A direct, SH2 domain-mediated interaction has been demonstrated between SHP-2 and FcεRI in RBL-2H3 cells.[53] Interestingly, this association depends upon aggregation of the receptor. The interaction between SHP-2 and the receptor is mediated by the SH2 domain of SHP-2 and the phosphorylated immunoreceptor tyrosine-based activation motif (ITAM) sequence of the β-chain of the receptor. Notably, SHP-2 interacted selectively with peptides based upon the phosphorylated ITAM sequence of the β-chain, failing to interact with the related ITAM sequence of the γ-chain. SHP-2 and FcεRIβ could be coimmunoprecipitated but only following 30 min of receptor aggregation. This interaction contrasted with that between SHP-1 and the receptor, which was indirect and appeared to be independent of receptor aggregation. Both SHP-1 and SHP-2 were shown in vitro to dephosphorylate the β- and γ-chains of the IgE receptor.[53] Notably, the rates of receptor dephosphorylation observed in studies with SHP-1 and SHP-2 fusion proteins[53] were markedly slower than those reported by Mao and Metzger[4] using extracts of RBL-2H3 cells. Nevertheless, that the receptor appears to interact with SHP-2 in an aggregation-dependent manner suggests a role in regulating signals generated by FcεRI. Given the differences in the nature of the interactions of SHP-1 and SHP-2 with the receptor, one might predict differences in the function of these two PTPs in regulating cell signaling via FcεRI. Further study to elucidate the functional consequences of under- or overexpression of these enzymes, as well as to identify their substrates in vivo, is necessary to understand how SHP-1 and SHP-2 impinge upon FcεRI signaling.

Other PTP Activities in RBL Cells

Several attempts have been made to characterize PTP activities in mast cells. Utilizing the RBL-2H3 cell line, Mao and Metzger[4] attempted to characterize PTP activities acting to dephosphorylate FcεRI. Monitoring the phosphorylation of the IgE receptor, they showed that up to 80% of the receptor-directed PTP activity present in whole-cell extracts was retained in the membrane component of fractionated cell sonicates. Furthermore, permeabilization of the cells with streptolysin-O, resulting in loss of approximately 90% of cytosolic proteins and 70% of two cytosolic PTPs (SHP-1 and SHP-2), had no effect on the kinetics of the receptor dephosphorylation that followed either receptor disaggregation with hapten or kinase inhibition by the addition of EDTA. In addition these authors demonstrated, using an in vitro receptor PTP assay, as well as permeabilized whole cells, that the rate at which receptor chains were dephosphorylated was independent of the state of receptor

aggregation. These data suggest that the accumulation of phosphotyrosine on the β- and γ-chains of the receptor is not the result of an aggregation-induced alteration in either the susceptibility to, or the catalytic activity of, PTPs. Rather they suggest that the aggregated and disaggregated receptors undergo continuous dephosphorylation by PTPs whose activity is not regulated by receptor aggregation.

Hampe and Pecht[58] presented data indicating that, in RBL-2H3 cells, receptor aggregation led to an enhancement in membrane-associated PTP activity toward a peptide substrate. The aggregation-sensitive activity was selective for phosphotyrosine over phosphoserine and phosphothreonine, although its activity against protein substrates such as the β- and γ-chains of the receptor or components of the FcεRI signaling cascade was not assessed. Thus, while the PTPs responsible for dephosphorylating the receptor appear to be insensitive to receptor aggregation,[4] other PTPs located in the membrane do appear to be responsive to FcεRI aggregation.[58] Interestingly, the aggregation- induced increase in membrane-associated PTP activity was demonstrated to be calcium-dependent. This observation could relate to the recent finding of Kimura et al. that the SH2 domain-containing PTP, SHP-1, becomes tyrosine phosphorylated in a calcium-dependent manner following receptor aggregation.[53]

SHIP

Recent data suggest that an SH2 domain-containing inositol-5 phosphatase, SHIP, may also play an important role in the regulation of FcεRI-generated signals.[59-61] SHIP appears to dephosphorylate position 5 of several inositol polyphosphate species including phosphatidylinositol 3,4,5-trisphosphate (PIP$_3$), the product of the action of phosphatidyl inositol 3-kinase (PI-3 kinase) upon phosphatidyl inositol 4,5-bisphosphate (PIP$_2$). Thus, at least one function of SHIP may be to attenuate PI-3 kinase-mediated signals. SHIP appears to inhibit the influx of calcium from the extracellular milieu that routinely accompanies the depletion of intracellular calcium stores.[62] The crucial molecular targets of SHIP, whose dephosphorylation interrupts calcium influx, have yet to be identified. However, a persuasive body of data suggest an important role for SHIP in negative signaling in diverse immunological settings.[63] More especially, several studies implicate SHIP in the regulation of cellular signals generated by FcεRI.

Two different groups have independently demonstrated an interaction between SHIP and FcεRI.[59,60] Using phosphorylated ITAM peptides derived from either the β- or γ-chain of FcεRI or from the CD3 ζ-chain of the T-cell receptor, Osborne et al.[59] were able to demonstrate an SH2-dependent interaction with SHIP. Similarly, Kimura et al.[60] identified a selective interaction between SHIP and the phosphorylated ITAM of the FcεRI β-chain but failed to see any interaction between SHIP and a phosphorylated peptide corresponding to the γ-chain ITAM sequence.

SHIP has only one SH2 domain and, the interaction between SHIP and the phosphorylated ITAM peptides was dependent upon only one of the two YxxL motifs found in either β- or γ-ITAM.[59] Mutation of the conserved Y or L residue of the N-terminal YxxL motif left the SHIP–ITAM interaction intact whereas similar modification of the C-terminal YxxL abolished the interaction. This is reminiscent of the interaction between Lyn and phosphorylated β-ITAM peptides.[64] Lyn kinase also has only one SH2 domain and has been shown to bind relatively efficiently to monophosphorylated β-ITAM peptides. In contrast, Syk, a dual SH2 domain-containing protein, requires both YxxL motifs to bind γ-chain ITAM peptides efficiently.[64]

Although an in vitro interaction between phosphorylated receptor ITAMs and SHIP could be demonstrated, neither group was able to detect a similar interaction in vivo. However, both studies demonstrated that, following FcεRI aggregation, the phosphotyrosine content of SHIP was increased and a constitutive association between SHIP and the Shc–Grb-2 complex was enhanced.[59,60] The functional significance of this finding as it relates to RBL cells remains to be clarified. However, data from the B-cell system suggest that the association of SHIP with Shc–Grb-2 may attenuate the activation of the Ras/MAP/ERK pathway normally mediated by the Shc–Grb-2–Sos complex.[65,66]

RBL cells express three forms of SHIP with molecular weights of 145, 140, and 105 kDa.[60] Notably, while phosphorylated β-chain ITAM peptides precipitated all three forms of SHIP from whole-cell lysates of RBL-2H3 cells, only the 145-kDa form was found associated with Shc–Grb-2. The functional ramifications of this selective interaction await evaluation.

Divergent, although not mutually exclusive, data regarding the nature of the interaction between Shc–Grb-2 and SHIP have been presented. One study demonstrated that the increase in Shc–SHIP association detected following receptor aggregation was dependent upon the concurrent increase in SHIP tyrosine phosphorylation.[60] These data therefore implied an SH2/phosphotyrosine-mediated interaction between SHIP and Shc–Grb-2. However, SHIP contains several proline-rich sequences that may serve as ligands for SH-3 domains. Osborne et al.[59] demonstrated that the in vitro interaction between SHIP and Shc–Grb-2 was dependent upon the SH-3 domain of Grb-2. Interactions between Shc–Grb-2 and other phosphatases have been observed, and these are dependent upon both SH2 and SH3 domains.[67] Further questions regarding the role of SHIP tyrosine phosphorylation are prompted by the finding that, at least in vitro, the tyrosine phosphorylation of SHIP by the Src family tyrosine kinase Lck strongly inhibits SHIP catalytic activity.[59]

Together, these data indicate the potential of SHIP to interact with FcεRI, identify SHIP as a target of tyrosine kinases downstream of receptor aggregation, and demonstrate an aggregation-sensitive association of SHIP with signaling molecules. In addition to its catalytic activity, SH2 domain and possible SH3 domain binding sites, SHIP contains two NxxY motifs, putative ligands for phosphotyrosine-binding (PTB) domains. These functional and structural characteristics strongly suggest that SHIP regulates FcεRI signaling, but the mechanistic details must still be determined.

Phosphatase-Dependent Inhibitory Receptors

A family of functionally related inhibitory receptors has recently emerged that specialize in downregulating stimulatory signals generated by ITAM-containing receptors. Rapidly accumulating data suggest that these receptors may allow external influences to modulate FcεRI-derived signals.[68,69]

The family of inhibitory receptors originally comprised the low-affinity receptor for IgG, FcγRIIB1,[70–72] and the human killer cell inhibitory receptors (KIRs) p58[73,74] and p70.[75,76] More recently the family has been expanded to include murine Ly49[77,78] receptors (functional homologs of human KIRs), the B-cell coreceptor CD22,[79] CTLA-4[80] (a T-cell receptor for B7.1 and B7.2), as well as the murine mast cell antigens gp49B1,[81] its human (HM18, HL9),[82] and murine (p91)[83] homologs (see Chapter 13), and the mast cell function-associated antigen, MAFA.[84]

Although structurally diverse, a conserved consensus motif (I/VxYxxL/V) has been identified and denoted the immune receptor tyrosine-based inhibitory motif (ITIM) on the basis of its similarity with the ITAM sequence.[72,85] This sequence appears to be necessary and sufficient to transduce inhibitory signals.[85] In addition to this shared sequence, inhibitory receptors retain important functional similarities. The inhibitory function of each receptor is dependent upon (i) cross-linkage to an ITAM-bearing receptor,[54,61,81,84,86–88] (ii) phosphorylation of the ITIM tyrosine by an ITAM-associated Src family tyrosine kinase, and (iii) the recruitment of an SH2 domain-containing phosphatase, for example, SHP-1 by KIR,[73–76] CD22,[79,88] and MAFA,[69] SHP-2 by CTLA-4,[80] and SHIP by FcγRIIB1.[54,61,72]

It has recently been demonstrated that the ITIM requirements of different SH2 domain-containing phosphatases differ.[89] Thus, using phosphorylated peptides corresponding to the ITIM sequences of FcγRIIB1 and p58KIR, a single mutation (I/V to A) at the tyrosine-2 position of the consensus ITIM sequence I/VxYxxL in FcγRIIB1 or KIR-derived ITIM peptides abolished the binding of SHP-1 and SHP-2 to either peptide. The same manipulation failed to interrupt the interaction between SHIP and the FcγRIIB1 ITIM peptide.[54] Although these data suggest a requirement for I or V at the tyrosine-2 position for a particular ITIM to recruit SHP-1 and SHP-2, this conclusion is undermined by the finding that the ITIM-like sequence found in CTLA-4 has a glycine, and MAFA, a serine, at the tyrosine-2 position and yet have been reported to bind SHP-2[80] and SHP-1 (CTLA-4), SHP-2 and SHIP (MAFA),[69] respectively. Furthermore, FcγRII, which selectively recruits SHIP in vivo,[61] binds efficiently to both SHP-1 and SHP-2 as well as SHIP in vitro,[89] suggesting that factors other than the ITIM sequence influence the phosphatase(s) utilized by a particular receptor.

The recruitment of SHP-1/SHP-2 or SHIP has important functional consequences. The co-cross-linking of FcεRI and FcγRIIB1 in transfected RBL cells (resulting in the recruitment of SHIP) attenuates positive signaling primarily at the level of calcium influx.[86] In this case, those downstream events dependent upon a prolonged increase in intracellular calcium ions (for example, the activation of certain transcription factors[90,91]) will be attenuated by SHIP, while other signaling events, particularly protein tyrosine phosphorylations, may proceed unabated. This residual signal may have profound effects on cell responses to future or concurrent stimuli.

By contrast, the association of FcεRI with transfected KIR in the same RBL cells, leads to the recruitment of SHP-1, which effectively blocks the aggregation-induced release of calcium from intracellular stores, suggesting a much more receptor-proximal site of action.[86] In the T cell, SHP-1 acts directly upon ZAP-70[92] (analogous to Syk in RBL-2H3) and upon the ζ-chain of the TCR[50] to block early tyrosine phosphorylation events. Collectively, these data combine to emphasize the potential significance of the nature and number of ITIM sequences found in the intracellular domains of inhibitory signaling receptors. These factors may not only have predictive value regarding the identity of the phosphatase species recruited but also impact importantly on the functional consequences of receptor engagement.

Phosphatase-Dependent Inhibitory Signaling in Mast Cells

In mast cells, FcγRIIB1,[54,85] gp49B1,[81] a member of the immunoglobulin superfamily, and the C-type lectin, MAFA,[84,93] have all been demonstrated to attenuate FcεRI-mediated responses. In several cases these intriguing functional observations have only recently been reported, and data regarding the mechanistic details have yet to be assessed.

Inhibitory signals generated by FcγRIIB1 were originally proposed to involve SHP-1.[71] However, a preponderance of data now indicate that SHP-1 is dispensable, and that SHIP is necessary and sufficient, for FcγRIIB1-mediated negative signaling in murine mast cells[54] as well as in B cells[94,95] and NK cells.[96] Co-cross-linking of FcεRI with FcγRIIB1 attenuates IgE-dependent mediator release from mouse mast cells,[54] RBL-2H3 cells,[72] and human basophils.[85] Furthermore, the importance of FcγRIIB1 in the regulation of mast cell function in vivo is underscored by the recent demonstration that passive cutaneous anaphylactic responses to IgG are enhanced in FcγRII-null mice.[97]

gp49B1 is a murine mast cell antigen[98] that appears also to be expressed upon murine NK cells.[99,100] Human homologs of gp49B1 have recently been described[82,83] and these also appear to be expressed on mast cells. gp49B1 has two Ig-like domains and two sequences closely resembling the consensus ITIM sequence.[85,101] Co-cross-linking of gp49B1 with FcεRI attenuates the release of inflammatory mediators from murine mast cells.[81] Both the murine and human proteins exhibit striking homology to the human KIRs.[82,83,101] KIRs are inhibitory receptors expressed on NK cells and some T cells that recognize particular subclasses of MHC-1 molecules and lead to the blockade of FcγRII1A-mediated cytolytic responses.[68,69] KIR function appears to be selectively dependent upon SHP-1.[86,95,96] Phosphorylation of the

gp49 ITIM and phosphatase recruitment to the ITIM in response to co-cross-linking with FcεRI have yet to be demonstrated. However, given their structural similarity to KIRs, including the presence of at least two consensus ITIM sequences, one might predict the involvement of SHP-1 in the inhibitory signals generated by gp49 and, potentially, in those generated by its homologs HL9 and p91.

MAFA is a C-type lectin, as are the members of the murine Ly49 family of inhibitory receptors that appear to be functional homologs of human KIRs. MAFA contains one YxxL motif related to part of the consensus ITIM sequence. MAFA is reported to associate with both aggregated[102] and unaggregated[103] FcεRI. The cytoplasmic domain of the protein is reported to be constitutively phosphorylated, and this is modified on FcεRI aggregation.[84,93] Aggregation of MAFA or FcεRI, but not coaggregation of MAFA with FcεRI, enhances tyrosine phosphorylation of the MAFA cytoplasmic domain, although whether the site of phosphorylation was the tyrosine of the YxxL motif was not assessed.[84] The aggregation of MAFA leads to inhibition of FcεRI-derived signals at a level upstream of the activation of PLCγ1. The FcεRI-mediated generation of IP_3 and the release of calcium from intracellular stores was attenuated in the presence of cross-linking anti-MAFA antibody.[102] These data suggest the involvement of SHP-1 rather than SHIP, which would be expected to leave IP_3 responses intact. However, it has been reported that phosphorylated peptides corresponding to the MAFA YxxL sequence bind SHP-1, SHP-2, and SHIP in vitro.[69] Elucidation of the physiologically important phosphatase will require further study. Although the antigen was characterized on murine mast cells, the same antibody appears to bind to human basophils suggesting the expression of a highly related protein. However, no functional studies were performed on the human cells.[104] MAFA exhibits several functional features suggesting a potentially important role in the regulation of FcεRI signaling. Characterization of the PTPs important for its function and identification of a ligand for the extracellular domain may clarify its physiological role.

Protein Serine/Threonine Phosphatases

The role played by protein serine/threonine phosphatases (PPs) in FcεRI-mediated signaling is still unclear. PPs are an expanding group of enzymes traditionally comprising PP1, PP2A, PP2B (calcineurin), and PP2C.[105,106] As summarized in Table 11.1, PPs have been classified on the basis of substrate preference, requirement for divalent metal cations, and sensitivity to endogenous inhibitor proteins. This classification system accounts for a large majority of known PP activities although the recent characterization of several novel PPs[107–109] indicates that it is not complete.

PP1 and PP2A

In addition to the endogenous inhibitors I-1 and I-2, several naturally occurring PP inhibitors have been characterized.[110,111] As summarized in Table 11.1, one inhibitor, okadaic acid, potently inhibits the activity of PP1 (IC_{50}, 10–20 nM) and PP2A (IC_{50}, 0.1–1 nM), whereas PP2B is affected only at much higher concentrations ($IC_{50} > 5\,\mu M$) and PP2C is insensitive to okadaic acid.

Pharmacological studies with okadaic acid suggest a role for PP1 and PP2A in the regulation of mast cell and basophil function. Pretreatment of human lung mast cells,[112] basophils,[113,114] and rat peritoneal mast cells[115] with okadaic acid attenuates the IgE-

dependent and IgE-independent release of a preformed inflammatory mediator, histamine; de novo synthesized mediators, leukotriene C_4 (LTC_4), and prostaglandin D_2 (PGD_2); and the de novo expression of a proinflammatory cytokine, interleukin-4 (IL-4), each with very similar IC_{50} values. In some of these studies less active analogs of okadaic acid, okadaol and okadaone, were shown to be less active inhibitors of mediator release. Notably, the same rank order of potency (okadaic acid > okadaol > okadaone) was observed for each mediator as well as in assays of PP activity in broken-cell extracts of purified HLMC[118] and basophils.[113,114] This order reflects the rank order of activity of these compounds as inhibitors of isolated PPs in vitro.[116,117] These and other studies confirm that PP activities with the functional characteristics of PP1 and PP2A are constitutively associated with extracts of HLMC[112,118] and basophils.[113] However, the molecular target(s) of okadaic acid have not been identified. The concentrations of okadaic acid required to attenuate mediator release are two to three orders of magnitude higher than those found to attenuate purified PP1 and PP2A in vitro.[111,119] Thus, no conclusion can be drawn, from these data alone, regarding the relative importance of PP1 and PP2A in mast cell function.

PP2B

The immunosuppressive agents cyclosporin A and FK506 selectively attenuate PP2B (calcineurin) activity, although neither compound in isolation has any effect on calcineurin activity. Rather, the inhibition of calcineurin results from the formation of a complex between the immunosuppressant and cognate receptor proteins, the immunophilins. The immunophilins, comprising cyclophilin, which binds to cyclosporin, and FK506-binding protein (FKBP), which binds to FK506, are essential for the inhibition of calcineurin by immunosuppressants.[120–123]

The essential role of immunophilins in mediating the inhibitory effects of immunosuppressants was illustrated in murine bone marrow-derived mast cells (BMMC), which lack

TABLE 11.1. Classification of protein serine/threonine phosphatases.

	PP1	PP2A	PP2B	PP2C
Substrate preference (subunit of phosphorylase kinase)	β	α	α	α
Divalent cation requirement	—	—	Ca^{2+}	Mg^{2+}
Sensitivity to I-1, I-2	Yes	No	No	No
Sensitivity to okadaic acid (IC_{50} nM)	10–20	0.1–1	>5000	No

Source: Adapted from ref. 105.

FKBP12, the immunophilin believed to mediate the effects of FK506.[124] While cyclosporin attenuated calcineurin PP activity and FcεRI-mediated generation of cytokines in these cells, FK506 had no effect on either PP activity or cytokine production.[125] In related murine mast cells also deficient in FKBP12, the inhibitory effect of FK506 on cytokine generation and calcineurin PP activity could be restored by the transfection and overexpression of FKBP12.[122]

The inhibitory effects of both cyclosporin and (in FKBP12-transfected cells) FK506 on FcεRI-mediated secretory responses were restricted to cytokine production. The exocytosis of granules containing preformed mediators was unaffected by either immunosuppressant.[122,125] In marked contrast, both FK506 and cyclosporin have been shown to inhibit the IgE-dependent generation of both preformed (histamine) and de novo synthesized (LTC$_4$) inflammatory mediators from human mast cells[126–129] and basophils.[130,131]

These data indicate that calcineurin may subserve distinct roles in mast cells derived from different sources. That the role of calcineurin in FcεRI-mediated activation in murine mast cells appears to be restricted to cytokine gene expression mirrors the situation in T cells in which both CsA and FK506 profoundly inhibit calcineurin activity and TCR-mediated IL-2 production.[132,133] In human mast cells and basophils, as well as in RBL-2H3 cells, calcineurin appears to have an important role in FcεRI-induced exocytosis of granules containing preformed mediators.

The mechanism of action of calcineurin in the regulation of cytokine production appears to involve the calcium-dependent transcription factor called nuclear factor of activated T cells (NF-AT).[134–137] Calcineurin is believed to mediate the calcium-dependent dephosphorylation of the cytoplasmic subunit of NF-AT (NF-ATc), which facilitates its translocation to the nucleus, whereupon it complexes with nuclear components including c-Jun and c-Fos to form a transcription-activating complex that regulates the expression of numerous cytokine genes.[138]

By contrast, the molecular targets of calcineurin that lead to the inhibition of exocytosis in human mast cells, basophils, and RBL-2H3 cells are unknown although it seems likely that these are distinct from NF-AT. Whatever the exocytosis-regulating substrates of calcineurin prove to be, it would appear either that they are redundant for exocytosis in BMMC or that, in these cells, other PPs can fulfill the role subserved selectively by calcineurin in human mast cells, basophils, and RBL cells. The characteristics of these unidentified calcineurin substrates (i.e., proteins involved in secretion that undergo calcium-dependent dephosphorylation) suggest that they may have important roles in granular exocytosis.

General Considerations

An important question regarding the role of phosphatases in FcεRI signaling is how phosphatases respond to the state of aggregation of the receptor. In some cells, receptor-mediated stimulation modulates the catalytic activity of protein phosphatases. For example, the catalytic activity of the SH2 domain-containing phosphatase SHP-1 is enhanced by its interaction with phosphorylated peptides corresponding to the ITIM sequence of FcγRIIB1.[71] In adrenal chromaffin cells, nicotine-induced catecholamine secretion is associated with the transient inhibition of a Src kinase-associated PTP activity.[139] In addition, the catalytic activity of PTPH1, a PTP related to a family of cytoskeleton-associated proteins, is substantially increased by removal of its N-terminal domain,[140] suggesting the possibility that its PTP activity might be regulated in vivo by the modulation of this intramolecular interaction. Thus, at a molecular level, both potential and precedent exist for the modulation of PTP catalytic activity in response to extracellular stimuli.

In mast cells, membrane-associated PTP activity was increased by receptor aggregation,[58] although whether this was the result of a recruitment of activity to the membrane or an enhancement of activity already located in the membrane was not addressed. Other data indicate that, at least for FcεRI itself, there is no alteration in the total receptor-directed PTP activity in whole-cell lysates before and after receptor aggregation.[4] It is possible that FcεRI is a special case, or that the activity of a relatively rare PTP enzyme is selectively regulated by FcεRI aggregation. However, an alternative explanation of these data is that receptor aggregation regulates the location rather than the catalytic activity of protein phosphatases.

Protein phosphatases have been regarded as relatively nonspecific although highly active enzymes, and this notion is, to some extent, borne out by in vitro data.[106,142] When considering the role of phosphatases in mast cells, this 'nonspecific' view of phosphatases colors any attempt to model the molecular consequences of receptor aggregation. Because there is no apparent increase in receptor-directed PTP activity following aggregation of FcεRI, one would have to argue that the receptor dephosphorylation that swiftly follows the disruption of receptor aggregates with hapten, or the interruption of kinase activity with EDTA,[4] comes about because the receptor is immersed in a 'sea' of relatively nonspecific phosphatase activity that immediately becomes dominant once receptors are disaggregated or kinase activity is diminished. In such a scheme, one could argue, the role of FcεRI aggregation in initiating intracellular signaling is to recruit sufficient kinase to ensure that tyrosine kinase activity temporarily exceeds the high basal levels of nonspecific phosphatase activity in the vicinity of the aggregate.

Thus, experimental observations relating to changes in receptor phosphorylation can be rationalized without invoking substrate targeting or specificity of PTP function if phosphatase activity is in large excess. However, accumulating data indicate that in vivo phosphatases actually exhibit a high level of substrate selectivity,[141,142] resulting in part from enzyme targeting.[143,144]

This selectivity is exemplified by the KIR inhibitory receptors. Several studies indicate that tyrosine phosphorylation of the KIR ITIM sequence, and the subsequent recruitment of phosphatase to the receptor, may take place in the absence of co-cross-linking with FcεRI[86] or other ITAM receptors.[74] However, the inhibitory effects of the receptor are only manifest when it is coassociated with FcεRI. It appears that only those targets in the immediate vicinity of the phosphatase recruited to the KIR are dephosphorylated. By extension, these observations suggest that substantial substrate specificity may be achieved in vivo by enzyme targeting. Indeed, numerous data indicate that phosphoproteins fall under the influence of PTPs targeted to the same subcellular site (e.g., the targeting of the protein serine/threonine phosphatase, PP1, to glycogen particles[106,143,145]), or to the same protein complex (e.g., numerous phosphatases have been found to associate with kinases or signaling components; PEP[146] and PTP-PEST[147] with Csk, SHIP,[59] and SHP-1[148] with Shc–Grb-2, SHP-2 with PI-3 kinase and Jak 3 (M. Gadina and J. O'Shea, personal communication), as well as SHP-1, SHP-2, and SHIP with FcεRI itself[53,59,60]).

Although enzyme targeting contributes to the substrate specificity of phosphatases in vivo, phosphatases, like kinases, show a preference for some substrates over others. For example, SHP-1 stimulates the kinase activity of Src kinase by dephosphorylating its negative regulatory tyrosine.[29] By contrast, the highly related kinase, Lyn, appears not to be a substrate for SHP-1 in vitro.[53] Similarly, quite distinct effects of CD45 on different members of the Src kinase family have been reported. In T cells the negative regulatory site of Lck is more sensitive to the expression of CD45 than is the similar site on Fyn.[31] In B cells, however, CD45 expression has a more profound effect on Lyn kinase activity than on Btk.[147] Although to some extent these data could also be explained by different members of the Src kinase family being targeted to distinct locations, these examples suggest that PTPs may distinguish between highly related substrates. Thus, it is probable that both structure-based substrate preference and enzyme targeting contribute to the specificity of phosphatase action in vivo.

Considerations of substrate specificity highlight the importance of assessing the regulation and function of phosphatses in an in vivo setting. One particularly promising technique in this regard, developed by Flint et al.,[141] involves the use of 'substrate-trapping' phosphatase mutants. The catalytic mechanism of protein tyrosine phosphatases, predicted from X-ray crystal structures of PTP–substrate complexes,[149,150] appears to begin with the formation of a thiophosphate intermediate between the incoming phosphotyrosine and a conserved cysteine residue within the active site.[150,153] The breakdown of this intermediate and the release of the dephosphorylated tyrosyl residue is brought about by a conserved aspartate residue.[149,150] Mutation of this aspartate generates an enzyme that is effectively catalytically inactive but retains its ability to bind substrate.[141] Indeed, the thiophosphate intermediate is sufficiently stable in aspartate mutants of PTP1B to allow in vivo substrates to be coprecipitated with the enzyme. Because the active site of PTP1B is common to all members of the protein tyrosine phosphatase family,[151,152] this technique may be applicable to a wide range of phosphatases. Thus, substrate-trapping mutants offer the potential to define the substrate specificity of different PTPs in myriad cellular settings.

The study of protein phosphatases in FcεRI signaling is still in its infancy. The application to the mast cell system of approaches such as 'substrate trapping', which allow the in situ assessment of diverse aspects of phosphatase function, may be necessary to understand properly how phosphatases regulate FcεRI signaling.

Acknowledgments. I am grateful to Dr. Henry Metzger for helpful discussion and critical reading of the manuscript.

References

1. Paolini R, Jouvin MH, Kinet JP. Phosphorylation and dephosphorylation of the high-affinity receptor for immunoglobulin E immediately after receptor engagement and disengagement. Nature (Lond) 1991;353:855–858.
2. Scharenberg AM, Kinet JP. Early events in mast cell signal transduction. Chem Immunol 1995;61:72–87.
3. Beaven MA, Metzger H. Signal transduction by Fc receptors: the Fc epsilon RI case. Immunol Today 1993;14:222–226.
4. Mao SY, Metzger H. Characterization of protein-tyrosine phosphatases that dephosphorylate the high affinity IgE receptor. J Biol Chem 1997;272:14067–14073.
5. Trowbridge IS, Thomas ML. CD45: an emerging role as a protein tyrosine phosphatase required for lymphocyte activation and development. Annu Rev Immunol 1994;12:85–116.
6. Chui D, Ong CJ, Johnson P, et al. Specific CD45 isoforms differentially regulate T cell receptor signaling. EMBO J 1994;13:798–807.
7. McKenney DW, Onodera H, Gorman L, et al. Distinct isoforms of the CD45 protein-tyrosine phosphatase differentially regulate interleukin 2 secretion and activation signal pathways involving Vav in T cells. J Biol Chem 1995;270:24949–24954.
8. Novak TJ, Farber D, Leitenberg D, et al. Isoforms of the transmembrane tyrosine phosphatase CD45 differentially affect T cell recognition. Immunity 1994;1:109–119.
9. Pingel JT, Thomas ML. Evidence that the leukocyte-common antigen is required for antigen-induced T lymphocyte proliferation. Cell 1989;58:1055–1065.
10. Koretzky GA, Picus J, Thomas ML, et al. Tyrosine phosphatase CD45 is essential for coupling T-cell antigen receptor to the phosphatidyl inositol pathway. Nature (Lond) 1990;346:66–68.
11. Byth KF, Conroy LA, Howlett S, et al. CD45-null transgenic mice reveal a positive regulatory role for CD45 in early thymocyte development, in the selection of CD4+CD8+ thymocytes, and B cell maturation. J Exp Med 1996;183:1707–1718.
12. Kishihara K, Penninger J, Wallace VA, et al. Normal B lymphocyte development but impaired T cell maturation in CD45-exon 6 protein tyrosine phosphatase-deficient mice. Cell 1993;74:143–156.
13. Justement LB, Campbell KS, Chien NC, et al. Regulation of B cell antigen receptor signal transduction and phosphorylation by CD45. Science 1991;252:1839–1842.
14. Justement LB, Brown VK, Lin J. Regulation of B-cell activation by CD45: a question of mechanism. Immunol Today 1994;15:399–406.
15. Cyster JG, Healy JI, Kishihara K, et al. Regulation of B-lymphocyte negative and positive selection by tyrosine phosphatase CD45. Nature (Lond) 1996;381:325–328.

16. Volarevic S, Niklinska BB, Burns CM, et al. Regulation of TCR signaling by CD45 lacking transmembrane and extracellular domains. Science 1993;260:541–544.

17. Hovis RR, Donovan JA, Musci MA, et al. Rescue of signaling by a chimeric protein containing the cytoplasmic domain of CD45. Science 1993;260:544–546.

18. Chan AC, Shaw AS. Regulation of antigen receptor signal transduction by protein tyrosine kinases. Curr Opin Immunol 1996;8:394–401.

19. Chan AC, Desai DM, Weiss A. The role of protein tyrosine kinases and protein tyrosine phosphatases in T cell antigen receptor signal transduction. Annu Rev Immunol 1994;12:555–592.

20. Cooper JA, Howell B. The when and how of Src regulation. Cell 1993;73:1051–1054.

21. Roussel RR, Brodeur SR, Shalloway D, et al. Selective binding of activated pp60$^{c\text{-}src}$ by an immobilized synthetic phosphopeptide modeled on the carboxyl terminus of pp60$^{c\text{-}src}$. Proc Natl Acad Sci USA 1991;88:10696–10700.

22. Chow LM, Veillette A. The Src and Csk families of tyrosine protein kinases in hematopoietic cells. Semin Immunol 1995;7:207–226.

23. Mustelin T, Coggeshall KM, Altman A. Rapid activation of the T-cell tyrosine protein kinase pp56^{lck} by the CD45 phosphotyrosine phosphatase. Proc Natl Acad Sci USA 1989; 86:6302–6306.

24. Mustelin T, Pessa-Morikawa T, Autero M, et al. Regulation of the p59^{fyn} protein tyrosine kinase by the CD45 phosphotyrosine phosphatase. Eur J Immunol 1992;22:1173–1178.

25. Ng DHW, Watts JD, Aebersold R, et al. Demonstration of a direct interaction between p56^{Lck} and the cytoplasmic domain of CD45 in vitro. J Biol Chem 1996;271:1295–1300.

26. Biffen M, McMichael-Phillips D, Larson T, et al. The CD45 tyrosine phosphatase regulates specific pools of antigen receptor-associated p59^{Fyn} and CD4-associated p56^{Lck} tyrosine kinases in human T-cells. EMBO J 1994;13:1920–1929.

27. Ostergaard HL, Shackelford DA, Hurley TR, et al. Expression of CD45 alters phosphorylation of the lck-encoded tyrosine protein kinase in murine lymphoma T-cell lines. Proc Natl Acad Sci U S A 1989;86:8959–8963.

28. McFarland ED, Hurley TR, Pingel JT, et al. Correlation between Src family member regulation by the protein-tyrosine-phosphatase CD45 and transmembrane signaling through the T-cell receptor. Proc Natl Acad Sci U S A 1993;90:1402–1406.

29. Somani AK, Bignon JS, Mills GB, et al. Src kinase activity is regulated by the SHP-1 protein-tyrosine phosphatase. J Biol Chem 1997;272:21113–21119.

30. Musci MA, Beaves SL, Ross SE, et al. Surface expression of hematopoietic cell phosphatase fails to complement CD45 deficiency and inhibits TCR-mediated signal transduction in a Jurkat T cell clone. J Immunol 1997;158:1565–1571.

31. Hurley TR, Hyman R, Sefton BM. Differential effects of expression of the CD45 tyrosine protein phosphatase on the tyrosine phosphorylation of the lck, fyn, and c-src tyrosine protein kinases. Mol Cell Biol 1993;13:1651–1656.

32. Brown VK, Ogle EW, Burkhardt AL, et al. Multiple components of the B cell antigen receptor complex associate with the protein tyrosine phosphatase, CD45. J Biol Chem 1994;269:17238–17244.

33. Berger SA, Mak TW, Paige CJ. Leukocyte common antigen (CD45) is required for immunoglobulin E-mediated degranulation of mast cells. J Exp Med 1994;180:471–476.

34. Hook WA, Berenstein EH, Zinsser FU, et al. Monoclonal antibodies to the leukocyte common antigen (CD45) inhibit IgE-mediated histamine release from human basophils. J Immunol 1991;147:2670–2676.

35. Bilwes AM, Den Hertog J, Hunter T, et al. Structural basis for inhibition of receptor protein-tyrosine phosphatase-a by dimerization. Nature (Lond) 1996;382:555–559.

36. Desai DM, Sap J, Schlessinger J, et al. Ligand-mediated negative regulation of a chimeric transmembrane receptor tyrosine phosphatase. Cell 1993;73:541–554.

37. Schneider H, Korn M, Haustein D. CD45-deficient RBL-2H3 cells: cellular response to FCεR- and ionophore-induced stimulation. Immunol Invest 1993;22:503–515.

38. Adamczewski M, Numerof RP, Koretzky GA, et al. Regulation by CD45 of the tyrosine phosphorylation of high affinity IgE receptor beta- and gamma-chains. J Immunol 1995;154:3047–3055.

39. Shen SH, Bastien L, Posner BI, et al. A protein-tyrosine phosphatase with sequence similarity to the SH2 domain of the protein-tyrosine kinases. Nature (Lond) 1991;352:736–739.

40. Yi TL, Cleveland JL, Ihle JN. Protein tyrosine phosphatase containing SH2 domains:

characterization, preferential expression in hematopoietic cells, and localization to human chromosome 12p12-p13. Mol Cell Biol 1992; 12:836–846.

41. Kozlowski M, Mlinaric-Rascan I, Feng GS, et al. Expression and catalytic activity of the tyrosine phosphatase PTP1C is severely impaired in motheaten and viable motheaten mice. J Exp Med 1993;178:2157–2163.

42. Shultz LD, Schweitzer PA, Rajan TV, et al. Mutations at the murine motheaten locus are within the hematopoietic cell protein-tyrosine phosphatase (Hcph) gene. Cell 1993;73:1445–1454.

43. Tsui HW, Siminovitch KA, de Souza L, et al. Motheaten and viable motheaten mice have mutations in the haematopoietic cell phosphatase gene. Nat Genet 1993;4:124–129.

44. Sidman CL, Shultz LD, Unanue ER. The mouse mutant "motheaten." II. Functional studies of the immune system. J Immunol 1978;121:2399–2404.

45. Shultz LD, Green MC. Motheaten, an immunodeficient mutant of the mouse. II. Depressed immune competence and elevated serum immunoglobulins. J Immunol 1976;116:936–943.

46. Cyster JG, Goodnow CC. Protein tyrosine phosphatase 1C negatively regulates antigen receptor signaling in B lymphocytes and determines thresholds for negative selection. Immunity 1995;2:13–24.

47. Pani G, Kozlowski M, Cambier JC, et al. Identification of the tyrosine phosphatase PTP1C as a B cell antigen receptor-associated protein involved in the regulation of B cell signaling. J Exp Med 1995;181:2077–2084.

48. Lorenz U, Ravichandran KS, Burakoff SJ, et al. Lack of SHPTP1 results in src-family kinase hyperactivation and thymocyte hyperresponsiveness. Proc Natl Acad Sci USA 1996; 93:9624–9629.

49. Pani G, Fischer KD, Mlinaric-Rascan I, et al. Signaling capacity of the T cell antigen receptor is negatively regulated by the PTP1C tyrosine phosphatase. J Exp Med 1996;184:839–852.

50. Raab M, Rudd CE. Hematopoietic cell phosphatase (HCP) regulates p56^{Lck} phosphorylation and ZAP-70 binding to T cell receptor zeta chain. Biochem Biophys Res Commun 1996; 222:50–57.

51. Plas DR, Johnson R, Pingel JT, et al. Direct regulation of ZAP-70 by SHP-1 in T cell antigen receptor signaling. Science 1996;272:1173–1176.

52. Swieter M, Berenstein EH, Swaim WD, et al. Aggregation of IgE receptors in rat basophilic leukemia 2H3 cells induces tyrosine phosphorylation of the cytosolic protein-tyrosine phosphatase HePTP. J Biol Chem 1995;270:21902–21906.

53. Kimura T, Zhang J, Segawa K, et al. Syk-independent tyrosine phosphorylation and association of the protein tyrosine phosphatase SHP-1 and SHP-2 with the high affinity IgE receptor. J Immunol 1997;159:4426–4434 (AU).

54. Ono M, Bolland S, Tempst P, et al. Role of the inositol phosphatase SHIP in negative regulation of the immune system by the receptor FcgammaRIIB. Nature (Lond) 1996;383:263–266.

55. Pazdrak K, Adachi T, Alam R. Src homology 2 protein tyrosine phosphatase (SHPTP2)/Src homology 2 phosphatase 2 (SHP2) tyrosine phosphatase is a positive regulator of the interleukin 5 receptor signal transduction pathways leading to the prolongation of eosinophil survival. J Exp Med 1997;186:561–568.

56. Bennett AM, Hausdorff SF, O'Reilly AM, et al. Multiple requirements for SHPTP2 in epidermal growth factor-mediated cell cycle progression. Mol Cell Biol 1996;16:1189–1202.

57. Yamauchi K, Milarski KL, Saltiel AR, et al. Protein-tyrosine-phosphatase SHPTP2 is a required positive effector for insulin downstream signaling. Proc Natl Acad Sci USA 1995;92:664–668.

58. Hampe CS, Pecht I. Protein tyrosine phosphatase activity enhancement is induced upon Fc epsilon receptor activation of mast cells. FEBS Lett 1994;346:194–198.

59. Osborne MA, Zenner G, Lubinus M, et al. The inositol 5′-phosphatase SHIP binds to immunoreceptor signaling motifs and responds to high affinity IgE receptor aggregation. J Biol Chem 1996;271:29271–29278.

60. Kimura T, Sakamoto H, Appella E, et al. The negative signaling molecule SH2 domain-containing inositol-polyphosphate 5-phosphata (SHIP) binds to the tyrosine-phosphorylated beta subunit of the high affinity IgE receptor. J Biol Chem 1997;272:13991–13996.

61. Fong DC, Malbec O, Arock M, et al. Selective in vivo recruitment of the phosphatidylinositol phosphatase SHIP by phosphorylated Fc gamma RIIB during negative regulation of IgE-dependent mouse mast cell activation. Immunol Lett 1996;54:83–91.

62. Hoth M, Penner R. Depletion of intracellular calcium stores activates a calcium current in mast cells. Nature (Lond) 1992;355:353–356.

63. Scharenberg AM, Kinet JP. The emerging field of receptor-mediated inhibitory signaling: SHP or SHIP? Cell 1996;87:961–964.

64. Kimura T, Kihara H, Bhattacharyya S, et al. Downstream signaling molecules bind to different phosphorylated immunoreceptor tyrosine-based activation motif (ITAM) peptides of the high affinity IgE receptor. J Biol Chem 1996; 271:27962–27968.

65. Tridandapani S, Kelley T, Cooney D, et al. Negative signaling in B cells: SHIP Grbs Shc. Immunol Today 1997;18:424–427.

66. Tridandapani S, Chacko GW, Van Brocklyn JR, et al. Negative signaling in B cells causes reduced Ras activity by reducing Shc-Grb2 interactions. J Immunol 1997;158:1125–1132.

67. Den Hertog J, Hunter T. Tight association of GRB2 with receptor protein-tyrosine phosphatase alpha is mediated by the SH2 and C-terminal SH3 domains. EMBO J 1996; 15:3016–3027.

68. Vivier E, Daeron M. Immunoreceptor tyrosine-based inhibition motifs. Immunol Today 1997;18:286–291.

69. Daeron M. ITIM-bearing negative coreceptors. Immunologist 1997;5:79–86.

70. Cambier JC. Inhibitory receptors abound? Proc Natl Acad Sci USA 1997;94:5993–5995.

71. D'Ambrosio D, Hippen KL, Minskoff SA, et al. Recruitment and activation of PTP1C in negative regulation of antigen receptor signaling by Fc gamma RIIB1. Science 1995;268:293–297.

72. D'Ambrosio D, Fong DC, Cambier JC. The SHIP phosphatase becomes associated with Fc gamma RIIB1 and is tyrosine phosphorylated during 'negative' signaling. Immunol Lett 1996;54:77–82.

73. Olcese L, Lang P, Vely F, et al. Human and mouse killer-cell inhibitory receptors recruit PTP1C and PTP1D protein tyrosine phosphatases. J Immunol 1996;156:4531–4534.

74. Binstadt BA, Brumbaugh KM, Dick CJ, et al. Sequential involvement of Lck and SHP-1 with MHC-recognizing receptors on NK cells inhibits FcR-initiated tyrosine kinase activation. Immunity 1996;5:629–638.

75. Burshtyn DN, Scharenberg AM, Wagtmann N, et al. Recruitment of tyrosine phosphatase HCP by the killer cell inhibitor receptor. Immunity 1996;4:77–85.

76. Campbell KS, Dessing M, Lopez-Botet M, et al. Tyrosine phosphorylation of a human killer inhibitory receptor recruits protein tyrosine phosphatase 1C. J Exp Med 1996;184:93–100.

77. Held W, Cado D, Raulet DH. Transgenic expression of the Ly49A natural killer cell receptor confers class I major histocompatibility complex (MHC) -specific inhibition and prevents bone marrow allograft rejection. J Exp Med 1996;184:2037–2041.

78. Moretta L, Mingari MC, Pende D, et al. The molecular basis of natural killer (NK) cell recognition and function. J Clin Immunol 1996;16:243–253.

79. Doody GM, Justement LB, Delibrias CC, et al. A role in B cell activation for CD22 and the protein tyrosine phosphatase SHP. Science 1995;269:242–244.

80. Marengere LE, Waterhouse P, Duncan GS, et al. Regulation of T cell receptor signaling by tyrosine phosphatase SYP association with CTLA-4. Science 1996;272:1170–1173.

81. Katz HR, Vivier E, Castells MC, et al. Mouse mast cell gp49B1 contains two immunoreceptor tyrosine-based inhibition motifs and suppresses mast cell activation when coligated with the high-affinity Fc receptor for IgE. Proc Natl Acad Sci USA 1996;93:10809–10814.

82. Arm JP, Nwankwo C, Austen KF. Molecular identification of a novel family of human Ig superfamily members that possess immunoreceptor tyrosine-based inhibition motifs and homology to the mouse gp49B1 inhibitory receptor. J Immunol 1997;159:2342–2349.

83. Hayami K, Fukuta D, Nishikawa Y, et al. Molecular cloning of novel murine cell-surface glycoprotein homologous to killer cell inhibitory receptors. J Biol Chem 1997;272;7320–7327.

84. Guthmann MD, Tal M, Pecht I. A secretion inhibitory signal transduction molecule on mast cells is another C-type lectin. Proc Natl Acad Sci USA 1995;92:9397–9401.

85. Daeron M, Latour S, Malbec O, et al. The same tyrosine-based inhibition motif, in the intracytoplasmic domain of Fc gamma RIIB, regulates negatively BCR-, TCR-, and FcR-dependent cell activation. Immunity 1995;3:635–646.

86. Blery M, Delon J, Trautmann A, et al. Reconstituted killer cell inhibitory receptors for major histocompatibility complex class I molecules control mast cell activation induced via immunoreceptor tyrosine-based activation motifs. J Biol Chem 1997;272:8989–8996.

87. Tooze RM, Doody GM, Fearon DT. Counterregulation by the coreceptors CD19 and CD22 of MAP kinase activation by membrane immunoglobulin. Immunity 1997;7:59–67.

88. Law CL, Sidorenko SP, Chandran KA, et al. CD22 associates with protein tyrosine phosphatase 1C, Syk, and phospholipase C-gamma (1) upon B cell activation. J Exp Med 1996;183:547–560.

89. Vély F, Olivero S, Olcese L, et al. Differential association of phosphatases with hematopoietic co-receptors bearing imunoreceptor tryosine-based inhibition motifs. Eur J Immunol 1997;27:1994–2000.

90. Healy JI, Dolmetsch RE, Timmerman LA, et al. Different nuclear signals are activated by the B cell receptor during positive versus negative signaling. Immunity 1997;6:419–428.

91. Dolmetsch RE, Lewis RS, Goodnow CC, et al. Differential activation of transcription factors induced by Ca^{2+} response amplitude and duration. Nature (Lond) 1997;386:855–858.

92. Plas DR, Johnson R, Pingel JT, et al. Direct regulation of ZAP-70 by SHP-1 in T cell antigen receptor signaling. Science 1996;272:1173–1176.

93. Guthmann MD, Tal M, Pecht I. A new member of the C-type lectin family is a modulator of the mast cell secretory response. Int Arch Allergy Immunol 1995;107:82–86.

94. Ono M, Okada H, Bolland S, et al. Deletion of SHIP or SHP-1 reveals two distinct pathways for inhibitory signaling. Cell 1997;90:293–301.

95. Nadler MJS, Chen B, Anderson JS, et al. Protein-tyrosine phosphatase SHP-1 is dispensable for Fc gamma RIIB-mediated inhibition of B cell antigen receptor activation. J Biol Chem 1997;272:20038–20043.

96. Gupta N, Scharenberg AM, Burshtyn DN, et al. Negative signaling pathways of the killer cell inhibitory receptor and Fc gamma RIIb1 require distinct phosphatases. J Exp Med 1997;186:473–478.

97. Takai T, Ono M, Hikida M, et al. Augmented humoral and anaphylactic responses in Fc gamma RII-deficient mice. Nature (Lond) 1996;379:346–349.

98. Castells MC, Wu X, Arm JP, et al. Cloning of the gp49B gene of the immunoglobulin superfamily and demonstration that one of its two products is an early-expressed mast cell surface protein originally described as gp49. J Biol Chem 1994;269:8393–8401.

99. Rojo S, Burshtyn DN, Long EO, et al. Type I transmembrane receptor with inhibitory function in mouse mast cells and NK cells. J Immunol 1997;158:9–12.

100. Wang LL, Mehta IK, LeBlanc PA, et al. Mouse natural killer cells express gp49B1, a structural homologue of human killer inhibitory receptors. J Immunol 1997;158:13–17.

101. Katz HR, Austen KF. A newly recognized pathway for the negative regulation of mast cell-dependent hypersensitivity and inflammation mediated by an endogenous cell surface receptor of the gp49 family. J Immunol 1997;158:5065–5070.

102. Ortega Soto E, Pecht I. A monoclonal antibody that inhibits secretion from rat basophilic leukemia cells and binds to a novel membrane component. J Immunol 1988;141:4324–4332.

103. Jurgens L, Arndt-Jovin D, Pecht I, et al. Proximity relationships between the type I receptor for Fc epsilon (FcεRI) and the mast cell function-associated antigen (MAFA) studied by donor photobleaching fluorescence resonance energy transfer microscopy. Eur J Immunol 1996;26:84–91.

104. Geller-Bernstein C, Berrebi A, Bassous Gedj L, et al. Antibodies specific to membrane components of rat mast cells are cross-reacting with human basophils. Int Arch Allergy Immunol 1994;105:269–273.

105. Cohen P. The structure and regulation of protein phosphatases. Annu Rev Biochem 1989;58:453–508.

106. Cohen P, Cohen PTW. Protein phosphatases come of age. J Biol Chem 1989;264:21435–21438.

107. Chen MX, McPartlin AE, Brown L, et al. A novel human protein serine/threonine phosphatase, which possesses four tetratricopeptide repeat motifs and localizes to the nucleus. EMBO J 1994;13:4278–4290.

108. Brewis ND, Street AJ, Prescott AR, et al. PPX, a novel protein serine/threonine phosphatase localized to centrosomes. EMBO J 1993;12:987–996.

109. Honkanen RE, Zwiller J, Daily SL, et al. Identification, purification, and characterization of a novel serine/threonine protein phosphatase from bovine brain. J Biol Chem 1991;266:6614–6619.

110. Haystead TAJ, Sim ATR, Carling D, et al. Effects of the tumor promoter okadaic acid on intracellular protein phosphorylation and metabolism. Nature (Lond) 1989;337:78–81.

111. Cohen P, Holmes CF, Tsukitani Y. Okadaic acid: a new probe for the study of cellular regulation. Trends Biochem Sci 1990;15:98–102.

112. Peachell PT, Munday MR. Regulation of human lung mast cell function by phosphatase inhibitors. J Immunol 1993;151:3808–3816.

113. Peirce MJ, Warner JA, Munday MR, et al. Regulation of human basophil function by phosphatase inhibitors. Br J Pharmacol 1996;119:446–453.

114. Botana LM, MacGlashan DW. Effect of okadaic acid on human basophil secretion. Biochem Pharmacol 1993;45:2311–2315.

115. Estevez MD, Vieytes MR, Louzao MC, et al. Effect of okadaic acid on immunologic and non-immunologic histamine release in rat mast cells. Biochem Pharmacol 1994;47:591–593.

116. Nishiwaki S, Fujiki H, Suganuma M, et al. Structure-activity relationship within a series of okadaic acid derivatives. Carcinogenesis (Oxf) 1990;11:1837–1841.

117. Takai A, Murata M, Torigoe K, et al. Inhibitory effect of okadaic acid-derivatives on protein phosphatases—a study on structure/affinity relationship. Biochem J 1992;284:539–544.

118. Peirce MJ, Cox SE, Munday MR, et al. Preliminary characterization of the role of protein serine/threonine phosphatases in the regulation of human lung mast cell function. Br J Pharmacol 1997;120:239–246.

119. Cohen P, Klumpp S, Schelling DL. An improved procedure for identifying and quantitating protein phosphatases in mammalian tissues. FEBS Lett 1989;250:596–600.

120. Liu J, Farmer JD, Lane WS, et al. Calcineurin is a common target of cyclophilin cyclosporine-A and FKBP-FK506 complexes. Cell 1991;66:807–815.

121. Fruman DA, Klee CB, Bierer BE, et al. Calcineurin phosphatase activity in T lymphocytes is inhibited by FK 506 and cyclosporin A. Proc Natl Acad Sci U S A 1992;89:3686–3690.

122. Fruman DA, Bierer BE, Benes JE, et al. The complex of FK506-binding protein 12 and FK506 inhibits calcineurin phosphatase activity and IgE activation-induced cytokine transcripts, but not exocytosis, in mouse mast cells. J Immunol 1995;154:1846–1851.

123. Swanson SK, Born T, Zydowsky LD, et al. Cyclosporin-mediated inhibition of bovine calcineurin by cyclophilins A and B. Proc Natl Acad Sci USA 1992;89:3741–3745.

124. Fruman DA, Wood MA, Gjertson CK, et al. FK506 binding protein 12 mediates sensitivity

125. Kaye RE, Fruman DA, Bierer BE, et al. Effects of cyclosporin A and FK506 on Fc epsilon receptor type I-initiated increases in cytokine mRNA in mouse bone marrow-derived progenitor mast cells: resistance to FK506 is associated with a deficiency in FK506-binding protein FKBP12. Proc Natl Acad Sci USA 1992;89:8542–8546.

126. de Paulis A, Stellato C, Cirillo R, et al. Anti-inflammatory effect of FK-506 on human skin mast cells. J Invest Dermatol 1992;99:723–728.

127. Stellato C, de Paulis A, Ciccarelli A, et al. Anti-inflammatory effect of cyclosporin A on human skin mast cells. J Invest Dermatol 1992;98:800–804.

128. de Paulis A, Cirillo R, Ciccarelli A, et al. Characterization of the anti-inflammatory effect of FK-506 on human mast cells. J Immunol 1991;147:4278–4285.

129. Triggiani M, Cirillo R, Lichtenstein LM, et al. Inhibition of histamine and prostaglandin-D_2 release from human-lung mast-cells by cyclosporine-A. Int Arch Allergy Appl Immunol 1989;88:253–255.

130. de Paulis A, Cirillo R, Ciccarelli A, et al. FK-506, a potent novel inhibitor of the release of proinflammatory mediators from human Fc epsilon RI+ cells. J Immunol 1991;146:2374–2381.

131. Cirillo R, Triggiani M, Siri L, et al. Cyclosporine A rapidly inhibits mediator release from human basophils, presumably by interacting with cyclophilin. J Immunol 1990;144:3891–3897.

132. Fruman DA, Burakoff SJ, Bierer BE. Immunophilins in protein folding and immunosuppression. FASEB J 1994;8:391–400.

133. Clipstone NA, Crabtree GR. Identification of calcineurin as a key signalling enzyme in T-lymphocyte activation. Nature (Lond) 1992;357:695–697.

134. Shaw KT, Ho AM, Raghavan A, et al. Immunosuppressive drugs prevent a rapid dephosphorylation of transcription factor NFAT1 in stimulated immune cells. Proc Natl Acad Sci USA 1995;92:11205–11209.

135. Luo C, Shaw KT, Raghavan A, et al. Interaction of calcineurin with a domain of the transcription factor NFAT1 that controls nuclear import. Proc Natl Acad Sci USA 1996;93:8907–8912.

136. Luo C, Burgeon E, Carew JA, et al. Recombinant NFAT1 (NFATp) is regulated by

calcineurin in T cells and mediates transcription of several cytokine genes. Mol Cell Biol 1996;16:3955–3966.

137. Loh C, Shaw KT, Carew J, et al. Calcineurin binds the transcription factor NFAT1 and reversibly regulates its activity. J Biol Chem 1996;271:10884–10891.

138. Rao A, Luo C, Hogan PG. Transcription factors of the NFAT family: regulation and function. Annu Rev Immunol 1997;15:707–747.

139. Van Hoek ML, Allen CS, Parsons SJ. Phosphotyrosine phosphatase activity associated with c-Src in large multimeric complexes isolated from adrenal medullary chromaffin cells. Biochem J 1997;326:271–277.

140. Zhang SH, Eckberg WR, Yang Q, Samatar AA, Tonks NK. Biochemical characterization of a human band 4.1-related protein-tyrosine phosphatase, PTPH1. J Biol Chem 1995;270:20067–20072.

141. Flint AJ, Tiganis T, Barford D, et al. Development of "substrate-trapping" mutants to identify physiological substrates of protein tyrosine phosphatases. Proc Natl Acad Sci USA 1997;94:1680–1685.

142. Garton AJ, Flint AJ, Tonks NK. Identification of p130(cas) as a substrate for the cytosolic protein tyrosine phosphatase PTP-PEST. Mol Cell Biol 1996;16:6408–6418.

143. Hubbard MJ, Cohen P. On target with a new mechanism for the regulation of protein phosphorylation. Trends Biochem Sci 1993;18:172–177.

144. Faux MC, Scott JD. More on target with protein phosphorylation: conferring specificity by location. Trends Biochem Sci 1997;21:312–315.

145. Klee C. Calcineurin as target for cyclosporin and FK 506. Proc Natl Acad Sci USA 1979;76:6270.

146. Cloutier JF, Veillette A. Association of inhibitory tyrosine protein kinase p50^{csk} with protein tyrosine phosphatase PEP in T cells and other hematopoietic cells. EMBO J 1996;15:4909–4918.

147. Davidson D, Cloutier J-F, Gregorieff A, et al. Inhibitory tyrosine protein kinase p50csk is associated with protein tyrosine phosphatase PTP-PEST in hematopoietic and non-hematopoietic cells. J Biol Chem 1997;272:23455–23462.

148. Kon-Kozlowski M, Pani G, Pawson T, et al. The tyrosine phosphatase PTP1C associates with Vav, Grb2, and mSos1 in hematopoietic cells. J Biol Chem 1996;271:3856–3862.

149. Jia Z, Barford D, Flint AJ, et al. Structural basis for phosphotyrosine peptide recognition by protein tyrosine phosphatase 1B. Science 1995;268:1754–1758.

150. Barford D, Flint AJ, Tonks NK. Crystal structure of human protein tyrosine phosphatase 1B. Science 1994;263:1397–1404.

151. Denu JM, Stuckey JA, Saper MA, et al. Form and function in protein dephosphorylation. Cell 1996;87:361–364.

152. Barford D, Jia Z, Tonks NK. Protein tyrosine phosphatases take off. Nat Struct Biol 1995;2:1043–1053.

153. Zhang ZY, Wang Y, Dixon JE. Dissecting the catalytic mechanism of protein-tyrosine phosphatases. Proc Natl Acad Sci USA 1994;91:1624–1627.

12
Protein Kinase C and Early Mast Cell Signals

Patrick G. Swann, Sandra Odom, and Juan Rivera

Protein kinase C (PKC) is a family of closely related serine/threonine kinases of which 11 distinct isoforms have been described to date.[1] It is commonly thought that the activity associated with this family of kinases is important for both the homeostasis of an organism and the regulation of its responses to environmental stimuli. The response of PKC isozymes to a stimulus can be beneficial (as in the activation of normal cellular functions) or detrimental (as in the induction of neoplasia). What determines whether a response is beneficial or detrimental is not understood, but recent studies on isozyme function and substrate specificity have begun to provide some clues as to how PKC is activated and what effect(s) this activity has on functional responses.

A variety of functional responses to PKC activation have been described.[2] It is likely that the diversity of response is mediated by the preferential activation of distinct isozymes, differences in the levels of PKC isozyme activity, targeting of specific substrates by specific isozymes, or a combination thereof. We now know that the intracellular targets of the isozymes may differ, and recent studies reveal that the differences in substrate specificity have specific functional consequences.[1] Furthermore, substrate specificity may be determined by a number of factors including substrate recognition, localization of enzyme and substrate, and presence of an appropriate microenvironment.[3] This latter criterion is gaining recognition as we learn more about the requirements for activation of the individual isozymes. Recent studies show that some isozymes are active in the presence of phosphatidylinositol 4,5-biphosphate (PIP$_2$), cis-unsaturated fatty acids, and other lipid metabolites in additon to the classic activator, diacylgycerol[4-6] (Table 12.1). This provides additional mechanisms for induction of PKC isozyme activity that may be important not only to the response mediated by a particular isozyme but also may impart a temporal component of activation that is independent of the generation of diacylglycerol (DAG).

In this chapter we focus on the role of PKC isozymes in early events in mast cell signaling. We also discuss how the activity of PKC is regulated and what effect this has on mast cell responses. Finally, we reflect on the appropriateness of considering PKC isozymes as potential targets for therapeutic intervention in mast cell-related diseases. In part, this latter topic is at best speculation because much of what we profess to know about PKC activation and function in the mast cell is inferred from more detailed studies on PKC activation and function in other systems. Thus, we warn the reader that this chapter should not be considered as the consummate review on PKC activity in early mast cell signals. Instead, it should be used as a source that consolidates what we know on PKC structure and function with the limited knowledge as to its role in early mast cell signals. We hope to provoke thought and induce further investigation in this challenging and provocative area.

TABLE 12.1. Protein kinase C (PKC) Isozyme "specific" activators.

Isozyme	Ca^{2+}	DAG	PtdIns-3,4-P$_2$[a] PtdIns-3,4,5-P$_3$	PtdIns-3,4,5-P$_3$[b]	LysoPC[c]	FFA[d]
PKC α	x	x			x	x
PKC β	x	x			x	x
PKC γ	x	x			x	x
PKC δ		x	x			
PKC ε		x	x			x
PKC η		x	x			
PKC ζ			—[e]	x		x

DAG, diacylglycerol.

[a] PtdIns-3,4-P$_2$, phosphotidylinositol-3,4-phosphate; PtdIns-3,4,5-P$_3$, phosphotidylinositol-3,4,5-phosphate. Baculovirus expressed and purified isozymes used in this study.[4]

[b] PKC isozyme was purified from bovine kidney.[5]

[c] LysoPC, lysophosphatidylcholine.[6]

[d] FFA, cis-unsaturated fatty acids.[6]

[e] PKC-ζ purified from baculovirus was constitutively active.

Molecular Structure of PKC Family Members

There are 12 isozymes of PKC, which fall into three subfamilies based on primary structure: conventional (α, βI, βII, γ), novel (δ, ε, η, θ, μ), and atypical (ζ, λ, ι).[1] Each PKC isotype has a regulatory domain and a catalytic domain consisting of conserved and variable regions.[6] The regulatory domain contains the first conserved region (C1), which binds DAG, and a second conserved region (C2) that, in the conventional (cPKC) isozymes contains the calcium-binding site. The novel (nPKC) isozymes contain a "C2-like" region[7] but do not require calcium for activation or translocation. The atypical (aPKC) isozymes contain half a "C1-like" region and do not bind diacylglycerol or calcium.[6] In the catalytic domain the C3 region contains the ATP-binding site and the catalytic site while the C4 region is necessary for substrate recognition.[6]

The C1 region of the regulatory domain contains the pseudosubstrate sequence that is thought to bind to the active site of the enzyme and inhibit its activity.[8] Immediately after the pseudosubstrate sequence are two cysteine-rich regions (also called zinc fingers as each coordinates with two zinc ions), which serve to bind DAG and phorbol esters.[9] Mutagenic studies on the second cysteine-rich domain have localized the binding site for DAG and phorbol esters to residues 107–113 and 121–128.[10] The crystal structure of the second cysteine-rich region in complex with phorbol ester was recently solved and shown to consist of two small β sheets with phorbol ester filling the gap between the two strands and forming a hydrophobic cap over polar main-chain groups.[11] This structure is thought to facilitate membrane binding.

Translocation and Activation of PKC Isozymes

Many of the studies on PKC function that have focused on isozyme subcellular redistribution speculate on the possible role(s) that changes in localization may play in cell function. In general, PKC isozymes are found predominantly in the cytosol in resting cells and subsequently translocate to the membrane or particulate fraction upon cell activation. In vitro measurement of PKC activity in the membrane or particulate fraction after cell activation usually reflects the increased presence of PKC in these fractions. This has promoted the view that translocation to the membrane or particulate fraction could be regarded as equivalent to PKC activation. This idea, however, is not supported by the fact that in certain cells some PKC isozymes are in the membrane fraction in an active or inactive form before cell activation.[12,13] Furthermore, various agents that effectively inhibit PKC activity are potent translocators of PKC.[13] One should also keep in mind that members of the aPKC isozymes reportedly do not translocate in response to certain stimuli, but these isozymes have increased activity when cells are stimulated.[14,15]

Translocation and activation of cPKC isozymes is modulated by calcium, diacylglycerol, and acidic phospholipids while that of nPKC isozymes is modulated by diacylglycerol and acidic phospholipids. The regulation of the aPKC isozymes is unclear, but PKC-ζ can be activated by phosphatidylinositol 3,4,5-trisphosphate in vitro[5] (Table 12.1); however, the requirements for translocation of aPKC isozymes are unknown.

Translocation

Protein kinase C isozymes exist primarily in the cytosol and require particular associations with cofactors to translocate. With an increase in intracellular calcium, Ca^{2+}-dependent isozymes bind with anionic phospholipid to form a preactivated form of the enzyme. An increase in DAG resulting from hydrolysis of phospholipids by phospholipase C (PLC) or by phospholipase D (PLD) and phosphatidic acid phosphomonoesterase increases the affinity of PKC for phosphatidylserine (PS) as well as activates both the Ca^{2+}-dependent and -independent PKC isozymes. Both cPKC and nPKC isozymes bind PS. Although PKC binding to PS is required, it does not necessarily lead to kinase activity.

The first step in the translocation of cPKC involves calcium binding to the C2 region of cPKC, which is rich in aspartic acid residues. Calcium forms a complex with negative charges of aspartic residues and negatively charged phospholipid in the membranes. Novel PKC isozymes contain a "C2-like" region that contains arginine, a positively charged residue, and may already be capable of binding negatively

charged phospholipids in the absence of an increase in intracellular Ca^{2+}.[9] Thus, for example, in rat basophilic leukemia (RBL-2H3) cells a greater percentage of the nPKC-δ and nPKC-ε isozymes are found associated with the membrane in nonstimulated cells than the cPKC-α or -β isoforms,[16,17] and the kinetics of nPKC translocation appear to precede that of cPKC.[17] This suggests the nPKC isozymes are able to interact with the membrane in the absence of a calcium response and may have a higher affinity for membrane interactions.

Activation

The binding of the physiological activator DAG to the C1 region of the regulatory domain reduces the affinity of the pseudosubstrate in the regulatory domain for the catalytic domain.[8,9] Diacylglycerol mediates its function by serving as a hydrophobic anchor that can recruit PKC to the membrane (translocation) by causing a selective increase in the affinity of cPKC and nPKC for PS.[9] Diacylglycerol may also stabilize the active conformation of PKC because DAG increases the catalytic efficiency of PKC bound to PS and reduces the concentration of Ca^{2+} required for PS-dependent cPKC activation. This apparent synergy of Ca^{2+} with DAG arises as both increasing Ca^{2+} and DAG, increase the affinity of PKC for membranes. This activation is mimicked by DAG analogs such as phorbol esters, although phorbol esters bind to PKC with an affinity greater than 1000 fold that of DAG in the presence of phospholipids. This high-affinity binding provides a more stable complex on the membrane and prolonged activation of PKC.[18]

Regulation of PKC Activity

The discovery that PKC is a family of isozymes[1] has expedited studies on specific cofactors and possible modulators of PKC activity in the cell. This evergrowing list of activators (see Table 12.1) provides evidence that multiple stimuli or signaling events may regulate PKC activity in many cell types. Of interest is that different isozymes may respond to different effectors. Thus, these isozymes may be activated by extracellular stimuli that activate the intracellular production of specific effectors. This topic has been extensively reviewed,[1,2,4–6] and thus we do not cover this topic in detail. However, we focus more extensively on the less-reviewed topic of phosphorylation of PKC and its effect on isozyme activity.

Serine/Threonine Phosphorylation of PKC: Effects on Enzyme Maturity and Activity

While PKC is acutely regulated by diacylglycerol, it is also known to be phosphorylated at multiple sites. Six autophosphorylation sites have been identified for the PKC-βII isozyme in vitro[19]; however, only one of these sites (Thr-641) was later identified from in vivo studies.[20] These observations emphasize the limitations of in vitro studies for identifying in vivo sites of phosphorylation. Using high pressure liquid chromatography and mass spectrometry, it was shown that there are three distinct sites of phosphorylation in PKC-βII in vivo: Thr-500, Thr-641, and Ser-660.[20,21] All these sites are in the C4 region of the catalytic domain. Thr-500 is in the activation loop and it is highly conserved among other PKC isozymes, while positions corresponding to Thr-641 and Ser-660 have conserved hydroxyl-containing or acidic residues in other PKC isozymes.[21] The phosphorylation of Thr-500 in PKC-βII is important for catalytic activity. The phosphorylation of Thr-500 is thought to render the kinase catalytically competent by aligning the residues involved in catalysis. Mutation of Thr-500 in PKC-βII, or its equivalent, Thr-497 in PKC-α, to an uncharged residue results in an inactive kinase. In contrast, mutation of the corresponding threonine (Thr 505) to alanine in PKC-δ did not inactivate the kinase, indicating that this isozyme differs in its phosphorylation requirements for activation.[22]

The functional significance of the other two phosphorylation sites at Thr-641 and Ser-660 in PKC-βII or their equivalents in other isozymes is the subject of debate. Newton has proposed the following model.[9] First, transphosphoryla-

tion at the activation loop (Thr-500 in PKC-βII) renders the kinase catalytically competent, followed by an autophosphorylation at the C-terminus (Thr-641 in PKC-βII) that stabilizes the catalytically competent conformation. The final step involves a second autophosphorylation (Ser 660) that releases PKC from a detergent-insoluble fraction to the cytosol. Parker and coworkers have proposed another model from mutational analysis of PKC-α.[23] Their studies show that phosphorylation of Thr-638 (corresponding to Thr-641 of PKC-βII) controls the rate of inactivation of kinase activity in vivo but is not required for catalytic activity. In addition, the phosphorylation of Ser-657 (corresponding to Ser-660 of PKC-βII) controls the accumulation of phosphate at other sites on PKC-α, as well as contributes to the maintenance of the phosphatase-resistant conformation. Thus, in this model, the initial phosphorylation at Thr-497 in the activation loop and Ser-657 at the carboxy terminus produces an intermediate form that is partially active and sensitive to phosphatases. In a subsequent step, Thr-638 is phosphorylated and PKC switches to a closed conformation that is fully active and resistant to phosphatases. In addition, Parker's group found no association between phosphorylation at Ser-657 and release of PKC from a detergent-insoluble fraction to the cytosol. Resolution of the observed differences in these two models of PKC maturation and competence awaits further studies.

Tyrosine Phosphorylation of PKC: Its Functional Implications

To the best of our knowledge, tyrosine phosphorylation has been reported for only two isozymes of PKC; namely, PKC-α and PKC-δ.[24,25] Protein kinase C-δ has been demonstrated to be an in vitro target of Src family tyrosine kinase activity.[26,27] However, whether this family of tyrosine kinases is solely responsible for PKC-δ phosphorylation in the cell remains to be defined. It is possible that other isozymes of PKC may also be phosphorylated on tyrosine residues and that this may depend on the stimulus or cell type in which they are expressed. We recently found that the overexpression of the PKC-βI isozyme and subsequent activation of the rat basophilic leukemia cell line (RBL-2H3) results in the tyrosine phosphorylation of PKC-βI (E.-Y. Chang and J. Rivera, unpublished observation) in addition to that of PKC-δ.[13] Since the original report[24] that PKC-δ is tyrosine phosphorylated, numerous studies have demonstrated that this modification of PKC-δ occurs in response to diverse stimuli.[13,24,28–30] The collective observations suggest an important role for this event although its functional significance is unclear. In fact, increased and decreased PKC-δ activity in response to this modification has been reported.[13,24,28] Recent observations in our laboratory are beginning to decipher the role of tyrosine phosphorylation of PKC-δ and its functional consequences, and these studies are detailed in the following sections. In summary of our findings, tyrosine phosphorylation of PKC isozymes appears to play a role in their ability to interact with other proteins and this may serve to regulate functional responses in a given cell type.

PKC, the Mast Cell, and the Basophil

The relationship between PKC activity and mast cell and basophil function has long been an area of study. At this time we can categorically state that PKC activity is important to the secretory response of mast cells.[31] Most early studies utilized phorbol esters as agents to provoke responses in secretory cells and analyzed the effect of this activating stimulus on the secretory response of these cells.[32,33] The discovery of PKC and the subsequent definition of this enzyme family as receptors for phorbol esters[34] initiated a new series of studies that permitted the analysis of translocation of this enzymatic activity to membranes as a measure of activation. Using this approach, several groups found that PKC activity in membrane fractions of mast cells is increased in response to stimulation of the high-affinity receptor for IgE (FcεRI),[35,36] and in the study from one of these groups the sustained translocation of PKC activity from the cytosol to the membrane

was found to be consistent with the secretory response.[36] This latter study also found an early and transient translocation of PKC activity that was detected within 15 s of receptor engagement. This finding suggests that PKC activity might be involved in early events in mast cell responses, and with subsequent analyses of the PKC isozymes found in mast cells and basophils, as detailed next, we can now begin to ask how the kinetics of isozyme response may regulate mast cell and basophil effector function.

Another definitive statement that can be made from the collective studies on PKC in mast cells is that PKC activity regulates more than one signaling pathway or response. This conclusion and the discovery of multiple PKC isoforms has advanced a new era of investigative effort focusing on whether PKC isozymes mediate responses in the mast cell by targeting specific substrates. This also raises the question of whether induction of a mast cell function is solely mediated by activation of a particular isozyme. Recent work has begun to delineate the role of individual PKC isozymes in mast cell and basophil function.

In RBL-2H3 cells, at least five PKC isozymes (α, β, δ, ε, ζ) have been identified.[16] The known functional consequences of specific isozyme activation are summarized in Table 12.2. The mouse mast cell line MC-9 contains PKC-α, -β, and -δ with PKC-δ being the most prevalent isozyme when measured by northern blot[37–40] and PKC-β the most prevalent when measured by western blot.[41] In human basophils, the presence of PKC-α, -β, and -δ has been described but little is known as to their function in these cells.[42] In a permeabilized RBL-2H3 cell model, reconstitution of a PKC-depleted cell with PKC-β or -δ isoforms was sufficient for reconstituting the secretory response of these cells.[16] Consistent with these findings, the overexpression of PKC-β in the RBL-2H3 cells resulted in an enhanced secretory response.[40] A negative regulatory role has been suggested for PKC-α and PKC-ε because these isozymes were found to inhibit the hydrolysis of phosphatidylinositol (PI) in the permeabilized and reconstituted cells.[39] However, in cells overexpressing these isozymes, negative regulation was observed on cPLA$_2$ activity rather

TABLE 12.2. PKC isozymes, concentrations, and function in FcεRI-activated RBL-2H3 cells.

	PKC-α	PKC-β	PKC-δ	PKC-ε	PKC-ζ
Serum deprived[37]	240 nM	30 nM	10 nM	20 nM	?
With serum[16]	85 nM	35 nM	60 nM	12.5 nM	?
Translocate to membrane with stimulation[16]	+	+	+	+	–
Associates with FcεRI β-chain[38]	–	–	+	–	?
Phosphorylates FcεRI γ-chain[38]	–	–	+	–	?
Inhibit PI hydrolysis[39]	+	–	–	+	–
Inhibit cPLA$_2$[40]	+	–	–	+	?
Induce c-Fos and c-Jun[37]	–	+	–	+	?
Enhance production of IL-2 and IL-6 mRNA[40]	–	+	–	–	?
Reconstitute secretion[16]	–	+	+	–	–

than on PI hydrolysis.[40] Although the target of negative regulation in these two studies differed, one is led to conclude that the function exerted by the activity of PKC-α and PKC-ε in the mast cell may be to negatively regulate one or more signaling pathways initiated by FcεRI activation of mast cells. What intracellular targets or cell function(s) may be regulated by PKC ζ remains to be determined, but one interesting observation is that phorbol myristate acetate (PMA) treatment followed by permeabilization and extensive washing of the RBL-2H3 cells allows depletion of all isozymes with the exception of ζ (P. Germano and J. Rivera, unpublished observation), suggesting that this isozyme may be physically trapped in the cell. Studies on the localization of this isozyme may shed some light as to its role in mast cell function.

Activation of PKC in the Mast Cell: Is Activation an Early Event?

Given that the focus of this chapter is on the role of PKC in early mast cell events, it seems reasonable to ask if the activation of PKC in mast cells is an early or late event. Using FcεRI aggregation in the RBL-2H3 cell as a paradigm, activation of PKC is classically thought to occur in the following manner.[44] (a) FcεRI aggregation allows the constitutively associated Lyn kinase to phosphorylate the receptor on tyrosine residues. (b) Tyrosine-phosphorylated FcεRI then binds the SH2 domains of Syk kinase, which may be subsequently activated by transphosphorylation or autophosphorylation or both. (c) Activated Syk kinase phosphorylates and activates PLC-γ. (d) Activated PLC-γ will hydrolyze the phosphatidylinositol lipid PIP$_2$, resulting in an increase in diacylglycerol (DAG) and inositol trisphosphate (IP$_3$). (e) IP$_3$ is then able to increase intracellular calcium by stimulating the release of calcium from intracellular stores. (f) This increase in DAG and intracellular calcium activates PKC isozymes. A sustained activation of PKC may be mediated by activation of PLD.

From this scenario, PKC is activated by the acute and reversible second messengers calcium and diacylglycerol. This suggests that activation of PKC activity is downstream of the initial and early events responsible for stimulus response. Nevertheless, several pieces of evidence suggest that activation of some PKC isozymes may actually precede or parallel the kinetics of the initial or early events best characterized by FcεRI phosphorylation. First, attempts to overexpress catalytically inactive forms of PKC in the RBL-2H3 cells or in other cells have been for the most part unsuccessful[40] suggesting the importance of constitutive PKC activity for cell viability. Second, Wolfe et al. reported that, following adhesion, RBL-2H3 cells display a constitutive activity of protein kinase C.[17] Third, prior studies suggest that a significant fraction of PKC-δ exists in the membrane in an active form and furthermore that receptor engagement causes a rapid translocation of a significant fraction of this isozyme within 5 to 15 s.[13] Fourth, a basal level of threonine phosphorylation of the FcεRI γ-chain has been described in the RBL-2H3 cells and in a variant of this line, and this phosphorylation is inhibited by PKC inhibitors.[38,45] Fifth, the kinetics of translocation described in a previous study by White and Metzger[36] suggested two distinct phases of PKC translocation occur in response to aggregation of the FcεRI. An early and transient phase occured within 15 to 30 s of receptor engagement, kinetics clearly suggestive of a role in early events. One must also consider that a growing list of effectors that do not require the activation of PLC-γ are capable of activating various PKC isozymes[4–6] (see Table 12.1), and the increased presence of these effectors may precede the activation of phospholipase C or D. It is likely that the constitutive PKC activity may be involved in more long-term cellular processes such as proliferation, multidrug resistance, apoptosis, and cellular transformation and differentiation.[46–48] However, we should seriously consider that redistribution of this constitutive activity in response to receptor engagement may occur as part of the earliest steps in mast cell activation and that PKC thus could participate in early events resulting from receptor engagement.

PKC and Receptor Phosphorylation

Phosphorylation of receptors by PKC activity has been described for diverse receptors in

TABLE 12.3. Receptor targets of PKC activity.

Angiotensin II receptor (type IA)[49]
Acetylcholine receptor (m1, muscarinic)[50]
β-Adrenergic receptor[51]
Cholecystokinin receptor (CCK)[52]
Complement receptor for C3d (CR2)[53]
Epidermal growth factor receptor[54]
FcεRI (γ subunit)[38]
Glutamate receptor (ionotropic NMDA) (metabotropic type 1α)[55,56]
Gamma-aminobutyric acid A (GABAA) receptor[57]
Glucagon-like peptide-1 receptor[58]
Glycine receptors (hippocampal)[59]
5-Hydroxytryptamine (type 1A)[60]
Insulin receptor[61]
c-Kit/stem cell factor receptor[62]
Natural cytotoxicity receptor (Nc-1.1)[63]
Prostacyclin receptor[64]
Receptor-like protein tyrosine phosphatase (RPTP)[65]
STa receptor (E. coli enterotoxin receptor)[66]

various cell types (Table 12.3). For those receptors in which a functional consequence of this phosphorylation has been studied, it appears that the modification of these receptors by PKC can result in either an increase or decrease in their ability to engage the signaling pathways that interpret their occupancy by a ligand. Early reports of PKC phosphorylation of receptors were provided by studies on the β-adrenergic receptor, which serves to activate adenylate cyclase in a variety of cells.[67] However, the consequence of this PKC-mediated phosphorylation was unclear because treatment of different cell types with phorbol esters led to increased or decreased cyclase responsiveness.[68,69] Nevertheless, three discernible effects could be seen after phorbol ester treatment: (a) an increase in the kinetics of activation (K_{act}) for β-agonists; (b) an increase in the maximal level of stimulation by β-agonists; and (c) a decrease in the G_i-mediated inhibition of forskolin stimulation.[70] Further studies using site-directed mutagenesis revealed that mutation of the site of PKC phosphorylation in the third intracellular loop eliminated the increase in K_{act} induced by phorbol ester treatment without affecting the maximal level of stimulation and the decrease in G_i-mediated inhibition. These results suggest that the phosphorylation of the PKC target site of this receptor mediates the increase in K_{act} and that the effect on maximal stimulation and G_i-mediated inhibition might be explained by an effect of PKC on G_i.

Receptors with intrinsic tyrosine activity can also be targets for PKC activity. Activation of PKC-α in CHO cells overexpressing this isozyme and transfected with varied members of the insulin receptor family demonstrated a dramatic reduction in the ability of these cells to activate PI-3 kinase.[71] Additionally, PKC-α was found to associate with the activated insulin receptor and could be tyrosine phosphorylated in response to insulin treatment of cells.[25] These experiments suggest a possible desensitization mechanism for the members of the insulin receptor family that is mediated by PKC-α. Table 12.3 lists receptors whose activity are known to be modified by the activation of PKC. In some cases it is still unclear as to whether the modification of receptor activity is a result of a direct or indirect phosphorylation of the receptor by PKC activity. However, in all cases activation of PKC results in either a gain or loss of receptor activity. Of particular interest, PKC activity may be important to the normal development of hematopoietic cells, melanoblasts, and germ cells by negatively regulating the Kit/stem cell factor receptor (Kit/SCF-R).[72] This is mediated by direct phosphorylation of serine-741 and -746,[61] which results in a weakened association with signaling proteins such as Grb2 and PI-3 kinase.[73] Inhibition of PKC activity increases SCF-induced tyrosine kinase activity and mitogenicity, but decreases motility.[73] As mast cells retain the Kit/SCF-R after differentiation, it is likely that the function of this receptor in mast cell growth is tightly regulated by PKC activity.

Phosphorylation of the FcϵRI by PKC Activity

The treatment of mast cells with phorbol 12-myristate 13-acetate (PMA) promotes gross morphological changes.[74] Mast cells that are normally rounded with a microvillus surface change to a flatter cell with large membrane folds; these changes mimic those observed when mast cells are stimulated via FcϵRI.[74,75] While it is likely that many of these changes occur as a consequence of PKC activation,

others may be mediated by activation of PMA-binding non-PKC proteins, such as n-chimaerins,[76] that can regulate cellular architecture by regulating the activity of small GTPases. One effect of PMA treatment of the mucosal mast cell analog RBL-2H3 is the induction of endocytosis of the FcϵRI. The kinetics of this response mimics that observed when receptors are stimulated by antigen.[75] However, the extent of PMA-mediated endocytosis was only half of that observed by antigen stimulation. These studies suggested the possibility that PKC activity might be part of the signals involved in the steps required for receptor recognition or in the machinery necessary for the endocytotic event. Prior studies had identified that both tyrosine and threonine residues on the FcϵRI γ-chain are phosphorylated in response to antigen stimulation of the RBL-2H3 cells. Further exploration, to define whether PMA might induce phosphorylation of FcϵRI, identified the threonine phosphorylation of the FcϵRI γ-chain as a consequence of PMA treatment.[38] In addition, PKC-δ was found to associate with the β-chain of the FcϵRI, and its activity appeared to be specifically directed toward the γ-chain both in vitro and in vivo. Inhibition of PKC activity by the PKC-specific inhibitor Ro 31-7549 resulted in complete inhibition of PMA-mediated endocytosis and a significant decrease in the kinetics of antigen-stimulated endocytosis with greater than 80% inhibition at 5 min of stimulation. These results demonstrate that the threonine phosphorylation of the FcϵRI γ-chain may play a role in marking the receptor for internalization.

Mapping of the Target Site of PKC Phosphorylation

A variety of methods can be employed to determine the site of protein phosphorylation. These include in vitro assays with purified substrates and mutagenic studies in which candidate residues are changed to residues that cannot be modified by phosphorylation. These methods have the disadvantage that the secondary or tertiary structure of a mutagenic protein in vitro or in vivo may not accurately reflect the wild-type structure. Alternatively, peptide

TABLE 12.4. Peptides used for in vitro kinase assay.[a]

FcεRI gamma chain: cytosolic region	Identifier
RLKIQVRKADIASREKS*DAVYTGLNTRNQETYETL*KHEKPPQ	Wild type
RLKIQVRKADIASREKS*DAVYAGLNARNQEAYEAL*KHEKPPQ	42T0
RLKIQVRKADIASREKS*DAVYTGLNARNQEAYEAL*KHEKPPQ	42T@1
RLKIQVRKADIASREKS*DAVYTGLNTRNQEAYEAL*KHEKPPQ	42T@1,2
RLKIQVRKADIASREKS*DAVYTGLNTRNQETYEAL*KHEKPPQ	42T@1,2,3
RLKIQVRKADIASREKS*DAVYAGLNARNQEAYETL*KHEKPPQ	42T@4

[a] Synthetic peptides (George Poy, NIDDK) were purified by preparative HPLC and the structure verified by amino acid analysis and mass spectrometry (Bill Lane, Harvard Microchemistry). Italicized region includes canonical tyrosines in the ITAM motif.

mapping of ^{32}P-labeled samples utilizing TLC[77] or HPLC[78] has been employed. These methods have the potential disadvantage of measuring only those phosphate groups that are incorporated within the experimental time frame. Finally, HPLC and mass spectrometry can determine sites of in vivo phosphorylation without using ^{32}P-labeled samples,[79] but these procedures are technically demanding.

FcεRI γ-Chain Phosphorylation

We sought to identify the in vitro site of threonine phosphorylation on the FcεRI γ-chain using recombinant PKC-δ and a set of γ-chain peptide homologs in which threonine residues were systematically replaced with alanine. This approach was chosen given the difficulty in obtaining sufficient labeling of threonine residues on the receptor with ^{32}P incorporation for two dimensional peptide mapping.[80] The cytosolic region of the FcεRI γ-chain is 42 amino acids long and contains 4 threonine residues. In addition to this wild-type peptide, we used peptides

in which either all the threonine residues were converted to alanine (42T0) or the threonines were limited to defined positions (42T@1, 42T@1,2, 42T@1,2,3, and 42T@4) corresponding to threonine positions in the wild-type peptide (Table 12.4). After incubation of PKC-δ with the individual peptides and the necessary cofactors including [^{32}P]-ATP, the peptides were isolated using reversed-phase HPLC. Collected peaks were then subjected to phosphoamino acid analysis using established methods.[81]

The results from this assay confirm previous reports,[38] which show that PKC-δ phosphorylates the FcεRI γ-chain wild-type peptide mainly on threonine residues (Table 12.5). There was a minor incorporation of phosphate to serine, but the ratio of threonine to serine phosphorylation was usually greater than 5 and never less than 2. When all threonines in the γ-chain were changed to alanine (42T0), the overall extent of phosphorylation was significantly diminished and phosphoamino acid analysis indicated that only serine residues were phosphorylated (Table 12.5). Gamma-chain peptides

TABLE 12.5. Phosphorylation of FcεRI γ-chain peptides by protein kinase Cδ.

Identifier	Radiolabel incorporation (Cerenkov counts from HPLC; relative measure normalized to 42T0)	Major residue phosphorylated (phosphoamino acid analysis)
Wild type	12+	Threonine
42T0	1+	Serine
42T@1	2+	Serine
42T@1,2	3+	Serine
42T@1,2,3	3+	Serine
42T@4	10+	Threonine

TABLE 12.6. Sequence comparison of FcεRI γ-chain with known protein kinase C-δ substrates.

FcεRI γ-chain, cytosolic region: RLKIQVRKADIASREKS*DAVYTGLNTRNQETYETL*KHEKPPQ		

protein kinase C-δ substrates		Reference
Pseudosubstrate	PTMNRRGA**I**KQAK	83
From peptide library	ARRKRKGS**F**FYGG	83
Elongation factor eEF-1α	RDMRQ**T**VAVGV	82

with progressively more threonine residues added (42T@1, 42T@1,2, and 42T@1,2,3) showed an increasing amount of threonine phosphorylation. However, the level remained far below the level of serine phosphorylation for these peptides. The FcεRI γ-chain variant that contained only one threonine residue in the position closest to the carboxy-terminus (42T@4) incorporated a similar amount of radiolabel compared to that seen for the wild-type peptide. In addition, as with the wild-type peptide, the amount of threonine phosphorylation was at least twofold greater than the amount of serine phosphorylation. We conclude from these data, using γ-chain peptide variants in an in vitro kinase assay, that PKC-δ phosphorylates the FcεRI γ-chain on the threonine residue closest to the carboxy terminus (T@4 = Thr-60).

To the best of our knowledge, this is the first time that a preferred site of threonine phosphorylation on the FcεRI γ-chain has been identified. However, the peptide region containing this site shows little correspondence to the sites on peptides that are thought to specifically interact with the catalytic domain of PKC-δ. Specifically, there is little correspondence to the pseudosubstrate peptide of PKC-δ, elongation factor eEF-1α,[82] or an optimized target sequence derived from an oriented peptide library[83] (Table 12.6).

Using an oriented peptide library, Nishikawa et al. found that PKC-δ selects for substrates with an arginine at p − 5, a basic residue at p − 2, a hydrophobic residue at p + 1, and a glycine at p + 4 (where residues are numbered relative to the phosphorylated residue).[83] Although eEF-1α meets all four of these criteria, the PKC-δ pseudosubstrate sequence meets only two criteria by containing a basic residue at p −

2 and a hydrophobic residue at p + 1. Of the FcεRI γ-chain threonine residues, only the threonine at the carboxy-terminus (Thr-60) matches any one of the criteria described by having a hydrophobic residue in the p + 1 location. An alternative interpretation can be found from the work of Lawrence and coworkers,[84] who have suggested that PKC isozymes lack stereospecificity and thus can bind substrates in either direction (i.e., C to N or N to C) in the substrate-binding cavity. If we apply Nishikawa's PKC-δ specificity in a C-to-N orientation, then only T4 (K at p − 2) or T1 (R at p − 5) satisfy at least one of the criteria. Finally, the Thr-60 (T4) site does exhibit some characteristics of general PKC substrates in that it falls into the general PKC substrate consensus sequence (S/T-X-K/R) described by Azzi et al.[85]

While we employed the entire cytosolic domain of the γ-chain peptide to more closely mimic the in vivo structure, interpretation of our studies (as well as the other specificity studies mentioned here) must be tempered by the fact that an in vitro system may not accurately reflect in vivo phosphorylation (see section on serine/threonine phosphorylation of PKC).

Implications of Serine/Threonine Phosphorylation of the FcεRI

The functional consequences of serine and threonine phosphorylation of the FcεRI are unclear. As briefly outlined in an earlier section, the inhibition of threonine phosphorylation of the FcεRI γ-chain decreases the rate of endocytosis of the receptor. However, the molecular mechanism for this effect is not defined, although from the preliminary experiments several scenarios are possible.

Early studies on endocytosis of the FcεRI demonstrated that aggregation of the receptor was a requirement for the initiation of this event.[86] With the discovery that the aggregation of receptors led to phosphorylation of receptors, it seemed reasonable to think that phosphorylation of the receptor might be a signal for endocytosis. Studies, in which mutant FcεRI receptors with deleted cytoplasmic domains were analyzed for the functional consequences of these mutations, failed to clearly define which of the cytoplasmic domains of the receptor is critical for endocytosis.[87] Nevertheless, some conclusions can be drawn from these studies because deletion of the carboxy-terminal (C-terminal) cytoplasmic domain of the β-chain significantly reduced the amount of receptor in intracellular vesicles, presumably through a reduction of the number of these receptors in coated pits. This suggests that these mutant receptors are either localized to the coated pits more slowly or cannot engage proteins in the coated pits that may serve to anchor them in these domains. Interestingly, later studies have suggested that both the src family tyrosine kinase Lyn[88] and the serine/threonine kinase PKC-δ[38] associate with the FcεRI via an interaction with the C-terminal domain of the β-chain. Therefore, if association with these kinases occurs at this domain of the receptor, one might extrapolate that this β-chain–C-terminal-deficient receptor is phosphorylation incompetent, suggesting that the threonine and/or tyrosine phosphorylation is necessary for movement to or engagement of proteins in the coated pits. To date, these studies along with those presented in the foregoing sections provide strong, but not definitive, evidence that phosphorylation of the receptor plays a role in the endocytotic event.

In molecular terms, how phosphorylation of the receptor subunits might serve as a signal for endocytosis is of interest because this may provide new areas of therapeutic promise. At the moment we can only speculate as to probable mechanisms. One attractive hypothesis, based in part on our results of the kinetic effects on endocytosis by inhibition of threonine phosphorylation of the FcεRI γ-chain and that the phosphorylated threonine of the γ-chain is lo-

calized within the ITAM, is that phosphorylation on threonine residues of the γ-chain may regulate the association of the receptor with interacting proteins like Syk kinase. Receptors where threonine phosphorylation is inhibited may lack these interactions or conversely interact with greater affinity, thus marking these receptors for longevity on the plasma membrane (an event potentially interpreted by a delay in endocytosis). Studies to determine if this scenario is possible are currently under way.

Molecular Associations of PKC In Vivo

PKC Receptors, Inhibitors, and Other Proteins

A growing list of proteins have been documented to associate with PKC (Table 12.7). These proteins are thought to be important in localizing PKC and regulating its activity. The receptors for activated PKC (RACKS) are the

TABLE 12.7. Proteins known to associate with PKC.

Protein	PKC isozyme(s)	References
Annexin I, II, VI	α	(89)
F-Actin	βII, ε	(90,91)
AKAP 79 (A kinase-anchoring protein)	α, βII	(92)
c-Abl tyrosine kinase	βI	(93)
Bruton's tyrosine kinase (BTK)	α, βI, βII, ε, ζ	(94)
FcεRI β-chain	δ	(38)
HIV nef protein	θ	(95)
Ina D protein	eye-PKC	(96)
p53/56 Lyn kinase	δ	(97)
MARCKS	α	(98)
Phosphatidylinositol 3-kinase (PI-3K)	δ	(99)
PICK-1	α	(100)
PKCI-1	β	(101)
Ras	ND	(102)
RACKS	cPKC	(103)
SRBC (sdr-related gene product that binds PKC)	δ	(104)
p60 Src kinase	δ	(97)
Talin	ND	(89)
Vinculin	ND	(89)
14-3-3 (tau)	θ	(105)

best characterized of these proteins whose sole known function to date is to bind activated PKC or other proteins with structural homology to PKCs.[103,106] The RACKS are comprised of a family of proteins of approximately 30 kDa. Of particular interest is that the protein encoded by the cDNA that has been cloned has high homology to the β-subunit of the heterotrimeric GTP-binding proteins and has stretches of homology with PKC and annexin I.[103] The homology of RACK with PKC is predominantly localized to the C2 region of PKC, a region on PKC that is thought to contain at least part of the binding site for RACK. Peptides of the RACK-binding domain were found to be effective inhibitors of PKC binding to RACKS. Other proteins that contain C2 regions may also interact with RACK proteins; one example is the binding of PLC-γ to the RACKS. This binding appears to be specific and requires the phosphorylation of PLC-γ, but unlike the binding of PKC to RACKS, the binding of PLC-γ does not require phospholipids and calcium.[106] Nevertheless, this in vitro finding suggests that PLC-γ may compete in vivo for RACK-binding sites and thus may serve to regulate PKC responses in vivo. The role of RACK proteins in PKC function is not yet clear. However, in vitro studies suggest that RACKs are not a good substrate for PKC, but it is possible that their presence may provide a stimulation of PKC activity.[107] Whether RACKS serve to regulate PKC activity in the cell remains to date unanswered. Numerous associations of PKC with cellular proteins have been described (Table 12.7); whether one might categorize any of these interactions as receptor–PKC interactions may ultimately depend on what we learn as to the role of these interacting proteins in the cell.

The existence of endogenous proteins that inhibit PKC activity in the cell has long been recognized.[108] However, because the functional role of these proteins in regulating PKC activity in the cell is not clear, we refer to these proteins as inhibitors, using this term as an operational definition. The endogenous inhibitors include a growing family of proteins termed PKC inhibitors (PKCI) that are characterized by a cysteine-rich zinc finger domain similar to that found in PKC.[100] For PKCI-1, a high degree of homology exists across species,[109,110] and southern blot analysis reveals the potential existence of multiple genes or psuedogenes that may be highly related. However, at present we know little of the mode of action of PKCI-1 inhibitory effect on PKC, although recent studies suggest that this PKC inhibitor could inhibit the norepinephrine-induced modulation of calcium currents in primary cultures of chicken neurons.[111]

Another family of proteins reported to inhibit the calcium phospholipid-dependent PKC isozymes are the members of the 14-3-3 family,[112] although one report suggests a stimulatory effect of these proteins on PKC.[113] The function of this family of proteins is not yet clear; however, it has been proposed that these proteins regulate enzymes involved in catecholamine and serotonin biosynthesis.[112] The binding of 14-3-3 proteins to many other proteins of the various signal transduction pathways and their effects on the activity of these pathways has been reported.[114] This brings into question the specificity of the effects of 14-3-3 members on PKC activity and may suggest a more general mechanism of modulating the activity of signaling proteins. One possible role is that of a chaperone protein that stabilizes the cytosolic conformation of signaling proteins, like PKC, as proposed by several groups.[114,115]

Other endogenous inhibitors of PKC activity include annexin V, which appears to be a potent and highly specific inhibitor with a half-maximal inhibition at 0.4 μM.[116] The family of annexins seem to be closely associated endogenously with PKC family members because annexin I has been characterized as highly homologous to a RACK and has some homology to the 14-3-3 family of proteins. Interestingly, annexin V is a potent inhibitor of PKC while annexin I, which is highly homologous to RACKS, stimulates its activity. This differential regulation of PKC by two proteins that have substantial homology may be of great importance in advancing our understanding as to how specific interactions may be important in the regulation of protein activity in vivo. Numerous other proteins have been suggested as endogenous inhibitors of PKC activity, although these systems have not been fully characterized,[112] nevertheless, it is probably safe to say that we have barely scratched the surface in the discovery of proteins that can both positively and negatively regulate the endogenous activity of PKC.

Many other proteins have been documented to interact with PKC (see Table 12.7). Of particular interest is the growing list of tyrosine kinases and substrates that are capable of interacting with the PKC family of kinases. For example, Bruton's tyrosine kinase (BTK) has been reported to interact with several PKC isozymes in mast cells.[94] It is likely that these associations have functional consequences in the responses mediated by these kinases. A glimpse of this is provided by the finding that the in vitro phosphorylation of BTK by PKC-β reduces the activity of BTK as measured by its ability to autophosphorylate.[94] Our previous studies[13] and those of others[24,28,29] have demonstrated that the tyrosine phosphorylation of PKC-δ may serve to regulate its kinase activity. Thus these results and the growing list of tyrosine kinases known to interact with PKC isozymes promote the notion that these interactions may be important to the regulation of in vivo kinase activity.

Association of PKC with Src Family Kinases in Mast Cells

Our prior observation that PKC-δ is a substrate of tyrosine kinase activity[13] has prompted us to further investigate the tyrosine kinase involved in this event. Several observations made in the initial studies served as a guide toward defining which tyrosine kinase might phosphorylate PKC-δ in response to aggregation of FcεRI. First, engagement of the receptor was required for detection of tyrosine phosphorylation of PKC-δ; second, this event occured in the membranes of cells and not in the cytosol; third, the phosphorylation of PKC-δ could be stimulated by pharmacological agents that caused PKC-δ translocation to membranes; and fourth, we demonstrated that tyrosine phosphorylation was not required for PKC-δ translocation. Collectively, these observations

provide strong evidence that the responsible tyrosine kinase is membrane localized and requires the presence of its substrate (PKC-δ) on the membrane to mediate its effect. One likely tyrosine kinase that fits the criteria mentioned is the p53/56 Lyn kinase (Lyn), which is predominantly membrane localized in the RBL-2H3 mast cell line used in these studies. Another possible candidate is p60 Src (Src), which can be found localized in membrane-associated focal adhesion plaques.[117]

Analysis of these possible candidates revealed the following.[97] First, both Lyn and Src were found to associate with PKC-δ, but Lyn association with PKC-δ required the engagement of the FcεRI while that of Src with PKC-δ was constitutive; second, immunoprecipitation of Src or Lyn from activated cells under conditions that maintained the protein–protein interactions revealed a tyrosine-phosphorylated species of PKC-δ associating only with Lyn following receptor engagement; third, receptor engagement did not significantly enhance the association of PKC-δ with Src but had a dramatic effect on its association with Lyn; fourth, we could reconstitute both the tyrosine phosphorylation and association of PKC-δ with Lyn in vitro; and fifth, mutation of the site of tyrosine phosphorylation on PKC-δ[27] resulted in the loss of interaction between PKC-δ and Lyn. These results provide evidence that following receptor engagement a fraction of PKC-δ becomes a substrate for Lyn and furthermore demonstrates that the phosphorylation of PKC-δ is an important step in its association with Lyn. The latter finding is reinforced by the fact that the kinetics of PKC-δ dephosphorylation correlated with the kinetics of disassociation of PKC-δ and Lyn.

One obvious question is this: What are the functional implications of the interactions between tyrosine and serine/threonine kinases? To date we have no clear answer; however, in preliminary experiments using a PKC-δ overexpressing cell line we have observed that the increased interaction of PKC-δ with Lyn or Src results in a dramatic increase of enzyme activity. In contrast, when FcεRI is engaged, a dramatic reduction in Src activity and a small reduction in Lyn activity are observed. Interestingly, our previous results suggested that the tyrosine phosphorylation of PKC-δ might serve to regulate its substrate selectivity because the enzyme was not inactivated by this event but was found incapable of recognizing its endogenous substrate, the FcεRI γ-chain.[38] From these experiments it seems evident that the cross talk of these kinases may play an important role in the regulation of their activities and in their substrate selection in vivo. It remains to be defined what role these events might play in the overall effector function of mast cells.

Protein Kinase C: A Suitable Target for Therapeutic Intervention?

As much as we who work in the field of signal transduction would like to justify our efforts based on their therapeutic potential, to date no substantial breakthroughs have resulted from the intensive efforts placed in defining receptor-activated pathways. Nevertheless, in avoidance of promoting ourselves as purveyors of doom and gloom, we can now point to studies in which a specific inhibitor of PKC-β activity has been used successfully in an animal model for the treatment of diabetes.[118] To the best of our knowledge this compound is now in clinical trials with many anxiously awaiting the results of these studies. Furthermore, novel parent compounds whose modes of action appear to be aimed at specific components of the signaling pathways, such as Syk kinase, are now beginning to be reported,[119] although the specificity and efficacy of these compounds in humans are far from being determined.

One might ask why are PKC isozymes or early and late events mediated by their activity reasonable targets for therapeutic intervention. It would be difficult to argue that targeting of PKC isozymes provides any significant advantage over targeting of other kinases. Nevertheless, in mast cells more than in other cells the targeting of PKC isozymes may provide a dis-

tinct advantage in the intervention of allergic responses. A relatively unique property of the mast cell is its apparent requirement for PKC activity in the process of degranulation.[31] Studies in which PKC activity is completely inhibited by PKC-specific pharmacological agents show that tyrosine phosphorylation proceeds without any effect although this treatment of the cells abrogates the secretory response.[38,119,120] This suggests that the development of isozyme-specific PKC inhibitors that ultimately target the secretory process may be a reasonable approach toward development of potent drugs that would inhibit responses regardless of the stimuli.

Conclusions

In summary, a persistent theme evolving from the described studies is that PKC isozymes play an important role in regulation of early and late mast cell signals. This concept is reflected in the studies on the phosphorylation of the FcεRI γ-chain by PKC and its role in endocytosis of the receptor. It is further reinforced by studies on the interaction of PKC isozymes and tyrosine kinases in which an increased interaction results in the regulation of tyrosine kinase activity. For the moment the only PKC isozyme that is a consistent positive effector is the β-isoform, whose role in regulating secretion and gene expression is well established. Thus, we might be led to believe that PKC-β is the major player in the induction of mast cell responses. It is likely that our current perception is limited by the paucity of knowledge, and future efforts should be focused toward defining whether PKC is the pivotal link in mast cell degranulation.

References

1. Dekker LV, Parker PJ. Protein kinase C—a question of specificity. Trends Biochem Sci 1994;19:73–77.
2. Nishizuka Y. Protein kinase C and lipid signalling for sustained cellular responses. FASEB J 1995;9:484–496.

3. Haller H, Lindschau C, Luft FC. Role of protein kinase C in intracellular signaling. Ann NY Acad Sci 1994;733:313–324.

4. Toker A, Meyer M, Reddy KK, et al. Activation of protein kinase C family members by the novel polyphosphoinositides PtdIns-3,4-P_2 and PtdIns-3,4,5-P_3. J Biol Chem 1994;269:32358–32367.

5. Nakanishi H, Brewer KA, Exton JH. Activation of the ζ isozyme of protein kinase C by phosphatidylinositol 3,4,5-trisphosphate. J Biol Chem 1993;268:13–16.

6. Nishizuka Y. Intracellular signaling by hydrolysis of phospholipids and activation of protein kinase C. Science 1992;258:607–614.

7. Sossin WS, Schwartz JH. Ca^{2+}-independent protein kinase C's contain an amino-terminal domain similar to the C2 consensus sequence. Trends Biochem Sci 1993;18:207–208.

8. House C, Kemp BE. Protein kinase C contains a pseudosubstrate prototype in its regulatory domain. Science 1987;238:1726–1728.

9. Newton AC. Protein kinase C: structure, function, and regulation. J Biol Chem 1995;270: 28495–28498.

10. Wang SM, Kazanietz MG, Blumberg PM, et al. Molecular modeling and site-directed mutagenesis studies of a phorbol ester-binding site in protein kinase C. J Med Chem 1996;39:2541–2553.

11. Zhang G, Kazanietz MG, Blumberg PM, et al. Crystal structure of the Cys2 activator-binding domain of protein kinase Cδ in complex with phorbol ester. Cell 1995;81:917–924.

12. Chakravarthy BR, Whitfield JF, Durkin JP. Inactive membrane protein kinase Cs: a possible target for receptor signalling. Biochem J 1994;304:809–816.

13. Haleem-Smith H, Chang EY, Szallasi Z, et al. Tyrosine phosphorylation of protein kinase C-δ in response to the activation of the high affinity receptor for immunoglobulin E modifies its substrate recognition. Proc Natl Acad Sci USA 1995;92:9112–9116.

14. Oliver AR, Kiley SC, Pears C, et al. The molecular biology of the nervous system-specific proteins. Biochem Soc Trans 1992;20: 603–607.

15. Ways DK, Cook PP, Webster C, et al. Effects of phorbol ester on protein kinase C ζ. J Biol Chem 1992;267:4799–4805.

16. Ozawa K, Szallasi Z, Kazanietz MG, et al. Ca^{2+}-dependent and Ca^{2+}-independent isozymes of protein kinase C mediate exocytosis in antigen-stimulated rat basophilic RBL-2H3 cells. J Biol Chem 1993;268:1749–1756.

17. Wolfe PC, Chang EY, Rivera J, et al. Differential effects of the protein kinase C activator phorbol 12-myristate 13-acetate on calcium responses and secretion in adherent and suspended RBL-2H3 mucosal mast cells. J Biol Chem 1996;271:6658–6665.

18. Quest AFG, Bell RM. The molecular mechanism of protein kinase C regulation by lipids. In: Kuo JF (ed) Protein Kinase C. New York: Oxford University Press, 1994:64–95.

19. Flint AJ, Paladini RD, Koshland DE Jr. Autophosphorylation of protein kinase C at three separated regions of its primary sequence. Science 1990;249:408–411.

20. Tsutakawa SE, Medzihradszky KF, Flint AJ, et al. Determination of in vivo phosphorylation sites in protein kinase C. J Biol Chem 1995;270:26807–26812.

21. Keranen LM, Dutil EM, Newton AC. Protein kinase C is regulated in vivo by three functionally distinct phosphorylations. Curr Biol 1995;5:1394–1403.

22. Stempka L, Girod A, Muller H-J, et al. Phosphorylation of protein kinase Cδ (PKCδ) at threonine 505 is not a prerequisite for enzymatic activity. Expression of rat PKCδ and an alanine 505 mutant in bacteria in a functional form. J Biol Chem 1997;272:6805–6811.

23. Bornancin F, Parker PJ. Phosphorylation of threonine 638 critically controls the dephosphorylation and inactivation of protein kinase Cα. Curr Biol 1996;6:1114–1123.

24. Denning MF, Dlugosz AA, Howett MK, et al. Expression of an oncogenic ras[HA] gene in murine keratinocytes induces tyrosine phosphorylation and reduced activity of protein kinase C δ. J Biol Chem 1993;268:26079–26081.

25. Liu F, Roth RA. Insulin stimulated tyrosine phosphorylation of protein kinase Cα: evidence for direct interaction of the insulin receptor and protein kinase Cα in cells. Biochem Biophys Res Commun 1994;200:1570–1577.

26. Gschwendt M, Kielbassa K, Kittstein K, et al. Tyrosine phosphorylation and stimulation of protein kinase C delta from porcine spleen by src in vitro. Dependence on the activated state of protein kinase C delta. FEBS Lett 1994;347:85–89.

27. Szallasi Z, Denning MF, Chang EY, et al. Development of a rapid approach to identification of tyrosine phosphorylation sites: application to PKC-δ phosphorylated upon activation of the

high affinity receptor for IgE in rat basophilic leukemia cells. Biochem Biophys Res Commun 1995;214:888–894.

28. Li W, Mischak H, Yu JC, et al. Tyrosine phosphorylation of protein kinase C-δ in response to its activation. J Biol Chem 1994;269:2349–2352.

29. Li W, Yu JC, Michieli P, et al. Stimulation of the platelet-derived growth factor β receptor signaling pathway activates protein kinase C-δ. Mol Cell Biol 1994;14:6727–6735.

30. Soltoff SP, Toker A. Carbachol, substance P, and phorbol ester promote the tyrosine phosphorylation of protein kinase C δ in salivary gland epithelial cells. J Biol Chem 1995;270:13490–13495.

31. Rivera J, Beaven MA. Regulation of secretion from secretory cells by protein kinase C. In: Parker PJ, Dekker DL (eds) Protein Kinase C. Austin, TX: RG Landes, 1997:131–164.

32. Schleimer RP, Gillespie E, Lichtenstein LM. Release of histamine from human leukocytes stimulated with tumor-promoting phorbol diesters. I. Characterization of the response. J Immunol 1980;126:570–575.

33. Heiman AS, Crews FT. Characterization of the effects of phorbol esters on rat mast cell secretion. J Immunol 1985;134:548–555.

34. Blumberg PM, Konig B, Sharkey NA, et al. Analysis of membrane and cytosolic phorbol ester receptors. IARC Sci Publ 1984;56:139–156.

35. White JR, Pluznik DH, Ishizaka K, et al. Antigen-induced increase in protein kinase C activity in plasma membrane of mast cells. Proc Natl Acad Sci USA 1985;82:8193–8197.

36. White KN, Metzger H. Translocation of protein kinase C in rat basophilic leukemic cells induced by phorbol ester or by aggregation of IgE receptors. J Immunol 1988;141:942–947.

37. Razin E, Szallasi Z, Kazanietz MG, et al. Protein kinase C-β and C-ε link the mast cell high-affinity receptor for IgE to the expression of c-*fos* and c-*jun*. Proc Natl Acad Sci USA 1994;91:7722–7726.

38. Germano P, Gomez J, Kazanietz MG, et al. Phosphorylation of the γ chain of the high affinity receptor for immunoglobulin E by receptor-associated protein kinase C-δ. J Biol Chem 1994;269:23102–23107.

39. Ozawa K, Yamada K, Kazanietz MG, et al. Different isozymes of protein kinase C mediate feedback inhibition of phospholipase C and stimulatory signals for exocytosis in rat baso-

philic RBL-2H3 cells. J Biol Chem 1993;268:2280–2283.

40. Chang E-Y, Szallasi Z, Acs P, et al. Functional effects of overexpression of protein kinase C-α, -β, -δ, -ε, and -η in the mast cell line RBL-2H3. J Immunol 1997;159:2624–2632.

41. Kim H-M, Hirota S, Chung H-T, et al. Differential expression of protein kinase C genes in cultured mast cells derived from normal and mast-cell-deficient mice and mast cell lines. Int Arch Allergy Immunol 1994;105:258–263.

42. Lewin I, Jacob-Hirsch J, Zang ZC, et al. Aggregation of the FcεRI in mast cells induces the synthesis of Fos-interacting protein and increases its DNA binding activity: the dependence on protein kinase C-β. J Biol Chem 1996;271:1514–1519.

43. MacGlashan DW Jr. Signal transduction issues in studies of human basophils and mast cells. In: Hamawy MM (ed) IgE receptor (FcεRI) function in mast cells and basophils. Austin, TX: RG Landes, 1997:265–286.

44. Scharenberg AM, Kinet J-P. Early events in mast cell signal transduction. Chem Immunol 1995;61:72–87.

45. Bingham BR, Monk PN, Helm BA, et al. Defective protein phosphorylation and Ca^{2+} mobilization in a low secreting variant of the rat basophilic leukemia cell line. J Biol Chem 1994;269:19300–19306.

46. Goodnight J, Mischak H, Mushinski JF. Selective involvement of protein kinase C isozymes in differentiation and neoplastic transformation. In: Van de Woude GF, Klein G (eds) Advances in Cancer Research. San Diego: Academic Press, 1994:159–209.

47. Blobe GC, Obeid LM, Hannun YA. Regulation of protein kinase C and role in cancer biology. Cancer Metastasis Rev 1994;13:411–431.

48. McCabe MJ Jr, Orrenius S. Protein kinase C: a key enzyme determining cell fate and apoptosis? In: Kuo JF (ed) Protein Kinase C. New York: Oxford University Press, 1994:290–304.

49. Oppermann M, Freedman NJ, Alexander RW, et al. Phosphorylation of the type 1A angiotensin II receptor by G protein-coupled receptor kinases and protein kinase C. J Biol Chem 1996;271:13266–13272.

50. Haga K, Kameyama K, Haga T, et al. Phosphorylation of human m1 muscarinic acetylcholine receptors by G protein-coupled receptor kinase 2 and protein kinase C. J Biol Chem 1996;271:2776–2782.

51. Yuan N, Friedman J, Whaley BS. cAMP-dependent protein kinase and protein kinase C consensus site mutations of the beta-adrenergic receptor. Effect on desensitization and stimulation of adenylcyclase. J Biol Chem 1994;269: 23032–23038.

52. Gates LK, Ulrich CD, Miller LJ. Multiple kinases phosphorylate the pancreatic cholecystokinin receptor in an agonist-dependent manner. Am J Physiol 1993;264:G840–G847.

53. Aquino A, Lisi A, Pozzi D, et al. EBV membrane receptor (CR2) is phosphorylated by protein kinase C (PKC) in the early stages of virus entry into lymphoblastoid cell line (Raji). Biochem Biophys Res Commun 1993;196:794–802.

54. Chen P, Xie H, Wells A. Mitogenic signaling from the EGF receptor is attenuated by a phospholipase C-gamma/protein kinase C feedback mechanism. Mol Biol Cell 1996;7:871–881.

55. Thomas KL, Davis S, Laroche S, et al. Regulation of the expression of NR1 NMDA glutamate receptor subunits during hippocampal LTP. Neuroreport 1994;6:119–123.

56. Alaluf S, Mulvihill ER, McIlhinney RA. Rapid agonist mediated phosphorylation of the metabotropic glutamate receptor 1 alpha by protein kinase C in permanently transfected BHK cells. FEBS Lett 1995;367:301–305.

57. Krishek BJ, Xie X, Blackstone C, et al. Regulation of GABAA receptor function by protein kinase C phosphorylation. Neuron 1994;12: 1081–1095.

58. Widmann C, Dolci W, Thorens B. Heterologous desensitization of the glucagon-like peptide-1 receptor by phorbol esters requires phosphorylation of the cytoplasmic tail at four different sites. J Biol Chem 1996;271:19957–19963.

59. Vaello ML, Ruiz-Gomez A, Lerma J, et al. Modulation of inhibitory glycine receptors by phosphorylation by protein kinase C and cAMP-dependent protein kinase. J Biol Chem 1994;269:2002–2008.

60. Raymond J. Protein kinase C induces phosphorylation and desensitization of the human 5-HT1A receptor. J Biol Chem 1991;266: 14747–14753.

61. Pillay TS, Xiao S, Olefsky JM. Glucose-induced phosphorylation of the insulin receptor. Functional effects and characterization of phosphorylation sites. J Clin Invest 1996;97:613–620.

62. Blume-Jensen P, Wernstedt C, Heldin CH, et al. Identification of the major phosphorylation

sites for protein kinase C in kit/stem cell factor receptor in vitro and in intact cells. J Biol Chem 1995;270:14192–14200.

63. Holmgreen SP, Wang X, Clarke GR, et al. Phosphorylation of the NC-1.1 receptor and regulation of natural cytotoxicity by protein kinase C and cyclic GMP-dependent protein kinase. J Immunol 1997;158:2035–2041.

64. Smyth EM, Nestor PV, FitzGerald GA. Agonist-dependent phosphorylation of an epitope-tagged human prostacyclin receptor. J Biol Chem 1996;271:33698–33704.

65. Tracy S, van der Geer P, Hunter T. The receptor-like protein-tyrosine phosphatase, RPTP alpha, is phosphorylated by protein kinase C on two serines close to the inner face of the plasma membrane. J Biol Chem 1995;270:10587–10594.

66. Crane JK, Shanks KL. Phosphorylation and activation of the intestinal guanylyl cyclase receptor for *Escherichia coli* heat-stable toxin by protein kinase C. Mol Cell Biochem 1996;165: 111–120.

67. Kelleher DJ, Pressin JE, Ruoho AE, et al. Phorbol ester induces desensitization of adenylate cyclase and phosphorylation of the beta-adrenergic receptor in turkey erythrocytes. Proc Natl Acad Sci USA 1984;81:4316–4320.

68. Katada T, Gilman AG, Watanabe Y, et al. Protein kinase C phosphorylates the inhibitory guanine-nucleotide-binding regulatory component and apparently suppresses its function in hormonal inhibition of adenylate cyclase. Eur J Biochem 1985;151:431–437.

69. Dixon BS, Breckon R, Burke C, et al. Phorbol esters inhibit adenylate cyclase activity in cultured collecting tubular cells. Am J Physiol 1988;254:C183–C191.

70. Johnson JA, Clark RB, Friedman J, et al. Identification of a specific domain in the beta-adrenergic receptor required for phorbol ester-induced inhibition of catecholamine-stimulated adenylyl cyclase. Mol Pharmacol 1990;38:289–293.

71. Danielsen AG, Liu F, Hosomi Y, et al. Activation of protein kinase Cα inhibits signaling by members of the insulin receptor family. J Biol Chem 1995;270:21600–21605.

72. Blume-Jensen P, Siegbahn A, Stabel S, et al. Increased kit/scf receptor induced mitogenicity but abolished cell motility after inhibition of protein kinase C. EMBO J 1993;12:4199–4209.

73. Blume-Jensen P, Ronnstrand L, Gout I, et al. Modulation of kit/stem cell factor receptor-induced signaling by protein kinase C. J Biol Chem 1994;269:21793–21802.

74. Pfeiffer JR, Seagrave JC, Davis BH, et al. Membrane and cytoskeletal changes associated with IgE-mediated serotonin release from rat basophilic leukemia cells. J Cell Biol 1985;101:2145–2155.

75. Ra C, Furuichi K, Rivera J, et al. Internalization of IgE receptors on rat basophilic leukemic cells by phorbol esters. Comparison with endocytosis induced by receptor aggregation. Eur J Immunol 1989;19:1771–1777.

76. Areces LB, Kazanietz MG, Blumberg PM. Close similarity of baculovirus-expressed n-chimaerin and protein kinase Cα as phorbol ester receptors. J Biol Chem 1994;269:19553–19558.

77. Boyle WJ, van der Geer P, Hunter T. Phosphopeptide mapping and phosphoamino acid analysis by two-dimensional separation on thin-layer cellulose plates. Methods Enzymol 1991;201:110–149.

78. Juhl H, Soderling TR. Peptide mapping and purification of phosphopeptides using high-performance liquid chromatography. Methods Enzymol 1983;99:37–42.

79. Hemling ME, Roberts GD, Johnson W, et al. Analysis of proteins and glycoproteins at the picomole level by on-line coupling of microbore high-performance liquid chromatography with flow fast atom bombardment and electrospray mass spectrometry: a comparative evaluation. Biomed Environ Mass Spectrom 1990;19:677–691.

80. Pribluda VS, Pribluda C, Metzger H. Biochemical evidence that the phosphorylated tyrosines, serines, and threonines on the aggregated high affinity receptor for IgE are in the immune receptor tyrosine-based activation motifs. J Biol Chem 1997;272:11185–11192.

81. Paolini R, Jouvin M-H, Kinet J-P. Phosphorylation and dephosphorylation of the high-affinity receptor for immunoglobulin E immediately after receptor engagement and disengagement. Nature (Lond) 1991;353:855–858.

82. Kielbassa K, Muller H-J, Meyer HE, et al. Protein kinase Cδ-specific phosphorylation of the elongation factor eEF-1α and an eEF-1α peptide at threonine 431. J Biol Chem 1995;270:6156–6162.

83. Nishikawa K, Toker A, Johannes F-J, et al. Determination of the specific substrate sequence motifs of protein kinase C isozymes. J Biol Chem 1997;272:952–960.

84. Kwon Y-G, Mendelow M, Lawrence DS. The active site substrate specificity of protein kinase C. J Biol Chem 1994;269:4839–4844.

85. Azzi A, Boscoboinik D, Hensey C. The protein kinase C family. Eur J Biochem 1992;208:547–557.

86. Isersky C, Rivera J, Segal DM, et al. The fate of IgE bound to rat basophilic leukemia cells. II. Endocytosis of IgE oligomers and effect on receptor turnover. J Immunol 1983;131:388–396.

87. Mao S-Y, Pfeiffer JR, Oliver JM, et al. Effects of subunit mutation on the localization to coated pits and internalization of cross-linked IgE-receptor complexes. J Immunol 1993;151:2760–2774.

88. Jouvin MH, Adamczewski M, Numerof R, et al. Differential control of the tyrosine kinases Lyn and Syk by the two signaling chains of the high affinity immunoglobulin E receptor. J Biol Chem 1994;269:5918–5925.

89. Hyatt SL, Liao L, Chapline C, et al. Identification and characterization of alpha-protein kinase C binding proteins in normal and transformed REF52 cells. Biochemistry 1994;33:1223–1228.

90. Blobe GC, Stribling DS, Fabbro D, et al. Protein kinase C beta II specifically binds and is activated by F-actin. J Biol Chem 1996;271:15823–15830.

91. Prekeris R, Mayhew MW, Cooper JB, et al. Identification and localization of an actin-binding motif that is unique to the epsilon isoform of protein kinase C and participates in the regulation of synaptic function. J Cell Biol 1996;132:77–90.

92. Klauck TM, Faux MC, Labudda K, et al. Coordination of three signaling enzymes by AKAP79, a mammalian scaffold protein. Science 1996;271:1589–1592.

93. Stribling S, Feng X, Blobe G, et al. Specific interactions of the protein kinase C β isozymes. Mol Biol Cell 1996;7:522a (abstract 3037).

94. Yao L, Kawakami Y, Kawakami T. Pleckstrin homology domain of Bruton tyrosine kinase interacts with protein kinase C. Proc Natl Acad Sci USA 1994;91:9175–9179.

95. Smith BL, Krushelnycky BW, Mochly-Rosen D, et al. The HIV nef protein associates with protein kinase C theta. J Biol Chem 1996;271:16753–16757.

96. Huber A, Sander P, Paulsen R. Phosphorylation of the Ina D gene product, a photoreceptor

membrane protein required for the recovery of visual excitation. J Biol Chem 1996;271:11710–11717.

97. Song JS, Swann PG, Szallasi Z, et al. Tyrosine phosphorylation-dependent and -independent associations of protein kinase C-δ with Src family kinases in the RBL-2H3 mast cell line: regulation of Src family kinase activity by protein kinase C-δ. Oncogene 1998;16:3357–3368.

98. Susannah L, Hyatt LL, Aderen A, et al. Correlation between protein kinase C binding proteins and substrates. Cell Growth Differ 1995;5:495–502.

99. Ettinger SL, Lauener RW, Duronio V, et al. Protein kinase C δ specifically associates with phosphatidylinositol 3-kinase following cytokine stimulation. J Biol Chem 1996;271:14514–14518.

100. Dong L, Chapline C, Mousseau B, et al. 35H, a sequence isolated as a protein kinase C binding protein, is a novel member of the adducin family. J Biol Chem 1995;270:25534–25540.

101. Lima CD, Klein MG, Weinstein IB, et al. Three-dimensional structure of human protein kinase C interacting protein 1, a member of the HIT family of proteins. Proc Natl Acad Sci USA 1996;93:5357–5362.

102. Marshall MS. Ras target proteins in eukaryotic cells. FASEB J 1995;9:1311–1318.

103. Ron D, Chen C-H, Caldwell J, et al. Cloning of an intracellular receptor for protein kinase C: a homolog of the β subunit of G proteins. Proc Natl Acad Sci USA 1994;91:839–843.

104. Izumi Y, Hirai S-I, Tamai Y, et al. A protein kinase Cδ-binding protein SRBC whose expression is induced by serum starvation. J Biol Chem 1997;272:7381–7389.

105. Meller N, Liu YC, Collins TL, et al. Direct interaction between protein kinase C theta (PKC theta) and 14-3-3 tau in T cells: 14-3-3 overexpression results in inhibition of PKC theta translocation and function. Mol Cell Biol 1996;16:5782–5791.

106. Disatnik M-H, Hernandez-Sotomayor SMT, Jones G, et al. Phospholipase C-γ1 binding to intracellular receptors for activated protein kinase C. Proc Natl Acad Sci USA 1994;91:559–563.

107. Ron D, Mochly-Rosen D. An autoregulatory region in protein kinase C: the psuedo-anchoring site. Proc Natl Acad Sci USA 1995;92:492–496.

108. Melner MH. Physiological inhibitors of protein kinase C. Biochem Pharmacol 1996;51:869–877.

109. Melner MH, Searles RP, Spindel ER, et al. Cloning and expression of an inhibitor of protein kinase C in humans. Endocrinology (Suppl) 1993;132:538.

110. Melner MH, Searles RP, Spindel ER, et al. Cloning of an inhibitor of protein kinase C in rats: high levels of expression in ovaries and testis. Biol Reprod (Suppl) 1993;48:189.

111. Rane SG, Walsh MP, McDonald JR, et al. Specific inhibitors of protien kinase C block transmitter induced modulation of sensory neuron calcium current. Neuron 1989;239–245.

112. Aitken A, Collinge DB, van Heusden BPH, et al. 14-3-3 proteins: a highly conserved, widespread family of eukaryotic proteins. Trends Biochem Sci 1992;17:498–501.

113. Tanji M, Horwitz R, Rosenfeld G, et al. Activation of protein kinase C by purified bovine brain 14-3-3: comparison with tyrosine hydroxylase activation. J Neurochem 1994;63:1908–1916.

114. Morrison D. 14-3-3: modulators of signaling proteins? Science 1994;266:56–57.

115. Alam R, Hachiya N, Sakaguchi M, et al. CDNA cloning and characterization of mitochondrial import factor (MSF) purified from rat liver cytosol. J Biochem (Tokyo) 1994;116:416–425.

116. Schlaepfer DD, Jones J, Haigler HT. Inhibition of protein kinase C by annexin V. Biochemistry 1992;31:1886–1891.

117. Hanks SK, Polte TR. Signaling through focal adhesion kinase. Bioessays 1997;19:137–145.

118. Ishii H, Jirousek MR, Koya D, et al. Amelioration of vascular dysfunction in diabetic rats by an oral PKC β inhibitor. Science 1996;272:728–731.

119. Moriya K, Rivera J, Odom S, et al. ER-27319, a novel acridone-related compound, inhibits release of antigen-induced allergic mediators from mast cells by selective inhibition of FcεRI-mediated activation of Syk. Proc Natl Acad Sci USA 1997;94:12539–12544.

120. Yamada K, Jelsema CL, Beaven MA. Certain inhibitors of protein serine/threonine kinases also inhibit tyrosine phosphorylation of phospholipase Cγ1 and other proteins and reveal distinct roles for tyrosine kinase(s) and protein kinase C in stimulated, rat basophilic RBL-2H3 cells. J Immunol 1992;149:1031–1037.

13

Mast Cell Inhibitory Receptors of the Immunoglobulin Superfamily

Howard R. Katz

Mast cells contain a panoply of pro-inflammatory mediators, which are released upon immunological activation through FcεRI or FcγRIII. Mast cell mediators can be classified into three groups, which differ in their origin inside the cell and in their time course of release. The secretory granule-derived mediators, including histamine, endoproteases, an exoprotease, and certain cytokines such as tumor necrosis factor-α, are released preformed from mast cells seconds to minutes after the initiation of activation. Lipid-derived mediators, such as leukotriene $(LT)C_4$ and prostaglandin $(PG)D_2$, are synthesized de novo from phospholipids by the concerted action of enzymes of the 5-lipoxygenase and cyclooxygenase pathways, respectively. The synthesis of these mediators is sufficiently rapid that they are released essentially as fast as the preformed, secretory granule-derived mediators. In contrast, certain mast cell-derived cytokines are released not only rapidly from the secretory granule, but during a delayed phase that occurs several to many hours after the initial activation of the mast cell. This delay is the result of the time required for cytokine gene transcription, generation of protein, post-translational processing of the protein, and transport of the cytokine out of the cell. Thus, activated mast cells release a diverse array of mediators that likely contribute to both early and delayed phases of inflammation.

In contrast to most other cells of the hematopoietic system, mast cells constitutively reside in certain tissues that are at interfaces between self and nonself. For example, mast cells are associated with the skin and the mucosal surfaces in the lung and intestine. Mast cells also have a perivenular localization, and hence are close to the interface between bloodborne leukocytes and those that have undergone extravasation into the tissues. The presence of mast cells at these sites leads to the hypothesis that mast cells may serve as early sentinels of foreign intrusion, and that they may play a beneficial role by releasing a discrete amount of mediators, for a limited time, that may play a key role in orchestrating the initiation of an inflammatory response. By contrast, in allergic diseases, mast cell mediator release is exaggerated and sufficiently prolonged to be detrimental.

It seems reasonable to speculate that because the mast cell armamentarium of proinflammatory mediators is so close to environmental signals, mast cells should be tightly regulated so as to achieve either no proinflammatory signal or only a modest and self-resolving proinflammatory signal. The failure to maintain either of these states may result in the manifestations of allergic disease. Historically, the regulation of mast cell activation has been primarily thought of in terms of the magnitude of the activating signal. Mast cell mediator release can be limited experimentally by decreasing the strength of the positive activation signal. For immunological activation, this can be achieved by decreasing the number of activating receptors at the mast cell surface, by reducing the concentration of the immunoglobulin (Ig) agonist, and/or by limiting the extent of cross-linking of Ig-bound receptors.

Recently, however, it has become clear that mast cells, like other cells of the hematopoietic system, bear as many, if not more, surface receptors that inhibit activation as receptors which stimulate it. It is possible to infer from these findings that the extent of net mast cell activation in vivo may be the result of an intricate combination of negative and positive signals sent from cell-surface receptors, with negative signaling playing a major role in maintaining beneficial mast cell homeostasis. In this scenario, pathological mast cell activation results from an imbalance between positive and negative signals. Hence, a deficiency in the amount of negative signals may be just as crucial to the genesis of allergic diseases as an abundance of positive signals.

In this chapter, I focus on the molecular and cellular biology of three inhibitory receptors of the Ig superfamily that are expressed on mouse mast cells: FcγRIIb, gp49B1, and p91/PIR-B1. In addition, I review the more limited knowledge about the expression of inhibitory receptors on human mast cells and basophils.

Mast Cell FcγRIIb

We and other investigators have shown that mouse mast cells express up to three species of FcγR.[1,2] Two of these, FcγRIIb1 and FcγRIIb2 (collectively referred to as FcγRIIb), arise from alternate splicing of mRNA from the FcγRIIβ gene.[3] FcγRIII is selectively expressed on certain mast cell populations, and upon cross-linking, initiates activation-secretion.[4-6] The cross-linking of FcγRIIb species does not elicit mast cell activation, however, and their function in mast cells remained a mystery for a number of years.

Clues as to this function came from studies of B cells. B cells can be activated by ligation of the B-cell antigen receptor (BCR) with F(ab')$_2$ fragments of antimembrane Ig antibody, but not by intact antibody.[7-10] The failure of intact antimembrane Ig to activate B cells is circumvented by the addition of monoclonal antibody

(mAb) 2.4G2, which is directed to the extracellular domains of FcγRIIb species and which blocks the binding of the Fc portion of IgG.[7,11] Hence, it has been hypothesized that intact antimembrane Ig fails to activate B cells because it coligates the BCR with FcγRIIb through its Fab and Fc ends, respectively.

Molecular evidence that FcγRIIb (specifically, FcγRIIb1) plays a key role in preventing B-cell activation was obtained with mutagenesis experiments. These studies revealed that a 13-amino-acid region of the cytoplasmic domain of FcγRIIb1 is sufficient for inhibition of B-cell stimulation through the BCR.[11] In fact, a subsequence, S/I/VxYxxL/V, termed the immunoreceptor tyrosine-based inhibitory motif (ITIM), appears to be sufficient to impart the suppressive activity, based on subsequent findings in natural killer (NK) cells (described here). Current thinking is that B-cell FcγRIIb1 may function in a negative feedback loop that regulates antibody production. As Ig levels rise during an immune or inflammatory response, the chances increase that a multivalent antigen bound by antigen-specific IgG will coligate a BCR and an FcγRIIb1.

FcγRIIb species are expressed on most cells of the mouse hematopoietic system, and they also function as inhibitory receptors for other activating receptors belonging to the Ig superfamily.[12] The coligation of FcγRIIb1 or FcγRIIb2 with FcεRI on mouse bone marrow culture-derived mast cells (mBMMC) results in inhibition of secretory granule mediator release.[13] Mouse serosal mast cells (mSMC), isolated from the peritoneal cavity, also express FcγRIIb species,[1] demonstrating that native mouse mast cells express inhibitory FcγR in vivo. In a rat mast cell line transfected with mouse FcγRIIb2, coligation of the mouse receptor with endogenous rat FcεRI inhibits both exocytosis and the secretion of tumor necrosis factor-α.[13] Inhibition is dependent upon the coligation of the two receptors to each other, and on the presence of the cytoplasmic domain of the FcγRIIb species. By analogy with the B cell, multivalent antigen bound to IgE in FcεRI may also be bound by antigen-specific IgG, resulting in coligation of FcεRI with FcγRIIb and inhibition of mast cell activation.

Although FcγRIIb species have been detected on all mouse mast cell populations, the human analogs (FcγRIIB1/FcγRIIB2) may not be ubiquitously expressed on human mast cells. Complicating the issue in human cells is the fact that they can also express FcγRIIA, which is an activating receptor[14] that does not have a homolog on mouse cells. Although FcγRIIA-specific mAbs exist, mAbs that react with FcγRIIB species also react with FcγRIIA.[15] Collectively, these receptors have been referred to as "CD32." It has been reported that human lung, foreskin, intestinal mucosa, and ascites mast cells do not express CD32,[16,17] but those studies were done using mAbs that were subsequently reported to be specific for FcγRIIA.[15] In contrast, uterine mast cells express an FcγRII that is recognized by anti-FcγRIIA-specific mAbs.[18] However, the mast cells were not activated by the binding of anti-FcγRIIA mAb. Overall, because of the mAbs used in these studies, it is still unclear whether human mast cells express inhibitory FcγR of the FcγRIIB type.

Human basophils bind anti-FcγRIIA-specific mAbs.[16,17] These cells also bind an FcγRIIA/IIB-reactive mAb,[19] but it is unknown whether the mAb recognizes FcγRIIA alone or FcγRIIB species as well. However, there is functional evidence that an inhibitory FcγRIIB species is expressed on human basophils, because coligation of an anti-FcγRIIA/IIB-reactive mAb with FcεRI on basophils inhibits FcεRI-directed exocytosis.[12]

Mouse gp49B1/Human HL9

Mouse gp49B1 is a member of a multigene family that includes at least three homologous forms.[20] The 5.6-kb gp49B gene encodes gp49B1 and an alternate splicing variant (gp49B2), which is identical to gp49B1 except for the deletion of exon 6 that encodes the transmembrane domain.[20] In addition to being expressed on mBMMC and mSMC, gp49B1 is also expressed on macrophages and NK cells, but not on T or B cells, neutrophils, eosinophils, or fibroblasts.[21–24] Although mRNA transcripts for gp49B2 are present in mast cells, it is

not known whether the transcripts are translated to protein, and if so, whether they are secreted or cytosolic proteins.

The first clue that gp49B1 is an inhibitory receptor arose from an amino acid sequence analysis, which revealed that gp49B1 has significant overall homology with killer inhibitory receptors (KIRs) belonging to the Ig superfamily.[25] KIRs are expressed on NK cells and some T lymphocytes and inhibit the cytotoxic activity of these cells upon recognition of appropriate MHC class I molecules on other cells.[26] Ig superfamily KIRs possess two ITIMs in their cytoplasmic domains that are required for inhibitory function. In addition to its overall homology with the Ig superfamily KIRs, gp49B1 also contains two ITIMs in its cytoplasmic domain.[25] Although the amino acid sequence of the extracellular domain of gp49B1 is 89% identical to another member of the gp49 family (gp49A), which is derived from a separate gene,[27] rat mAb B23.1 recognizes a gp49B1-specific epitope.[25] The cross-linking of mAb B23.1 bound to gp49B1 and rat IgE bound to FcεRI on the surface of mBMMC inhibits the release of secretory granule mediators and the generation of LTC$_4$.[25] Thus, gp49B1 is a mast cell inhibitory receptor with distinctive features compared with FcγRIIb species, including a low overall amino acid sequence homology and the presence of two, rather than one, ITIMs in the cytoplasmic domain. The latter difference has important implications for differences in the mechanisms by which FcεRI and gp49B1 inhibit mast cell activation, as explained below.

Three gp49B1-like human cDNAs have been isolated by low-stringency, cross-species cloning using a mouse gp49A cDNA probe.[28] HM18, cloned from a cultured monocyte library, resembles gp49B1 most closely, having two Ig-like domains (Fig. 13.1). The same cDNA has been cloned from a B-cell-derived library.[29] HL9, derived from a human lung library, has four Ig domains, in which domains 1 and 4 are most homologous to domains 1 and 2 of mouse gp49B1, respectively. HM43, also derived from the cultured monocyte library, is similar to HL9 in having four Ig-like domains, but contains a stop codon before the predicted

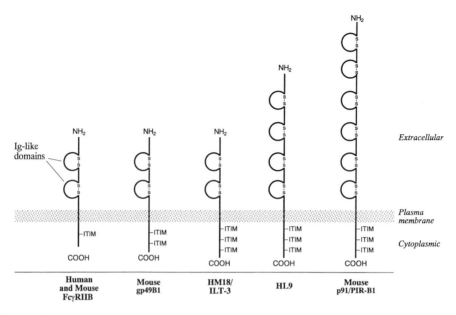

FIGURE 13.1. Members of the ITIM-bearing receptor superfamily expressed by mast cells.

transmembrane domain, allowing for the possibility of a soluble, secreted protein. The HM18 and HL9 cDNAs predict cytoplasmic domains with two ITIM motifs and an ITIM-like motif with a serine, rather than an isoleucine or valine, at the −2 position relative to the tyrosine. RT-PCR analysis with primers specific for HM18, HL9, and HM43 revealed that human lung mast cells of 96% purity contain mRNA for HL9, whereas mRNA for the other molecules is not detected.[28] Hence, human lung mast cells may express on their surface an ITIM-bearing receptor related to mouse gp49B1.

Mast Cell p91 and PIR-B1

In addition to its homology with human Ig superfamily KIRs, mouse gp49B1 also has significant amino acid sequence homology with the human FcαR.[25] In attempting to clone the mouse homolog of human FcαR, two groups of investigators have cloned two gp49B1-homologous molecules, p91 and PIR-B1, that are 98% and 97% identical to each other at the

nucleotide and amino acid levels, respectively. These differences may reflect allelic variation, because p91 and PIR-B1 were cloned from B10.A and BALB/c mouse-derived cDNA libraries, respectively. mRNA transcripts encoding p91 and PIR-B1 are expressed in mBMMC and the P815 mouse mastocytoma cell line, respectively, making p91/PIR-B1 the third potentially inhibitory molecule expressed by mouse mast cells. However, the expression of the transcripts as cell-surface proteins on mast cells has not been established, although p91 protein has been detected in macrophage extracts by immunoblotting.[30] In addition, it is not known whether two truncated mRNA forms of p91, which are present at a minimum in macrophages, are translated.

In contrast to gp49B1, p91 and PIR-B1 contain six Ig-like domains, each of which bears roughly 45% amino acid sequence homology with each of the two Ig-like domains of gp49B1. Notably, the cytoplasmic domains of p91 and PIR-B1 possess 53% amino acid sequence homology with that of gp49B1, allowing for gaps to maximize the alignment (Fig. 13.2). In contrast to mouse gp49B1, p91 and PIR-B1 each contain three ITIMs, plus one ITIM-like

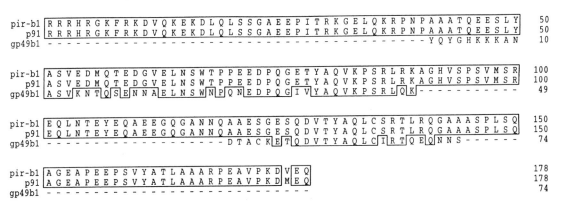

```
pir-b1  R R R H R G K F R K D V Q K E K D L Q L S S G A E E P I T R K G E L Q K R P N P A A A T Q E E S L Y    50
p91     R R R H R G K F R K D V Q K E K D L Q L S S G A E E P I T R K G E L Q K R P N P A A A T Q E E S L Y    50
gp49b1  - - - - - - - - - - - - - - - - - - - - - - - - - - - - - - - - - - - - - - Y Q Y G H K K K A N    10

pir-b1  A S V E D M Q T E D G V E L N S W T P P E E D P Q G E T Y A Q V K P S R L R K A G H V S P S V M S R   100
p91     A S V E D M Q T E D G V E L N S W T P P E E D P Q G E T Y A Q V K P S R L R K A G H V S P S V M S R   100
gp49b1  A S V K N T Q S E N N A E L N S W N P Q N E D P Q G I V Y A Q V K P S R L Q K - - - - - - - - - -     49

pir-b1  E Q L N T E Y E Q A E E G Q G A N N Q A A E S G E S Q D V T Y A Q L C S R T L R Q G A A A S P L S Q   150
p91     E Q L N T E Y E Q A E E G Q G A N N Q A A E S G E S Q D V T Y A Q L C S R T L R Q G A A A S P L S Q   150
gp49b1  - - - - - - - - - - - - - - - - - - - D T A C K E T Q D V T Y A Q L C I R T Q E Q N N S - - - - -    74

pir-b1  A G E A P E E P S V Y A T L A A A R P E A V P K D V E Q                                              178
p91     A G E A P E E P S V Y A T L A A A R P E A V P K D M E Q                                              178
gp49b1  - - - - - - - - - - - - - - - - - - - - - - - - - -                                                 74
```

FIGURE 13.2. Amino acid sequence alignment of the cytoplasmic domains of PIR-B1, p91, and gp49B1. Identical amino acids are *boxed*; gaps are indicated by *dashes*.

motif with glutamic acid at the -2 position. Interestingly, the regions of greatest sequence homology between the cytoplasmic domains of gp49B1 and p91/PIR-B1 occur around the two gp49B1 ITIM sequences. In particular, the second gp49B1 ITIM region is identical to one of the p91/PIR-B1 ITIMs.

Another contrast between gp49B1 and p91/PIR-B1 is that the expression of gp49B1 mRNA is essentially limited to mast cells, macrophages, and NK cells, whereas mRNA for p91 and/or PIR-B1 has been detected not only in those cell types but also in certain B cells and myeloid cells (e.g., granulocytes). Thus, it seems reasonable to hypothesize that gp49B1 and p91/PIR-B1 may have unique functions or ligands, based not only on their different numbers of Ig-like domains, but because of their overlapping, but nonidentical, profiles of cell expression. With regard to ligands, p91 does not appear to be an FcR,[30] and thus like gp49B1, its ligand(s) is unidentified.

Functional Implications of Multiple ITIM-Bearing Inhibitory Receptors in Mast Cells

Mutational studies of the ITIMs of FcγRIIb and KIRs have demonstrated that inhibitory function is dependent on the tyrosine in the motif.[11] The ITIM tyrosine becomes phosphorylated under conditions of functional inhibition. For FcγRIIb1 in B cells, the phosphorylation event is associated with the binding of the *src* homology-2 domain-containing tyrosine phosphatase (SHP) -1.[31] However, a *src* homology-2 domain-containing inositol polyphosphate 5-phosphatase (SHIP) also binds to BCR-coligated FcγRIIb1.[32] In contrast, only the latter enzyme binds to tyrosine phosphorylated, FcεRI-coligated FcγRIIb in mast cells, and SHIP appears to be the relevant transducer of FcγRIIb-mediated inhibition of mast cell activation.[32] The KIRs associate with SHP-1, which is capable of mediating KIR-dependent inhibition of cell activation.[33]

The participation of SHP-1 versus SHIP has important implications for the pathway of inhibition that is undertaken by the cell. SHP-1-mediated suppression of cell activation is associated with the inhibition of early tyrosine phosphorylation events, the release of calcium from intracellular stores, and the secondary influx of extracellular free calcium. In contrast, SHIP inhibits only the calcium influx step.[32] As noted previously, inhibition of mast cell activation by FcγRIIb proceeds via SHIP. Thus, the inhibition of the influx of extracellular calcium is sufficient to prevent exocytosis of secretory granule mediators in native mast cells and, in cross-species transfectants, the release of tumor necrosis factor-α (TNF-α). Preliminary data indicate that the inhibition of mast cell activation by gp49B1 proceeds predominantly via a SHP-1-dependent mechanism. Hence, mouse mast cells are likely to utilize at least two patterns of inhibition that can be signaled from surface receptors of the Ig superfamily.

As noted in the introduction, it is conceivable that under certain conditions, partial inhibition of mast cell activation is desirable to achieve a limited proinflammatory stimulus that is beneficial to the host. For that scenario to occur, inhibition of mast cell activation via FcγRIIb might be most appropriate, although it is not yet clear which function(s) of activated mast cells is resistant to FcγRIIb inhibition. By contrast, the complete resolution of mast cell-dependent inflammation, or the constitutive homeostatic inhibition of mast cell activation, would be most effectively achieved through a receptor that inhibits activation from the earliest positive signal transduction step, namely, tyrosine phosphorylation. The more global inhibition of activation may be a predominant role for ITIM-bearing gp49 family members in mouse and human mast cells. Ultimately, the different inhibitory capabilities of ITIM-bearing receptors may provide new strategies for fine-tuned interventions in the treatment of mast cell-dependent allergic diseases.

References

1. Katz HR, Arm JP, Benson AC, et al. Maturation-related changes in the expression of FcγRII and FcγRIII on mouse mast cells derived in vitro and in vivo. J Immunol 1990;145:3412–3417.
2. Benhamou M, Bonnerot C, Fridman WH, et al. Molecular heterogeneity of murine mast cell Fcγ receptors. J Immunol 1990;144:3071–3077.
3. Hogarth PM, Witort E, Hulett MD, et al. Structure of the mouse βFcγ receptor II gene. J Immunol 1991;146:369–376.
4. Katz HR, Raizman MB, Gartner CS, et al. Secretory granule mediator release and generation of oxidative metabolites of arachidonic acid via Fcγ bridging in mouse mast cells. J Immunol 1992;148:868–871.
5. Daëron M, Bonnerot C, Latour S, et al. Murine recombinant FcγRIII, but not FcγRII, trigger serotonin release in rat basophilic leukemia cells. J Immunol 1992;149:1365–1373.
6. Alber G, Kent UM, Metzger H. Functional comparison of FcεRI, FcγRII, and FcγRIII in mast cells. J Immunol 1992;149:2428–2436.
7. Cambier JC, Ransom JT. Molecular mechanisms of transmembrane signaling in B lymphocytes. Annu Rev Immunol 1987;5:175–199.
8. Amigorena S, Bonnerot C, Drake JR, et al. Cytoplasmic domain heterogeneity and functions of IgG Fc receptors in B lymphocytes. Science 1992;256:1808–1812.
9. Wilson HA, Greenblatt D, Taylor CW, et al. The B lymphocyte calcium response to anti-Ig is diminished by membrane immunoglobulin cross-linkage to the Fcγ receptor. J Immunol 1987;138:1712–1718.
10. Bijsterbosch MK, Klaus GG. Crosslinking of surface immunoglobulin and Fc receptors on B lymphocytes inhibits stimulation of inositol phospholipid breakdown via the antigen receptors. J Exp Med 1985;162:1825–1836.
11. Muta T, Kurosaki T, Misulovin Z, et al. A 13-amino-acid motif in the cytoplasmic domain of FcγRIIB modulates B-cell receptor signalling. Nature (Lond) 1994;368:70–73.
12. Daëron M, Latour S, Malbec O, et al. The same tyrosine-based inhibition motif, in the intracytoplasmic domain of FcγRIIb, regulates negatively BCR-, TCR-, and FCR-dependent cell activation. Immunity 1995;3:635–646.
13. Daëron M, Malbec O, Latour S, et al. Regulation of high-affinity IgE receptor-mediated mast cell activation by murine low-affinity IgG receptors. J Clin Invest 1995;95:577–585.
14. Odin JA, Edberg JC, Painter CJ, et al. Regulation of phagocytosis and [Ca^{2+}]$_i$ flux by distinct regions of an Fc receptor. Science 1991;254:1785–1788.
15. Warmerdam PA, Van Den Herik-Oudijk IE, Parren PW, et al. Interaction of a human FcγRIIb1 (CD32) isoform with murine and human IgG subclasses. Int Immunol 1993;5(3):239–247.
16. Valent P, Majdic O, Maurer D, et al. Further characterization of surface membrane structures expressed on human basophils and mast cells. Int Arch Allergy Appl Immunol 1990;91:198–203.
17. Valent P, Ashman LK, Hinterberger W, et al. Mast cell typing: demonstration of a distinct hematopoietic cell type and evidence for immunophenotypic relationship to mononuclear phagocytes. Blood 1989;73:1778–1785.
18. Guo CB, Kagey-Sobotka A, Lichtenstein LM, et al. Immunophenotyping and functional analysis

of purified human uterine mast cells. Blood 1992;79:708–712.

19. Anselmino LM, Perussia B, Thomas LL. Human basophils selectively express the FcγRII (CDw32) subtype of IgG receptor. J Allergy Clin Immunol 1989;84:907–914.

20. Castells MC, Wu X, Arm JP, et al. Cloning of the gp49B gene of the immunoglobulin superfamily and demonstration that one of its two products is an early-expressed mast cell surface protein originally described as gp49. J Biol Chem 1994; 269:8393–8401.

21. LeBlanc PA, Russell SW, Chang S-MT. Mouse mononuclear phagocyte heterogeneity detected by monoclonal antibodies. J Reticuloendothel Soc 1982;32:219–231.

22. LeBlanc PA, Biron CA. Mononuclear phagocyte maturation: a cytotoxic monoclonal antibody reactive with postmonoblast stages. Cell Immunol 1984;83:242–254.

23. Wang LL, Mehta IK, LeBlanc PA, et al. Mouse natural killer cells express gp49B1, a structural homology of human killer inhibitory receptors. J Immunol 1997;158:13–17.

24. Rojo S, Burshtyn DN, Long EO, et al. Type I transmembrane receptor with inhibitory function in mouse mast cells and NK cells. J Immunol 1997;158:9–12.

25. Katz HR, Vivier E, Castells MC, et al. Mouse mast cell gp49B1 contains two immunoreceptor tyrosine-based inhibition motifs and suppresses mast cell activation when coligated with the high-affinity Fc receptor for IgE. Proc Natl Acad Sci USA 1996;93:10809–10814.

26. Colonna M. Specificity and function of immunoglobulin superfamily NK cell inhibitory and stimulatory receptors. Immunol Rev 1997;155: 127–133.

27. Castells MC, McCormick MJ, Austen KF, et al. Cloning of the mouse gp49A gene and expression of gp49A protein on mast cells. J Allergy Clin Immunol 1997;99:S90 (abstract).

28. Arm JP, Nwankwo C, Austen KF. Molecular identification of a novel family of human immunoglobulin superfamily members that possess immunoreceptor tyrosine-based inhibition motifs and homology to the mouse gp49B1 inhibitory receptor. J Immunol 1997;159:2342–2349.

29. Cella M, Dohring C, Samaridis J, et al. A novel inhibitory receptor (ILT3) expressed on monocytes, macrophages, and dendritic cells involved in antigen processing. J Exp Med 1997;185(10): 1743–1751.

30. Hayami K, Fukuta D, Nishikawa Y, et al. Molecular cloning of novel murine cell-surface glycoprotein homologous to killer cell inhibitory receptors. J Biol Chem 1997;272:7320–7327.

31. D'Ambrosio D, Hippen KL, Minskoff SA, et al. Recruitment and activation of PTP1C in negative regulation of antigen receptor signaling by FcγRIIB1. Science 1995;268:293–297.

32. Ono M, Bolland S, Tempst P, et al. Role of the inositol phosphatase SHIP in negative regulation of the immune system by the receptor FcγRIIb. Nature (Lond) 1996;383(6597):263–266.

33. Binstadt BA, Brumbaugh KM, Dick CJ, et al. Sequential involvement of Lck and SHP-1 with MHC-recognizing receptors on NK cells inhibits FcR-initiated tyrosine kinase activation. Immunity 1996;5:629–638.

Section III

14
Signaling Pathways That Regulate Effector Function: Perspectives

Reuben P. Siraganian

The β- and γ-subunits of FcεRI transduce the signals generated by aggregation of this receptor.[1] The cytoplasmic domains of both these subunits contain a motif with the amino acid sequence (D/E)x_2Yx_2Lx_{6-8}Yx_2(L/I) that is essential for cell activation.[2-5] This immunoreceptor tyrosine-based activation motif (ITAM) is also present in the subunits of the T-cell and B-cell antigen receptors.[4,6-10] The aggregation of these receptors results in phosphorylation of the tyrosine residues within the ITAM, which is critical for cell signaling.[6-9]

This overview chapter presents (1) a model of the pathways that are activated after receptor aggregation, (2) tyrosine phosphorylation of regulatory molecules independent of Syk, and (3) the regulation of degranulation by cell adhesion. The molecules and the early steps in cell activation have been outlined in Chapter 8 by Henry Metzger.

Model of Pathways Activated by FcεRI Aggregation

Protein tyrosine phosphorylation is an early and critical signal for FcεRI-induced degranulation.[11,12] FcεRI associates or activates several nonreceptor protein tyrosine kinases. Lyn, a member of the *src* family of kinases, is associated with the inner surface of the plasma membrane and with the β-subunit of FcεRI in nonstimulated cells.[13-15] Aggregation of the receptor results in the tyrosine phosphorylation of the ITAMs of the β- and γ-subunits of the receptor.[16,17] This phosphorylation is probably by Lyn or another Src family kinase that associates with the receptor.[13,18,19] The tyrosine-phosphorylated ITAMs in the COOH-terminal cytoplasmic domain of FcεRIβ and the cytoplasmic domain of FcεRIγ function as scaffolds for binding of additional signaling molecules.[20] These include adaptor molecules that have no catalytic activity, but have multiple binding sites and can recruit additional molecules to the receptor, and a second group of molecules with catalytic domains that regulates more distal signaling events. Among these molecules with enzymatic activity is Syk, which is critical for propagating the downstream signals.

The association of Syk with FcεRI, especially after cell activation,[21,22] results from the interaction of the two SH-2 domains of Syk with the two phosphorylated tyrosines in the ITAM of the γ-subunit and less from that of the β-subunit.[14,20,23] Analysis of the crystal structure of the tandem SH-2 domains of ZAP-70 bound to the diphosphorylated ITAM peptide suggests that the carboxy-terminal SH-2 domain binds to the amino-terminal YxxL sequence of the ITAM whereas the amino-terminal SH-2 domain binds the carboxy-terminal YxxL of the ITAM.[24] An unusual aspect of this interaction is that the two SH-2 domains do not function independently but cooperate to form one of the binding sites for pTyr.

The clustering of Syk to FcεRI at the membrane is probably critical for the subsequent downstream signaling events. The binding of Syk to the tyrosine-phosphorylated ITAM results in a conformational change in Syk, with an increase in its enzymatic activity.[25-27] This change results in autophosphorylation of Syk, which may be caused by intermolecular or intramolecular reactions or both. Syk recruited by the tyrosine-phosphorylated ITAM of the γ-subunit of FcεRI would be close to Lyn, which is associated with the β-subunit. This location of Syk in the membrane together with the change in its conformation may also allow it to be further phosphorylated by Lyn or other kinases, resulting in further activation of Syk.[28] In transfection experiments, Syk is activated when only the γ-subunit is contransfected, but this activation is enhanced by the presence of the β-subunit.[15,29,30] Therefore, Syk activation may involve several mechanisms that include conformational changes from ITAM binding, autophosphorylation, and phosphorylation by other kinases. However, whatever the mechanism, the phosphorylation of the tyrosines in the activation loop of Syk is important for the capacity of Syk to phosphorylate other molecules and to induce signal transduction in cells (Figure 14.1).

Syk bound to FcεRI may then phosphorylate membrane proteins in the vicinity of the receptor. Some of these proteins could also have been recruited to the receptor after it is aggregated or tyrosine phosphorylated. For example, the tyrosine-phosphorylated ITAM of FcεRIβ

binds not only Syk but also other proteins that have SH-2 domains, including Shc, PLC-γ1, SHIP, and SHP-2, whereas the similar peptide based on the ITAM of γ precipitates only Syk.[14,20,23] The tyrosine phosphorylation of Syk could also recruit other proteins such as Lyn or other Src family kinases that bind to it through their own SH-2 domains.[14,31,32] Such Src family kinases associated with Syk may phosphorylate Syk and further regulate its activity. The function of Syk may also be regulated by other molecules such as p120[c-cbl] (Cbl) and protein tyrosine phosphatases, which may modulate the kinase activity of Syk by controlling the level of its tyrosine phosphorylation.[33,34]

In a Syk-negative variant of the RBL-2H3 cell line, FcεRI aggregation induced minimal increase in protein tyrosine phosphorylations of total cellular proteins.[35] Although some proteins are tyrosine phosphorylated independent of Syk (see next section), most of the FcεRI-induced cellular protein tyrosine phosphorylations require Syk. The phosphorylation of the following proteins is downstream of Syk: PLC-γ1, PLC-γ2, Vav, SLP-76, Shc, MAP kinase, JNK kinase, the focal adhesion kinase pp125[FAK] (FAK), the protein tyrosine kinase Pyk-2, paxillin, the adhesion molecule PECAM-1 (CD31), Cbl, ras-GAP, and HS-1. These biochemical changes result in cytoskelatal reorganization, activation of genes, and degranulation. However, these results do not im-

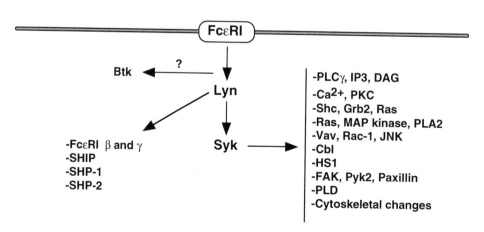

FIGURE 14.1. Schematic diagram of signaling pathways in mast cells.

ply that Syk is directly phosphorylating these proteins. The tyrosine phosphorylation and activation of Btk probably depends on a Src family kinase, although whether Syk is also required is unknown. These different pathways are discussed in detail in this volume.

Binding of Negatively Regulating Molecules to FcεRI and Their Tyrosine Phosphorylation Independent of Syk

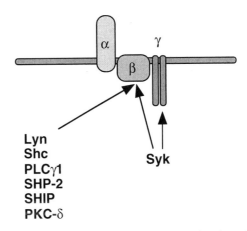

FIGURE 14.2. Interaction of different molecules with subunits of FcεRI.

The FcεRI-induced tyrosine phosphorylation of a number of proteins is independent of the presence of Syk in cells (Figure 14.2). These include the following: the β- and γ-subunits of FcεRI, the SH-2-domain containing inositol phosphate polyphosphate 5-phosphatase (SHIP), and the SH-2-domain containing protein tyrosine phosphatases SHP-1 and SHP-2.[35,37] Some of these molecules are negative regulators of signaling in B cells and mast cells.[38–41] Both SHP-1 and SHP-2 can dephosphorylate several proteins including the receptor subunits. SHIP hydrolyzes inositol(1,3,4,5)phosphate with the formation of inositol(1,3,4)phosphate and also catalyzes the hydrolysis of phosphatidylinositol(3,4,5)phosphate to phosphatidylinositol(3,4)phosphate.[42] The decrease in the concentration of inositol(1,3,4,5)phosphate would limit the receptor-induced calcium influx. Therefore, these molecules may regulate signal transduction: SHIP by controlling the level of inositol phosphates and SHP-1/SHP-2 by regulating the extent of the tyrosine phosphorylation of the receptor and other associated signaling molecules. This suggests that the control of the molecules that regulate signaling is independent of Syk. Therefore, the pathway that generates signals which induce secretion and the pathway for regulating those signals diverge at an early step after receptor aggregation.

Interestingly, several of the molecules that are tyrosine phosphorylated independent of

Syk bind to tyrosine-phosphorylated peptides based on the ITAM of the β- but not γ-subunit of FcεRI. For example, a synthetic tyrosine-diphosphorylated peptide based on the ITAM sequence of the β-subunit of FcεRI precipitates SHIP and SHP-2, in addition to some other signaling molecules. Therefore, tyrosine phosphorylation of the ITAM of the γ-subunit recruits Syk, which is crucial in downstream activating signals. In contrast, tyrosine phosphorylation of the β-subunit recruits molecules such as SHIP and SHP-2 that regulate or modulate signal transduction.

The binding of these negative signaling molecules to the β-subunit of FcεRI is probably caused by the immunoreceptor tyrosine-based inhibiting motif (ITIM), which is important for downregulating signals from immune receptors.[43,45] This ITIM consensus sequence of I/V-X-pY-X2-L/V is similar to the ITAM of the β-subunit of FcεRI. The ITIM sequence is found on the cytoplasmic domain of several molecules, including FcγRIIB in B cells and mast cells, CD22 in B cells, the killer-inhibitory receptors in NK cells, and gp49B1 in mast cells.[46,47] These molecules are thought to downregulate receptor function: for example, in B cells or mast cells, the coligation of the immune receptor to FcγRIIB results in generating inhibitory signals.[41,48] Synthetic tyrosine-phosphorylated peptides based on this motif bind SHP-1, SHP-2, and SHIP.[39,41] Therefore, phosphorylation of the tyrosine in the ITIM may recruit molecules that downregulate the function of immune receptors.

These observations suggest that the different subunits of FcεRI recruit both activating and regulatory molecules; the γ-subunit recruits Syk, which is crucial for downstream activating signals, whereas the β-subunit recruits molecules such as SHIP and SHP-2 that regulate signal transduction. Therefore, unlike B cells in which the inhibitory molecules such as SHP-2 or SHIP are recruited by coreceptors to downregulate B-cell antigen receptor- (BCR-) initiated signals, subunits of the FcεRI recruit molecules that both activate and inhibit signal transduction.

Cell Adhesion, Focal Adhesion Kinases, and Signaling

Basophils and mast cells have surface adhesion receptors (e.g., integrins) that are involved in the binding of these cells to other cells or to the extracellular matrix (Figure 14.3).[49–51] At these sites of binding of cells, the cytoplasmic domains of the integrins form focal adhesion complexes that contain cytoskeletal proteins such as talin, vinculin, α-actinin, filamin, and the focal adhesion kinase pp125[FAK] (FAK), and the related kinase Pyk-2.[52–57] Adherence of cells also results in aggregation of adhesion receptors with the generation of intracellular signals,

which includes protein tyrosine phosphorylations, phosphatidylinositol hydrolysis, changes in intracellular pH or calcium, and the expression of new genes.[58] Some of the proteins that are tyrosine phosphorylated are protein tyrosine kinases FAK and Pyk-2 and the cytoskeletal protein paxillin.[52,59–61] RBL-2H3 and mast cells bind through integrin receptors to surfaces coated with fibronectin, resulting in changes in the cytoskeleton, cell spreading, and a redistribution of the granules to the periphery of the cells.[50,58,62] These changes induced by cell adherence may play an important role in modulating the intracellular signaling that leads to degranulation. For example, there is enhanced secretion from RBL-2H3 cells that are adherent to surfaces coated with fibronectin.[58,62]

The aggregation of FcεRI results in tyrosine phosphorylation of the protein tyrosine kinases FAK and Pyk-2 in RBL-2H3 cells.[59,61] This receptor-induced phosphorylation of FAK and Pyk-2 is dramatically enhanced by adherence of the cells to fibronectin and, as noted, this adherence also results in enhanced secretion.[59,62] The tyrosine phosphorylation and activation of FAK and Pyk-2 is downstream of Syk and of the rise in intracellular Ca^{2+}. Stimulation of the cells by adenosine or thrombin that activate G protein-coupled receptors also results in the tyrosine phosphorylation of FAK and Pyk-2. In contrast to the FcεRI-mediated phosphorylation, the G protein-coupled receptor phosphorylation is independent of Syk.

FAK and Pyk-2 are expressed in many different cells and localize with integrins and several

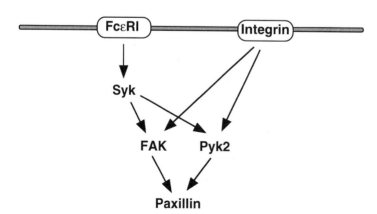

FIGURE 14.3. Activation of FAK and Pyk-2 by FcεRI and integrins.

proteins to intracellular sites where cells are in contact with the extracellular matrix.[52,63–65] These kinases link signals from cell-surface receptors to downstream events. The stimulation of many different cell-surface receptors including immune receptors and G protein-coupled receptors results in the tyrosine phosphorylation and activation of FAK and Pyk-2.[59,66–69] FAK and Pyk-2 are members of a family of cytoplasmic protein tyrosine kinases that have a central catalytic domain and amino- and carboxyl-terminal regions which lack SH-2 and SH-3 domains and membrane localization signals. The activation of these kinases results in autophosphorylation, which creates a binding site for the SH-2 domains of Src family kinases.[52] The Src family kinase bound to these molecules, then further phosphorylates FAK or Pyk-2 and other proteins such as paxillin, tensin, and p130^{cas} that accumulate at focal adhesion sites.[52,70] FAK and Pyk-2 also interact with several signaling molecules. For example, the cytoskeletal protein paxillin associates with the C-terminal region of both FAK and Pyk-2, and the tyrosine-phosphorylated FAK or Pyk-2 has binding sites for the SH-2 domains of the p85 subunit of phosphatidylinositol 3'-kinase, Grb-2, Crk, and Csk.

The FcεR-induced tyrosine phosphorylation of FAK and Pyk-2 may be important for further propagation of signaling in the late steps of degranulation. In vitro, FAK and Pyk-2 can phosphorylate paxillin, a cytoskeletal protein that accumulates at focal adhesion sites.[71] In vivo, FAK, Pyk-2, and paxillin become coordinately phosphorylated on tyrosine in response to FcεRI aggregation and, like FAK, the FcεRI-induced tyrosine phosphorylation of paxillin is downstream of Syk and is enhanced by adherence of the cells.[60,71,72] Paxillin has multiple binding sites for SH-2 and SH-3 domains and forms complexes with many proteins including protein tyrosine kinases such as Lyn and Csk, adaptor molecules such as Crk, and cytoskeletal proteins such as FAK, vinculin, α-actinin, and talin.[73] Paxillin may help direct actin filament interactions with the membrane. The tyrosine phosphorylation of paxillin, FAK, and Pyk-2 may be important in the development of actin stress fibers and the reorganiza-

tion of F-actin and thus may play a role in signal transduction from receptors to the cytoskeleton.[74,75]

Experiments suggest that FAK may play an important role during the late stages of signaling leading to degranulation. A monoclonal antibody to an FcεRI-associated ganglioside induces tyrosine phosphorylation of FcεRIβ, FcεRIγ, Lyn, PLC-γ1, and Syk, but not of FAK.[76,77] This antibody, however, does not induce significant degranulation, suggesting a role for FAK in secretion from mast cells.[78] Further evidence for an important function of FAK in degranulation are recent studies with a FAK-deficient variant of the RBL-2H3 cells. FcεRI but not ionophore-mediated secretion was defective in these variant cells that had a decreased level of FAK. The stable transfection of FAK markedly enhanced the FcεRI-mediated secretion. This enhancement does not require the catalytic activity or the autophosphorylation site of FAK. These results strongly suggest that FAK plays an important role in FcεRI-induced secretion by functioning as an adaptor or linker molecule. However, the catalytically inactive FAK is still tyrosine phosphorylated after FcεRI aggregation. The tyrosine phosphorylation of the kinase-inactive FAK may still allow for binding to it of other proteins mediated by SH-2 domains, and therefore the mutated FAK can still function to transduce FcεRI-initiated signals.

Tyrosine phosphorylation and activation of FAK and Pyk-2 are by themselves not sufficient signals for degranulation but still require other signals generated by FcεRI aggregation. For example, both Pyk-2 and FAK are tyrosine-phosphorylated by phorbol esters, but that does not result in degranulation. In FAK-transfected cells, the overexpression and the tyrosine phosphorylation of FAK still require FcεRI aggregation to result in degranulation.

FcεRI aggregation of cells also results in enhanced cell adherence, indicating that it regulates adhesion molecules. This regulation could be secondary to signals such as tyrosine phosphorylation that are induced by receptor aggregation. For example, FcεRI aggregation results in the tyrosine phosphorylation of the cell-surface adhesion molecule platelet/endothelial cell adhesion molecule-1 (PECAM-1 or CD31).[79] PECAM-1 is an integral membrane glycoprotein member of the Ig superfamily of cell adhesion molecules that functions in the transmigration of cells across the endothelium.[80–82] The FcεRI-induced phosphorylation of PECAM-1 is an early event, downstream of Syk but independent of Ca^{2+} influx or of the activation of protein kinase C. PECAM-1 is also transiently tyrosine phosphorylated after adherence of RBL-2H3 cells to fibronectin.[83] The protein tyrosine phosphatase SHP-2, but not the related protein tyrosine phosphatase SHP-1, associates with PECAM-1. This association of the two proteins correlates with the extent of the tyrosine phosphorylation of PECAM-1 and is mediated by the SH-2 domains of SHP-2 binding to phosphorylated PECAM-1. This interaction can lead to the dephosphorylation of PECAM-1. Therefore, integrin and immune receptor activation results in tyrosine phosphorylation of PECAM-1 and the binding of the protein tyrosine SHP-2, which could regulate receptor-mediated signaling in cells. PECAM-1 is required for the transmigration of leukocytes across the endothelium into sites of inflammation and also can control the function of β1 and β2 integrins. Furthermore, the function of PECAM-1 as an adhesion molecule is regulated by its cytoplasmic domain.[84,85] Therefore, the level of the tyrosine phosphorylation of PECAM-1 may modulate its interaction with other molecules, thereby regulating the migration of basophils into inflammatory sites. PECAM-1 also regulates the adhesive properties of β1 and β2 integrins.[86–88] Therefore, tyrosine phosphorylation of PECAM-1 could also influence its capacity to regulate the adhesive activity of integrins.

Conclusions

Remarkable progress has been made in the last few years in unraveling the signaling pathways in basophils and mast cells. What has emerged is the observation that signal transduction pathways initiated by immune recep-

tors on T cells, B cells, and mast cells are very similar. Activation signals in all these cells critically requires Syk or ZAP-70. Secretion, however, is also regulated by adhesion receptors, which can also activate other kinases such as FAK.

Although Syk is critical for initiating signal transduction in basophils, several regulatory proteins are tyrosine phosphorylated independent of Syk. Interestingly, the activating Syk binds predominantly to the γ-subunit, whereas these inhibitory molecules bind to the β-subunit of FcεRI. Therefore, unlike B cells in which the inhibitory molecules such as SHP-2 or SHIP are recruited by coreceptors to downregulate BCR-initiated signals, different subunits of the FcεRI recruit molecules that both activate and regulate signal transduction.

References

1. Alber G, Miller L, Jelsema CL, Varin-Blank N, Metzger H. Structure-function relationships in the mast cell high affinity receptor for IgE. Role of the cytoplasmic domains and of the β subunit. J Biol Chem 1991;266:22613–22620.
2. Reth M. Antigen receptor tail clue. Nature (Lond) 1989;338:383–384.
3. Chan AC, Desai DM, Weiss A. The role of protein tyrosine kinases and protein tyrosine phosphatases in T cell antigen receptor signal transduction. Annu Rev Immunol 1994;12:555–592.
4. Cambier JC, Pleiman CM, Clark MR. Signal transduction by the B cell antigen receptor and its coreceptors. Annu Rev Immunol 1994;12:457–486.
5. Reth M, Wienands J. Initiation and processing of signals from the B cell antigen receptor. Annu Rev Immunol 1997;15:453–479.
6. Letourneur F, Klausner RD. T-cell and basophil activation through the cytoplasmic tail of T-cell-receptor ζ family proteins. Proc Natl Acad Sci USA 1991;88:8905–8909.
7. Romeo C, Amiot M, Seed B. Sequence requirements for induction of cytolysis by the T cell antigen/Fc receptor ζ chain. Cell 1992;68:889–897.
8. Irving BA, Chan AC, Weiss A. Functional characterization of a signal transducing motif present in the T cell antigen receptor ζ chain. J Exp Med 1993;177:1093–1103.

9. Gauen LKT, Zhu Y, Letourneur F, et al. Interactions of p59[fyn] and ZAP-70 with T-cell receptor activation motifs: defining the nature of a signalling motif. Mol Cell Biol 1994;14:3729–3741.

10. Daeeron M. Fc receptor biology. Annu Rev Immunol 1997;15:203–234.

11. Benhamou M, Gutkind JS, Robbins KC, Siraganian RP. Tyrosine phosphorylation coupled to IgE receptor-mediated signal transduction and histamine release. Proc Natl Acad Sci USA 1990;87:5327–5330.

12. Benhamou M, Siraganian RP. Protein-tyrosine phosphorylation: an essential component of FcεRI signaling. Immunol Today 1992;13:195–197.

13. Eiseman E, Bolen JB. Engagement of the high-affinity IgE receptor activates *src* protein-related tyrosine kinases. Nature (Lond) 1992;355:78–80.

14. Kihara H, Siraganian RP. Src homology 2 domains of Syk and Lyn bind to tyrosine-phosphorylated subunits of the high affinity IgE receptor. J Biol Chem 1994;269:22427–22432.

15. Jouvin MH, Adamczewski M, Numerof R, Letourneur O, Valle A, Kinet JP. Differential control of the tyrosine kinases Lyn and Syk by the two signaling chains of the high affinity immunoglobulin E receptor. J Biol Chem 1994;269:5918–5925.

16. Paolini R, Jouvin MH, Kinet JP. Phosphorylation and dephosphorylation of the high-affinity receptor for immunoglobulin E immediately after receptor engagement and disengagement. Nature (Lond) 1991;353:855–858.

17. Pribluda VS, Pribluda C, Metzger H. Biochemical evidence that the phosphorylated tyrosines, serines, and threonines on the aggregated high affinity receptor for IgE are in the immunoreceptor tyrosine-based activation motifs. J Biol Chem 1997;272:11185–11192.

18. Pribluda VS, Pribluda C, Metzger H. Trans-phosphorylation as the mechanism by which the high-affinity receptor for IgE is phosphorylated upon aggregation. Proc Natl Acad Sci USA 1994;91:11246–11250.

19. Yamashita T, Mao S-Y, Metzger H. Aggregation of the high-affinity IgE receptor and enhanced activity of p53/56[lyn] protein-tyrosine kinase. Proc Natl Acad Sci USA 1994;91:11251–11255.

20. Kimura T, Kihara H, Bhattacharyya S, Sakamoto H, Appella E, Siraganian RP. Downstream signaling molecules bind to different phosphorylated immunoreceptor tyrosine-based

21. Hutchcroft JE, Geahlen RL, Deanin GG, Oliver JM. FcεRI-mediated tyrosine phosphorylation and activation of the 72-kDa protein-tyrosine kinase, PTK72, in RBL-2H3 rat tumor mast cells. Proc Natl Acad Sci USA 1992;89:9107–9111.

22. Benhamou M, Ryba NJP, Kihara H, Nishikata H, Siraganian RP. Protein-tyrosine kinase p72[syk] in high affinity IgE receptor signaling. Identification as a component of pp72 and association with the receptor γ chain. J Biol Chem 1993;268:23318–23324.

23. Shiue L, Green J, Green OM, et al. Interaction of p72[syk] with the γ and β subunits of the high-affinity receptor for immunoglobulin E, FcεRI. Mol Cell Biol 1995;15:272–281.

24. Hatada MH, Lu X, Laird ER, et al. Molecular basis for interaction of the protein tyrosine kinase ZAP-70 with the T-cell receptor. Nature (Lond) 1995;377:32–38.

25. Kimura T, Sakamoto H, Appella E, Siraganian RP. Conformational changes induced in the protein tyrosine kinase p72[syk] by tyrosine phosphorylation or by binding of phosphorylated immunoreceptor tyrosine-based activation motif peptides. Mol Cell Biol 1996;16:1471–1478.

26. Rowley RB, Burkhardt AL, Chao HG, Matsueda GR, Bolen JB. Syk protein-tyrosine kinase is regulated by tyrosine-phosphorylated Igα/Igβ immunoreceptor tyrosine activation motif binding and autophosphorylation. J Biol Chem 1995;270:11590–11594.

27. Shiue L, Zoller MJ, Brugge JS. Syk is activated by phosphotyrosine-containing peptides representing the tyrosine-based activation motifs of the high affinity receptor for IgE. J Biol Chem 1995;270:10498–10502.

28. El-Hillal O, Kurosaki T, Yamamura H, Kinet JP, Scharenberg AM. Syk kinase activation by a *src* kinase-initiated activation loop phosphorylation chain reaction. Proc Natl Acad Sci USA 1997;94:1919–1924.

29. Scharenberg AM, Lin S, Cuenod B, Yamamura H, Kinet JP. Reconstitution of interactions between tyrosine kinases and the high affinity IgE receptor which are controlled by receptor clustering. EMBO J 1995;14:3385–3394.

30. Lin SQ, Cicala C, Scharenberg AM, Kinet JP. The FcεRIβ subunit functions as an amplifier of FcεRIγ-mediated cell activation signals. Cell 1996;85:985–995.

activation motif (ITAM) peptides of the high affinity IgE receptor. J Biol Chem 1996;271:27962–27968.

phosphorylated immunoreceptor tyrosine-based

31. Sidorenko SP, Law CL, Chandran KA, Clark EA. Human spleen tyrosine kinase p72^{syk} associates with the Src-family kinase p53/56^{lyn} and a 120-kDa phosphoprotein. Proc Natl Acad Sci USA 1995;92:359–363.

32. Aoki Y, Kim YT, Stillwell R, Kim TJ, Pillai S. The SH2 domains of Src family kinases associate with Syk. J Biol Chem 1995;270:15658–15663.

33. Ota Y, Beitz LO, Scharenberg AM, Donovan JA, Kinet JP, Samelson LE. Characterization of Cbl tyrosine phosphorylation and a Cbl-Syk complex in RBL-2H3 cells. J Exp Med 1996;184:1713–1723.

34. Ota Y, Samelson LE. The product of the proto-oncogene c-*cbl*: a negative regulator of the Syk tyrosine kinase. Science 1997;276:418–420.

35. Zhang J, Berenstein EH, Evans RL, Siraganian RP. Transfection of Syk protein tyrosine kinase reconstitues high affinity IgE receptor mediated degranulation in a Syk negative variant of rat basophilic leukemia RBL-2H3 cells. J Exp Med 1996;184:71–79.

36. Kimura T, Sakamoto H, Appella E, Siraganian RP. The negative signaling molecule inositol polyphosphate 5-phosphatase, SHIP, binds to tyrosine phosphorylated β subunit of the high affinity IgE receptor. J Biol Chem 1997;272:13991–13996.

37. Kimura T, Zhang J, Sagawa K, Sakaguchi K, Appella E, Siraganian RP. Syk independent tyrosine phosphorylation and association of the protein tyrosine phosphatase SHP-1 and SHP-2 with the high affinity IgE receptor. J Immunol 1997;159:4426–4434.

38. DeFranco AL, Law DA. Tyrosine phosphatases and the antibody response. Science 1995;268: 263–264.

39. D'Ambrosio D, Hippen KL, Minskoff SA, et al. Recruitment and activation of PTP1C in negative regulation of antigen receptor signaling by FcγRIIB1. Science 1995;268:293–297.

40. Chacko GW, Tridandapani S, Damen JE, Liu L, Krystal G, Coggeshall KM. Negative signaling in B lymphocytes induces tyrosine phosphorylation of the 145-kDa inositol polyphosphate 5-phosphatase, SHIP. J Immunol 1996;157: 2234–2238.

41. Ono M, Bolland S, Tempst P, Ravetch JV. Role of the inositol phosphatase SHIP in negative regulation of the immune system by the receptor FcγRIIB. Nature (Lond) 1996; 383:263–266.

42. Damen JE, Liu L, Rosten P, et al. The 145-kDa protein induced to associate with Shc by mul-

tiple cytokines is an inositol tetraphosphate and phosphatidylinositol 3,4,5-trisphosphate 5-phosphatase. Proc Natl Acad Sci USA 1996; 93:1689–1693.

43. Muta T, Kurosaki T, Misulovin Z, Sanchez M, Nussenzweig MC, Ravetch JV. A 13-amino-acid motif in the cytoplasmic domain of FcγRIIB modulates B-cell receptor signalling. Nature (Lond) 1994;368:70–73.

44. Katz HR, Austen KF. A newly recognized pathway for the negative regulation of mast cell-dependent hypersensitivity and inflammation mediated by an endogenous cell surface receptor of the gp49 family. J Immunol 1997;158:5065–5070.

45. Cambier JC. Inhibitory receptors abound? Proc Natl Acad Sci USA 1997;94:5993–5995.

46. Doody GM, Justement LB, Delibrias CC, et al. A role in B cell activation for CD22 and the protein tyrosine phosphatase SHP. Science 1995;269:242–244.

47. Katz HR, Vivier E, Castells MC, McCormick MJ, Chambers JM, Austen KF. Mouse mast cell gp49B1 contains two immunoreceptor tyrosine-based inhibition motifs and suppresses mast cell activation when coligated with the high-affinity Fc receptor for IgE. Proc Natl Acad Sci USA 1996;93:10809–10814.

48. Daeron M, Latour S, Malbec O, et al. The same tyrosine-based inhibition motif, in the intracytoplasmic domain of FcγRIIB regulates negatively BCR-, TCR-, and FcR-dependent cell activation. Immunity 1995;3:635–646.

49. Valent P, Bettelheim P. Cell surface structures on human basophils and mast cells: biochemical and functional characterization. Adv Immunol 1992;52:333–423.

50. Hamawy MM, Mergenhagen SE, Siraganian RP. Adhesion molecules as regulators of mast cell and basophil function. Immunol Today 1994;15:62–66.

51. Bochner BS, Schleimer RP. The role of adhesion molecules in human eosinophil and basophil recruitment. J Allergy Clin Immunol 1994;94:427–438.

52. Clark EA, Brugge JS. Integrins and signal transduction pathways: the road taken. Science 1995;268:233–239.

53. Li JZ, Avraham H, Rogers RA, Raja S, Avraham S. Characterization of RAFTK, a novel focal adhesion kinase, and its integrin-dependent phosphorylation and activation in megakaryocytes. Blood 1996;88:417–428.

54. Sasaki H, Nagura K, Ishino M, Tobioka H, Kotani K, Sasaki T. Cloning and characterization of cell adhesion kinase β, a novel protein-tyrosine kinase of the focal adhesion kinase subfamily. J Biol Chem 1995;270:21206–21219.

55. Schwartz MA, Schaller MD, Ginsberg MH. Integrins: emerging paradigms of signal transduction. Annu Rev Cell Dev Biol 1995;11:549–599.

56. Chothia C, Jones EY. The molecular structure of cell adhesion molecules. Annu Rev Biochem 1997;66:823–862.

57. Burridge K, Chrzanowska-Wodnicka M, Zhong CL. Focal adhesion assembly. Trends Cell Biol 1997;7:342–347.

58. Juliano RL, Haskill S. Signal transduction from the extracellular matrix. J Cell Biol 1993;120:577–585.

59. Hamawy MM, Mergenhagen S, Siraganian RP. Tyrosine phosphorylation of pp125^{FAK} by the aggregation of high affinity immunoglobulin E receptors requires cell adherence. J Biol Chem 1993;268:6851–6854.

60. Hamawy MM, Swaim WD, Minoguchi K, de Feijter AW, Mergenhagen SE, Siraganian RP. The aggregation of the high affinity IgE receptor induces tyrosine phosphorylation of paxillin, a focal adhesion protein. J Immunol 1994;153:4655–4662.

61. Okazaki H, Zhang J, Hamawy MM, Siraganian RP. Activation of protein-tyrosine kinase Pyk2 is downstream of Syk in FcεRI signaling. J Biol Chem 1997;272:32443–32447.

62. Hamaway MM, Oliver C, Mergenhagen SE, Siraganian RP. Adherence of rat basophilic leukemia (RBL-2H3) cells to fibronectin-coated surfaces enhances secretion. J Immunol 1992;149:615–621.

63. Schaller MD, Borgman CA, Cobb BS, Vines RR, Reynolds AB, Parsons JT. pp125^{FAK}, a structurally distinctive protein-tyrosine kinase associated with focal adhesions. Proc Natl Acad Sci USA 1992;89:5192–5196.

64. Hanks SK, Calalb MB, Harper MC, Patel SK. Focal adhesion protein-tyrosine kinase phosphorylated in response to cell attachment to fibronectin. Proc Natl Acad Sci USA 1992;89:8487–8491.

65. Lev S, Moreno H, Martinez R, et al. Protein tyrosine kinase PYK2 involved in Ca^{2+}-induced regulation of ion channel and MAP kinase functions. Nature (Lond) 1995;376:737–745.

66. Hanks SK, Polte TR. Signaling through focal adhesion kinase. BioEssays 1997;19:137–145.

67. Tokiwa G, Dikic I, Lev S, Schlessinger J. Activation of Pyk2 by stress signals and coupling with JNK signaling pathway. Science 1996;273:792–794.

68. Dikic I, Tokiwa G, Lev S, Courtneidge SA, Schlessinger J. A role for Pyk2 and Src in linking G-protein-coupled receptors with MAP kinase activation. Nature (Lond) 1996;383:547–550.

69. Ganju RK, Hatch WC, Avraham H, et al. RAFTK, a novel member of the focal adhesion kinase family, is phosphorylated and associates with signaling molecules upon activation of mature T lymphocytes. J Exp Med 1997;185:1055–1063.

70. Harte MT, Hildebrand JD, Burnham MR, Bouton AH, Parsons JT. p130^{Cas}, a substrate associated with v-Src and v-Crk, localizes to focal adhesions and binds to focal adhesion kinase. J Biol Chem 1996;271:13649–13655.

71. Salgia R, Avraham S, Pisick E, et al. The related adhesion focal tyrosine kinase forms a complex with paxillin in hematopoietic cells. J Biol Chem 1996;271:31222–31226.

72. Hiregowdara D, Avraham H, Fu YG, London R, Avraham S. Tyrosine phosphorylation of the related adhesion focal tyrosine kinase in megakaryocytes upon stem cell factor and phorbol myristate acetate stimulation and its association with paxillin. J Biol Chem 1997;272:10804–10810.

73. Minoguchi K, Kihara H, Nishikata H, Hamawy MM, Siraganian RP. Src family tyrosine kinase Lyn binds several proteins including paxillin in rat basophilic leukemia cells. Mol Immunol 1994;31:519–529.

74. Burridge K, Turner CE, Romer LH. Tyrosine phosphorylation of paxillin and pp125^{FAK} accompanies cell adhesion to extracellular matrix: a role in cytoskeletal assembly. J Cell Biol 1992;119:893–903.

75. Pfeiffer JR, Oliver JM. Tyrosine kinase-dependent assembly of actin plaques linking FcεRI cross-linking to increased cell substrate adhesion in RBL-2H3 tumor mast cells. J Immunol 1994;152:270–279.

76. Minoguchi K, Swaim WD, Berenstein EH, Siraganian RP. Src family tyrosine kinase p53/56^{lyn}, a serine kinase and FcεRI associate with α-galactosyl derivatives of ganglioside GD1b in rat basophilic leukemia RBL-2H3 cells. J Biol Chem 1994;269:5249–5254.

77. Swaim WD, Minoguchi K, Oliver C, et al. The anti-ganglioside monoclonal antibody AA4 in-

duces protein tyrosine phosphorylations, but not degranulation, in rat basophilic leukemia cell. J Biol Chem 1994;269:19466–19473.

78. Hamawy MM, Swieter M, Mergenhagen SE, Siraganian RP. Reconstitution of high affinity IgE receptor-mediated secretion by transfecting protein tyrosine kinase pp125^{FAK}. J Biol Chem 1997;272:30498–30503.

79. Sagawa K, Swain SL, Zhang J, Unsworth E, Siraganian RP. Aggregation of the high affinity IgE receptor results in the tyrosine phosphorylation of the surface adhesion protein PECAM-1 (CD31). J Biol Chem 1997; 272:13412–13418.

80. Newman PJ. The biology of PECAM-1. J Clin Invest 1997;99:3–7.

81. DeLisser HM, Newman PJ, Albelda SM. Molecular and functional aspects of PECAM-1/CD31. Immunol Today 1994;15:490–495.

82. Muller WA. The role of PECAM-1 (CD31) in leukocyte emigration: studies in vitro and in vivo. J Leukocyte Biol 1995;57:523–528.

83. Sagawa K, Kimura T, Swieter M, Siraganian RP. The protein tyrosine phosphatase SHP-2 associates with tyrosine-phosphorylated adhesion molecule PECAM-1 (CD31). J Biol Chem 1997;272:31086–31091.

84. DeLisser HM, Chilkotowsky J, Yan HC, Daise ML, Buck CA, Albelda SM. Deletions in the cytoplasmic domain of platelet-endothelial cell adhesion molecule-1 (PECAM-1, CD31) result in changes in ligand binding properties. J Cell Biol 1994;124:195–203.

85. Yan HC, Baldwin HS, Sun J, Buck CA, Albelda SM, DeLisser HM. Alternative splicing of a specific cytoplasmic exon alters the binding characteristics of murine platelet/endothelial cell adhesion molecule-1 (PECAM-1). J Biol Chem 1995;270:23672–23680.

86. Tanaka Y, Albelda SM, Horgan KJ, et al. CD31 expressed on distinctive T cell subsets is a preferential amplifier of β_1 integrin-mediated adhesion. J Exp Med 1992;176:245–253.

87. Piali L, Albelda SM, Baldwin HS, Hammel P, Gisler RH, Imhof BA. Murine platelet endothelial cell adhesion molecule (PECAM-1)/CD31 modulates β_2 integrins on lymphokine-activated killer cells. Eur J Immunol 1993;23:2464–2471.

88. Berman ME, Muller WA. Ligation of platelet/endothelial cell adhesion molecule 1 (PECAM-1/CD31) on monocytes and neutrophils increases binding capacity of leukocyte CR3 (CD11b/CD18). J Immunol 1995;154:299–307.

15
Phosphoinositide-Derived Second Messengers in FcεRI Signaling: PI-3 Kinase Products Control Membrane Topography and the Translocation and Activation of PLC-γ1 in Antigen-Stimulated Mast Cells

Bridget S. Wilson, Sheryll A. Barker, Timothy E. Graham, Janet R. Pfeiffer, and Janet M. Oliver.

In mast cells and basophils, cross-linking the high-affinity IgE receptor, FcεRI, leads to the secretion of histamine and other granule constituents, to changes in adhesive properties, cell shape, and surface topography, and to the synthesis of lipid mediators of inflammation and of cytokines. Key features of this signaling cascade were summarized by contributors to an earlier volume[1] and are updated in other chapters in this volume. In brief, cross-linking IgE-bound receptors with multivalent ligands activates receptor-associated Lyn, a member of the Src tyrosine kinase family whose principal mast cell substrates are immunoreceptor tyrosine-based activation motifs (ITAMs) found in the FcεRI γ- and β-subunit cytoplasmic domains.[2-4] Lyn-mediated receptor subunit phosphorylation in turn permits the recruitment and activation of Syk tyrosine kinase[5] and the phosphorylation of multiple protein substrates.[6] These early events activate signaling pathways leading within minutes to secretion and within hours to cytokine production.

FcεRI-coupled Lyn and Syk activation also leads within minutes to a series of membrane and cytoskeletal responses, including the polymerization of actin and its redistribution into ruffles on the dorsal cell surface (Fig. 15.1A, B)[7] and into characteristic adhesion structures, called actin plaques, at sites of cell–substrate interaction.[8] The transformation of surface topography from a microvillous to a lamellar or ruffled appearance is accompanied by a substantial increase in fluid uptake by macropinocytosis.[7,9] The assembly of actin-plaques is associated with increased spreading and adhesion[8] and with increased activity of the β_1 integrin, VLA-4.[10] The role of membrane/cytoskeletal responses in mast cell and basophil function is incompletely understood. Their contribution to the control of cell migration and tissue localization seems obvious. Additionally, receptor–cytoskeletal interactions have been implicated in the termination of signaling responses.[10] The hypothesis that membrane/cytoskeletal responses help to determine the localization and activity of macromolecular signaling complexes is developed in this chapter.

This review focuses on the roles of phosphatidylinositol-3 kinase (PI-3 kinase) and the γ1- and γ2-isoforms of phospholipase C (PLC-γ1, PLC-γ2) in FcεRI signaling. These enzymes are activated in response to FcεRI cross-linking.[6,11-13] Their substrates

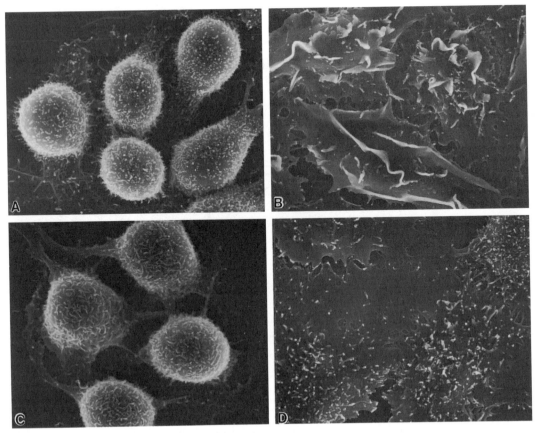

FIGURE 15.1. FcεRI cross-linking changes mast cell shape, adhesion, and surface topography. RBL-2H3 cells were cultured on glass coverslips and primed with DNP-specific IgE. (A) Resting cells are rounded, weakly adherent, and have a microvillous surface. (B) After 10-min stimulation with DNP-BSA, cells flatten, adhere tightly, and develop a highly ruffled surface. (C) Treatment with 10 nM wortmannin (15 min) has little effect on the appearance of resting cells. (D) A + 10 nM, wortmannin abolishes the membrane ruffling response following 10-min stimulation with DNP-BSA. However, the antigen-stimulated increase in adhesion and flattening is unaffected by wortmannin. (From Barker et al.,[13] with permission.)

are membrane phosphoinositides, particularly PtdIns(4,5)P$_2$. Their products are key second messengers in FcεRI signaling pathways. We particularly emphasize recent results that localize PLC-γ1 and PLC-γ2 to different subcellular compartments in activated mast cells and that implicate PI-3 kinase activation in the pathway coupling FcεRI cross-linking to the formation of membrane ruffles and to the translocation and activation of PLC-γ species. Most of the work described here was performed using the RBL-2H3 rat mast cell model. However, where possible, we also report data obtained in mouse and human bone marrow-derived basophils and mast cells and in primary human basophils. All these cells express high levels of the FcεRI and are expected to use the same or similar signaling pathways to generate functional responses.

Phosphatidylinositol 3-Kinase

PI-3 Kinase Expression and Function in Mast Cells

Mast cells contain a form of PI-3 kinase that is composed of a p85 adaptor subunit and a 110-kDa catalytic subunit (Fig. 15.2A).

This dual-specificity enzyme phosphorylates phosphatidylinositols in the D3 position of the inositol ring[14–16] and is also capable of acting on proteins as a serine kinase.[13,17,18] Its preferred substrate is the minor membrane phosphoinositide, PtdIns(4,5)P$_2$, but it also phosphorylates PtdIns and PtdIns(4)P, which are more abundant in plasma membranes. It is a member of a family of kinases that includes the yeast lipid kinase, VPS34p, that utilizes only PtdIns as a substrate, Cpk which uses only PtdIns and PtdIns(4)P as substrate, and several serine kinases that function in primarily in the nucleus.[16,19] A closely related PI-3 kinase heterodimer, PI-3 kinase γ, is regulated by G protein-coupled receptors.[20] The

phosphoinositide products of PI-3 kinase control membrane architecture and determine the localization and activation of a wide range of intracellular signaling molecules.[15,16]

In RBL-2H3 cells, FcεRI crosslinking induces a severalfold increase in the lipid kinase activity of PI-3 kinase (Fig. 15.3A). FcεRI cross-linking also stimulates the serine kinase activity of anti-PI-3 kinase immune complexes.[13] Studies with two chemically unrelated inhibitors of the p85/110 form of PI-3 kinase, wortmannin and LY294002, have established the importance of this enzyme activation in FcεRI signaling. These inhibitors block secretion from human basophils, rat basophilic leukemia cells, and murine bone marrow-derived

A) Phosphatidylinositol 3-Kinase

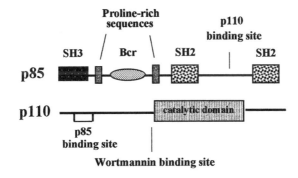

B) Phospholipase Cγ Isotypes

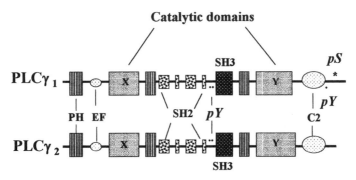

FIGURE 15.2. Diagrams of PI-3 kinase (A) and PLC-γ (B) isoforms. (From Barker et al.,[13] with permission.)

FcεR1 and Syk Chimera Crosslinking Activates PI 3-K

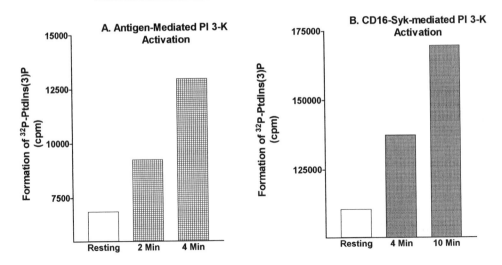

Wortmannin Inhibits PI 3-K Activity

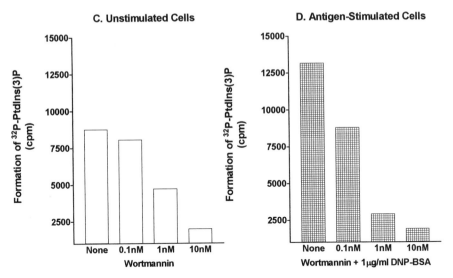

FIGURE 15.3. FcεRI and Syk chimera cross-linking activates wortmannin-sensitive PI-3 kinase activity. Lipid kinase activity was measured using phosphatidylinositol as a substrate for in vitro phosphorylation by anti-p85 immunoprecipitates. (A,B) Time-dependent increase in PI-3 kinase activity induced by the addition of 0.1 μg/ml DNP-BSA to IgE-primed wild-type RBL-2H3 cells (A) or by the addition of anti-CD16 antibodies to Rbl- 2H3 transfectants expressing a transmembrane form of Syk (CD16-CD8-Syk) (B). Transfected cells were a generous gift of J. Brugge and V. Rivera, Ariad Pharmaceuticals.[58] (C,D) Wortmannin causes a concentration-dependent inhibition of PI-3 kinase activity in both unstimulated (C) and antigen-activated (0.1 μg/ml DNP-BSA, 2 min) (D) RBL-2H3 cells. Results are the average of triplicate measurements in a single experiment.

mast cells (BMMC).[12,21-23] Wortmannin also blocks the antigen-induced transformation of the cell surface from a microvillous to a lamellar or ruffled topography (Fig. 15.1C,D), it blocks macropinocytosis, and it attenuates JNK-1 activation in RBL-2H3 cells.[13,24,25] These effects are generally attributed to the inhibition of PI-3 kinase-mediated D3 phosphoinositide synthesis that occurs when wortmannin is added to either resting or antigen-stimulated cells (Fig. 15.3C,D). However, wortmannin also inhibits the protein kinase activity of anti-PI-3 kinase immune complexes prepared from antigen-activated mast cells,[13] and this inhibition may contribute to the functional properties of wortmannin-treated cells.

Importantly, wortmannin has no effect on the activities of Lyn and Syk in immune complex kinase assays.[13] Furthermore, it inhibits only a subset of receptor-mediated responses. In particular, wortmannin fails to prevent FcεRI-induced spreading and actin plaque assembly in antigen-stimulated RBL-2H3 cells.[13] Additionally, wortmannin does not inhibit the activation of the Erk-1/Erk-2 mitogen-activated protein (MAP) kinases[13] or the induction of tumor necrosis factor-α (TNF-α) expression[26] in antigen-stimulated cells. These pharmacological studies place PI-3 kinase downstream of Lyn and Syk and at a relatively early branch point in the FcεRI signaling cascade.

How Is PI-3 Kinase Activated?

We found recently that antibody-induced aggregation of a chimeric, transmembrane form of Syk is sufficient to activate PI-3 kinase in transfected RBL-2H3 cells (Fig. 15.3B). This result places PI-3 kinase activation downstream of Syk in the FcεRI signaling cascade but does not establish the activation mechanism. Although others have suggested that Syk may interact directly with PI-3 kinase,[27] we have failed to detect any stably associated tyrosine kinase in p85 immune complexes using sensitive in vitro kinase assays.[13] We are presently exploring the possibility that an adaptor protein may link PI-3 kinase to Syk in mast cells. Cbl is a likely candidate: this adaptor protein was

shown recently to coprecipitate with Syk in RBL-2H3 cells[28] and to bind a p85β/p110 form of PI-3 kinase in Jurkat T cells and A20 B cells.[29]

How Does PI-3 Kinase Activation Transduce Signals?

Recent studies have identified motifs within signaling and structural proteins that specifically recognize and interact with the D3 phosphoinositide products of activated PI-3 kinase, resulting in changes in protein distribution and function. The best-studied of these motifs are src homology-2 (SH-2) and pleckstrin homology (PH) domains, which bind to specific inositol phospholipids, providing a mechanism for protein association with membranes.[30,31] The PH module in particular may serve primarily as a membrane association motif,[32] although it can also bind to counterstructures within certain proteins such as the βγ-subunits of G proteins[33] and to soluble inositol phosphate head groups.[34] While there has been significant progress in the structural analysis of PH domains,[32] it is not yet possible to predict the preferred binding partners for individual PH domains. For example, the PH domains of PLC-δ1, pleckstrin, dynamin, and mSos1 bind specifically to PtIns(4,5)P$_2$,[34,35] while the PH domains of other proteins have been demonstrated to bind exclusively to D3-phosphorylated phosphoinositols. Signaling proteins that specifically bind PI-3 kinase products include neurogranin, synaptotagmin, Akt, Bruton's tyrosine kinase, several PKC isoforms, GRP-1, cytohesin-1, and others.[16,36,37] Additionally, D3 phosphoinositides bind to PH domains in a number of cytoskeletal proteins that regulate actin polymerization, including profilin and gelsolin.[16]

The consequences of protein binding to D3-phosphorylated phosphoinositols are being studied in many systems. In some cases, protein–lipid interaction appears to improve protein presentation to enzymes. Thus, Lu and Chen[36] reported that binding a neurogranin-derived peptide to PtdIns(3,4,5)P$_3$-containing vesicles markedly enhanced its phosphorylation by PKC. Protein–lipid interaction may also directly alter enzymatic activity. For example, binding the pleckstrin homology (PH) domain of the serine kinase, Akt, to PtdIns(3)P or PtdIns(3,4)P$_2$ results in the activation of Akt, a serine/threonine kinase that appears to have antiapoptotic activity.[38–40] RBL-2H3 cells express Akt (also known as RAC-PK or PKB), and we have noted its wortmannin-sensitive activation following FcεRI cross-linking. D3-phosphorylated phosphoinositides are also known to directly activate protein kinase C types ε,η, and δ[41]; of these, RBL-2H3 cells express protein kinase C (PKC-ε and PKC-δ).[42] New results, described next, additionally implicate PI-3 kinase in the localization and activation of specific PLC isoforms.

How Does PI-3 Kinase Stimulate Membrane Ruffling?

We and others are exploring two, not mutually exclusive, hypotheses to explain how PI-3 kinase products control membrane architecture. First, it is possible that PI-3 kinase activation generates locally D3-phosphoinositide-rich membrane regions that recruit target proteins through SH-2 or PH domain interactions. These target proteins may be cytoskeletal proteins such as gelsolin or profilin that could in turn direct actin polymerization to stabilize ruffles. They may alternatively be adaptor proteins that recruit specific cytoskeletal proteins responsible for the control of cell shape. The latter model is supported by Ma et al.,[43] who showed that overexpression of pleckstrin in Cos cells led to its association with the membrane and the formation of membrane ruffles. These responses depended on an intact NH$_2$-terminal PH domain, very likely needed for the association of pleckstrin with membrane phosphoinositides. They also required the serine/threonine phosphorylation of the pleckstrin, perhaps needed for its interactions with actin and other cytoskeletal elements that might stabilize the ruffles. Second, work by Hall and others has strongly implicated members of the Rho family of small GTP-binding proteins in ruffling and spreading responses.[44] In particular, cells transfected with constitutively active Rac typically show exaggerated

membrane ruffles. Recent evidence that transfectants expressing constitutively active PI-3 kinase also show exaggerated ruffling implicates PI-3 kinase in the pathway linking receptors to Rac-dependent ruffling.[45] The mechanism is unknown, but may involve interactions between D3 phosphoinositides and a PH-domain-containing GTP exchange factor for Rac.[16] Such an interaction could activate Rac and also localize Rac-induced actin polymerization to membrane regions defined by PI-3 kinase.

Phospholipase C

PLC Expression and Function in Mast Cells

PLC-γ1 and PLC-γ2 are monomeric enzymes that contain an amino-terminal Ca^{2+}-binding region, one intact and one split pleckstrin homology (PH) domain, two SH-2 domains, one SH-3 domain, a C2 domain, and a split catalytic domain (Fig. 15.2B).[46] PLC-γ1 is ubiquitously expressed, and its absence leads to embryonic lethality.[47] The distribution of PLC-γ2 is more restricted, with expression primarily limited to cells of hematopoetic lineage.

RBL-2H3 cells contain both PLC-γ1 and PLC-γ2, with PLC-γ2 being the more abundant species.[48] Both PLC-γ isoforms are phosphorylated on tyrosine and serine following antigen stimulation.[6,11,48] Anti-PLC-γ1 and anti-PLC-γ2 immune complexes from antigen-activated cells both show increased activity over control in immune complex phospholipase assays, indicating that both isoforms are activated by FcεRI cross-linking (Fig. 15.4A,B).[48] Both isoforms are translocated to particulate fractions that include both plasma membrane and organellar membranes in response to FcεRI cross-linking[49]; however, as described next, PLC-γ1 associates primarily with the plasma membrane of RBL-2H3 cells, whereas PLC-γ2 associates primarily with Golgi membranes.

The antigen-induced stimulation of PLC activity is a particularly well-known and important feature of the FcεRI signaling cascade.[50–52]

FIGURE 15.4. FcεRI cross-linking activates both isoforms of phospholipase C (PLC-γ). Phospholipase activity was measured in anti-PLC-γ immune complexes prepared from resting and antigen-stimulated RBL-2H3 cells from the hydrolysis of [^{3}H]-PtdIns(4,5)P$_2$ in a Triton-based mixed micelle assay. *Solid bars* show activation of PLC-γ1 (A) and PLC-γ2 (B) at 2 and 10 min after addition of antigen (0.1 μg/ml DNP-BSA) to control cells; *hatched bars* show that the antigen-stimulated activation of PLC-γ1, but not PLC-γ2, is significantly blocked by 15-min preincubation with 10 nM wortmannin. (From Barker et al.,[48] with permission.)

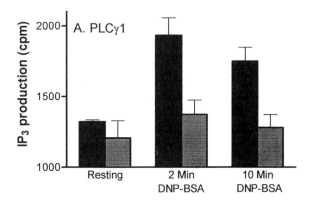

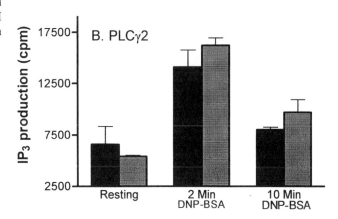

The PLC product, inositol 1,4,5-trisphosphate [Ins(1,4,5)P$_3$], is directly responsible for the release of cytoplasmic Ca^{2+} stores, and, because Ca^{2+} influx occurs by the capacitative pathway, is also indirectly responsible for the influx of Ca^{2+} in antigen-activated basophils and mast cells. Antigen-stimulated Ca^{2+} influx is the principal event in the pathways leading to the activation of Ca^{2+}-dependent kinases such as p110, and to functional responses including secretion and the production of selected cytokines such as IL-4. The other main PLC product, diacylglycerol (DAG), activates PKC and may additionally activate proteins such as Vav that also contain the C1 homology domain which binds DAG.[53] Based on studies with the DAG analog, phorbol myristate acetate (PMA), PLC-γ-mediated DAG production has been linked to the stimulation of membrane ruffling, spreading, F-actin assembly, and macropinocytosis in antigen-activated cells. PKC isoforms are also implicated in the serine phosphorylation and downregulation of PLC-γ.[54]

Role of Tyrosine Phosphorylation in PLC-γ Activation

Early pharmacological studies established that FcεRI-mediated tyrosine kinase activation is required for antigen-stimulated phospholipase activity. Thus the addition of broad-spectrum tyrosine kinase inhibitors, including herbimycin and tyrphostin AG490, to intact RBL-2H3 cells or mouse BMMC markedly inhibited tyrosine phosphorylation of PLC-γ1 and inositol phosphate formation.[11,55,56] Another general tyrosine kinase inhibitor, genistein, blocked Ins(1,4,5)P$_3$ release in streptolysin O-permeabilized RBL-2H3 cells.[57]

Syk was specifically implicated in antigen-induced PLC activation by studies with piceatannol, a more selective kinase inhibitor that blocked antigen-induced Syk but not Lyn activation and concomitantly blocked both Ins(1,4,5)P$_3$ release and calcium mobilization in antigen-stimulated cells.[3] Supporting a role for Syk in PLC-γ activation, Rivera and Brugge[58] showed that cross-linking chimeric, transmembrane forms of Syk are sufficient to stimulate both tyrosine phosphorylation of PLC-γ1 and calcium mobilization. Conversely, Siraganian and colleagues showed that a variant of RBL-2H3 cells lacking detectable Syk expression could not couple FcεRI cross-linking to either histamine release or phosphorylation of PLC-γ proteins.[59] Reintroduction of Syk expression by stable transfection restored tyrosine phosphorylation of both isoforms of PLC-γ as well as Ca^{2+} mobilization and degranulation. Additionally, mast cells have been generated from precursors in the embryonic livers from Syk knockout mice. Like RBL-2H3 cells exposed to the Syk-selective inhibitor, piceatannol, these cells respond to FcεRI cross-linking by activating Lyn and phosphorylating receptor subunits. However, they demonstrate no further tyrosine phosphorylation, no Ins(1,4,5)P$_3$ synthesis, no Ca^{2+} mobilization, and no functional responses.[60] Thus, although both Lyn and Syk immunoprecipitates from antigen-activated RBL-2H3 cells can induce the phosphorylation and activation of highly purified preparations of PLC-γ1 and PLC-γ2 (our unpublished studies using purified enzymes kindly provided by S. G. Rhee, NIH), it is likely that Syk is the key enzyme activating both PLC-γ isoforms in the intact mast cell.

The mechanism of Syk-mediated PLC-γ activation is not yet clear. It was demonstrated recently that immobilized, diphosphorylated FcεRIβ- and -γ ITAM peptides can precipitate a complex mixture of signaling molecules from RBL-2H3 lysates, including PLC-γ1, Shc, Grb-2, Syk, and Lyn.[61] However, Fukamachi et al.[62] failed to detect FcεRI subunits or kinases that comigrated with Syk or Src family members in anti-PLC-γ1 immune complexes isolated from BMMC. Other recent studies have implicated the 78-kDa Bruton's tyrosine kinase (Btk) in the coupling of Syk to PLC-γ activation in immunoreceptor signaling pathways. Although not yet studied in mast cells, analyses of B-cell-receptor (BCR) signaling

in wild-type and genetically modified DT40 chicken B cells showed that PLC-γ2 phosphorylation is abolished in Syk-deficient cells and markedly reduced in Btk-deficient cells, building a model for the sequential roles of these two tyrosine kinases.[63,64] Unfortunately, only PLC-γ2 is expressed in DT40 cells, so the relationship of Btk to PLC-γ1 activation remains unknown.

Tyrosine Phosphorylation Is Necessary But Not Sufficient to Activate PLC-γ

Both isoforms of PLC-γ can be activated in vitro by incubation with ATP and selected tyrosine kinases, as first described using epidermal growth factor (EGF) receptor preparations.[65] However, the relationship of phosphorylation to activation is complex, even in the case of the well-studied growth factor receptor systems. For example, substitution of phenylalanine for tyrosine at amino acid 783 of PLC-γ1 results in a protein that can associate with the platelet-derived growth factor (PDGF) receptor cytoplasmic domain and can become phosphorylated on other tyrosine sites, but is not activated, while similar substitutions at Y1254 partially inhibit, and at Y771 enhance, the PDGF-induced activation of PLC-γ1.[66,67] Furthermore, a C-terminal mutant of the EGF receptor, which lacks Y922 as a binding site for one of the PLC-γ-SH-2 domains, is capable of tyrosine-phosphorylating PLC-γ1 without stimulating PtdIns(4,5)P$_2$ hydrolysis.[68]

In immune system cells, there are now several recent reports in BCR and T-cell-receptor (TCR) signaling cascades where tyrosine phosphorylation of PLC-γ is seen in the absence of measurable Ins(1,4,5)P$_3$ production. Pao et al.[69] showed that PLC-γ1 and PLC-γ2 could be tyrosine phosphorylated in CD45– plasmacytoma cells following BCR ligation despite very low levels of inositol phosphate production and a complete block in calcium mobilization. Motto et al.[70] and Jensen et al.,[71] working with chimeric membrane proteins to model TCR receptor signaling, have also reported experimental conditions in which PLC-γ1 tyrosine phosphorylation can be induced in the absence of Ins(1,4,5)P$_3$ production.

Different Distributions of PLC-γ1 and PLC-γ1 in Resting and Activated Cells

We hypothesized that PLC-γ activation in antigen-stimulated RBL-2H3 cells may depend on translocation as well as tyrosine phosphorylation and tested this hypothesis by immunofluorescence and immunoelectron microscopy.[48] We discovered that a small percentage of PLC-γ1 associates with the plasma membrane of resting RBL-2H3 cells and, where present at the membrane, is primarily associated with membrane projections such as the leading edges of lamellae (illustrated by immunofluorescence labeling in Fig. 15.5A). Following FcεRI cross-linking, PLC-γ1 was strongly associated with membrane ruffles (Fig. 15.5B). The results of immunogold electron microscopy confirmed that PLC-γ1 is translocated to the ruffles of antigen-activated cells. Thin sections of LR White-embedded RBL-2H3 cells were incubated with anti-PLC-γ1 antibodies, followed by 15-nm protein A-gold particles, and the number and distribution of particles localizing PLC-γ isoforms was analyzed in the resulting micrographs. We showed first that the numbers of anti-PLC-γ1-gold particles at the plasma membrane increase approximately twofold following antigen stimulation.[48] Closer analysis showed that 70% of the plasma membrane-associated PLC-γ1 was specifically found in the membrane ruffles of activated cells. We speculate that the targeting of activated PLC-γ1

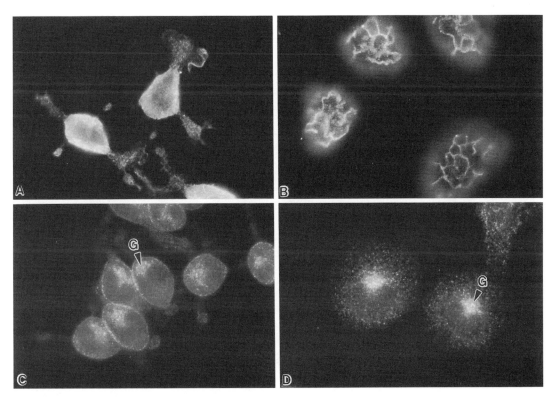

FIGURE 15.5. Immunofluorescence studies of fixed, permeabilized RBL-2H3 show distinct intracellular distributions for PLC-γ isoforms. (A) Isoform-specific antibody to PLC-γ1 shows a generally diffuse distribution in resting RBL-2H3 cells, with some label found on leading edges. (B) After 10-min stimulation with antigen, PLC-γ1 is strongly associated with membrane ruffles. In contrast, PLC-γ2 is found in the Golgi region (*arrowheads*) and in small clusters associated with the plasma membrane of both resting (C) and antigen-stimulated cells (D). (From Barker et al.,[48] with permission.)

to membrane ruffles places it ideally for access to phospholipid substrate in the membrane of antigen-stimulated mast cells. In contrast with PLC-γ1, the majority of PLC-γ2 was concentrated in the Golgi region of both resting and activated cells (Fig. 15.5C,D). The substantial proportion of PLC-γ2 that resided near the Golgi complex provided the first suggestion that this isoform may be limited in its ability to contribute to antigen-stimulated Ins(1,4,5)IP$_3$ production.

Our discovery that PLC-γ1 associates with membrane ruffles in RBL-2H3 cells contrasts with data obtained in fibroblasts, where a truncated form of PLC-γ1, containing only the SH-3 domain, localized to stress fibers.[72] RBL-2H3 cells make actin plaques, not stress fibers, following antigen stimulation. However, even when they are induced to form stress fibers by incubation with PMA, no association of PLC-γ1 with these structures was observed. Importantly, PLC-γ1 also failed to associate with ruffles that are induced by PMA treatment. These results suggest that PMA-induced ruffles

are different from antigen-induced ruffles, and in particular may lack D3-phosphoinositides needed for PLC-γ1 recruitment. Alternatively, ruffling alone may be an insufficient signal for PLC-γ1 translocation, just as tyrosine phosphorylation is an insufficient signal for full PLC-γ activation.

PI-3 Kinase Controls the Translocation and Activation of PLC-γ1

Differential Sensitivity of PLC-γ1 and PLC-γ2 to Wortmannin Inhibition

Very recent studies from this and other laboratories implicate PI-3 kinase in the regulation of PLC-γ activity in immunoreceptor signaling pathways. We showed that nanomolar wortmannin concentrations inhibit Ins(1,4,5)P$_3$ synthesis following FcεRI cross-linking by 50% to 70% in RBL-2H3 cells (Fig. 15.6).[13] Similar results were reported by Vossebeld et al.[73] in

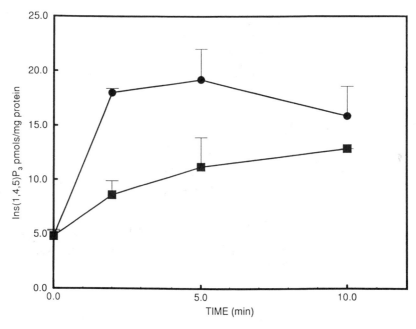

FIGURE 15.6. Wortmannin inhibits Ins(1,4,5)P$_3$ production in antigen-stimulated RBL-2H3 cells. Anti-DNP-IgE-Primed control cells (*circles*) or wortmannin-treated (*squares*) cells were activated with DNP-BSA for indicated times. Ins(1,4,5)P$_3$ levels were measured in cell extracts by a radioreceptor assay. (From Barker et al.,[13] with permission.)

human neutrophils stimulated via Fcγ receptors. These results were unexpected because the inhibition of PI-3 kinase might be expected to increase the availability of PtdIns(4,5)P$_2$ as a substrate for PLC-γ and were therefore pursued by use of in vitro immune complex phospholipase assays.[48] Figure 15.4A,B shows the results of these further studies. It is clear that wortmannin specifically blocks the antigen-stimulated activation of PLC-γ1 by tyrosine phosphorylation while having no effect on PLC-γ2 activity.[48] These data support evidence from the localization studies that PLC-γ1 and PLC-γ2 may be regulated separately.

As described, wortmannin inhibits antigen-induced membrane ruffling. Not surprisingly, the inhibition of ruffling was associated not only with the inhibition of PLC-γ1 activation but also with the inhibition of PLC-γ1 translocation to the plasma membrane.

How Might PI-3 Kinase Or Its Lipid Products Control PLC-γ1 Distribution and Activity?

We recently found that PI-3 kinase contributes to the activation of both PLCγ isoforms through interaction of its lipid products, the 3-phosphorylated phosphoinositides, with the PLC-γ SH-2 or PH domains (Barker et al., in press). This is consistent with data that the PH domain-mediated binding of PLC-δ to PtsIns(4,5)P$_2$ stimulates enzymatic activity.[74] Additionally, Abrams et al.[75] showed that expression of platelet pleckstrin in Cos-1 or HEK-293 cells markedly inhibited phosphatidylinositol hydrolysis mediated by both PLC-β and PLC-γ; this inhibition was dependent upon the presence of the PH domain at the amino-terminus of pleckstrin that may compete for binding to phosphoinositides at the plasma membrane.

The observation that 3-phosphorylated phosphoinositides both directly activate the PLC-γ's and also are a requirement for PLC-γ1 tyrosine phosphorylation suggests a complex relationship between PI 3-kinase and PLCγ regulation. We speculate that PLC-γ1 recruitment could occur indirectly via a membrane-associated platform–adaptor complex whose

assembly is controlled by PI-3 kinase and its metabolites. Once at the membrane, PLC-γ1 would be in close proximity to PtdIns(4,5)P$_2$ and its other substrates, PtdIns and PtdIns(4)P, as well as to membrane-associated tyrosine kinases. Consistent with the adaptor hypothesis, Motto et al.[70] transfected Jurkat T cells with a chimeric membrane phosphatase engineered to bind the adaptor protein, Grb-2, via its SH-2 domain and cause its dephosphorylation. This manipulation blocked inositol phosphate hydrolysis despite the persistence of TCR-coupled PLC-γ1 tyrosine phosphorylation, presumably because phosphorylated Grb-2 is required for the membrane localization of PLC-γ.

It remains possible that the serine kinase activity of PI-3 kinase contributes to PLC-γ redistribution and activation. A large fraction of ^{32}P incorporated into PLC-γ isoforms following IgE receptor stimulation is on serine,[6,48] and we found that wortmannin specifically inhibits antigen-stimulated PLC-γ1 phosphorylation of both serine and tyrosine residues in RBL-2H3 cells.[48] Previous studies have strongly implicated PKA- and PKC-mediated serine phosphorylation in PLC-γ1 inactivation.[42,54,76,77] On the other hand, there is also evidence that treating RBL-2H3 cells with serine/threonine kinase inhibitors reduces, rather than enhances, both the antigen-stimulated tyrosine phosphorylation of PLC-γ1 and the hydrolysis of total inositol phospholipids.[78,79] This result suggests that a serine kinase is upstream of the translocation of PLC-γ1 to the membrane, or, alternatively, that PLC-γ possesses other serine phosphorylation sites with positive effects on PLC-γ activity. It is thus conceivable that PI-3 kinase directly phosphorylates PLC-γ1 on a stimulatory serine residue. Alternatively, the lipid products of PI-3 kinase may directly stimulate another serine kinase, such as Akt,[80,81] that in turn phosphorylates either PLC-γ1 or binding partners which contribute to its translocation.

Conclusions

The importance of phosphoinositide metabolites as second messengers in receptor-activated signaling pathways has been recognized for several decades. Nevertheless, our understanding of how the metabolism of a minor membrane lipid, PtdIns(4,5)P$_2$, contributes to the activation of diverse signaling pathways remains incomplete. Here, we have summarized current data from our own and other laboratories on the basic mechanisms controlling the activation of PI-3 kinase and PLC-γ isoforms in mast cells and basophils and the consequences of their activation for FcεRI-mediated signaling. New features of PI-3 kinase regulation and function include the ability of the activated enzyme to phosphorylate proteins on serine residues and the contribution of its D3 phosphoinositide products to the transformation of the cell surface from a microvillous to a lamellar topography in antigen-stimulated RBL-2H3 cells. It was previously known that both PLC-γ isoforms are tyrosine phosphorylated in response to FcεRI cross-linking. However, tyrosine phosphorylation alone is unlikely to control their activities. New features of PLC-γ regulation include evidence that PLC-γ1 and PLC-γ2 have distinctly different distributions in mast cells and that both isoforms can be directly activated by the lipid products of PI-3 kinase. Importantly, we show that PI-3 kinase activation is required for the phosphorylation of PLC-γ1, but not PLC-γ2, in antigen-stimulated mast cells. The mechanism is not fully defined, but very likely involves a process of PI-3 kinase-mediated membrane remodeling that creates binding sites for PLC-γ1 and permits its activation by tyrosine phosphorylation. Previously, the membrane/cytoskeletal and secretory responses to FcεRI cross-linking appeared to be regulated separately and serve separate purposes. The discovery that PLC-γ1, critical to Ca^{2+} influx and secretion, is activated in part by its translocation to membrane ruffles provides a new level of integration between these response pathways.

We emphasize that PLC-γ1 is not alone in showing a marked redistribution following FcεRI cross-linking in mast cells. Previous work has demonstrated that high-affinity IgE receptors are distributed randomly in the plasma membrane of resting RBL-2H3 cells.[82] Following cross-linking, the receptors rapidly redistribute away from areas of membrane curvature and are restricted to the relatively flat, or planar, regions of the membrane at the base of ruffles. Whereas high-affinity IgE receptors are excluded from membrane projections of antigen-stimulated cells, other signaling and structural proteins are markedly enriched within lamellae and microvilli. In addition to PLC-γ1, our immunofluorescence microscopy studies have revealed the specific association of PI-3 kinase, inositol 5-phosphatases, lyn, and csk with ruffles in activated RBL-2H3 cells. Work in fibroblasts has documented a similar localization of phospholipase A$_2$,[83] Ras, and Grb-2[72] to microvilli and membrane ruffles; the transmembrane protein, E-selectin, is localized to the tips of microvilli in resting neutrophils.[84] A large array of actin-binding proteins also localize to membrane ruffles.[85] Thus, membrane ruffles, stabilized by association with cytoskeletal elements, may represent important sites of assembly for macromolecular signaling complexes.

Acknowledgments. Studies in the authors' laboratories were supported in part by NIH grants GM-50562 (B.S.W.), GM-49814 (J.M.O.), and HL56384 (J.M.O. and B.S.W.) and by developmental funds from a Howard Hughes Medical Institute grant to the University of New Mexico Medical School (B.S.W.). S.B. and TG are predoctoral (S.B.) and medical student (T.G.) fellows of the Howard Hughes Medical Institute. This chapter is dedicated to our late colleague, Grace Deanin, who shared with us her enthusiastic quest to understand phosphoinositide regulation. The authors thank Kevin Caldwell for invaluable help in establishing assays for PLC-γ and PI-3 kinase activity.

References

1. Hamawy MM (ed). The High Affinity IgE Receptor (FcεRI): structure and Function. Austin, TX: RG Landes, 1997.
2. Eiseman E, Bolen JB. Engagement of the high affinity IgE receptor activates src protein-related tyrosine kinases. Nature (Lond) 1992; 355:78–80.

3. Oliver JM, Burg DL, Wilson BS, et al. Inhibition of mast cell FcεRI-mediated signaling and effector function by the Syk-selective inhibitor, piceatannol. J Biol Chem 1994;269:29697–29703.

4. Cambier JC. Antigen and Fc receptor signaling. J Immunol 1995;15:3281–3285.

5. Hutchcroft JE, Geahlen RL, GG Deanin, et al. FcεRI-mediated tyrosine phosphorylation and activation of the 72-kDa protein-tyrosine kinase, PTK72, in RBL-2H3 rat tumor mast cells. Proc Natl Acad Sci USA 1992;89:9107–9111.

6. Li W, Deanin GG, Margolis B, et al. FcεRI-mediated tyrosine phosphorylation of multiple proteins, including phospholipase Cγ1 and the receptor $\beta\gamma_2$ complex, in RBL-2H3 rat basophilic leukemia cells. Mol Cell Biol 1992;12:3176–3182.

7. Pfeiffer JR, Deanin GG, Seagrave JC, et al. Membrane and cytoskeletal changes associated with IgE-mediated serotonin release in rat basophilic leukemia cells. J Cell Biol 1985;101:2145–2155.

8. Pfeiffer JR, Oliver JM. Tyrosine kinase-dependent assembly of actin plaques linking FcεRI cross-linking to increased cell-substrate adhesion in RBL-2H3 tumor mast cells. J Immunol 1994;152:270–279.

9. Wilson BS, Deanin GG, Standefer JC, et al. Depletion of guanine nucleotides with mycophenolic acid suppresses IgE receptor-mediated degranulation in rat basophilic leukemia cells. J Immunol 1989;143:259–265.

10. Oliver JM, Pfeiffer JR, Wilson BS. Regulation and roles of the membrane, cytoskeletal and adhesive responses of RBL-2H3 rat tumor mast cells to FcεRI cross-linking. In: MM Hamawy, ed. The High-Affinity IgE Receptor (FcεRI): Structure and Function. Austin, TX: RG Landes, 1997:139–172.

11. Park DJ, Min HK, Rhee SG. IgE-induced tyrosine phosphorylation of phospholipase C-γ1 in rat basophilic leukemia cells. J Biol Chem 1991;266:24237–24240.

12. Yano H, Nakanishi S, Kimura K, et al. Inhibition of histamine secretion by wortmannin through the blockade of phosphatidylinositol 3-kinase in RBL-2H3 cells. J Biol Chem 1993;268:25846–25856.

13. Barker SA, Caldwell KK, Hall A, et al. Wortmannin blocks lipid and protein kinase activities associated with PI-3 kinase and inhibits a specific subset of responses induced by FcεRI cross-linking. Mol Biol Cell 1995;6:825–839.

14. Stephens LR, Jackson TR, Hawkins PT. Ago-nist-stimulated synthesis of phosphatidylinositol 3,4,5-trisphosphate: a new intracellular signaling system? Biochim Biophys Acta 1993;1179:27–75.

15. Kapeller R, Cantley LC. Phosphatidylinositol 3-kinase. Bioessays 1994;16:565–576.

16. Toker A, Cantley L. Signaling through the lipid products of phosphoinositide 3-OH kinase. Nature (Lond) 1997;387:673–676.

17. Dhand R, Hiles I, Panayotou G, et al. PI-3 kinase is a dual enzyme: autoregulation by an intrinsic protein-serine kinase activity. EMBO J 1994;13:522–533.

18. Lam K, Carpenter CL, Ruderman NB, et al. The phosphatidylinositol 3-kinase serine kinase phosphorylates IRS-1. J Biol Chem 1994; 269:20648–20652.

19. Abraham RT. Phosphatidylinositol 3-kinase related kinases. Curr Opin Immunol 1996;8:412–418.

20. Stephens LR, Eguinoa A, Erdjument-Bromage H, et al. The G βγ sensitivity of a PI3K is dependent upon a tightly associated adaptor, p101. Cell 1997;89:105–114.

21. Knol EF, Koenderman L, Jul FPJ, et al. Differential activation of human basophils by anti-IgE and formyl-methionyl-leucyl-phenylalanine: indications for protein kinase C-dependent and -independent activation pathways. Eur J Immunol 1991;21:881–885.

22. Kitani S, Teshima R, Morita Y, et al. Inhibition of IgE-mediated histamine release by myosin light chain kinase inhibitors. Biochem Biophys Res Commun 1992;183:48–54.

23. Marquardt DL, Alongi JL, Walker JL. The phosphatidylinositol 3-kinase inhibitor wortmannin blocks mast cell exocytosis but not IL-6 production. J Immunol 1996;156:1942–1945.

24. Yamada K, Jelsema CL, Beaven MA. Certain inhibitors of protein serine/threonine kinases also inhibit tyrosine phosphorylation of phospholipase Cγ1 and other proteins and reveal distinct roles for tyrosine kinase(s) and protein kinase C in stimulated, rat basophilic RBL-2H3 cells. J Immunol 1991;149:1031–1037.

25. Ishizuka T, Oshiba A, Sakata N, et al. Aggregation of the FcεRI on mast cells stimulates c-Jun amino-terminal kinase activity. J Biol Chem 1996;271:12762–12766.

26. Pendl GG, Prieschl EE, Thumb W, et al. Effects of phosphatidylinositol-3-kinase inhibitors on degranulation and gene induction in allergically triggered mouse mast cells. Int Arch Allergy Immunol 1997;112:392–399.

27. Yanagi S, Sada K, Tohyama Y, et al. Transloca-
    tion, activiation and association of protein-
    tyrosine kinase (p72syk) with phosph-
    atidylinositol 3-kinase are early events during
    platelet activation. Eur J Biochem 1994;224:329–
    333.
28. Ota Y, Beitz LO, Scharenberg AM, et al. Char-
    acterization of Cbl tyrosine phosphorylation and
    a Cbl-Syk complex in RBL-2H3 cells. J Exp Med
    1996;184:1713–1723.
29. Hartley D, Meisner H, Corvera S. Specific
    association of the β isoform of the p85 subunit of
    phosphatidylinositol-3 kinase with the proto-
    oncogene c-cbl. J Biol Chem 1995;270:18260–
    18263.
30. Harlan JE, Hajduk PJ, Yoon HS, et al.
    Pleckstrin homology domains bind to
    phosphatidylinotisol-4,5-bisphosphate. Nature
    (Lond) 1994;371:168–170.
31. Rameh LE, Chen S-S, Cantley LC.
    Phosphatidylinositol 3,4,5P$_3$ interacts with SH-2
    domains and modulates PI-3 kinase association
    with tyrosine-phosphorylated proteins. Cell
    1995;83:821–830.
32. Hemmings BA. PH domains—a universal mem-
    brane adaptor. Science 1997;275:1899.
33. Touhara K, Inglese J, Pitcher JA. Binding of
    G protein βγ subunits to pleckstrin homology
    domains. J Biol Chem 1994;269:10217–10220.
34. Lemmon MA, Ferguson KM, O'Brien R, et al.
    Specific and high affinity binding of inositol
    phosphates to an isolated pleckstrin homology
    domain. Proc Natl Acad Sci USA
    1995;92:10472–10476.
35. Kubiseski TJ, Chook YM, Parris WE, et al.
    High affinity binding of the pleckstrin homology
    domain of mSos1 to phosphatidylinositol
    4,5-bisphosphate. J Biol Chem 1997;272:1799–
    1804.
36. Lu P-J, Chen C-S. Selective recognition of
    phosphatidylinositol 3,4,5-trisphosphate by a
    synthetic peptide. J Biol Chem 1997;272:466–
    472.
37. Klarlund JK, Builherme A, Holik JJ, et al.
    Signaling by phosphoinositide-3,4,5-trisph-
    osphate through proteins containing pleckstrin
    and Sec7 homology domains. Science
    1997;275:1927–1930.
38. Franke TF, Yang S, Chan TO, et al. The
    protein kinase encoded by the Akt proto-
    oncogene is a target of the PDGF-activated
    phosphatidylinositol 3-kinase. Cell 1995;81:727–
    736.
39. Klippel A, Kavanaugh WM, Pot D, et al. A

40. specific product of phosphatidylinositol 3-kinase
    directly activates the protein kinase Akt through
    its pleckstrin homology domain. Mol Cell Biol
    1997;17:338–344.
40. Frech M, Andjelkovic M, Ingley E, et al. High
    affinity binding of inositol phosphates and
    phosphoinositides to the pleckstrin homology
    domain of RAC/protein kinase B and their
    influence on kinase activity. J Biol Chem
    1997;272:8474–8481.
41. Toker A, Meyer M, Reddy KK, et al. Activation
    of protein kinase C family members by the
    novel polyphosphoinositides PtdIns-3,4-P$_2$ and
    PtdIns-3,4,5-P$_3$. J Biol Chem 1994;269:32358–
    32367.
42. Ozawa K, Yamada K, Kazanietz MG, et al.
    Different isozymes of protein kinase C mediate
    feedback inhibition of phospholipase C and
    stimulatory signals for exocytosis in rat RBL-
    2H3 cells. J Biol Chem 1993;268:2280–2283.
43. Ma AD, Brass LF, Abrams CS. Pleckstrin
    associates with plasma membranes and induces
    the formation of membrane projections:
    requirements for phosphorylation and the NH$_2$-
    terminal PH domain. J Cell Biol 1997;136:1071–
    1079.
44. Hall A. Small GTP-binding proteins and the
    regulation of the actin cytoskeleton. Annu Rev
    Cell Biol 1994;10:31–54.
45. Reif K, Nobes CD, Thomas G, et al.
    Phosphatidylinositol 3-kinase signals activate a
    selective subset of Rac/Rho-dependent effector
    pathways. Curr Biol 1996;6;11:1445–1455.
46. Lee SB, Rhee SG. Significance of PIP$_2$ hydrolysis
    and regulation of phospholipase C isozymes.
    Curr Biol 1995;7:183–189.
47. Ji Q-S, Winnier GE, Niswender KD, et al.
    Essential role of the tyrosine kinase substrate
    phospholipase C-γ1 in mammalian growth
    and development. Proc Natl Acad Sci USA
    1997;94:2999–3003.
48. Barker SA, Caldwell KK, Pfeiffer JR, et al.
    Wortmannin-sensitive phosphorylation, translo-
    cation and activation of PLC-γ1, but not PLC-γ2,
    in antigen-stimulated RBL-2H3 mast cells. Mol
    Biol Cell 1997;9:483–496.
49. Atkinson TP, Lee C-W, Rhee SG, et al.
    Orthovanadate induces translocation of phos-
    pholipase C-γ1 and -γ2 in permeabilized mast
    cells. J Immunol 1993;151:1448–1455.
50. Beaven MA, Moore JP, Smith GA, et al. The
    calcium signal and phosphatidylinositol break-
    down in 2H3 cells. J Biol Chem 1984;259:7137–
    7142.

51. Cunha-Melo JR, Dean NM, Moyer JD, et al. The kinetics of phosphoinositide hydrolysis in rat basophilic leukemia (RBL-2H3) cells varies with the type of IgE receptor cross-linking agent used. J Biol Chem 1987;262:11455–11463.

52. Ali H, Cunha-Melo JR, Beaven MA. Receptor-mediated release of inositol 1,4,5-trisphosphate and inositol 1,4-bisphosphate in rat basophilic leukemia RBL-2H3 cells permeabilized with streptolysin O. Biochim Biophys Acta 1989;1010:88–99.

53. Hurley JH, Newton AC, Parker PJ, et al. Taxomomy and function of C1 protein-kinase-C homology domains. Protein Sci 1997;6:477–480.

54. Rhee SG, Lee C-W, Jhon D-Y. Phospholipase C isozymes and modulation by cAMP-dependent protein kinase. In: Brown BL, Dobson RM, eds. Advances in Second Messenger and Phosphoprotein Research. Vol 23. New York: Raven Press, 1993:57–64.

55. Kawakami T, Inagaki N, Takaei M, et al. Tyrosine phosphorylation is required for mast cell activation by FcεRI cross-linking. J Immunol 1992;148:3513–3519.

56. Schneider H, Cohen-Dayag A, Pecht I. Tyrosine phosphorylation of phospholipase Cγ1 couples the Fcε receptor mediated signal to mast cell secretion. Int Immunol 1992;4:447–453.

57. Deanin GG, Martinez AM, Pfeiffer JR, et al. Tyrosine kinase-dependent phosphatidylinositol turnover and functional responses in the FcεRI signaling pathway. Biochem Biophys Res Commun 1991;179:551–557.

58. Rivera VM, Brugge JS. Clustering of Syk is sufficient to induce tyrosine phosphorylation and release of allergic mediators from rat basophilic leukemia cells. Mol Cell Biol 1995;15:1582–1590.

59. Zhang J, Berenstein EH, Evans RL, et al. Transfection of syk tyrosine kinase reconstitutes high affinity IgE receptor-mediated degranulation in a syk-negative variant of rat basophilic leukemia RBL-2H3 cells. J Exp Med 1996;184:71–79.

60. Costello PS, Turner M, Walters AE, et al. Critical role for the tyrosine kinase Syk in signaling through the high affinity IgE receptor of mast cells Oncogene 1996;13:2595–605.

61. Kimura T, Kihara H, Bhattacharyya S, et al. Downstream signaling molecules bind to different phosphorylated immunoreceptor tyrosine-based activation motif (ITAM) peptides of the high affinity IgE receptor. J Biol Chem 1996;271:27962–27968.

62. Fukamachi H, Kawakami Y, Takei M, et al. Association of protein-tyrosine kinase with phospholipase C-γ1 in bone marrow-derived mouse mast cells. Proc Natl Acad Sci USA 1992;89:9524–9528.

63. Takata M, Sabe H, Hata A, et al. Tyrosine kinases Lyn and Syk regulate B cell receptor-coupled Ca^{2+} mobilization through distinct pathways. EMBO J 1994;13:1341–1349.

64. Takata M, Kurosaki T. A role for Bruton's tyrosine kinase in B cell antigen receptor-mediated activation of phospholipase C-γ2. J Exp Med 1996;184:31–40.

65. Nishibe S, Wahl MI, Hernandez-Sotomayor SMT, et al. Increase in the catalytic activity of phospholipase C-γ1 by tyrosine phosphorylation. Science 1990;250:1253–1255.

66. Kim JW, Sim SS, Kim U-H, et al. Tyrosine residues in bovine phospholipase C-γ phosphorylated by the epidermal growth factor receptor in vitro. J Biol Chem 1990;265:3940–3943.

67. Kim HK, Kim JW, Zilberstein A, et al. PDGF stimulation of inositol phospholipid hydrolysis requires PLC-γ1 phosphorylation on tyrosine residues 783 and 1254. Cell 1991;65:435–441.

68. Vega QC, Cochet C, Filhol O, et al. A site of tyrosine phosphorylation in the C terminus of the epidermal growth factor receptor is required to activate phospholipase C. Mol Cell Biol 1992;12:28–135.

69. Pao LI, Bedzyk WD, Persin C. Molecular targets of CD45 in B cell antigen receptor signal transduction. J Immunol 1997;158:1116–1124.

70. Motto DG, Musci MA, Ross SE, et al. Tyrosine phosphorylation of Grb-2-associated proteins correlates with phospholipase Cγ1 activation in T cells. Mol Cell Biol 1996;16:2823–2829.

71. Jensen WA, Peiman CM, Beaufils P, et al. Qualitatively distinct signaling through T cell antigen receptor subunits. Eur J Immunol 1997;27:707–716.

72. Bar-Sagi D, Rotin D, Batzer A, et al. SH-3 domains direct cellular localization of signaling molecules. Cell 1993;74:83–91.

73. Vossebeld PJM, Homburg CHE, Schweizer RC, et al. Tyrosine phosphorylation-dependent activation of phosphatidylinositide 3-kinase occurs upstream of Ca^{2+}-signaling induced by Fcγ receptor cross-linking in human neutrophils. Biochem J 1997;323:87–94.

74. Lomasney JW, Cheng H-F, Wang L-P, et al. Phosphatidylinositol 4,5-bisphosphate binding to the pleckstrin homology domain of phospholipase C-δ1 enhances enzyme activity. J Biol Chem 1996;271:25316–25326.

75. Abrams CS, Wu H, Zhao W, et al. Pleckstrin inhibits phosphoinositide hydrolysis initiated by G-protein-coupled and growth factor receptors. J Biol Chem 1995;270:14485–14492.

76. Park DJ, Min HK, Rhee SG. Inhibition of CD3-linked phospholipase C by phorbol ester and by cAMP is associated with decreased phosphotyrosine and increased phosphoserine contents of PLC-γ1. J Biol Chem 1992;267:1496–1501.

77. Sidorenko SP, Law C-L, Klaus SJ, et al. Protein kinase Cμ (PKCμ) associates with the B cell antigen receptor complex and regulates lymphocyte signaling. Immunity 1996;5:353–363.

78. Yamada K, Jelsema CL, Beaven MA. Certain inhibitors of protein serine/threonine kinases also inhibit tyrosine phosphorylation of phospholipase Cγ1 and other proteins and reveal distinct roles for tyrosine kinases and protein kinase C in stimulated rat basophilic RBL-2H3 cells. J Biol Chem 1992;149:1031–1037.

79. Apgar JP. Polymerization of actin in RBL-2H3 cells can be triggered through either the IgE receptor or the adenosine receptor but different signaling pathways are used. Mol Biol Cell 1994;5:313–322.

80. Burgering BM, Coffer PJ. Protein kinase B (c-Akt) in phosphatidylinositol-3-OH kinase signal transduction. Nature (Lond) 1995;376:599 602.

81. Bos JL. A target for phosphoinositide 3-kinase: Akt/PKB. Trends Biochem Sci 1995;20:441–442.

82. Seagrave JC, Pfeiffer JR, Wofsy C, et al. Relationship of IgE receptor topography to secretion in RBL-2H3 cells. J Cell Physiol 1991;148:139–151.

83. Bar-Sagi D, Suhan JP, McCormick F, et al. Localization of phospholipase A₂ in normal and ras-transformed cells. J Cell Biol 1988;106:1649–1658.

84. Erlandsen SL, Hasslin SR, Nelson RD. Detection and spatial distribution of the β₂ integrin Mac-1 and L-selectin LECAM-1 adherence receptors on human neutrophils by high resolution field emission SEM. J Histochem Cytochem 1993;41:327–334.

85. Ridley AJ. Membrane ruffling and signal transduction. Bioessays 1994;16:321–327.

16
Phospholipase D and Its Role in Mast Cells

David S. Cissel, Paul F. Fraundorfer, and Michael A. Beaven

Phospholipase D (PLD) is widely distributed among prokaryotic and eukaryotic organisms, and in mammalian cells it is regulated by a variety of receptor agonists and chemical stimulants. Its major substrate is phosphatidylcholine (PC), although in some organisms phosphatidylethanolamine (PE) and phosphatidylinositol (PI) can serve as substrates.[1,2] The hydrolysis of PC yields choline and phosphatidic acid (PA), which is then rapidly dephosphorylated by phosphatidate phosphohydrolase (PPH) to form 1,2-*sn*-diacylglycerol (DAG) (Fig. 16.1). PA may also be converted to *lyso*-phosphatidic acid (LPA) through the catalytic action of a PA-specific phospholipase A_2. PA, DAG, and LPA modify the activities of numerous proteins including protein kinase C (PKC) and, for this reason, PLD is thought to regulate multiple signaling pathways in cells. In addition to catalyzing the hydrolysis of PC, PLD in the presence of a primary alcohol catalyzes a phosphatidyl transfer reaction in which the primary alcohol acts as a nucleophilic acceptor instead of H_2O (Fig. 16.1). Although the physiological significance of this transphosphatidylation reaction is unknown, the reaction yields a unique product, phosphatidyl alcohol, which can be used as a measure of PLD activity.

PLD activity has been found in all mammalian tissues so far examined. It predominates in plasma membrane, Golgi, and other membrane fractions, although some activity is located in the cytosolic fractions. Biochemical studies with highly or partially purified enzyme preparations have revealed two broad categories of PLD, namely oleate-activated and oleateinhibited forms. The latter form is dependent on phosphatidylinositol-4,5-bisphosphate (PIP_2) for activity and is further activated by small monomeric G proteins such as the ADP-ribosylation factor (ARF) and Rho A when these proteins are converted to their active forms by guanosine 5′-*O*-(3-thiotriphosphate) (GTPγ-S). Other differences in properties of these enzyme preparations suggested the existence of isoforms of PLD,[2] as is now evident from the recent cloning of PLD isoforms from bacteria, yeast, plant, and animal sources. The cloned PLDs and certain synthases from *E. coli* share sequence homologies as well as the ability to catalyze the transphosphatidylation reaction, and thus it seems likely that the PLDs are representatives of a broad family of enzymes that has early evolutionary origins.[2–4]

With respect to the biological roles of PLD, the localization of PLD in Golgi and its stimulation by ARF have focused interest on the possible role of PLD in regulating vesicle trafficking by ARF in Golgi. This interest is reinforced by the findings that brefeldin A, an inhibitor of ARF activation, suppresses Golgi trafficking and PLD activation.[5,6] PLD has been implicated in other cellular functions, among them cell proliferation, degranulation in several types of cells including mast cells, the generation of oxygen radicals in neutrophils, and regulation of the cycling of synaptic vesicles in

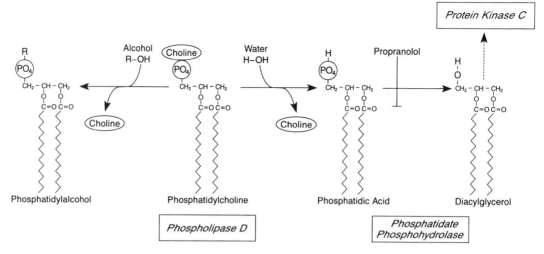

FIGURE 16.1 Reactions catalyzed by phosphatidyl-choline-specific phospholipase D and the link to protein kinase C. The figure depicts the conversion of phosphatidylcholine to phosphatidic acid or, in the presence of a primary alcohol, to phosphatidyl alcohol (the transphosphatidylation reaction) and the subsequent dephosphorylation of phosphatidic acid by phosphatidate phosphohydrolase to form diacylglycerol. The latter reaction is inhibited by propranolol. The activation of protein kinase C by diacylglycerol is also shown.

neurons.[1,2] PLD is the primary source of DAG in stimulated mast cells where DAG regulates degranulation through activation of PKC and may serve as a substrate for DAG lipase to generate arachidonic acid.[7-9] These roles are attributed to PLD largely on the basis of studies with inhibitors, as, for example, primary alcohols for inhibition of PLD and propranolol for inhibition of PPH (see Fig. 16.1). Unfortunately, none of these inhibitors are highly specific. For example, wortmannin, first described as an inhibitor of PLD, is also a potent inhibitor of phosphoinositide 3-kinase (PI-3 kinase).

Most recent advances in the field have come from studies with partially purified and cloned forms of PLD. In this chapter, we first discuss the enzymology, assay, cloning, and putative roles of PLD in general terms to set the stage for the subsequent discussion of the literature on mast cells. Our hope is that recent advances in PLD research will rekindle interest in the enzyme in the mast cell.

Catalytic Reactions and Assays of PLD: Phospholipid Hydrolysis and Transphosphatidylation

The actions catalyzed by PLD with PC as substrate are illustrated in Fig. 16.1. In the presence of a primary alcohol, the transphosphatidylation reaction is favored such that 200 mM alcohol will usually ensure substantial to complete suppression of formation of PA. A detectable proportion of PC is converted to the phosphatidyl alcohol in the presence of 20 mM alcohol. The phosphatidyl group may be transferred to methanol, ethanol, propanol, butanol, or even glycerol.[10] As the alcohols rapidly penetrate cell membranes, these agents can be used in vivo to suppress formation of PA with the resultant production of phosphatidyl alcohol. The phosphatidyl alcohols, which are metabolically stable, accumulate in tissues,[11] and this accumulation is considered a specific indicator of PLD activity. In PLD-enriched preparations, PA and phosphatidylethanol are generated to the same extent, and activation of PLD is equally apparent by measurement of either product.[12,13]

Measurement of the hydrolytic (i.e., PA and choline) and transphosphatidylation (i.e., phosphatidyl alcohol) products has been utilized for the assay of PLD activity.[11] The release of radiolabeled choline from choline-labeled phospholipids is a convenient method for the assay of PLD. For assays in vivo, the choline can then be separated from protein-free extracts by ion-exchange chromatography or thin-layer chromatography.[14] In vitro, the assay is performed with mixed lipid vesicles and release of labeled choline into the water-soluble fraction is measured directly.[12] An alternative approach is to incorporate radiolabeled fatty acids such as myristic, palmitic, or oleic acids into intracellular phospholipids and measure formation of radiolabeled PA. This procedure, however, lacks specificity and may underestimate the actual amounts of PA formed because PA is rapidly metabolized by PPH and other enzymes or can be generated by DAG kinase-catalyzed phosphorylation of phospholipase C- (PLC-) derived DAG.[11] Because of this, the preferred method is to stimulate the radiolabeled cells or tissues in the presence of low concentrations of ethanol, propanol, or butanol that yield measurable amounts of radiolabeled phosphatidyl alcohol but do not disrupt other signaling mechanisms.[11] Typically, lipid-soluble radiolabeled products are then isolated by the Bligh and Dyer solvent extraction procedure[15] and then separated by thin-layer chromatography before quantitation of the radioactive phosphatidyl alcohol[16–18] or qualitative analysis of acyl/alkyl components of the phosphatidyl alcohol or PA fractions.[17,19,20] A fluorescent derivative of PC has also been used for the measurement of phosphatidyl alcohol formation in vitro.[21]

PLD can hydrolyze either alkyl- (i.e., ether) or acyl-linked species of PC and PE. As the acylated and alkylated species of DAG have different effects on PKC, preferential formation of one or the other species of DAG (i.e., via PPH) might have functional consequences. Formation of one species of DAG over the other has been reported for Madine-Darby canine kidney cells[22] but not other[23,24] types of cells, and the significance of this phenomenon is unclear.

Other potential messenger molecules, namely DAG and LPA, are formed from PA by the actions of PPH (Fig. 16.1)[25–28] and a PA-specific phospholipase A_2,[29] respectively. Neither enzyme has been extensively characterized. PPH exists in two forms, one cytosolic and one plasma membrane associated, which differ in physical and enzymatic properties. Initial reports indicated that both forms of PPH are inhibited by propranolol, chlorpromazine, sphingosine, and spermine but that they did not appear to be regulated over the short term by hormones.[26,27] A recent report, however, indicated otherwise in that PPH dissociated from the epidermal growth factor (EGF) receptor and associated with PKCε following treatment with EGF.[28] The PA-specific phospholipase A_2 (58 kDa), which has been purified from rat brain, does not hydrolyze PC or PE, is calcium independent, and also differs in other respects from previously described forms of phospholipase A_2.[29] Although formation of LPA from PA appears likely, the extent to which LPA is formed by this route in stimulated cells is difficult to determine because LPA is a transient intermediate in the triacylglycerol cycle.[30]

Biochemical Characteristics and Regulation of PLD

Studies with partially purified preparations of PLD indicate that different types of PLD exist in tissues. One type, with an estimated mass of 95 kDa, from porcine brain membranes, is stimulated by PIP_2 and the monomeric G proteins, ARF and Rho A.[12] Another of estimated size 190 kDa from porcine lung microsomes is stimulated by oleate but is only weakly stimulated by PIP_2.[31] Other partially purified preparations of PLD differ in their responses to Ca^{2+}, PIP_2, oleate, and detergents.[1,2] In general, these studies have revealed at least two forms of PLD

TABLE 16.1. Characteristics of the cloned and partially purified phospholipase D (PLD) enzymes.

	Recombinant proteins		Partially purified proteins	
	Human PLD-1[a]	Mouse/rat PLD-2[b]	Oleate dependent[c]	Oleate independent[d]
Homology	100%	50%	—	—
Size	~120 kDa	~97 kDa (mouse) ~106 kDa (rat)	—	—
Activity	Regulated	Constituitively Active	Unknown	Regulated
Substrate	PC	PC	PC	PC
Transphosphatidylation reaction	Yes	Yes	Yes	Yes
Requirements				
PIP_2	Yes	Yes[e]	No	Yes
Ca^{2+}/Mg^{2+}	Yes	Yes	Yes	Yes
Stimulants	GTPγ-S, PKCα[f]; ARF > Rho A > Rac 1 > cdc42	None	Oleic acid	GTPγ-S, ARF, Rho A
Inhibitors	Oleic acid, synaptojanin, AP3	18-kDa protein (mouse), oleic acid (rat)	—	Oleic acid, 30-kDa inhibitor (bovine), AP3 synaptojanin
Localization	Perinuclear (ER, Golgi vesicles, late endosomes)[g]	Plasma membrane > cytosol[g]	Membrane fraction (bovine) Nucleus (liver)[h]	Membrane fraction (bovine) Nucleus (liver)[h]

[a] *Source:* Hammond et al.[72,73]

[b] *Source:* Colley et al.[71]; Kodaki and Yamashita.[74]

[c,d] *Source:* Massen berg et al.[32]; Newton.[94]

[e] PIP_2 and PE are necessary for Rat PLD-2 activation.

[f] The effect of PKCα on PLD-1 is ATP independent.

[g] Distribution when recombinant proteins were expressed in mammalian cells.

[h] Additional studies of the subcellular distribution of native PLD are described in the text.

(Table 16.1), one that is oleate activated and another that is inhibited by oleate but activated by PIP_2, ARF, Rho A, and PKC.[13,32–35] In addition to these stimulatory factors, there are indications from studies in intact and permeabilized cells that PLD may be regulated by tyrosine kinases and calcium.[1]

The stimulation of PLD by PIP_2, first described by Brown and coworkers,[36] has been noted for several preparations of enzyme,[1] and this effect is mimicked by PI-3,4,5 trisphosphate[21] but not by other acidic phospholipids including PA, phosphatidylserine, PI, PI-3,4-bisphosphate, and PI-4-monophosphate.[21,36] An indication that the effects of PIP_2 are physiologically relevant is that neomycin, which binds acidic phospholipids, inhibits activity of membrane-associated PLD as well as the stimulation of partially purified PLD by PIP_2 without affecting the basal activity of the par-

tially purified enzyme preparation.[21] Neomycin also inhibits GTPγ-S stimulation of PLD in permeabilized cells.[37] Other studies indicate that PIP_2 is required for the stimulation of PLD activity by ARF, Rho, and PKC.[6,12,32,35,36,38] Not all sources of PLD require PIP_2 for activation,[31,39,40] and an oleate-dependent form of PLD from rat brain nuclei is inhibited by PIP_2.[40] Such observations suggest that PIP_2 is an essential cofactor for some but not all forms of PLD.

The question arises as to whether or not changes in membrane levels of PIP_2 in vivo regulate activity of PLD. PIP_2 levels are normally dependent on the sequential phosphorylation of PI by magnesium/ATP-dependent PI-4 kinase and PI-4 phosphate 5-kinase. Reduction of PIP_2 levels by antibody blockade of PI-4 kinase in permeabilized cells or, as noted earlier, the binding of PIP_2 with neomycin, suppresses activation of PLD by GTPγ-S and phorbol ester.[37] With respect to PI-4-phosphate 5-kinase, because this enzyme is activated by the PLD product, PA, it is speculated that regulation of this kinase by PA and in turn the regulation of PLD by PIP_2 constitute a positive communicating loop between the two enzymes.[21,37] In addition, PI-4-phosphate 5-kinase, like PLD, can be regulated by monomeric G proteins of the Rho family,[41] and it is possible that the stimulatory effects of GTPγ-S on PLD in permeabilized cell systems may be exerted indirectly through stimulation of PIP_2 synthesis by PI-4-phosphate 5-kinase.[1] PIP_2 may have also indirect input into PLD. The reported-activation of several ARF guanine nucleotide exchange factors by PIP_2[42] and its metabolite, PI-3,4,5-trisphosphate,[43] may result in stimulation of ARF and, as a consequence, PLD.

Negative regulation of PLD by PIP_2 is also possible[44] because PIP_2 activates ARF guanine nucleotide-activating protein (ARF-GAP) and thereby inactivates ARF.[45] Another possible feedback inhibitory loop is through the inositol polyphosphate 5-phosphatase, synaptojanin.[46] Synaptojanin, a nerve terminal protein implicated in endocytosis of spent synaptic vesicle membrane, also hydrolyzes PIP_2 and thus inhibits PIP_2-dependent ARF activation of rat brain PLD and a cloned PLD (hPLD-1; see later). It is proposed that this set of reactions counteracts the positive signals emanating from PLD production of PA and thereby promote endocytosis of synaptic vesicles.[46]

Membrane PIP_2 levels are also dependent on PI transfer proteins, which transfer PI and PC from sites of synthesis to acceptor membranes[47] and to enzymes such as PI-4 kinase that utilize these phospholipids.[48–50] A topic of current interest is the possible involvement of PI transfer proteins, PI-4 kinase, and PLD in Golgi-mediated vesicle trafficking.[48,51,52]

The regulation of PLD by monomeric G proteins was first described in studies with GTPγ-S in cell lysates and membrane preparations. ARF, whose role in Golgi-mediated vesicle trafficking and formation of clathrin-coated vesicles is now well recognized,[52] was shown to mediate GTPγ-S stimulation of PLD first by the laboratories of Sternweis[36] and Cockcroft[53] and

then by others.[1] All subtypes of ARF tested (i.e., ARF-1, -3, -5, and -6), and in particular the myristoylated forms of ARF, exhibit this activity,[12,32] which is further enhanced by PKCα[13] and calmodulin.[54] The stimulatory effects of ARF are inhibited by synaptojanin (see earlier) and the clathrin assembly protein-3 (AP-3). AP-3, unlike synaptojanin, inhibits PLD directly by binding to the enzyme and, presumably, could act as an additional negative regulator of PLD for control of synaptic membrane recycling.[55] An important aspect of the ARF/Golgi model is that Golgi-associated PLD activity is stimulated by ARF,[5,6,56,57] and this stimulation is blocked by the ARF inhibitor, brefeldin A.[5,6] Brefeldin A was also shown to inhibit muscarinic (M3) receptor-induced activation of PLD in vivo.[58] This and another study[59] suggest the possibility of receptor-mediated regulation of PLD through ARF.

Other G proteins that regulate PLD include those of the Rho family of GTPases, which have been implicated in various activities associated with actin/cytoskeletal rearrangements. As with ARF, the effects of these proteins on PLD were discovered in studies with GTPγ-S. Stimulation of membrane-associated PLD by GTPγ-S is blocked by Rho-GDP dissociation inhibitor, a protein that inhibits dissociation of GDP from Rho and prevents Rho activation.[60] In dialyzed preparations, the response to GTPγ-S can be restored by addition of Rho proteins with a rank order of potency of Rho A > Rac-1 > Cdc42.[61–63] Rho proteins may participate in receptor-mediated stimulation of PLD, as indicated from the alterations in PLD response to EGF by expression of wild-type Rac-1 or a dominant negative Rac-1 in fibroblasts.[64] Additional indications come from studies with bacterial toxins that either ADP-ribosylate (i.e., the C3 exoenzyme from *Clostridium botulinum*) or glucosylate (i.e., toxin B from *Clostridium difficile*) Rho A. Activation of PLD by LPA,[65] carbachol,[66] or FcεRI aggregation[67] is blocked by one or the other toxin.

PLD appears to be regulated by PKC in many types of cells, including mast cells, as indicated by numerous studies with phorbol esters and PKC inhibitors.[1] As is discussed in later sections, many of these inhibitors do not target PKC exclusively nor do they inhibit all actions of PKC. Indeed, PKCα and PKCβ,[13,34] but not other isoforms of PKC,[34] were shown to activate PLD preparations in the absence of ATP, and this activation is unaffected by the PKC inhibitor, staurosporine.[13] Evidently, PKCα and PKCβ interact with PLD through their regulatory rather than catalytic domains.[13] As discussed by Exton,[2] these findings do not exclude additional regulation of PLD by PKC in intact cell systems through phosphorylation of PLD itself or of other proteins that stimulate PLD activity or PIP$_2$ synthesis.

Calcium ions and tyrosine kinases have also been implicated in the regulation of PLD on the basis of the effects of calcium reagents and inhibitors of tyrosine kinases and phosphatases. A number of studies indicate that the calcium ionophore-induced rise in cytosolic calcium concentration ($[Ca^{2+}]_i$) is associated with stimulation of PLD and that activation of PLD via G protein-coupled receptors is blocked by calcium chelation. The complicating issues in studies with these types of agents are discussed by Exton[1] and, as is discussed later, these issues are of concern in studies of PLD in mast cells (see Activation of PLD in Mast Cells).

Two points should be emphasized in regard to the regulation of PLD. The first is that the Rho proteins, ARF, and PKCα markedly synergize each other in their activation of PLD from various sources,[1] including recombinant PLD (see later). The second point is that other proteins can regulate PLD activity. These include the actin-binding protein, gelsolin,[68] other unidentified cytosolic proteins,[2] sphingosine, and sphingosine 1-phosphate,[69] all of which either stimulate or enhance PLD activity. Inhibitory factors include synaptojanin and AP-3 (see earlier), fodrin,[70] a protein in brain that is selective for PLD-2,[71] and ceramides.[2] As is noted later, ceramides have been implicated in the regulation of PLD in mast cells.

Cloned PLD Isoforms

The cloned mammalian PLDs include PLD-1a,[72] PLD-1b (a spliced variant that lacks a stretch of 38 amino acids),[73] and PLD-2.[71,74]

Each variant of PLD-1 resembles the ARF-regulated PLD that has been characterized in tissues and cells (see Table 16.1). Both are inhibited by oleate and are synergistically regulated by ARF-1, Rho A, Rac-1, Cdc42, and PKCα in the presence of PIP$_2$.[73] PLD-2, in contrast, is constitutively active in the presence of PIP$_2$, and this activity is not enhanced by PKCα or by members of the ARF or Rho family of proteins, either individually or in combination.[71] This isoform is inhibited by oleate and has no apparent tissue cognate.

PLD-1a (1072 amino acids; estimated size, 120 kDa) and PLD-1b (1034 amino acids) have been cloned from human[72,73] and rat[2] cDNA. Partial sequences for the PLD-1 variants have been reported also for human tissues[75] and C6 glioma cells.[76] PLD-2 has been cloned from mouse embryonic (932 amino acids; estimated size, 96 kDa) and rat brain (933 amino acids; estimated size, 106 kDa) cDNA.[71,74] PLD-1 and PLD-2 are 51% identical, with greatest similarity in regions that are thought to be essential for catalytic activity.[71] The most notable difference between the two isoforms is the presence of a unique segment of about 120 amino acids in PLD-1, which is absent in PLD-1 and other cloned nonmammalian PLDs. The prokaryote and eukaryote PLD homologs otherwise possess five conserved regions.[71]

It is generally assumed that the PLD-1 variants contain separate but interacting sites for the ARF and Rho proteins as well as for PKCα because of the synergistic interactions between these different regulatory factors. However, no particular sequence motifs have been noted to account for these interactions, and neither PLD-1 or PLD-2 contains recognizable Src homology (SH-2 and SH-3) or pleckstrin homology (PH) domains.

Recombinant PLD-1 and PLD-2 exhibit different patterns of distribution when expressed in rat embryo fibroblasts. PLD-2 associates primarily with the plasma membrane whereas PLD-1 associates with perinuclear membranes such as Golgi, endoplasmic reticulum, and late endosomes.[71] This finding is consistent with the subcellular distribution of PLD activity which, in many tissues, is present in the plasma membrane, Golgi, nuclei and to a lesser extent in cytosolic fractions[1] (see Table 16.1).

Pharmacological Probes for Study of PLD Function

We have mentioned the use of various inhibitors in studies of PLD function including the clostridial toxins for inhibition of Rho, brefeldin A for inhibition of ARF, staurosporine for inhibition of PKC, wortmannin and primary alcohols for inhibition of PLD, and various agents for inhibition of PPH. As mentioned earlier, most of these inhibitors have other actions, and studies based on these agents alone provide insufficient evidence to attribute specific roles to PLD.

A caveat with the use of clostridial toxins is that the toxins could interfere with other signaling mechanisms especially those dependent on Rho proteins such as the previously mentioned activation of PI-4-phosphate 5-kinase by Rho.[41] The consequences of inhibiting this pathway could be a decline of PIP_2 to levels that are inadequate to support activation of PLD or formation of inositol phosphates through PLC. As discussed elsewhere,[8] the problem with inhibitors of PKC is that many of them inhibit other protein kinases. For example, one widely used PKC inhibitor, staurosporine, suppresses antigen-induced responses in RBL-2H3 cells, not through its action on PKC, but by blocking early tyrosine kinase-dependent phosphorylation events.[77] The concentrations of butanol[78] and propranolol[27,79,80] that are used to inhibit PLD and PPH, respectively, also inhibit PKC to a similar extent.[81,82] Wortmannin inhibits one form of PI-4 kinase,[83] as well as PI-3 kinase,[84] at nanomolar concentrations and could therefore impede the action of PLC. Interestingly at these concentrations, wortmannin also inhibits thapsigargin-induced activation of PLD[148] and secretion in RBL-2H3 cells.[85] These observations raise the question as to whether or not PI-3 kinase is involved in the activation of PLD, although such a relationship has not been reported.

Physiological Roles of PLD

It is now apparent that PLD-derived PA increases rapidly in a monophasic fashion in most stimulated cells. DAG levels, in contrast, usually exhibit a biphasic increase with an initial transient increase, attributable to hydrolysis of inositol phospholipids via PLC, followed by a sustained increase, attributable to dephosphorylation of PA by PPH. The formation of LPA via PA-specific PLA_2 is especially pronounced in platelets, although it probably occurs in other types of cells as well.[1] As noted earlier, each of these lipid metabolites is a potential messenger molecule. Also, because PC is by far the most prominent phospholipid in all membranes, the extent of PLD-induced hydrolysis of PC in most cell systems is likely to have profound changes, at least locally, on the physical properties of the membrane through alteration in levels of PC and PA.

The immediate PLD product, PA, may influence binding of membrane-associated proteins because of its negative charge and may induce conformational changes in the membrane because of changes in ratio of PA to PC. Studies with GTPγ-S and rat liver membranes suggest that much of the PC in the cytosolic side of the membrane is hydrolyzed by PLD,[86] and this may be true also for Golgi membranes.[1] Much of the interest in PLD has focused on its possible role in mediating the ARF-regulated binding of β-COP, a component of coatomer protein complexes, to Golgi vesicles.[51,52] Because ethanol inhibits intra-Golgi vesicular transport and receptor-mediated secretion[87] as well as PA formation (see previous section), PA may facilitate this transport. It has also been suggested that PA serves in this role as a membrane fusogen.[88] Other actions of PA[1] include stimulation of PKCζ, other protein kinases, a protein tyrosine phosphatase, n-chimaerin, PLCγ, and inhibition of the Rac guanine nucleotide-activating protein and consequently the interaction of Rac with Rho. A well-documented role of PA is the activation of NADPH oxidase and the resulting respiratory burst in stimulated neutrophils. There is evidence that PA acts through a protein kinase to activate a component of NADPH oxidase, $p47^{phox}$.[89] The possibility that PA acts as a mitogen has been suggested, but experimental limitations in studies with PA have also been noted.[1]

LPA, one of the two major metabolites of PA, is known to act as an extracellular messenger by interacting with a G protein-coupled receptor to initiate PLC- and PLD- mediated events in various types of cells. LPA is also mitogenic, acting through a pathway that may involve G_i, PI-3 kinase, and mitogen-activated protein (MAP) kinase.[90] No well-defined intracellular function has been assigned to LPA at this time although a number of interesting effects of LPA have been noted.[1,90] Most research effort has been directed toward the other PA metabolite, DAG. Its major function is the activation of PKC isozymes that have well-defined roles in the exocytotic secretion of preformed granules[8] and receptor desensitization as demonstrated in a mast cell model.[91] Less well defined roles include myosin phosphorylation and cytoskeletal reorganization in RBL-2H3 cells,[92] regulation of nuclear events,[1] nuclear transcription, and regulation of cell growth among other functions.[93] The interaction of DAG with PKC isoforms and the mechanisms of action of PKC have been discussed in detail in recent reviews.[93,94]

In addition to functions inferred from studies of PLD-derived products, changes induced by overexpression of cloned PLDs in rat embryo fibroblasts suggest that PLD-2 may regulate cortical reorganization and endocytosis.[71] In these studies, overexpression of PLD-2 resulted in cortical reorganization and actin polymerization and, following serum stimulation of these cells, PLD-2 translocated from the plasma membrane to submembranous vesicles. PLD-1, in contrast, remained confined to perinuclear membranes where it was assumed, on the basis of previously mentioned studies on Golgi, to regulate intravesicular membrane trafficking. A possible candidate for mediating actin polymerization is PLD-derived PA.[95,96]

In summary, the potential roles of PLD in cell function are many but only some are being actively investigated. Interest has focused largely on the possible roles of PLD in cell proliferation, cytoskeletal rearrangements, and vesicular trafficking. The latter two subjects are discussed further in the context of mast cells (see Other Potential Roles for PLD in Mast Cells).

PLD in Mast Cells

Activation of PLD in Mast Cells

PLD activity is present in rat peritoneal mast cells[97–105] and cultured RBL-2H3 mast cells,[67,106–114] where it appears to enhance stimulatory signals for degranulation.[149] Kennerly and his associates first recognized and precisely documented the substantial contribution of PLD in stimulating phospholipid metabolism in activated mast cells.[17,97–99] Subsequent studies, mainly from the laboratories of Gilfillan[114] and Nozawa,[67] have provided additional information on the activation of PLD but, in general, the mechanisms of activation and role of PLD in mast cells are not entirely clear. As in other types of cells, PLD is stimulated after a rise in $[Ca^{2+}]_i$ and activation of PKC[113] to produce sustained increases in levels of PA and DAGs.[106,110,113,115] As is explained later, it has not been unequivocally determined whether PLD is directly coupled to early receptor-mediated events or activated exclusively through changes in $[Ca^{2+}]_i$ and PKC. The PLD-derived DAGs probably contribute to the activation of PKC and thereby degranulation, but they may also serve as substrates for DAG lipase for release of arachidonic acid.[7–9]

PLD is activated in rat peritoneal mast cells and RBL-2H3 cells by a broad spectrum of stimulants. These include calcium-mobilizing agents such as calcium ionophores[98,104–106,109,113] and thapsigargin,[113] activators of protein kinase C such as phorbol 12-myristate 13-acetate (hereafter referred to as phorbol ester),[67,103,108–110,113] phenylarsine oxide,[110,112] the mast cell secretagogue compound 48/80,[98] and ligands for various receptors that are expressed in mast cells. The ligands include FcεRI cross-linking agents,[67,99,100,114,116] stem cell factor,[101,102] agonists of adenosine receptors,[117] and in RBL-2H3 cells transfected with the muscarinic m1 receptor, carbacho.[149] FcεRI-mediated stimulation of PLD is enhanced twofold in the presence of substrates that permit cell adhesion and spreading.[118] In permeabilized RBL-2H3 cells, PLD is activated by GTPγ-S.[67] The stimulation of PLD by calcium-mobilizing agents and phorbol ester is consistent with the notion that PLD is regulated by signals generated through either calcium or PKC. These findings highlight an important difference between PLD and PLC in RBL-2H3 cells in that PKC is a positive regulator of PLD and a negative regulator of PLC.[119]

PC is the major substrate for PLD in IgE-stimulated mast cells[100] and RBL-2H3 cells[80] and, consistent with studies in other types of cells[48,120] PLD appears to be responsible for the bulk of the DAGs that are generated in mast cells, particularly at late stages of stimulation. In rat peritoneal mast cells, FcεRI ligation causes rapid production of PA and choline (within 15 s), followed shortly thereafter by an increase in levels of DAG.[100] Quantitative measurements of PLC- and PLD- derived metabolites[100] as well as comparison of the fatty acid composition of DAG and phospholipids[98] indicate that DAG is derived primarily from PC via PLD. In antigen-stimulated RBL-2H3 cells, production of DAG is biphasic. A transient increase in DAG, which reaches a maximum within 60 s of addition of antigen, is followed by an increase in DAG that lasts for at least 30 min. The initial increase is attributed to a phospholipase other than PLD, possibly PLC,[9] whereas the sustained increase is attributed to stimulation of PC-specific PLD by PKC. This sustained increase is blocked by the PPH inhibitor, propranolol, and by the PKC inhibitor, staurosporine,[116] and is accompanied by formation of PA and, in the presence of ethanol, phosphatidyl ethanol.[106]

The pathways for activation of PLD have been studied in detail in RBL-2H3 cells. Here a rise in $[Ca^{2+}]_i$ and activation of PKC appear to be two important mechanisms for activation of PLD. In addition to the ability of calcium-mobilizing agents and phorbol ester to individually stimulate PLD in RBL-2H3 cells (see earlier), we find that phorbol ester and calcium-mobilizing agents markedly synergize each other in this stimulation (unpublished data). Antigen-mediated activation of PLD is virtually blocked by chelation of extracellular or

intracellular calcium,[113] by various PKC inhibitors,[108,113] and by downregulation of PKC through prolonged exposure of cells to phorbol ester.[113] The PLD response to antigen thus appears to be totally dependent on a rise in $[Ca^{2+}]_i$ and activation of PKC, which might indicate potent synergistic interaction between these two signals.

Mast cells are normally activated through recruitment of the cytosolic tyrosine kinases, Lyn and Syk, by FcεRI and the tyrosine phosphorylation of proteins, including PLC-γ.[121] Tyrosine phosphorylation of PLD has been suggested as a possible mechanism of activation,[107,114] although the data are equivocal. It has been shown, for example, that activation of PLD is blocked by tyrosine kinase inhibitors in antigen-stimulated RBL-2H3 cells,[107,114] but the inhibitors tested (genistein, lavendustin A, and tyrphostin) also inhibit other kinases including PKC.[8] Surprisingly, these inhibitors failed to block the initial transient rise in DAG that has been attributed to hydrolysis of phosphatidylinositol lipids by PLC instead of PC by PLD.[9,114] This result is paradoxical in that FcεRI-mediated activation of other phospholipases, such as PLC and PLA$_2$, is clearly dependent on the tyrosine kinase, Syk, in RBL-2H3 cells.[123,124] It has also been reported that immunoprecipitated tyrosine-phosphorylated proteins show significant increases in PC-PLD activity upon antigen stimulation of RBL-2H3 cells.[114] These studies were performed, however, before the description of the cloned isoforms of PLD and the interaction of PLD with other proteins. It is not clear from these studies, therefore, whether the immunoprecipitated material contained tyrosine-phosphorylated PLD or tyrosine-phosphorylated proteins that might have associated with PLD isoforms.

The difficulty in interpreting the pharmacological data is that, on the one hand, PLD appears to be activated by increases in $[Ca^{2+}]_i$ or stimulation of PKC alone and, on the other hand, the activation of PLD by antigen is totally dependent on a rise in $[Ca^{2+}]_i$, stimulation of PKC, and tyrosine phosphorylation of protein, presumed to be PLD. In the light of present knowledge, the simplest interpretation of the data is that the rise in $[Ca^{2+}]_i$ and stimulation of PKC are both dependent on the tyrosine kinase, Syk,[123] and that both provide primary synergistic signals for the activation of PLD. This interpretation, nevertheless, does not exclude additional synergistic signals including tyrosine phosphorylation of PLD itself. Preliminary evidence has also been presented from studies in RBL-2H3 cells for the regulation of PLD by calcium/calmodulin,[108] PKCα and β isozymes,[108] ARF,[111] Rho proteins,[67] and ceramide,[109] as follows. Antigen-mediated stimulation of PLD is inhibited by calcium deprivation, as previously noted by Gilfillan and coworkers,[113] by the calmodulin inhibitors W-7 and trifluoperazine,[108] and by selectively depleting cells of PKCα and β, through short exposure to phorbol ester.[108] In addition, antigen-induced translocation of PKCα and β, as well as stimulation of PLD and calcium influx, can be blocked by treatment of RBL-2H3 cells with the membrane-permeable ceramide analog, N-acetylsphingosine.[109] This analog does not block translocation of the calcium-independent isoforms of PKCδ and ε. It appears to act indirectly on PLD by suppressing influx of extracellular calcium and, as a consequence, activation of PKCα and β. The ARF inhibitor, brefeldin A, partially inhibits the PLD response to antigen (by ~50% to ~75%) and calcium ionophore (by 27%) but not to phorbol ester.[111] Contrary to these results, however, we find that brefeldin A has no effect on PLD responses even at concentrations that completely disrupt Golgi structures in RBL-2H3 cells (our unpublished data). The PLD[111] and secretory responses[111,124] to antigen stimulation are effectively blocked by prior treatment of RBL-2H3 cells with the *Clostridium difficile* toxin B, which as noted earlier glucosylates and inactivates Rho proteins,[67] among them Rho A and Cdc42.[124] Of these two proteins, Cdc42 seems the most likely candidate for regulating PLD, either directly or indirectly, because the selective inactivation of Rho A by ADP ribosylation with C3 transferase from *Clostridium botulinum* does not lead to loss of secretory responses in RBL-2H3 cells.[124] While these findings point to several potential regulators of PLD in mast cells, the effects of these inhibitory agents on other signaling systems have not been adequately explored and the information adds little to that already obtained from studies with other cell systems.

In addition to a rise in $[Ca^{2+}]_i$ and activation of PKC, it is apparent that PLD is regulated, directly or indirectly, by monomeric and trimeric G proteins. The stimulation of PLD activity in permeabilized mast cells by GTPγ-S[67] is impaired after extensive permeabilization but the response can be reconstituted by addition of ARF (Fraundorfer et al., submitted for publication). Nevertheless, additional activity remains in permeabilized cells that is not dependent on ARF but is sensitive to calcium, PKC, and possibly subunits of trimeric G proteins (see final section). Regulation of PLD by trimeric G proteins is also evident from studies of the adenosine A_3 receptor that is present in RBL-2H3 cells.[125] Stimulation through this receptor results in an initial transient activation of PLC-mediated hydrolysis of inositol phospholipids and a transient increase in $[Ca^{2+}]_i$, followed by sustained activation of both PLD and PKC.[117] All responses are blocked by treatment with pertussis toxin,[117,126] which inhibits dissociation of trimeric G_i proteins, and are enhanced by treatment with dexamethasone,[127] which is associated with a severalfold increase in expression of the G_i proteins.[128] A model currently being examined in our laboratory is that dissociation of G_i results in activation of PLC and PLD by βγ-subunits of G_i and that subsequent activation of PKC reinforces activation of PLD and negatively regulates PLC.

The c-Kit receptor, a tyrosine kinase receptor that is required for mast cell maturation, utilizes signaling pathways that overlap with those activated by FcεRI. The c-Kit ligand (stem cell factor) stimulates PLD activity,[102] degranulation,[102] and release of arachidonic acid[101] in rat peritoneal mast cells. The PLD response to the c-Kit ligand is blocked by the tyrosine kinase inhibitor genistein.[102] An unexpected finding is that c-Kit ligand failed to stimulate production of inositol 1,4,5 trisphosphate[102] which seems to conflict with previous reports that it induces rapid increases in $[Ca^{2+}]_i$[129] and calcium-dependent degranulation[130] in human mast cells. With the possible exception of PLC, c-Kit appears to communicate with the same effector enzymes as FcεRI, including PLD and the MAP kinases (see also Chapter 18).

PLD and Degranulation

In RBL-2H3 cells, secretion of granules requires both an increase in $[Ca^{2+}]_i$ and the translocation of certain isoforms of PKC,[131] in a DAG-dependent manner.[132] As noted throughout this review, a significant proportion of DAG is generated through the activation of PLD, although PLC may provide an additional source of DAG. The dilemma in designing studies to investigate whether or not PLD is necessary for adequate activation of PKC for secretion is that the same signals, namely, elevation of $[Ca^{2+}]_i$ and translocation of PKC, activate PLD and secretion in RBL-2H3 cells. Disruption of either signal will block both responses.[113,131] The problem is confounded by the likelihood that PKC serves as both upstream regulator and downstream effector of PLD and thereby provides a reinforcing loop for sustained activation of PKC. Moreover, early studies were undertaken before the recognition of the nonspecific effects of PKC or tyrosine kinase inhibitors. To cite a few examples, the inhibition of secretion by propranolol[80,116,133] or genistein[102] and the reduction in DAG levels by staurosporine[116] could be caused, at least in part, by inhibition of PKC by propranolol[81] and tyrosine kinase by staurosporine.[77] While correlations between PLD activity and degranulation have been noted under a variety of conditions in mast cells,[67,80,109,111,113,133] there is no direct evidence that PLD is an absolute requirement for secretion of granules in antigen-stimulated cells.

Previous work has established that activation of PLD by phorbol ester is an insufficient stimulus for secretion[111,113] and that activation of PLD via the adenosine A_3 receptor (see previous section) is not associated with secretion in the absence of a calcium signal.[117] In the presence of a calcium signal, PLD appears to provide a necessary signal for secretion, at least

when RBL-2H3 cells are stimulated with phorbol ester (our data, unpublished) or an adenosine analog along with low concentrations of calcium ionophore.[117] A criticism of these studies, however, is that even ionophores such as A23187 and ionomycin stimulate some hydrolysis of inositol phospholipids[134,135] and activate PKC in RBL-2H3 cells.[92] To address this concern, studies have been conducted with thapsigargin (Cissel et al., submitted for publication). In RBL-2H3 cells, thapsigargin induces substantial increases in $[Ca^{2+}]_i$, release of arachidonic acid,[135] and activation of PLD[113] with minimal stimulation of PLC.[135] The PLD response, however, could be manipulated independently of these other responses by use of cholera toxin (see final section) and PKC inhibitors, and this response was shown to correlate closely with secretion. Of particular note, the PLD and secretory responses to thapsigargin, which selectively stimulates the calcium-independent PKCε in RBL-2H3 cells,[136] are markedly inhibited by Ro31-7549 but are resistant to the inhibitory actions of Gö6976.[148] The former compound inhibits most isozymes of PKC[137] while the latter suppresses primarily calcium-dependent isozymes of PKC.

Collectively, these studies provide strong evidence that PLD can regulate secretion in mast cells, especially in response to stimulation with calcium-mobilizing agents such as the calcium ionophores and thapsigargin. Determination of the exact contribution of PLD to FcεRI-mediated secretion, however, must await molecular identification of PLD isozymes in mast cells.

PLD and Generation of Arachidonic Acid

The possibility that DAG is a source of free arachidonic acid was first suggested by Kennerly and associates,[138] who noted increased levels of DAG in stimulated mast cells and that DAG itself was rapidly hydrolyzed to form free arachidonic acid, monoglycerides, and other lipids in the presence of mast cell homogenates, possibly through the action of DAG lipase. Arachidonic acid released from stimulated cells is derived primarily from PC.[139] In antigen-stimulated RBL-2H3 cells[80] and rat peritoneal mast cells[140] that have been labeled with [^{14}C]arachidonic acid, this release is preceded by increased labeling of first the PA and then the DAG pool. This time course is consistent with the notion that a PLD/PPH/ DAG lipase pathway is involved in the release of arachidonic acid from PC in these cells.[9] The essential supporting data are that the presence of ethanol or propranolol suppresses the increase in DAG levels as well as the subsequent release of arachidonic acid and its metabolites in stimulated RBL-2H3 cells[80,116,133] and rat peritoneal mast cells[101,104,105] that have been stimulated with antigen, calcium ionophores, or c-Kit ligand. In rat peritoneal mast cells, ethanol suppressed release of arachidonic acid by 40% but the combination of ethanol and the cytosolic PLA_2 (see following) inhibitor, p-bromophenylacyl bromide, inhibited release by 90% in an additive manner. This result suggested that release of arachidonic acid was mediated in part by DAG lipase and in part by PLA_2.[104,105] A discrepancy in these two papers, however, is that propranolol is reported to enhance arachidonic acid release in one report[104] and to suppress this release in the other.[105]

An alternative mechanism for release of arachidonic acid is the activation and translocation of cytosolic PLA_2 through the action of MAP kinases and calcium.[9] As noted elsewhere in this volume (see Chapter 18), inhibition of the MAP kinases is associated with similar inhibition of PLA_2 activity and arachidonic acid release. The effects of primary alcohols and propranolol on the MAP kinase/ PLA_2 pathway and, in turn, of the MAP kinase inhibitors on the PLD/PPH/DAG lipase pathway are unknown. Consequently, the relative roles of PLD and PLA_2 in the release of arachidonic acid (and arachidonic acid-derived inflammatory mediators) cannot be evaluated at this time. Again, molecular biological rather than pharmacological approaches may help further advance the research in this field.

Other Potential Roles for PLD in Mast Cells

Additional roles for PLD such as those described earlier (Physiological Roles of PLD) have not been rigorously examined in the mast cell. This cell and its cognates have ideal attributes for examining the roles of PLD in exocytosis, vesicle trafficking, and actin-dependent cortical rearrangements when the appropriate molecular biological tools become available. Much is known about the RBL-2H3 cell in molecular terms, especially in relation to FcεRI-mediated events.[141] In addition to the possible roles of PLD in degranulation and release of arachidonic acid, another potential, but unexplored, role is in the regulation of secretion of proteins via Golgi. Stimulation of RBL-2H3 cells induces production of the cytokine, tumor necrosis factor-α (TNF-α) which is subsequently secreted in a Golgi/PKC-dependent manner.[142] Others have noted PKC/ARF-regulated secretion of proteins via Golgi in RBL-2H3 cells.[143,144]

RBL-2H3 cells also adhere to fibronectin via β1 integrin (mast cells in general express several types of integrins) with the resultant tyrosine phosphorylation of certain proteins including pp125^{FAK} and paxillin.[145] The cell adhesion by itself does not induce degranulation or release of arachidonic acid but enhances such responses to antigen and ionophores. Cell adhesion and spreading is associated with reorganization of microfilaments and microtubules. When cells are subsequently stimulated, additional polymerization and reorganization of the actin cytoskeleton occurs,[118,146] along with phosphorylation of myosin chains by PKC and calcium-activated myosin light chain kinase.[92,147] Although cytoskeletal rearrangements are undoubtedly intrinsic to cell adhesion and spreading, the role of these rearrangements and the participation of actin and myosin in degranulation is suspected but not understood. As noted earlier, antigen-stimulated activation of PLD and other responses are enhanced severalfold in adherent RBL-2H3 cells compared to nonadherent cells.[118] The critical observation in the context of this review is that cell adhesion does not stimulate basal PLD or PLC activities. Thus, it would appear cytoskeletal rearrangements that are necessary for cell adhesion and spreading are not dependent on PLD. Furthermore, if PLD regulates cytoskeletal function, this regulation is restricted to cells in the activated state, for example during the process of degranulation.

Unanswered Questions and Conclusions

The studies to date indicate that PLD is responsible for a substantial, if not the major, part of the metabolic changes in phospholipids in stimulated mast cells and that its primary substrate is PC, the most abundant phospholipid in cell membranes. Although the metabolic products PA, LPA, and DAG may all function as messenger molecules, at present only DAG has established credentials as a messenger molecule for the stimulation of PKC. In this role, PLD-derived DAG is essential for the actions of some secretagogues, for example, thapsigargin, and probably contributes substantially to the secretagogue activities of other stimulants. There is evidence, although not conclusive, that PLD regulates, at least in part, release of arachidonic acid and the unexplored possibility that PLD regulates secretion of cytokines and other proteins via Golgi. In view of the lack of specific inhibitors of PLD, identification of the precise roles of PLD in mast cells will most likely require molecular biological approaches.

The studies in mast cells also indicate that increases in $[Ca^{2+}]_i$ and activation of PKC synergize each other to activate PLD and that together they can activate PLD sufficiently to account for PLD-related changes observed in antigen-stimulated cells. Preliminary evidence has been presented for further regulation of PLD by tyrosine kinase(s) and monomeric GTP-binding proteins, but receptor-mediated pathways for PLD activation, other than through calcium and PKC, are unclear. At this point, one essential task is the identification of the PLD isoforms in mast cells. RBL-2H3 cells contain the homologous gene for human PLD-1

(Holbrook, unpublished data) and the presence of this PLD would account for the observations with brefeldin A[111] and *Clostridium difficile* toxin B[111] in RBL-2H3 cells, as noted earlier. It is apparent, however, that a brefeldin A- and ARF-insensitive PLD activity exists in RBL-2H3 cells and that this activity is responsible for most of the PLD-derived metabolites in stimulated RBL-2H3 cells (Fraundorfer et al., submitted for publication). Our studies with bacterial toxins also indicate that this activity represents a previously unrecognized form of PLD or a novel pathway of activation of PLD1 via trimeric G proteins (Fraundorfer et al., submitted for publication). Because of the recent characterization of cloned PLDs and the substantial advances in our understanding of FcεRI-mediated signaling processes, it seems an opportune time to revisit PLD in mast cells.

Addendum

Since submission of this manuscript, a report appeared of the tyrosine phosphorylation of human PLD-1a and PLD-1b, and their association with tyrosine-phosphorylated proteins, after exposure of human granulocytes to peroxides of vanadate (see Marcil J, Harbour D, Naccache PH, Bourgoin S. Human phospholipase D1 can be tyrosine phosphorylated in HL-60 granulocytes. J Biol Chem 1997; 272:20660–20664).

References

1. Exton JH. Phospholipase D: enzymology, mechanisms of regulation, and function. Physiol Rev 1997;77:303–320.
2. Exton JH. New developments in phospholipase D. J Biol Chem 1997;272:15579–15582.
3. Ponting CP, Kerr ID. A novel family of phospholipase D homologues that includes phospholipid synthases and putative endonucleases: identification of duplicated repeats and potential active site residues. Protein Sci 1996;5:914–922.
4. Morris AJ, Engebrecht J, Frohman MA. Structure and regulation of phospholipase D. Trends Pharmacol Sci 1996;17:182–185.

5. Ktistakis NT, Brown AH, Sternweis PC, Roth MG. Phospholipase D is present on Golgi-enriched membranes and its activation by ADP ribosylation factor is sensitive to brefeldin A. Proc Natl Acad Sci USA 1995; 92:4952–4956.

6. Provost JJ, Fudge J, Israelit S, Siddiqi AR, Exton JH. Tissue-specific distribution and subcellular distribution of phospholipase D in rat: evidence for distinct RhoA- and ADP-ribosylation factor (ARF) -regulated isoenzymes. Biochem J 1996;319:285–291.

7. Beaven MA, Kassessinoff T. Role of phospholipases, kinases and calcium in FcεRI-induced secretion. In: Hamaway MM, ed. IgE Receptor (FcεRI) Function in Mast Cells and Basophils. Austin, TX: RG Landes, 1997:55–73.

8. Rivera J, Beaven MA. Role of protein kinase C isozymes in secretion. In: Parker P, Dekker L, eds. Protein Kinase C. (Molecular Biology Intelligence Unit series.) Austin, TX: RG Landes, 1997:131–164.

9. Gilfillan AM. Signal transduction pathways regulating arachidonic acid metabolite generation following FcεRI aggregation. In: Hamawy MM, ed. IgE Receptor (FcεRI) Function in Mast Cells and Basophils. Austin, TX: RG Landes, 1997:181–208.

10. Kanfer JN. The base exchange enzymes and phospholipase D of mammalian tissue. Can J Biochem 1980;58:1370–1380.

11. Billah MM, Anthes JC. The regulation and cellular functions of phosphatidylcholine hydrolysis. Biochem J 1990;269:281–291.

12. Brown HA, Gutowski S, Kahn RA, Sternweis PC. Partial purification and characterization of Arf-sensitive phospholipase D from porcine brain. J Biol Chem 1995;270:14935–14943.

13. Singer WD, Brown HA, Jiang X, Sternweis PC. Regulation of phospholipase D by protein kinase C is synergistic with ADP-ribosylation factor and independent of protein kinase activity. J Biol Chem 1996;271:4504–4510.

14. Cook SJ, Wakelam MJ. Analysis of the water-soluble products of phosphatidylcholine breakdown by ion-exchange chromatography. Bombesin and TPA(12-O-tetradecanoylphorbol 13-acetate) stimulate choline generation in Swiss 3T3 cells by a common mechanism. Biochem J 1989;263:581–587.

15. Bligh EG, Dyer WJ. A rapid method of total lipid extraction and purification. Can J Biochem Physiol 1959;37:911–917.

16. Shukla SD, Halenda SP. Phospholipase D in cell signalling and its relationship to phospholipase C. Life Sci 1991;48:851–866.

17. Kennerly DA. Diacylglycerol metabolism in mast cells. Analysis of lipid metabolic pathways using molecular species analysis of intermediates. J Biol Chem 1987;262:16305–16313.

18. Cook SJ, Briscoe CP, Wakelam MJ. The regulation of phospholipase D activity and its role in sn-1,2-diradylglycerol formation in bombesin- and phorbol 12-myristate 13-acetate-stimulated Swiss 3T3 cells. Biochem J 1991;280:431–438.

19. Holbrook PG, Pannell LK, Murata Y, Daly JW. Molecular species analysis of a product of phospholipase D activation. Phosphatidylethanol is formed from phosphatidylcholine in phorbol ester- and bradykinin-stimulated PC12 cells. J Biol Chem 1992;267:16834–16840.

20. Kennerly DA. Quantitative analysis of water-soluble products of cell-associated phospholipase C- and phospholipase D-catalyzed hydrolysis of phosphatidylcholine. Methods Enzymol 1991;197:191–197.

21. Liscovitch M, Chalifa V, Pertile P, Chen CS, Cantley LC. Novel function of phosphatidylinositol 4,5-bisphosphate as a cofactor for brain membrane phospholipase D. J Biol Chem 1994;269:21403–21406.

22. Huang C, Wykle RL, Daniel LW. Phospholipase D hydrolyzes ether- and ester-linked glycerophospholipids by different pathways in MDCK cells. Biochem Biophys Res Commun 1995;213:950–957.

23. Hii CS, Kokke YS, Pruimboom W, Murray AW. Phorbol esters stimulate a phospholipase D-catalysed reaction with both ester- and ether-linked phospholipids in HeLa cells. FEBS Lett 1989;257:35–37.

24. Lauritzen L, Hansen HS. Differential phospholipid-labeling suggests two subtypes of phospholipase D in rat Leydig cells. Biochem Biophys Res Commun 1995;217:747–754.

25. Day CP, Yeaman SJ. Physical evidence for the presence of two forms of phosphatidate phosphohydrolase in rat liver. Biochim Biophys Acta 1992;1127:87–94.

26. Martin A, Gomez-Munoz A, Jamal Z, Brindley DN. Characterization and assay of phosphatidate phosphatase. Methods Enzymol 1991;197:553–563.

27. Jamal Z, Martin A, Gomez-Munoz A, Brindley DN. Plasma membrane fractions from rat liver contain a phosphatidate phosphohydrolase distinct from that in the endoplasmic reticulum

and cytosol. J Biol Chem 1991;266:2988–2996.

28. Jiang Y, Lu Z, Zang Q, Foster DA. Regulation of phosphatidic acid phosphohydrolase by epidermal growth factor. Reduced association with the EGF receptor followed by increased association with protein kinase C-ε. J Biol Chem 1996;271:29529–29532.

29. Thomson FJ, Clark MA. Purification of a phosphatidic-acid-hydrolysing phospholipase A₂ from rat brain. Biochem J 1995;306:305–309.

30. Moolenaar WH, Jalink K, van Corven EJ. Lysophosphatidic acid: a bioactive phospholipid with growth factor-like properties. Rev Physiol Biochem Pharmacol 1992;119:47–65.

31. Okamura S, Yamashita S. Purification and characterization of phosphatidylcholine phospholipase D from pig lung. J Biol Chem 1994;269:31207–31213.

32. Massenberg D, Han J-S, Liyanage M, et al. Activation of rat brain phospholipase D by ADP-ribosylation factors 1, 5, and 6: separation of ADP-ribosylation factor-dependent and oleate-dependent enzymes. Proc Natl Acad Sci USA 1994;91:11718–11722.

33. Vinggaard AM, Provost JJ, Exton JH, Hansen HS. Arf and RhoA regulate both the cytosolic and the membrane-bound phospholipase D from human placenta. Cell Signal 1997;9:189–196.

34. Conricode KM, Smith JL, Burns DJ, Exton JH. Phospholipase D activation in fibroblast membranes by the α and β isoforms of protein kinase C. FEBS Lett 1994;342:149–153.

35. Ohguchi K, Banno Y, Nakashima S, Nozawa Y. Regulation of membrane-bound phospholipase D by protein kinase C in HL60 cells. Synergistic action of small GTP-binding protein RhoA. J Biol Chem 1996;271:4366–4372.

36. Brown AH, Gutowski S, Moomaw CR, Slaughter C, Sternweis PC. ADP-ribosylation factor, a small GTP-dependent regulatory protein, stimulates phospholipase D activity. Cell 1993;75:1137–1144.

37. Pertile P, Liscovitch M, Chalifa V, Cantley LC. Phosphatidylinositol 4,5-bisphosphate synthesis is required for activation of phospholipase D in U937 cells. J Biol Chem 1995;270:5130–5135.

38. Kuribara H, Tago K, Yokozeki T, et al. Synergistic activation of rat brain phospholipase D by ADP-ribosylation factor and rhoA p21, and its inhibition by Clostridium botulinum C3 exoenzyme. J Biol Chem 1995;270:25667–25671.

39. Nakamura S, Kiyohara Y, Jinnai H, et al. Mammalian phospholipase D: phosphatidylethanolamine as an essential component. Proc Natl Acad Sci USA 1996;93:4300–4304.

40. Kanfer JN, McCartney DG, Singh IN, Freysz L. Acidic phospholipids inhibit the phospholipase D activity of rat brain neuronal nuclei. FEBS Lett 1996;383:6–8.

41. Chong LD, Traynor-Kaplan A, Bokoch GM, Schwartz MA. The small GTP-binding protein Rho regulates a phosphatidylinositol 4-phosphate 5-kinase in mammalian cells. Cell 1994;79:507–513.

42. Chardin P, Paris S, Antonny B, et al. A human exchange factor for ARF contains Sec7- and pleckstrin-homology domains. Nature (Lond) 1996;384:481–484.

43. Klarlund JK, Guilherme A, Holik JH, Virbasius JV, Chawla A, Czech MP. Signaling by phosphoinositide-3,4,5-trisphosphate through proteins containing pleckstrin and Sec7 homology domains. Science 1997;275:1927–1930.

44. Liscovitch M, Cantley LC. Signal transduction and membrane traffic: the PITP/phosphoinositide connection. Cell 1995;81:659–662.

45. Randazzo PA. Functional interaction of ADP-ribosylation factor 1 with phosphatidylinositol 4,5-bisphosphate. J Biol Chem 1997;272:7688–7692.

46. Chung J-K, Sekiya F, Kang H-S, et al. Synaptojanin inhibition of phospholipase D activity by hydrolysis of phosphatidylinositol 4,5-bisphosphate. J Biol Chem 1997;272:15980–15985.

47. Wirtz KWA. Phospholipid transfer proteins. Annu Rev Biochem 1991;60:73–99.

48. Liscovitch M, Cantley LC. Lipid second messengers. Cell 1994;77:329–334.

49. Kauffmann-Zeh A, Thomas GM, Ball A, et al. Requirement for phosphatidylinositol transfer protein in epidermal growth factor signaling. Science 1995;268:1188–1190.

50. Cunningham E, Thomas GMH, Ball A, Hiles I, Cockcroft S. Phosphatidylinositol transfer protein dictates the rate of inositol trisphosphate production by promoting the synthesis of PIP₂. Curr Biol 1995;5:775–783.

51. Kahn RA, Yucel JK, Malhotra V. ARF Signaling: a potential role for phospholipase D in membrane traffic. Cell 1993;75:1045–1048.

52. Moss J, Vaughan M. Structure and function of ARF proteins: activators of cholera toxin and critical components of intracellular vesicular transport processes. J Biol Chem 1995; 270:12327–12330.

53. Cockcroft S, Thomas GM, Fensome A, et al. Phospholipase D: a downstream effector of ARF in granulocytes. Science 1994;263:523–526.

54. Takahashi K-I, Tago K, Okano H, Ohya Y, Katada T, Kanaho Y. Augmentation by calmodulin of ADP-ribosylation factor-stimulated phospholipase D activity in permeabilized rabbit peritoneal neutrophils. J Immunol 1996;156:1229–1234.

55. Lee C, Kang H-S, Chung J-K, et al. Inhibition of phospholipase D by clathrin assembly protein 3 (AP3). J Biol Chem 1997;272:15986–15992.

56. Ktistakis NT, Brown HA, Waters MG, Sternweis PC, Roth MG. Evidence that phospholipase D mediates ADP ribosylation factor-dependent formation of Golgi coated vesicles. J Cell Biol 1996;134:295–306.

57. Liscovitch M, Chalifa-Caspi V. Enzymology of mammalian phospholipases D: *in vitro* studies. Chem Phys Lipids 1996;80:37–44.

58. Rumenapp U, Geiszt M, Wahn F, Schmidt M, Jakobs KH. Evidence for ADP-ribosylation-factor-mediated activation of phospholipase D by m3 muscarinic acetylcholine receptor. Eur J Biochem 1995;234:240–244.

59. Houle MG, Kahn RA, Naccache PH, Bourgoin S. ADP-ribosylation factor translocation correlates with potentiation of GTPγ S-stimulated phospholipase D activity in membrane fractions of HL-60 cells. J Biol Chem 1995;270:22795–22800.

60. Bowman EP, Uhlinger DJ, Lambeth JD. Neutrophil phospholipase D is activated by a membrane-associated Rho family of small molecular weight GTP-binding protein. J Biol Chem 1993;268:21509–21512.

61. Malcolm KC, Ross AH, Qui R-G, Symons M, Exton JH. Activation of rat liver phospholipase D by the small GTP-binding protein RhoA. J Biol Chem 1994;269:25951–25954.

62. Siddiqi AR, Smith JL, Ross AH, Qui R-G, Symons M, Exton JH. Regulation of phospholipase D in HL60 cells. Evidence for a cytosolic phospholipase D. J Biol Chem 1995;270:8466–8473.

63. Kwak J-Y, Lopez I, Uhlinger DJ, Ryu SH, Lambeth JD. RhoA and cytosolic 50-kDa factor reconstitute GTPγ-dependent phospholipase D activity in human neutrophil subcellular fractions. J Biol Chem 1995;270:27093–27098.

64. Hess JA, Ross AH, Qiu RG, Symons M, Exton JH. Role of Rho family proteins in phospholipase D activation by growth factors. J Biol Chem 1997;272:1615–1620.

65. Malcolm KC, Elliott CM, Exton JH. Evidence for Rho-mediated agonist stimulation of phospholipase D in rat1 fibroblasts. Effects of *Clostridium botulinum* C3 exoenzyme. J Biol Chem 1996;271:13135–131139.

66. Schmidt M, Rumenapp U, Bienek C, Keller J, von Eichel-Streiber C, Jakobs KH. Inhibition of receptor signaling to phospholipase D by *Clostridium difficile* toxin B. Role of Rho proteins. J Biol Chem 1996;271:2422–2426.

67. Ojio K, Banno Y, Nakashima S, et al. Effect of *Clostridium difficile* toxin B on IgE receptor-mediated signal transduction in rat basophilic leukemia cells: inhibition of phospholipase D activation. Biochem Biophys Res Commun 1996;224:591–596.

68. Steed PM, Nagar S, Wennogle LP. Phospholipase D regulation by a physical interaction with the actin-binding protein gelsolin. Biochemistry 1996;35:5229–5237.

69. Spiegel S, Milstien S. Sphingoid bases and phospholipase D activation. Chem Phys Lipids 1996;80:27–36.

70. Lukowski S, Lecomte M-C, Mira J-P, et al. Inhibition of phospholipase D activity by fodrin. An active role for the cytoskeleton. J Biol Chem 1996;271:24164–24171.

71. Colley WC, Sung T-C, Roll R, et al. Phospholipase D2, a distinct phospholipase D isoform with novel regulatory properties that provokes cytoskeletal reorganization. Curr Biol 1997;7:191–201.

72. Hammond SM, Altshuller YM, Sung T, et al. Human ADP-ribosylation factor-activated phosphatidylcholine-specific phospholipase D defines a new and highly conserved gene family. J Biol Chem 1995;270:29640–29643.

73. Hammond SM, Jenco JM, Nakashima S, et al. Characterization of two alternately spliced forms of phospholipase D1. Activation of the purified enzymes by phosphatidylinositol 4,5-bisphosphate, ADP-ribosylation factor, and Rho family monomeric GTP-binding proteins and protein kinase C-α. J Biol Chem 1997;272:3860–3868.

74. Kodaki T, Yamashita S. Cloning, expression, and characterization of a novel phospholipase

D complementary DNA from rat brain. J Biol Chem 1997;272:11480–114813.

75. Ribbes G, Henry J, Cariven C, Pontarotti P, Chap H, Record M. Expressed sequence tags identify human isologs of the ARF-dependent phospholipase D. Biochem Biophys Res Commun 1996;224:206–211.

76. Yoshimura S, Nakashima S, Ohguchi K, et al. Differential mRNA expression of phospholipase D (PLD) isozymes during cAMP-induced differentiation in C6 glioma cells. Biochem Biophys Res Commun 1996;225:494–499.

77. Yamada K, Jelsema CL, Beaven MA. Certain inhibitors of protein serine/threonine kinases also inhibit tyrosine-phosphorylation of phospholipase Cγ 1 and other proteins and reveal distinct roles for tyrosine kinase(s) and protein kinase C in stimulated, rat basophilic RBL-2H3 cells. J Immunol 1992;149:1031–1037.

78. Yang SF, Freer S, Benson AA. Transphosphatidylation by phospholipase D. J Biol Chem 1967;242:477–484.

79. Koul O, Hauser G. Modulation of rat brain cytosolic phosphatidate phosphohydrolase: effect of cationic amphiphilic drugs and divalent cations. Arch Biochem Biophys 1987;253:453–461.

80. Lin P, Wiggan GA, Gilfillan AM. Activation of phospholipase D in a rat mast (RBL-2H3) cell line: a possible unifying mechanism for IgE-dependent degranulation and arachidonic acid metabolite release. J Immunol 1991;146:1609–1616.

81. Sozzani S, Agwu DE, McCall CE, et al. Propranolol, a phosphatidate phosphohydrolase inhibitor, also inhibits protein kinase C. J Biol Chem 1992;267:20481–204818.

82. Slater SJ, Cox KJ, Lombardi JV, et al. Inhibition of protein kinase C by alcohols and anaesthetics. Nature (Lond) 1993;364:82–84.

83. Downing GJ, Kim S, Nakanishi S, Catt KJ, Balla T. Characterization of a soluble adrenal phosphatidylinositol 4-kinase reveals wortmannin sensitivity of type III phosphatidylinositol kinases. Biochemistry 1996; 35:3587–3594.

84. Yano H, Nakanishi S, Kimura K, et al. Inhibition of histamine secretion by wortmannin through the blockade of phosphatidylinositol 3-kinase in RBL-2H3 cells. J Biol Chem 1993;268:25846–25856.

85. Hirasawa N, Sato Y, Yomogida S-I, Mue S, Ohichi K. Role of phosphatidylinositol 3-

86. Bocckino SB, Blackmore PF, Wilson PB, Exton JH. Phosphatidate accumulation in hormone-treated hepatocytes via a phospholipase D mechanism. J Biol Chem 1987;262:15309–15315.

87. Stutchfield J, Cockcroft S. Correlation between secretion and phospholipase D activation in differentiated HL60 cells. Biochem J 1993;293:649–655.

88. Eastman SJ, Hope MJ, Wong KF, Cullis PR. Influence of phospholipid asymmetry on fusion between large unilamellar vesicles. Biochemistry 1992;31:4262–4268.

89. Waite KA, Wallin R, Qualliotine-Mann D, McPhail LC. Phosphatidic acid-mediated phosphorylation of the NADPH oxidase component p47^{phox}. Evidence that phosphatidic acid may activate a novel protein kinase. J Biol Chem 1997;272:15569–15578.

90. Moolenaar WH. Lysophosphatidic acid, a multifunctional phospholipid messenger. J Biol Chem 1995;270:12949–12952.

91. Ali H, Fisher I, Haribabu B, Richardson RM, Snyderman R. Role of phospholipase Cβ phosphorylation in the desensitization of cellular responses to platelet-activating factor. J Biol Chem 1997;272:11706–11709.

92. Choi OH, Adelstein RS, Beaven MA. Secretion from rat basophilic RBL-2H3 cells is associated with phosphorylation of myosin light chains by myosin light chain kinase as well as phosphorylation by protein kinase C. J Biol Chem 1994;269:536–541.

93. Nishizuka Y. Protein kinase C and lipid signaling for sustained cellular responses. FASEB J 1995;9:484–496.

94. Newton AC. Protein kinase C: structure, function, and regulation. J Biol Chem 1995; 270:28495–28498.

95. Ha K-S, Exton JH. Activation of actin polymerization by phosphatidic acid derived from phosphatidylcholine in IIC9 fibroblasts. J Cell Biol 1993;123:1789–1796.

96. Cross MJ, Roberts S, Ridley AJ, et al. Stimulation of actin stress fibre formation mediated by activation of phospholipase D. Curr Biol 1996;6:588–597.

97. Murray JJ, Dinh TT, Truett AP, Kennerly DA. Isolation and enzymic assay of choline and phosphocholine present in cell extracts

with picomole sensitivity. Biochem J 1990;
270:63–68.

98. Kennerly DA. Phosphatidylcholine is a quantitatively more important source of increased
1,2-diacylglycerol than is phosphatidylinositol
in mast cells. J Immunol 1990;144:3912–3919.

99. Gruchalla RS, Dinh TT, Kennerly DA. An
indirect pathway of receptor-mediated 1,2-
diacylglycerol formation in mast cells. I. IgE
receptor-mediated activation of phospholipase
D. J Immunol 1990;144:2334–23420.

100. Dinh TT, Kennerly DA. Assessment of receptor-dependent activation of phosphatidylcholine hydrolysis by both phospholipase D and
phospholipase C. Cell Regul 1991;2:299–309.

101. Koike T, Mizutani T, Hirai K, Morita Y,
Nozawa Y. SCF/c-kit receptor-mediated
arachidonic acid liberation in rat mast cells. Involvement of PLD activation-associated tyrosine phosphorylation. Biochem Biophys Res
Commun 1993;197:1570–1577.

102. Koike T, Hirai K, Morita Y, Nozawa Y. Stem
cell factor-induced signal transduction in rat
mast cells. Activation of phospholipase D but
not phosphoinositide-specific phospholipase C
in c-*kit* receptor stimulation. J Immunol 1993;
151:359–366.

103. Yamada K, Kanaho Y, Miura K, Nozawa Y.
Antigen-induced phospholipase D activation in
rat mast cells is independent of protein kinase
C. Biochem Biophys Res Commun 1991;
175:159–164.

104. Sato T, Ishimoto T, Akiba S, Fujii T. Enhancement of phospholipase A_2 activation by phosphatidic acid endogenously formed through
phospholipase D action in rat peritoneal mast
cell. FEBS Lett 1993;323:23–26.

105. Ishimoto T, Akiba S, Sato T, Fujii T. Contribution of phospholipases A_2 and D to arachidonic
acid liberation and prostaglandin D_2 formation
with increase in intracellular Ca^{2+} concentration
in rat peritoneal mast cells. Eur J Biochem
1994;219:401–406.

106. Nakashima S, Fujimiya H, Miyata H, Nozawa
Y. Antigen-induced biphasic diacylglycerol formation in RBL-2H3 cells: the late sustained
phase due to phosphatidylcholine hydrolysis is
dependent on protein kinase C. Biochem
Biophys Res Commun 1991;177:336–342.

107. Kumada T, Miyata H, Nozawa Y. Involvement
of tyrosine phosphorylation in IgE receptor-
mediated phospholipase D activation in rat
basophilic leukemia (RBL-2H3) cells. Biochem
Biophys Res Commun 1993;191:1363–1368.

108. Kumada T, Nakashima S, Nakamura Y, Miyata
H, Nozawa Y. Antigen-mediated phospholipase D activation in rat basophilic leukemia
(RBL-2H3) cells: possible involvement of
calcium/calmodulin. Biochim Biophys Acta
1995;1258:107–114.

109. Nakamura Y, Nakashima S, Ojio K, Banno Y,
Miyata H, Nozawa Y. Ceramide inhibits IgE-
mediated activation of phospholipase D, but
not of phospholipase C, in rat basophilic leukemia (RBL-2H3) cells. J Immunol 1996;156:256–
262.

110. Kumada T, Nakashima S, Miyata H, Nozawa Y.
Potent activation of phospholipase D by
phenylarsine oxide in rat basophilic leukemia
(RBL-2H3) cells. Biochem Biophys Res
Commun 1994;199:792–798.

111. Nakamura Y, Nakashima S, Kumada T, Ojio K,
Miyata H, Nozawa Y. Brefeldin A inhibits anti-
gen- or calcium ionophore-mediated but not
PMA-induced phospholipase D activation in
rat basophilic leukemia (RBL-2H3) cells.
Immunobiology 1996;195:231–242.

112. Kumada T, Nakashima S, Nakamura Y, Miyata
H, Nozawa Y. Phenylarsine oxide (PAO)-
mediated activation of phospholipase D in rat
basophilic leukemia (RBL-2H3) cells: possible
involvement of calcium and protein kinase C.
Immunobiology 1996;195:347–359.

113. Lin P, Gilfillan AM. The role of calcium and
protein kinase C in the IgE-dependent activation of phosphatidylcholine-specific phospholipase D in a rat mast (RBL-2H3) cell line. Eur J
Biochem 1992;207:163–168.

114. Lin P, Fung SJ, Li S, et al. Temporal regulation
of the IgE-dependent 1,2-diacylglycerol production by tyrosine kinase activation in a rat
(RBL 2H3) mast-cell line. Biochem J 1994;
299:109–114.

115. Marcotte GV, Millard PJ, Fewtrell C. Release
of calcium from intracellular stores in rat basophilic leukemia cells monitored with the fluorescent probe chlortetracycline. J Cell Physiol
1990;142:78–88.

116. Lin P, Fung WJ, Gilfillan AM. Phosphatidylcholine-specific phospholipase D-derived 1,2-
diacylglycerol does not initiate protein kinase C
activation in the RBL 2H3 mast-cell line.
Biochem J 1992;287:325–331.

117. Ali H, Choi OH, Fraundorfer PF, Yamada K,
Gonzaga HMS, Beaven MA. Sustained activation of phospholipase D via adenosine A_3 receptors is associated with enhancement of
antigen- and Ca^{2+}-ionophore-induced secretion

in a rat mast cell line. J Pharmacol Exp Ther 1996;276:837–845.

118. Apgar JR. Increased degranulation and phospholipase A₂, C, and D activity in RBL cells stimulated through FcεRI is due to spreading and not simply adhesion. J Cell Sci 1997; 110:771–780.

119. Ozawa K, Yamada K, Kazanietz MG, Blumberg P, Beaven MA. Different isozymes of protein kinase C mediate feed-back inhibition of phospholipase C and stimulatory signals for exocytosis in rat RBL-2H3 cells. J Biol Chem 1993;268:2280–2283.

120. Billah MM. Phospholipase D and cell signaling. Curr Opin Immunol 1993;5:114–123.

121. Beaven MA, Metzger H. Signal transduction by Fc receptors: the Fcε case. Immunol Today 1993;14:222–226.

122. Zhang J, Berenstein EH, Evans RL, Siraganian RP. Transfection of Syk protein tyrosine kinase reconstitutes high affinity IgE receptor-mediated degranulation in a Syk-negative variant of rat basophilic leukemia RBL-2H3 cells. J Exp Med 1996;184:71–80.

123. Hirasawa N, Scharenberg A, Yamamura H, Beaven MA, Kinet J-P. A requirement for Syk in the activation of the MAP kinase/phospholipase A₂ pathway by FcεRI is not shared by a G protein-coupled receptor. J Biol Chem 1995; 270:10960–10967.

124. Prepens U, Just I, von Eichel-Streiber C, Aktories K. Inhibition of FcεRI-mediated activation of rat basophilic leukemia cells by *Clostridium difficile* toxin B (monoglucosyltransferase). J Biol Chem 1996; 271:7324–7329.

125. Ramkumar V, Stiles GL, Beaven MA, Ali H. The A₃R is the unique adenosine receptor which facilitates release of allergic mediators in mast cells. J Biol Chem 1993;268:16887–16890.

126. Ali H, Cunha-Melo JR, Saul WF, Beaven MA. The activation of phospholipase C via adenosine receptors provides synergistic signals for secretion in antigen-stimulated RBL-2H3 cells: evidence for a novel adenosine receptor. J Biol Chem 1990;265:745–753.

127. Collado-Escobar D, Cunha-Melo JR, Beaven MA. Treatment with dexamethasone down-regulates IgE-receptor mediated signals and up-regulates adenosine-receptor mediated signals in a rat mast cell (RBL-2H3) line. J Immunol 1990;144:244–250.

128. Ramkumar V, Wilson M, Dhanraj DN, Gettys TW, Ali H. Dexamethasone upregulates A₃ ad-

enosine receptors in rat basophilic leukemia (RBL-2H3) cells. J Immunol 1995;154:5436–5443.

129. Columbo M, Horowitz EM, Botana LM, et al. The human recombinant c-kit receptor ligand, rhSCF, induces mediator release from human cutaneous mast cells and enhances IgE-dependent mediator release from both skin mast cells and peripheral blood basophils. J Immunol 1992;149:599–608.

130. Sperr WR, Czerwenka K, Mundigler G, et al. Specific activation of human mast cells by the ligand for *c-kit*: comparison between lung, uterus and heart mast cells. Int Arch Allergy Immunol 1993;102:170–175.

131. Ozawa K, Szallasi Z, Kazanietz MG, et al. Ca²⁺-Dependent and Ca²⁺-independent isozymes of protein kinase C mediate exocytosis in antigen-stimulated rat basophilic RBL-2H3 cells: reconstitution of secretory responses with Ca²⁺ and purified isozymes in washed permeabilized cells. J Biol Chem 1993;268:1749–1756.

132. Beaven MA, Ozawa K. Role of calcium, protein kinase C, and MAP kinase in the activation of mast cells. Allergol Int 1996;45:73–84.

133. Lin P, Wiggan GA, Welton AF, Gilfillan AM. Differential effects of propranolol on the IgE-dependent, or calcium ionophore-stimulated, phosphoinositide hydrolysis and calcium mobilization in a mast (RBL 2H3) cell line. Biochem Pharmacol 1991;41:1941–1948.

134. Lo TN, Saul W, Beaven MA. The actions of Ca²⁺ ionophores on rat basophilic (2H3) cells are dependent on cellular ATP and hydrolysis of inositol phospholipids. A comparison with antigen stimulation. J Biol Chem 1987; 262:4141–4145.

135. Ali H, Maeyama K, Sagi-Eisenberg R, Beaven MA. Antigen and thapsigargin promote influx of Ca²⁺ in rat basophilic RBL-2H3 cells by ostensibly similar mechanisms that allow filling of inositol 1,4,5-trisphosphate-sensitive and mitochondrial Ca²⁺ stores. Biochem J 1994;304:431–440.

136. Wolfe PD, Chang E, Rivera J, Fewtrell C. Differential effects of the protein kinase C activator phorbol 12-myristate 13-acetate on calcium responses and secretion in adherent and suspended RBL-2H3 mucosal mast cells. J Biol Chem 1996;271:6658–6665.

137. Wilkinson SE, Parker PJ, Nixon JS. Isoenzyme specificity of bisindolylmaleimides, selective in-

hibitors of protein kinase C. Biochem J 1993;294:335–337.

138. Kennerly DA, Sullivan TJ, Sylwester P, Parker CW. Diacylglycerol metabolism in mast cells: a potential role in membrane fusion and arachidonic acid release. J Exp Med 1979;150:1039–1044.

139. Garcia-Gil M, Siraganian RP. Source of the arachidonic acid released on stimulation of rat basophilic leukemia cells. J Immunol 1986;136:3825–3828.

140. Yamada K, Okano Y, Miura K, Nozawa Y. A major role for phospholipase A_2 in antigen-induced arachidonic acid release in rat mast cells. Biochem J 1987;247:95–99.

141. Beaven MA. Calcium signalling: sphingosine kinase versus phospholipase C. Curr Biol 1996;6:798–801.

142. Baumgartner RA, Yamada K, Deramo VA, Beaven MA. Secretion of tumor necrosis factor (TNF) from a rat mast cell line is a brefeldin A-sensitive and a calcium/protein kinase C-regulated process. J Immunol 1994;153:2609–2617.

143. Matteis MAD, Santini G, Kahn RA, Tullio GD, Luini A. Receptor and protein kinase C-mediated regulation of ARF binding to the Golgi complex. Nature (Lond) 1993;364:818–821.

144. Luini A, Matteis MAD. Receptor-mediated regulation of constitutive secretion. Trends Cell Biol 1993;3:290–292.

145. Hamawy MM, Mergenhagen SE, Siraganian RP. Adhesion molecules as regulators of mast-cell and basophil function. Immunol Today 1994;15:62–66.

146. Pfeiffer JR, Oliver JM. Tyrosine kinase-dependent assembly of actin plaques linking Fc epsilon RI cross-linking to increased cell substrate adhesion in RBL-2H3 tumor mast cells. J Immunol 1994;152:270–279.

147. Choi OH, Park C-S, Itoh K, Adelstein RS, Beaven MA. Cloning of the cDNA encoding rat myosin heavy chain-A and evidence for the absence of myosin heavy chain-B in cultured rat mast (RBL-2H3) cells. J Muscle Res Cell Motil 1996;17:69–77.

148. Cissel DS, Fraundorfer PF, Beaven MA. Thapsigargin-induced secretion is dependent on activation of a cholera-toxin sensitive phospholipase D in a mast cell line. J Pharm Exper Ther 1998;285:110–118.

17

New Perspectives on Ca^{2+} Influx in Mast Cells

Michael A. McCloskey

The importance of extracellular calcium for anaphylactic release of histamine was suggested by experiments of Mongar and Schild in 1958, years before the discovery of IgE or its high-affinity receptor, the FcεRI.[1] Many subsequent studies using radiotracer flux and fluorescent Ca^{2+} indicators revealed that multivalent binding of antigen to the IgE–FcεRI complex elicits the release of internal as well as the influx of extracellular Ca^{2+}, and there appears to be a causal relation between these two events. For the RBL-2H3 mast cell line used in many of these studies,[2] it is generally agreed that Ca^{2+} influx is crucial for antigen-induced secretion of preformed inflammatory mediators.[3] But it was not until 1995 that antigen-evoked Ca^{2+} influx into mast cells was directly observed,[4,5] and the mechanistic details of this process as well as the molecular nature of the presumed Ca^{2+} channel are yet unclear. Moreover, the functional implications of FcεRI-elicited Ca^{2+} entry may extend far beyond those envisioned by Bill Douglas in stimulus–secretion coupling[6,7] or James Putney in refilling of intracellular Ca^{2+} stores.[8] In fact, some investigators still question the relevance of Ca^{2+} influx to degranulation of peritoneal mast cells, in which antigen-evoked Ca^{2+} entry has eluded direct observation. In this chapter, I describe our detection in RBL-2H3 cells of an antigen-evoked Ca^{2+} conductance that appears to regulate secretion,[4] consider some practical matters that complicate this line of inquiry, and address a few current topics on FcεRI-linked Ca^{2+} influx. I also outline our recent evidence

pointing to a new and potentially important function of antigen-driven Ca^{2+} influx, that is, activation of the nuclear factor of activated T cells (NF-AT) transcription factor via cross-linkage of the FcεRI.[9]

Historical Frame of Reference

Seminal experiments conducted in the 1970s and early 1980s supported the concept that mast cell secretagogues mobilize Ca^{2+} from extracellular or intracellular sources, the resulting elevation of cytosolic free [Ca^{2+}] being a *necessary* and *sufficient* condition for exocytotic release of histamine. This stimulus–secretion coupling model evolved from studies of chromaffin cells,[6,7] in which the Ca^{2+} signal is rapidly generated by conspicuous voltage-gated Ca^{2+} currents. Although its applicability to mast cells has been attacked on multiple fronts, the model has emerged less bruised than is commonly perceived. Parallel signals initiated by FcεRI cross-linkage may contribute to antigen-evoked secretion,[10] but here I focus exclusively on the Ca^{2+} signal.*

---

* An issue that I have not touched upon is the enhancement of GTP-γS-induced exocytosis by elevated [Ca^{2+}]$_i$, a phenomenon of as yet uncertain relevance to FcεRI-mediated secretion by intact cells.[36] It is unclear whether G proteins play other than housekeeping roles in *antigen*-evoked secretion (but see Chapters 20, 21).

Among the early supportive data were the findings that rat peritoneal mast cells (RPMC) can be induced to degranulate with noncyto-toxic concentrations of Ca^{2+}-selective iono-phores,[11-13] by intracellular iontophoresis of Ca^{2+} but not Mg^{2+} or K^+ salts,[14] by fusion with lipid vesicles containing Ca^{2+} but not Mg^{2+} salts,[15] or by readdition of Ca^{2+} to cells whose internal Ca^{2+} stores had been depleted by incubation in low-Ca^{2+} medium.[16,17] Extracellular $[Ca^{2+}]$ of 1 to 3 mM is required for maximal mediator release from antigen-stimulated RPMC,[18] and indeed, cross-linkage of the IgE receptor enhances uptake of external $^{45}Ca^{2+}$ and $^{89}Sr^{2+}$ ions.[11,19,20] Similar results were obtained with a secreting rat basophilic leukemia (RBL) cell line[21] and its RBL-2H3 variant.[22,23]

Measurements of $^{45}Ca^{2+}$ uptake did not yield information on a key variable, $[Ca^{2+}]_i$, but Tsien's development of fluorescent Ca^{2+} indicators[24] soon permitted direct measurement of $[Ca^{2+}]_i$ in cell populations. Initial studies with quin2 showed that FcεRI cross-linkage elevates $[Ca^{2+}]_i$ in both RPMC[25,26] and RBL-2H3 cells,[27] consonant with previous suggestions by Pearce and others[28] that antigen mobilizes Ca^{2+} in RPMC through a combination of release and influx. Maintenance of the Ca^{2+} signal in RBL-2H3 cells was shown to be an active process, dependent upon external Ca^{2+}, but it remained for Mohr and Fewtrell to demonstrate that antigen also mobilizes Ca^{2+} from intracellular stores in these cells.[29,30] The dynamic nature of the Ca^{2+} signal in single RBL-2H3 cells was revealed dramatically by dual excitation imaging of cells loaded with the new Ca^{2+} indicator, fura-2.[31] Even in the absence of external Ca^{2+}, antigen was found to induce sharp $[Ca^{2+}]_i$ oscillations, the persistence of which required external calcium.[32]

The necessity of Ca^{2+} influx for antigen-evoked secretion by RBL-2H3 cells was supported by the discovery that certain metal ions inhibit with similar potency the antigen-evoked $[Ca^{2+}]_i$ rise, $^{45}Ca^{2+}$ uptake, and mediator secretion[3,27]; lanthanides also had been shown to inhibit IgE-mediated secretion by RPMC.[3,33] As Kanner and Metzger first demonstrated, these physiological responses of RBL-2H3 cells are also blocked by high external K^+, which by depolarizing the plasma membrane reduces the driving force for Ca^{2+} entry.[3,23,30] That a high-K^+ solution did not directly raise $[Ca^{2+}]_i$ or elicit $^{45}Ca^{2+}$ uptake indicated that conventional voltage-gated Ca^{2+} channels were absent from these cells. Although most of these findings were consistent with Douglas' hypothesis, direct evidence for the putative FcεRI-activated Ca^{2+} channel was lacking.

Enter the biophysicists in the mid-1980s with their powerful gigaseal patch-clamp technology. In agreement with the previous work, initial patch-clamp studies of RPMC and RBL-2H3 cells did not detect voltage-activated Ca^{2+} channels.[34,35] Yet despite their introduction of a new recording method that preserves cytoplasmic integrity, these studies also failed to identify antigen-evoked Ca^{2+} currents. These and various other findings were misinterpreted by many as a death knell for the Ca^{2+} signal hypothesis, it being pronounced that $[Ca^{2+}]_i$ elevation and Ca^{2+} influx were *neither* necessary *nor* sufficient to initiate exocytosis.[26,36] Absence of evidence is not necessarily evidence of absence, and we sought conditions that would permit reliable observation of the presumed Ca^{2+} channel. Our interest was sparked by a growing awareness that independent of its contested role in secretion by RPMC, Ca^{2+} influx makes essential contributions to other mast cell functions, among them activation of the NF-AT transcription factor.[9]

A further impetus came from the discovery in T cells and mast cells of a Ca^{2+} current, dubbed I_{CRAC} (for Ca^{2+} release-activated Ca^{2+} current), that is activated by the depletion of intracellular Ca^{2+} stores[37-40]; CRAC channels* are now viewed as the prototype of a set of store-operated or capacitative calcium entry channels.[41-43] Fueling speculation that I_{CRAC} might be induced by FcεRI cross-

---

* Nonstationary fluctuation analysis of T cell I_{CRAC}[39] predicts a unitary Ca^{2+} current at -80 mV of just 3.7 fA (110 mM Ca^{2+}), smallest of any ion channel. Classification as a gated channel is supported by the Lorentzian shape of the power spectrum,[39,107] permeation by monovalent cations in the absence of Ca^{2+},[46,107,108] and a predicted unitary Na^+ current of 230 fA (-80 mV).[107]

linkage, depletion of internal Ca^{2+} stores in RBL-2H3 cells with thapsigargin, which inhibits the Ca^{2+} pump of the endoplasmic reticulum,[44] was found to elicit $^{45}Ca^{2+}$ uptake via a pathway with similar ionic selectivity and pharmacological sensitivity to that induced by antigen.[45] But there were apparent inconsistencies, as in the report that I_{CRAC} is carried poorly by Ba^{2+} and Sr^{2+} ions[37,46] whereas Ba^{2+} and Sr^{2+} permeate the antigen-induced pathway quite well.[3,20] RPMC also contain a 50-pS nonselective cation channel that makes a minor contribution to the plateau phase of elevated $[Ca^{2+}]_i$ elicited by compound 48/80,[47] but there is no convincing evidence linking it to the FcεRI. Speculation notwithstanding, the critical experiment remained to be done.

Practical Lessons from the Past

In considering possible reasons for the previous failures to observe antigen-evoked Ca^{2+} influx currents, two major factors came to mind. First, signal transduction via the FcεRI in RBL-2H3 cells is highly temperature sensitive, and below 20°C almost no inositol phospholipid hydrolysis or elevation of $[Ca^{2+}]_i$ occurs in response to FcεRI aggregation.[48] This may not present a problem in Gôttingen in January, where a nonfunctional thermostat could pose a health risk, but as I discovered it can make a huge difference in Irvine (California), where winters are on the mild side. During 2 days of Ca^{2+} imaging experiments with RBL-2H3 cells in the winter of 1989, I noticed that the fraction of cells exhibiting an increase in $[Ca^{2+}]_i$ following antigen addition jumped from nil to about 60% as the day wore on. This correlated with an increase in ambient temperature from about 19° to 23°C as heat generated by equipment warmed the small room. Upon preheating the buffers to 37°C, literally every cell responded. The problem is not just one of a threshold below which detection is impossible: in terms of the population average calcium signal, there is a 14-fold increase between 20° and 37°C.[48] As noted later, temperature control proved to be an important factor for reliable detection of

antigen-induced Ca^{2+} currents in RBL-2H3 cells.

Second, the experiments with RBL-2H3 cells were carried out with standard whole-cell recording (WCR), in which electrical continuity between cytoplasm and pipette is achieved by breaking the patch of membrane covering the pipette opening. Given that diffusible cell components, for example, secondary or tertiary messengers, can escape rapidly into the effectively infinite volume of the pipette,[49,50] and that antigen fails to trigger degranulation of RPMC during conventional WCR,[34] preservation of cytoplasmic integrity seemed crucial. Lindau and Fernandez skirted this problem in their study of RPMC by using a pipette solution containing a very modest concentration of ATP^{4-}. RPMC contain a P_{2Z} (or P_{2X7}) receptor for ATP^{4-}, ligation of which induces the formation of electrically conductive pores in the membrane, obviating the need for breakin to achieve WCR. But in these experiments, the first to employ perforated-patch recording, the electrical access resistance was so high (Ra = 0.2–5 GΩ) as to hinder detection of small Ca^{2+} currents now considered sufficient to support secretion by mast cells. The antimycotic nystatin was later found to be more versatile than ATP^{4-}, affording dramatically reduced access resistance and high-resolution WCR even with cells that do not express P_{2z} receptors.[51] As discussed next, nystatin perforated-patch recording at 36°–37°C permits highly consistent detection of antigen-evoked Ca^{2+} currents in RBL-2H3 cells.

Properties of an Antigen-Evoked Ca^{2+} Current in RBL-2H3 Cells

The current–voltage (I/V) curves in Fig. 17.1A show the time-dependent appearance of an inwardly rectifying current in RBL-2H3 cells following the addition of a multivalent antigen (TNP_{15}-BSA, 50 ng/ml) cognate to the sensitizing antibody (anti-TNP IgE). In our initial study, antigen induced this current in 47 of 48 cells not containing exogenous Ca^{2+} chelators,

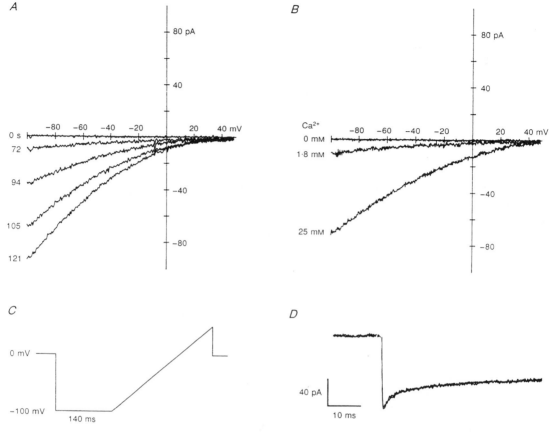

FIGURE 17.1. Time-dependent appearance of Ca^{2+}-selective inward current on antigenic stimulation of IgE-sensitized RBL-2H3 cells. (A) I/V curves recorded at the indicated times (*left* of each trace) after addition of polyvalent antigen (TNP-BSA, 50 ng/ml). Voltage ramps were applied every 10 s from the holding potential of 0 mV, according to the protocol shown in (C). (B) Increase in extracellular $[Ca^{2+}]$ increases magnitude of antigen-induced inward current. (C) Voltage clamp protocol used to obtain instantaneous I/V curves. (D) Voltage-independent activation and partial inactivation of Ca^{2+} current during voltage steps from 0 to −100 mV. Two traces are shown, indicating that recovery from inactivation was complete in ≤5 s. (From Zhang and McCloskey,[4] with permission.)

and since then we have used the same method to record from several hundred cells with the same reliability of induction. The current is carried by Ca^{2+} ions, as indicated by its extrapolated reversal potential than exceeds +40 mV and the fact that we observe it when the composition of the extracellular buffer is chosen to isolate Ca^{2+} current, that is, when it contains no K^+ or Na^+ but does contain tetraethylammonium to block outward K^+ current. Further evidence of its Ca^{2+} selectivity is shown in Fig. 17.1B, in which elevation of external $[Ca^{2+}]$ increases the magnitude of the

inward current at all voltages. In cells bathed in 10 mM extracellular Ca^{2+}, antigen evoked an average steady-state current (see following) at −80 mV of −25.7 ± 4.7 pA ($n = 25$).

As shown from the response of membrane current to voltage steps in Fig. 17.1D, the Ca^{2+} current does not appear to be voltage activated, but it does inactivate rapidly in response to hyperpolarization. This rapid inactivation is partial, reaching an average steady state value of about 50% within 100 ms. Complete recovery from rapid inactivation occurs within 5 s of returning the membrane potential to 0 mV (Fig.

17.1D). Although we have not systematically studied either the voltage or $[Ca^{2+}]_i$ dependence of rapid inactivation, loading cells with a sufficiently high concentration of the Ca^{2+} chelator BAPTA (as BAPTA-AM) to block the antigen-induced $[Ca^{2+}]_i$ rise did not affect the extent of rapid inactivation. In addition to rapid inactivation, in most cells the Ca^{2+} current also decays by a slower process. Typically, addition of antigen is followed by a variable lag phase; the current then appears rather abruptly, reaches a maximal value, and decays within minutes to values of 2 pA or less. This "slow inactivation" is usually monotonic, but in 8 of 91 cells the induced Ca^{2+} current oscillated dramatically over a 5- to 10-mm period.[4]

As with voltage-gated Ca^{2+} channels, at sufficiently negative potentials the antigen-evoked current is carried even better by Ba^{2+} and Sr^{2+} than it is by Ca^{2+} (Fig. 17.2A,B). For example, at −80 mV the Ba^{2+} and Sr^{2+} currents are on average about 2.5 fold larger than is the corresponding Ca^{2+} current measured in the same cells. This is true for peak as well as steady-state currents,[4] indicating a greater conductance to Ba^{2+} or Sr^{2+} rather than just reduced inactivation by these ions. It is remarkable that the shape of the I/V plot for Ba^{2+} is much steeper, that is, more strongly rectifying, than it is for Ca^{2+} ions. The basis for this remains to be worked out, but it resembles relief from ionic block by hyperpolarization. Also in analogy with voltage-gated Ca^{2+} channels, the antigen-induced current is carried by Na^+ in the nominal absence of extracellular calcium.[4]

These observations on the ionic selectivity of the antigen-evoked current are consistent with its possible physiological role in antigen-induced degranulation. Thus Sr^{2+} can replace Ca^{2+} in support of degranulation by RPMC and RBL cells,[18,52] antigen enhances $^{89}Sr^{2+}$ uptake by RPMC[20] and Ba^{2+} and Sr^{2+} both compete with $^{45}Ca^{2+}$ for entry into antigen-stimulated RBL-2H3 cells.[3] Moreover, that Na^+ can carry the antigen-induced current parallels the finding of Kanner and Metzger that FcεRI cross-linkage induces the uptake of $^{22}Na^+$ by RBL cells in the absence but not the presence of extracellular calcium.[23] Thus, one conductance regulator, ion channel or ion carrier, may control perme-

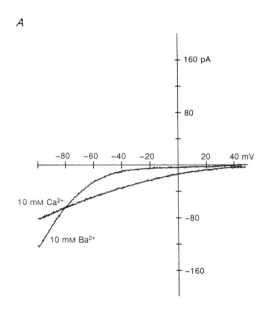

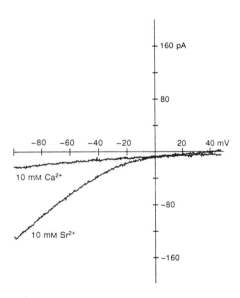

FIGURE 17.2. Antigen-evoked current is carried by Ba^{2+} and Sr^{2+}. After induction of current in the presence of 10 mM Ba^{2+} (A) or 10 mM Sr^{2+} (B), these ions were replaced by 10 mM Ca^{2+}, resulting in a decrease in inward current; addition of Ba^{2+} or Sr^{2+} (plus EGTA) after induction in the presence of Ca^{2+} did not change the result. Note the distinctly different shapes of the I/V curves for Ba^{2+} and Ca^{2+}. (From Zhang and McCloskey,[4] with permission.)

ability to both Na^+ and Ca^{2+} ions. One can exclude the electrogenic Na/Ca exchanger from consideration,[53] because the direct effect of membrane potential on the Ca^{2+} current would be opposite to that seen in the macroscopic I/V curve of Fig. 17.1.

The pharmacological sensitivity of the induced current also suggests that it identifies the primary conduit for antigen-stimulated $^{45}Ca^{2+}$ uptake and for the Ca^{2+} influx that drives secretion. Thus, at concentrations equal to their IC_{50} for blockade of $^{45}Ca^{2+}$ uptake,[3] La^{3+} (1.4 μM) and Zn^{2+} (42 μM) block the induced current by 52% and 57%, respectively. At concentrations in the micromolar range these ions also block antigen-driven secretion.[3] As might be expected on the basis of previous work as reviewed earlier, nitrendipine (5 μM), an organic antagonist of L-type voltage-gated Ca^{2+} channels, does not directly block the antigen-induced Ca^{2+} current. This suggests that the inhibition of mediator secretion from RBL-2H3 cells by nitrendipine[54] is not caused by a direct effect on Ca^{2+} influx, but is consistent with an indirect effect via membrane potential, the critical drug target being the outwardly rectifying K^+ channel.[54] The antiasthmatic drug cromolyn sodium (0.5 mM) has no effect on the induced current, indicating that this Ca^{2+} current is not mediated by the cromolyn-sensitive Ca^{2+} channel ostensibly purified from these cells.[55]

What can be said about the mechanism of induction of this Ca^{2+} conductance? First, it requires persistent cross-linkage of the FcεRI. Compelling evidence for this is shown in Fig. 17.3, where, following its maximal induction by TNP_{15}-BSA, the Ca^{2+} conductance (g_{Ca}) is abrogated by addition of a cross-reactive univalent hapten (20 μM DNP-lysine). Note that g_{Ca} can be reinduced multiple times (≥4 in some cells) through cyclical addition of polyvalent antigen and univalent hapten. These observations provide further evidence for the physiological relevance of the induced Ca^{2+} conductance, as uptake of $^{45}Ca^{2+}$ by RBL-2H3 cells also requires persistent cross-linkage of the FcεRI.[23,56] Although necessary, FcεRI cross-linkage is not sufficient to maintain the initial peak conductance, which as noted usually decays by 10 fold

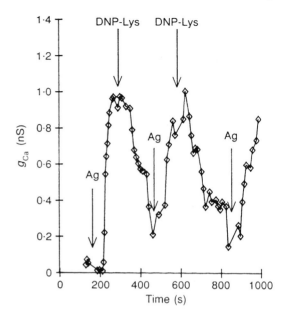

FIGURE 17.3. Hapten-reversible induction of Ca^{2+} conductance (g_{Ca}) by antigen. Cyclical activation and deactivation of g_{Ca} upon addition of polyvalent antigen TNP-BSA (50 ng/ml) (*Og, arrows*) followed by competitive univalent happen DNP-lysine (20 μM). (From Zhang and McCloskey,[4] with permission.)

within minutes even in the absence of monovalent hapten. In more than 90% of cells, g_{Ca} does not increase again, and monovalent hapten will not fully restore g_{Ca} if added after "slow inactivation" has begun. Whether this failure might reflect internalization of the cross-linked receptor or some other factors, such as protein kinase C- (PKC-) dependent phosphorylation,[5] remains to be determined. Regardless, the initial biochemical events that couple FcεRI cross-linkage to the Ca^{2+} current are readily reversible by disruption of antigen–antibody bonds.

Further insight into the mechanism of induction stems from the similarity of the antigen-evoked current to I_{CRAC} in T cells, RPMC, and RBL-2H3 cells. Like the antigen-induced Ca^{2+} current in RBL-2H3 cells, among other factors I_{CRAC} is characterized by an inwardly rectifying I/V curve, high Ca^{2+} selectivity, permeation by Na^+ in the absence of Ca^{2+}, block by micromolar La^{3+}, and voltage-independent gating.[37–40] Also, we found that under the same recording conditions as described that thapsigargin induces a Ca^{2+} current with properties very similar, although not identical, to those of the antigen-evoked current. This induction appears to depend upon Ca^{2+} store depletion per se rather than elevated $[Ca^{2+}]_i$, because BAPTA-AM at sufficient concentration (20 μM for 30 min) to block the calcium signal did not inhibit the induction (Fig. 17.4).

The question is, does the IgE receptor recruit calcium channels from the same pool as those recruited by thapsigargin? Our tentative answer is yes, given that addition of antigen following thapsigargin did not cause an additive induction (Fig. 17.4). This conclusion should be tested by experiments conducted in the opposite sequence, and is tempered by the fact that the thapsigargin-elicited current is carried better by Ca^{2+} than it is by Ba^{2+} at voltages down to –90 mV, behavior distinct from that of the antigen-evoked current. However, intracellular buffering with BAPTA reverts the order of thapsigargin-induced Ba^{2+} and Ca^{2+} currents to that of the antigen-induced currents. Barring these uncertainties, the induced current appears to be I_{CRAC}.

CRAC channels in mast cells initially were reported to conduct Ba^{2+} and Sr^{2+} currents very poorly,[37,46] but as previously noted Ba^{2+} and Sr^{2+} permeate the antigen-induced pathway. This apparent conflict appears in part to result from the previous measurement of Ba^{2+} current at 0 mV, a potential at which antigen-stimulated RBL-2H3 cells do not do well,[57,58] and where as indicated by the I/V plot the Ba^{2+} current is much smaller than the Ca^{2+} current (Fig. 17.2A); it may also reflect time-dependent reversal of Ca^{2+}-dependent potentiation of CRAC conductance[59] after substitution of Ba^{2+} or Sr^{2+} for Ca^{2+}. In a systematic study of the relative abilities of Ca^{2+} and Ba^{2+} to carry charge through CRAC channels, Hoth confirmed and extended our observations, and found the same steeply rectifying I/V curve for Ba^{2+} ions as shown in Fig. 17.2A.[60]

BAPTA-AM has multiple effects on the antigen-induced current, the origin(s) of which are not yet apparent. In addition to effects mentioned, pretreatment of cells with BAPTA-AM allowed the induction of I_{CRAC} by antigen in about 50% of cells during perforated-patch recording at room temperature, compared to

A

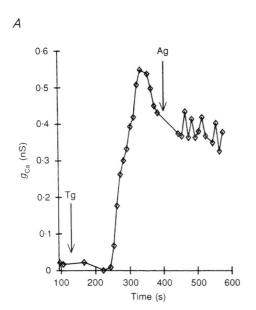

B

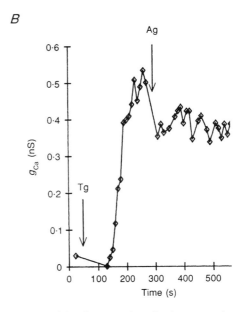

FIGURE 17.4. Thapsigargin and antigen recruit the same pool of Ca^{2+} channels. Typical inductions of g_{Ca} by 0.1 mM thpasigargin (*Tg*) in control (A) and BAPTA-AM-treated cells (B). Antigen (*Ag*) added at *arrows* neither increased g_{Ca} further nor enhanced its decay, suggesting that one set of Ca^{2+} channels is activated by both stimuli. (From Zhang and McCloskey,[4] with permission.)

0% without BAPTA-AM. About 50% of untreated cells also exhibit an antigen-induced $[Ca^{2+}]_i$ rise at 23°C. As its effect on rapid inactivation is modest, perhaps BAPTA amplifies the Ca^{2+} current by facilitating $InsP_3$-triggered store depletion. Penner has urged caution in the interpretation of patch-clamp studies using BAPTA and dibromo-BAPTA, pointing to possible Ca^{2+}-independent effects.[5,61]

Measurements of Ca^{2+} Current Versus $^{45}Ca^{2+}$ Influx

In principle, patch clamping is a useful adjunct to Ca^{2+} imaging and $^{45}Ca^{2+}$ flux measurements for studies of FcεRI-linked Ca^{2+} entry. The practical challenge is to find ways to make electrical measurements without overtly perturbing the system. Consider as an example the following problem posed by previous tracer flux studies. One can measure the initial rate of unidirectional uptake of $^{45}Ca^{2+}$ into antigen-stimulated RBL-2H3 cells after buffering their cytosol with quin2 at concentrations that permit a normal secretory response.[56] At a Ca^{2+} concentration (10mM) required to saturate $^{45}Ca^{2+}$ uptake, the maximal influx rate corresponds to a Ca^{2+} current of just 2.9pA per cell, nearly equivalent to that calculable from quin2 data in reference.[27] Calcium influx is thought to mediate antigen-induced depolarization of these cells,[57,62] but it is not clear how an inward current of merely 3pA could accomplish this. The constitutively active K^+ conductance of RBL cells passes 10 to 20pA or more of outward current between −70 and −40mV.[35,63] Barring inactivation of this K^+ conductance by antigen, and we find no evidence of this, the induced inward current(s) in this voltage region must be greater than 10 to 20pA to significantly depolarize the cell.

Because our data suggest that the maximal Ca^{2+} currents in this voltage region are much larger than 3pA, depolarization is not such a mystery. Studies of $^{45}Ca^{2+}$ uptake may not reveal this detail because they are conducted on asynchronously responding cell populations, the inferred value of 3pA being a population average of cells with variable lag periods and perhaps unsynchronized oscillations of Ca^{2+} current. On the other hand, the artificial buffers and voltage-clamp routine we use to assay Ca^{2+} current do not faithfully mimic the natural extracellular milieu and voltage excursions of antigen-stimulated mast cells. As does $[Ca^{2+}]_i$ itself, it is possible, perhaps probable, that under less perturbing conditions the Ca^{2+} influx rate and membrane potential oscillate in single cells. When Lewis and Cahalan first observed the Ca^{2+} current now called I_{CRAC}, they found that during perforated-patch recording from T cells the lectin phytohemagglutinin induced regular oscillations of Ca^{2+} current.[38] The pronounced oscillations of Ca^{2+} current that we observed in 9% of cells might typify the CRAC channel activity one would observe under recording conditions more closely physiological than those yet employed.

Benefits of Noninvasive Measurements at Warm Temperatures

There are some great advantages of conducting WCR the standard way (breakin) at ambient temperature. One can control the ionic composition of the cytosol, introduce magic molecules by dialysis, and data can be acquired much more quickly and with less frustration than when using perforated-patch WCR while heating the sample. But there are some clear disadvantages that are often ignored, not the least of which is that stimulus–secretion coupling via the FcεRI is abrogated in RPMC[34] and apparently in RBL cells.[35,63] Thus, one cannot measure antigen-evoked Ca^{2+} currents and secretion simultaneously, as is necessary for directly probing the route of Ca^{2+} entry critical for antigen-induced secretion. In the following, I discuss other reasons why I think it is important to use recording conditions that mimic the native situation with more fidelity than is commonly done.

A major problem with conventional WCR is that I_{CRAC} spontaneously appears when T cells or mast cells are dialyzed with pipette solutions containing a high concentration of BAPTA or EGTA and relatively low [Ca^{2+}]$_i$, that is, conditions that facilitate detection of Ca^{2+} current.[37,38] We found this to be especially problematic at physiological temperatures. Presumably Ca^{2+} store depletion contributes to spontaneous activation of the Ca^{2+} current, but what are the critical experimental parameters that provoke this? Spontaneous induction apparently is reduced if pipette free [Ca^{2+}] is buffered at 60 to 90nM with 10mM EGTA or BAPTA,[5,40] but in our experience it is not eliminated. Using a BAPTA-based pipette solution,[40] we found that independent of free [Ca^{2+}] from 10 to 350nM (checked with fura-2), large I$_{CRAC}$ consistently appeared in RBL-2H3 cells during conventional WCR at 35°–37°C. Spontaneous induction also occurred at room temperature, and although slower it was appreciable in size, for example, 10pA at −80mV (Zhang and McCloskey, unpublished observations). Anant Parekh was kind enough to ship us some of the same buffer (60–90nM [Ca^{2+}]) that he used to study regulation of I$_{CRAC}$ in RBL-2H3 cells,[5] and we found that modest I$_{CRAC}$ still appeared spontaneously in our cells during conventional WCR at room temperature. Zweifach and Lewis also find that buffering pipette free [Ca^{2+}] at 100 or 200nM does not prevent spontaneous activation of I$_{CRAC}$ in Jurkat T cells (R.S. Lewis, personal communication).

Because low [Ca^{2+}]$_i$ cannot fully account for spontaneous induction, what other factors are likely to be important? Perhaps rates of constitutive InsP$_3$ production or passive Ca^{2+} leak differ with temperature, culture conditions, or between different cell populations, creating differential susceptibility to "spontaneous" induction by strong Ca^{2+} buffering. It will be a significant advance for this line of research when the critical variable is brought to light. In the meantime, perforated patching seems to forestall spontaneous induction of I$_{CRAC}$ in both RBL-2H3[64] and Jurkat T cells.[38]

Obviously, spontaneous induction of the Ca^{2+} current complicates investigation into the mechanism of its coupling to the FcεRI. If one's purpose is to characterize the activity of a putative "calcium-influx factor"[65] or to block pro-

duction of a second messenger that purportedly links FcεRI cross-linkage to Ca^{2+} release,[66] it just will not do to have I$_{CRAC}$ appear spontaneously. Furthermore, it is a post hoc fallacy to conclude that the appearance of I$_{CRAC}$ after antigen addition is a consequence of antigen addition if it is impossible to say whether or when the current will appear spontaneously in any given cell. One has to resort to population-average behavior on a day-by-day basis, something that requires heroic effort in the case of multiple experimental treatments.

There are further disadvantages of standard WCR in studies of FcεR-linked Ca^{2+} influx. As implied, it is critical to establish cause and effect if I$_{CRAC}$ appears after antigen addition, as we did by demonstrating reversibility with excess univalent hapten (see Fig. 17.3).[4] We suspect that the hapten-induced decay of the Ca^{2+} current mirrors the refilling of Ca^{2+} stores after the releasing stimulus is shut off. However, dialysis with internal solutions of high Ca^{2+} buffer capacity prevents store replenishment and prolongs I$_{CRAC}$, at least in T cells.[67] In addition to complicating the demonstration of specificity, replacement of endogenous with artificial Ca^{2+} buffers confounds studies of the temporal evolution of [Ca^{2+}]$_i$ and Ca^{2+} current under conditions in which calcium homeostasis in RBL-2H3 cells has been well studied.

The importance of temperature control for studies of IgE-mediated signal transduction was presaged by studies of Mongar and Schild in 1958 that showed a steep, biphasic temperature dependence of anaphylactic release of histamine from guinea pig lung.[68] In their words, "The limits of temperature within which the anaphylactic reaction can function are remarkably narrow and any departure from the physiological range impairs . . . histamine release." By interpolating their data one estimates that the percent of release increases from 2% to 48% in going from 20° to 35.5°C! Over this same range of temperature, there is also a very steep increase (~14 fold) in magnitude of the [Ca^{2+}]$_i$ rise and inositol phosphates production in antigen-stimulated RBL-2H3 cells.[48] Whether this reflects an increase in frequency of responsive cells or peak [Ca^{2+}]$_i$ per cell, it makes sense to patch clamp at temperatures that facilitate consistent measurements of Ca^{2+} current.

Clearly, not every signal transduction event in mast cells is compromised during conventional WCR at room temperature. For instance, ligation of G_i-linked P_2 purinoceptors elicits a latent K^+ conductance in RBL-2H3 cells and in rat BMMC, and this coupling is readily reversible through multiple cycles of agonist addition and removal during standard WCR at room temperature.[54,69] Activation of this K^+ conductance appears to be membrane delimited, however, whereas evidence suggests this is not the case for the FcεRI-linked Ca^{2+} conductance. Even channel systems activated by Ca^{2+} release can be studied effectively with standard WCR.[70] My point is that one cannot assume a priori that simplified systems mimic with high fidelity what happens in the warm, intact cell. Intriguingly, Innocenti et al.[71] recently reported that intracellular ADP regulates $InsP_3$-induced I_{CRAC} in RBL-1 cells in a highly temperature-dependent manner.

Future Directions

Two of the most pressing questions regarding IgE-mediated Ca^{2+} influx are these: What is the molecular structure of the putative IgE-linked Ca^{2+} channel and how is its activity regulated by the FcεRI? Solution of the first problem may well provide clues and tools to answer the second question. Progress toward the cloning of putative store-operated Ca^{2+} channels was recently covered,[42,72,73] and research into the coupling between store depletion and influx of extracellular Ca^{2+} has also been reviewed.[41] We are pursuing several observations that suggest there are two routes from the FcεRI to I_{CRAC}, only one of which depends directly upon Ca^{2+} store depletion. The second pathway is not mimicked by store depletion alone nor is it observed during conventional WCR.

Regarding upstream signals from the FcεRI, a provocative report appeared last year suggesting that sphingosine-1-phosphate is the pivotal antigen-induced Ca^{2+} releaser in RBL-2H3 cells,[66] this despite 10 years of evidence implicating $InsP_3$ as the key second messenger

linking antigen binding to Ca^{2+} release. Of possible relevance is the existence in RBL microsomes of a Ba^{2+}-permeable channel ostensibly gated by lyso-sphingomyelin.[74] It seems that even the principal architect of the $InsP_3$ linkage hypothesis has embraced the sphingosine-1-phosphate story with open arms.[75] Yet if this new model is accurate, it is curious that blockade of inositol phosphates production completely abrogates the induction of I_{CRAC} by antigen without affecting I_{CRAC} induction by thapsigargin (L. Zhang, master's thesis, Iowa State University, 1995).

Although 5 years have elapsed since the presence of I_{CRAC} in RPMC was reported,[37] there is no published evidence that antigen elicits I_{CRAC} therein. Given the controversial role of Ca^{2+} entry in antigen-driven secretion by RPMC, direct observation of this presumptive coupling would seem an important goal. The use of perforated-patch recording at elevated temperature may facilitate reaching this goal, not just as it has done with RBL-2H3 cells, but with rat bone marrow-derived mast cells (BMMC) as well (Zhang and McCloskey, unpublished data). Interestingly, the specific Ca^{2+} conductance is at least twice as great in these rat BMMC as it is in RBL-2H3 cells. In any case, the link between I_{CRAC} and the FcεRI is not restricted to immortalized mast cells.

Does Ca^{2+} influx drive exocytosis by sustaining global $[Ca^{2+}]_i$ above some threshold for secretion, or is the putative Ca^{2+} sensor tuned to respond to a local high $[Ca^{2+}]$ near the CRAC channels? Depolarization of excitable cells generates steep $[Ca^{2+}]$ gradients beneath the mouth of voltage-gated Ca^{2+} channels,[76,77] and Ca^{2+} entering the cell via such channels triggers exocytosis more effectively than do other stimuli that elevate global $[Ca^{2+}]_i$ to the same extent.[78] In chromaffin cells and neurons, Ca^{2+} channels are thought to be strategically placed adjacent to (or may define) the release sites where the low-affinity Ca^{2+} sensor can respond to an increase in $[Ca^{2+}]_i$ of perhaps 200 to 300 μM.[79] Because of the tiny throughput of CRAC channels the $[Ca^{2+}]$ gradients are predicted to be much less steep,[80] which together with other observations might suggest that in mast cells the main function of Ca^{2+} influx in secretion is to sustain globally elevated $[Ca^{2+}]_i$ in the face of active Ca^{2+} extrusion mechanisms. Predictions aside, it may be possible to measure $[Ca^{2+}]_i$ just beneath the plasma membrane of mast cells using near-field microscopy.[81]

Beyond the mechanisms of antigen-evoked Ca^{2+} entry, one might ask what functions it serves other than maintaining Ca^{2+} homeostasis or driving regulated secretion. As has become apparent since the mid-1980s, mast cells are multifunctional, exchanging various signals with other components of the immune and nervous systems. In addition to secreting preformed mediators, mast cells respond to antigen by elaborating and releasing cytokines and lipid mediators, and by dividing or undergoing cell-cycle arrest. Interestingly, Beaven and colleagues find that secretion of newly synthesized tumor necrosis factor-α (TNF-α) by RBL-2H3 cells is blocked by chelation of extracellular calcium,[82] and in T cells transcription of the TNF-α gene is Ca^{2+} dependent.[83] Also, removal of extracellular calcium completely blocks the production and release of IL-6 by RPMC stimulated with anti-IgE,[84] but whether this is caused by an effect on synthesis or release is not known. We have recently identified two further Ca^{2+} influx-dependent responses of mast cells, one of which is discussed next.

Ca²⁺-Dependent Activation of NF-AT by the FcεRI

Nuclear factor of activated T cells (NF-AT) is a transcription factor complex essential for production of the autocrine growth factor interleukin-2 (IL-2) by antigen-stimulated T lymphocytes.[85,86] NF-AT binds to two recognition sequences within an approximately 300-bp enhancer region of the of IL-2 promoter, and together with Oct-1, Oct-2, NFκB, and AP-1 promotes transcription of the IL-2 gene. According to the prevailing model, T-cell receptor (TCR) stimulation induces the assembly of nuclear NF-AT complexes from newly made AP-1 components (Fos and Jun)[87] and a preexisting cytosolic protein, either NFATc or NFATp.[85,86] The cytosolic proteins NFATc and

NFATp are distinct gene products, 73% identical within a domain of about 300 residues that bears similarity to the DNA-binding and dimerization regions of Dorsal/Rel/NF-κB family members.[88–90]

As originally defined in T cells, induction of AP-1 and translocation of NFATc/p are mediated by PKC activation and elevation of $[Ca^{2+}]_i$, respectively. Transcription of the IL-2 gene and commitment to DNA synthesis require a sustained Ca^{2+} signal, hence Ca^{2+} influx, during a 2 h period.[85] IL-2 expression cannot be induced by transient Ca^{2+} elevation in Jurkat T-cell lines containing mutations that prevent a sustained Ca^{2+} signal or on addition of extracellular EGTA at earlier times.[91,92] Redistribution of cytosolic NFATc/p is driven by the Ca^{2+}-regulated phosphatase, calcineurin, itself a target of the potent immunosuppressants cyclosporin A (CsA) and FK506.[93,94] These drugs block transcription of the genes for IL-2 and several other cytokines produced by activated T cells and mast cells.[64,83]

Several clues summarized in Hutchinson and McCloskey[9] led us to suspect that mast cells might contain NF-AT, which at the time was thought restricted to T and B lymphocytes. We used electrophoretic mobility shift assays (EMSA) to assay for NF-AT-specific DNA-binding activity in nuclear extracts from RBL-2H3 cells and rat BMMC. The radiolabeled oligonucleotide probe used matches a sequence from the distal NF-AT site of the human IL-2 enhancer. As shown in Fig. 17.5, stimulation of RBL-2H3 cells or Jurkat T cells with the combination of ionomycin and phorbol ester caused the nuclear appearance of NF-AT-selective DNA-binding activity. Antigenic stimulation of IgE-sensitized RBL-2H3 cells and rat BMMC (not shown) also induced NF-AT DNA-binding activity.

Evidence on the DNA-binding specificity and subunit composition of the presumed NF-AT complex in rat mast cells was obtained with antibody supershift experiments and by cross-competition with oligonucleotides representing recognition sequences for other transcription factors. As indicated in Fig. 17.6a, of five different binding sites, only those for NF-AT and AP-1 competed with the probe from the distal

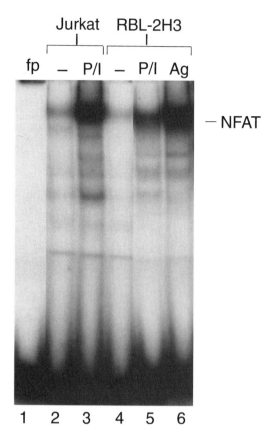

Jurkat   RBL-2H3

fp   –   P/I   –   P/I   Ag

— NFAT

1   2   3   4   5   6

FIGURE 17.5. Induction of nuclear factor of activated T cells (NF-AT) DNA-binding activity in Jurkat T cells and RBL-2H3 cells. Electrophoretic mobility shift assay (EMSA) of nuclear extracts prepared from cells stimulated for 2 h either with phorbol myristate acetate (32 nM) and ionomycin (2 μM) (*P/I*) or with 50 ng/ml TNP-BSA (antigen) as indicated. Equal amounts (20 μg) of nuclear proteins from each treatment were incubated with double-stranded ^{32}P-labeled oligonucleotide containing the distal NF-AT site from the human IL-2 promoter. Retarded bands containing NF-AT complexes are indicated by a *line*. The *first lane* shows free probe (*fp*) without nuclear extract. Data represent five experiments. (From Hutchinson and McCloskey,[9] with permission.)

NF-AT site of the human IL-2 enhancer. We interpret the ablation of the NF-AT band by the AP-1 oligonucleotide as evidence for the presence of one or more AP-1 components in the NF-AT complex.[87] Although AP-1 does not bind directly to the radiolabeled NF-AT probe from the murine IL-2 enhancer,[95] its association

with NFATc/p markedly enhances their DNA-binding affinity.[87,96] Tight binding of this oligo to AP-1 within the complex weakens the interaction of NFATc/p with DNA, either by displacing AP-1 or altering its interaction with NFATc/p. Preincubation of nuclear extracts from stimulated cells with a combination of antibodies against Fos and Jun proteins ablated the NF-AT band seen by EMSA (Fig. 17.6B). This suggests the presence of both Fos and Jun in the putative NF-AT complex from RBL-2H3 cells.

The time course of induction was followed out to 2 h, and throughout this period NF-AT avtivity continued to accumulate in the nucleus (Fig. 17.7a). A similar time course was observed for the induction of AP-1 (Fig. 17.7b), reminiscent of that for antigen-induced increases in the mRNA for c-*jun* and junB in RBL cells, but more sustained than expression of c-*fos* mRNA.[97] Somewhat greater variability in the basal AP-1 activity has appeared in subsequent experiments, and this may relate to variability in constitutive PKC activity. The cells are eluted from flasks with EDTA and allowed to recover in "suspension" for 90 min before stimulation. During this period of recovery and especially after stimulation, a variable fraction of cells adheres to the polypropylene incubation tubes. As adherent RBL-2H3 cells appear to contain a greater constitutive PKC activity than do suspended cells,[98] the basal AP-1 activity may also vary.

Variations on the aforementioned scheme for assembly of NF-AT have been described, including a CsA-resistant pathway[99] and the involvement of ras in AP-1 induction.[100,101] But in RBL-2H3 cells, using the distal site of the human IL-2 promoter as a probe, CsA inhibits the activation of NF-AT by antigen in a concentration-dependent manner (Fig. 17.8). The variability in concentration response from day to day was significant, perhaps reflecting the limited water solubility of this hydrophobic drug. If calcineurin activity is required for induction of mast cell NF-AT, is elevation of $[Ca^{2+}]_i$ also required, and if so, will transient elevation caused by Ca^{2+} release suffice? As shown in Fig. 17.9, pretreatment of cells with sufficient

BAPTA-AM to severely truncate the antigen-induced Ca^{2+} rise completely inhibited the NF-AT induction. Dramatic inhibition also resulted from addition of extracellular EGTA at a concentration calculated to yield free $[Ca^{2+}]$, approximately 80 nM. Together, these observations suggest that the induction of nuclear NF-AT DNA-binding activity in RBL-2H3 cells is dependent upon sustained elevation of $[Ca^{2+}]_i$.

The question arises as to which cytokine genes are regulated by mast cell NF-AT. Of the cytokines known to be produced by mast cells in response to FcεRI cross-linkage, several have putative NF-AT-binding sites in the regulatory regions of their genes.[86] The intergenic enhancer between the closely linked IL-3 and GM-CSF genes contains DNAse I-hypersensitive sites that bind to Jurkat NF-AT.[95]

Although previous work suggests that IL-3 levels in the PB-3c mast cell line and its immortalized derivatives are regulated by Ca^{2+} posttranscriptionally,[102,103] it might be useful to determine the generality of this finding. It also would be interesting if the constitutive transcription of the IL-3 gene by these cells requires NF-AT, as "basal" NF-AT activity has been detected in nuclear extracts from T cells and mast cells. A second example is the gene for TNF-α, known to be synthesized de novo in antigen-stimulated RBL-2H3 cells[98] and transcribed in T cells in a CsA-sensitive fashion.[83] CsA also inhibits the antigen-stimulated production of TNF-α and IL-6 by mouse BMMC at IC_{50}s of 72 and 143 ng/ml, respectively,[64] a similar effective concentration to that indicated in Fig. 17.7. This indirect evidence suggests the possible involvement of calcineurin and per-

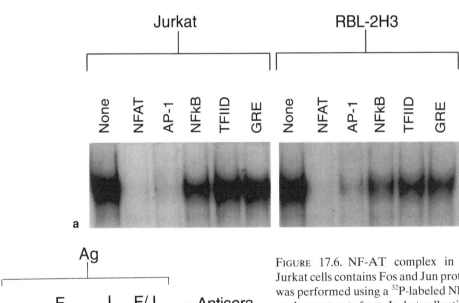

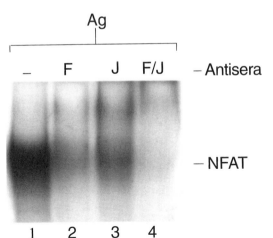

FIGURE 17.6. NF-AT complex in RBL-2H3 and Jurkat cells contains Fos and Jun proteins. (a) EMSA was performed using a ³²P-labeled NF-AT probe and nuclear extracts from Jurkat cells stimulated with P/I for 2 h or RBL cells triggered with antigen for 60 min. As indicated, either no competitor DNA or a 200-fold excess of unlabeled *NF-AT, AP-1, NF-κB, TFIID,* or *GRE* oligonucleotide was added to reaction mixture. (b) EMSA was performed with nuclear extract from antigen-stimulated RBL-2H3 cells. Either no antibody (−) or 1 μg or anti-Fos antibody (*F*), 1 μg of anti-Jun antibody (*J*), or 1 μg each of these two antibodies (*F/J*) were added to the binding reaction. Data represent three experiments. (From Hutchinson and McCloskey,[9] with permission.)

0   10   30   60   120

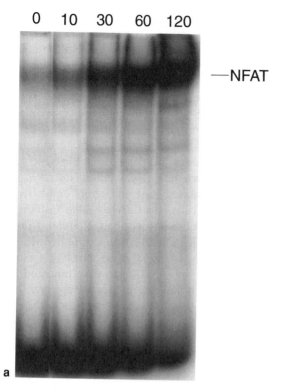

—NFAT

a

0   10   30   60   120

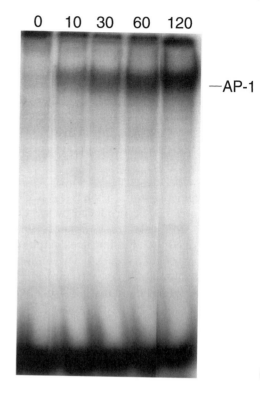

—AP-1

b

FIGURE 17.7. Rate of appearance of nuclear NF-AT and AP-1 in antigen-stimulated RBL-2H3 cells. Nuclear extracts were prepared from RBL cells at the indicated times (min) after addition of antigen (50 ng/ml TNP-BSA). Equivalent amounts of nuclear proteins (20 µg, a; 10 µg, b) from each time point were incubated with ^{32}P-labeled double-stranded oligonucleotide containing the *NF-AT-* or *AP-1*-binding sites as indicated. Retarded bands representing NF-AT (a) and AP-1 (b) complexes are indicated by *arrows*. Data represent three experiments. (From Hutchinson and McCloskey,[9] with permission.)

haps NF-AT in the production of TNF-α and IL-6 by mast cells, but it does not establish that NF-AT is transcriptionally active therein.

Subsequent to our publication, Baumruker's group reported that cross-linkage of the FcεRI in the murine mast cell line CPII induces NF-AT DNA-binding activity, and evidence from transfected reporter constructs suggests that

Ag

—   0   1   10   100 1000   —CsA

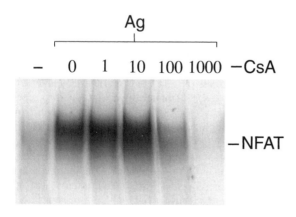

—NFAT

FIGURE 17.8. Cyclosporin A (CsA) inhibits NF-AT induction in RBL-2H3 cells. EMSA was performed with an NF-AT probe and nuclear proteins (20 µg) from RBL-2H3 cells unstimulated or stimulated with antigen (*Ag*) for 2 h in the presence of 1, 10, 100, or 1000 ng/ml CsA. Data represent three experiments. (From Hutchinson and McCloskey,[9] with permission.)

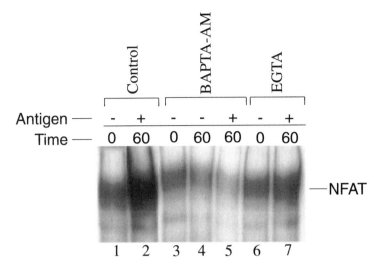

FIGURE 17.9. Calcium-dependent activation of mast cell NF-AT. Nuclear extracts (20 μg protein) from RBL-2H3 cells unstimulated or stimulated with antigen for 60 min as shown: basal medium, after incubation with 20 μM *BAPTA-AM* or in the presence of excess extracellular *EGTA* (extracellular free [Ca²⁺] ~80 nM). Data represent three experiments. (From Hutchinson and McCloskey,[9] with permission.)

NF-AT regulates transcription of the MARC chemokine and IL-5 genes in these cells.[104] The authors argue that NF-AT functions independently of AP-1 or PKC in mast cells. This is an intriguing speculation, but it is clearly premature to generalize given the limited number of mast cell subsets, animal species, and NF-AT-binding sites tested so far and the fact that DNA sequence alone can determine which of the two NF-AT forms (±AP-1) bind to DNA.[96] Using the distal site from the human IL-2 promoter, we observe just one apparent DNA–protein complex (NF-AT plus AP-1) by EMSA of nuclear extracts from antigen-stimulated RBL-2H3 cells. The identical nuclear extract probed with an oligonucleotide from the murine IL-2 promoter yields two bands on EMSA, one containing NF-AT plus AP-1 and the other NF-AT alone (Hutchinson and McCloskey, unpublished data). Great caution must also be exercised in the interpretation of PKC downregulation experiments: Not only is the ζ-group of isozymes insensitive to phorbol esters, but residual activity from the α through θ isozymes that are degraded may be sufficient to induce AP-1,[105] particularly when the reagent used to downregulate is also a potent activator of these isozymes. It is diffi-

cult to reach meaningful conclusions based upon a negative result unless PKC activity has been measured, and even then the question as to how much activity is sufficient is nontrivial.

The list of cytokines known to be regulated by mast cell NF-AT is likely to expand, and as it does, so may our understanding of the different roles this protein plays in the biology of mast cells. Given the multiplicity of forms of cytosolic NF-AT, and the recent discovery that transgenic mice with a deletion of NFAT1 (NFATp) have an enhanced immune response,[106] the possibility looms that combinatorial association of different NFATc/p family members may yield complexes that differentially regulate cytokine production. As the signal transduction pathways from the FcεRI to nuclear NF-AT(s) come into focus more clearly, it will be important to define the exact route of Ca²⁺ influx required for sustained NF-AT activity. Given the precedent for differential regulation of CaM kinase II by repetitive versus uniform Ca²⁺ signals, it will be of interest to determine whether the oscillatory form of the Ca²⁺ signal (and I_{CRAC}?) in RBL-2H3 cells similarly modulates calcineurin, hence NF-AT.

Conclusions

It has been 36 years since Douglas and Rubin formulated their stimulus-secretion coupling hypothesis, and longer still since the first indication that extracellular Ca^{2+} is necessary for the anaphylactic release of histamine. Since then we have come to appreciate additional ways that stimulated Ca^{2+} influx may regulate mast cell responses to antigen, and their number is likely to increase in the near future. Extraordinary parallels with the T-cell system, from homologous antigen receptors to CRAC channels to nuclear transcription factors, have accelerated this understanding. Now, with a reliable assay in hand to directly monitor IgE-mediated Ca^{2+} influx, the path is clear to real-time studies of the mechanism(s) coupling FcεRI cross-linkage to Ca^{2+} channel activation. In such effort the simulation of physiological recording conditions, cumbersome though it may be, should complement studies with more simplified systems and may yield big dividends.

Acknowledgments. The work described here was supported by an NIH grant (GM48144). I am grateful to Dr. Paul Ross and Dr. Phil Haydon for helpful comments on the manuscript.

References

1. Mongar JL, Schild HO. The effect of calcium and pH on the anaphylactic reaction. J Physiol (Camb) 1958;140:272–284.
2. Barsumian EL, Isersky C, Petrino MB, et al. IgE-induced histamine release from rat basophilic leukemia cell lines: isolation of releasing and nonreleasing clones. Eur J Immunol 1981;11:317–323.
3. Hide M, Beaven MA. Calcium influx in a rat mast cell (RBL-2H3) line. Use of multivalent metal ions to define its characteristics and role in exocytosis. J Biol Chem 1991;266:15221–15229.
4. Zhang L, McCloskey MA. Immunoglobulin E receptor-activated calcium conductance in rat mast cells. J Physiol (Camb) 1995;483:59–66.
5. Parekh AB, Penner R. Depletion-activated calcium current is inhibited by protein kinase in RBL-2H3 cells. Proc Natl Acad Sci USA 1995;92:7907–7911.
6. Douglas WW, Rubin RP. The role of calcium in the secretory response of the adrenal medulla to acetylcholine. J Physiol (Camb) 1961;159:40–57.
7. Douglas WW. Stimulus-secretion coupling: the concept and clues from chromaffin and other cells. Br J Pharmacol 1968;34:451–474.
8. Putney JW. Capacitative calcium entry revisited. Cell Calcium 1990;11:611–624.
9. Hutchinson LE, McCloskey MA. FcεRI-mediated induction of nuclear factor of activated T cells. J Biol Chem 1995;270:16333–16338.
10. Ozawa K, Szallasi Z, Kazanietz MG, et al. Ca^{2+}-dependent and Ca^{2+}-independent isozymes of protein kinase C mediate exocytosis in antigen-stimulated rat basophilic RBL-2H3 cells. J Biol Chem 1993;268:1749–1756.
11. Foreman JC, Mongar JL, Gomperts BD. Calcium ionophores and movement of calcium ions following the physiological stimulus to a secretory process. Nature (Lond) 1973;245:249–251.
12. Cochrane DE, Douglas WW. Calcium-induced extrusion of secretory granules (exocytosis) in mast cells exposed to 48/80 or the ionophores A-23187 and X-537A. Proc Natl Acad Sci USA 1974;71:408–412.
13. Siraganian RP, Kulczycki A Jr, Mendoza G, et al. Ionophore A-23187 induced histamine release from rat mast cells and rat basophil leukemia (RBL-1) cells. J Immunol 1975;115:1599–1602.
14. Kanno T, Cochrane DE, Douglas WW. Exocytosis (secretory granule extrusion) induced by injection of calcium into mast cells. Can J Physiol Pharmacol 1973;51:1001–1004.
15. Theoharides TC, Douglas WW. Secretion in mast cells induced by calcium entrapped within phospholipid vesicles. Science 1978;201:1143–1145.
16. Douglas WW, Kagayama M. Calcium and stimulus-secretion coupling in the mast cell: stimulant and inhibitory effects of calcium-rich media on exocytosis. J Physiol (Camb) 1977;270:691–703.
17. WoldeMussie E, Moran NC. Histamine release by compound 48/80: evidence for the depletion and repletion of calcium using

chlortetracycline and ^{45}calcium. Agents Actions 1984;15:268–272.

18. Foreman JC, Mongar JL. The role of the alkaline earth ions in anaphylactic histamine secretion. J Physiol (Camb) 1972;224:753–869.

19. Foreman JC, Hallett MB, Mongar JL. The relationship between histamine secretion and ^{45}calcium uptake by mast cells. J Physiol (Camb) 1977;271:193–214.

20. Foreman JC, Hallett MB, Mongar JL. Movement of strontium ions into mast cells and its relationship to the secretory response. J Physiol (Camb) 1977;271:233–251.

21. Taurog JD, Mendoza GR, Hook WA, et al. Noncytotoxic IgE-mediated release of histamine and serotonin from murine mastocytoma cells. J Immunol 1977;119:1757–1761.

22. Crews FT, Morita Y, McGivney A, et al. IgE-mediated histamine release in rat basophilic leukemia cells: receptor activation, phosph-olipid methylation, Ca^{2+} flux, and release of arachidonic acid. Arch Biochem Biophys 1981;212:561–571.

23. Kanner BI, Metzger H. Initial characterization of the calcium channel activated by the cross-linking of the receptors for immunoglobulin E. J Biol Chem 1984;259:10188–10193.

24. Tsien RY. New calcium indicators and buffers with high selectivity against magnesium and protons: design, synthesis, and properties of prototype structures. Biochemistry 1980; 19:2396–2404.

25. White JR, Ishizaka T, Ishizaka K, et al. Direct demonstration of increased intracellular concentration of free calcium as measured by quin-2 in stimulated rat peritoneal mast cell. Proc Natl Acad Sci USA 1984;81:3978–3982.

26. Neher E, Almers W. Fast calcium transients in rat peritoneal mast cells are not sufficient to trigger exocytosis. EMBO J 1986;5:51–53.

27. Beaven MA, Rogers J, Moore JP, et al. The mechanism of the calcium signal and correlation with histamine release in 2H3 cells. J Biol Chem 1984;259:7129–7136.

28. Pearce FL, Ennis AT, White JR. Role of intra- and extracellular calcium in histamine release from rat peritoneal mast cells. Agents Actions 1981;11:51–54.

29. Mohr FC, Fewtrell C. The relative contributions of extracellular and intracellular calcium to secretion from tumor mast cells. Multiple effects of the proton ionophore carbonyl cyanide m-chlorophenylhydrazone. J Biol Chem 1987;262:10638–10643.

30. Mohr FC, Fewtrell C. Depolarization of rat basophilic leukemia cells inhibits calcium uptake and exocytosis. J Cell Biol 1987; 104:783–792.

31. Grynkiewicz G, Poenie M, Tsien RY. A new generation of Ca^{2+} indicators with greatly improved fluorescence properties. J Biol Chem 1985;260:3440–3450.

32. Millard PJ, Ryan TA, Webb WW, et al. Immunoglobulin E receptor cross-linking induces oscillations in intracellular free ionized calcium in individual tumor mast cells. J Biol Chem 1989;264:19730–19739.

33. Pearce FL, White JR. Effect of lanthanide ions on histamine secretion from rat peritoneal mast cells. Br J Pharmacol 1981; 72:341–347.

34. Lindau M, Fernandez JM. IgE-mediated degranulation of mast cells does not require opening of ion channels. Nature (Lond) 1986;319:150–153.

35. Lindau M, Fernandez JM. A patch-clamp study of histamine-secreting cells. J Gen Physiol 1986;88:349–368.

36. Neher E. The influence of intracellular calcium concentration on degranulation of dialysed mast cells from rat peritoneum. J Physiol (Camb) 1988;395:193–214.

37. Hoth M, Penner R. Depletion of intracellular calcium stores activates a calcium current in mast cells. Nature (Lond) 1992;355:353–356.

38. Lewis RS, Cahalan MD. Mitogen-induced oscillations of cytosolic Ca^{2+} and transmembrane Ca^{2+} current in human leukemic T cells. Cell Regul 1989;1:99–112.

39. Zweifach A, Lewis RS. Mitogen-regulated Ca^{2+} current of T lymphocytes is activated by depletion of intracellular Ca^{2+} stores. Proc Natl Acad Sci USA 1993;90:6295–6299.

40. Fasolato C, Hoth M, Penner R. A GTP-dependent step in the activation mechanism of capacitative calcium influx. J Biol Chem 1993; 268:20737–20740.

41. Berridge MJ. Capacitative calcium entry. Biochem J 1995;312:1–11.

42. Fanger CM, Hoth M, Crabtree GR, et al. Characterization of T cell mutants with defects in capacitative calcium entry: genetic evidence for the physiological roles of CRAC channels. J Cell Biol 1995;131:655–667.

43. Fasolato C, Innocenti B, Pozzan T. Receptor-activated Ca^{2+} influx: how many mechanisms for how many channels? Trends Pharmacol Sci 1994;15:77–83.

44. Thastrup O, Cullen PJ, Drobak BK, et al. Thapsigargin, a tumor promoter, discharges intracellular Ca^{2+} stores by specific inhibition of the endoplasmic reticulum Ca^{2+}-ATPase. Proc Natl Acad Sci USA 1990;87:2466–2470.

45. Ali H, Maeyama K, Sagi-Eisenberg R, et al. Antigen and thapsigargin promote influx of Ca^{2+} in rat basophilic RBL-2H3 cells by ostensibly similar mechanisms that allow filling of inositol 1,4,5-trisphosphate-sensitive and mitochondrial Ca^{2+} stores. Biochem J 1994;304:431–440.

46. Hoth M, Penner R. Calcium release-activated calcium current in rat mast cells. J Physiol (Camb) 1993;465:359–386.

47. Fasolato C, Hoth M, Matthews G, et al. Ca^{2+} and Mn^{2+} influx through receptor-mediated activation of nonspecific cation channels in mast cells. Proc Natl Acad Sci USA 1993;90:3068–3072.

48. Wolde Mussie E, Maeyama K, Beaven MA. Loss of secretory response of rat basophilic leukemia (2H3) cells at 40°C is associated with reversible suppression of inositol phospholipid breakdown and calcium signals. J Immunol 1986;137:1674–1680.

49. Pusch M, Neher E. Rates of diffusional exchange between small cells and a measuring patch pipette. Pflügers Arch 1988;411:204–211.

50. Penner R, Pusch M, Neher E. Washout phenomena in dialyzed mast cells allow discrimination of different steps in stimulus-secretion coupling. Biosci Rep 1987;7:313–321.

51. Horn R, Marty A. Muscarinic activation of ionic currents measured by a new whole-cell recording method. J Gen Physiol 1988;92:145–159.

52. Fewtrell C, Kessler A, Metzger H, Weissmann G, eds. Advances in Inflammation Research. New York: Raven Press, 1979.

53. Stump RF, Oliver JM, Cragoe EJ Jr, et al. The control of mediator release from RBL-2H3 cells: roles for Ca^{2+}, Na^+, and protein kinase C. J Immunol 1987;139:881–886.

54. Qian Y-X, McCloskey MA. Activation of mast cell K^+ channels through multiple G protein-linked receptors. Proc Natl Acad Sci USA 1993;90:7844–7848.

55. Mazurek N, Schindler H, Schurholz T, et al. The cromolyn binding protein constitutes the Ca^{2+} channel of basophils opening upon immu-

nological stimulus. Proc Natl Acad Sci USA 1984;81:6841–6845.

56. Fewtrell C, Sherman E. IgE receptor-activated calcium permeability pathway in rat basophilic leukemia cells: measurement of the unidirectional influx of calcium using Quin2-buffered cells. Biochemistry 1987;26:6995–7003.

57. Kanner BI, Metzger H. Crosslinking of the receptors for immunoglobulin E depolarizes the plasma membrane of rat basophilic leukemia cells. Proc Natl Acad Sci USA 1983;80:5744–5748.

58. Sagi-Eisenberg R, Pecht I. Membrane potential changes during IgE-mediated histamine release from rat basophilic leukemia cells. J Membr Biol 1983;75:97–104.

59. Zweifach A, Lewis RS. Calcium-dependent potentiation of store-operated calcium channels in T lymphocytes. J Gen Physiol 1996;107:597–610.

60. Hoth M. Calcium and barium permeation through calcium release-activated calcium (CRAC) channels. Pflügers Arch 1995;430:315–322.

61. Penner R, Neher E. Secretory responses of rat peritoneal mast cells to high intracellular calcium. FEBS Lett 1988;226:307–313.

62. Mohr FC, Fewtrell C. IgE receptor-mediated depolarization of rat basophilic leukemia cells measured with the fluorescent probe bis-oxonol. J Immunol 1987;138:1564–1570.

63. McCloskey MA, Cahalan MD. G protein control of potassium channel activity in a mast cell line. J Gen Physiol 1990;95:205–227.

64. Hultsch T, Albers MW, Schreiber SL, et al. Immunophilin ligands demonstrate common features of signal transduction leading to exocytosis or transcription. Proc Natl Acad Sci USA 1991;88:6229–6233.

65. Randriamampita C, Tsien RY. Emptying of intracellular Ca^{2+} stores releases a novel small messenger that stimulates Ca^{2+} influx. Nature (Lond) 1993;364:809–814.

66. Choi OH, Kim J-H, Kinet J-P. Calcium mobilization via sphingosine kinase in signalling by the FcεRI antigen receptor. Nature (Lond) 1996;380:634–636.

67. Zweifach A, Lewis RS. Slow calcium-dependent inactivation of depletion-activated calcium current. J Biol Chem 1995;270:14445–14451.

68. Mongar JL, Schild HO. Effect of temperature on the anaphylactic reaction. J Physiol (Camb) 1958;135:320–338.

69. McCloskey MA, Qian Y-X. Selective expression of K⁺ channels during mast cell differentiation. J Biol Chem 1994;269:14813–14819.

70. Fan Y, McCloskey MA. Dual pathways for GTP-dependent regulation of chemoattractant-activated K⁺ conductance in murine J774 monocytes. J Biol Chem 1994;269:31533–31543.

71. Innocenti B, Pozzan T, Fasolato C. Intracellular ADP modulates the Ca²⁺ release-activated Ca²⁺ current in a temperature- and Ca²⁺-dependent way. J Biol Chem 1996;271:8582–8587.

72. Birnbaumer L, Zhu X, Jiang M, et al. On the molecular basis and regulation of cellular capacitative calcium entry: roles for Trp proteins. Proc Natl Acad Sci USA 1996;93:15195–15202.

73. Hardie RC. Calcium signaling: setting store by calcium channels. Curr Biol 1996;6:1371–1373.

74. Kindman LA, Kim S, McDonald TV, et al. Characterization of a novel intracellular sphingolipid-gated Ca²⁺-permeable channel from rat basophilic leukemia cells. J Biol Chem 1994;269:13088–13091.

75. Beaven MA. Calcium signaling: sphingosine kinase versus phospholipase C? Curr Biol 1996;6:798–801.

76. Simon SM, Llinas RR. Compartmentalization of the submembrane calcium activity during calcium influx and its significance in transmitter release. Biophys J 1985;48:485–498.

77. Llinas R, Sugimori M, Silver RB. Microdomains of high calcium concentration in a presynaptic terminal. Science 1992;256:677–679.

78. Kim K-T, Westhead EW. Cellular responses to Ca²⁺ from extracellular and intracellular sources are different as shown by simultaneous measurements of cytosolic Ca²⁺ and secretion from bovine chromaffin cells. Proc Natl Acad Sci USA 1989;86:9881–9885.

79. Robinson IM, Finnegan JM, Monck JR, et al. Colocalization of calcium entry and exocytotic release sites in adrenal chromaffin cells. Proc Natl Acad Sci USA 1995;92:2474–2478.

80. Zweifach A, Lewis RS. Rapid inactivation of depletion-activated caclium current (I_CRAC) due to local calcium feedback. J Gen Physiol 1995;105:209–226.

81. Haydon PG, Marchese-Ragona S, Basarsky TA, et al. Near-field confocal optical spectroscopy (NCOS): subdiffraction optical resolution for biological systems. J Microsc (Oxf) 1996;182:208–216.

82. Baumgartner RA, Yamada K, Deramo VA, et al. Secretion of TNF from a rat mast cell line is a brefeldin A-sensitive and a calcium/protein kinase C-regulated process. J Immunol 1994;153:2609–2617.

83. Goldfeld AE, Tsai E, Kincaid R, et al. Calcineurin mediates tumor necrosis factor-α gene induction in stimulated T and B cells. J Exp Med 1994;180:763–768.

84. Leal-Berumen I, Conlon P, Marshall JS. IL-6 production by rat peritoneal mast cells is not necessarily preceded by histamine release and can be induced by bacterial lipopolysaccharide. J Immunol 1994;152:5468–5476.

85. Crabtree GR, Clipstone NA. Signal transmission between the plasma membrane and nucleus of T lymphocytes. Annu Rev Biochem 1994;63:1045–1083.

86. Rao A. NF-AT_p: a transcription factor required for the co-ordinate induction of several cytokine genes. Immunol Today 1994;15:274–281.

87. Jain J, McCaffrey PG, Valge-Archer VE, et al. Nuclear factor of activated T cells contains Fos and Jun. Nature (Lond) 1992;356:801–804.

88. Northrop JP, Ho SN, Chen L, et al. NF-AT components define a family of transcription factors targeted in T-cell activation. Nature (Lond) 1994;369:497–502.

89. McCaffrey PG, Luo C, Kerppola TK, et al. Isolation of the cyclosporin-sensitive T cell transcription factor NFATp. Science 1993;262:750–754.

90. Li X, Ho SN, Luna J, et al. Cloning and chromosomal localization of the human and murine genes for the T-cell transcription factors NFATc and NFATp. Cytogenet Cell Genet 1995;68:185–191.

91. Goldsmith MA, Weiss A. Early signal transduction by ther antigen receptor without committment to T cell activation. Science 1988; 240:1029–1031.

92. Timmerman LA, Clipstone NA, Ho SN, et al. Rapid shuttling of NF-AT in discrimination of Ca²⁺ signals and immunosuppression. Nature (Lond) 1996;383:837–840.

93. Clipstone NA, Crabtree GR. Identification of calcineurin as a key signalling enzyme in T-lymphocyte activation. Nature (Lond) 1992;357:695–697.

94. Loh C, Shaw KT-Y, Carew J, et al. Calcineurin binds the transcription factor NFAT1 and reversibly regulates its activity. J Biol Chem 1996; 271:10884–10891.

95. Cockerill PN, Shannon MF, Bert AG, et al. The granulocyte-macrophage colony-simulating factor/interleukin 3 locus is regulated by an inducible cyclosporin A-sensitive enhancer. Proc Natl Acad Sci USA 1993;90:2466–2470.

96. Jain J, Miner Z, Rao A. Analysis of the preexisting and nuclear forms of nuclear factor of activated T cells. J Immunol 1993;151:837–848.

97. Baranes D, Razin E. Protein kinase C regulates proliferation of mast cells and the expression of the mRNAs of fos and jun proto-oncogenes during activation by IgE-Ag or calcium ionophore A23187. Blood 1991; 78:2354–2364.

98. Wolfe PC, Chang E-Y, Rivera J, et al. Differential effects of the protein kinase C activator phorbol-12-myristate 13-acetate on calcium responses and secretion in adherent and suspended RBL-2H3 mucosal mast cells. J Biol Chem 1996;271:6658–6665.

99. Ghosh P, Sica A, Cippitelli M, et al. Activation of nuclear factor of activated T cells in a cyclosporin A-resistant pathway. J Biol Chem 1996;271:7700–7704.

100. Woodrow MA, Rayter S, Downward J, et al. p21^{ras} function is important for T cell antigen receptor and protein kinase C regulation of nuclear factor of activated T cells. J Immunol 1993;150:3853–3861.

101. Woodrow M, Clipstone NA, Cantrell D. p21^{ras} and calcineurin synergize to regulate the nuclear factor of activated T cells. J Exp Med 1993;178:1517–1522.

102. Wodnar-Filipowicz A, Moroni C. Regulation of interleukin 3 mRNA expression in mast cells occurs at the posttranscriptional level and is mediated by calcium ions. Proc Natl Acad Sci USA 1990;87:777–781.

103. Nair APK, Hahn S, Banholzer R, et al. Cyclosporin A inhibits growth of autocrine tumor cell lines by destabilizing interleukin-3 mRNA. Nature (Lond) 1994;369:239–242.

104. Prieschl EE, Gouilleux-Gruart V, Walker C, et al. A nuclear factor of activated T cell-like transcription factor in mast cells is involved in IL-5 gene regulation after IgE plus antigen stimulation. J Immunol 1995;154:6112–6119.

105. Razin E, Szallasi Z, Kazanietz MG, et al. Protein kinases C-β and C-ε link the mast cell high-affinity receptor for IgE to the expression of c-*fos* and c-*jun*. Proc Natl Acad Sci USA 1994;91:7722–7726.

106. Xanthoudakis S, Viola JPB, Shaw KTY, et al. An enhanced immune response in mice lacking the transcription factor NFAT1. Science 1996;272:892–895.

107. Lepple-Wienhues A, Cahalan MD. Conductance and permeation of monovalent cations through depletion-activated Ca^{2+} channels (I_{CRAC}) in Jurkat T cells. Biophys J 1996;71:787–794.

108. Premack BA, McDonald TV, Gardner P. Activation of Ca^{2+} current in Jurkat T cells following the depletion of Ca^{2+} stores by microsomal Ca^{2+}-ATPase inhibitors. J Immunol 1994;152:5226–5240.

18
The MAP Kinases and Their Role in Mast Cells and Basophils

Cheng Zhang and Michael A. Beaven

The available information on the relationships of mitogen-activated protein (MAP) kinases to mast cell function, and more so on the relationship to FcεRI-mediated function, is insufficient for meaningful review. However, meaningful scenarios can be discussed when comparisons are made with cells that have features in common with the mast cell and with other multimeric immune receptors that employ similar signaling systems to FcεRI. In this review, we provide an overview of the MAP kinases and then discuss their activation in lymphocytes and blood granulocytes before discussion of the mast cell. We have chosen for comparison with the mast cell, T and B lymphocytes, eosinophils and neutrophils because these cells express multimeric immune receptors and share some of the features of mast cells. Our hope is that these comparisons will stimulate research on similar systems in mast cells, in particular the effects of costimulants on FcεRI-mediated responses.

As noted in other chapters in this volume, mast cells of various phenotypes express FcεRI, in addition to low-affinity IgG-binding receptors (FcγRII and FcγRIII) and receptors for the complement-derived anaphylatoxins, C3a and C5a.[1] Stimulation through these receptors causes release of preformed inflammatory mediators from granules, synthesis of additional inflammatory mediators from arachidonic acid through activation of phospholipase A_2 and the generation of cytokines through gene transcription. In addition to these receptors, other receptors are essential for mast cell differentia-

tion and proliferation.[2,3] Immature mast cells from pluripotential stem cells differentiate into distinct subtypes depending upon the tissue microenvironment. Proliferation and differentiation of rodent connective tissue mast cells require fibroblast-derived stem cell factor (SCF), which acts through a tyrosine kinase growth-promoting receptor (Kit) that is encoded by c-kit. Another rodent phenotype, the mucosal mast cell, requires T-cell-derived cytokines, interleukin-3 (IL-3) and IL-4 for proliferation. Mast cells also express adhesion receptors (integrins) that bind to extracellular matrix glycoproteins and initiate cytosolic signals to synergize, for example, FcεRI-mediated degranulation and IL-3-induced proliferation.

The studies to date indicate that the MAP kinases are activated in mast cells via FcεRI, Kit and IL-3 receptors. Other mast cell receptors have not been studied with respect to the MAP kinases, although, as is noted, activation of MAP kinases have been reported for analogous receptors in lymphocytes and granulocytes. We also note examples of synergistic receptor interactions in the activation of MAP kinases in lymphocytes and granulocytes that have yet to be examined in the mast cell.

The rapid advances in the field of MAP kinases, often marked by the simultaneous discovery of the same enzymes by several groups, have left a legacy of confusing nomenclature and inconsistent acronyms. We have tried to minimize confusion by providing alternate identifying names initially and then adopting a

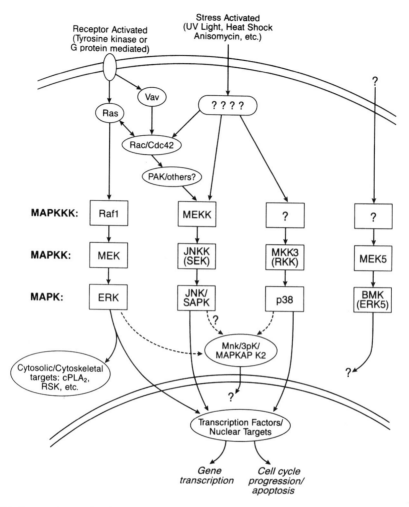

FIGURE 18.1. Pathways for activation of the MAP kinases in mammalian cells. The acronyms and designations for MAP kinases (*MAPK*), MAP kinase kinases (*MAPKK*), MAP kinase kinase kinases (*MAPKKK*), upstream regulators, and downstream targets are defined in the text; p38 denotes p38 MAP kinase. With the exception of BMK, which has not been studied in mast cells, all MAP kinases are activated by cross-linking FcεRI.

single appellation (Fig. 18.1) for the remainder of the chapter.

The MAP Kinases

Family of Mitogen-Activated Protein Kinases

MAP kinases are a multigene family of serine/ threonine kinases whose activation is associated with stimulation of cellular proliferation,

differentiation, apoptosis and other cellular processes.[4,5] A distinguishing feature of MAP kinases is the presence of a threonine/tyrosine phosphorylation motif, TXY, where X can be Glu (E), Pro (P), or Gly (G). The kinases are activated through dual phosphorylation of the threonine and tyrosine in this motif by a unique group of dual-specificity kinases (MAP kinase kinases), and they are inactivated through dephosphorylation of either or both amino acids by protein phosphatases. The pathways that lead to the activation of several MAP kinases have been defined in yeast and in mammalian cells.[6] To date, four classes of MAP kinases have been described in mammalian cells (Fig. 18.1). These include the extracellular signal-regulated kinases (ERK-1 and ERK-2, also called p44 and p42 MAP kinases); the c-*Jun N*-amino-terminal kinase (JNK), also referred to as the stress-activated protein kinase (SAPK); p38 MAP kinase (a mammalian homolog of the yeast high-osmolarity glycerol response-1 kinase, HOG-1); and, most recently, ERK-5,[7] renamed as big MAP kinase-1 (BMK-1) to distinguish it from the ERKs.[8] We shall refer to these kinases as ERKs, JNKs, p38 MAP kinase, and BMK-1, respectively. Work in this field has rapidly unraveled the extraordinary complexity of the MAP kinase pathways but, from this complexity, there has emerged a basic hierarchy that is common to all MAP kinases in yeast as well as vertebrates. We sometimes refer to this generic hierarchy of kinases (i.e., MAPK for MAP kinase, MAPKK for MAP kinase kinase, and MAPKKK for MAP kinase kinase kinase; see left-hand side of Fig. 18.1) where further clarification of nomenclature is required.

As depicted in Fig. 18.1, the MAP kinases are the terminal enzymes of a series of protein kinases that activate one another through phosphorylation to propagate a linear cascade of phosphorylation signals. The signaling pathways for the ERKs, p38 MAP kinase, and JNKs are discussed in detail in the next section. Although the mechanisms of communication between membrane receptors and the individual MAP kinase pathways vary, within the individual MAP kinase pathways, the kinases are highly specific. These kinases serve to amplify[9] and maintain linear fidelity of the signaling cascade such that the pathways can be activated in parallel.[10] Recent evidence points to additional kinases, downstream of the MAP kinases, that allow convergence and perhaps coordination of signals for activation of transcription factors (see Fig. 18.1). These downstream regulators include MAP kinase signal-integrating kinases (Mnk-1 and -2),[11,12] MAP kinase-activated protein kinase-3 (MAPKAP-K3, 3pK), and the closely related MAPKAP-K2.[13]

The MAP kinases are selective in their interaction with both upstream activators and substrates. This selectivity is predicated by the precise sequence of the dual phosphorylation motif and is influenced by another structural feature of the MAP kinases, a linker loop (L12), as is discussed in a later section (Structural Motifs and Downstream Substrates).[14] Other common features are that the MAP kinases and their associated kinases are cytosolic. Following cell activation, the ERKs, JNKs, and the downstream regulators may translocate to the nucleus and, thus poised, phosphorylate nuclear transcription factors.[6] The MAP kinase pathways are tightly regulated by phosphatases, that may help to determine whether signals propagated through individual MAP kinase pathways are short- or long-lived.[15]

The foregoing features permit signaling from receptors to the nucleus for the activation of transcription factors and gene transcription while the repertoire of MAP kinases confer pleiotropic potential. Nevertheless, we have little comprehension of how the cell coordinates signals through the MAP kinase pathways and responds appropriately to a given stimulus. Although a given stimulus may mediate activation of more than one MAP kinase, the best-studied systems suggest that the pattern of activation of individual MAP kinases, both in extent and duration, is determined by the type of receptor or stimulus. As noted earlier, MAP kinases participate in a wide range of cellular responses to stimulants. The ERKs play an important role in regulating responses to growth factors as well as ligands to G protein-coupled and multimeric immune receptors. The JNKs and p38 MAP kinases are activated by proinflammatory cytokines, bacterial pathogens, and adverse environmental changes. The topic has been discussed extensively in recent reviews,[4–6,10,15–19] but a small sampling of the literature illustrates the point. The ERKs, for example, are activated in response to growth factors including the epidermal[20] and the platelet-derived[21] growth factors (EGF and PDGF) or to differentiating factors such as nerve growth factor (NGF).[22] This activation leads to phosphorylation of such cytosolic substrates as cytosolic phospholipase (PL) A_2,[23] p90[rsk],[24] the EGF receptor,[25] and, after translocation of the ERKs to the nucleus, transcription factors such as Elk-1.[26] Activation of the ERKs by growth-promoting cytokines or cell adhesion via integrin receptors is associated with specific gene expression, DNA synthesis, cell growth, and suppression of apoptosis.[6,27] In contrast to the ERKs, JNKs and p38 MAP kinase[18,28] are activated by a variety of environmental stress signals, such as UV light and osmotic shock, and pathological stimuli such as proinflammatory cytokines and microbial pathogens.[6,29] Activation of these two kinases result in growth arrest, apoptosis, or stimulation of immune cells.

MAP kinases do not act exclusively in the nucleus nor do they have the exclusive role in regulating fundamental cellular processes. The phosphorylation of cytosolic substrates has been noted. Other regulatory systems, such as the protein kinase C,[30] the JAK/STAT (for Janus kinase/signal transducers and activators of transcription) pathway,[18] and phosphoinositide-3 kinase (PI-3 kinase),[31] which can impinge on the MAP kinase pathways (Fig. 18.2), also regulate cellular processes in their own right.

Pathways of Activation

Generic Patterns of Activation of ERK-1 and ERK-2 via Ras

The best-studied MAP kinase pathway is that for ERK-1 and ERK-2. Several mechanisms are available for input into the ERK pathway from receptors of diverse types (Fig. 18.2). In this section we discuss the best-studied mechanisms; those initiated by receptors that possess intrinsic tyrosine kinase activity (e.g., growth factor receptors), receptors that recruit or associate with soluble tyrosine kinases (e.g. cytokine receptors, integrins, and multimeric immune receptors), and G protein-coupled re-

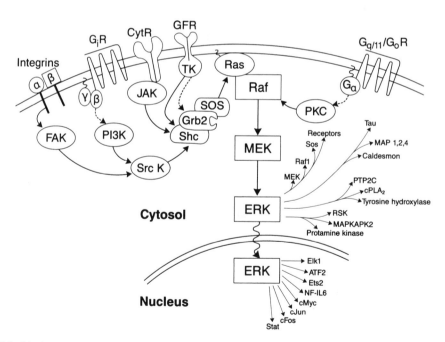

FIGURE 18.2. Various receptor-mediated pathways for the activation of the ERK cascade via Ras/Raf and substrates targeted by ERK in cytosol and nucleus. G_iR, $G_{q/11}/G_oR$, CytR, and GFR designate receptors; i.e., G_i-coupled, $G_{q/11}/G_o$-coupled, cytokine, and growth factor receptors. G_α and $\gamma\beta$ represent the α- and $\beta\gamma$-subunits of the relevant trimeric G proteins G_i, $G_{q/11}$, and G_o. Other acronyms and abbreviations are as defined in the text. The *squiggly lines* designate lipid tethers to the plasma membrane.

ceptors. Multimeric immune receptors are discussed in further detail in our review of lymphocytes and blood granulocytes (see later sections).

An important focal point is the small GTP-binding protein, Ras (see Fig. 18.2). Communication between Ras and tyrosine kinase-dependent or G_i-coupled receptors occurs through the formation of multimeric complexes of adaptor proteins, other intermediary proteins, and guanine nucleotide exchange factors. For example, phosphorylation of the cytoplasmic domains of growth factor receptors results in recruitment of the adaptor protein, Grb-2 (growth factor receptor-bound protein-2), which itself is constitutively bound to the guanine-nucleotide exchange factor, SOS (for son of sevenless, the gene product of *Drosophila*). Grb-2, like other adaptor proteins, has no intrinsic enzymatic activity but it contains one Src homology-2 (SH-2) and two Src homology-3 (SH-3) binding domains. The SH-2 domain recognizes specific motifs that contain the phosphotyrosine residue of the receptors while the SH-3 domains recognize sequences enriched in proline and hydrophobic amino acids (PH domains) that are present in SOS. Binding of the Grb-2/SOS complex to the phosphorylated receptor, via Grb-2, causes translocation of SOS to the membrane-bound Ras, which is converted to its active GTP-bound state by SOS. The Grb-2/SOS complex is utilized in a similar manner for communication between adhesion receptors (i.e., integrins) and the ERK pathway (see later in this section).

Stimulation through G_i-coupled receptors (other G proteins are discussed in the next section) and some growth factor receptors leads to the tyrosine phosphorylation of a third intermediary protein, Shc, which can then bind to the SH-2 domain of Grb-2 (i.e., as the Grb-2/SOS complex) to permit engagement of SOS with Ras (Fig. 18.2). Here, the release of βγ-subunits from G_i leads to the phosphorylation of Shc through a G protein-activated form of PI-3 kinase (PI-3 kinase g) and a Src-like kinase.[32,33] Shc, of which there are three known isoforms (p46, p52, and p68), may function as a docking protein whose function is facilitated by the presence of a phosphotyrosine binding (PTB) domain that binds to tyrosine-phosphorylated motifs in growth factor receptors and other proteins. We revisit Shc in the context of docking proteins (see next section) and multimeric immune receptors.

The cytokine receptors and integrins, like the multimeric immune receptors, have no intrinsic tyrosine kinase activity but they rely instead on specific soluble tyrosine kinases for initiating signals (see Fig. 18.2). The cytokine and certain other (growth hormone and erythropoietin) receptors associate with JAKs, which were first identified as upstream activators of STATs and, through STATs, interferon-responsive genes. Most ligands that activate this pathway activate the ERK pathway as well. It now seems from deletion experiments that JAKs may activate both pathways. Activation of the ERK pathway may occur by Ras-regulatory proteins interacting directly with phosphorylated receptor or indirectly through binding of multimeric complexes of Shc/Grb-2/SOS to the receptor-associated phosphorylated JAK.[34]

Integrins consist of transmembrane α- and β-subunits of different subtypes to form a family of at least 20 different heterodimeric integrin receptors, some of which are present in mast cells.[2] These receptors promote formation of multimeric complexes of cytoskeletal and protein kinases including FAK (for focal adhesion protein tyrosine kinase). FAK localizes with integrin receptors at sites of cell attachment to the extracellular matrix (i.e., focal adhesion sites) and is activated by cross-linking of the integrin receptor with fibronectin or antiintegrin antibodies. Such activation induces transient association of FAK with Src tyrosine kinase and as a consequence tyrosine phosphorylation of FAK (Fig. 18.2). Various mutational disruptions suggest that phosphorylated FAK engages the Grb-2/SOS complex and thereby activate the ERK pathway via Ras.[27]

The convergence of signals through Ras allow recruitment of both membranal and cytosolic components of the ERK activation pathway (Fig. 18.1).[6,15] The activated GTP-bound form of Ras binds to the N-terminal portion of Raf-1, a serine-threonine protein kinase, to bring this kinase in close proximity to the membrane. Tyrosine phosphorylation of Raf-1, by an as yet unidentified kinase, leads to its activation. Raf activity may be further modulated by phosphorylation via protein kinase C and protein kinase A. Thereafter, Raf-1 selectively phosphorylates two regulatory serine residues on MAP kinase kinase, which exists in two isoforms (MEK-1 and MEK-2, for MAP kinase/ERK kinase). The MEKs are highly selective activators of the ERKs. They do so by phosphorylating Thr-183 and Tyr-185 in the dual-phosphorylation motif of ERK-2, for example, in a nonprocessive manner in which the MEK–ERK complex dissociates and then reassociates for each phosphorylation event.[35] Thr-183 and Tyr-185 are present in the linker L12 and form a lip[36] (see section on Structural Motifs and Downstream Substrates). Conformational changes induced by this dual phosphorylation[37] are thought to transform ERKs into their activated states. ERK-1 and ERK-2 are considered by some as functionally redundant because they are 90% identical as compared to 40% to 50% identity with the JNKs.[38]

Other Ras-Dependent and Ras-Independent Mechanisms for Activation of the ERKs

These scenarios are probably incomplete. Recent reports indicate additional pathways and signaling components that are not shown in Fig. 18.2. One report, for example, describes the cloning of a myristoylated membrane-anchored docking protein, FRS-2 (for FGF receptor substrate-2), which migrates as a doublet (92 and 95 kDa) on SDS-PAGE. Like Shc, FRS-2 contains a PTB domain, is tyrosine phosphorylated, and forms ternary complexes with Grb-2/SOS when cells are stimulated via tyrosine kinase receptors for nerve growth factor (NGF) and epidermal growth factor (EGF).[39] It now appears that growth factor receptors can activate the ERK pathway by more than one mechanism. Thus, the EGF receptor recruits Grb-2/SOS directly and indirectly through tyrosine phosphorylation of Shc,[40] the fibroblast factor (FGF) receptor recruits Grb-2/SOS indi-

rectly via Shc and FRS-2,[39] and the insulin receptor recruits Grb-2/SOS through phosphorylation of Shc and the insulin receptor substrate-1 (IRS-1).[41] Although the docking proteins share certain features, FRS-2 and IRS-1 are engaged by a more limited number of receptors than Shc and may activate, therefore, a more limited repertoire of responses than Shc.

Other candidates for regulating Ras in hematopoietic cells include the guanine nucleotide exchange factor (GEF), p95^{vav}, and the p120 GTPase-activating protein (GAP). Vav[42] has an interesting assortment of domains that include, among others, one SH-2 flanked by two SH-3 domains, a cysteine-rich zinc-binding domain analogous to that in protein kinase C, a pleckstrin domain, and a GEF domain. The latter domain is highly homologous to the GEF domain of the Rho exchange protein Dbl and is sometimes referred to as the Dbl homology, or DH domain. Vav is tyrosine phosphorylated and so activated in response to a wide variety of stimuli, as well as FcεRI ligation, and thereafter associates with a disparate array of cytoplasmic, cytoskeletal and nuclear proteins that include Shc/Grb-2 and, potentially, SOS.[43] These features provide potential mechanisms for divergent signals through the Ras/ERK pathway and through GTPases of the Rho family and potentially the JNK pathway (see next section). In addition, the presence of the cysteine-rich domain makes Vav a target for phorbol 12-myristate 13-acetate in hematopoietic cells and a point of convergence of signals generated by tyrosine kinases and protein kinase C. The Ras exchange activity of Vav is stimulated in vivo and in vitro by the phorbol ester, diacylglycerol, and ceramide.[44] Such activation is not observed after expression of a cysteine-deleted mutated form of Vav.

Although GAP was initially viewed as a negative regulator of Ras, there are indications that a Ras/GAP complex may relay necessary signals for cellular responses that are not fulfilled by signals transduced through Raf and ERK alone.[45] As we note later in the context of lymphocytes, phosphorylation of Ras/GAP leads to accretion of the complex through recruitment of other proteins, particularly the p62 and p190 proteins. Apart from the carboxy-terminal catalytic domain, which is necessary for Ras binding, the amino-terminal domain of GAP contains an SH-3 domain flanked by two SH-2 domains as well as PH, calcium-dependent lipid binding, and phospholipid-binding domains.[45] This array of domains raises the suspicion that GAP, like Vav, has the flexibility for multiple but still largely undetermined roles in cell signaling.

In addition to the aforementioned proteins, other candidates have been invoked in the activation of the ERKs. Stimulation of fibroblasts via PDGF (but not EGF) receptors and transfected CHO cells via a G_i-coupled serotonergic receptor (5-HT_{1A}) results in increased phosphatidylcholine hydrolysis by phospholipase (PL) D and phosphatidylcholine-specific phospholipase (PL) C as well as activation of the ERKs. Studies with tricyclodecan-9-yl-xanthogenate (D609), an inhibitor of phosphatidylcholine-specific PLC, and wortmannin, an inhibitor of PI-3 kinase, suggest that both enzymes provide overlapping pathways for activation of Ras.[46,47] Also, there is evidence for Ras-independent pathways of activation. On the basis of pharmacological studies it has been suggested that PI-3 kinase and protein kinase C may promote sustained signals for activation independently of MEK and that activation may involve more than one simple linear pathway.[48] Receptors that employ $G_{q/11}$ can activate the ERKs independently of Ras through stimulation of phospholipid metabolism, protein kinase C, and thence Raf.[49] Activation of the ERKs though G_o-coupled receptors also requires protein kinase C, but pathways remain undefined.[50]

JNK p38 MAP Kinase, and BMK

Two human kinases, JNK-1 (46kDa) and JNK-2 (55kDa), were initially characterized and one (JNK-1) was subsequently cloned.[28] At the same time, three rat genes (referred to as SAPK-α, -β, and -γ) were cloned.[51,52] SAPK-α and SAPK-β encode proteins of similar size to JNK-2 while SAPK-γ encodes a protein of similar size to JNK-1. Alternative mRNA splicing leads to further diversification into at least 12 isoforms, although some of these isoforms may be functionally redundant.[53] JNKs are activated via a MEK kinase (a JNK kinase kinase; see following) and a JNK kinase (also called SEK-1 for SAPK/ERK kinase-1) (see Fig. 18.1). The confused nomenclature arose because in vitro studies and the effects of overexpression indicated that MEK could be phosphorylated by MEK kinase. Subsequent studies have established that MEK kinase selectively phosphorylates and activates JNK kinase in mammalian cells to indicate separate activation pathways for ERK and JNK.[54-56] As noted earlier, JNK kinase activates JNK by phosphorylation of threonine and tyrosine residues of the phosphorylation motif (-Thr-Pro-Tyr-) of JNK.

Like ERK, JNK is activated through tyrosine kinase-dependent and G protein-coupled receptors but, unlike ERK, JNK is also activated by stress factors such as UV light and osmotic shock and by chemical stimulants such as anisomycin and arsenite. The events initiated by stress stimuli and chemicals that lead to activation of the JNK kinase kinase/JNK kinase/JNK pathway remain undefined. Some receptor ligands, however, are known to activate the pathway through Ras (EGF), or independently of Ras (the cytokine, tumor necrosis factor-α, TNF-α), by engagement of the Rho-like small GTP-binding proteins, Rac-1 and Cdc42[57,58] and possibly the p21-activated serine/threonine kinase, PAK[57] (Fig. 18.1). Other ligands (PDGF) may engage both Ras and Rac-1, but apparently not Rho A or Cdc42, in a PI-3 kinase-dependent manner for the activation of JNK.[59] As depicted in Fig. 18.1, a linkage has been established in transfection experiments between tyrosine phosphorylation of Vav and enhanced nucleotide exchange on Rac-1 and subsequently activation of JNK in vivo.[60] As is discussed later, the Vav/Rac-1/JNK connection is utilized by FcεRI[61] and presumably by other multimeric immune receptors that are known to activate Vav.

Least is known about the mechanisms of activation of p38 MAP kinases and BMK-1. Three members of the p38 MAP kinase are alternatively spliced forms of a single gene[62-65] and a recently described homolog, p38β.[66] The primary[67] and tertiary (at 2.1 Å resolution)[68,69] structures of p38 MAP kinase have been determined. It contains the dual phosphorylation motif -Thr-Gly-Tyr-, and its upstream kinase, MAP kinase kinase (MKK-3, recently referred to as RKK for reactivating kinase kinase) has been identified.[70] The preceding kinase is unknown. Unlike the ERKs and JNKs, p38 MAP kinase is weakly activated through tyrosine kinase-linked receptors. p38 MAP kinase is activated, however, by the same stress stimuli as JNK and via Rac and Cdc42.[29] Therefore, the physiological significance of p38 MAP kinase is unclear. The dual-specificity kinase for BMK-1 (ERK-5) has been cloned and called MEK-5,[7] but the upstream kinases have not been defined. Because little is known about BMK-1, although it is activated by H_2O_2 in some types of cells,[8] this MAP kinase is not discussed further here.

Structural Motifs and Downstream Substrates

The crystallographic studies of ERK-2[36] and p38 MAP kinase[68,69] as well as the amino acid sequence comparisons of members of the MAP kinase families[51,62,63,68] point to structural features that determine the substrate specificity of the MAP kinases.[14] Some of the major similarities and differences are highlighted by the structure of the previously mentioned linker

loop-12 (L12) (Fig. 18.3). L12 extends from a conserved region in domain VII through much of domain VIII. An intermediate region (highlighted in Fig. 18.3) of variable sequence and length contains the dual phosphorylation motif and forms a phosphorylation lip. The structure and topography of this lip and the intervening amino acid of the dual phosphorylation motif may account for the specific interactions of MAP kinases with substrates but, surprisingly, not with upstream activators.[14,71] For the latter interactions, the N-terminal and other domains may be involved.[14,72]

In view of the functional abilities of ERKs in regulating, at least in part, cell cycle proliferation and differentiation along with the attendant changes in cell biochemistry and morphology, it is not surprising that these kinases phosphorylate a large array of substrates. Until some semblance of order can be established among the various processes, the substrates are best listed as cytosolic, membranal, cytoskeletal and nuclear. Known cytosolic substrates of the ERKs[6,38] are p90[RSK], which can, in turn, phosphorylate the transcription factor c-Fos, MAP kinase-activated protein kinase-2 (MAPKAP-K2), protamine kinase, Raf-1, SOS and MEK (see Fig. 18.2). Phosphorylation of Raf-1, SOS and MEK may provide negative feedback signals for the ERK cascade while phosphorylation of c-Fos by p90[RSK] is

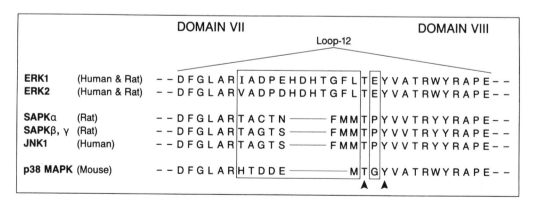

FIGURE 18.3. Comparison of amino acid sequences in the linker loop-12 (L12) in domains *VII* and *VIII* of the various MAP kinases. L12 contains the dual phosphorylation motif as indicated by the sites of phosphorylation (*arrowheads*) by the MAP kinase kinases. Variable regions (see text) are *boxed*. The SAPKs and JNK are referred to collectively as JNK in the text.

thought to regulate transcriptional activity of the AP-1 complex. Several cytosolic enzymes (e.g., the protein tyrosine phosphatase, PTP-2C) are phosphorylated and, as noted earlier, phosphorylation of cytosolic PLA$_2$ results in activation of this enzyme. Membranal substrates include the NGF and EGF receptors. Known cytoskeletal substrates are the microtubule-associated proteins (MAP-1, -2, and -3), Tau and caldesmon whose phosphorylation may regulate cytoskeletal changes.

Potential ERK substrates in the nucleus are the transcription factors Elk-1, c-Myc, NF-IL6, the STAT proteins and c-Jun (Fig. 18.2). The phosphorylation and activation of Elk-1 results in its binding to serum response element (SRE) and serum response factor (SRF) and thereby activates the c-*fos* gene. The Fos protein in combination with Jun protein induces transcriptional activity of a variety of genes (see next paragraph). ERK, however, phosphorylates the inhibitory carboxy-terminal site (Ser-243) of c-Jun,[73] and activation of both ERK (for induction of Fos synthesis) and JNK (for phosphorylation and activation of Jun) is necessary for Fos/Jun dimerization (see next paragraph). As the serine phosphorylation of STATs by ERK leads to increased binding of STATs to DNA, it is suggested that ERK may be involved in the activation of cytokine genes by the JAK/STAT signaling cascade.[74]

Some biological substrates for JNK and p38 MAP kinase have been identified, but the body of information is insufficient to link most of these substrates to the presumed physiological functions of these MAP kinases. The emerging picture is that both kinases inhibit cell growth and promote apoptotic or necrotic cell death of nonlymphoid cells while they positively regulate differentiation and activation of lymphoid cells. Like ERK, both MAP kinases phosphorylate the transcription factor, ATF-2.[6,29] In addition, JNK appears capable of initiating gene transcription through phosphorylation of its primary targets, the N-terminal Ser-63 and Ser-73 sites of c-Jun; this is in contrast to the inhibitory phosphorylation of c-Jun by ERK noted earlier.[73] The N-terminal phosphorylation of c-Jun promotes formation of Jun/Fos heterodimers and c-Jun homodimers, that bind

to AP-1 sites in the promotor regions of numerous genes. As noted, induction of the c-*fos* gene and production of Fos protein is dependent primarily on the ERK/Elk-1 pathway. p38 MAP kinase was initially identified as the binding site of SB 203573 and related cytokine-suppressive antiinflammatory drugs (see next section) and as a participant of a cascade that culminates in the activation of MAPKAP-K2 and phosphorylation of the heat-shock proteins, Hsp-25 and Hsp-27.[62,75] In addition to AFT-2 and MAPKAP-K2, a protein called Max that is essential for the DNA-binding and *trans*-activating activity of the early gene, c-*myc*, is also phosphorylated by p38 MAP kinase.[65]

Chemical Stimulants and Inhibitors

As noted earlier, the ERKs are activated through diverse receptors and appear to respond to signals generated by tyrosine kinases and hydrolysis of membrane phospholipids. Chemical activators of intracellular signaling pathways such as guanosine 5'-(2-*O*-thio)triphosphate (GTPγS; an activator of GTP-binding proteins), phorbol 12-myristate 13-acetate (an activator of protein kinase C and possibly other proteins), calcium ionophores, and tyrosine phosphatase inhibitors also activate the ERKs, at least transiently. Stimulation of protein kinase C with phorbol 12-myristate 13-acetate and elevation of calcium alone activate the ERK pathway via, respectively, Raf-1[76] (see Fig. 18.2) and a calmodulin-dependent protein kinase cascade,[77] although calcium/protein kinase C-independent pathways clearly exist for activation of the ERKs. Studies with knockout and constitutively active kinase mutant cells indicate that Raf-dependent stimulation of MEK and ERK by phorbol 12-myristate 13-acetate occurs through protein kinase C and does not require Ras.[78] The possibility exists, however, that phorbol ester-binding proteins such as Vav[44] and n-chimaerin, a Rac-GTPase activating protein,[79] could also stimulate the Raf/MEK/ERK pathway through Ras (or Rac; see Fig. 18.2) in some types of cells.[78]

With respect to the stress-activated JNKs and p38 MAP kinases, there is limited evidence that specific signaling pathways are recruited by ostensibly "nonspecific" stimuli such as reactive oxygen species, misfolded or modified proteins, and free radicals.[29] Agents that have been employed as experimental stimulants include tunicamycin and inhibitors of protein synthesis (cycloheximide and anisomycin). These compounds may act by generating the required "nonspecific" stimuli, but the current information is sketchy. There are also indications that in some circumstances the stress-activated MAP kinases may respond to the "conventional" signaling pathways. For example, the combination of calcium ionophore and phorbol ester (but not phorbol ester alone) will activate JNK (SAPK) in T cells,[53] and there is evidence that the proinflammatory cytokines, TNF-α and IL-1β, act by stimulating hydrolysis of membrane sphingomyelin.[29]

Various pharmacological probes have been used to delineate pathways upstream of Ras/Raf (see Fig. 18.2) and thereby demonstrate the participation of tyrosine kinases (with herbimycin A, tyrphostin, genistein, etc.), PI-3 kinase (wortmannin and LY294002), $\beta\gamma$ of trimeric G proteins (pertussis toxin to prevent dissociation of $\beta\gamma$-subunits from G_i and βARK-1 peptide to prevent $\beta\gamma$ binding to PI-3 kinase), and protein kinase C (phorbol 12-myristate 13-acetate to downregulate protein kinase C).

Two useful inhibitors of the ERK and p38 MAP kinase pathways have been described. One, the flavone PD 098059,[80] selectively inhibits activation of MEK without significantly inhibiting ERK itself or other protein kinases.[81] The compound inhibits phosphorylation and activation of MEK by Raf but does not inhibit MEK once it is activated by Raf.[81] This feature and the lack of competition with ATP[80] suggest that the compound is not an inhibitor of MEK activity but rather impairs the interaction of Raf with MEK. PD 098059 is active in vitro (IC$_{50}$, 2–7μM) and in vivo (at 50μM with partial to complete inhibition of MEK downstream targets). Its activity in vivo varies with the potency and concentration of the agonist; weak stimulants of the ERK pathway being inhibited the most and strong stimulants the least.[81] Initial studies with PD 098059 showed that, in addition to suppression of the MAP kinase cascade, it suppresses Ras-induced

phenotypic transformation[80] as well as NGF-induced cell differentiation,[82] growth arrest[83] and cell-cycle-dependent kinases[83] in specific cell lines.

The other inhibitor, SB 203573, was first identified as a member of a novel group of pyridinyl imidazole antiinflammatory drugs that inhibited lipopolysaccharide-induced release of TNF-α and IL-1β from macrophages.[84] Subsequently, these drugs were found to bind to and directly inhibit p38 MAP kinase, thus suggesting a role for this kinase in release of these cytokines.[64] At submicromolar concentrations, SB 203580 inhibits p38 MAP kinase in vitro (IC$_{50}$, 0.6μM)[85] and in vivo as well as downstream reactions, namely, the activation of MAPKAP-K2 and -K3 and the phosphorylation of heat-shock protein, Hsp-27.[85–87] The actions of SB 203580 appear to be specific in that it does not inhibit other MAP and protein kinases in vitro.[85] This specificity has been attributed to reversible binding of the compound to the catalytic residues in the ATP pocket and locking of the enzyme in the inactive conformation. The compound binds equally well to both the inactive and activated forms of p38 MAP kinase, but it does not prevent the phosphorylation of the enzyme by endogenous MAP kinase kinases.[88] Like PD 098059, the efficacy of SB 203580 in intact cell systems appears to be related to potency of the stimulant. Studies with SB 203580 in vivo suggest roles for p38 MAP kinase in platelet aggregation in response to low but not high concentrations of collagen,[89] in the synthesis of IL-6 induced by TNF in a manner that excludes involvement of the transcription factor NF-κB,[87] and in the synthesis of IL-8 in IL-1-stimulated or osmotically shocked blood mononuclear cells.[90] Significant antiinflammatory and immunosuppressive activities of SB 203580 are also apparent in animal models.[91]

MAP Kinases in Lymphocytes and Blood Granulocytes

T Cells

Stimulation of the T cell through its antigen receptor (TCR) will evoke responses that depend on the stage of maturity of the T cell and the nature of costimulatory signals through receptors such as CD28. TCR stimulation results in either maturation or death (i.e., positive or negative selection) of immature T cells and proliferation, apoptosis, or refractoriness (i.e., anergy) of mature cells. The interest here is whether the MAP kinases define the nature of the response in T cells. Some (i.e., c-*kit* ligand, colony-stimulating factor-1, IL-3, and phorbol 12-myristate 13-acetate) but not all (i.e., IL-4) mitogenic stimulants enhance Ras/ERK-1 and -2 activity in T-cell lines.[92] Engagement of the TCR complex results in rapid and sustained activation of the Ras/Raf/MEK/ERK pathway,[93,94] but no activation of JNK.[53] Stimulation of T cells via CD28 activates neither ERK nor JNK, but when costimulated via CD28 and TCR both kinases are activated, the ERKs primarily by TCR and the JNKs synergistically by both receptors. These activations can be mimicked by Ca^{2+} ionophore and phorbol 12-myristate 13-acetate. Calcium ionophore by itself is inactive while the phorbol ester induces modest stimulation of the ERKs (1 and 2) and JNKs (1 and 2). In combination, both drugs elicit synergistic activation of JNK. The latter activation is blocked by cyclosporin A to indicate the possible involvement of the calcium-sensitive phosphatase, calcineurin.[53] p38 MAP kinase and to a much lesser extent ERK, were found to be constitutively active in freshly isolated thymocytes. These activities decay after removal of cells from the thymus but the p38 MAP kinase activity alone could be rescued by TNF-α or IL-1α.[95] High p38 MAP kinase activity is found in both immature (i.e., CD4- and CD8-negative) and mature (i.e., CD4- and CD8-positive) thymocytes. Freshly isolated thymocytes exhibit no detectable JNK activity.[95]

Although these reports suggest that T cell mitogenesis may often be associated with ERK activation, this may not be true for all growth factors (e.g., IL-4).[92] Indeed, studies in transgenic mice suggest that the ERK pathway is not essential for TCR-induced proliferation but rather this pathway may play an essential role in positive selection of thymocytes.[94,96] Thymocyte maturation and survival within the thymus may be dependent instead on p38

MAP kinase.[95] However, in the context of costimulation of T cells via TCR and CD28, signals mediated through ERK/Elk-1/Fos and calcineurin/JNK/Jun are probably necessary signals for induction of IL-2 production and T cell activation.[53] Binding of Fos and Jun to AP-1 could provide a point of convergence for these signals (see previous sections). Moreover, findings from recent studies imply that the shutdown of IL-2 production in anergic T cells might result from diminished activities of enzymes in both the ERK and JNK pathways and as a consequence of the failure of AP-1 to switch on the IL-2 gene.[97,98] As the anergic block lay in the pathway between TCR and Ras, Ras may mediate TCR-dependent activation of the ERK pathway and synergize CD28-dependent activation of the JNK pathway through Rac-1 (see Fig. 18.1).

An issue of relevance to FcεRI signaling in mast cells is the linkage between TCR and Ras. In Jurkat cells the linkage appears to be through tyrosine phosphorylation of Vav rather than of Shc.[99,100] An important caveat is that cells of other lineage and maturity and other modes of stimulation may engage different mechanisms. The role of GAP, if any, in TCR regulation of Ras activity is undetermined, although phorbol 12-myristate 13-acetate inhibits GAP activity in T cells, and the possibility exists that Ras activity might be regulated by protein kinase C via GAP.[101]

B Cells

As with T cells, stimulation of B cells through its antigen receptor (BCR) can induce proliferation, apoptosis, or anergy depending on the state of cell maturation and nature of costimulatory signals, in particular those generated through another B cell receptor, CD40, whose ligand (CD40L) is expressed on activated T cells. The initial studies with B lymphoma-cell lines showed that the Ras/Raf-1/MEK/ERK pathway as well as p90[rsk] are activated after stimulation of B cells with anti-IgM antibody (for cross-linking BCR), phorbol ester[102] and NGF,[103] and that stimulation through BCR markedly activates ERK-2 but weakly so ERK-1, JNK, and p38 MAP kinase.[104] Costimulation through CD40 suppresses BCR-induced apoptosis and anergy while synergizing BCR-induced B-cell proliferation, differentiation and Ig production. Activation of CD40 by itself causes robust activation of JNK and p38 MAP kinase (and MAPKAP-K2)[104] and little[104] or only transient[105] activation of ERK-2. Costimulation through both receptors results in unimpaired activation of all three MAP kinases and synergistic activation of JNK.[104] While the stimulation of all three MAP kinases in B cells correlates with cell survival, and stimulation of ERK correlates with apoptosis, this situation is the reverse of that observed in PC12 cells[106] and nonlymphoid cells in general.

With respect to the linkage of BCR to Ras, in some B-cell lines BCR ligation results in tyrosine phosphorylation of Vav[107] and GAP as well as the association of GAP with p62 and p120.[108,109] These associations may inhibit GAP activity and thereby increase levels of Ras-GTP. Again, the caveat is that mechanisms may differ among cell lines and stimulants. In fact, Ramos cells, a cell line of mature B-cell lineage, exhibit increased tyrosine phosphorylation of Shc and not of Vav after BCR ligation.[110]

Blood Granulocytes

Neutrophils express Ig (Fc), cytokine, CD15 (the surface lipopolysaccharide-binding protein), and G protein-linked receptors. Examples of the latter receptors include those for chemotactic formylated-peptides (e.g., f-Met.Leu.Phe), platelet-activating factor (PAF), complement-derived C5a and prostenoids. As neutrophils are committed differentiated cells, the property of relevance to the MAP kinases is the ability these cells to assume a "primed" state; that is, cells can be primed for activation by "dedicated primers" such as lipopolysaccharide and the cytokines, TNF-α, and granulocyte-macrophage colony-stimulating (GM-CSF) factor, or by low concentrations of primary stimulants.[111] Possible targets of MAP kinases include cytosolic PLA$_2$ for release of arachidonic acid and the NADPH oxidase complex for the generation of oxygen radicals. One

component of the NADPH oxidase complex, p47-*phox*, is a substrate for ERK-2 and p38 MAP kinase, but not for JNK.[112]

A commonly used primary neutrophil stimulant, f-Met.Leu.Phe, activates ERK-1[113-115] and ERK-2[116] via the classical Ras/Raf/MEK pathway[117,118] as well as p38 MAP kinase.[112,119] A role for ERKs in cell adhesion and chemotaxis is suspected. Elevation of intracellular cAMP is associated with inhibition of neutrophil adhesion and chemotaxis, and studies with pharmacological probes suggest that cAMP inhibits formylpeptide-induced activation of ERK-1 and ERK-2, possibly through protein kinase A-mediated phosphorylation of Raf-1.[116] Some have proposed a role for the activation of ERK in the generation of oxygen radicals on the basis of similarities in the kinetics of the two responses,[113-115,117] while others have noted that, under certain conditions, the two responses could be dissociated.[120,121] In contrast to formylpeptide, GM-CSF stimulates the ERKs without affecting p38 MAP kinase activity (identified as p40^{hera}),[122] whereas lipopolysaccharide induces slow stimulation of p38 MAP kinase without directly activating the ERK and JNK pathways[123] although it potentiates activation of ERKs by PAF.[124] These selective activations may be related to the "priming" ability of lipopolysaccharide and GM-CSF, which by themselves do not fully activate neutrophils.[123,124] For example, both stimulants induce phosphorylation and activation of cytosolic PLA$_2$, presumably via MAP kinases, while causing minimal release of arachidonic acid unless cells are costimulated with a calcium-mobilizing agent.[124,125] The cytokine TNF-α also selectively activates p38 MAP kinase, and this activation has been linked to phosphorylation and activation of cytosolic PLA$_2$.[119] At present, it is not clear whether different ligands can activate cytosolic PLA$_2$ through different MAP kinases, but the studies do illustrate the fact that MAP kinases can be selectively activated especially by agents that evoke a limited repertoire of responses.

Eosinophils like neutrophils, exhibit a similar array of responses to inflammatory stimuli, but a distinguishing feature is that eosinophils can be programmed for survival or apoptosis by certain cytokines. Survival and function of eosinophils is dependent primarily on IL-5, although GM-CSF and IL-3 can subserve this function. IL-5 stimulates the Ras/Raf-1/MEK/ERK-1 pathway via Lyn[126,127] and the JAK2/STAT1[127,128] pathway without detectable activation of ERK-2.[126,127] Activation of both pathways is suppressed by transforming growth factor-β (TGF-β), which abrogates the survival-promoting properties of IL-5.[129] While the effects of IL-5 and TGF-β are highly suggestive, a causal relationship between these pathways and growth promotion has not been established. Interestingly, leukotriene B$_4$[130] and 5-oxo-eicosatraenoate,[131] the two chemotaxins investigated so far, transiently activate both ERK-1 and ERK-2 in contrast to the selective and more sustained activation of ERK-1 by IL-5. The effects mediated through the MAP kinases are unclear because 5-oxo-eicosatraenoate is an extraordinary potent chemotactic agent compared to leukotriene B$_4$.[131]

The presence of Fc receptors on blood granulocytes should provide instructive comparisons with Fc-mediated signaling in mast cells and antigen-mediated signals in lymphocytes. Unfortunately, there are no reports of activation of the MAP kinases in basophils and only one pertinent report in neutrophils. This report noted that the time course of phagocytosis of IgG-opsinized erythrocytes correlated with phosphorylation of ERK-1 and ERK-2, an effect attributed to IgG cross-linking of FcRII and FcRIII.[132] Ceramide, which is produced during phagocytosis, as well as the MEK inhibitor PD 098059, suppressed ERK phosphorylation and phagocytosis to indicate possible association of these events and negative regulation of these events by ceramide.

MAP Kinases in Mast Cells

MAP Kinases Activated by FcεRI and Other Receptors

ERK, JNK, and p38 MAP kinase are activated by different mechanisms in antigen-stimulated mast cells. At this time, studies are limited to cultured cell lines including mouse bone marrow-derived mast cells (BMMC). Activation through FcεRI results in activation of ERK-2 in RBL-2H3,[133–138] PT18,[139] MC/9,[140,141] and BMMC[142,143] cell lines, in addition to p38 MAP kinase[137,141,143] and JNK[140,141,143] in all four cell lines. Of these activations, that of JNK (70-fold increase in activity) and ERK-2 (40-fold increase) are robust and that of p38 MAP kinase the least dramatic (12-fold increase) but still substantial.[141] In RBL-2H3 cells, a shift in electrophoretic migration of ERK-1 has been noted, but this shift is not accompanied by increased tyrosine phosphorylation of ERK-1 to suggest that this MAP kinase is not strongly activated in these cells. In BMMC, however, ERK-1 and ERK-2 appear to be equally activated by FcεRI cross-linking as were JNK-1 and JNK-2.[143] One technical problem with some early studies was insufficient inhibition of phosphatases in cell extracts, which diminished the apparent activation of the MAP kinases.[138] In general, the phosphorylation of MAP kinases is detectable by an alteration in electrophoretic migration of the MAP kinase (gel shift assay), increased tyrosine phosphorylation of the MAP kinase on immunoblotting, and increased in vitro kinase activity in immunoprecipitated MAP kinase. All procedures should show a correlation but all are susceptible to the action of cellular phosphatases.

The responses of the MAP kinases to FcεRI cross-linking are of long duration. The activation of ERK-2, although apparent within 60 s,[134] persists for 30 min or even longer,[138,140,142] except in an early study where activation was said to be transient.[133] Activation of JNK is of slow onset but it is apparent within 5 min and reaches a maximum at 15 to 30 min. By 60 min, JNK activity has decayed considerably but is still elevated.[140] Activation of MEKK-1 clearly precedes that of JNK and reaches a maximum within 3 min of addition of antigen.[140] Like ERK-2, p38 MAP kinase responds rapidly and reaches maximal activity within 1 to 5 min of stimulation. This activity then slowly declines but is still significantly elevated at 60 min.[141] The extent of these activations is dependent on the concentration of antigen, and these show a general correlation with functional responses in the few instances where such correlations have been investigated.[133,141]

c-Kit ligand also promotes relatively sustained increases in ERK[92,142,144] and JNK[143] activities in BMMC. The effect on p38 MAP kinase has not been reported. The time course of increase in ERK activity is similar to that induced by antigen.[142] Stimulation through c-Kit is associated also with activation of Ras GDP/GTP exchange,[92,144] PI-3 kinase,[144] p90[RSK], and p70-S6 kinase,[142] as well as expression of c-fos, c-jun, c-myc, and c-myb mRNA.[144] The relationship of these various activations to the MAP kinases is discussed later.

Of the other mast cell growth-promoting hormones (see Introduction), IL-3 stimulates JNK activity to the same extent as c-Kit ligand (i.e., stem cell factor) in IL-3-deprived BMMC.[143] IL-4 apparently does not stimulate Ras, ERK-1, or ERK-2 in IL-4-dependent BMMC although effects on other MAP kinases were not investigated.[92] An important point, however, is that deprivation alone may have significant effect on MAP kinase activities. Removal of IL-3 from BMMC cultures causes modest increases in JNK and p38 MAP kinase activities, without detectable changes in ERK activity, and eventually apoptosis.[143] Therefore, in this instance, the study revealed a possible link of apoptosis to JNK/p38 MAP kinase activation in the absence of ERK activation.

Finally, numerous other stimulants activate ERK in mast cells. These stimulants include calcium-mobilizing agents such as thapsigargin and calcium ionophores, the protein kinase C agonist phorbol 12-myristate 13-acetate, the phosphatase inhibitor phenylarsine oxide, the G protein stimulant GTPγS in permeabilized cells, and carbachol in cells transfected with the muscarinic m1 receptor.[133–135] Of particular note, small physiologically relevant increases in free calcium ($[Ca^{2+}]_i$) in permeabilized RBL-2H3 cells,[135] and low nonsecretory concentrations of thapsigargin in intact RBL-2H3 cells[138] cause significant enhancement of ERK activity. In both cases increases in $[Ca^{2+}]_i$ correlate with ERK activity. Also of note, the stimulation of ERK by phorbol 12-myristate 13-acetate is not associated with secretion, stimulation of PLC or increase in $[Ca^{2+}]_i$[135,138] (and our unpublished

observations). The effects of "chemical stressors" on JNK and p38 MAP kinases have not been systematically tested, but anisomycin does selectively activate p38 MAP kinase in RBL-2H3 cells.[137]

ERK-Activating Pathways in Mast Cells

The aforementioned studies with calcium in permeabilized cells or thapsigargin and phorbol ester in intact RBL-2H3 cells strongly suggest that increases in $[Ca^{2+}]_i$ or activation of protein kinase C alone stimulate the ERK pathway. Consistent with this conclusion, the protein kinase C inhibitor Ro31-7549 selectively blocks activation by phorbol ester but not the activation by thapsigargin.[138] Kinetic analysis of responses to antigen suggest that ERK-2 is transiently activated by protein kinase C and by a more sustained signal that is protein kinase C independent.[138] Candidates for transducing this signal include Shc and Vav, which are downstream of the tyrosine kinases, Lyn and Syk. Both Shc and Vav are rapidly phosphorylated following cross-linking of FcεRI in a Syk-dependent[136,145,146] and protein kinase C-independent[138] manner. Current evidence, as discussed in the next paragraph, points to separate functions for Shc and Vav, each mediating the ERK and JNK cascades, respectively.

The interactions of Lyn and Syk with FcεRI and the role of immunoreceptor tyrosine-based activation motifs (ITAMs) in these interactions are described in other chapters of this volume. At least three cascades are activated through Syk: the calcium/protein kinase C cascade via PLCγ,[147,148] the ERK cascade via Shc/Grb-2/SOS and Ras,[61,136] and JNK via Vav and Rac-1.[61] These have been established in RBL-2H3 cells by expression of a dominant negative Syk that blocks calcium signaling,[147] Shc and Vav phosphorylation and activation of ERK-2[136]; in studies with a Syk-negative RBL-2H3 variant in which antigen-induced phosphorylation of PLCγ1 and γ2 and calcium response can be reconstituted by expression of Syk[148]; and by transfection of a chimeric Tac receptor (a hybrid of the cytosolic domain of FcεRIγ and extracellular domain of biotinylated IL-2 receptor), a membrane-targeted SOS, and an onco-Vav in COS-7 cells.[61] To elaborate further on these studies, a Shc/Grb-2/SOS/Ras/ERK pathway is probable because expression of the dominant negative Syk in RBL-2H3 cells not only blocks Shc phosphorylation but disrupts the association of Shc with Grb-2 and Grb-2 with SOS.[146] Also, expression of the membrane-targeted SOS in COS-7 cells stimulates Ras GTPase activity without stimulating the GTPase activity of other small G proteins.[61] The alternative candidate, Vav, which as noted is tyrosine phosphorylated in stimulated RBL-2H3 cells[145] in a Syk-dependent manner,[136] associates with phosphorylated β- and γ-chains of FcεRI as a multimeric signaling complex that includes Grb-2, Raf-1, and ERK-2.[149]

There is considerable overlap in the responses evoked by the tyrosine kinase receptor c-Kit and FcεRI. Stimulation of mast cells via FcεRI suppresses c-Kit-mediated proliferation whereas c-Kit ligand enhances FcεRI-mediated release of inflammatory mediators and by itself stimulates such release.[150] The c-Kit-mediated pathway(s) for activation of ERK is uncertain but the general picture is as follows. In several types of cells, the c-Kit ligand induces dimerization and autophosphorylation of the Kit receptor, which then interacts with SH-2-containing proteins that include the p85 subunit of PI-3 kinase, PLCγ1, and the protein tyrosine phosphatase Shp-1, but not Ras-GAP. Shc, Vav, Raf-1, and Ras are phosphorylated or activated as well.[144,151]

We noted earlier the Kit-mediated activation of PI-3 kinase and Ras in mast cells. Therefore, potential models for activation of ERK in mast cells could be the association of phosphorylated Shc or Vav with Grb-2/SOS complexes and subsequently activation of the Ras/Raf/MEK/ERK cascade. Studies in BMMC from mice with spontaneous mutations in genes that encode the c-Kit receptor tyrosine kinase and Shp-1 indicate that this tyrosine phosphatase negatively regulates activation of ERK-1 and tyrosine phosphorylation of Shc.[151] The study demonstrates that c-Kit-mediated cell differentiation is dependent on a finely tuned balance between tyrosine kinase and tyrosine phosphatase activities. The conundrum from this study, however, was that formation of the Shc/Grb-2 complex was not apparent during ERK-1 activation (Vav was not investigated). As demonstrated by genetic manipulations in c-Kit-negative BMMC, most c-Kit-mediated events are dependent on phosphorylation of either or both Tyr^{719} and Tyr^{821} in the kinase domain of c-Kit.[144] Point mutations (i.e., phenylalanine substitution) of these tyrosine sites indicate overlapping (for cell adhesion) and divergent (for adhesion and cell proliferation) signals emanating from c-Kit.[144,152] This study raised another conundrum in that neither mutation impaired activation of ERK-1 and -2 while the Tyr^{719} mutation partially reduced Ras activation. The mechanisms for stimulating Ras and its role in stimulating the ERKs are, therefore, unclear with respect to c-Kit.

JNK and p38 MAP Kinase-Activating Pathways

The information on these pathways is incomplete, although the connections, PI-3 kinase >> MEKK > JNKK > JNK, have been established in MC/9 cells,[141] and FcεRI > Syk > Vav > Rac1 >>> JNK has been established by cotransfection experiments in COS-7 cells.[61] The former connection was demonstrated in studies with the PI-3 kinase inhibitor wortmannin (see following). The latter connection was demonstrated by expression of onco-Vav, which stimulated GTPase activity of Rac-1 specifically without affecting the activities of Ras and other small G proteins (see also previous section).

The mechanisms of activation of JNK and p38 MAP kinase are clearly different from those of ERK in mast cells. One difference is that wortmannin suppresses activation of JNK, but not of ERK, and weakly suppresses activation of p38 MAP kinase in antigen-stimulated MC/9 cells.[141] Suppression of JNK activation occurs at exactly the concentrations of wortmannin that would be expected to inhibit PI-3 kinase. Another difference is that BMMC from Bruton's tyrosine kinase (Btk) -defective immunodeficient mice or btk-deficient mice exhibit severely defective JNK activation and partially defective p38 activation in response to FcεRI cross-linking, IL-3, or c-Kit ligand. These activations were restored by transfection of defective cells with wild-type *btk*. Activation of the ERKs are unaffected by these mutations.[143]

Btk is a member of the Tec family tyrosine kinases, all of which contain, in progression from the carboxy- to amino-termini, Pleckstrin, Tec homology, three SH-3, SH-2, and kinase domains. Various conditions related to defects in the *btk* gene (X-linked agammaglobulinemia and immunodeficiency *xid*) and to *btk* knockout mice indicate that Btk is essential for B cell function and development.[143] Upon FcεRI or BCR cross-linking, Btk is tyrosine phosphorylated[153] by Lyn[154,155] and subsequently undergoes autophosphorylation and activation.[156] Btk is negatively regulated by protein kinase C in BMMC.[157]

The mechanisms by which PI-3 kinase and Btk apparently regulate JNK, and partially so p38 MAP kinase, are uncertain. Upsteam activators of MEKK include the Rho family GTPases, Rac-1 and Cdc42, which bind to PI-3 kinase.[158] FcεRI-mediated activation of PAK65 is enhanced in the *btk*-mutated cells, and the speculation is that Btk works upstream of Rac-1 and Cdc42 to activate PAK-65 (see Fig. 18.1) and thereafter MEKK, JNKK, and JNK.[143]

Roles of MAP Kinases in Mast Cells

Degranulation

The correlations between activation of ERK-2 (then designated p42^{mapk}) and secretion of granules following stimulation of RBL-2H3 cells with various secretagogues led to early speculation that both events were related.[133,134] Subsequent studies revealed discrepancies that make this view untenable. The most striking discrepancies were noted with quercetin[135] and the glucocorticoid dexamethasone,[159] both of which inhibit the ERK pathway at the level of Raf-1 or Ras at relatively low concentrations (see following section: Potential Therapeutic Targets). Concentrations of quercetin (30 μM) and dexamethasone (10 nM) that completely suppressed ERK-2 activation only partially

suppressed secretion. Other discrepancies were noted upon comparing these two responses to various stimulants[135] or by selectively blocking these responses with the MEK inhibitor, PD 098059,[137] and the protein kinase C inhibitor, Ro31-7549.[138] PD 098059 has minimal effects on secretion while Ro31-7549 totally blocks secretion and transiently delays ERK-2 activation in antigen-stimulated RBL-2H3 cells.

The published data indicate that, at least in RBL-2H3 cells, elevation of $[Ca^{2+}]_i$ and activation of protein kinase C are necessary and sufficient signals for secretion[160] but that elevation of $[Ca^{2+}]_i$ and activation of ERK alone are insufficient signals for secretion.[138] The latter condition can be achieved with low non-secretory concentrations of thapsigargin. With respect to other MAP kinases, SB 203580 neither impairs nor enhances antigen-induced secretion even though this compound inhibits stimulation of p38 MAP kinase and enhances stimulation of ERK.[137] Finally, wortmannin only partially suppresses antigen-induced secretion at concentrations that block PI-3 kinase and JNK activations (Cissel and Beaven, unpublished data).

Activation of Cytosolic PLA$_2$ and Release of Arachidonic Acid

In contrast to the story on degranulation, studies with quercetin,[135] dexamethasone,[159] PD 098059, and SB 203580[137] indicate close correlations between activation of ERK-2 and release of arachidonic acid in RBL-2H3 cells. These correlations are associated with ERK-dependent phosphorylation of cytosolic PLA$_2$, which is known to increase intrinsic cytosolic PLA$_2$ activity.[23] The only other necessary signal for the release of arachidonic acid is an elevation of $[Ca^{2+}]_i$, which does not activate cPLA$_2$ directly but rather promotes association of cPLA$_2$ with the membrane fraction and, presumably, its access to membrane substrates.[135] Others have shown that the cytosolic PLA$_2$ binds reversibly to the nuclear membrane in a calcium-dependent manner in RBL-2H3 cells.[161] As further support for the notion that ERK phosphorylation of PLA$_2$ and calcium-induced translocation to the membrane regulate release of arachidonic acid, calcium-mobilizing agents such as A23187 and thapsigargin stimulate substantial release of arachidonic acid whereas phorbol 12-myristate 13-acetate does not, even though all agents are potent stimulants of ERK.[135]

As previously noted, the ERK pathway can be regulated by protein kinase C, which may further potentiate activation of the ERK pathway in cells stimulated with antigen (via PLC) or calcium-mobilizing agents (via PLD). The transient inhibition of ERK activation by the protein kinase C inhibitor, Ro31-7549, in antigen-stimulated RBL-2H3 cells is accompanied by an equally transient delay in release of arachidonic acid.[138] The potential pathways for regulating PLA_2 activity in RBL-2H3 cells are depicted in Fig. 18.4.

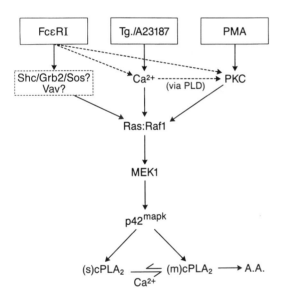

FIGURE 18.4. Potential pathways for the activation of cytosolic PLA_2 through the ERK cascade by FcεRI, thapsigargin (*Tg.*), or calcium ionophore (*A23187*), and phorbol 12-myristate 13-acetate (*PMA*) in RBL-2H3 cells. The scheme depicts convergent pathways to *Ras/Raf1*, a common pathway via p42[mapk] (otherwise referred to as ERK-2 in text), and an equilibrium between soluble (*s*) and membrane-associated (*m*) cytosolic PLA_2 (cPLA_2), under the regulation of calcium, for release of arachidonic acid (*A.A.*). Phorbol 12-myristate 13-acetate, while activating ERK, does not increase cytosolic calcium levels and does not stimulate arachidonic acid release, as discussed in the text.

While the evidence for ERK regulation of arachidonic acid release in RBL-2H3 cells appears compelling, several points should be noted. There is an increasing body of evidence that although MAP kinase may be essential, it is not sufficient for the catalytic action of cytosolic PLA_2 on membrane lipids. Some authors note additional effects through calcium,[162] while others report that the PLD pathway is involved in FcεRI- and c-Kit-mediated arachidonic acid release and eicosanoid production,[163,164] possibly as a consequence of the action of diglyceride lipase on PLD-generated diglycerides.[165] The presence of other forms of PLA_2 (secretory and calcium-independent) in mast cells[166] also obscures the issues, although these forms do not appear to be involved in FcεRI-mediated arachidonic release in RBL-2H3 cells.[138] Finally, various cell types may utilize different MAP kinases or even other kinases to phosphorylate cytosolic PLA_2. Thrombin-induced release of arachidonic acid from platelets, for example, is not inhibited by PD 098059[167] or SB 203580,[168] even though both ERK-2 and p38 MAP kinase phosphorylate cytosolic PLA_2 in vitro and in vivo.

Activation of Transcription Factors and Gene Transcription

Of the various regulators of gene transcription that are thought to be regulated by MAP kinases (see Structural Motifs and Downstream Substrates), relatively few have been linked directly to MAP kinases in mast cells. Also, no coherent picture has yet emerged in studies of mast cells. JNK has been linked to the TNF-α gene promotor in MC/9 cells.[141] Expression of an inhibitory mutant of JNK-2 or exposure to wortmannin significantly suppressed FcεRI-mediated stimulation of JNK and TNF-α promotor-regulated luciferase gene expression, whereas both were strongly stimulated by expression of a constitutively active MEKK-1 in MC/9 cells. On the basis of studies with inhibitors, ERK and p38 MAP kinase did not appear to regulate TNF-α promotor activity in these cells. The TNF-α promotor contains AP-1, AP-2, NF-AT, Ets, NF-κB, and AP-1/ATF-related elements, and, as noted earlier, the JNK pathway is known to regulate activation of c-Jun,

ATF-2, and NF-κB. A role for NF-κB was considered unlikely, however, as expression of a dominant negative NF-κB did not significantly affect stimulation of the promotor-driven luciferase expression system in MC/9 cells.[141] It is possible, nevertheless, that different cytokine genes are regulated by different MAP kinases. Activation of a recently described NF-AT family member that regulates IL-5[169] and chemokine[170] gene expression in a cultured mast cell line (CPII) appears to be Ras dependent.[170] The overexpression of Raf-1[171,172] or ERK-1[173] results in enhanced expression of a variety of cytokine genes in T cells and macrophages,[171,173] the inactivation of the inhibitory transcription factor, IκB,[172] and the enhanced DNA-binding activity of NF-κB, NF-AT, and AP-1.[173]

The studies with c-Kit deficient in BMMC and induction of early response genes[144] indicate that c-Kit ligand induction of these genes (c-*myc*, c-*myb*, c-*fos*, and c-*jun*B), as well as the activation of Ras and ERK-1 and -2 (identified as p44^{mapk} and p42^{mapk}, respectively), is dependent on c-Kit. Interestingly, ERK activation was not affected by the mutations tested (Tyr^{719} and Tyr^{821}; see ERK-Activating Pathways in Mast Cells), while substitution of Tyr^{719} by phenylalanine reduced expression of c-*fos* and c-*jun*B mRNA, and the Tyr^{821} substitution had no effect on any of the early response genes. Thus, induction of c-*fos* and c-*jun*B genes are likely regulated through mechanisms other than, or in addition to, the ERK system, whereas induction of c-*myc* and c-*myb* might be related to the ERKs.

Production of Cytokines

Stimulated mast cells produce a variety of cytokines, which include the interleukins-1,-3, -4,-5, and -6, as well as TNF-α and GM-CSF.[174,175] Typically, increased expression of cytokine mRNA and protein is detectable 30 min to several hours after addition of stimulant.[176] These cytokines, particularly TNF-α, are thought to mediate pathological inflammatory reactions[175] and protective responses to bacterial infection.[177] The production and release of TNF-α are regulated through signals

transduced by calcium and protein kinase C, although there are indications that additional FcεRI-mediated signals may operate for optimal production of TNF-α in cultured RBL-2H3 mast cells.[178] The MAP kinases are enticing candidates for mediating these additional signals, but data so far are limited.[137,141]

In RBL-2H3 cells the activation of the ERK-2 pathway[136,146] persists through the period when production of TNF-α would be most apparent,[138] and inhibition of this pathway by PD 098059 suppresses TNF-α production to the same extent.[137] The p38 MAP kinase inhibitor, SB 203580, has no effect on TNF-α production,[137] and wortmannin has minimal effect on this production (Baumgartner et al., unpublished results). These observations suggest that TNF-α production is ERK-2 dependent. In MC/9 cells, in contrast, TNF-α promotor activity is regulated by JNK (see previous section) rather than by the ERKs or p38 MAP kinases, as this activity is not diminished by the MEK or p38 MAP kinase inhibitors.[141] The role of MAP kinases in cytokine regulation clearly requires further scrutiny. An intriguing question, of course, is whether this regulation varies for different cytokines and cell phenotypes.

Cell Growth, Differentiation, and Apoptosis

Results from several studies that have been mentioned earlier provide circumstantial evidence of participation of the MAP kinases in mast cell differentiation and apoptosis. We have noted the selective activation of JNK and p38 MAP kinase by IL-3 withdrawal and apoptosis in BMMC.[143] The studies with BMMC from *xid* or *btk*-deficient mice also support the view that JNK and p38 MAP kinase regulate BMMC apoptosis.[143] With respect to the ERKs, the studies with BMMC from mice with mutations in genes that encode c-Kit and the tyrosine phosphatase, SHP-1, suggest that the ERKs are associated with mast cell differentiation and survival rather than proliferation.[151] These findings are consistent with the general scenarios that were outlined earlier in this chapter but, here again, these topics warrant further scrutiny in different types of mast cells.

Potential Therapeutic Targets

It should be apparent from the foregoing remarks that several agents have utility in investigating the roles of MAP kinases in cell function. Some of them are established (glucocorticoids) or potential (SB 203580) therapeutic agents or have known antiinflammatory activity (quercetin). On this basis, it could be argued that the MAP kinases might provide a rich field for development of new therapeutic agents for, among others, immunological and inflammatory disorders. The multiplicity of actions of some of these agents (e.g., glucocorticoids and quercetin) is well recognized, but the fact that these agents were not designed originally as MAP kinase inhibitors, but were found later to target specific components of the MAP kinase pathways, makes it probable that even better inhibitors could be designed. For mast cells and basophils, the interest is likely to focus on the role of MAP kinases and the effects of inhibitors on stimulated production of cytokines. A widely held view is that mast cell-derived cytokines are an important component of mast cell-related inflammatory disorders.

mast cells. The ability of strong stimuli to override the inhibitory actions of PD 098059 and SB 203580 means that dose–response curves for stimulant and inhibitor are necessary for full interpretation of the data. Other considerations include the fact that effects of stimulants may vary with the stage of maturity of cells or cell lines and nature of costimulants. With respect to the mast cell, the potential interactions of integrins, cytokine receptors, and c-Kit on FcεRI-mediated signaling remains largely unexplored territory.

It should be apparent from this review that information on MAP kinases in mast cells is fragmentary. There is little question at this point that most, if not all, ligands that promote growth and differentiation of mast cells also stimulate the MAP kinases. The notion that the MAP kinases are linked in some way to mast cell survival and differentiation is probably correct, on the basis of studies in other types of cells, but information on the linkages is lacking.

The evidence that these kinases also regulate activation cytosolic PLA_2 and generation of some cytokines is also enticing, but questions remain as to which MAP kinases are linked to these events. Nevertheless, the information gained so far provide us with good prospects for answers to some of these questions.

Conclusions

We hope that lessons learned from this broad review of MAP kinases can be applied to studies of the mast cell. As early studies were restricted to one (e.g., ERK) or two (e.g., ERK and JNK) MAP kinases, these studies provide incomplete pictures. Even now, there may be as yet undiscovered MAP kinases and current studies may need reinterpretation. It is now clear with respect to cell growth and differentiation, for example, that ERK does not represent the whole story. Full accounting of the MAP kinases also requires kinetic data, for example, whether responses were short- or long lived. In this regard, technical problems such as inadequate suppression of phosphatases were not fully appreciated in the early studies on

References

1. Fureder W, Agis H, Willheim M, et al. Differential expression of complement receptors on human basophils and mast cells: evidence for mast cell heterogeneity and CD88/C5aR expression on skin mast cells. J Immunol 1995;155:3152–3160.
2. Smith TJ, Weis JH. Mucosal T cells and mast cells share common adhesion receptors. Trends Immunol 1996;17:60–63.
3. Beaven MA, Baumgartner RA. Downstream signals initiated in mast cells by FcεRI and other receptors. Curr Opin Immunol 1996;8:766–772.
4. Marshall CJ. Specificity of receptor tyrosine kinase signaling: transient versus sustained extracellular signal-regulated kinase activation. Cell 1995;80:179–185.

5. Campbell JS, Seger R, Graves J, et al. The MAP kinase cascade. Recent Prog Horm Res 1995;50:131–159.

6. Bokemeyer D, Sorokin A, Dunn MJ. Multiple intracellular MAP kinase signaling cascades. Kidney Int 1996;49:1187–1198.

7. Zhou G, Bao ZQ, Dixon JE. Components of a new human protein kinase signal transduction pathway. J Biol Chem 1995;270:12665–12669.

8. Abe J, Kusuhara M, Ulevitch RJ, et al. Big mitogen-activated protein kinase 1 (BMK1) is a redox-sensitive kinase. J Biol Chem 1996;271:16586–16590.

9. Ferrell JE. Tripping the switch fantastic: how a protein kinase can convert graded inputs into switch-like outputs. Trends Biochem Sci 1996;21:460–466.

10. Cano E, Mahadevan LC. Parallel signal processing among mammalian MAPKs. Trends Biochem Sci 1995;20:117–122.

11. Fukunaga R, Hunter T. MNK1, a new MAP kinase-activated protein kinase, isolated by a novel expression screening method for identifying protein kinase substrates. EMBO J 1997;16:1921–1933.

12. Waskiewicz AJ, Flynn A, Proud CG, et al. Mitogen-activated protein kinases activate the serine/threonine kinases Mnk1 and Mnk2. EMBO J 1997;16:1909–1920.

13. Ludwig S, Engel K, Hoffmeyer A, et al. 3pK, a novel mitogen-activated protein (MAP) kinase-activated protein kinase, is targeted by three MAP kinase pathways. Mol Cell Biol 1996;16:6687–6697.

14. Jiang Y, Li Z, Schwarz EM, et al. Structure-function studies of p38 mitogen-activated protein kinase. Loop 12 influences substrate specificity and autophosphorylation, but not upstream kinase selection. J Biol Chem 1997;272:11096–11102.

15. Hunter T. Protein kinases and phosphatases: the yin and yang of protein phosphorylation and signaling. Cell 1995;80:225–236.

16. L'Allemain G. Deciphering the MAP kinase pathway. Prog Growth Factor Res 1994;5:291–334.

17. Keyse SM. An emerging family of dual specificity MAP kinase phosphatases. Biochim Biophys Acta 1995;1265:152–160.

18. Karin M, Hunter T. Transcriptional control by protein phosphorylation: signal transmission from the cell surface to the nucleus. Curr Biol 1995;5:747–757.

19. Denhardt DT. Signal-transducing protein phosphorylation cascades mediated by Ras/Rho proteins in the mammalian cell: the potential for multiplex signalling. Biochem J 1996;318:729–747.

20. Ahn NG, Weiel JE, Chan CP, et al. Identification of multiple epidermal growth factor-stimulated protein serine/threonine kinases from Swiss 3T3 cells. J Biol Chem 1990;265:11487–11494.

21. Rossomando AJ, Payne DM, Weber MJ, et al. Evidence that pp42, a major tyrosine kinase target protein, is a mitogen-activated serine/threonine protein kinase. Proc Natl Acad Sci USA 1989;86:6940–6943.

22. Miyasaka T, Chao MV, Sherline P, et al. Nerve growth factor stimulates a protein kinase in PC-12 cells that phosphorylates microtubule-associated protein-2. J Biol Chem 1990;265:4730–4735.

23. Lin LL, Wartmann M, Lin AY, et al. cPLA$_2$ is phosphorylated and activated by MAP kinase. Cell 1993;72:269–278.

24. Sturgill TW, Ray LB, Erikson E, et al. Insulin-stimulated MAP-2 kinase phosphorylates and activates ribosomal protein S6 kinase II. Nature (Lond) 1988;334:715–718.

25. Takishima K, Griswold-Prenner I, Ingebritsen T, et al. Epidermal growth factor (EGF) receptor T669 peptide kinase from 3T3-l1 cells is an EGF-stimulated "MAP" kinase. Proc Natl Acad Sci USA 1991;88:2520–2524.

26. Gille H, Kortenjann M, Thomae O, et al. ERK phosphorylation potentiates Elk-1-mediated ternary complex formation and transactivation. EMBO J 1995;14:951–962.

27. Schlaepfer DD, Hunter T. Signal transduction from the extracellular matrix—a role for the focal adhesion protein-tyrosine kinase FAK. Cell Struct Funct 1996;21:445–450.

28. Derijard B, Hibi M, Wu IH, et al. JNK1: a protein kinase stimulated by UV light and Ha-Ras that binds and phosphorylates the c-Jun activation domain. Cell 1994;76:1025–1037.

29. Kyriakis JM, Avruch J. Sounding the alarm: protein kinase cascades activated by stress and inflammation. J Biol Chem 1996;271:24313–24316.

30. Rivera J, Beaven MA. Role of protein kinase C isozymes in secretion. In: Parker P, Dekker L, eds. Protein Kinase C. (Molecular Biology Intelligence Unit Series.) Austin, TX: RG Landes, 1997:131–164.

31. Hunter T. When is a lipid kinase not a lipid kinase? When it is a protein kinase. Cell 1995;83:1–4.

32. van Biesen T, Hawes BE, Luttrell DK, et al. Receptor-tyrosine-kinase- and Gβγ-mediated MAP kinase activation by a common signalling pathway. Nature (Lond) 1995;376:781–784.

33. Touhara K, Hawes BE, van Biesen T, et al. G protein βγ subunits stimulate phosphorylation of Shc adapter protein. Proc Natl Acad Sci USA 1995;92:9284–9287.

34. Winston LA, Hunter T. Intracellular signalling: putting JAKs on the kinase MAP. Curr Biol 1996;6:668–671.

35. Burack WR, Sturgill TW. The activating dual phosphorylation of MAPK by MEK is non-processive. Biochemistry 1997;36:5929–5933.

36. Zhang F, Strand A, Robbins D, et al. Atomic structure of the MAP kinase ERK2 at 2.3 Å resolution. Nature (Lond) 1994;367:704–711.

37. Cobb MH, Goldsmith EJ. How MAP kinases are regulated. J Biol Chem 1995;270:14843–14846.

38. Seger R, Krebs EG. The MAPK signaling cascade. FASEB J 1995;9:726–735.

39. Kouhara H, Hadari YR, Spivak-Kroizman T, et al. A lipid-anchored Grb2-binding protein that links FGF-receptor activation to the Ras/MAPK signaling pathway. Cell 1997;89:693–702.

40. Batzer AG, Rotin D, Urena JM, et al. Hierarchy of binding sites for Grb2 and Shc on the epidermal growth factor receptor. Mol Cell Biol 1994;14:5192–5201.

41. Skolnik EY, Batzer A, Li N, et al. The function of GRB2 in linking the insulin receptor to Ras signaling pathways. Science 1993;260:1953–1955.

42. Katzav S, Martin-Zanca D, Barbacid M. *vav*, a novel human oncogene derived from a locus ubiquitously expressed in hematopoietic cells. EMBO J 1989;8:2283–2290.

43. Collins TL, Deckert M, Altman A. Views on Vav. Immunol Today 1997;18:221–225.

44. Gulbins E, Coggeshall KM, Baier G, et al. Direct stimulation of Vav guanine nucleotide exchange activity for Ras by phorbol esters and diglycerides. Mol Cell Biol 1994;14:4749–4758.

45. Tocque B, Delumeau I, Parker F, et al. Ras-GTPase activating protein (GAP): a putative effector for Ras. Cell Signal 1997;9:153–158.

46. van Dijk MCM, Muriana FJG, de Widt J, et al. Involvement of phosphatidylcholine-specific phospholipase C in platelet-derived growth factor-induced activation of the mitogen-activated protein kinase pathway in rat-1 fibroblasts. J Biol Chem 1997;272:11011–11016.

47. Cowen DS, Sowers RS, Manning DR. Activation of a mitogen-activated protein kinase (ERK2) by the 5-hydroxytryptamine1A receptor is sensitive not only to inhibitors of phosphatidylinositol 3-kinase, but to an inhibitor of phosphatidylcholine hydrolysis. J Biol Chem 1996;271:22297–22300.

48. Grammer TC, Blenis J. Evidence for MEK-independent pathways regulating the prolonged activation of the ERK-MAP kinases. Oncogene 1997;14:1635–1642.

49. Hawes BE, van Biesen T, Koch WJ, et al. Distinct pathways of Gi- and Gq-mediated mitogen-activated protein kinase activation. J Biol Chem 1995;270:17148–17153.

50. van Biesen T, Hawes BE, Raymond JR, et al. G$_o$-protein α-subunits activate mitogen-activated protein kinase via a novel protein kinase C-dependent mechanism. J Biol Chem 1996; 271:1266–1269.

51. Kyriakis JM, Banerjee P, Nikolakaki E, et al. The stress-activated protein kinase subfamily of c-Jun kinases. Nature (Lond) 1994;369:156–160.

52. Sluss HK, Barrett T, Derijard B, et al. Signal transduction by tumor necrosis factor mediated by JNK protein kinases. Mol Cell Biol 1994; 14:8376–8384.

53. Su B, Jacinto E, Hibi M, et al. JNK is involved in signal integration during costimulation of T lymphocytes. Cell 1994;77:727–736.

54. Yan M, Dai T, Deak JC, et al. Activation of stress-activated protein kinase by MEKK1 phosphorylation of its activator SEK1. Nature (Lond) 1994;372:798–800.

55. Minden A, Lin A, McMahon M, et al. Differential activation of ERK and JNK mitogen-activated protein kinases by Raf-1 and MEKK. Science 1994;266:1719–1723.

56. Lin A, Minden A, Martinetto H, et al. Identification of a dual specificity kinase that activates the Jun kinases and p38-Mpk2. Science 1995; 268:286–290.

57. Minden A, Lin A, Claret FX, et al. Selective activation of the JNK signaling cascade and c-Jun transcriptional activity by the small GTPases Rac and Cdc42Hs. Cell 1995;81:1147–1157.

58. Coso OA, Chiariello M, Yu JC, et al. The small GTP-binding proteins Rac1 and Cdc42 regu-

late the activity of the JNK/SAPK signaling pathway. Cell 1995;81:1137–1146.

59. Lopez-Ilasaca M, Li W, Uren A, et al. Requirement of phosphatidylinositol-3 kinase for activation of JNK/SAPKs by PDGF. Biochem Biophys Res Commun 1997;232:273–277.

60. Crespo P, Schuebel KE, Ostrom AA, et al. Phosphotyrosine-dependent activation of Rac-1 GDP/GTP exchange by the *vav* proto-oncogene product. Nature (Lond) 1997;385: 169–172.

61. Teramoto H, Salem P, Robbins KC, et al. Tyrosine phosphorylation of the *vav* proto-oncogene product links FcεRI to the Rac1-JNK pathway. J Biol Chem 1997;272:10751–10755.

62. Rouse J, Cohen P, Trigon S, et al. A novel kinase cascade triggered by stress and heat shock that stimulates MAPKAP kinase-2 and phosphorylation of the small heat shock proteins. Cell 1994;78:1027–1037.

63. Han J, Lee JD, Bibbs L, et al. A MAP kinase targeted by endotoxin and hyperosmolarity in mammalian cells. Science 1994;265:808–811.

64. Lee JC, Laydon JT, McDonnell PC, et al. A protein kinase involved in the regulation of inflammatory cytokine biosynthesis. Nature (Lond) 1994;372:739–746.

65. Zervos AS, Faccio L, Gatto JP, et al. Mxi2, a mitogen-activated protein kinase that recognizes and phosphorylates Max protein. Proc Natl Acad Sci USA 1995;92:10531–10534.

66. Jiang Y, Chen C, Li Z, et al. Characterization of the structure and function of a new mitogen-activated protein kinase (p38β). J Biol Chem 1996;271:17920–17926.

67. Li Z, Jiang Y, Ulevitch RJ, et al. The primary structure of p38γ: a new member of p38 group of MAP kinases. Biochem Biophys Res Commun 1996;228:334–340.

68. Wang Z, Harkins PC, Ulevitch RJ, et al. The structure of mitogen-activated protein kinase p38 at 2.1-Å resolution. Proc Natl Acad Sci USA 1997;94:2327–2332.

69. Wilson KP, Fitzgibbon MJ, Caron PR, et al. Crystal structure of p38 mitogen-activated protein kinase. J Biol Chem 1996;271:27696–27700.

70. Derijard B, Raingeaud J, Barrett T, et al. Independent human MAP kinase signal transduction pathways defined by MEK and MKK isoforms. Science 1995;267:682–685.

71. Robinson MJ, Cheng M, Khokhlatchev A, et al. Contributions of the mitogen-activated protein (MAP) kinase backbone and phosphorylation loop to MEK specificity. J Biol Chem 1996; 271:29734–29739.

72. Brunet A, Pouyssegur J. Identification of MAP kinase domains by redirecting stress signals into growth factor responses. Science 1996;272:1652–1655.

73. Minden A, Lin A, Smeal T, et al. c-Jun N-terminal phosphorylation correlates with activation of the JNK subgroup but not the ERK subgroup of mitogen-activated protein kinases. Mol Cell Biol 1994;14:6683–6688.

74. David M, Petricoin E III, Benjamin C, et al. Requirement for MAP kinase (ERK2) activity in interferon alpha- and interferon beta-stimulated gene expression through STAT proteins. Science 1995;269:1721–1723.

75. Freshney NW, Rawlinson L, Guesdon F, et al. Interleukin-1 activates a novel protein kinase cascade that results in the phosphorylation of Hsp27. Cell 1994;78:1039–1049.

76. Kolch W, Heidecker G, Kochs G, et al. Protein kinase C alpha activates RAF-1 by direct phosphorylation. Nature (Lond) 1993;364:249–252.

77. Enslen H, Tokumitsu H, Stork PJS, et al. Regulation of mitogen-activated protein kinases by a calcium/calmodulin-dependent protein kinase cascade. Proc Natl Acad Sci USA 1996;93: 10803–10808.

78. Ueda Y, Hirai SI, Osada SI, et al. Protein kinase C activates the MEK-ERK pathway in a manner independent of Ras and dependent on Raf. J Biol Chem 1996;271:23512–23519.

79. Ahmed S, Lee J, Kozma R, et al. A novel functional target for tumor-promoting phorbol esters and lysophosphatidic acid. The p21^{rac}-GTPase activating protein n-chimaerin. J Biol Chem 1993;268:10709–10712.

80. Dudley DT, Pang L, Decker SJ, et al. A synthetic inhibitor of the mitogen-activated protein kinase cascade. Proc Natl Acad Sci USA 1995;92:7686–7689.

81. Alessi DR, Cuenda A, Cohen P, et al. PD 098059 is a specific inhibitor of the activation of mitogen-activated protein kinase kinase in vitro and in vivo. J Biol Chem 1995;270:27489–27494.

82. Pang L, Sawada T, Decker SJ, et al. Inhibition of MAP kinase kinase blocks the differentiation of PC-12 cells induced by nerve growth factor. J Biol Chem 1995;270:13585–13588.

83. Pumiglia KM, Decker SJ. Cell cycle arrest mediated by the MEK/mitogen-activated protein kinase pathway. Proc Natl Acad Sci USA 1997;94:448–452.

84. Lee JC, Adams JL. Inhibitors of serine/threonine kinases. Curr Opin Biotechnol 1995;6:657–661.

85. Cuenda A, Rouse J, Doza YN, et al. SB 203580 is a specific inhibitor of a MAP kinase homologue which is stimulated by cellular stresses and interleukin-1. FEBS Lett 1995;364:229–233.

86. McLaughlin MM, Kumar S, McDonnell PC, et al. Identification of mitogen-activated protein (MAP) kinase-activated protein kinase-3, a novel substrate of CSBP p38 MAP kinase. J Biol Chem 1996;271:8488–8492.

87. Beyaert R, Cuenda A, Vanden Berghe W, et al. The p38/RK mitogen-activated protein kinase pathway regulates interleukin-6 synthesis response to tumor necrosis factor. EMBO J 1996;15:1914–1923.

88. Young PR, McLaughlin MM, Kumari S, et al. Pyridinyl imidazole inhibitors of p38 mitogen-activated protein kinase bind in the ATP site. J Biol Chem 1997;272:12116–12121.

89. Saklatvala J, Rawlinson L, Waller RJ, et al. Role for p38 mitogen-activated protein kinase in platelet aggregation caused by collagen or a thromboxane analogue. J Biol Chem 1996; 271:6586–6589.

90. Shapiro L, Dinarello CA. Osmotic regulation of cytokine synthesis in vitro. Proc Natl Acad Sci USA 1995;92:12230–12234.

91. Badger AM, Bradbeer JN, Votta B, et al. Pharmacological profile of SB 203580, a selective inhibitor of cytokine suppressive binding protein/p38 kinase, in animal models of arthritis, bone resorption, endotoxin shock and immune function. J Pharmacol Exp Ther 1996;279:1453–1461.

92. Welham MJ, Duronio V, Schrader JW. Interleukin-4-dependent proliferation dissociates p44erk-1, p42erk-2, and p21ras activation from cell growth. J Biol Chem 1994;269:5865–5873.

93. Franklin RA, Tordai A, Patel H, et al. Ligation of the T cell receptor complex results in activation of the Ras/Raf-1/MEK/MAPK cascade in human T lymphocytes. J Clin Invest 1994; 93:2134–2140.

94. Alberola-Ila J, Forbush KA, Segar R, et al. Selective requirement for MAP kinase activation in thymocyte differentiation. Nature (Lond) 1995;373:620–623.

95. Sen J, Kapeller R, Fragoso R, et al. Intrathymic signals in thymocytes are mediated by p38 mitogen-activated protein kinase. J Immunol 1996;156:4535–4538.

96. Levin SD, Anderson SJ, Forbush KA, et al. A dominant-negative transgene defines a role for p56lck in thymopoiesis. EMBO J 1993;12:1671–1680.

97. Li W, Whaley CD, Mondino A, et al. Blocked signal transduction to the ERK and JNK protein kinases in anergic CD4[+] T cells. Science 1996;271:1272–1276.

98. Fields PE, Gajewski TF, Fitch FW. Blocked Ras activation in anergic CD4[+] T cells. Science 1996;271:1276–1278.

99. Gulbins E, Coggeshall KM, Baier G, et al. Tyrosine kinase-stimulated guanine nucleotide exchange activity of Vav in T cell activation. Science 1993;260:822–825.

100. Gupta S, Weiss A, Kumar G, et al. The T-cell antigen receptor utilizes Lck, Raf-1, and MEK-1 for activating mitogen-activated protein kinase. Evidence for the existence of a second protein kinase C-dependent pathway in an Lck-negative Jurkat cell mutant. J Biol Chem 1994;269:17349–17357.

101. Downward J, Graves JD, Warne PH, et al. Stimulation of p21[ras] upon T-cell activation. Nature (Lond) 1990;346:719–723.

102. Tordai A, Franklin RA, Patel H, et al. Cross-linking of surface IgM stimulates the Ras/Raf-1/MEK/MAPK cascade in human B lymphocytes. J Biol Chem 1994;269:7538–7543.

103. Franklin RA, Brodie C, Melamed I, et al. Nerve growth factor induces activation of MAP-kinase and p90rsk in human B lymphocytes. J Immunol 1995;154:4965–4972.

104. Sutherland CL, Health AW, Pelech SL, et al. Differential activation of the ERK, JNK, and p38 mitogen-activated protein kinases by CD40 and the B cell antigen receptor. J Immunol 1996;157:3381–3390.

105. Kashiwada M, Kaneko Y, Yagita H, et al. Activation of mitogen-activated protein kinases via CD40 is distinct from that stimulated by surface IgM on B cells. Eur J Immunol 1996;26:1451–1458.

106. Xia Z, Dickens M, Raingeaud J, et al. Opposing effects of ERK and JNK-p38 MAP kinases on apoptosis. Science 1995;270:1326–1331.

107. Gulbins E, Langlet C, Baier G, et al. Tyrosine phosphorylation and activation of Vav GTP/ GDP exchange activity in antigen receptor-triggered B cells. J Immunol 1994;152:2123–2129.

108. Gold MR, Crowley MT, Martin G, et al. Targets of B lymphocyte antigen receptor signal transduction include the p21ras GTPase-activating protein (GAP) and two GAP-associated proteins. J Immunol 1993;150:377–386.

109. Lazarus AH, Kawauchi K, Rapoport MJ, et al. Antigen-induced B lymphocyte activation involves the p21ras and ras-GAP signaling pathway. J Exp Med 1993;178:1765–1769.

110. Kumar G, Wang S, Gupta S, et al. The membrane immunoglobulin receptor utilizes a Shc/ Grb2/hSOS complex for activation of the mitogen-activated protein kinase cascade in a B-cell line. Biochem J 1995;307:215–223.

111. Hallett MB, Lloyds D. Neutrophil priming: the cellular signals that say "amber" but not "green". Immunol Today 1995;16:264–268.

112. El Benna J, Han J, Park JW, et al. Activation of p38 in stimulated human neutrophils: phosphorylation of the oxidase component p47^{phox} by p38 and ERK but not by JNK. Arch Biochem Biophys 1996;334:395–400.

113. Grinstein S, Furuya W. Chemoattractant-induced tyrosine phosphorylation and activation of microtubule-associated protein kinase in human neutrophils. J Biol Chem 1992;267: 18122–18125.

114. Thompson HL, Marshall CJ, Saklatvala J. Characterization of two different forms of mitogen-activated protein kinase kinase induced in polymorphonuclear leukocytes following stimulation by N-formylmethionyl-leucyl-phenylalanine or granulocyte-macrophage colony-stimulating factor. J Biol Chem 1994;269:9486–9492.

115. Torres M, Hall FL, O'Neill K. Stimulation of human neutrophils with formyl-methionyl-leucyl-phenylalanine induces tyrosine phosphorylation and activation of two distinct mitogen-activated protein-kinases. J Immunol 1993;150:1563–1577.

116. Pillinger MH, Feoktistov AS, Capodici C, et al. Mitogen-activated protein kinase in neutrophils and enucleate neutrophil cytoplasts: evidence for regulation of cell-cell adhesion. J Biol Chem 1996;271:12049–12056.

117. Worthen GS, Avdi N, Buhl AM, et al. FMLP activates Ras and Raf in human neutrophils.

Potential role in activation of MAP kinase. J Clin Invest 1994;94:815–823.

118. Obel D, Rasmussen LH, Christiansen NO. Protein kinase C subtypes in human neutrophils. Scand J Clin Lab Invest 1991;51:299–302.

119. Waterman WH, Molski TF, Huang CK, et al. Tumour necrosis factor-α-induced phosphorylation and activation of cytosolic phospholipase A$_2$ are abrogated by an inhibitor of the p38 mitogen-activated protein kinase cascade in human neutrophils. Biochem J 1996;319:17–20.

120. Yu H, Suchard SJ, Nairn R, et al. Dissociation of mitogen-activated protein kinase activation from the oxidative burst in differentiated HL-60 cells and human neutrophils. J Biol Chem 1995;270:15719–15724.

121. el Benna J, Faust LP, Babior BM. The phosphorylation of the respiratory burst oxidase component p47^{phox} during neutrophil activation. Phosphorylation of sites recognized by protein kinase C and by proline-directed kinases. J Biol Chem 1994;269:23431–23436.

122. Waterman WH, Sha'afi RI. Effects of granulocyte-macrophage colony-stimulating factor and tumour necrosis factor-α on tyrosine phosphorylation and activation of mitogen-activated protein kinases in human neutrophils. Biochem J 1995;307:39–45.

123. Nick JA, Avdi NJ, Gerwins P, et al. Activation of a p38 mitogen-activated protein kinase in human neutrophils by lipopolysaccharide. J Immunol 1996;156:4867–4875.

124. Fouda SI, Molski TF, Ashour MS, et al. Effect of lipopolysaccharide on mitogen-activated protein kinases and cytosolic phospholipase A$_2$. Biochem J 1995;308:815–822.

125. Nahas N, Waterman WH, Sha'afi RI. Granulocyte-macrophage colony-stimulating factor (GM-CSF) promotes phosphorylation and an increase in the activity of cytosolic phospholipase A$_2$ in human neutrophils. Biochem J 1996;313:503–508.

126. Pazdrak K, Schreiber D, Forsythe P, et al. The intracellular signal transduction mechanism of interleukin 5 in eosinophils: the involvement of Lyn tyrosine kinase and the Ras-Raf-1-MEK-microtubule-associated protein kinase pathway. J Exp Med 1995;181:1827–1834.

127. Bates ME, Bertics PJ, Busse WW. IL-5 activates a 45-kilodalton mitogen-activated protein (MAP) kinase and Jak-2 tyrosine kinase in human eosinophils. J Immunol 1996;156:711–718.

128. Pazdrak K, Stafford S, Alam R. The activation of the Jak-STAT 1 signaling pathway by IL-5 in eosinophils. J Immunol 1995;155:397–402.

129. Pazdrak K, Justement L, Alam R. Mechanism of inhibition of eosinophil activation by transforming growth factor-β. Inhibition of Lyn, MAP, Jak2 kinases and STAT1 nuclear factor. J Immunol 1995;155:4454–4458.

130. Araki R, Komada T, Nakatani K, et al. Protein kinase C-independent activation of Raf-1 and mitogen-activated protein kinase by leukotriene B_4 in guinea pig eosinophils. Biochem Biophys Res Commun 1995;210:837–843.

131. O'Flaherty JT, Kuroki M, Nixon AB, et al. 5-Oxo-eicosatetraenoate is a broadly active, eosinophil-selective stimulus for human granulocytes. J Immunol 1996;157:336–342.

132. Suchard SJ, Mansfield PJ, Boxer LA, et al. Mitogen-activated protein kinase activation during IgG-dependent phagocytosis in human neutrophils: inhibition by ceramide. J Immunol 1997;158:4961–4967.

133. Offermanns S, Jones SVP, Bombien E, et al. Stimulation of mitogen-activated protein kinase activity by different secretory stimuli in rat basophilic leukemia cells. J Immunol 1994;152:250–261.

134. Santini F, Beaven MA. Tyrosine phosphorylation of a mitogen activated protein (MAP) kinase-like protein occurs at a late step in exocytosis: studies with tyrosine phosphatase inhibitors and various secretagogues in RBL-2H3 cells. J Biol Chem 1993;268:22716–22722.

135. Hirasawa N, Santini F, Beaven MA. Activation of the mitogen-activated protein kinase/cytosolic phospholipase A_2 pathway in a rat mast cell line: indications of different pathways for release of arachidonic acid and secretory granules. J Immunol 1995;154:5391–5402.

136. Hirasawa N, Scharenberg A, Yamamura H, et al. A requirement for Syk in the activation of the MAP kinase/phospholipase A_2 pathway by FcεRI is not shared by a G protein-coupled receptor. J Biol Chem 1995;270:10960–10967.

137. Zhang C, Baumgartner RA, Yamada K, et al. Mitogen activated protein (MAP) kinase regulates production of TNF-α and release of arachidonic acid in mast cells: indications of communication between p38 and p42 MAP kinases. J Biol Chem 1997;272:13397–13402.

138. Zhang C, Hirasawa N, Beaven MA. Antigen activation of mitogen-activated protein kinase in mast cells through protein kinase C dependent and independent pathways. J Immunol 1997;158:4968–4975.

139. Fukamachi H, Takei M, Kawakami T. Activation of multiple protein kinases including a MAP kinase upon FcεRI cross-linking. Int Arch Allergy Immunol 1993;102:15–25.

140. Ishizuka T, Oshiba A, Sakata N, et al. Aggregation of the FcεRI on mast cells stimulates c-Jun amino-terminal kinase activity. A response inhibited by wortmannin. J Biol Chem 1996; 271:12762–12766.

141. Ishizuka T, Terada N, Gerwins P, et al. Mast cell tumor necrosis factor-α production is regulated by MEK kinases. Proc Natl Acad Sci USA 1997;94:6358–6363.

142. Tsai M, Chen RH, Tam SY, et al. Activation of MAP kinases, pp90rsk and pp70-S6 kinases in mouse mast cells by signaling through the c-kit receptor tyrosine kinase or FcεRI: rapamycin inhibits activation of pp70-S6 kinase and proliferation in mouse mast cells. Eur J Immunol 1993;23:3286–3291.

143. Kawakami Y, Miura T, Bissonnette R, et al. Bruton's tyrosine kinase regulates apoptosis and JNK/SAPK kinase activity. Proc Natl Acad Sci USA 1997;94:3939–3942.

144. Serve H, Yee NS, Stella G, et al. Differential roles of PI3-kinase and Kit tyrosine 821 in Kit receptor-mediated proliferation, survival and cell adhesion in mast cells. EMBO J 1995;14:473–483.

145. Margolis B, Hu P, Katzav S, et al. Tyrosine phosphorylation of *vav* proto-oncogene product containing SH2 domain and transcription factor motifs. Nature (Lond) 1992;356:71–74.

146. Cuenod B, Zhang C, Scharenberg AM, et al. Syk-dependent phosphorylation of Shc: a potential link between FcεRI and the Ras/MAP kinase signaling pathway through Sos and Grb2. J Biol Chem 1996;271:16268–16272.

147. Scharenberg AM, Lin S, Cuenod B, et al. Reconstitution of interactions between tyrosine kinases and the high affinity IgE receptor which are controlled by receptor clustering. EMBO J 1995;14:3385–3394.

148. Zhang J, Berenstein EH, Evans RL, et al. Transfection of Syk protein tyrosine kinase reconstitutes high affinity IgE receptor-mediated degranulation in a Syk-negative variant of rat basophilic leukemia RBL-2H3 cells. J Exp Med 1996;184:71–80.

149. Song JS, Gomez J, Stancato LF, et al. Association of a p95 Vav-containing signaling complex with the FcεRI chain in the RBL-2H3 mast cell

line. Evidence for a constitutive *in vivo* association of Vav with Grb2, Raf-1 and ERK2 in an active complex. J Biol Chem 1996;271:26962–26970.

150. Murakami M, Austen KF, Arm JP. The immediate phase of c-kit ligand stimulation of mouse bone marrow-derived mast cells elicits rapid leukotriene C_4 generation through posttranslational activation of cytosolic phospholipase A_2 and 5-lipoxygenase. J Exp Med 1995;182:197–206.

151. Paulson RF, Vesely S, Siminovitch KA, et al. Signalling by the W/Kit receptor tyrosine kinase is negatively regulated in vivo by the protein tyrosine phosphatase Shp1. Nat Genet 1996;13:309–315.

152. Kinashi T, Escobedo JA, Williams LT, et al. Receptor tyrosine kinase stimulates cell-matrix adhesion by phosphatidylinositol 3 kinase and phospholipase Cγ 1 pathways. Blood 1995;86: 2086–2090.

153. Kawakami Y, Yao L, Miura T, et al. Tyrosine phosphorylation and activation of Bruton tyrosine kinase upon FcεRI cross-linking. Mol Cell Biol 1994;14:5108–5113.

154. Rawlings DJ, Scharenberg AM, Park H, et al. Activation of BTK by a phosphorylation mechanism initiated by SRC family kinases. Science 1996;271:822–825.

155. Mahajan S, Fargnoli J, Burkhardt AL, et al. Src family protein tyrosine kinases induce autoactivation of Bruton's tyrosine kinase. Mol Cell Biol 1995;15:5304–5311.

156. Park H, Wahl MI, Afar DE, et al. Regulation of Btk function by a major autophosphorylation site within the SH3 domain. Immunity 1996;4: 515–525.

157. Yao L, Kawakami Y, Kawakami T. The pleckstrin homology domain of Bruton's tyrosine kinase interacts with protein kinase C. Proc Natl Acad Sci USA 1994;91:9175–9179.

158. Tolias KF, Cantley LC, Carpenter CL. Rho family GTPases bind to phosphoinositide kinases. J Biol Chem 1995;270:17656–17659.

159. Rider LG, Hirasawa N, Santini F, et al. Activation of the mitogen-activated protein kinase cascade is suppressed by low concentrations of dexamethasone in mast cells. J Immunol 1996; 157:2374–2380.

160. Ozawa K, Szallasi Z, Kazanietz MG, et al. Ca^{2+}-dependent and Ca^{2+}-independent isozymes of protein kinase C mediate exocytosis in antigen-stimulated rat basophilic RBL-2H3 cells: reconstitution of secretory responses with Ca^{2+} and purified isozymes in washed permeabilized cells. J Biol Chem 1993;268:1749–1756.

161. Glover S, de Carvalho MS, et al. Translocation of the 85-kDa phospholipase A_2 from cytosol to the nuclear envelope in rat basophilic leukemia cells stimulated with calcium ionophore or IgE/antigen. J Biol Chem 1995;270:15359–15367.

162. Ishimoto T, Arisato K, Akiba S, et al. Requirement of calcium influx for hydrolytic action of membrane phospholipids by cytosolic phospholipase A_2 rather than mitogen-activated protein kinase activation in FcεRI-stimulated rat peritoneal mast cells. J Biochem (Tokyo) 1996;120:1247–1252.

163. Lin P, Wiggan GA, Gilfillan AM. Activation of phospholipase D in a rat mast (RBL-2H3) cell line: a possible unifying mechanism for IgE-dependent degranulation and arachidonic acid metabolite release. J Immunol 1991;146: 1609–1616.

164. Koike T, Mizutani T, Hirai K, et al. SCF/c-kit receptor-mediated arachidonic acid liberation in rat mast cells. Involvement of PLD activation associated tyrosine phosphorylation. Biochem Biophys Res Commun 1993;197: 1570–1577.

165. Ishimoto T, Akiba S, Sato T, et al. Contribution of phospholipases A_2 and D to arachidonic acid liberation and prostaglandin D_2 formation with increase in intracellular Ca^{2+} concentration in rat peritoneal mast cells. Eur J Biochem 1994;219:401–406.

166. Murakami M, Kudo I, Umeda M, et al. Detection of three distinct phospholipases A_2 in cultured mast cells. J Biochem (Tokyo) 1992;111:175–181.

167. Borsch-Haubold AG, Kramer RM, Watson SP. Inhibition of mitogen-activated protein kinase kinase does not impair primary activation of human platelets. Biochem J 1996;318:207–212.

168. Kramer RM, Roberts EF, Um SL, et al. P38 mitogen-activated protein kinase phosphorylates cytosolic phospholipase A_2 (cPLA$_2$) in thrombin-stimulated platelets. Evidence that proline-directed phosphorylation is not required for mobilization of arachidonic acid release by cPLA$_2$. J Biol Chem 1996;271: 27723–27729.

169. Prieschl EE, Gouilleux V, Walker C, et al. A nuclear factor of activated T cell-like transcription factor in mast cells is involved in IL-5 gene regulation after IgE plus antigen stimulation. J Immunol 1995;154:6112–6119.

170. Prieschl EE, Pendl GG, Harrer NE, et al. P21^{ras} links FcεRI to NF-AT family member in mast cells. The AP3-like factor in this cell type is an NF-AT family member. J Immunol 1995; 155:4963–4970.

171. Reimann T, Buscher D, Hipskind RA, et al. Lipopolysaccharide induces activation of the Raf-1/MAP kinase pathway. A putative role for Raf-1 in the induction of the IL-1β and the TNF-α genes. J Immunol 1994;153:5740–5749.

172. Li S, Sedivy JM. Raf-1 protein kinase activates the NF-κB transcription factor by dissociating the cytoplasmic NF-κB-IκB complex. Proc Natl Acad Sci USA 1993;90:9247–9251.

173. Park JH, Levitt L. Overexpression of mitogen-activated protein kinase (ERK1) enhances T-cell cytokine gene expression: role of AP1, NF-AT, and NF-κB. Blood 1993;82:2470–2477.

174. Baumgartner RA, Beaven MA. Mediator release by mast cells and basophils. In: Herzenberg LA, Herzenberg L, Weir DM, Blackwell C, eds. Weir's Handbook of Experimental Immunology, vol. 4. 5th Ed. Cambridge: Blackwell, 1996:213.1–213.8.

175. Galli SJ. New concepts about the mast cell. New Engl J Med 1993;328:257–265.

176. Gordon JR, Burd PR, Galli SJ. Mast cells as a source of multifunctional cytokines. Immunol Today 1990;11:458–464.

177. Galli SJ, Wershil BK. The two faces of the mast cell. Nature (Lond) 1996;381:21–22.

178. Baumgartner RA, Yamada K, Deramo VA, et al. Secretion of tumor necrosis factor (TNF) from a rat mast cell line is a brefeldin A-sensitive and a calcium/protein kinase C-regulated process. J Immunol 1994;153:2609–2617.

19
Tec Family Protein Tyrosine Kinases and Their Interaction with Protein Kinase C

Toshiaki Kawakami, Libo Yao, and Yuko Kawakami

"The completion of the budding yeast genome sequence project has made it possible to determine not only the total number of genes, but also the exact number of genes of a particular type. As a consequence, we now know exactly how many protein kinases are encoded by the yeast genome, a number of considerable interest because of the importance of protein phosphorylation in the control of so many cellular processes."[1] Activation of the immune cells, including mast cells, is a typical cellular process that is under the control of the intricate phosphorylation network composed of both protein tyrosine kinases (PTKs) and serine/threonine kinases, among many types of signaling molecules. Since Siraganian and associates first demonstrated the activation of PTKs upon cross-linking of the high-affinity IgE receptor (FcεRI) in 1990,[2] several groups including ourselves have provided convincing evidence for the essential role of PTKs in mast cell activation.[3–5]

Members of three PTK families have emerged as main players in the initiation of mast cell activation. A current model[6] based on numerous studies suggests that the following series of events take place. The β-subunit of FcεRI is physically associated with Lyn, a Src family PTK abundantly expressed in mast cells, in a poorly defined mechanism. Upon cross-linking of FcεRI, Lyn becomes activated and phosphorylates the tyrosine residues in the immunoreceptor tyrosine-based activation motifs (ITAMs) of the cytoplasmic domains of both β- and γ-subunits of FcεRI. Phosphory-

lated ITAMs of β- and γ-subunits now recruit new Lyn molecules and Syk, respectively. Syk is a member of another PTK family containing two tandemly arranged Src homology (SH) 2 domains that are N-terminal to the catalytic domain. The newly recruited Syk, bound to the γ-subunit via the SH-2–phosphotyrosine interactions, becomes activated.[7,8] Syk activation seems to be dependent, at least partly, on the conformational change impinged on by the SH-2—phospho–ITAM binding.[9] Because Lyn and other Src family PTKs can activate Bruton's tyrosine kinase (Btk),[10] a Tec family PTK, the members of these three PTK families are implicated in the early signaling events. Indeed, *lyn* gene-targeted mice failed to exhibit IgE-dependent anaphylactic response.[11] However, mast cells derived from another *lyn* knockout mouse exhibited normal abilities to degranulate, to adhere to a fibronectin-coated substratum, and to produce cytokines despite a generalized decrease in tyrosine phosphorylation and attenuated Ca²⁺ response.[12] RBL-2H3 variant cells devoid of Syk expression showed neither increased protein tyrosine phosphorylation, Ca²⁺ response, nor degranulatory activity upon FcεRI cross-linking.[13] Furthermore, we have data showing that mast cells derived from *btk* knockout mice have impaired abilities to degranulate and secrete cytokines (Hata et al., unpublished data).

Budding yeast has 113 conventional protein kinase genes, corresponding to about 2% of the total genes.[1] If this number can be extrapolated to higher eukaryotes, we probably need to con-

274

sider possible involvement of many more protein kinases in mast cell activation. Therefore, it would be safe to say that we have just started to understand the exquisite signaling machinery for mast cell activation, even when the topic is limited to the early tyrosine phosphorylation-related aspect. We are also in our infancy in understanding the other side of the coin: almost nothing is known about the roles of the phosphorylation-antagonizing enymes, that is, phosphatases, except for the requirement of CD45, a transmembrane phosphatase, for mast cell activation.[14,15]

In contrast with the enormous number of protein kinase genes in mammals expected from the yeast genome study, previous studies suggested that relatively small numbers of PTKs are expressed in mouse mast cells[16] and K-562 chronic myelogenous leukemia cells.[17] Because of technical limitations it was (and it is even now) impossible to estimate the number of PTKs expressed in a particular type of cells. Nonetheless, we searched for new PTK genes expressed in mast cells using a reverse transcription-polymerase chain reaction (RT-PCR) strategy.[18] The obtained PCR products amplified between the 3′-ends of mRNAs and two degenerate oligonucleotides, corresponding to the conserved PTK catalytic domain sequences, included two novel, highly related sequences, *emb* and *emt*, as well as seven known protein kinase sequences, *lyn, hck, jak-1, jak-2, tyk-2,* IGF-I receptor, and B-*raf. emb* turned out to be

the same gene referred to as *atk*[19] or *BPK*,[20] which was isolated as the gene responsible for the disease X-linked agammaglobulinemia (XLA). This gene, now termed as Bruton's tyrosine kinase (Btk), was later demonstrated to be mutated in an X-linked immunodeficient (*xid*) mouse.[21,22] *emt* cDNA was also isolated by other groups with the different names *itk,*[23] *Tsk,*[24] or *emt.*[25]

Structure of Tec Family PTKs

The overall Btk (and Emt) molecule deduced from its cDNA sequence exhibits a prominent multidomain structure with a number of distinct features[18] (Fig. 19.1). First, they lack hydrophobic amino acid stretches characteristic of the transmembrane domains found in growth factor receptor PTKs. Second, they are similar to the Src family PTKs in displaying SH-3, SH-2, and a kinase (SH-1) domain that has a consensus "autophosphorylation" site corresponding to Tyr-416 in the activation segment in p60[c-src]. Nevertheless, Btk and Emt differ from Src family PTKs in that they lack, first, a glycine residue at the second position that serves as a myristylation and membrane-anchoring site and, second, a negative regulatory tyrosine residue (corresponding to Tyr-527 in p60[c-src]), whose phosphorylation suppresses the kinase activity of Src family PTKs. These differences suggest a different regulatory mode

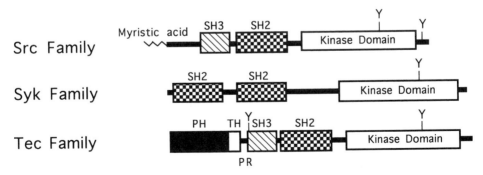

FIGURE 19.1. Structure of three protein tyrosine kinase (PTK) families involved in mast cell activation. The domain structures of the PTKs are shown by *blocks*. Tyrosine residues important for regulation are also shown. The Tec homology (TH) domain in Tec family PTKs includes both regions, denoted by *TH* and *PR* (proline rich).

for the kinase activity of Btk and Emt relative to members of the Src family (see following). Third, the kinase domains reveal relatively high homologies (55%–64%) among Btk, Emt, TecII, a hematopoietic PTK,[26] and Dsrc28,[27] a Src-related PTK isolated from *Drosophila melanogaster*. The homology level between any of these four kinases and other PTKs is significantly lower (<49%). Fourth, Btk and Emt share a unique, extensively homologous amino-terminal region. Within this region, the amino-terminal 110–130 amino acids form a pleckstrin homology (PH) domain. Downstream of the PH domain, Btk and Emt share a short sequence, termed the Tec homology (TH) domain,[28] containing a short proline-rich (PR) motif [KKPLPPTPE(E/D)]. Btk has another proline-rich motif [KKPLPPEPTA] 7 residues downstream. These PR sequences of Btk were shown to interact with the SH-3 domains of several Src family PTKs (Fyn, Hck and Lyn).[29,30] The equivalent region of Emt was recently shown to interact with the SH-3 domain in an intramolecular fashion (see following). SH-3-binding proteins identified to date include p120[c-cbl] (with Btk),[31] Sam68, and Grb-2 (with Emt).[32] Finally, the Btk-specific sequence, PERQIPRRGEESSEME, which resides in the PH domain, has a high degree of homology (11 identities and 3 conserved substitutions over the 16-residue stretch) with a portion of the cytoplasmic domain of CD22, suggesting that the same or similar protein(s) might interact with CD22 and Btk through this hydrophilic sequence motif. Based on these structural features, Btk, Emt, and Tec II along with a more recently cloned member, BMX,[33] constitute a novel family of PTKs, termed the Tec family, that may perform similar functions and share a unique mode of regulation. TXK/Rlk[34,35] and Dsrc28 may form a subtype of Tec family PTKs because these kinases lack the PH domain.

Functions of Tec Family PTKs

Activation of Btk and Emt on FcεRI Cross-Linking

The molecular masses of Btk (77 kDa) and Emt (72 kDa) were close to those of the tyro-sine-phosphorylated proteins whose phosphorylation was induced strongly by FcεRI stimulation.[2,3] Because of this similarity and relatively specific expression of these PTKs in mast cells, effects of FcεRI cross-linking on them were examined.[36,37] Mouse bone marrow-derived cultured mast cells (BMMC) and RBL-2H3 rat basophilic leukemia cells exhibited increased tyrosine phosphorylation and enzymatic activity of Btk. Emt was also tyrosine phosphorylated and enzymatically activated upon FcεRI cross-linking in BMMC, whereas RBL-2H3 cells expressed little, if any, Emt protein. The increase in tyrosine phosphorylation of these proteins was detected within 1 min and attained the maximal level about 1 to 3 min after stimulation. Phosphoamino acid analysis of ^{32}P metabolically labeled proteins demonstrated that these PTKs are phosphoproteins with phosphoserine as the most abundant phosphorylated residue before mast cell activation and that phosphorylation in activated cells was induced not only on tyrosine, but also on serine and threonine.

As shown for several PTKs involved in signal transduction pathways for FcεRI as well as other antigen receptors, tyrosine phosphorylation of PTKs is accompanied by their activation. Btk activity was increased as rapidly as its tyrosine phosphorylation, and reached its peak about 3 min after antigen stimulation. Similar kinetics was observed for Emt activity. Interestingly, the increased activity of Btk and Emt lasted significantly longer than the duration of their tyrosine phosphorylation. The mechanism of this phenomenon has remained unsolved, although a detailed study on their phosphorylation is expected to give a clue on this enigma. By contrast, Tec was shown not to be activated by FcεRI cross-linking, whereas it was activated by stem cell factor.[38]

Because the two PTKs (Lyn and Syk) were shown to be associated with FcεRI upon receptor cross-linking, it was interesting to test whether Btk (and Emt) were also associated with the receptor. However, extensive experiments using numerous lysis conditions failed to detect significant association between Btk (and Emt) and either β- or γ-subunits of FcεRI. On the other hand, we purified the PTK from RBL-2H3 cells that phosphorylates a peptide modeled on the sequence of the cytoplasmic region of the γ-subunit.[39] The purified PTK was identified as Btk, suggesting that Btk might phosphorylate the ITAM motif in the γ-subunit in vivo. Experiments testing this possibility are under way.

Regulation of Tec Family PTKs

Cotransfection studies showed that Btk is tyrosine phosphorylated on Tyr-551 by Lyn and other Src family PTKs.[10] Once phosphorylated at Tyr-551, Btk exhibits an increased kinase activity (5–10 fold) and autophosphorylates at Tyr-223 in its SH-3 domain.[40] This activation mechanism for Btk is distinctly different from that of Src family PTKs, as expected from structural differences. c-Src phosphorylated at Tyr-527 by Csk (c-Src kinase) has an inactive conformation impinged on by the intramolecular interaction between its SH-2 domain and phosphotyrosine at 527. Intramolecular interactions between the SH-3 domain and the SH-2-kinase linker and between the SH-3 and the small lobe of the kinase domain further stabilize the inactive c-Src molecule in a conformation that simultaneously disrupts the kinase active site and sequesters the binding surfaces of the SH-2 and SH-3 domains.[41,42] However, Btk and other Tec family PTKs lack the C-terminal tyrosine residue corresponding to Tyr-527 of c-Src.

A recent nuclear magnetic resonance study on the fragments of Emt demonstrated the intramolecular interaction between the SH-3 domain and the PR region of the TH domain.[43] Binding of the SH-3 domain ligands to variously truncated SH-3 fusion proteins also suggested the possible interaction between the N-terminal portion of the TH domain and the SH-2 domain. Furthermore, a substitution mutant (E41K) in the PH domain of Btk results in a constitutively active form.[44] This mutant Btk protein tends to be localized to the membrane compartment and transforms the NIH/3T3 fibroblasts. This finding, together with the fact that numerous XLA patients have mutations in the Btk PH domain,[45,46] suggests that the PH domain might interact with the kinase domain.

Based on these observations, we propose a model for the activation of Btk and Emt (Fig.

19.2). This model is also applicable to other Tec family PTKs. In unstimulated cells, Btk and Emt should have an inactive closed form with extensive intramolecular interactions, that is, one between the PR region and the SH-3 domain, one between the TH and SH-2 domains, and one between the PH and kinase domains. Intramolecular interactions observed for inactive c-Src probably help to further stabilize the locked Btk (and Emt) conformation. However, the PR region would be a favored interaction partner for the SH-3 domain over the SH-2-kinase linker.[43] Abundant serine phosphorylation may also contribute to the stability of the intramolecular interactions. When activated Lyn comes close to inactive Btk (or Emt)

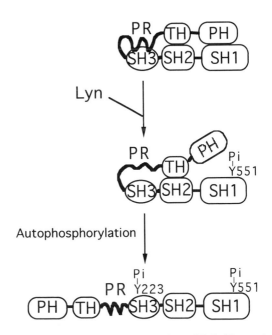

FIGURE 19.2. Enzymatic activation of Btk (Bruton's tyrosine kinase. The two-step activation process is hypothesized in this model. In unstimulated mast cells, Btk is in an inactive conformation impinged upon by extensive intramolecular interactions. Upon FcεRI cross-linking, active Lyn or other Src family PTKs phosphorylate Btk at Tyr-551 and at least partially activate Btk. Btk in this conformation can autophosphorylate (probably in *trans*) at Tyr-223 in the SH-3 domain. The resulting, doubly tyrosine-phosphorylated Btk assumes a fully active conformation with an ability to phosphorylate its target proteins such as phospholipase C (PLC-γ1 and PLC-γ2).

upon FcεRI cross-linking, Btk is phosphory-lated at Tyr-551 by Lyn. Then, Btk takes a partially open configuration, allowing it to trans-autophosphorylate Tyr-223. Therefore, at this step the PR region–SH-3 interaction should be loose enough for trans-autophosphorylation to take place. Btk phos-phorylated at Tyr-551 and Tyr-223 is thought to have a fully active, open configuration to phos-phorylate its target proteins. This model re-quires the functions of the individual domains to be intact for normal functions of Btk protein. This notion is supported by the presence of mutations over all the individual domains in XLA patients.

Biological Functions of Tec Family PTKs

As amply exemplified in human and mouse immunodeficiencies, Btk plays a pivotal role in B-cell differentiation and activation. Btk seems to have a broader role in various signaling path-ways, such as those for IL-5,[47] IL-6,[48] IL-10,[49] CD38,[50,51] and CD40[52] in addition to that for FcεRI. The availability of btk-mutated mice (xid and btk gene-targeted mice[53]) has facili-tated the studies on the functions of Btk in these and other events. Chicken lymphoma DT-40 cells with targeted btk alleles exhibited a reduced tyrosine phosphorylation of phospho-lipase C (PLC-γ_2) and no Ca^{2+} response upon anti-μ stimulation.[54] By contrast, a study using XLA-derived B cells demonstrated that Btk restores deficient Ca^{2+} influx in XLA B cells although Btk exerted little effect on Ca^{2+} release from intracellular stores.[55] Although the mechanism must be worked out to recon-cile these two observations, it is clear that Btk plays an essential role in regulation of intracellular Ca^{2+} concentrations. Our recent study demonstrated that Btk plays roles in de-granulation and cytokine production/secretion induced by FcεRI cross-linking (unpublished data).

In our previous study, we showed that Emt is expressed not only in T cells but also in mast cells.[18] We also showed that CD28, a major T-cell costimulatory receptor, can be associated with Emt and that this association with Emt is enhanced by CD28 stimulation.[56] Furthermore, CD3 stimulation induced the activation of Emt in T cells.[57] Because the CD3-induced Emt acti-vation requires the expression of Lck, a Src family PTK, Emt activity seems to be regulated by Lck just like Btk that is activated by Lyn. These studies prompted us to investigate whether CD28 is expressed in mast cells. We demonstrated the expression of CD28 by cytofluorometric analysis of BMMC and its de-rivative line, MCP-5.[58] The authenticity of the expressed CD28 was verified by RT-PCR tech-niques and sequence analysis of the PCR prod-ucts. Our analysis of early and late signaling events generated by cross-linking of CD28 with anti-CD28 antibodies revealed that CD28 on mast cells are functional: tyrosine phosphoryla-tion of several cellular proteins was induced, including Emt, Btk, Syk, p120^{c-cbl}, Shc, and Vav. The kinase activities of Emt, Btk, and Syk were enhanced. Furthermore, anti-CD28 stimulation induced secretion of tumor necrosis factor-α (TNF-α) and interleukin-2 (IL-2). More impor-tantly, TNF-α secretion induced by FcεRI cross-linking was enhanced by concurrent stimulation of CD28. Because TNF-α secreted from activated mast cells is at least partly responsible for late-phase reactions of allergy, this costimulation-induced enhancement of TNF-α secretion is implicated in the late-phase reactions and other disease states.

Our recent study indicated that Btk is re-quired for apoptosis induced by growth factor withdrawal from the culture of mast cells.[59] We had been puzzled for years by the observation that bone marrow cells derived from xid mice yield about two times more mast cells after 4 weeks of culture in IL-3-containing medium than the wild-type control, despite no differ-ence in mitogenic response to IL-3 between the two types of mast cells. Similar differences in mast cell yields were found between btk knock-out and control mice. Since the culture medium of bone marrow cells was changed weekly, it was suspected that IL-3 levels might not be high enough to sustain the growth of mast cells, es-pecially those from wild-type mice. Then, the survival response to IL-3 depletion was com-pared between xid mast cells and wild-type mast cells. xid mast cells turned out to be more

FIGURE 19.3. Mast cell activation pathways initiated by FcεRI cross-linking. FcεRI cross-linking induces activation of the members of the three PTK families. PTK activation leads to the activation of PLC-γ and three major mitogen-activated protein kinase (MAPK) pathways. The catalytic products of PLC, diacylglycerol (DAG) and IP$_3$, act as second messengers to activate PKC and to mobilize Ca^{2+}, respectively. These events are believed to lead to degranulation. MAPK (ERKs, JNKs, and p38) activation is linked to transcriptional activation of cytokine genes, while ERKs are required for PLA$_2$ activation.

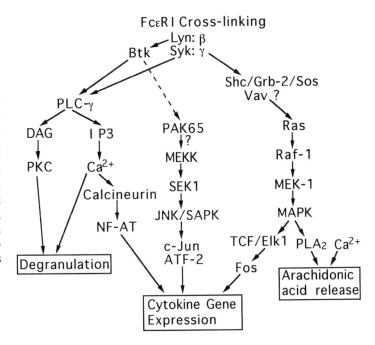

resistant to IL-3-deprived culture conditions than the control cells. A number of methods indicated that growth factor deprivation induces the rapid activation of apoptosis in wild-type cells but not in *xid* mast cells. Similar observations were made on *btk* knockout mast cells compared to control cells. Because apoptosis is induced by activation of stress-activated MAP kinases (MAPKs) such as JNK/SAPK and p38/RK, the effects of the absence of Btk on the activity of these MAPKs were examined. Growth factor depletion induced a sustained activation of JNK (and p38 to a lesser extent) in wild-type mast cells while the same treatment did not induce activation of either ERK-1 or ERK-2. FcεRI cross-linking also induced a transient activation of all three major classes of MAPKs, that is, ERK-1 and ERK-2, JNK-1 and JNK-2, and p38, in wild-type mast cells. However, *btk* knockout mast cells exhibited a very low level of JNK activation and a slightly reduced activation of p38. The specificity of *btk* mutations on apoptosis and JNK activation was confirmed by reconstitution of the responses in *btk* knockout mast cells transfected by the wild-type *btk* cDNA, but not vector control, kinase-dead *btk* mutant, or other kinase cDNAs. Therefore, Btk regulates apoptosis and the activity of JNK and p38. These results are consistent with the fact that γ-irradiation- and anti-μ-induced apoptosis in DT-40 chicken B lymphoma cells are dependent on Btk. These data, together with our unpublished results on Btk-regulated transcription of cytokine (IL-2 and TNF-α) genes, are summarized in Fig. 19.3.

PH Domains

Recent work has identified several protein motifs that mediate intermolecular and intramolecular interactions in a wide variety of cellular contexts. These include SH-2, SH-3, and phosphotyrosine-binding (PTB) domains. SH-2 and PTB domains bind to phosphotyrosine residues in the context of their flanking sequences[60,61] while SH-3 as well as the recently identified WWP/WW domains bind to proline-rich peptides.[61,62]

The PH domain is another short protein motif composed of loosely conserved sequences of about 100 amino acid residues. Originally recognized as repeated sequences in the platelet protein pleckstrin, a prominent substrate for protein kinase C (PKC), PH domains have been found in numerous proteins, most of which are signal-transducing or cytoskeletal proteins.[63–66] These include protein tyrosine kinases, serine/threonine kinases, lipid kinases, GTPase-activating proteins (GAPs), guanine nucleotide exchange factors (GEFs), and phospholipase Cs (PLCs). Tertiary structures of several PH domains have been solved by nuclear magnetic resonance spectroscopic or x-ray crystallographic techniques.[67–73] The core of the compact domain structures shared by the PH domains of pleckstrin, β-spectrin, dynamin, and PLC-δ1 is a β-sandwich formed by two nearly orthogonal antiparallel β-sheets of four and three strands respectively. One corner of the β-sandwich is capped by the C-terminal α-helix. So far, three classes of molecules have been shown to interact with these apparently multifunctional domains.

First, studies by Lefkowitz and coworkers[74,75] established that the C-terminal portion and their downstream sequences of the PH domains of the β-adrenergic receptor kinase (β-ARK) and several other proteins bind to the βγ-complexes of heterotrimeric GTP-binding proteins (G proteins). The WD40 repeats of the β-subunit of G protein (Gβ) are the PH domain-interacting counterparts.[76] Because the downstream sequences abutting the PH domains are involved in binding to Gβ, a doubt is cast on whether Gβ is an authentic PH domain ligand. Second, several studies[73,77,78] showed that various PH domains bind to phosphatidylinositol 4,5-bisphosphate (PIP$_2$) and inositol 1,4,5-trisphosphate (IP$_3$) through their positively charged residues in the amino-terminal halves. Low-affinity binding of PIP$_2$ ($K_D \cong 30$ [pleckstrin N-terminal PH domain] and $1.7\,\mu M$ [PLC-δ1]) and Gβγ led to the hypothesis that PH domains function as membrane-localizing surfaces for the PH domain-containing proteins. More recently, phosphatidylinositol 3,4,5-trisphosphate was shown to be a specific ligand for the Btk PH domain under conditions in which the dynamin PH domain specifically interacts with PIP$_2$ and phosphatidylinositol 4-phosphate.[79] Third, our previous studies[37,80] demonstrated that multiple

isoforms of PKC interact directly with several PH domains, including those of Btk and Emt. Both Btk and Emt physically associate with PKC. Btk in mast cells and B lymphocytes coimmunoprecipitates with PKC-βI and PKC-βII while Emt in mast cells coimmunoprecipitates with many more PKC isoforms, including α, βI, βII, ε, ζ, and θ. Purified Btk was shown to be phosphorylated and enzymatically downregulated by PKC in vitro. Downregulation of PKC expression by an overnight treatment of mast cells induced an enhanced tyrosine phosphorylation of Btk upon FcεRI cross-linking. Similarly, pretreatment of mast cells with PKC inhibitors enhanced FcεRI-induced tyrosine phosphorylation of Btk.

These in vivo effects of downregulation or enzymatic inhibition of PKC confirmed the in vitro inhibitory activity on the Btk kinase activity of Btk phosphorylation by PKC. The physiological relevance of the Btk–PKC-β interaction was suggested by a recent study of the *PKC*-β gene knockout mice.[81] The phenotype of *PKC*-β$^{-/-}$ mice was quite similar to those of *btk*$^{-/-}$ or *xid* mice. (1) B-1 B-cell numbers were drastically reduced, and splenic B cell numbers were modestly reduced. (2) IgM and IgG-3 levels in the sera were reduced. (3) No immune responses were observed against a T-cell-independent antigen. (4) Proliferative responses to anti-IgM, anti-CD40, and lipopolysaccharide were reduced. By contrast, Emt appears to be phosphorylated and upregulated by PKC. More recently, it was shown that the Btk PH domain interacts with PKC with high affinity ($K_D = 39\,nM$).[82] Unlike other tested phospholipids, PIP$_2$ competed with PKC for binding to the PH domain, apparently because their binding sites on the amino-terminal portion of the PH domains overlap. The minimal PKC-binding sequence within the Btk PH domain was found to correspond roughly to the second and third β-sheets of the PH domains of known tertiary structures.

On the other hand, the C-1 regulatory region of PKC-ε containing the pseudosubstrate and zinc finger-like sequences was found to be sufficient for strong binding to the Btk PH domain. A potent activator of PKC, phorbol

12-myristate 13-acetate (PMA), that interacts with the C-1 region of PKC inhibited the PKC–PH domain interaction whereas the bioinactive PMA, 4-α-PMA, was ineffective. The ζ-isoform of PKC, which has a single zinc finger-like motif instead of the two tandem zinc finger-like sequences present in conventional and novel PKC isoforms, does not bind PMA. As expected, PMA did not interefere with PH domain binding to PKC-ζ. Further, inhibitors that are known to attack the catalytic domains of serine/threonine kinases did not affect this PKC–PH domain interaction. In contrast, the presence of physiological concentrations of Ca^{2+} induced less than twofold increase in PKC–PH domain binding. These results indicate that PKC binding to PH domains involves the β2–β3 region of the Btk PH domain and the C-1 region of PKC, and that agents which interact with either of these regions (i.e., PIP_2 binding to the PH domain and PMA binding to the C-1 region of PKC) might act to regulate PKC–PH domain binding. Our preliminary study provides evidence that some, but not all, tested PH domains bind to the filamentous form of actin (F-actin) (Yao et al., unpublished data). Therefore, PH domains seem to provide to signaling and cytoskeletal proteins a stage for intermolecular interactions for their functions in cell activation.

Perspectives

Recent studies have unraveled the important roles of Btk not only in B-cell differentiation and activation but also in fundamental functions in mast cells. The pace of major findings on functions of other Tec family PTKs will be kept along with that on Btk. Future studies are expected to reveal whether each Tec family member has unique functions or redundant functions among the members. Since the functions of Btk in apoptosis, degranulation, and cytokine gene expression have been revealed, the next phase of the study on the Tec family PTKs will be shifted to deciphering the signaling pathways in which they are involved. The two major issues in this field will be how Btk regulates Ca^{2+} response and

how Btk signals the activation of stress-activated MAPKs.

As described, several specific ligands can interact with PH domains. PH domains might function as a membrane-tethering device because of their binding to phospholipids and Gβ. However, their functions seem to be broader as shown for the Btk–PKC interaction. The presence of multiple ligands on the small protein–protein or protein–lipid interaction motifs such as PH domains present an interesting possibility that multiple ligands might interact with each other, giving an additional level of complexity in terms of regulatory functions of the signaling complexes. This kind of interaction is quite likely to take place, because all the PH domain ligands identified to date are also known as key signal transducers in cell activation, growth, and differentiation. Because Btk and PKC-β seem to work in the same signaling pathways leading to degranulation and gene regulation in mast cells, the interaction of these protein kinases mediated by the Btk PH domain and the C1 region of PKC-β as well as the kinase activities of these proteins could be targets of pharmacological interference.

Acknowledgments. Studies in our laboratory are supported in part by NIH grants RO1 AI33617 and RO1 AI38348 (to T.K.). This is Publication 282 from the La Jolla Institute for Allergy and Immunology.

References

1. Hunter T, Plowman GD. The protein kinases of budding yeast: six score and more. Trends Biochem Sci 1997;22:18–22.
2. Benhamou M, Gutkind JS, Robbins KC, et al. Tyrosine phosphorylation coupled to IgE receptor-mediated signal transduction and histamine release. Proc Natl Acad Sci USA 1990;87:5327–5330.
3. Kawakami T, Inagaki N, Takei M, et al. Tyrosine phosphorylation is required for mast cell activation by FcεRI cross-linking. J Immunol 1992;148:3513–3519.
4. Yu K-T, Lyall R, Jariwala N, et al. Antigen- and ionophore-induced signal transduction in rat basophilic leukemia cells involves protein tyrosine

phosphorylation. J Biol Chem 1991;266:22564–22568.

5. Stephen V, Benhamou M, Gutkind JS, et al. FcεRI-induced protein tyrosine phosphorylation of pp72 in rat basophilic leukemia cells (RBL-2H3). Evidence for a novel signal transduction pathway unrelated to G protein activation and phosphatidylinositol hydrolysis. J Biol Chem 1992;267:5434–5441.

6. Jouvin M-H, Adamczewski M, Numerof R, et al. Differential control of the tyrosine kinases Lyn and Syk by the two signaling chains of the high affinity immunoglobulin E receptor. J Biol Chem 1994;269:5918–5925.

7. Shiue L, Zoller MJ, Brugge JS. Syk is activated by phosphotyrosine-containing peptides representing the tyrosine-based activation motifs of the high affinity receptor for IgE. J Biol Chem 1995;270:10498–10502.

8. Rowley RB, Burkhardt AL, Chao H-G, et al. Syk protein-tyrosine kinase is regulated by tyrosine-phosphorylated Igα/Igβ immunoreceptor tyrosine activation motif binding and autophosphorylation. J Biol Chem 1995;270:11590–11594.

9. Kimura T, Sakamoto H, Appella E, et al. Conformational changes induced in the protein tyrosine kinase p72^{syk} by tyrosine phosphorylation or by binding of phosphorylated immunoreceptor tyrosine-based activation motif peptides. Mol Cell Biol 1996;16:1471–1478.

10. Rawlings DJ, Scharenberg AM, Park H, et al. Activation of BTK by a phosphorylation mechanism initiated by SRC family kinases. Science 1996;271:822–825.

11. Hibbs ML, Tarlinton DM, Armes J, et al. Multiple defects in the immune system of Lyn-deficient mice, culminating in autoimmune disease. Cell 1995;83:301–311.

12. Nishizumi H, Yamamoto T. Impaired tyrosine phosphorylation and Ca^{2+} mobilization, but not degranulation, in Lyn-deficient bone marrow-derived mast cells. J Immunol 1997;158:2350–2355.

13. Zhang J, Berenstein EH, Evans RL, et al. Transfection of Syk protein tyrosine kinase reconstitutes high affinity IgE receptor-mediated degranulation in a Syk-negative variant of rat basophilic leukemia RBL-2H3 cells. J Exp Med 1996;184:71–79.

14. Hook WA, Berenstein EH, Zinsser FU, et al. Monoclonal antibodies to the leukocyte common antigen (CD45) inhibit IgE-mediated histamine release from human basophils. J Immunol 1991;147:2670–2676.

15. Berger SA, Mak TW, Paige CJ. Leukocyte common antigen (CD45) is required for immunoglobulin E-mediated degranulation of mast cells. J Exp Med 1994;180:471–476.

16. Fukamachi H, Takei M, Kawakami T. Activation of multiple protein kinases including a MAP kinase upon FcεRI cross-linking. Int Arch Allergy Immunol 1993;102:15–25.

17. Partanen J, Makela TP, Alitalo R, et al. Putative tyrosine kinases expressed in K-562 human leukemia cells. Proc Natl Acad Sci USA 1990; 87:8913–8917.

18. Yamada N, Kawakami Y, Kimura H, et al. Structure and expression of novel protein tyrosine kinases, Emb and Emt, in hematopoietic cells. Biochem Biophys Res Commun 1993;192:231–240.

19. Vetrie DJ, Vorechovsky I, Sideras P, et al. The gene involved in X-linked agammaglobulinemia is a member of the src family of protein-tyrosine kinase. Nature (Lond) 1993;361:226–233.

20. Tsukada S, Saffran DC, Rawlings DJ, et al. Deficient expression of a B cell cytoplasmic tyrosine kinase in human X-linked agammaglobulinemia. Cell 1993;72:279–290.

21. Thomas JD, Sideras P, Smith CIE, et al. Colocalization of X-linked agammaglobulinemia and X-linked immunodeficiency genes. Science 1993;261:355–358.

22. Rawlings DJ, Saffran DC, Tsukada S, et al. Mutation of unique region of Bruton's tyrosine kinase in immunodeficient XID mice. Science 1993;261:358–361.

23. Siliciano JD, Morrow TA, Desiderio SV. itk, a T-cell-specific tyrosine kinase gene inducible by interleukin 2. Proc Natl Acad Sci USA 1992; 89:11194–11198.

24. Heyeck SD, Berg LJ. Developmental regulation of a murine T-cell-specific tyrosine kinase gene, Tsk. Proc Natl Acad Sci USA 1993;90:669–673.

25. Gibson S, Leung B, Squire JA, et al. Identification, cloning, and characterization of a novel human T-cell-specific tyrosine kinase located at the hematopoietin complex on chromosome 5q. Blood 1993;83:1561–1572.

26. Mano H, Mano K, Tang B, et al. Expression of a novel form of Tec kinase in hematopoietic cells and mapping of the gene to chromosome 5 near kit. Oncogene 1993;8:417–424.

27. Gregory RJ, Kammermeyer KL, Vincent WS III, et al. Primary sequence and developmental expression of a novel Drosophila melanogaster src gene. Mol Cell Biol 1987; 7:2119–2127.

28. Vihinen M, Nilsson L, Smith CIE. Tec homology (TH) adjacent to the PH domain. FEBS Lett 1994;350:263–265.

29. Cheng G, Ye Z-S, Baltimore D. Binding of Bruton's tyrosine kinase to Fyn, Lyn, or Hck through a Src homology 3 domain-mediated interaction. Proc Natl Acad Sci USA 1994;91:8152–8155.

30. Yang W, Malek SN, Desiderio S. An SH3-binding site conserved in Bruton's tyrosine kinase and related tyrosine kinases mediates specific protein interactions *in vitro* and *in vivo*. J Biol Chem 1995;270:20832–20840.

31. Cory GOC, Lovering RC, Hinshelwood S, et al. The protein product of the *c-cbl* protooncogene is phosphorylated after B cell receptor stimulation and binds the SH3 domain of Bruton's tyrosine kinase J Exp Med 1995;182:611–615.

32. Bunnell SC, Henry PA, Kolluri R, et al. Identification of Itk/Tsk Src homology 3 domain ligands. J Biol Chem 1996;271:25646–25656.

33. Tamagnone L, Lahtinen I, Mustonen T, et al. *BMX*, a novel nonreceptor tyrosine kinase gene of the *BTK/ITK/TEC/TXK* family located in chromosome Xp22.2. Oncogene 1994;9:3683–3688.

34. Haire RN, Ohta Y, Lewis JE, et al. TXK, a novel human tyrosine kinase expressed in T cells shares sequence identity with Tec family kinases and maps to 4p12. Hum Mol Genet 1994;3:897–901.

35. Hu Q, Davidson D, Schwartzberg PL, et al. Identification of Rlk, a novel protein tyrosine kinase with predominant expression in the T cell lineage. J Biol Chem 1995;270:1928–1934.

36. Kawakami Y, Yao L, Tsukada S, et al. Tyrosine phosphorylation and activation of Bruton tyrosine kinase upon FcεRI cross-linking. Mol Cell Biol 1994;14:5108–5113.

37. Kawakami Y, Yao L, Tashiro M, et al. Activation and interaction with protein kinase C of a cytoplasmic tyrosine kinase, Itk/Tsk/Emt, upon FcεRI cross-linking on mast cells. J Immunol 1995;155:3556–3562.

38. Tang B, Mano H, Yi T, et al. Tec kinase associates with c-kit and is tyrosine phosphorylated and activated following stem cell factor binding. Mol Cell Biol 1994;14:8432–8437.

39. Price DJ, Kawakami Y, Kawakami T, et al. Purification of a major tyrosine kinase from RBL 2H3 cells phosphorylating FcεRI γ-cytoplasmic domain and identification as the Btk tyrosine kinase. Biochim Biophys Acta 1995;1265:133–142.

40. Park H, Wahl MI, Afar DEH, et al. Regulation of Btk function by a major autophosphorylation site within the SH3 domain. Immunity 1996;4:515–525.

41. Xu W, Harrison SC, Eck MJ. Three-dimensional structure of the tyrosine kinase c-Src. Nature (Lond) 1997;385:595–602.

42. Sicheri F, Moarefi I, Kuriyan J. Crystal structure of the Src family tyrosine kinase Hck. Nature (Lond) 1997;385:602–609.

43. Andreotti AH, Bunnell SC, Feng S, et al. Regulatory intramolecular association in a tyrosine kinase of the Tec family. Nature (Lond) 1997;385:93–97.

44. Li T, Tsukada S, Satterthwaite A, et al. Activation of Bruton's tyrosine kinase (BTK) by a point mutation in its pleckstrin homology (PH) domain. Immunity 1995;2:1–20.

45. Vihinen M, Cooper MD, de Saint Basile G, et al. BTKbase: a database of XLA-causing mutations. Immunol Today 1995;16:460–465.

46. Vihinen M, Iwata T, Kinnon C, et al. BTKbase, mutation database for X-linked agammaglobulinemia (XLA). Nucleic Acids Res 1996;24:160–165.

47. Sato S, Katagiri T, Takaki S, et al. IL-5 receptor-mediated tyrosine phosphorylation of SH2/SH3-containing proteins and activation of Bruton's tyrosine and Janus 2 kinases. J Exp Med 1994;180:2101–2111.

48. Matsuda T, Takahashi-Tezuka M, Fukada T, et al. Association and activation of Btk and Tec tyrosine kinases by gp130, a signal transducer of the interleukin-6 of cytokines. Blood 1995;85:627–633.

49. Go NF, Castle BE, Barrett R, et al. Interleukin 10, a novel B cell stimulatory factor: inresponsiveness of X-chromosome-linked immunodeficiency B cells. J Exp Med 1990;172:1625–1631.

50. Santos-Argumedo L, Lund FE, Heath AW, et al. CD38 unresponsiveness of *xid* B cells implicates Bruton's tyrosine kinase (*btk*) as a regulator of CD38 induced signal transduction. Int Immunol 1995;7:163–170.

51. Yamashita Y, Miyake K, Kikuchi Y, et al. A monoclonal antibody against a murine CD38 homologue delivers a signal to B cells for prolongation of survival and protection against apoptosis *in vitro*: unresponsiveness of X-linked immunodeficient B cells. Immunology 1995;85:248–255.

52. Hasbold J, Klaus GGB. B cells from CBA/N mice do not proliferate following ligation of CD40. Eur J Immunol 1994;24:152–157.

53. Khan WN, Alt FW, Gerstein RM, et al. Defective B cell development and function in Btk-deficient mice. Immunity 1995;3:283–299.

54. Takata M, Kurosaki T. A role for Bruton's tyrosine kinase in B cell antigen receptor-mediated activation of phospholipase C-γ2. J Exp Med 1996;184:31–40.

55. Rawlings DJ, Li Z, Fluckiger A-C, et al. Biphasic regulation of B cell receptor mediated Ca^{2+} flux by Bruton's tyrosine kinase. J Allergy Clin Immunol 1997;99:S469.

56. August A, Gibson S, Kawakami Y, et al. CD28 is associated with and induces the immediate tyrosine phosphorylation of the *tec* family kinase *itk/tsk/emt*. Proc Natl Acad Sci USA 1994; 91:9347–9351.

57. Gibson S, August A, Kawakami Y, et al. The EMT/ITK/TSK (EMT) tyrosine kinase is activated during TCR signaling. LCK is required for optimal activation of EMT. J Immunol 1996;156:2716–2722.

58. Tashiro M, Kawakami Y, Abe R, et al. Increased secretion of TNF-α by costimulation of mast cells via CD28 and FcεRI. J Immunol 1997;158:2382–2389.

59. Kawakami Y, Miura T, Bissonnette R, et al. Bruton's tyrosine kinase regulates apoptosis and JNK/SAPK kinase activity. Proc Natl Acad Sci USA. 1997;94:3938–3942.

60. Koch CA, Anderson D, Moran MF, et al. SH2 and SH3 domains: elements that control interactions of cytoplasmic signaling proteins. Science 1991;252:668–674.

61. Cohen GB, Ren R, Baltimore D. Modular binding domains in signal transduction proteins. Cell 1995;80:237–248.

62. Sudol M. The WW module competes with the SH3 domain? Trends Biochem Sci 1996;21:161–163.

63. Haslam RJ, Koide HB, Hemmings BA. Pleckstrin domain homology. Nature (Lond) 1993;363:309–310.

64. Mayer BJ, Ren R, Clark KL, et al. A putative modular domain present in diverse signaling proteins. Cell 1993;73:629–630.

65. Musacchio A, Gibson T, Rice P, et al. The PH domain: a common piece in the structural patchwork of signalling proteins. Trends Biochem Sci 1993;18:343–348.

66. Gibson TJ, Hyvonen M, Musacchio A, et al. PH domain: the first anniversary. Trends Biochem Sci 1994;19:349–353.

67. Yoon HS, Hajduk PJ, Petros AM, et al. Solution structure of a pleckstrin-homology domain. Nature (Lond) 1994;369:672–675.

68. Macias MJ, Musacchio A, Ponstingl H, et al. Structure of the pleckstrin homology domain from β-spectrin. Nature (Lond) 1994;369:675–677.

69. Ferguson KM, Lemmon MA, Schlessinger J, et al. Crystal structure at 2.2 Å resolution of the pleckstrin homology domain from human dynamin. Cell 1994;79:199–209.

70. Timm D, Salim K, Gout I, et al. Crystal structure of the pleckstrin homology domain from dynamin. Nat Struct Biol 1994;1:782–788.

71. Fushman D, Cahill S, Lemmon MA, et al. Solution structure of pleckstrin homology domain of dynamin by heteronuclear NMR spectroscopy. Proc Natl Acad Sci USA 1995; 92:816–820.

72. Ferguson KM, Lemmon MA, Schlessinger J, et al. Structure of the high affinity complex of inositol trisphosphate with a phospholipase C pleckstrin homology domain. Cell 1995;83:1037–1046.

73. Hyvonen M, Macias MJ, Nilges M, et al. Structure of the binding site for inositol phosphates in a PH domain. EMBO J 1995;14:4676–4685.

74. Koch WJ, Inglese J, Stone WC, et al. The binding site for the βγ subunits of the heterotrimeric G proteins on the β-adrenergic receptor kinase. J Biol Chem 1993;268:8256–8260.

75. Touhara K, Inglese J, Pitcher JA, et al. Binding of G protein βγ-subunits to pleckstrin homology domains. J Biol Chem 1994;269:10217–10220.

76. Wang D-S, Shaw R, Winkelmann JC, et al. Binding of PH domains of β-adrenergic receptor kinase and β-spectrin to WD40/β-transducin repeat containing regions of the β-subunit of trimeric G-proteins. Biochem Biophys Res Commun 1994;203:29–35.

77. Harlan JE, Hajduk PJ, Yoon HS, et al. Pleckstrin homology domains bind to phosphatidylinositol-4,5-bisphosphate. Nature (Lond) 1994;371:168–170.

78. Lemmon MA, Ferguson KM, O'Brien R, et al. Specific and high-affinity binding of inositol phosphates to an isolated pleckstrin homology domain. Proc Natl Acad Sci USA 1995;92: 10472–10476.

79. Salim K, Bottomley MJ, Querfurth E, et al. Distinct specificity in the recognition of phosphoinositides by the pleckstrin homology domains of dynamin and Bruton's tyrosine kinase. EMBO J 1996;22:6241–6250.

80. Yao L, Kawakami Y, Kawakami T. The pleckstrin homology domain of Btk tyrosine kinase interacts with protein kinase C. Proc Natl Acad Sci USA 1994;91:9175–9179.

81. Leitges M, Schmedt C, Guinamard R, et al. Immunodeficiency in protein kinase Cβ-deficient mice. Science 1996;273:788–791.

82. Yao L, Suzuki H, Ozawa K, et al. Interactions between protein kinase C and the Pleckscrin homology domains: inhibition by hosphatidylinositol 4,5-bisphosphate and phorbol 12-myristate 13-acetate. J Biol Chem 1997;272:13033–13039.

20
Activation of Heterotrimeric GTP-Binding Proteins

Ronit Sagi-Eisenberg

The Supergene Family of GTP-Binding Proteins

Guanosine triphosphate- (GTP-) binding proteins constitute a supergene family of proteins that utilize GTP binding and hydrolysis as a chemical switch. These proteins are activated by exchanging bound guanosine diphosphate (GDP) with GTP and are subsequently inactivated by hydrolyzing bound GTP to GDP.[1-3] Hence, by shuttling between an inactive, GDP-bound to an active, GTP-bound conformation, these proteins couple ligand binding to effector activation. This superfamily includes several subclasses of proteins, which share structural and functional homology in the core structure of their GTP-binding domains, but maintain distinct functions. Included are the small monomeric "Ras-like" GTPases, which play a major role in controlling cell proliferation,[4] protein traffic,[5] and cytoskeleton organization[6]; the heterotrimeric G proteins, originally believed to mediate the coupling of membrane surface receptors to their effector systems[1] but recently shown also to control membrane trafficking,[7] and finally the "large" GTP-binding proteins such as the signal recognition particle (SRP) and the SRP receptor, that control protein translocation across the endoplasmic reticulum (ER),[8] and dynamin, which controls endocytosis.[9] In this chapter, I focus on the role of the heterotrimeric G proteins in the control of mast cell exocytosis. For further information on the role of the monomeric GTPases, the reader is referred to Chapter 21 (this volume).

The heterotrimeric G proteins comprise a large family of structurally related proteins, which are all composed of three subunits, α, β, and γ. Based on structural similarities, they can be divided into four major subfamilies (Table 20.1). Up to date, 20 Gα-subunits have been identified. In addition, there are at least 5 distinct β-subunits and 6 distinct γ-subunits. Overall, these can generate more than 600 different forms of G proteins.

The Gα-subunit includes the GTP-binding site as well as the binding sites for the activating receptor and the downstream effector system. Upon binding of the ligand to its specific G protein-coupled receptor, an exchange reaction is facilitated whereby the G protein exchanges GDP for GTP. This exchange reaction causes subunit dissociation, thus allowing the GTP-bound Gα-subunit to bind and activate its specific effector system. The Gα-subunit thus acts as a shuttle between two membrane proteins, the G protein-coupled receptor and the effector (Fig. 20.1).

The Gα-subunit also functions as a GTPase that hydrolyzes bound GTP. Recent studies have demonstrated that this activity is subjected, at least in the case of a number of Gα-subunits, to tight regulation by the regulators of G-protein signaling (RGS; Fig. 20.1), a newly discovered family of GTPase activating proteins (GAPs).[10] Evidence for the existence of RGS first emerged from studies by De Vries et al.,[11] who made use of the two-hybrid system in an attempt to identify proteins that bind to Gi$_3$. The protein isolated by this approach, named

TABLE 20.1. The major families of heterotrimeric G proteins.

Family	Member	Modifying toxin
Gs	Gαs, Gαolf	Cholera toxin
Gi	Gαi₁, Gαi₂, Gαi₃, Gαo, Gαt₁, Gαt₂, Gαz, Gαgust	Pertussis toxin (except for Gαz)
Gq	Gαq, Gα₁₁, Gα₁₄, Gα₁₅, Gα₁₆	No effect
G12	Gα₁₂, Gα₁₃	No effect

GAIP for G-alpha interacting protein, shared a great deal of homology with a number of other proteins whose function was largely unknown. There were, however, several hints that helped unveil the secret. Ss2p, a yeast homolog of GAIP, was shown by Dohlman et al.[12] to prevent yeast cells from recovering from pheromone-induced growth arrest, thus raising the intruiging possibility that Ss2p binds the one of the G-protein subunits to control pheromone signaling. The finding that EGL-10, a *Caenorhabditis elegans* homolog of GAIP, also suppresses G protein-mediated signaling lent further support to the hypothesis that members of the RGS family regulate G-protein signaling. Indeed, subsequent studies by Berman et al.,[13]

Hepler et al.,[14] and Hunt et al.[15] have demonstrated that GAIP, RGS-4, and RGS-10, additional mammalian RGS, act as GAPs for the Gi or Gq subfamilies of Gα-subunits. The existence of RGS has important implications on G-protein functions. Variations among the rate of GTP hydrolysis could allow different effectors under the control of a single G protein to display different patterns of regulation. Gα-subunits under the control of a strong GAP activity will generate short-lived signals, as opposed to Gα-subunits that are under little or no GAP regulation and which will effect prolonged generation of second messengers.

The βγ-dimers were initially thought to participate mainly in the anchorage of the α-subunit to the membrane. However, it is well established now that both the α- and the βγ-subunits are involved in effector activation (see Fig. 20.1). Effectors activated by Gβγ include phospholipases C (PLC) and A₂, phosphatidylinositol-3 kinase, and K⁺ and Ca²⁺ ion channels.[16] More recent studies have indicated that mitogenic agonists that act via G protein-coupled receptors, such as thrombin, LPA, and bradykinin, stimulate protein tyrosine phosphorylation resulting in signaling interactions that are similar to those initiated by growth factor receptors, which possess an intrinsic tyrosine kinase activity or associate with cytosolic protein tyrosine kinases. Although the mechanism by which G proteins activate protein tyrosine kinases is as yet unknown, an extensive body of evidence[17-24] indicates that it involves the βγ-subunits of the activated G protein.

Tools to Study G-Protein Functions

Several reagents have been proven useful in studies aimed at exploring the possible involvement of a G protein in mediating a cellular function (Fig. 20.2). First are the non-hydrolyzable analogs of GTP such as GTP-γS and Gpp-NHp. These analogs can bypass the need for a ligand and therefore result in the direct activation of a G protein-mediated function. The toxins produced by the cholera and

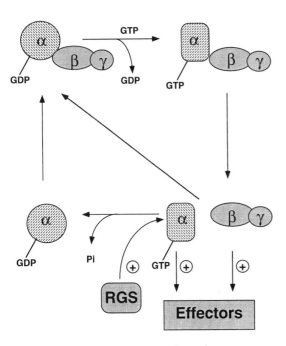

FIGURE 20.1. G-protein cycle.

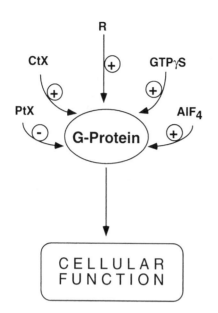

FIGURE 20.2. Tools to study G-protein functions.

Direct Evidence for the Involvement of GTP-Binding Proteins in Mast Cell Exocytosis

Gp: a G Protein Coupled to Phospholipase C

The first indication that activation of GTP-binding proteins could lead to mast cell activation and degranulation came from studies utilizing permeabilized mast cells. The finding, made by Gomperts,[27] that introduction of nonhydrolyzable analogs of GTP (Gpp-NHp, GTP-γS) into the cytosol of ATP^{-4} permeabilized and resealed cells leads to their degranulation in response to external Ca^{2+} has implicated a role for a GTP-binding protein in the gating of Ca^{2+}. Barrowman et al.[28] subsequently demonstrated that neomycin, an aminoglycoside antibiotic known to bind and inhibit phosphatidylinositol 4,5-bisphosphate (PIP_2) breakdown,[29] when cointroduced into the cytosol of the permeabilized mast cells inhibits Ca^{2+}-dependent and GTP-γS-induced secretion. This finding has indicated that mast cell degranulation triggered by GTP-γS, under these experimental conditions, is dependent on the activation of a PIP_2-hydrolyzing PLC. Furthermore, the direct demonstration by these researchers[28] that GTP-γS can activate PIP_2 breakdown when added to isolated neutrophil membranes has provided the first direct evidence for a role of a G protein in activating a PIP_2-hydrolyzing PLC, thus yielding the concept of Gp, a G protein that activates PLC. Indeed, we[30] have subsequently demonstrated that entrapment of GTP-γS in the ATP^{-4}-permeabilized and -resealed mast cells stimulates not only histamine secretion but also diacylglycerol (DAG) formation, measured by generation of its phosphorylation product phosphatidic acid (PA).

That the source of the PA formed was indeed from PIP_2 metabolism was indicated by the observation that GTP-γS-induced PA formation was inhibited when neomycin was introduced into the cells during their permeabilization. The stimulatory effect of GTP-γS on both secretion

pertussis bacteria comprise two additional powerful tools.[25] These toxins interfere with normal G-protein function by irreversibly ADP-ribosylating the Gα-subunits. For example, ADP ribosylation of the GTP-bound form of Gαs by cholera toxin (Ctx) inhibits the intrinsic GTPase activity of the α-subunit. ADP-ribosylated Gαs therefore remains permanently activated, resulting in a long-lived activation of its effector system. In contrast, ADP ribosylation by pertussis toxin (Ptx) of the α-subunits of the Gi, Go, and transducin classes of G proteins inhibits the G-protein function by preventing the dissociation of the G-protein subunits. ADP ribosylation by Ptx thus abrogates those cellular functions that are mediated by Ptx-sensitive G proteins. The complex of AlF_4^- binds to Gα-GDP, conferring on the α-subunit a GTP-bound conformation.[26] Thus, like GTP-γS, binding of AlF_4^- effects persistent activation of G proteins. Hence, activation of a cellular function by either nonhydrolyzable analogs of GTP or by AlF_4^-, or its modulation by Ctx or Ptx, would strongly suggest the involvement of a heterotrimeric G protein in the affected response (Fig. 20.2).

and PA production could be inhibited by the simultaneous introduction of excess of GDP-βS (1 mM) into the cells, as is expected if a G protein is involved. Similar results were also obtained by Ali et al.[31] using the rat basophilic leukemia (RBL-2H3) cells, a tumor analog of mucosal mast cells, where the introduction of GTP-γS into streptolysin-O (SLO) -permeabilized cells resulted in PIP_2 breakdown.

The original idea of Gomperts[27] that Gp is involved in Ca^{2+} gating only recently gained direct support when the close relationship between PIP_2 hydrolysis and Ca^{2+} influx became apparent. In agreement with the hypothesis, first proposed by Putney et al.[32] and referred to as the capacitance entry, Ca^{2+} influx into mast cells was shown by Hoth and Penner[33] to switch on when the intracellular pools of Ca^{2+} are depleted. This depletion follows the generation of inositol trisphosphate (IP_3), the breakdown product of PIP_2, which binds to its intracellular receptor thus resulting in Ca^{2+} release from internal stores. Hence, by activating the PIP_2-hydrolyzing PLC enzyme and stimulating production of IP_3, the putative Gp protein causes the mobilization of Ca^{2+} from intracellular stores, thereby facilitating the opening of a plasma membrane Ca^{2+} channel and regulating Ca^{2+} gating.

Although the G protein designated Gp has not been identified as a molecular species, the current functional evidence indicates that Gp constitutes several distinct entities, all belonging to the heterotrimeric family of GTP-binding proteins. As noted by Nakamura and Ui,[34] activation of PIP_2 hydrolysis by the class of positively charged agonists collectively known as the basic secretagogues of mast cells[35] is sensitive to Ptx. This implicates a Ptx-sensitive G protein in fulfilling the role of Gp in these agonist signaling pathways. As shown by Ali et al.,[36] a Ptx-sensitive G protein also couples adenosine receptor, presumably A_3 to PIP_2 hydrolysis in the RBL cells. In contrast, the Ptx-insensitive Gq protein presumably mediates the activation of PLC brought about by the muscarinic M1 receptor, transfected into RBL cells.[37]

G_E: a G Protein Regulating Late Steps in Exocytosis

Evidence for the involvement of yet another GTP protein in the control of exocytosis came from studies by Neher[38] and Barrowman et al.[28] showing that, when introduced into either patch-clamped or SLO-permeabilized mast cells, GTP-γS could also stimulate exocytosis independently of PLC. Under these conditions, GTP-γS served as a sufficient stimulus, with Ca^{2+} serving merely to accelerate the response. On the basis of these observations, the involvement of a late-acting GTP-binding protein (G_E), presumably activated by a Ca^{2+}-binding protein (C_E), which has the characteristics of an intracellular pseudoreceptor, has been suggested.[39]

Recent studies have indicated the existence of a functional G_E protein in exocytosis of a variety of other secretory cells. These include eosinophils, neutrophils, melanotrophs, and the insulin-secreting RINm5F β-cells.[39] In the latter case, as shown by Lang et al.,[40] yet another G protein appears to negatively regulate this exocytotic process. In a similar fashion, Vitale et al.[41] have noted that two distinct Ptx-sensitive G proteins act in series to control exocytosis in chromaffin cells. One G protein negatively regulates the ATP-priming step, whereas the second stimulates the late Ca^{2+}-dependent and ATP-independent step (see following).

Characterization of G_E

In attempting to identify the G_E protein one needs to consider both the monomeric small Ras-like GTPases as well as the heterotrimeric G proteins. Ras-related proteins have been implicated as mediators of vesicle fusion at a number of stages in the pathway leading to both constitutive and regulated secretion.[5] Moreover, in mast cells, Oberhauser et al.[42] have shown that synthetic peptides corresponding to the putative effector domain of the small GTPase Rab-3A can stimulate exocytosis, thus supporting the notion that Rab-3 homologs are likely to be key signaling molecules in the control exocytosis. Nevertheless, based on the characteristics of G_E, Lillie and Gomperts[43] have proposed that G_E belongs to the heterotrimeric class of G proteins. This proposal was based mainly on the observation that when mast cells are deprived of Mg^{2+}, GTP becomes equipotent with GTP-γS, as is expected from a heterotrimeric G protein whose GTPase activity requires the presence of Mg^{2+}. This idea gained ultimate support from our studies on the mechanism of action of the basic secretagogues of mast cells.

G Proteins as Mediators of the Peptidergic Pathway of Mast Cell Activation

The major pathways of activating exocytosis in mast cells include the immunological trigger, which as was initially shown by Segal et al.[44] involves the aggregation of the cell high-affinity receptors for immunoglobulin E (FcϵRI) by

their corresponding antigens, and the peptidergic pathway, achieved by polycationic compounds, including positively charged peptides such as substance P and bradykinin, various amines such as the synthetic compound 48/80 (c48/80), and naturally occurring polyamines (the basic secretagogues of mast cells.[35] As shown by us[45,46] and by Katakami et al.,[47] mast cells can also be activated, pharmacologically, by the combination of an ionophore for Ca^{2+} and a phorbol ester such as TPA that activates protein kinase C (PKC). Indeed, an early event in the stimulation of mast cells through both the immunological and the peptidergic trigger is the activation of PLC, resulting in the breakdown of PIP_2 and the subsequent activation of PKC and rise in cytosolic Ca^{2+} (Fig. 20.3).

That G proteins indeed mediate exocytosis, stimulated by physiological secretagogues of mast cells, came from the observations by Nakamura et al.[34,48] that pretreatment with Ptx

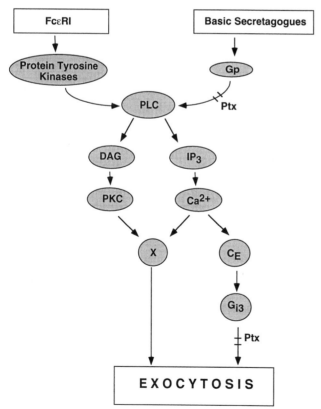

FIGURE 20.3. Mechanisms involved in activation of mast cell exocytosis.

inhibited histamine release as well as the activation of phospholipases C and A_2 brought about by basic secretagogues such as c48/80 or peptides such as substance P (shown by Colombo et al.[49]), bradykinin (shown by Bueb et al.[50]), or mastoparan (shown by Mousli et al.[51] and Aridor et al.[52]). In contrast, neither histamine release nor the activation of phospholipases by the immunological IgE-mediated trigger are sensitive to this toxin. These findings have implicated a Ptx-sensitive G-protein in mediating the peptidergic but not the immunological trigger for mast cell exocytosis (Fig. 20.3). Indeed, we[52] and others[53] have shown that basic secretagogue-induced secretion, from permeabilized, cells is strictly dependent on the provision of GTP. This response is washed out if the intracellular concentration of GTP is reduced. Hence, although both IgE- and basic secretagogue-induced exocytosis are accompanied by PIP_2 breakdown, there are a number of important differences in their mode of action. FcεRI-mediated exocytosis is relatively slow, extending over a period of about 2 min. It is enhanced by phosphatidylserine, is strictly dependent on external Ca^{2+}, and as previously mentioned is PtX insensitive. In contrast, basic secretagogue-induced secretion is far more rapid (10 s), is independent of external Ca^{2+}, and is PtX sensitive. Hence, basic secretagogues appear to induce secretion by a common mechanism, which is different from that mediated by FcεRI. Therefore, additional signals, which are presuambly activated in parallel to PLC, must also contribute to exocytosis.

G Proteins Are the Cellular Receptors of Basic Secretagogues

Substance P belongs to the family of tachykinins, whose receptors have been cloned by Masu et al.[54] However, several observations have strongly suggested that activation of mast cells by substance P may not be mediated by its conventional cell-surface receptor. For example, the activation of tachykinin receptors is dependent on the carboxy-terminal end of the tachykinins. In contrast, it is the N-terminal

domain of substance P (Arg-Pro-Lys-Pro-) that is required for mast cell activation. As was shown by Repke et al.,[55,56] the carboxy-terminus can be replaced by a simple 12-carbon chain, and not only does this modification not impair the stimulatory activity of substance P, it increases its potency. Also the concentrations of substance P required to activate mast cells are different from those needed for receptor activation. Although the tachykinin receptor is activated by nanomolar concentrations of substance P, concentrations in the micromolar range are required to activate mast cells. Finally, as shown by Krumins and Broomfield,[57] known antagonists of the receptor for substance P do not antagonize the ability of substance P to trigger mast cell secretion but they do in fact act as agonists themselves. Taken together with the facts that the basic secretagogues of mast cells include a large number of members, all sharing similar structural features that are essential for their activity (e.g., a combination of positive charges associated with a hydrophobic domain, like a stretch of hydrophobic amino acids in the carboxy-terminus of substance P or aromatic rings in the case of c48/80; see Fig. 20.4), the existence of classical receptors for these secretagogues on mast cells has been questioned. Therefore, the finding by Higashijima et al.[58] that the wasp venom-derived peptide mastoparan, a potent basic secretagogue, acts as a receptor mimetic agent which directly activates purified G proteins raised the intriguing possibility that histamine secretion induced by all polybasic molecules may result from their direct interaction and subsequent activation of G proteins. Indeed, working independently, both our group[52,59] and others[60,61] have demonstrated that basic secretagogues such as c48/80, neomycin, substance P, mast cell-degranulating peptide (MCD), and a variety of kinins can stimulate the GDP–GTP exchange reaction on the Gα-subunit, activate GTPase activity, and inhibit PtX-catalyzed ADP ribosylation.

Collectively, these studies have established that basic secretagogues activate mast cells in a receptor-independent manner by directly activating the G proteins that control excytosis. This activation presumably involves the inser-

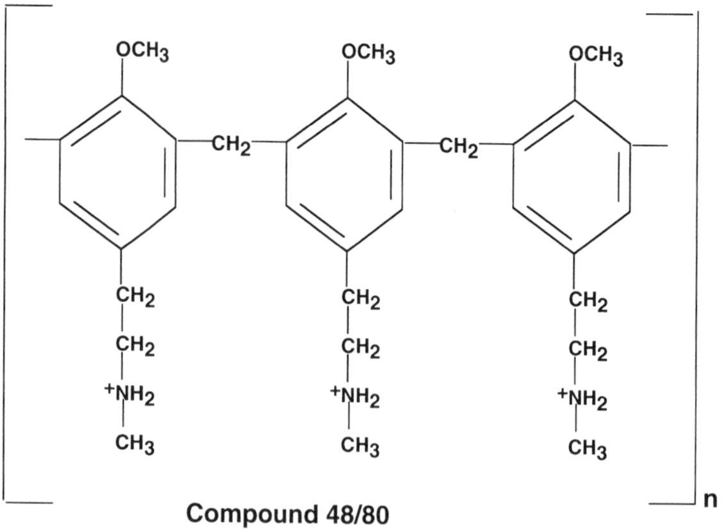

Compound 48/80

Arg-Pro-Lys-Pro-Gln-Gln-Phe-Phe-Gly-Leu-Met

Substance P

FIGURE 20.4. Structure of the basic secretagogues of mast cells: compound 48/80 and substance P.

tion of the basic secretagogue hydrophobic moiety into the membrane, which then enables the positively charged domain to interact and directly activate the G protein. Indeed, spectroscopic studies by Ortner and Chingell[62] on the binding properties of spin-labeled c48/80 have indicated that c48/80 penetrates the membrane up to a point where it is no longer exposed to the external aqueous environment. In this context, it should be noted that positively charged clusters, rather than an amphiphilic structure, are required for G-protein activation. In studies by Voss et al.[63] in which the ability of peptides, derived from the third cytoplasmic loops of the dopamine (D1 and D2) or the β1-adrenergic receptors, to activate directly G-proteins was examined, no correlation was found to exist between the α-helical structure of the peptides and their potency to stimulate G-protein activation. Rather, the most potent peptide did not show any α-helical structure but it contained the highest charge density. This observation agrees nicely with our structure–function relationship studies on the stimulatory activity of neomycin and several of its analogs (geneticin, gentamicin, and paromomycin), which act as basic secretagogues when added to intact mast

cells. We could demonstrate a close correlation between the positive charge of the analogue and its effectiveness in inducing the secretory response.[59]

That the basic secretagogues indeed function as receptor mimetic agents was further illustrated by the findings, by Mousli et al.[64] and Weingarten et al.,[65] that in a fashion similar to the interactions between Gα-subunits and their cognate receptors basic secretagogues also interact with the carboxy-terminus end of Gα. Taken together, these results therefore defined the basic secretagogues of mast cells as receptor mimetic agents that can serve as powerful tools to study the possible involvement of G proteins in various cellular functions.

Activation of Mast Cell Exocytosis by the Heterotrimeric G Protein Gi$_3$

G$_E$ Is a Ptx Substrate

The ability of basic secretagogues to directly activate G proteins suggested to us that they may also activate directly the putative G$_E$ protein, thus evoking exocytosis independently of PLC. In this case the overall sensitivity of the exocytotic process to PtX may reflect the fact that the putative G_E protein serves as a substrate for PtX-catalyzed ADP ribosylation. Indeed, our dose–response analysis studies have revealed that c48/80 can evoke exocytosis at concentrations that do not activate PLC.[52] This result suggested that the functionally defined Gp and G_E proteins are distinct entities, which are both activated by c48/80, but with different dose–response relationships. This notion was further supported by the finding of Muhlen et al.[66] that different stereoisomers of GTP-γS differentially activate the two functions. We could further demonstrate that basic secretagogues can induce exocytosis also in the presence of internally applied neomycin, unveiling their ability to directly activate the late-acting G_E protein.[59,67] The finding that treatment with Ptx still inhibits exocytosis proceeding under these conditions provided the first indication that a Ptx-sensitive heterotrimeric G protein controls late steps in regulated exocytosis.

Gi_3 Serves as G_E in Basic Secretagogue-Induced Secretion

Toward the identification of G_E, we have analyzed the Ptx-sensitive G proteins present in mast cells and have identified them as Gi_2 and Gi_3.[68] To determine which of the two fulfills the function of G_E, we have adopted the strategy described by Simonds et al.[68] whereby synthetic peptides that correspond to the C-terminal sequences of Gα-subunits, or antibodies directed against these peptides, prevent coupling of G proteins to their respective receptors. Taking this approach, we were able to show that the introduction into ATP^{-4}-permeabilized mast cells, of a synthetic peptide that corresponds to the sequence of the C-terminal end of Gαi_3, specifically blocked basic secretagogue-induced secretion.[67] Similar results were obtained when introducing antibodies directed against this peptide (EC antibodies). Moreover, these antibodies could block Ca^{2+}-dependent and GTP-γS-activated secretion from SLO-permeabilized cells.[67] Taken together, our re-

sults have indicated that Gi_3 functions as the G_E protein that controls late steps in regulated exocytosis triggered by either the basic secretagogues or Ca^{2+} and GTP-γS.

A G_E protein has been implicated in regulating exocytosis also in other cell types, including the chromaffin cells (see earlier). That Gi_3 also fulfills this role in chromaffin cells has emerged from a recent study by Vitale et al.,[69] which demonstrated the sequential involvement of G_o in the control of the ATP-dependent priming reaction and of Gi_3 in the control of the late Ca^{2+}-dependent step. Hence, in a similar fashion to their inhibitory effects on mast cell exocytosis, a synthetic peptide comprising the C-terminal end of Gαi_3 and anti-Gαi_3 antibodies were found to inhibit Ca^{2+}-dependent secretion from the chromaffin cells.

Cellular Localization of Gi_3

We have shown that in connective tissue mast cells (CTMC), exemplified by the rat peritoneal mast cells, Gαi_3 is localized to both the Golgi and the plasma membrane compartments.[67] Interestingly, the subcellular localization of Gi_3 appears to be cell type specific. In studies by Stow et al.,[70,71] Gαi_3 has been localized to both the Golgi and the apical plasma membrane border in renal epithelial cells, but it is almost exclusively restricted to the Golgi in LLC-Pk1 cells or fibroblasts. The carboxy-terminal end of Gαi_3 appears to include a sequence responsible for the targeting of the protein to the Golgi membrane. This targeting may be a saturable process, because, as shown by Hermouet et al.,[72] overexpression of Gαi_3 targets the protein to the plasma membrane.

The distinct localizations of Gαi_3 are apparently associated with distinct regulatory functions of this protein. The membrane-bound form of Gαi_3 was shown by us[67] and by Vitale et al.[69] to be responsible for stimulating regulated exocytosis in mast and chromaffin cells, respectively, it is the Golgi-bound form that regulates constitutive traffic from the ER to Golgi, as shown by Wilson et al.,[73] and from the Golgi to the plasma membrane, as shown by Stow et al.[74]

Possible Effectors for Gi₃

The intracellular effector(s) that is controlled by Gi₃ has not been defined yet. The finding by Cantiello et al.[75] that Gαi₃ activates PLA₂ in the A6 epithelial cells makes PLA₂ an attractive candidate to consider. Indeed, as shown by Nakamura and Ui,[34] basic secretagogues activate PLA₂ in a Ptx-sensitive manner. Moreover, McGivney et al.[76] have shown that inhibitors of PLA₂ inhibit both arachidonic acid release and histamine secretion, thus suggesting a possible role for PLA₂ in the stimulation of exocytosis. In support of this notion, it has recently been reported by Vitale et al.[69] that in chromaffin cells the inhibitory Gαi₃ peptide, which inhibits exocytosis, also inhibits the generation of arachidonic acid, induced by mastoparan. Against this hypothesis, however, stand the observations by Churcher et al.[77] that despite the general similarity of the conditions causing exocytosis and arachidonic acid release, under some circumstances it is possible to obtain exocytosis without measurable release of arachidonic acid.

Another effector system that might be regulated by Gi₃ is the cytoskeleton network. As noted by Norman et al.,[78] treatment of intact mast cells with c48/80 produces cytoskeletal changes that are very similar to those induced by GTP-γS in permeabilized cells. These changes are characterized by the disassembly of F-actin in the cortical region and the appearance of actin filaments in the cell interior. The disassembly of the cortex was found to be regulated by an AlF₄⁻-sensitive GTP-binding protein. This finding therefore raises the question as to whether this protein might be Gi₃.

The Involvement of Ctx-Sensitive G Proteins in Mast Cell Signaling

While resistant to Ptx treatment, there is evidence for the involvement of GTP-binding proteins also in FcεRI-mediated release. For example, Wilson et al.[79] have shown that pretreatment of RBL cells with mycophenolic acid, which reduces the endogenous levels of GTP,

results in the reversible inhibition of FcεRI-mediated secretion and Ca²⁺ influx. Ali et al.[31] have shown that the introduction of GDP-βS into SLO-permeabilized RBL cells inhibits both FcεRI-mediated PLC activation and exocytosis. Also, shown by Ali et al.,[36] activation of the A3 adenosine receptors present on RBL cells markedly potentiates FcεRI-induced secretion in both a PtX- and a CtX-sensitive manner. Both McCloskey[80] and Narasimhan et al.[81] have demonstrated that treatment of RBL cells with Ctx, which normally ADP-ribosylates and activates Gαs, increases antigen-induced ⁴⁵Ca²⁺ uptake and increase in the cytosolic concentrations of Ca²⁺, antigen-induced PIP₂ hydrolysis, and the rate of antigen-induced secretion. These Ctx stimulatory effects are not mimicked by agents that elevate cAMP levels, indicating that they do not result from the activation of adenylyl cyclase by the ADP-ribosylated and activated Gs protein. Thus, the simplest explanation would be that a Ctx-sensitive G protein, possibly Gs itself, regulates FcεRI signaling. Such an interaction between the FcεRI and a G protein has however not been demonstrated as yet.

The Role of Gαz in Mediating FcεRI-Induced Responses

Dexamethasone was shown by Collado et al.[82] to downregulate FcεRI-induced responses. Because dexamethasone was also found by Hide et al.[83] to decrease the amount of mRNA encoding the Ptx-insensitive G protein Gαz, it was suggested that Gαz is a potential candidate for coupling the FcεRI to effector systems that mediate exocytosis. Evidence in favor of this notion was recently obtained in a study by Baumgartner et al.[84] utilizing RBL-2H3 cells, which were stably transfected to overexpress Gαz. In antigen-stimulated cells, overexpression of Gαz significantly enhanced the production of TNF-α, one of the cytokines produced and released by activated mast cells. This effect was restricted to this cytokine synthesis, as neither antigen-induced PIP₂ breakdown nor secretion of hexosaminidase was affected.

Future Perspectives

The studies described in this chapter provide clear indications for the involvement of heterotrimeric G proteins in a variety of mast cell functions. They do however also raise several intriguing questions. Gi_3, for example, is definitely involved in the propagation of the peptidergic signaling pathways. Hence, activation of Gi_3 provides a suffcient signal for exocytosis in CTMC, but is this activation also essential? The fact that, in the same cells, the immunological trigger as well as secretion triggered by an ionophore for Ca^{2+} or the combination of Ca^{2+} ionophore and TPA are insensitive to Ptx treatment indicates that mast cells possess multiple signaling pathways, at least one of which is mediated by the PLC-derived second messengers Ca^{2+} and PKC and does not involve the participation of the Ptx-sensitive Gi_3 protein (Fig. 20.3). What is then the relative contribution of the Gi_3 signaling pathway to the overall secretory process elicited by the basic secretagogues under physiological conditions? This question surely has an enormous clinical impact for the design of future therapeutics, considering the fact that the peptidergic signaling pathway is responsible for a large number of diseases (e.g., psychogenic asthma, psoriasis, interstitial cystitis, bowel diseases, migraines, multiple sclerosis) associated with neurogenic inflammation, a type of allergy initiated by the chemical mediators released from substance P-activated mast cells.

This brings us to yet another question, which is why does Gi_3 not play any role in FcεRI- or Ca^{2+} ionophore-induced secretion in intact cells? The answer to this question may be related to the characteristics of the physiological activator of Gi_3. This putative receptor has not been identified as yet, but it has been suggested by Howell et al.[85] to be a Ca^{2+}-binding protein, designated C_E. Thus, if C_E is a low-affinity Ca^{2+}-binding protein, then the rise in cytosolic Ca^{2+}, which occurs during antigen-induced stimulation, may be too low to allow its activation. This would explain why the contribution of Gi_3 becomes apparent only if directly activated by the receptor mimetic basic secretagogues, which bypass the need for Ca^{2+} (see Fig. 20.3).

However, although this argument could provide an explanation as to why Gi_3 is not involved in the FcεRI-induced response, it could not explain why Gi_3 is activated by Ca^{2+} in permeabilized cells, but it is not activated by an ionophore for Ca^{2+} in intact cells. The reason for this discrepancy is presently unknown, but it may be related to the fact that the GTPase activity of Gi_3 is regulated by GAIP, a member of the RGS family.[11] It should be noted that activation of Gi_3 by Ca^{2+} was demonstrated in permeabilized cells using the combination of Ca^{2+} and nonhydrolyzable analogs of GTP, such as GTP-γS. A plausible explanation would then be that because of the strong GTPase activity manifested by Gi_3/GAIP, the signal elicited by Ca^{2+} is too short lived to constitute a significant contribution. It would then appear that for Gi_3 to be able to activate an effector system, when by itself activated by Ca^{2+} (but not by the receptor mimetic agents), its intrinsic GTPase activity must be blocked, for example, by the use of nonhydrolyzable analogs of GTP. Future studies using GTPase-defective mutants of Gi_3 or dominant negative mutants of GAIP will be able to test this hypothesis.

Another open question deals with the cell-type-specific activation of Gi_3 by the basic secretagogues. These agonists trigger secretion from CTMC but not from the less mature mucosal mast cells (MMC). Because the cellular localization of Gi_3 is cell type specific, an interesting possibilty might be that the lack of responsiveness of MMC to basic secretagogues may result from a different cellular localization of Gi_3 in these cells. For example, if Gi_3 resides in the Golgi compartment in the MMC, this would explain why these cells do not respond to the peptidergic pathway. In this context, it is interesting to note that Swieter et al.[86] have shown that MMC cocultured with fibroblasts acquire responsiveness to basic secretagogues. Whether this process is accompanied by a change in the cellular localization of Gi_3 is undoubtedly a possibility that is worth testing.

Also awaiting to be resolved are questions regarding the downstream effector(s) of Gi_3, the identity of the putative Ca^{2+} receptor C_E, and the identity of Gp in the peptidergic pathway. Additional questions concern the role of Gz and the mechanism by which this G protein enhances the FcεRI-stimulated synthesis of TNF-α. Future studies should provide answers to these questions and shed more light on our understanding of the multiple signaling pathways that control mast cell exocytosis.

References

1. Gilman A. G proteins: tranducers of receptor generated signals. Annu Rev Biochem 1987; 56:615–649.
2. Neer EJ, Smith TF. G protein heterodimers: new structures propel new questions. Cell 1996;84:175–178.
3. Hamm HE, Gilchrist. Heterotrimeric G proteins. Curr Opin Cell Biol 1996;8:189–196.
4. Marshall MS. Ras target proteins in eukaryotic cells. FASEB J 1995;9(13):1311–1318.
5. Balch WE. Small GTP-binding proteins in vesicular transport. Trends Biol Sci 1990;15:473–477.
6. Tapon N, Hall A. Rho, Rac and Cdc42 GTPases regulate the organization of the actin cytoskeleton. Curr Opin Cell Biol 1997;9(1):86–92.
7. Bomsel M, Mostov K. Role of heterotrimeric G proteins in membrane traffic. Mol Biol Cell 1992;3:1317–1328.
8. Lutcke H. Signal recognition particle (SRP), a ubiquitous initiator of protein translocation. Eur J Biochem 1995;228(3):531–550.
9. McClure SJ, Robinson PJ. Dynamin, endocytosis and intracellular signalling. Mol Membr Biol 1996;13(4):189–215.
10. Koelle MR. A new family of G-protein regulators—the RGS proteins. Curr Opin Cell Biol 1997;9(2):143–147.
11. De Vries L, Mousli M, Wurmser A, et al. GAIP, a protein that specifically interacts with the trimeric G protein G alpha-i_3, is a member of a protein family with a highly conserved core domain. Proc Natl Acad Sci USA 1995;92:11916–11920.
12. Dohlman HG, Song J, Ma D, et al. Sst2, a negative regulator of pheromone signalling in the yeast *Saccharomyces cerevisiae*: expression, localization, and genetic interaction and physical association with Gpa₁ (the G-protein alpha subunit). Mol Cell Biol 1996;16(9):5194–5209.
13. Berman DM, Wilkie TM, Gilman AG. GAIP and RGS4 are GTPase-activating proteins for the Gi subfamily of G protein alpha subunits. Cell 1996;86:445–452.
14. Hepler JR, Berman DM, Gilman AG, et al. RGS4 and GAIP are GTPase-activating proteins for Gq alpha and block activation of phospholipase C beta by gamma-thio-GTP-Gq alpha. Proc Natl Acad Sci USA 1997;94(2):428–432.
15. Hunt TW, Fields TA, Casey PJ, et al. RGS10 is a selective activator of G alpha$_i$ GTPase activity. Nature (Lond) 1996;383(6596):175–177.
16. Clapham DE, Neer EJ. New roles for G-protein beta gamma-dimers in transmembrane signalling. Nature (Lond) 1993;365(6445):403–406.
17. Coso OA, Teramoto H, Simonds WF, et al. Signaling from G protein-coupled receptors to c-Jun kinase involves beta gamma subunits of heterotrimeric G proteins acting on a Ras and Racl-dependent pathway. J Biol Chem 1996; 271:3963–3966.
18. Dikic I, Tokiwa G, Lev S, et al. A role for Pyk2 and Src in linking G-protein-coupled receptors with MAP kinase activation. Nature (Lond) 1996;383:547–550.
19. Hawes BE, van Biesen T, Koch WJ, et al. Distinct pathways of Gi- and Gq-mediated mitogen-activated protein kinase activation. J Biol Chem 1995;270(29):17148–17153.
20. Hawes BE, Luttrell LM, van Biesen T, et al. Phosphatidylinositol 3-kinase is an early intermediate in the G beta gamma-mediated mitogen-activated protein kinase signaling pathway. J Biol Chem 1996;271(21):12133–12136.
21. Koch WJ, Hawes BE, Allen LF, et al. Direct evidence that Gi-coupled receptor stimulation of mitogen-activated protein kinase is mediated by G beta gamma activation of p21ras. Proc Natl Acad Sci USA 1994;91(26):12706–12710.
22. Lev S, Moreno H, Martinez R, et al. Protein tyrosine kinase PYK2 involved in Ca²⁺-induced regulation of ion channel and MAP kinase functions. Nature (Lond) 1995;376(6543):737–745.
23. Lopez-llasaka M, Crespo P, Giuseppe P, et al. Linkage of G-protein-coupled receptors to the MAPK signaling pathway through PI 3-kinase. Science 1997;275:394–397.
24. Touhara K, Hawes BE, van Biesen T, et al. G protein beta gamma subunits stimulate phosphorylation of Shc adapter protein. Proc Natl Acad Sci USA 1995;92(20):9284–9287.
25. Moss J. Signal transduction by receptor-responsive guanyl nucleotide-binding proteins: modulation by bacterial toxin-catalyzed ADP-ribosylation. Clin Res 1987;35(5):451–458.

26. Yatani A, Brown AM. Mechanism of fluoride activation of G protein-gated muscarinic atrial K$^+$ channels. J Biol Chem 1991;266(34):22872–22877.

27. Gomperts BD. Involvement of guanine nucleotide-binding protein in gating of calcium by receptors. Nature (Lond) 1983;306:64–66.

28. Barrowman MM, Cockroft S, Gomperts BD. Two roles for guanine nucleotides in the stimulus-secretion sequence of neutrophils. Nature (Lond) 1986;319:504–507.

29. Schacht J. Purification of polyphosphoinositides by chromatography on immobilized neomycin. J Lipid Res 1978;19:1063–1067.

30. Aridor M, Traub LM, Sagi-Eisenberg, R. Exocytosis in mast cells by basic secretagogues: evidence for direct activation of GTP-binding proteins. J Cell Biol 1990;111(3):909–917.

31. Ali H, Cunha-Melo JR, Beaven M. Receptor-mediated release of inositol 1,4,5-trisphosphate and inositol 1,4-bisphosphate in rat basophilic leukemia RBL-2H3 cells permeabilized with streptolysine O Biochem Biophys Acta 1989;1010:88–99.

32. Putney JJ, Takemura H, Hughes AR, et al. How do inositol phosphates regulate calcium signaling? FASEB J 1989;3(8):1899–1905.

33. Hoth M, Penner R. Depletion of intracellular calcium stores activates a calcium current in mast cells. Nature (Lond) 1992;355:353–355.

34. Nakamura T, Ui M. Simultaneous inhibitions of inositol phospholipid breakdown, arachidonic acid release, and histamine secretion in mast cells by islet-activating protein, pertussis toxin. J Biol Chem 1985;260:3584–3593.

35. Lagunoff D, Martin TW, Read G. Agents that release histamine from mast cells. Annu Rev Pharmacol Toxicol 1983;23:331–351.

36. Ali H, Cunha-Melo JR, Saul WF, et al. Activation of phospholipase C via adenosine receptor provides synergistic signals for secretion in antigen-stimulated RBL-2H3 cells. J Biol Chem 1990;265:745–753.

37. Choi OH, Lee JH, Kassessinoff T, et al. Antigen and carbachol mobilize calcium by similar mechanisms in a transfected mast cell line (RBL-2H3 cells) that expresses ml muscarinic receptors. J Immunol 1993;151(10):5586–5595.

38. Neher E. Inositol 1,4,5-triphosphate and GTP-γS induce calcium transients in isolated rat peritoneal mast cells. J Physiol 1986;381:71.

39. Gomperts BD. GE: a GTP-binding protein mediating exocytosis. Annu Rev Physiol 1990; 52:591–606.

40. Lang J, Nishimoto I, Okamoto T, et al. Direct control of exocytosis by receptor-mediated activation of the heterotrimeric GTPases Gi and Go or by the expression of their active Gα subunits. EMBO J 1995;14:3635–3644.

41. Vitale N, Aunis D, Bader MF. Distinct heterotrimeric GTP-binding-proteins act in series to control the exocytotic machinery in chromaffin cells. Cell Mol Biol 1994;40(5):707–715.

42. Oberhauser AF, Monck JR, Balch WE, et al. Exocytotic fusion is activated by Rab3a peptides. Nature (Lond) 1992;360(6401):270–273.

43. Lillie TH, Gomperts BD. Kinetic characterization of guanine-nucleotide-induced exocytosis from permeabilized rat mast cells. Biochem J 1993;290:389–394.

44. Segal DM, Taurog J, Metzger H. Dimeric immunoglobulin E serves as a unit signal for mast cell degranulation. Proc Natl Acad Sci USA 1977;74: 2993–2997.

45. Sagi-Eisenberg R, Pecht I. Protein kinase C, a coupling element between stimulus and secretion of basophils. Immunol Lett 1984;8:237–241.

46. Sagi-Eisenberg R, Lieman H, Pecht I. Protein kinase C regulation of the receptor-coupled calcium signal in histamine-secreting rat basophilic leukemia cells. Nature (Lond) 1985; 313:59–60.

47. Katakami Y, Kaibuchi K, Sawamura Y, et al. Synergistic action of protein kinase C and calcium for histamine release from rat peritoneal mast cells. Biochem Biophys Res Commun 1984;121:573–578.

48. Nakamura T, Ui M. Islet activating protein, pertussis toxin, inhibits calcium-induced and guanine nucleotide dependent releases of histamine and arachidonic acid from rat mast cells. FEBS Lett 1984;173:414–418.

49. Columbo M, Horowitz EM, Kagey SA, et al. Substance P activates the release of histamine from human skin mast cells through a pertussis toxin-sensitive and protein kinase C-dependent mechanism. Clin Immunol Immunopathol 1996;81(1):68–73.

50. Bueb JL, Mousli M, Landry Y, et al. A pertussis toxin-sensitive G protein is required to induce histamine release from rat peritoneal mast cells by bradykinin. Agents Actions 1990;30(1–2):98–101.

51. Mousli M, Bronner C, Bueb JL, et al. Activation of rat peritoneal mast cells by substance P and mastoparan. Eur J Pharmacol 1989;250:329–335.

52. Aridor M, Sagi-Eisenberg R. The role of GTP-binding proteins in the control of mast cell exo-

cytosis. Cell Cytokine Networks Tissue Immun 1991;11:169–175.

53. Penner R, Pusch M, Neher E. Washout phenomena in dialyzed mast cells allow discrimination of different steps in stimulus-secretion coupling. Biosci Rep 1987;7:313–321.

54. Masu Y, Nakayama K, Tamaki H, et al. cDNA cloning of bovine substance-K receptor through oocyte expression system. Nature (Lond) 1987;329:836–838.

55. Repke H, Bienert M. Mast cell activation—a receptor-independent mode of substance P action? FEBS Lett 1987;221:236–240.

56. Repke H, Piotrowski W, Bienert M, et al. Histamine release induced by Arg-Pro-Lys-Pro (CH2)11CH3 from rat peritoneal mast cells. J Pharmacol Exp Therap 1987;243:317–321.

57. Krumins SA, Broomfield CA. C-terminal substance P fragments elicit histamine release from a murine mast cell line. Neuropeptides 1993;24(1):5–10.

58. Higashijima T, Uzu S, Nakajima T, et al. Mastoparan, a peptide toxin from wasp venom, mimics receptors by activating GTP-binding regulatory proteins (G-proteins). J Biol Chem 1988;263:6491–6494.

59. Aridor M, Sagi-Eisenberg R. Neomycin is a potent secretagogue of mast cells that directly activates a GTP-binding protein involved in exocytosis. J Cell Biol 1990;111:2885–2891.

60. Beub JL, Mously M, Bronner C, et al. Activation of Gi-like proteins, a receptor-independent effect of kinins in mast cells. Mol Pharmacol 1990;38:816–822.

61. Mousli M, Bronner C, Landry Y, et al. Direct activation of GTP-binding regulatory proteins (G-proteins) by substance P and compound 48/80. FEBS Lett 1990;259:260–262.

62. Ortner MJ, Chingell CF. Spectroscopic studies of rat mast cells, mouse mastocytoma cells and compound 48/80. Biochem Pharmacol 1981;30:283–288.

63. Voss T, Wallner E, Czernilofsky AP, et al. Amphipathic alpha-helical structure does not predict the ability of receptor-derived synthetic peptides to interact with guanine nucleotide-binding regulatory proteins. J Biol Chem 1993;268(7):4637–4642.

64. Mousli M, Bronner C, Bockaert J, et al. Interaction of substance P, compound 48/80 and mastoparan with the alpha-subunit C-terminus of G protein. Immunol Lett 1990;25(4):355–357.

65. Weingarten R, Ransna L, Mueller H, et al. Mastoparan interacts with the carboxy terminus of the alpha subunit of Gi. J Biol Chem 1990;265:11044–11049.

66. Muhlen FVZ, Eckstein F, Penner R. Guanosine 5′-[b-thio] triphosphate selectively activates calcium signaling in mast cells. Proc Natl Acad Sci USA 1991;88:926–930.

67. Aridor M, Rajmilevich G, Beaven MA, et al. Activation of exocytosis by the heterotrimeric G protein Gi3. Science 1993;262(5139):1569–1572.

68. Simonds WF, Goldsmith PK, Woodard CJ, et al. Receptor and effector interactions of Gs. Functional studies with antibodies to alpha s carboxy-terminal decapeptide. FEBS Lett 1989;249:189–194.

69. Vitale N, Gensse M, Chasserot-Golaz S, et al. Trimeric G proteins control regulated exocytosis in bovine chromaffin cells: sequential involvement of Go associated with secretory granules and Gi3 bound to the plasma membrane. Eur J Neurosci 1996;8:1275–1285.

70. Stow JL, Sabolic I, Brown D. Heterogeneous localization of G protein alpha-subunits in rat kidney. Am J Physiol 1991;261:F831–F840.

71. Stow JL, de Almeida JB. Distribution and role of heterotrimeric G proteins in the secretory pathway of polarized epithelial cells. J Cell Sci Suppl 1993;17(33):33–39.

72. Hermouet S, de Mazancourt P, Spiegel AM, et al. High level expression of transfected G protein alpha-i3 is required for plasma membrane targeting and adenylyl cyclase inhibition in NIH 3T3 fibroblasts. FEBS Lett 1992;312:223–228.

73. Wilson BS, Palade GE, Farquhar MG. Endoplasmic reticulum-through-Golgi transport assay based on O-glycosylation of native glycophorin in permeabilized erythroleukemia cells: role for Gi3. Proc Natl Acad Sci USA 1993;90:1681–1685.

74. Stow JL, De Almeida JB, Narula N, et al. A heterotrimeric G protein, Gαi3, on Golgi membranes regulates the secretion of heparan sulfate proteoglycan in LLC PK1 epithelial cells. J Cell Biol 1991;114:1113–1124.

75. Cantiello HF, Patenaude CR, Codina J, et al. G alpha i-3 regulates epithelial Na+ channels by activation of phospholipase A2 and lipoxygenase pathways. J Biol Chem 1990;265(35):21624–21628.

76. McGivney A, Morita Y, Crews FT, et al. Phospholipase activation in the IgE-mediated and calcium ionophore A23187-induced release of histamine from rat basophilic leukemia

cells. Arch Biochem Biophys 1981;212:572–580.

77. Churcher Y, Allan D, Gomperts BD. Relationship between arachidonate generation and exocytosis in permeabilized mast cells. Biochem J 1990;266:157–163.

78. Norman JC, Price LS, Ridley AJ, et al. Actin filament organization in activated mast cells is regulated by heterotrimeric and small GTP-binding proteins. J Cell Biol 1994;126:1065–1075.

79. Wilson BS, Deanin CG, Stndefer JC, et al. Depletion of guanine nucleotides with mycophenolic acid suppresses IgE receptor-mediated degranulation in rat basophilic leukemia cells. J Immunol 1989;143:259–265.

80. McCloskey MA. Cholera toxin potentiates IgE-coupled inositol phospholipid hydrolysis and mediator secretion by RBL-2H3 cells. Proc Natl Acad Sci USA 1988;85:7260–7264.

81. Narasimhan V, Holowka D, Fewtrell C, et al. Cholera toxin increases the rate of antigen-stimulated calcium influx in rat basophilic leukemia cells. J Biol Chem 1988;263:19626–19632.

82. Collado ED, Cunha MJ, Beaven MA. Treatment with dexamethasone down-regulates IgE-receptor-mediated signals and up-regulates adenosine-receptor-mediated signals in a rat mast cell (RBL-2H3) line. J Immunol 1990;144(1):244–250.

83. Hide M, Ali H, Price SR, et al. GTP-binding protein G alpha Z: its down-regulation by dexamethasone and its credentials as a mediator of antigen-induced responses in RBL-2H3 cells. Mol Pharmacol 1991;40(4):473–479.

84. Baumgartner RA, Hirasawa N, Ozawa K, et al. Enhancement of TNF-alpha synthesis by overexpression of G alpha z in a mast cell line. J Immunol 1996;157(4):1625–1629.

85. Howell TW, Cockcroft S, Gomperts BD. Essential synergy between calcium and guanine nucleotides in exocytotic secretion from permeabilized rat mast cells. J Cell Biol 1987;105:191–198.

86. Swieter M, Midura RJ, Nishikata H, et al. Mouse 3T3 fibroblasts induce rat basophilic leukemia (RBL-2H3) cells to acquire responsiveness to compound 48/80. J Immunol 1993;150(2):617–624.

21
Activation of Small GTP-Binding Proteins

Anna Koffer and Richard Sullivan

Guanosine triphosphate- (GTP-) binding proteins act as molecular switches: active in GTP-bound form and inactivated when bound GTP is hydrolyzed. Switching on these proteins initiates a wide range of cellular responses and is controlled by multiple factors. In particular, GEFs (guanine nucleotide exchange factors or guanine nucleotide-releasing proteins) activate GTP-binding proteins by promoting exchange of the bound GDP for GTP, while GAPs (GTPase-activating proteins) and GDIs (guanine nucleotide dissociation inhibitors) inactivate GTP-binding proteins by promoting hydrolysis of GTP and inhibiting dissociation of the bound GDP, respectively. In general, as is the case with many other signaling molecules, proteins that regulate GTPases contain various consensus domains (e.g., SH-2, SH-3, PH, DH, IQ) that provide sites for input and output from and into other signaling pathways. In this way, a fine tuning of cellular responses to a variety of receptors can be achieved.

GTP-binding proteins include two main superfamilies: (1) the heterotrimeric GTP-binding proteins (G proteins), which consist of α-, β-, and γ- subunits and are discussed in Chapter 20, and (2) the small GTP-binding proteins, monomers of 20 to 30 kDa related to the 21-kDa protein, Ras. In addition there is a separate family of large monomeric GTPases, related to dynamin (~100 kDa), that are involved in membrane protein trafficking and controlled primarily by phosphorylation and self-assembly.[1-3] Dynamin forms rings at the neck of invaginated coated pits and a confor-

mational change in such rings, dependent on GTP hydrolysis, is essential for vesicle fission.[1,2] The superfamily of Ras-related GTPases includes more than 50 members[4] divided according to their primary sequence into six subgroups: Ras, Rho, Rab, Ran, Rad, and Arf. Ran is a nuclear GTPase required for the transport of proteins and ribonucleoproteins across the nuclear pore complex and for the maintenance of nuclear structures.[5] Rad is the product of a gene that is overexpressed in the skeletal muscle of humans with type II diabetes. Rad displays some unique features: first, it is phosphorylated by both a serine/threonine kinase and protein kinase A (PKA), and second, it interacts with tropomyosin and calmodulin.[6,7] This chapter discusses the Ras, Rab, Arf, and Rho subfamilies with respect to their reported or anticipated participation in mast cell signal transduction. The emphasis is on the most recent findings so as to update our recent report on the role of GTP-binding proteins in FcεRI signaling.[8]

Ras

In the "classical" Ras signaling pathway the activated (tyrosine-phosphorylated) protein tyrosine kinase (PTK) receptor recruits an adapter protein, Grb-2, either directly or indirectly via another adapter protein, p52 Shc (refer to the scheme in Fig. 21.1). Grb-2 is complexed with a Ras-GEF, Sos, and mediates both translocation of Sos to the plasma mem-

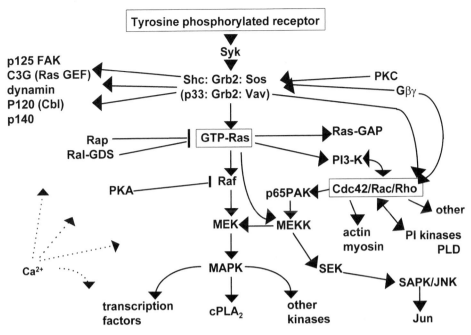

FIGURE 21.1. Upstream and downstream of Ras and Rho. Receptor-dependent activation of protein tyrosine kinase (PTK) Syk leads to recruitment of multiple adapter proteins (Grb-2, Shc, p33, and others) and consequently of various signaling molecules including guanine nucleotide exchange factors (GEFs) for Ras (Sos) and Rho (Vav?). These GTPases are coregulated by G proteins, protein kinase C (PKC), and phospholipids. The signaling cascade leading to MAPK forms an axis with diverse inputs and outputs. Ras activation leads to upregulation of Rho GTPases while the latter in turn may share the same downstream targets with Ras (e.g., MEKK). Calcium contributes to the regulation of many components of this network.

brane and its activation.[9–11] The recruitment/ activation of Sos may be additionally regulated by PKC or mitogen-activated protein kinase (MAPK), which phosphorylate Sos on multiple sides, and thus may modulate its interaction with the adapter proteins.[12,13] The activated Sos consequently converts Ras to its active GTP-bound state and this initiates MAP kinase, PI-3 kinase, phospholipase A_2 (PLA_2), and other signaling pathways.[9,10,14] Previously, the adapter protein Shc was found to be phosphorylated constitutively in RBL cells and another adapter protein (p33) was implicated in the recruitment of Grb-2–Sos complex to the FcεRI receptor.[15] More recent data have shown that at low serum concentrations the classical pathway to Ras activation (via the Shc–Grb-2–Sos complex) takes place in RBL cells after FcεRI cross-linking and is dependent on PTK Syk.[13,16] However, multiple alternative adapter proteins may exist

in mast cells and lymphocytes, related to both Shc (p33, p36/38) or Grb-2 (Grap).[17] Other phosphoproteins associate with Shc–Grb-2–Sos complexes (e.g., p120, p140, dynamin, Vav), and again this implies multiple levels of control and divergence of Ras signaling.[13,18,19] Ras activation is also initiated by C-kit, interleukins, and other hemopoietin receptors.[20] IL-3 and stem cell factor (SCF) were found to stimulate phosphorylation of Shc and activation of the Ras/Raf-1/MAPK pathway in a synergistic manner.[21] Signaling from G protein-coupled receptors to Ras seems to occur through Gβγ subunits via PI-3 kinase-γ.[22,23]

In mast cells, Ras activation leads to activation of MAPK and $cPLA_2$ but secretory responses are independent of these two events.[24–26] The pathways to Ras-induced transcriptional regulation in mast cells are divergent. For example, recent results have shown that activation of a transcription factor NF-AT (nuclear factor of activated T cells) is controlled by Rac-1 in parallel with or via Ras, while the Ras/Raf-1/MAPK cascade alone is sufficient for activation of the Elk-1 transcription factor.[27] Among the substrates for tyrosine kinases that are activated after aggregation of the FcεRI is a protein Vav, whose phosphorylation is dependent on PTK Syk.[19,24] Vav is the product of a proto-oncogene *vav*, which is expressed specifically in hematopoietic cells.[28] Vav shares sequence homology, designated Dbl homology (DH) domain, with the products of the human *dbl* gene and the yeast *cdc24* gene. The protein products possess GEF activity for Rho-related GTPases.[29,30] In addition, Vav contains other sequence motifs including one SH-2 and two SH-3 domains, and the pleckstrin homology (PH) domain, as well as sequences similar to those found in some transcription factors: helix-loop-helix/leucine zipper and zinc finger (diacylglycerol-binding domain).[29] More than 15 proteins containing a DH-like domain (a family of Dbl-like GEFs) have now been identified. Interestingly Sos and another Ras-GEF (Ras-GRF) were also found to contain DH domains, suggesting that they may act as GEFs for both Ras and Rho families.[29] Vav, however, does not contain the Ras-GEF motif and its role as a Ras-GEF is still controversial. Experiments with the COS-7 cell transient expression system have shown that oncogenic Vav could activate JNK/SAPK (stress-activated PK) but not MAPK and that this effect was Rac dependent.[31] On the other hand, in RBL cells Vav was found to be constitutively associated with the components of the MAPK pathway.[19] Figure 21.1 shows a scheme based on this discussion and that in our previous review.[8]

Rab

Rab proteins are localized to distinct cellular compartments and regulate vesicular traffic, that is, budding, targeting, docking, and fusion of membrane organelles. First identified in mammalian cells, their homologs have now been identified in yeast and plants and include more than 30 members.[32-34] Distribution and function of the individual Rabs is shown in Table 21.1. Posttranslational C-terminus prenylation and geranylgeranylation allow Rabs to insert into their respective membrane compartments, while cytosolic Rabs are associated with GDI proteins. GDIs bind preferentially to diprenylated Rabs in GDP-bound form and deliver Rab-GDP to its correct membrane target (Fig. 21.2). This step is followed by nucleotide exchange, catalyzed by Rab-GEF (such as MSS-4 protein), and following vesicle budding, fission, transport, docking, and fusion, Rab-bound GTP is hydrolyzed and Rab-GDP is again retrieved by GDI and recycled.[34] The nucleotide exchange is crucial for the vesicle budding step because Rab-GTP seems to be required for the proper formation of transport vesicles.[34] Rab-GTP hydrolysis, on the other hand, may be associated with the fusion step.[34,35]

Correct targeting and docking depends on the binding of specific vesicle proteins, V-SNARES (i.e., SNAP receptors or VAMPs or synaptobrevins), to specific partners on target membranes, T-SNARES (syntaxins), and on the cytosolic proteins SNAP (soluble NSF-attachment protein) and NSF (N-ethylmaleimide-sensitive factor).[36] Recent evidence indicates that Rab proteins promote or stabilize the assembly of SNARE complexes.[37,38] Another protein that seems to associate with and regulate these complexes is a synaptic vesicle protein, synaptotagmin.[39,40] Synaptotagmin contains (in its cytoplasmic C-terminal region) two domains of protein kinase C homology (C2A and C2B) and consequently binds phospholipids in a calcium-dependent manner. Its N-terminal region contains the Rab-binding site. This protein has an important function in endo- and exocytosis, acting as a Ca^{2+} sensor.

Rab-3 proteins are associated with synaptic vesicles and implicated in the docking and regulation of exocytosis in neuronal cells.[41] Rab-3 isoforms are also associated with other types of secretory organelles, including pancreatic zymogen and mast cell secretory granules.[42,43] In chromaffin cells, GTP-bound form of Rab3a is a component of the prefusion complex and may

TABLE 21.1 Currently identified Rab proteins: their localization and possible functions.

Rab subfamily	Intracellular location	Transport step	References
Rat-1	Smooth ER to Golgi	ER to Golgi	58, 59
Ypt-1 (yeast homolog)	Golgi apparatus	ER to Golgi	60
Rab-2	Intermediate compartment between ER and Golgi	ER to Golgi	61
Rab-3a	Synaptic vesicles	Regulated exocytosis	33, 43, 62, 63
Rab-4	Early endosomes	Early endosomes, recycling vesicles	64
Rab-5	Early endosomes, clathrin-coated plasma membrane	Early endosome lateral fusion, plasme membrane to early endosome	56
Rab-6	Golgi; medial and *trans*-cisternae	Early step in secretory pathway, probably intra-Golgi transport	64, 65
Ypt-6	Golgi; medial and *trans*-cisternae	Early step in secretory pathway, probably intra-Golgi transport	65
Rab-7	Late endosomes	Late endosomes to lysosomes	66, 67
Rab-8	Golgi	*trans*-Golgi network to plasme membrane	57, 68, 69
Rab-9	*trans*-Golgi network and late endosomes	Late endosomes to *trans*-Golgi; TGN, recycling	70, 71
Sec-4 (yeast homolog)	Secretory vesicles and plasme membrane	Golgi to plasme membrane	
Rab-10	Perinuclear membrane	Unknown	72
Rab-11	Pericentriolar recycling compartment	Recycling endosome	73, 74
Rab-12	Golgi	Unknown	73

Source: Modified from Chavrier et al.[32]

act as a negative fusion clamp, inhibiting fusion of granule and plasma membranes.[44,45] Rab-3 binds, in a GTP-dependent manner, to a synaptotagmin-like protein, Rabphilin-3a, and recruits it to the synaptic vesicle membrane.[46,47] Rabphilin may participate in the docking step because it interacts specifically with a 115-kDa plasma membrane protein.[48] Like synaptotagmin, Rabphilin-3a is composed of the N-terminal Rab-3a-binding domain and the C-terminal calcium- and phospholipid-binding domain.[49] The middle, regulatory region is phosphorylated both by calmodulin-dependent protein kinase II and by PKA, offering a potential for further calcium and cAMP regulation of its activity.[50–52] Expression of various Rabphilin-3a mutants in PC-12 or chromaffin cells has shown that this protein is a positive regulator of calcium-dependent exocytosis from these neuroendocrine cells.[53,54] Rabphilin inhibits the Rab-GAP-stimulated GTPase activity of Rab-3a,[55] and its function may be to maintain Rab in

a GTP-bound form at low calcium concentrations. After calcium-induced membrane fusion and exocytosis, both Rab-3 and Rabphilin were found to dissociate from synaptic vesicles.[47] The precise mechanism of individual steps is still unknown. Figure 21.2 represents our attempt to summarize current understanding of Rab participation in vesicle traffic.

As well as regulated exocytosis, mast cells carry out constitutive and receptor-mediated endocytosis and must have a full repertoire of Rabs that direct vesicle traffic. Rab-3a is present[43] and, together with a rabphilin-like protein, may be expected to participate in the control of regulated exocytosis. Constitutive endocytosis may be under the control of, perhaps, Rab-5, which has been localized to early endosomes, clathrin-coated vesicles, and the plasma membrane.[56] A novel member of the Rab family, Rab-8b, has been cloned from RBL cells.[57] When overexpressed, Rab-8b induced outgrowths of the plasma membrane of

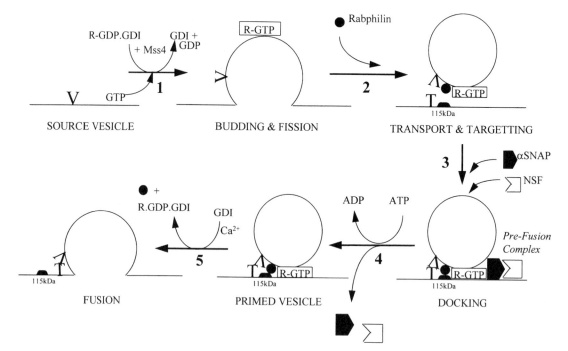

FIGURE 21.2. Rab participation in vesicle traffic. (*1*) Guanosine nucleotide dissociation inhibitor (GDI) delivers Rab-GDP (R-GDP) to the membrane. Nucleotide exchange, catalyzed by Rab-GEF (Mss4), is necessary for vesicle budding. (*2*) GTP-Rab promotes or stabilizes the assembly of **V**- and **T**-SNARE complex and recruits a calcium sensor, Rabphilin (●). Rabphilin associates with a 115-kDa protein on the target membrane (▬). (*3*) α-SNAP (◗) and consequently NSF (⊐) bind to the SNARE complex, forming a prefusion complex. (*4*) ATP hydrolysis by the ATPase NSF primes the vesicle for fusion. (*5*) calcium binds to Rabphilin, thus releasing its inhibitory effect on Rab-GTP hydrolysis and allowing fusion.

these cells.[57] Little more is known about the function of Rabs in mast cells and indeed about whether and how they are activated in response to receptor stimulation. Although most Rabs have defined positions in the endocytotic and exocytotic pathways,[32] it is unlikely that they regulate in isolation; redundancy will have been built into their repertoire. The goal is to identify redundant and key elements to this regulation as well as the upstream events that control the constitutive and inducible levels of activated Rabs.

ARF

ADP-ribosylation factors (ARFs) were originally identified by their ability to enhance the ADP-ribosyltransferase activity of cholera toxin toward $G_s\alpha$ and were recently recognized as participants in vesicular trafficking pathways and phospholipase D activation.[75] There are at least six isoforms, ubiquitously present in all tissues, with different functions and localization.[76] For example, ARF-1 is required for the assembly of the coat protein complex (COP) that initiates budding of vesicles from Golgi and the endoplasmic reticulum (ER),[77] and ARF-6 is localized primarily to the plasma membrane and controls receptor-mediated endocytosis.[78] Recent evidence indicates that, rather than ARF-6 itself, it is the ARF-induced activation of phospholipase D (PLD) and the subsequent production of phosphatidic acid (an intracellular second messenger) which is crucial for the formation of coatamer-coated vesicles (Fig. 21.3).[75,79] Brefeldin A (BFA) is a fungal metabolite that acts as an antagonist of

ARF activities; it interacts with and inhibits ARF-GEFs.[80,81] The nucleotide exchange on ARF is also strictly dependent on phosphatidy linositol 4,5-biphosphate (PIP$_2$), which acts as a cofactor.[82] It now appears that the local activation of PLD may induce conformational changes in the lipid bilayer that may potentiate coat protein assembly and play an important role in a whole range of other cellular functions, including exocytosis.[83,84] In addition to ARF and PIP$_2$, PLD is also regulated (in a synergistic manner) by Rho A as well as by PKC.[75,85,86] Inhibition of PKC kinase activity did not affect PKC ability to stimulate PLD, sug-

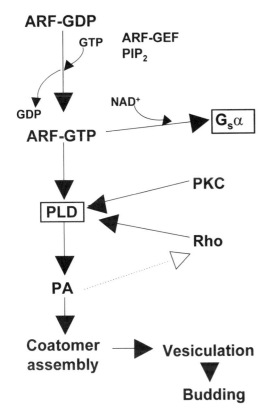

FIGURE 21.3. Phospholipase D (PLD) is the main identified target for ADP-ribosylation factors (ARF). ARF-GEFs and PIP$_2$ are required for ARF activation. ARF-GTP upregulates PLD in synergy with PKC and Rho. Local activation of PLD and consequent production of phosphaditic acid (PA) induces conformational changes in the lipid bilayer that potentiate coat protein assembly. ARF also acts as a cofactor of cholera toxin, facilitating its ribosyltransferase activity toward G$_s$α (mechanism unknown).

gesting that direct interaction of PLD with PKC may be responsible.[86]

In RBL and mast cells, ARF has been shown to regulate both PLD activity and secretion.[87,88] BFA inhibited the antigen- and calcium ionophore A23187-induced activation of PLD, but failed to inhibit PLD activation in response to phorbol myristate acetate (PMA),[87] again indicating ARF-dependent and-independent pathways to PLD activation. BFA inhibited antigen-induced serotonin release at lower concentrations than those needed for the inhibition of PLD. Reconstitution experiments with permeabilized RBL cells have shown that antigen-induced secretion is enhanced by addition of (myristoylated) recombinant ARF-1 (S. Cockcroft group, unpublished results). In parallel, rARF-1 stimulated PLD activity of these cells. Similarly to their results obtained with HL-60 cells,[83] secretion (but not PLD) was also enhanced by phosphatidylinositol transfer protein (PITP), acting in an additive manner. Prostaglandin D$_2$ (PGD$_2$) production and arachidonic acid release were also found to be controlled by ARF. Thus, in permeabilized rat peritoneal mast cells, addition of ARF increased arachidonic acid and PGD$_2$ release in response to GTP-γ-S, and this effect was inhibited by ethanol[88] (in the presence of ethanol, phosphatidylethanol is generated instead of the phosphatidic acid).

Rho

The family of Rho-related GTPases includes Rho A, B, C, D, E, G, Rac-1 and-2, Cdc42, and TC10. The best established role of these proteins is to mediate signals from cell-surface receptors to the cytoskeleton and the assembly of focal adhesion complexes.[89] As discussed earlier, Dbl-related proteins, containing a DH domain in tandem with a PH domain, exhibit GEF activity toward Rho-related GTPases.[29] In addition, there are alternative pathways to activation of these GTPases. PIP$_2$ interacts with the C-terminal domain of Cdc42 and Rho (a site distinct from the Dbl-binding site), stimulates GDP dissociation, and stabilizes the nucleotide-depleted state.[90] On the other hand, enzymes involved in phospholipid metabolism are controlled by Rho-GTPases,[89] adding complexity to their regulation. Both Rho[91] and Rac[92] interact with PIP$_5$ kinase, but this interaction is independent of the nucleotide state. In platelets, thrombin-stimulated activity of phosphoinositide-3 kinase (PI-3 kinase) is dependent on Rho.[93] Cdc42 binds to and activates the p85 subunit of PI-3 kinase,[94] but Rac has been shown to be downstream of PI-3 kinase.[95] PLD activity is stimulated in a synergistic manner by Rho/Cdc42 together with Arf and PIP2.[82,96] On the other hand, phosphatidic acid, which is generated as a product of PLD activity, stimulates Rho.[97] Thus, there are still many unresolved questions with respect to the flow of control to and from these GTPases (see Fig. 21.1).

Many other new potential targets for the members of this family have been identified in recent years. These include a number of kinases such as serine threonine kinases PKN, Rho-kinase, and p65PAK (analog of yeast Ste20 protein) and tyrosine kinases p120ACK and MLK3.[89] Some of these (e.g., p65PAK, MLK3) are implicated in the control of transcription factors.[98] The kinase p65-PAK also phosphorylates a component of the NADPH oxidase system, p47phox, while another component of this system, p67phox, interacts with Rac directly.[89] Rho-kinase was found to phosphorylate and so inactivate the myosin-binding subunit (MBS) of myosin light chain phosphatase, leading to an increase in the level of phosphorylated myosin II light chain (MLC).[99] In addition, MBS of MLC phosphatase was found to bind to Rho directly.[99] A further level of complexity is added by the finding that Rho-kinase also phosphorylates MLC on Ser-19, the same site as that which is phosphorylated by MLC kinase.[100] Other isoforms of myosin may also be controlled by Rho-GTPases. Myosin-1 (minimyosin) may be regulated by Cdc42/Rac via PAK (Ste20) -related kinases, which phosphorylate its heavy chain.[101] Another isoform, myosin myr-5, contains a Rho-GAP region in its tail.[102] This indicates that the control of actin organization by Rho-GTPases may be exerted via their regulation of myosin activity. However, there are many other potential targets, including the

proteins of focal adhesion complexes[103] and of the ezrin/radixin/moesin family.[104]

The number of cellular functions that are known to be controlled by these proteins is growing at an alarming rate. In addition to cell motility, morphology, and adhesion, Rho-GTPases control neuronal development, cell cycle, and gene transcription.[89] Discussion of the mechanisms is beyond the scope of this chapter, which is focused on the role of small GTPases in mast cell signaling. Worth mentioning, however, at this point is the new evidence for the participation of Rho-GTPases in the regulation of vesicular transport in general. In *Xenopus laevis* oocytes, activity of Rho is required for constitutive (clathrin-independent) fluid-phase endocytosis, which is independent of actin polymerization.[105] On the other hand, receptor-mediated endocytosis in clathrin-coated pits, measured in intact (transiently transfected) HeLa cells and in permeabilized A4312 cells, was inhibited by active mutants of Rho and Rac and stimulated by Rho-GDI and C3 transferase.[106] This suggests that these GTPases exert negative control over clathrin-coated vesicle formation. A new member of the Rho family, the GTPase Rho D, has dramatic effects on the distribution and motility of early endosomes in addition to its effects on the cytoskeleton.[107] Cdc42 was shown to be localized to the Golgi apparatus and dispersed on exposure of cells to the drug brefeldin A.[108] This effect of brefeldin was prevented in cells expressing a GTPase-deficient form of ARF. Such evidence points to a major role for Rho-GTPases in membrane traffic.

Evidence for the participation of Rho-GTPases in signaling pathways initiated after FcεRI cross-linking is accumulating. We have shown that inhibitors of endogenous Rac and Rho, N17Rac-1 (a dominant negative mutant), and C3 transferase (which selectively inactivates Rho by ADP ribosylation) reduce secretory responses of permeabilized rat peritoneal mast cells.[109,110] On the other hand, secretion is enhanced by constitutively active mutant proteins, V14-Rho A and V12-Rac-1.[109] Immunoblotting has shown both Rac and Rho to be present in mast cells, and a major part of these proteins (~60%–70%) remained associated with intracellular structures after permeabilization. Using the recombinant proteins as standards, the intracellular content of Rho and Rac was estimated to be 1 and 0.5 µM, respectively.[109,111] After stimulation of both intact and permeabilized mast cells, a centripetal redistribution of filamentous actin is observed as a consequence of disassembly of the cortical F-actin, polymerization of actin in the cell interior, and relocalization of cortical actin filaments.[111] This effect can be mimicked or inhibited by the activators or inhibitors of Rho-related proteins.[111] Figure 21.4 shows V14-Rho A-induced increase in the level of F-actin in permeabilized mast cells. Newly polymerized filaments appear despite the loss of the soluble monomeric actin from permeabilized cells, suggesting the existence of a membrane-bound G-actin pool. Another function of Rho may thus be to initiate the release of this membrane-sequestered G-actin.

The effect shown in Fig. 21.4 is not accompanied by an increase in secretion unless calcium is also present. Both Rho and Rac enhance calcium-induced secretion in the presence of cytochalasin, which abrogates the F-actin increase. Secretory response is therefore independent of actin polymerization. However, cytoskeletal reorganization and secretion are strongly correlated: secreting cells preferentially display both relocalization and polymerization of actin, indicating a common upstream regulator of these two functions.[110] Calcium and ATP induce loss of cortical F-actin.[112,113] This loss is enhanced in the presence of V14Rho, suggesting another, calcium-dependent, function of Rho (our unpublished results). Phalloidin prevents the F-actin loss without any effect on secretion and without diminishing the secretion-enhancing ability of Rho.[110] Again, this suggests that Rho controls cortical F-actin disassembly and secretion by divergent pathways. Both V14-Rho A and V12-Rac-1 reduce calcium requirements of permeabilized mast cells for secretion, mimicking the effects of GTP-γ-S (Fig. 21.5). Like GTP-γ-S, both active mutants also abolish the delay in the onset of calcium-induced secretion (Fig. 21.5).

Calmodulin inhibitors (specific calmodulin-binding peptides or W7) as well as MLCK inhibitor ML7 greatly reduced F-actin disassembly also without affecting secretion from permeabilized mast cells (our unpublished

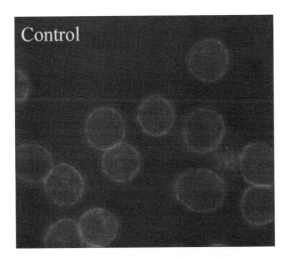 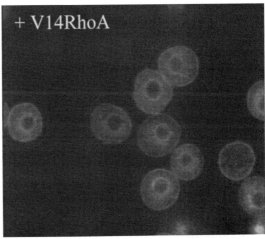

FIGURE 21.4. Rho induces polymerization of actin in permeabilized mast cells. The active mutant of Rho, V14-Rho A, was introduced into streptolysin-O-permeabilized cells in the presence of 3 mM EGTA; after 15 min, cells were stained with rhodamine-phalloidin. F-actin increase is independent of calcium and ATP, although in their presence additional effects of V14Rho A are observed: enhancement of both secretion and cortical F-actin disassembly. *Left*, control; *right*, with V14Rho A.

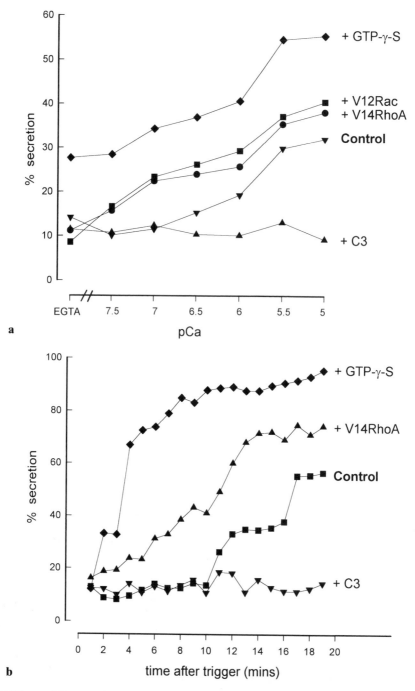

FIGURE 21.5. Rho and Rac act in synergy with calcium. (a) Rho and Rac reduce calcium requirements
for secretion. Permeabilized mast cells were exposed
to increasing concentrations of calcium in the presence or absence of the indicated inhibitors or activators (0.1 μg/ml C3, 8 μg/ml V14-Rho A, 16 μg/ml
V12-Rac-1, 50 μM GTP-γ-S) and 3 mM MgATP. Released hexosaminidase was assayed after 20 min. (b)
Rho and Rac eliminate the delay in the onset of
calcium-induced secretion: time course of secretion
from cells exposed to 10 μM calcium, 3 mM MgATP,
and the indicated inhibitors or activators.

results). These results indicate participation of myosin in cortical F-actin disassembly and a possibility that Rho controls this function via its regulation of MLCK activity. In RBL cells, phosphorylation of both myosin heavy and light chains by protein kinases PKC and MLCK is correlated with antigen-or calcium/PMA-induced secretion. MLC phosphorylation occurs on residues Ser-19, Thr-18, and Ser-1 or Ser-2.[114–116] Inhibition of any of the two kinases blocked phosphorylation as well as secretion,[115] but it appears that secretion correlates best with the PKC-induced phosphorylation.[116] It is very likely that the Rho-kinase participates in the overall control of myosin in these cells, but the relationship of the myosin control to the control of secretion is unclear. W7 (100 μM) and other cell-permeable calmodulin antagonists did inhibit secretion from intact mast cells responding to antigen, indicating that calmodulin may have a role in an upstream event, mediating signaling from the FcεRI receptor. On the other hand, ML7 had no effect on intact cells except at much higher and probably unspecific concentrations.

The importance of Rac in the signaling pathway to secretion has recently been confirmed. Fractions from a brain extract have been examined for their ability to reconstitute secretion from permeabilized mast cells, and the active fraction contained Rac coupled to Rho-GDI. Interestingly, native Rac by itself did not have any effect on secretion while purified Rho-GDI was inhibitory.[117] The difference between activities of native and recombinant Rac proteins may result from the lack of posttranslational modification of the latter. Native Rho-GTPases require Rho-GDI to remain in solution, and Rho-GDI will solubilize membrane-bound GTPases.[118] Indeed, inhibitory effects of Rho-GDI were observed when exocytosis was monitored by recording cell capacitance using a patch pipette.[119] Use of toxins that specifically inhibit Rho-GTPases provided further evidence for their participation in mast cell signaling. Thus, C3 transferase was found to inhibit antigen as well as calcium- or GTP-γ-S-induced secretion from permeabilized RBL cells.[120] The *Clostridium difficile* toxin B, which monoglucosylates and inactivates Rho-

GTPases, inhibited phospholipase D activation in both intact and permeabilized RBL cells responding to antigen, GTP-γ-S, or PMA.[121] This toxin also strongly inhibited serotonin release from intact RBL cells responding to antigen, carbachol, mastoparan, and ionophore A23187.[122] Immunoblotting indicated Rho A and Cdc42 as its protein substrates. As in this report C3 transferase was found to be without any effects on secretion, the authors concluded that Cdc42 is the GTPase that participates in regulated exocytosis.[122] In fact, reconstitution experiments with permeabilized peritoneal mast cells have now shown that a GTP-γ-S-loaded Cdc42 has secretion-enhancing effects while the dominant negative mutant (N17-Cdc42) is inhibitory (A. Brown et al., unpublished results). Finally dominant negative mutant forms of Cdc42 and Rac-1, when expressed in RBL-2H3 cells, had distinct effects on cell morphology but both inhibited degranulation.[123]

Conclusions

Activation of Rho-GTPases as a direct consequence of FcεRI activation and its most probable dependence on the PTK Syk and on Vav phosphorylation still remains to be established. These proteins are controlled by multiple mechanisms and other receptors, such as integrins, G protein-coupled chemokine receptors, or c-*kit* receptors are very likely to upregulate them as well. Moreover, cross-talk and overlapping functions of Rho-related GTPases are common, and a hierarchical, sequential activation of Cdc42 > Rac > Rho has been observed in fibroblasts.[89]

Vesicular traffic seems to be controlled by mechanisms that include ARF and Rab and Rho family members, converging onto the same or interdependent targets. Therefore the task of dissecting the individual role for each participant will be difficult. It is interesting that in mast cells, although the cytoskeletal effects of Rac, Rho, and Cdc42 are distinct,[111] the three GTPases seem to have a common target in the signaling pathway to exocytosis (Fig. 21.6).

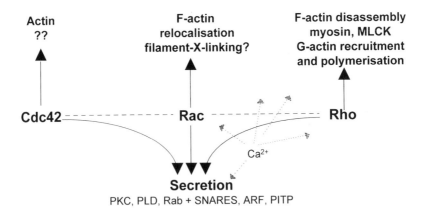

FIGURE 21.6. Rho-GTPases regulate actin organization and secretion in mast cells. Cytoskeletal effects of Rho, Rac, and Cdc42 are distinct; there seems to be a common target on the signaling pathway to secretion. Exocytosis is regulated by a complex mechanism involving the enzymes of phospholipid metabolism, PKC, calcium, and other GTP-binding proteins, including Rabs and ARFs.

The relationship of the signaling pathways that lead from Cdc42/Rac/Rho to secretion and to the cytoskeleton may be more manifest for mast cells that are, in physiological conditions, exposed to a gradient of stimuli. Our results have shown that cytoskeletal responses are much more resistant to GDP-β-S than the secretory ones, and we have postulated an additional, GDP-β-S-sensitive, GTP-binding protein downstream of Rac/Rho.[110] However, the results could also be explained by differential requirements of the two functions for the strength of the stimuli. Cytoskeletal responses may be induced by low concentrations of stimuli leading to migration, while stronger signals may be necessary to elicit secretion. Participation of Rho in mediating adhesion in response to activation of G protein-linked receptors of the chemoattractant subfamily has already been demonstrated.[124] Thus mast cells may respond to chemokines and other stimuli in a specific, graded, and probably hierarchical manner in this order: adhesion > kinesis > chemotaxis > secretion (Fig. 21.7).

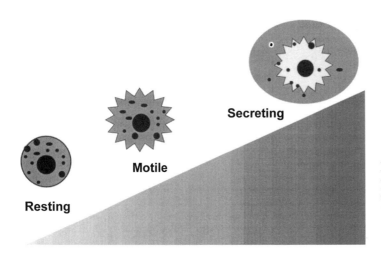

FIGURE 21.7. Mast cells are exposed to stimuli gradients. Responses may be elicited in a hierarchical manner in this order: adhesion > kinesis > chemotaxis > secretion.

References

1. de Camilli P, Takei K, McPherson PS. The function of dynamin in endocytosis. Curr Opin Neurobiol 1995;5:559–565.

2. Warnock DE, Schmid SL. Dynamin GTPase, a force-generating molecular switch. BioEssays 1996;18:885–893.

3. Pullar CE, Repetto B, Gilfillan AM. Rapid dephosphorylation of the GTPase dynamin after FcepsilonRI aggregation in a rat mast cell line. J Immunol 1996;157:1226–1232.

4. Hall A. Ras-related proteins. Curr Opin Cell Biol 1993;5:265–268.

5. Rush MG, Drivas G, D'Eustachio P. The small nuclear GTPase Ran: how much does it run? BioEssays 1996;18:103–112.

6. Fischer R, Wei Y, Anagli J, et al. Calmodulin binds to and inhibits GTP binding of the ras-like GTPase Kir/Gem. J Biol Chem 1996;271:25067–25070.

7. Zhu J, Bilan PJ, Moyers JS, et al. Rad, a novel Ras-related GTPase, interacts with skeletal muscle beta-tropomyosin. J Biol Chem 1996;271:768–773.

8. Koffer A. Role of GTP-binding proteins in FcεRI signalling. In: Hamawy-MM, ed. IgE Receptor (FcεRI) Function in Mast Cells and Basophils. Vol. 5. 1st Ed. Austin, TX: RG Landes, 1997:75–105.

9. McCormick F. How receptors turn ras on. Nature (Lond) 1993;363:15–17.

10. Downward J. Control of ras activation. Cancer Surv 1996;27:87–100.

11. Byrne JL, Paterson HF, Marshall CJ. p21Ras activation by the guanine nucleotide exchange factor Sos, requires the Sos/Grb2 interaction and a second ligand-dependent signal involving the Sos N-terminus. Oncogene 1996;13:2055–2065.

12. Corbalan Garcia S, Yang SS, Degenhardt KR, et al. Identification of the mitogen-activated protein kinase phosphorylation sites on human Sos1 that regulate interaction with Grb2. Mol Cell Biol 1996;16:5674–5682.

13. Jabril Cuenod B, Zhang C, Scharenberg AM, et al. Syk-dependent phosphorylation of Shc. A potential link between FcepsilonRI and the Ras/mitogen-activated protein kinase signaling pathway through SOS and Grb2. J Biol Chem 1996;271:16268–16272.

14. Quilliam LA, Khosravi Far R, Huff SY, et al. Guanine nucleotide exchange factors: activa-

tors of the Ras superfamily of proteins. BioEssays 1995;17:395–404.

15. Turner H, Reif K, Rivera J, et al. Regulation of the adapter molecule Grb2 by the FcεRI in the mast cell line RBL2H3. J Biol Chem 1995;270:9500–9506.

16. Beaven MA, Baumgartner RA. Downstream signals initiated in mast cells by Fc epsilon RI and other receptors. Curr Opin Immunol 1996;8:766–772.

17. Trub T, Frantz JD, Miyazaki M, et al. The role of a lymphoid-restricted, Grb2-like SH3-SH2-SH3 protein in T cell receptor signaling. J Biol Chem 1997;272:894–902.

18. Smit L, van der Horst G, Borst J. Formation of Shc/Grb2- and Crk adaptor complexes containing tyrosine phosphorylated Cbl upon stimulation of the B-cell antigen receptor. Oncogene 1996;13:381–389.

19. Song JS, Gomez J, Stancato LF, et al. Association of a p95 Vav-containing signaling complex with the FcepsilonRI gamma chain in the RBL-2H3 mast cell line. Evidence for a constitutive in vivo association of Vav with Grb2, Raf-1, and ERK2 in an active complex. J Biol Chem 1996;271:26962–26970.

20. Duronio V, Welham MJ, Abraham S, et al. p21ras activation via hemopoietin receptors and c-kit requires tyrosine kinase activity but not tyrosine phosphorylation of p21ras GTPase-activating protein. Proc Natl Acad Sci USA 1992;89:1587–1591.

21. O'Farrell AM, Ichihara M, Mui AL, et al. Signaling pathways activated in a unique mast cell line where interleukin-3 supports survival and stem cell factor is required for a proliferative response. Blood 1996;87:3655–3668.

22. Garnovskaya MN, van Biesen T, Hawe B, et al. Ras-dependent activation of fibroblast mitogen-activated protein kinase by 5-HT1A receptor via a G protein beta gamma-subunit-initiated pathway. Biochemistry 1996;35:13716–13722.

23. Lopez Ilasaca M, Crespo P, Pellici PG, et al. Linkage of G protein-coupled receptors to the MAPK signaling pathway through PI 3-kinase gamma. Science 1997;275:394–397.

24. Hirasawa N, Scharenberg A, Yamamura H, et al. A requirement for Syk in the activation of the microtubule-associated protein kinase/phospholipase A2 pathway by FcεRI is not shared by a G protein-coupled receptor. J Biol Chem 1995;270:10960–10967.

25. Hirasawa N, Santini F, Beaven MA. Activation of the mitogen-activated protein kinase/cytosolic PLA2 pathway in a rat mast cell line. J Immunol 1995;154:5391–5402.

26. Rider LG, Hirasawa N, Santini F, et al. Activation of the mitogen-activated protein kinase cascade is suppressed by low concentrations of dexamethasone in mast cells. J Immunol 1996;157:2374–2380.

27. Turner H, Cantrell DA. Distinct Ras effector pathways are involved in Fc epsilon RI regulation of the transcriptional activity of Elk-1 and NFAT in mast cells. J Exp Med 1997; 185:43–53.

28. Katzav S, Martin Zanca D, Barbacid M. vav, a novel human oncogene derived from a locus ubiquitously expressed in hematopoietic cells. EMBO J 1989;8:2283–2290.

29. Cerione RA, Zheng Y. The Dbl family of oncogenes. Curr Opin Cell Biol 1996;8:216–222.

30. Crespo P, Schuebel KE, Ostrom AA, et al. Phosphotyrosine-dependent activation of Rac-1 GDP/GTP exchange by the vav proto-oncogene product. Nature (Lond) 1997; 385:169–172.

31. Crespo P, Bustelo XR, Aaronson DS, et al. Rac-1 dependent stimulation of the JNK/SAPK signaling pathway by Vav. Oncogene 1996;13:455–460.

32. Chavrier P, Gorvel JP, Bertoglio J. An immunologist's look at the Rho and Rab GTP-binding proteins. Immunol Today 1993;14:440–444.

33. Madison DL, Kruger WH, Kim T, et al. Differential expression of rab3 isoforms in oligodendrocytes and astrocytes. J Neurosci Res 1996;45:258–268.

34. Pfeffer SR. Rab GTPases: master regulators of membrane trafficking. Curr Opin Cell Biol 1994;6:522–526.

35. Bock JB, Scheller RH. A fusion of new ideas. Nature (Lond) 1997;387:133–135.

36. Rothman JE, Warren G. Implications of the SNARE hypothesis for intracellular membrane topology and dynamics. Curr Biol 1994; 4(3):220–233.

37. Lian JP, Stone S, Jiang Y, et al. Ypt1p implicated in v-SNARE activation. Nature (Lond) 1994;372:698–701.

38. Sogaard M, Tani K, Ye RR, et al. A rab protein is required for the assembly of SNARE complexes in the docking of transport vesicles. Cell 1994;78:937–948.

39. Schiavo G, Stenbeck G, Rothman JE, et al. Binding of the synaptic vesicle v-SNARE, synaptotagmin, to the plasma membrane t-SNARE, SNAP-25, can explain docked vesicles at neurotoxin-treated synapses [see comments]. Proc Natl Acad Sci USA 1997;94: 997–1001.

40. Shao X, Li C, Fernandez I, et al. Synaptotagmin-syntaxin interaction: the C2 domain as a Ca^{2+}-dependent electrostatic switch. Neuron 1997;18:133–142.

41. Lledo PM, Johannes L, Vernier P, et al. Rab3 proteins: key players in the control of exocytosis. Trends Neurosci 1994;17:426–432.

42. Jena BP, Gumkowski FD, Konieczko EM, et al. Redistribution of a Rab3-like GTP-binding protein from secretory granules to the Golgi complex in pancreatic acinar cells during regulated exocytosis. J Cell Biol 1994;124:43–53.

43. Oberhauser AF, Balan V, Fernandez Badilla CL, et al. RT-PCR cloning of Rab3 isoforms expressed in peritoneal mast cells. FEBS Lett 1994;339:171–174.

44. Ludger J, Lledo P, Roa M, et al. The GTPase rab3A negatively controls calcium-dependent exocytosis in neuroendocrine cells. EMBO J 1994;13:2029–2037.

45. Holz RW, Brondyk WH, Senter RA, et al. Evidence for the involvement of rab3A in Ca^{2+}-dependent exocytosis from adrenal chromaffin cells. J Biol Chem 1994;269:10229–10234.

46. Li B, Warner JR. Mutation of the Rab6 homologue of *Saccharomyces cerevisiae*, YPT6, inhibits both early Golgi function and ribosome biosynthesis. J Biol Chem 1996;271:16813–16819.

47. Stahl B, Chou JH, Li C, et al. Rab3 reversibly recruits rabphilin to synaptic vesicles by a mechanism analogous to raf recruitment by ras. EMBO J 1996;15:1799–1809.

48. Miyazaki M, Kaibuchi K, Shirataki H, et al. Rabphilin-3A binds to a M(r) 115,000 polypeptide in a phosphatidylserine- and Ca(2+)-dependent manner. Brain Res Mol Brain Res 1995;28:29–36.

49. Yamaguchi T, Shirataki H, Kishida S, et al. Two functionally different domains of rabphilin-3A, Rab3A p25/smg p25A-binding and phospholipid- and Ca(2+)-binding domains. J Biol Chem 1993;268:27164–27170.

50. Kato M, Sasaki T, Imazumi K, et al. Phosphorylation of Rabphilin-3A by calmodulin-dependent protein kinase II. Biochem Biophys Res Commun 1994;205:1776–1784.

51. Numata S, Shirataki H, Hagi S, et al. Phosphorylation of Rabphilin-3A, a putative target protein for Rab3A, by cyclic AMP-dependent protein kinase. Biochem Biophys Res Commun 1994;203:1927–1934.

52. Fykse EM, Li C, Sudhof TC. Phosphorylation of rabphilin-3A by Ca^{2+}/calmodulin- and cAMP-dependent protein kinases in vitro. J Neurosci 1995;15:2385–2395.

53. Komuro R, Sasaki T, Orita S, et al. Involvement of rabphilin-3A in Ca^{2+}-dependent exocytosis from PC12 cells. Biochem Biophys Res Commun 1996;219:435–440.

54. Chung SH, Takai Y, Holz RW. Evidence that the Rab3a-binding protein, rabphilin3a, enhances regulated secretion. Studies in adrenal chromaffin cells. J Biol Chem 1995;270:16714–16718.

55. Fujita Y, Sasaki T, Araki K, et al. GDP/GTP exchange reaction-stimulating activity of Rabphilin-3A for Rab3A small GTP-binding protein. FEBS Lett 1994;353:67–70.

56. Rybin V, Ullrich O, Rubino M, et al. GTPase activity of Rab5 acts as a timer for endocytic membrane fusion. Nature (Lond) 1996;383:266–269.

57. Armstrong J, Thompson N, Squire JH, et al. Identification of a novel member of the Rab8 family from the rat basophilic leukaemia cell line, RBL.2H3. J Cell Sci 1996;109:1265–1274.

58. Peter F, Nuoffer C, Pind SN, et al. Guanine nucleotide dissociation inhibitor is essential for Rab1 function in budding from the endoplasmic reticulum and transport through the Golgi stack. J Cell Biol 1994;126:1393–1406.

59. Pind SN, Nuoffer C, McCaffery JM, et al. Rab1 and Ca^{2+} are required for the fusion of carrier vesicles mediating endoplasmic reticulum to Golgi transport. J Cell Biol 1994;125:239–252.

60. Jedd G, Richardson C, Litt R, et al. The Ypt1 GTPase is essential for the first two steps of the yeast secretory pathway. J Cell Biol 1995;131:583–590.

61. Tisdale EJ, Balch WE. Rab2 is essential for the maturation of pre-Golgi intermediates. J Biol Chem 1996;271:29372–29379.

62. Masumoto N, Sasaki T, Tahara M, et al. Involvement of Rabphilin-3A in cortical granule exocytosis in mouse eggs. J Cell Biol 1996;135:1741–1747.

63. Senbonmatsu T, Shirataki H, Jin no Y, et al. Interaction of Rabphilin3 with synaptic vesicles through multiple regions. Biochem Biophys Res Commun 1996;228:567–572.

64. Bottger G, Nagelkerken B, van der Sluijs P. Rab4 and Rab7 define distinct nonoverlapping endosomal compartments. J Biol Chem 1996;271:29191–29197.

65. Mayer T, Touchot N, Elazar Z. Transport between *cis* and medial Golgi cisternae requires the function of the Ras-related protein Rab6. J Biol Chem 1996;271:16097–16103.

66. Meresse S, Gorvel JP, Chavrier P. The rab7 GTPase resides on a vesicular compartment connected to lysosomes. J Cell Sci 1995;108:3349–3358.

67. Papini E, Satin B, Bucci C, et al. The small GTP binding protein rab7 is essential for cellular vacuolation induced by *Helicobacter pylori* cytotoxin. EMBO J 1997;16:15–24.

68. Ren M, Zeng J, De Lemos Chiarandini C, et al. In its active form, the GTP-binding protein rab8 interacts with a stress-activated protein kinase. Proc Natl Acad Sci USA 1996;93:5151–5155.

69. Peranen J, Auvinen P, Virta H, et al. Rab8 promotes polarized membrane transport through reorganization of actin and microtubules in fibroblasts. J Cell Biol 1996;135:153–167.

70. Soldati T, Rancano C, Geissler H, et al. Rab7 and Rab9 are recruited onto late endosomes by biochemically distinguishable processes. J Biol Chem 1995;270:25541–25548.

71. Riederer MA, Soldati T, Shapiro AD, et al. Lysosome biogenesis requires Rab9 function and receptor recycling from endosomes to the trans-Golgi network. J Cell Biol 1994;125:573–582.

72. Chen YT, Holcomb C, Moore HP. Expression and localization of two low molecular weight GTP-binding proteins, Rab8 and Rab10, by epitope tag. Proc Natl Acad Sci USA 1993;90:6508–6512.

73. Olkkonen VM, Dupree P, Killisch I, et al. Molecular cloning and subcellular localization of three GTP-binding proteins of the rab subfamily. J Cell Sci 1993;106:1249–1261.

74. Oishi H, Sasaki T, Takai Y. Interaction of both the C2A and C2B domains of rabphilin3 with Ca^{2+} and phospholipid. Biochem Biophys Res Commun 1996;229:498–503.

75. Frohman MA, Morris AJ. Rho is only Arf the story. Curr Biol 1996;6:945–947.

76. Hosaka M, Toda K, Takatsu H, et al. TI: structure and intracellular localization of mouse

ADP-ribosylation factors type 1 to type 6 (ARF1-ARF6). J Biochem (Tokyo) 1996; 120(4):813–819.

77. Orci L, Palmer DJ, Ravazzola M, et al. Budding from golgi membranes requires the coatomer complex of non-clathrin coat proteins. Nature (Lond) 1993;362:648–652.

78. D'Souza Schorey C, Li G, Colombo MI, et al. A regulatory role for ARF6 in receptor-mediated endocytosis. Science 1995;267:1175–1178.

79. Ktistakis NT, Brown HA, Waters MG, et al. Evidence that phospholipase D mediates ADP ribosylation factor-dependent formation of Golgi coated vesicles. J Cell Biol 1996;134:295–306.

80. Peyroche A, Paris S, Jackson CL. Nucleotide exchange on ARF mediated by yeast Gea1 protein. Nature (Lond) 1996;384:479–481.

81. Morinaga N, Tsai SC, Moss J, et al. Isolation of a brefeldin A-inhibited guanine nucleotide-exchange protein for ADP ribosylation factor (ARF) 1 and ARF3 that contains a Sec7-like domain. Proc Natl Acad Sci USA 1996;93: 12856–12860.

82. Kahn RA, Terui T, Randazzo PA. Effects of acid phospholipids on ARF activities: potential roles in membrane traffic. J Lipid Mediat Cell Signal 1996;14:209–214.

83. Fensome A, Cunningham E, Prosser S, et al. ARF and PITP restore GTP gamma S-stimulated protein secretion from cytosol-depleted HL60 cells by promoting PIP$_2$ synthesis. Curr Biol 1996;6:730–738.

84. Galas MC, Helms JB, Vitale N, et al. Regulated exocytosis in chromaffin cells. A potential role for a secretory granule-associated ARF6 protein. J Biol Chem 1997;272:2788–2793.

85. Kanaho Y, Yokozeki T, Kuribara H. Regulation of phospholipase D by low molecular weight GTP-binding proteins. J Lipid Mediat Cell Signal 1996;14:223–227.

86. Singer WD, Brown HA, Jiang X, et al. Regulation of phospholipase D by protein kinase C is synergistic with ADP-ribosylation factor and independent of protein kinase activity. J Biol Chem 1996;271:4504–4510.

87. Nakamura Y, Nakashima S, Kumada T, et al. Brefeldin A inhibits antigen- or calcium ionophore-mediated but not PMA-induced phospholipase D activation in rat basophilic leukemia (RBL-2H3) cells. Immunobiology 1996;195:231–242.

88. Ishimoto T, Akiba S, Sato T. Importance of the phospholipase D-initiated sequential pathway for arachidonic acid release and prostaglandin D2 generation by rat peritoneal mast cells. J Biochem (Tokyo) 1996;120:616–623.

89. Tapon N, Hall A. Rho, Rac and Cdc42 GTPases regulate the organisation of the actin cytoskeleton. Curr Opin Cell Biol 1997;9:86–92.

90. Zheng Y, Glaven JA, Wu WJ, et al. Phosphatidylinositol 4,5-bisphosphate provides an alternative to guanine nucleotide exchange factors by stimulating the dissociation of GDP from Cdc42Hs. J Biol Chem 1996;271:23815–23819.

91. Ren XD, Bokoch GM, Traynor Kaplan A, et al. Physical association of the small GTPase Rho with a 68-kDa phosphatidylinositol 4-phosphate 5-kinase in Swiss 3T3 cells. Mol Biol Cell 1996;7:435–442.

92. Tolias KF, Cantley LC, Carpenter CL. Rho family GTPases bind to phosphoinositide kinases. J Biol Chem 1995;270:17656–17659.

93. Zhang J, Benovic JL, Sugai M, et al. Sequestration of a G-protein beta gamma subunit or ADP-ribosylation of Rho can inhibit thrombin-induced activation of platelet phosphoinositide 3-kinases. J Biol Chem 1995;270:6589–6594.

94. Zheng Y, Bagrodia. Activation of phosphoinositide 3-kinase activity by Cdc42Hs binding to p85. J Biol Chem 1994;269:18727–18730.

95. Hawkins PT, Eguinoa A, Qiu RG, et al. PDGF stimulates an increase in GTP-Rac via activation of phosphoinositide 3-kinase. Curr Biol 1995;5:393–403.

96. Singer WD, Brown HA, Bokoch GM, et al. Resolved phospholipase D activity is modulated by cytosolic factors other than Arf. J Biol Chem 1995;270:14944–14950.

97. Cross MJ, Roberts S, Ridley AJ, et al. Stimulation of actin stress fibre formation mediated by activation of phospholipase D. Curr Biol 1996;6:588–597.

98. Teramoto H, Coso OA, Miyata H, et al. Signaling from the small GTP-binding proteins Rac1 and Cdc42 to the c-Jun N-terminal kinase/stress-activated protein kinase pathway. A role for mixed lineage kinase 3/protein-tyrosine kinase 1, a novel member of the mixed lineage kinase family. J Biol Chem 1996;271:27225–27228.

99. Kimura K, Ito M, Amano M, et al. Regulation of myosin phosphatase by Rho and Rho-associated kinase (Rho-kinase). Science 1996; 273:245–248.

100. Amano M, Ito M, Kimura K, et al. Phosphorylation and activation of myosin by Rho-

associated kinase (Rho-kinase). J Biol Chem 1996;271:20246–20249.

101. Wu C, Lee SF, Furmaniak Kazmierczak E, et al. Activation of myosin-I by members of the Ste20p protein kinase family. J Biol Chem 1996;271:31787–31790.

102. Reinhard J, Scheel AA, Diekmann D, et al. A novel type of myosin implicated in signalling by rho family GTPases. EMBO J 1995;14:697–704.

103. Parsons JT. Integrin mediated signalling: regulation by protein tyrosine kinases and small GTP-binding proteins. Curr Opin Cell Biol 1996;8:146–152.

104. Tsukita S, Yonemura S. ERM (ezrin/radixin/moesin) family: from cytoskeleton to signal transduction. Curr Opin Cell Biol 1997;9:70–75.

105. Schmalzing G, Richter HP, Hansen A, et al. Involvement of the GTP binding protein Rho in constitutive endocytosis in *Xenopus laevis* oocytes. J Cell Biol 1995;130:1319–1332.

106. Lamaze C, Chuang TH, Terlecky LJ, et al. Regulation of receptor-mediated endocytosis by Rho and Rac. Nature (Lond) 1996;382:177–179.

107. Murphy C, Saffrich R, Grummt M, et al. Endosome dynamics regulated by a Rho protein. Nature (Lond) 1996;384:427–432.

108. Erickson JW, Zhang CJ, Kahn RA, et al. Mammalian Cdc42 is a brefeldin A-sensitive component of the Golgi apparatus. J Biol Chem 1996;271:26850–26854.

109. Price LS, Norman JC, Ridley AJ, et al. Small GTPases, rac and rho, as regulators of secretion in mast cells. Curr Biol 1995;5(1):68–73.

110. Norman JC, Price LS, Ridley AJ, et al. The small GTP-binding proteins, Rac and Rho, regulate cytoskeletal organisation and exocytosis in mast cells by parallel pathways. Mol Biol Cell 1996;7(9):1429–1442.

111. Norman JC, Price LS, Ridley AJ, et al. Actin filament organisation in activated mast cells is regulated by heterotrimeric and small GTP-binding proteins. J Cell Biol 1994;126(4):1005–1015.

112. Koffer A, Tatham PER, Gomperts BD. Changes in the state of actin during the exocytotic reaction of permeabilised rat mast cells. J Cell Biol 1990;111:919–927.

113. Borovikov YS, Norman JC, Price LS, et al. Secretion from permeabilised mast cells is enhanced by addition of gelsolin: contrasting effects of endogenous gelsolin. J Cell Sci 1995; 108:657–666.

114. Ludowyke RI, Peleg I, Beaven MA, et al. Antigen-induced secretion of histamine and protein kinase C phosphorylation of myosin in rat basophilic leukemia cells. J Biol Chem 1989;264:12492–12501.

115. Choi OH, Adelstein RS, Beaven MA. Secretion from rat basophilic RBL-2H3 cells is associated with diphosphorylation of myosin light chain kinase as well as phosphorylation by protein kinase C. J Biol Chem 1994;269:536–541.

116. Ludowyke RI, Scurr LL, McNally CM. Calcium ionophore-induced secretion from mast cells correlates with myosin light chain phosphorylation by protein kinase C. J Immunol 1996; 157:5130–5138.

117. O'Sullivan AJ, Brown AM, Freeman HNM, et al. Purification and identification of FOAD-II, a cytosolic protein that regulates secretion in streptolysin-O permeabilised mast cells, as a Rac/RhoGDI complex. Mol Cell Biol 1996; 7(3):397–408.

118. Takai Y, Kaibuchi K, Kikuchi A, et al. Small GTP-binding proteins. Int Rev Cytol 1992;133:187–230.

119. Mariot P, O'Sullivan AJ, Brown AM, et al. Rho guanine nucleotide dissociation inhibitor protein (RhoGDI) inhibits exocytosis in mast cells. EMBO J 1996;15(23):6476–6482.

120. Yonei SG, Oishi K, Uchida MK. Regulation of exocytosis by the small GTP-binding protein Rho in rat basophilic leukemia (RBL-2H3) cells. Genetics 1995;1583–1589.

121. Ojio K, Banno Y, Nakashima S, et al. Effect of *Clostridium difficile* toxin B on IgE receptor-mediated signal transduction in rat basophilic leukemia cells: inhibition of phospholipase D activation. Biochem Biophys Res Commun 1996;224:591–596.

122. Prepens U, Just I, von Eichel Streiber C, et al. Inhibition of FcεRI-mediated activation of RBL cells by *Clostridium difficile* toxin B (monoglucosyltransferase). J Biol Chem 1996; 271:7324–7329.

123. Guillemot JC, Montcourrier P, Vivier E, Davoust J, Chavrier P. Selective control of membrane ruffling and actin plaque assembly by the Rho GTPases Rac1 and CDC42 in FcepsilonRI-activated rat basophilic leukemia (RBL-2H3) cells. J Cell Sci 1997;110:2215–2225.

124. Laudanna C, Campbell JJ, Butcher EC. Role of Rho in chemoattractant-activated leukocyte adhesion through integrins. Science 1996; 271:981–983.

Section IV

22
Regulation of Mediator Synthesis and Secretion: Overview

Richard L. Stevens

In this section of the volume, Dr. Susan MacDonald (Chapter 29) describes some of the progress that has been made recently in the identification and cloning of the non-immunoglobulin histamine-releasing factors (HRF), Dr. Donald MacGlashan (Chapter 28) discusses how secretion is regulated in human basophils, and Dr. Mathes and coworkers (Chapter 27) discuss how calcium release-activated calcium current (I_{CRAC}) channels regulate secretion in different populations of mast cells. Drs. Hovav Nechushtan and Ehud Razin (Chapter 23) point out how protein kinase C (PKC) and certain induced transcription factors regulate mast cell growth, and Drs. Thomas Baumruker, Eva Prieshl, Melanie Sherman, and Melissa Brown (Chapters 24 and 25) discuss how nuclear factor of activated T cells (NF-AT) and other transcription factors regulate the FcεRI-mediated expression of cytokines in mast cells. In Chapter 26, Dr. Clifton Bingham and coworkers provide insight as to how arachidonic acid is metabolized in mast cells that have been exposed to specific combinations of cytokines.

Because all mast cells and basophils that have been examined so far express large numbers of high-affinity receptors for IgE (FcεRI) on their surfaces and because these two cell types play such a prominent role in allergic reactions, it is generally perceived that mast cells and basophils degranulate in vivo predominantly via IgE/antigen-dependent mechanisms. As described in Dr. MacDonald's chapter, additional ways of inducing the degranulation of a mast cell and basophil have been uncovered during the last decade. Certain cytokines and chemokines can induce basophils and mast cells to release a modest amount of histamine. However, the major HRF in the body that is formed during the late-phase reaction in an allergic response appears to be a protein that was previously called p21 in mice and p23 in humans. In her chapter, Dr. MacDonald describes some of the recent advances made in the cloning, expression, and function of this novel HRF.

FcεRI-mediated degranulation of mast cells has been studied in depth because of its importance in allergic inflammation. Nevertheless, it has been known for some time that certain populations of mast cells can be induced to degranulate when they are exposed to a diverse array of immunoregulatory proteins (e.g., p21, galectin-3, and certain subclasses of IgG) and peptides (e.g., C5a). As pointed out elsewhere in this volume, it is now apparent that mast cells are quite heterogenous in terms of their granule constituents. Moreover, a mast cell that resides in a specific location can rapidly and reversibly change its phenotype to respond to changes in its microenvironment. Thus, an important and challenging area of future investigation will be understanding the physiological importance of these alternate mechanisms for inducing mast cell degranulation.

When mast cells and basophils are exposed to a secretagogue, complex signal transduction events are initiated that result in the generation of second messengers which ultimately cause the intracellular granules to move and fuse with a specific region on the cell plasma membrane. In their chapter, Dr. Mathes and coworkers address one aspect of the exocytosis process in activated mast cells, namely, the role of I_{CRAC} channels. Within seconds after mast cells are exposed to a secretagogue, there is a dramatic influx of Ca^{2+} into the cell. Dr. Mathes gives insight as to how the influx of calcium through I_{CRAC} channels regulates the secretion of granule mediators from mast cells.

Dr. MacGlashan focuses attention on the intrinsic and extrinsic regulation of histamine secretion from human basophils, and then the subsequent generation and release of leukotriene C_4 (LTC_4) and cytokines from the activated cells. Although rodent mast cells such as the rat basophil leukemia-2H3 cell line have contributed greatly to our understanding of the mechanisms involved in cellular activation of histamine-containing cells, human basophils isolated from unrelated donors can differ considerably in their ability to undergo FcεRI-mediated degranulation. One of the major challenges in the next decade will be to understand at the molecular level why human basophils (and presumably mast cells) from different donors respond so differently when exposed to the same amount of IgE and antigen. Dr. MacGlashan points out what is known about the activation–secretion response in human basophils and then discusses some of the factors that might contribute to the variable outcome among the donors.

Unlike mast cells, normal human basophils contain very little tryptase, chymase, and carboxypeptidase in their granules. Human basophils also have segmented/multilobular nuclei and tend to reside in the peripheral blood, whereas mast cells have nonsegmented nuclei and reside in connective tissues. Based on these and other differences, it has been concluded that basophils and mast cells are developmentally unrelated. Nevertheless, recent studies have revealed that mouse mast cells can have segmented nuclei and can even reside in the peripheral blood for a short period of time after they have outlived their usefulness. Mouse mast cells can rapidly and reversibly change the panel of protease mediators they store in their granules. Because it is possible that a periph-

eral blood human basophil also can alter its granule constituents, more studies need to be carried out to understand the developmental and functional relationship between mast cells and basophils.

Drs. Marshall Plaut and Paris Burd and their coworkers made the seminal observations in 1989 that different populations of rodent mast cells release a diverse array of cytokines and chemokines when they are activated through their high-affinity IgE receptors. Because of the global importance of these newly generated mast cell factors in the regulation of different aspects of the immune response, numerous investigators have focused their attention during the subsequent years on understanding how the transcription of cytokine and chemokine genes are regulated in resting and FcεRI-activated mast cells. The identification of the transcription factors that a mast cell expresses when it is exposed to a specific combination of cytokines or is exposed to IgE and antigen is also relevant to our understanding how this cell type regulates its expression of lipid and granule mediators.

Nuclear factor of activated T cells (NF-AT) was originally described as an activity required for the transcription of the interleukin-2 (IL-2) gene in T cells. Because IL-2 is selectively expressed by activated T cells, it had been thought that NF-AT is a single transcription factor that also is selectively expressed in T cells. Drs. Baumruker and Prieschl note that a number of NF-AT-like transcription factors have been identified and cloned in the past couple of years. Although mast cells contain proteins that are recognized by antibodies raised against NF-ATc and NF-ATp, the mast cell-derived NF-AT probably is a novel member of this family of transcription factors. In FcεRI-activated mast cells, the NF-AT-like factor regulates the transcription of the genes that encode the cytokines IL-4, IL-5, and tumor necrosis factor-α and the chemokine MARC by forming complexes with other transcription factors inside the cell. Formation of these complexes is dependent on MAP-kinases and Ca^{2+}-activated phosphatase calcineurin but not on the varied members of the PKC family. Although PI-3 kinase seems to play an important role in the FcεRI-mediated exocytosis of granular proteins, the PI-3 kinase pathway does not appear to be play a major role in the activation of these transcription factors.

Because IL-4 is required for IgE production by B cells and because IL-4 influences the differentiation of T-helper cell subsets, Drs. Sherman and Brown focused their attention on understanding how NF-AT and numerous other *trans*-acting factors regulate the transcription of the IL-4 gene in mast cells. In addition to the *cis*-acting element in the promoter of the IL-4 gene that is recognized by NF-AT, these investigators discovered that transcription of a reporter gene in mast cells can be enhanced in transient transfection experiments by another *cis*-acting element that surprisingly resides in the second intron of the IL-4 gene. It has been known for some time that the GATA family of transcription factors regulate the expression of the genes that encode a number of the neutral proteases which are stored in the mast cell granule. Drs. Sherman and Brown discuss how GATA-1, GATA-2, and the ets family member PU.1 probably control transcription of the IL-4 gene in mast cells and the expression of an alternate IL-4 transcript that lacks exons 1 and 2. Because the 5'-flanking region of every gene that has been examined has a *cis*-acting element that regulates its transcription, investigators tend to focus their attention on the nucleotide sequences that flank the 5'- and 3'-ends of the gene of interest. Based on the emerging data that show how transcription of the IL-4 gene is regulated in mast cells, it is apparent that mast cell-derived transcription factors sometimes can recognize *cis*-acting elements that reside in introns.

During the early phases of *Trichinella spiralis* infection before the helminth is expelled, the number of mast cells in the jejunal mucosa increase more than 50 fold. Thus, mature mast cells or their immature progenitors are able to respond to changes in the microenvironment to quickly increase their numbers in tissues to respond to noxious or infectious agents. Certain transcription factors are selectively expressed shortly after cultured mast cells are exposed to growth-enhancing cytokines or are activated

via their high-affinity IgE receptors. In their chapter, Drs. Nechushtan and Razin focus their attention on how the transcription factors c-*fos*, c-*jun*, USF2, and MITF interact with one another in FcεRI-activated mast cells and in cytokine-activated mast cells to regulate the proliferation of this effector cell. The role of PKC in signal transduction pathways induced by cytokine exposure is also addressed by these investigators.

Surprisingly, most transcription factors are not very specific in terms of the nucleotide sequences they recognize in vitro. For example, the GATA family of transcription factors all bind to (A/T)GATA(A/G) motifs in the promoter and enhancer elements of genes. Nearly every gene in the body has multiple GATA-binding sites somewhere in its nucleotide sequence. Yet, GATA-1, GATA-2, and GATA-3 somehow are able to selectively induce the transcription of certain protease genes in mast cells. One of the mysteries in the transcription field is how a generic transcription factor like GATA-2 induces a multipotential stem cell to develop into a mature mast cell that expresses a limited number of genes. Studies on the bHLH family of transcription factors have given insight as to how this occurs. The number of mast cells in the *mi/mi* mouse is substantially less than in wild-type mice because of a genetic defect in the bHLH transcription factor designated MITF. The bHLH transcription factors fall into classes (e.g., type A and B). They are only active when they are dimers, and they recognize the so-called E-boxes in promoters having the consensus sequence of "CANNTG." The class A factors tend to be ubiquitous, but when they bind to the class B factors certain tissue-specific differentiation pathways are allowed to proceed.

Recent data have indicated that the Id family of proteins control the levels and ratios of homotypic and heterotypic dimers of the class A and B bHLH transcription factors. Four Id proteins have been identified. While Id proteins are related to the bHLH family of transcription factors, they cannot bind to DNA because they have a defective amino-terminus. Nevertheless, because Id proteins can bind to class A factors, they can modulate the steady-state levels of class A/B heterotypic complexes and the class A homotypic complexes in cells. Thus, Id proteins can act as dominant negative or positive regulators of differentiation pathways. MITF recognizes the "CANNTG" motif and has been proposed to control the transcription of the mMCP-6 gene in mast cells. Besides events that modify the mMCP-6 gene (e.g., methylation of sequences around "CANNTG" motifs) or the transcription factor (e.g., phosphorylation of MITF), one now must consider the role of Id factors in the regulated expression of mMCP-6.

Although cytokine expression is transcriptionally regulated in all cell types, it has been known for more than a decade that repetitive "AUUUA" motifs in the 3′-untranslated regions of most cytokine transcripts dramatically influence their stability, which in turn dominantly controls how much cytokine protein a cell can express. IL-3 expression in certain FcεRI-activated mouse mast cells is regulated in part by this posttranscriptional mechanism, and recent studies have revealed that a posttranscriptional mechanism dominantly controls the expression of at least three chymases in BALB/c mouse mast cells. Interestingly, these three chymase transcripts have repetitive "UGXCCCC" motifs in their 3′-untranslated regions. It appears that cells use multiple mechanisms to control the steady-state levels of proteins that exhibit potent bioactivities such as cytokines and proteases. If the transcription of a cytokine or protease gene gets out of control in a mast cell, this cell can use its fail-safe posttranscriptional mechanisms to rapidly and selectively catabolize the transcript before too much product is expressed.

Mast cells that have been exposed to IgE and antigen not only release cytokines and preformed granule mediators, but they also metabolize arachidonic acid via the 5-lipoxygenase pathway to LTC_4 or via the cyclooxygenase pathway to prostaglandin D_2 (PGD_2). Dr. Bingham and his coworkers discuss how specific combinations of cytokines induce mast cells to metabolize arachidonic acid to different biologically active mediators. These investigators show that IL-1, IL-3, IL-9, IL-10, and c-*kit* ligand regulate the biosynthesis of PGD_2 and LTC_4 in mouse mast cells at a number of levels in a time-dependent manner whether or not these cells are subsequently exposed to IgE and antigen. For many years, an "activated mast cell" was thought to be a degranulated cell. The ability of cytokines to induce the spontaneous release of LTC_4 and PGD_2 from mast cells that have not been exposed to IgE and antigen dramatically changes our concept of what is an "activated" mast cell. It is now apparent that a mast cell that has been exposed to a particular cytokine environment can profoundly influence the other cell types in the tissue even if it has not been induced to release its preformed granule mediators.

23
Early-Response Genes in Mast Cell Activation

Hovav Nechushtan and Ehud Razin

Cytokines and lymphokines have been implicated in having a vital influence on the regulation, maturation, activation, proliferation, and specific functions of mast cells. The intercellular events resulting from mast cell–cytokine interactions are of considerable interest for the understanding and ultimate management of mast cell-associated disease states and for identifying new targets for drug development.

Among the predominant cytokines that regulate the function of murine mast cells are interleukin-3 (IL-3), interleukin-4 (IL-4), and c-*kit* ligand. IL-3, a T-cell-derived growth factor that has been extensively characterized in both mouse and human,[1] is known to stimulate the proliferation and differentiation of a broad spectrum of hemopoietic cells, including pluripotential stem cells, mature megakaryocytes, macrophages, and mouse mast cells. IL-4 was the first mast cell growth factor shown to have an IL-3 synergistic growth regulatory capacity.[2] This cytokine plays a pivotal role in allergic reactions, not only via its direct effects on mast cells but also by its role in immunoglobulin switching in B cells to IgE production. The almost total lack of mast cells in mice with a defect in either c-*kit* or c-*kit* ligand already pointed to the crucial importance of this growth factor for mast cell growth and survival.[3,4]

PKC and Mast Cell Proliferation

In the past few years we and others have extensively investigated the signal transduction pathways induced by IL-3 and IL-4 in murine mast cells. We have produced evidence that protein kinase C (PKC) is a major component in the regulation of cytokine-mediated mast cell proliferation.[5] PKC is a family of regulatory enzymes that play a major role in the control of a wide variety of physiological processes including tumor promotion, membrane receptor function, differentiation, and proliferation.[6]

We studied the role played by PKC in the induction of mast cell growth by either IL-3 or IL-4.[4,7] Both IL-3 and IL-4 induced PKC activity in the particulate fractions of mouse bone marrow-derived mast cells (BMMC) as well as in the murine fetal liver-derived mast cell line (MC-9). PKC depletion in these cells by long-term treatment with phorbol myristate acetate (PMA) led to an increase in the proliferative response of mast cells to either IL-3 or IL-4.

Mast cells have been reported to generate and secrete lymphokines in response to immunological stimulation with IgE-Ag.[8] Our group observed that antibodies against IL-3 and IL-4 strictly inhibited the respective enhancing effects of each lymphokine, while the enhancement of these chronically phorbol ester-treated cells was not affected by either antibody. This indicates that the proliferation in response to IL-3 and IL-4 in PKC-depleted mast cells is caused by a direct effect of these agents rather then secondary phorbol ester-induced secretion of IL-3 or IL-4, revealing an inhibitory role played by PKC in cytokine-mediated mast cell proliferation.[4]

AP-1 in Activated Mast Cells

A similar approach to the identification of phosphorylated proteins, following stimulation of cells, in signal transduction pathway research is the one that characterizes the expression patterns of the early-response genes to gain insight into proliferation mechanisms. The expression of early-response genes is significantly elevated following various cellular stimuli, including those coding for the c-*fos* and c-*jun* transcription factor protein families.

Aberrant expression of many of these genes may lead to uncontrolled cellular proliferation, suggesting an important role for these proteins in cellular growth control.[9] Two studies showed that c-*jun* and c-*fos* mRNA levels are increased in mast cells after stimulation by c-*kit* ligand or IL-3.[10,11] However, stimulation of transformed murine mast cells by IgE-Ag, which is known to cause short-term downregulation of mast cell proliferation rates, also led to similar increases in c-*jun* and c-*fos* levels.[12] Thus, because both growth-promoting and growth-inhibiting external stimuli induce similar patterns of early-response gene transcripts, the fact that one of the early-response genes is induced in mast cells is not concrete evidence for the involvement of this gene product in mast cell growth regulation. The c-*Fos* and c-*Jun* proteins can dimerize to form the transcription factor AP-1.[9] Similar protein complexes are formed by dimerization of other members of the Jun family proteins (c-*jun* heterodimer, *jun-D*, c-*jun* heterodimer, etc.) and are also known as AP-1.[9] However, c-*Fos* does not form a homodimer and does not bind DNA on its own without interacting with other proteins. Furthermore, both Fos and Jun may dimerize with nonfamily members such as upstream stimulating factor-2 (USF-2) or cAMP response element binding protein (CREB).[9]

PKC and AP-1 in Activated Mast Cells

In an attempt to recognize different effects of PKC on c-*jun* and c-*fos* protein levels, PKC activity was depleted by 72-h treatment with PMA or inhibited pharmacologically in IgE-Ag- or IL-3-stimulated mast cells and the expression of these genes was then determined.[5,11,12] IgE-Ag or IL-3 induction of the c-*jun* protein accumulation was inhibited by lowering the activity of PKC, while the same treatment had hardly any effect on the induction of the c-*fos* protein. Because the inhibition of c-*jun* induction by PKC depletion was associated with an increased positive proliferative response to IL-3 in mast cells, or almost prevented the IgE-Ag-induced repression of mast cell proliferation, we postulated that c-*jun* induction might be associated with repression rather than induction of the mast cell proliferative response.

To test this assumption, our group utilized antisense (AS) oligodeoxynucleotides to specifically reduce the levels of c-*Jun*. Reduction of *jun* mRNA levels in IL-3-treated cells significantly enhanced IL-3-induced mast cell proliferation rates.[11] Therefore, c-*jun* is apparently an endogenous suppressive factor of mast cell growth in cells, this contrasts with the role of c-*Jun* as a growth promoter in other cell types such as fibroblasts and erythroleukemia cells. In these cell types, inhibition of expression either by insertion of antibody against c-*Jun* or by applying mRNA antisense inhibited cellular proliferation.[11] Thus, the influence of c-*jun* on mast cell growth regulation is opposite to its effect on the regulation of growth of other cell types. Elucidating the mechanisms whereby c-*jun* negatively controls cellular proliferation may provide significant insight into the intracellular molecular events associated with mast cell proliferation. Moreover, based on these results, it is logical to assume that at least in part the autonomous proliferation of mast cells of mastocytosis patients may be caused by an interruption in c-Jun function.

An interesting finding in this respect is the lack of the induction of AP-1 DNA-binding activity by IL-4 in mast cells.[7] It is known that differences exist in the signal transduction pathways of this growth factor as compared to IL-3.[1] Whether the lack of AP-1 DNA-binding activity induction is one of the reasons for the difference in the induction of mast cell growth by these two growth factors is currently unknown.

USF-2 in Mast Cells

To obtain information on the specific inter-actions of c-*fos* and c-*jun* proteins following IgE-Ag stimulation of mast cells, immuno-precipitations of c-Jun-containing protein complexes were carried out.[13] The c-Fos content in these protein complexes was then compared to total c-Fos content. An intriguing finding from our studies was the observation that the newly synthesized c-Fos did not associate with c-Jun in the AP-1 complex but rather was found to associate with USF-2, also termed Fos-interacting protein.[14]

The ability of c-Fos to interact with USF-2 may allow receptor-specific regulation of gene expression by increasing the diversity of DNA-binding protein complexes that regulate gene transcription. One method of determining how receptors regulate gene expression and to what extent particular genes may be activated is by following the synthesis, assembly, and phos-phorylation of transcription factor complexes in response to activation of receptors. For ex-ample, the formation of a USF-2–Fos bimo-lecular complex might lead to activation of specific genes while at the same time attenuate the activity of the Fos–Jun complex. This pos-sibility would serve to indirectly control the transcription of genes that require AP-1 for expression.

The USF-2 sequence analysis reveals that the protein contains both the consensus for the basic helix-loop-helix (HLH) and the leucine zipper domains.[15] The leucine zipper (Zip) and the HLH dimerization motifs exist in a large group of transcription factors that are de-scribed as members of three different families: the basic region-Zip (bZip), the basic region-HLH (bHLH), and the bHLH-Zip families. Members of the latter recognize the DNA se-quence of CACGTG.[15] This family of bHLH-Zip includes TFE-3, USF-1, USF-2, TFEB, AP-4, Myc, and Max.[15] USF2 −/− mice were recently produced.[16] By using these knockout mice, it was shown that USF-2 is required in vivo for a normal transcriptional response of L-type pyruvate kinase and Spot-14 genes to glucose in the liver.

We cloned the murine mast cell cDNA for USF-2 and have prepared rabbit polyclonal antibodies specific for this protein.[14] This al-lowed us to show in vivo induction of USF-2 synthesis by aggregation of the mast cell FcεRI and increase in its DNA-binding activity.[14] Moreover, we showed that downregulation of the PKC-β by a specific antisense phosphorothioate oligonucleotide resulted in profound inhibition of USF-2 expression and its DNA-binding activity.[14] Thus, aggrega-tion of the FcεRI on mast cells elicits a PKC-β-dependent signaling pathway that regulates USF-2 expression and function. Recently, we also observed that IL-3, which is a major growth factor for mast cells, induces USF-2 protein synthesis in murine mast cells.[17] Sur-prisingly, it does not significantly affect the level of USF-2 mRNA; in using polysomal frac-tionation and RNA analysis we demonstrated that USF-2 is translationally regulated. More-over, PKC inhibitors prevented both the induc-tion of USF-2 protein synthesis and increase in USF-2 translational efficiency in IL-3-activated mast cells.

The physiological function of USF-2 in hematopoietic-derived cells in general, and in mast cells in particular, is the focus of our re-cent investigative effort. Considering the regu-latory role of other transcription factors such as Myc and Max that bind to the same DNA-binding motif as does USF-2, and our observa-tions that regulation of USF-2 expression is tightly linked to growth factor stimulation, it is highly likely that USF-2 is involved in cell growth regulation. More direct evidence for the involvement of USF-2 in cell division has been recently described.[18] It was shown that USF-2 may serve as a negative regulator of cell prolif-eration by antagonizing the transforming func-tion of Myc and also through a more general inhibitory effect on growth.

Microphthalmia Transcription Factor (*mi*) in Mast Cells

In mast cells, the microphthalmia transcription factor (*mi*) plays a major role in the regulation of growth and differentiation.[19] Mutations in the gene encoding for this transcription factor were shown to cause the mouse microphthalmia phenotype.[20,21] The mouse phenotype is now known to contain more than 10 allelic variations.[21] Homozygous mutations at this locus cause small unpigmented eyes, skin and inner ear melanocyte depletion, microphthalmia, deafness, a defect in osteoclasts, and a major decrease in mast cell number.[21] Kitamura and his coworkers showed that although *mi/mi* mice are mast cell deficient, splenic cells from these mice develop into mast cells when cultured with conditioned medium.[22,23] However, these cells are different from mast cells grown from normal mice in many important aspects. For example, when mast cells from *mi/mi* mice are cocultured with fibroblasts, they fail to enter the S-phase and gradually disappear from the tissue culture plates.[21] Furthermore, they express only low levels of c-Kit and nerve growth factor (140-kDa) receptors, two growth factor receptors with important roles in mast cell development.[1]

The isolation of the gene causing the *mi* mutation was recently achieved by two groups as a result of the fortunate observation of homozygous transgenic mice strains displaying phenotypes similar to that of the *mi/mi* mice.[20,21] The integration site of the transgene from these mice was then cloned and analyzed. Computer analysis of the nucleotide sequence revealed that the *mi* gene encodes a novel member of the basic helix-loop-helix leucine zipper (bHLH-ZIP) protein family of transcription factors. Because *mi* was shown to have an essential role in mast cell function and development, we decided to characterize the expression patterns of this transcription factor in mast cells.

We investigated the regulation of the synthesis of *mi* in murine mast cells activated by various physiological stimuli.[24] Using a specific rabbit polyclonal anti-*mi* antibody, we found that IL-3, IL-4, or aggregation of the mast cell FcεRI induced the synthesis of *mi* protein in these cells. None of these stimuli significantly affected the level of *mi* mRNA in the mast cells at any of the time points tested. Also, using this specific anti-*mi* antibody an increase in *mi* protein synthesis was demonstrated during differ-

entiation of mast cells from their bone marrow cell precursors. Surprisingly, nondifferentiated mast cells can be grown in vitro from the bone marrow of *mi/mi* mice, and so *mi* is not an essential gene for the initial cellular commitment to the mast cell lineage. Therefore, *mi* might play an important role in the differentiation, proliferation, and survival of committed mast cells as a key molecule in response of these cells to various external stimuli. Recently, several studies have provided us with information regarding the mast cell progenitor population.[25] Study of *mi* expression in this cell population could provide important insights into the role of *mi* in early mast cell differentiation events. Moreover, a complex containing *mi* bound to USF-2 was detected only in activated mast cells.

Thus, *mi* and USF-2 participate in the coordination of gene expression in mast cells either directly, by binding to DNA-binding sites, or indirectly, through its effects on the activities of other transcription factors.

References

1. Nechushtan H, Razin E. Regulation of mast cell growth and proliferation. Crit Rev in Oncol Hematol 1996;23:131–150.
2. Boulay JL, Paul WE. The interleukin-4 family of lymphokines. Curr Opin Immunol 1992;4:294–298.
3. Kitamura Y, Go S. Decreased production of mast cells in S1/S1d anemic mice. Blood 1979;53:492–497.
4. Kitamura Y, Go S, Hatanaka K. Decrease of mast cells in W/Wv mice and their increase by bone marrow transplantation. Blood 1978;52:447–452.
5. Chaikin E, Ziltener H, Razin E. Protein kinase C plays an inhibitory role in IL-3 and IL-4 mediated mast cell proliferation. J Biol Chem 1990;265:22109–22116.
6. Nishizuka Y. Studies and prespectives of protein kinase C. Science 1992;258:607–614.
7. Chaikin E, Hakeem I, Razin E. The incapability of IL-4 to induce AP-1 activity in murine mast cells. Int Arch Allergy Immunol 1995;107:57–59.
8. Galli SJ. New concepts about the mast cell. N Engl J Med 1993;328:257–256.
9. Angel P, Karin M. The role of Jun, Fos and the AP-1 complex in cell proliferation and transfor-

mation. Biochim Biophys Acta 1991;1072:129–157.

10. Tsai M, Tam SY, Galli SJ. Distinct patterns of early response gene expression and proliferation in mouse mast cells stimulated by stem cell factor, interleukin-3, or IgE and antigen. Eur J Immunol 1993;23:867–872.

11. Chaikin E, Hakeem I, Razin E. Enhancement of interleukin-3-dependent mast cell proliferation by supression of c-*jun* expression. J Biol Chem 1994;269:8498–8503.

12. Baranes D, Razin E. PKC regulates proliferation of mast cells and the expression of Fos and Jun proto-oncogene mRNAs during activation by IgE-Ag or calcium ionophore A23187. Blood 1991;78:2354–2366.

13. Lewin I, Nechushtan H, Qingen K, Razin E. Regulation of AP-1 expression and activity in antigen-stimulated mast cells: the role played by protein kinase C and the possible involvement of Fos interacting protein. Blood 1993;82:3745–3751.

14. Lewin I, Jacob-Hirsch J, Zang ZC, et al. Aggregation of the FcεRI in mast cells induces the synthesis of *fos*-interacting protein and increases its DNA-binding activity: the dependency on PKC-β. J Biol Chem 1996;271:1514–1519.

15. Sirito M, Lin Q, Maity T, Sawadogo M. Ubiquitous expression of the 43- and 44-kDa forms of transcription factor USF in mammalian cells. Nucleic Acids Res 1994;22:427–429.

16. Vallet SV, Henrion AA, Bucchini D, et al. Glucose-dependent liver gene expression in upstream stimulatory factor 2 –/– mice. J Biol Chem 1997;272:21944–21949.

17. Zhang ZC, Nechushtan H, Jacob-Hirsch J, Avni D, Meyuhas O, Razin E. Growth-dependent and PKC-mediated translational regulation of the upstream stimulating factor 2 (USF2) mRNA in hematopoietic cells. Oncogene 1997;16:763–769.

18. Luo X, Sawadogo M. Antiproliferative properties of the USF family of helix-loop-helix transcription factors. Proc Natl Acad Sci USA 1996;93:1308–1311.

19. Ebi Y, Kasugai T, Seino Y, Onoue H, Kanemoto T, Kitamura Y. Mechanism of mast cell deficiency in mutant mice of *mi/mi* genotype: an analysis by co-culture of mast cells and fibroblasts. Blood 1990;75:1247–1258.

20. Hodgkinson CA, Moore KJ, Nakayama A, Steingrimsson E, Copeland NG, Jenkins NA. Mutations at the mouse microphthalmia locus are associated with defects in a gene encoding a novel basic-helix-loop-helix-zipper protein. Cell 1993;74:395–407.

21. Hughes MJ, Lingrel JB, Krakowsky JM, Anderson KP. A DNA insertional mutation results in microphthalmia in transgenic mice. J Biol Chem 1993;268:20687–20699.

22. Jippo-Kanemoto T, Adachi S, Ebi Y, et al. BALB/3T3 fibroblast-conditioned medium attracts cultured mast cells derived from W/W but not from *mi/mi* mutant mice, both of which are deficient in mast cells. Blood 1992;80:1933–1946.

23. Jippo T, Ushio H, Hirota S, et al. Poor response of cultured mast cells derived from *mi/mi* mutant mice to nerve growth factor. Blood 1994;84:2977–2989.

24. Nechushtan H, Zhang ZC, Razin E. Microphthalmia (*mi*) in murine mast cells: regulation of its stimuli-mediated expression on the translational level. Blood 1997;89:2999–3008.

25. Rodewald HR, Dessing M, Dvorak AN, Galli SJ. Identification of a committed precursor for the mast cell lineage. Science 1996;271:881–885.

24
FcεRI-Mediated Activation of NF-AT

Thomas Baumruker and Eva E. Prieschl

The term nuclear factor of activated T-cells (NF-AT) appeared in the scientific literature in 1988 to describe a protein visualized in gel shift assays binding to the ARRE (antigen receptor response element) of the IL-2 promoter and forming a complex called NF-IL2-E.[1] In contrast to many other binding activities and transcription factors that were first described in one specific cell type and later found to be of broader significance, NF-AT has remained as a more or less T-cell-induction-specific DNA-binding activity.[1-5] It escaped a detailed molecular characterization because all efforts to clone the corresponding gene or cDNA failed. An explosion in our knowledge came in the last 3 years with the cloning of NF-ATp[6] and NF-ATc,[7] the realization that it is a family of proteins,[8-10] and the finding that a number of other cell types besides T-cells express NF-AT family members for the regulation of gene transcription.[10-17]

Transcription Factors in Mast Cells

In the late 1980s it became clear that mast cells are a major source of cytokines if triggered at the FcεRI via IgE and antigen.[18-22] However, neither the role of these mast cell mediators in health and disease nor their regulation at the transcriptional level was clarified further in the following years. Although numerous investigations in T-cells were fostered by the link of the gel shift binding activity of NF-AT to IL-2 gene expression and the availability of immunosuppressive drugs indirectly targeting the activation of this factor with clinical relevance in transplantation surgery,[6,23,24] no such development took place in mast cells. A number of transcription factors have been described in mast cells, such as AP 1 (*fos* and *jun*),[25] FIP (Fos-interacting protein),[26] *mi* (microphthalmia),[27,28] SCL/TAL-1,[29,30] EKLF (erythroid Kruppel-like factor),[31] GATA-1 and GATA-2,[32,33] and B-1 (identical to proto-oncogene Spi-1/Pu.1),[34] but none of these has subsequently been linked to FcεRI triggering and cytokine gene induction.

NF-AT-Like Factors in Mast Cells

Two publications in 1995 in the *Journal of Immunology* (using the mouse mast cell line CPII)[11,12] and one in the *Journal of Biological Chemistry* (using the rat basophilic leukemia cell line RBL-2H3)[25] described for the first time the existence of an NF-AT-like binding activity in mast cells/basophils and linked cell activation and cytokine gene induction to this activity. Further publications in the meantime by our group as well as by the group of M. Brown clearly demonstrated the requirement of NF-AT-like factors in mast cells/basophils for the transcriptional activation of IL-4,[35] IL-5,[11] MARC (a chemokine),[12] and tumor necrosis factor-α (TNF-α)[36] after IgE plus antigen as well as after phorbol myristate acetate (PMA) plus ionomycin triggering. These results are supported by the finding that antiserum generated against NF-ATp from murine T-cells stained primary human bone marrow-derived mast cells,[37] indicating the expression of an antigenically-related protein in this cell type. The molecular nature of the NF-AT protein(s) in these allergic effector cells, however, is currently unknown because no cDNA clones have been isolated so far. All experimental evidence has been derived using gel shift assays, transient transfections, the use of antisera raised against NF-AT from T-cells, and a limited polymerase chain reaction (PCR) analysis. Initial data by M. Brown using these antisera/antibodies gave rise to the assumption that various members of the NF-AT family are expressed in mast cells and basophils. An anti-NF-ATc antibody detects two proteins of 120 to 160 kDa and a 56-kDa protein, while an anti-NF-ATp antibody detects a 145- and a 41-kDa protein in a western blot.[35] In CPII mouse mast cells, using an antiserum against NF-ATp, we also detected two proteins of about 110 kDa and 130 kDa in a western blot analysis. In a southwestern blot analysis, using the distal NF-AT binding site of the murine IL-2 promoter as a radioactive probe, similar findings were made.[38] PCR analysis, however, demonstrated that these factors most likely are distinct from the NF-ATp of T-cell origin. Although a fragment of approximately 450 bp from the DNA-binding domain of this factor could be amplified (see later), primers outside this region failed to give any amplification product from CPII mouse mast cell mRNA (Prieschl, unpublished data). All these findings further strengthen the results obtained concerning different transcription cofactors interacting with (potentially different) NF-AT factor(s) in the regulation of these cytokines in mast cells.

NF-AT Factor(s) and Cofactors in Mast Cells

Currently, from mouse and man and using different cell sources, six members of the NF-AT transcription factor family have been cloned and partially characterized (see Table 24.1).[6–10]

TABLE 24.1. Overview of currently cloned members of the NF-AT family.

NF-AT family member	mRNA size	Cloned from cell type
NF-ATp[6]	8 kb	murine T cells
NF-ATc[7]	2.7 and 4.5 kb	Jurkat and human PBL
NF-AT3[8]	Three species between 3 and 5 kb	Jurkat
NF-AT4 a, b, c[8]	6.5 kb	Jurkat (a, c), skeletal muscle (b, c)
NF-ATx[10]	7 kb	Jurkat
NF-ATc3[9]	n.d. 3619 bp cDNA clone	T-cell-receptor transgenic murine thymus

As also seen in the case of the NF-κB transcription factor family they all share good homology in the DNA-binding region, reflecting the fact that their DNA sequence recognition is, as far as is known today, identical (the exception is NF-AT 3c).[9] It is noteworthy that in this region there is also a distant homology between the NF-AT and the NF-κB family (17% at the amino acid level, called the relB homology domain). Outside this approximately 187-amino-acid region, which A. Rao identified as the minimal DNA-binding domain of NF-ATp,[39] no or a much lesser degree of conservation is found (also a finding seen in the various NF-κB family members). This divergence obviously determines the different biological profiles of the NF-AT transcription factor family members, especially their interactions with other transcription cofactors. According to current knowledge in T-cells, AP 1 is the dominant cofactor described for NF-AT, using a number of different promoters as a model system such as TNF-α, GM-CSF, IL-2, and IL-3.[40] More recent publications also outline cAMP response element binding protein (CREB) as a potential cofactor in this cell type, and a few findings point toward other possible interaction partners such as SP 1, EGR-1, and ATF 2.[41]

The broader expression of NF-AT-like factors, which are now described also in B-cells,[16] NK cells,[42] and endothelial cells,[43] with an indis-

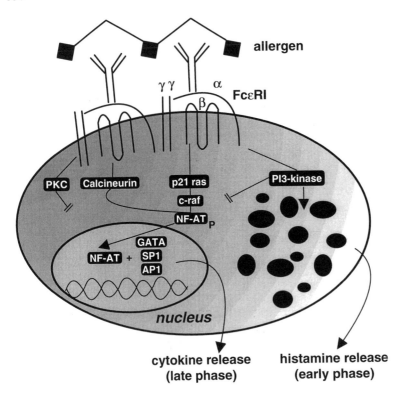

tinguishable DNA recognition capability raises the question of how the cell type specificity in gene expression of all the NF-AT-dependent cytokines is mediated (clearly not at the expression level of NF-AT or the primary DNA recognition site). Again, an explanation is offered from our knowledge about the NF-κB proteins and strongly strengthened by the finding of different cofactors for the NF-AT molecules in mast cells.[11,36,38] So-called supracomplexes[44] of the NF-κB factor(s) with various additional transcription cofactors are assembled at a given promoter, and the specific composition of these supracomplexes determines which cytokine genes are expressed in a given cell type. In this respect, mast cells offer a suitable experimental system to address this question for NF-AT(s). For IL-5 gene induction in mast cells, NF-AT was shown to cooperate with a GATA transcription factor family member.[11] In mast cells the existence and function of several GATA family members has been shown before,[45] with GATA-2 specifically found in proliferating and differentiating mast cells.[33] At the MARC chemokine pro-

moter, which confers a relatively mast cell-specific expression pattern, we did not detect GATA but instead found SP 1.[12] At the TNF-α promoter, AP 1, as in T-cells, was shown to interact with NF-AT.[36] This also holds true for IL-4 gene expression, as described by McClosky's group in RBL cells.[25] These findings together lead to the hypothesis that the different assembly of cofactors with various NF-AT family members determines the cell type specificity of cytokine gene expression (see Figure 24.1).

NF-AT DNA-Binding Sites in Mast Cells

Binding sites for NF-AT-like factors in mast cells for the respective promoters (IL-5, TNF-α, MARC) were mostly determined functionally in transient transfection assays with successive 5′-promoter deletions and confirmed by using the corresponding oligonucleotides in gel shift assays.[11,12,46] A comparison of these NF-

AT sites (including the IL-4 site described by M. Brown et al.) shows that they all share the so-called common half site motif TTTCC[47] (for TNF-α, the last C is changed to a T) originally outlined as being of importance for the binding of the three factors NF-κB, NF-AT, and CK-1. This similar recognition capability is reflected by the weak homology in the DNA-binding region (17% at the amino acid level) of the corresponding proteins of NF-AT and NF-κB, as described here (CK-1 has not been cloned so far). The distance of the NF-AT-binding sites relative to the TATA box in the corresponding promoters is more or less conserved (MARC, 88 N; IL-5, 82 N; TNF-α, 64 N; IL-4, 69 N). However, the binding sites are found in both orientations.

NF-AT Activation in Mast Cells via the MAP Kinase Pathway

D. Cantrell and her colleagues have extensively elucidated over the last few years the signaling pathways involved in gene activation in T-cells after T-cell receptor triggering (CD3 stimulation) including the Ca^{2+} signal (see following).[48,49] Her work and the similarity of the FcεRI γ-chain to the T-cell receptor ζ-chain prompted us to investigate the pathways from the FcεRI to the gene level in a similar experimental setting.[12] Earlier publications had already shown that the MAP kinases are activated after allergic triggering as well as after SCF (stem cell factor) treatment in mast cells.[50] The activation (phosphorylation) of the 42- and 44-kDa MAP kinases after various stimuli such as the calcium ionophore A23187, carbachol, and IgE plus antigen was also reported in the rat basophilic leukemia cell line (RBL-2H3).[51] To investigate the relevance of the p21[ras]/pp74[raf]/MAPK signaling pathway for cytokine gene induction in mast cells, transdominant negative mutants of p21[ras] and pp74[raf] (ras N17 and raf 301)[52,53] were used in transient cotransfection assays together with a cytokine reporter gene. A clear inhibition was seen for the expression of the cotransfected cytokine reporter gene if the cells were stimulated with IgE plus antigen. This inhibition was linked to

the function of the NF-AT-like transcription factor, as reporter gene constructs devoid of this binding site were not affected by the transdominant mutants. In a reverse experimental setting using an activated p21[ras] expression plasmid (v-ras) in cotransfections with cytokine reporter gene constructs containing the NF-AT-binding site, activation was already observed in nonstimulated cells. These results are identical to the picture seen in T cells after T-cell receptor (TCR) stimulation. They are supported by the finding that certain commercially available MAP kinase pathway inhibitors (i.e., Apigenin, $IC_{50} = 25\,\mu M$) prevent the FcεRI-mediated cytokine induction in the expected dose range (Csonga and Baumruker, unpublished). Taken together, these data strongly suggest an involvement of p21[ras], the raf kinase, and subsequent MAP kinases in the signal transmission for cytokine gene induction and NF-AT activation in mast cells (see Figure 24.1). In this respect it is interesting that p21[ras] was also recently postulated to be part of the signaling pathways in B-cells using an identical set of experiments.[54] Also supportive of this claim is the finding that p21[ras] is in complex with the SH 3 domain of Grb 2 after FcεRI crosslinking in mast cells.[55]

NF-AT Activation via Calcineurin

As in T cells, a second signal involving Ca^{2+} is necessary for cytokine gene activation in our mast cell line CPII. This is in contrast to some other mast cell lines in which the Ca^{2+} signal alone induces lymphokines such as IL-3 and GM-CSF.[56,57] However, in those cell lines it was reported that gene induction is primarily caused by posttranscriptional stabilization of the mRNA and does not involve promoter (NF-AT, transcriptional) activation. This difference might be the result of different mast cell lines reflecting the various types found also in vivo or the generation of the cell line (certain cell lines, especially factor-independent ones, might already have p21[ras] activated constitutively, providing a permanent first signal). The sensitivity of the CPII mouse mast cell line to immunosuppressive drugs of the cyclic undecapeptide and macrolide classes and the finding that an NF-AT family member is present in this cell type strongly suggested that the Ca^{2+}-dependent phosphatase calcineurin is involved in this signal transmission. This was also concluded previously from experiments with bone marrow-derived mouse mast cells, which are not sensitive to inhibition by FK506 because of lack of expression of FK506 binding protein (FKBP) 12.[58] Messenger RNA expression of calcineurin is detectable by PCR analysis in our mast cell line. In analogy to the experiments with the constitutively active v-*ras* expression plasmid, a constitutively active calcineurin mutant, cotransfected with a cytokine reporter gene plasmid, led to an increased expression of the reporter gene in nonstimulated cells (Pendl, unpublished data). These findings imply that the second signal for NF-AT activation, as in T-cells, is mediated via the Ca^{2+}-dependent phosphatase calcineurin (see Figure 24.1).

Signal Pathways That Do Not Link the FcεRI to NF-AT, the PKC Pathway, and the PI-3 Kinase Pathway

Several studies have described the expression pattern of protein kinase C (PKC) isozymes in mast cells and basophils,[26,59–65] and very recently the phosphorylation of a member of the Tec subfamily of tyrosine kinases (Emt) after crosslinking the FcεRI on mast cells by PKC was implied.[66] PKC activation was shown to be involved in the expression of c-*fos* and c-*jun*,[60] and PKC-β was specifically linked to the regulation of FIP (Fos-interacting protein).[26] Contrary to all these data, depletion of PMA-dependent PKCs in our mast cell line CPII by 48-h treatment with PMA did not abolish the subsequent induction of NF-AT-dependent reporter gene plasmids by IgE plus antigen, but induction by PMA plus ionomycin was abolished.[12] From these data, a picture emerges in which activation of PMA-dependent PKCs is not an absolute requirement for the activation

of NF-AT and its cofactors in cytokine gene induction.

Investigations regarding PI-3 kinase involvement in signaling pathways are mostly done using Wortmannin as an inhibitor of the catalytic subunit p110 of this enzyme,[67] which blocks the lipid as well as the protein kinase activity at nanomolar to low micromolar concentrations. Effects of this drug on mast cell activation were previously seen in the stem cell factor- (SCF) triggered histamine release in rat peritoneal mast cells and in the *kit*-induced adhesion of bone marrow-derived mast cells (BMMC).[68] In addition, IgE plus antigen-provoked degranulation of RBL-2H3 cells was reported to be inhibited also by this compound.[69] These data in general suggested that PI-3 kinase is used for signaling in mast cells and basophils as well as being necessary for the degranulation reaction after activation of the FcεRI. Using Wortmannin as well as the related but more potent compound Demethoxyviridin, we were able to confirm that degranulation after IgE plus antigen triggering is dose dependently inhibited in our CPII mouse mast cells by both compounds, although no effect was seen on cytokine gene induction and release (see Figure 24.1). This effect was measured via transient transfections of reporter genes such as MARC, TNF-α, and IL-5, as well as by an ELISA for TNF-α.[69] Similar data were recently reported for IL-6 in mast cells.[70] However, contrary to these data, leukotriene synthesis (LTC_4) was also affected by the two inhibitors in our CPII mouse model system. On the basis of these data, although an essential involvement of PI-3 kinase in the activation of the NF-AT factors and the corresponding cofactors for lymphokine gene regulation can be ruled out, this molecule clearly participates in the degranulation reaction and most likely in the activation of arachidonic acid metabolism.

Acknowledgment. We thank all our colleagues from the Novartis Research Institute who over the years have contributed to our work. In particular we thank Penelope Andrew for critical reading of the manuscript.

References

1. Shaw J, Utz PJ, Durand DB, et al. Identification of a putative regulator of early T cell activation genes. Science 1988;241:202–205.

2. Jain J, McCaffrey PG, Miner Z, et al. The T-cell transcription factor NFATp is a substrate for calcineurin and interacts with fos and jun. Nature (Lond) 1993;365:352–355.

3. Masuda ES, Tokumitsu H, Tsuboi A, et al. The granulocyte-macrophage colony-stimulating factor promoter cis-acting element CLEO mediates induction signals in T cells and is recognized by factors related to AP1 and NFAT. Mol Cell Biol 1993;13:7399–7407.

4. Ullman KS, Northrop JP, Verweij CL, et al. Transmission of signals from the T lymphocyte antigen receptor to the genes responsible for cell proliferation and immune function: the missing link. Annu Rev Immunol 1990;8:421–452.

5. Crabtree GR, Clipstone NA. Signal transmission between the plasma membrane and nucleus of T lymphocytes. Annu Rev Biochem 1994;63:1045–1083.

6. McCaffrey PG, Luo C, Kerppola TK, et al. Isolation of the cyclosporin-sensitive T cell transcription factor NFATp. Science 1993;262:750–753.

7. Li X, Ho SN, Luna J, et al. Cloning and chromosomal localization of the human and murine genes for the T-cell transcription factors NFATc and NFATp. Cytogenet Cell Genet 1995;68:185–191.

8. Hoey T, Sun YL, Williamson K, et al. Isolation of two new members of the NF-AT gene family and functional characterization of the NF-AT proteins. Immunity 1995;2:461–472.

9. Ho SN, Thomas DJ, Timmerman LA, et al. NFATc3, a lymphoid-specific NFATc family member that is calcium-regulated and exhibits distinct DNA binding specificity. J Biol Chem 1995;270:19898–19907.

10. Masuda ES, Naito Y, Tokumitsu H, et al. NFATx, a novel member of the nuclear factor of activated T cells family that is expressed predominantly in the thymus. Mol Cell Biol 1995;15:2697–2706.

11. Prieschl EE, Gouilleux-Gruart V, Walker C, et al. An NFAT-like transcription factor on mast cells is involved in IL5 gene regulation after IgE plus Ag stimulation. J Immunol 1995;154:6112–6119.

12. Prieschl EE, Pendl GG, Harrer NE, et al. p21^{ras} links FcεRI to NF-AT family member in mast cells. The AP3-like factor in this cell type is an NF-AT family member. J Immunol 1995;155:4963–4970.

13. Park J, Yaseen NR, Hogan PG, et al. Phosphorylation of the transcription factor NFATp inhibits its DNA binding activity in cyclosporin A-treated human B and T cells. J Biol Chem 1995;270:20653–20659.

14. Ruff VA, Leach KL. Direct demonstration of NFATp dephosphorylation and nuclear localization in activated HT-2 cells using a specific NFATp polyclonal antibody. J Biol Chem 1995;270:22602–22607.

15. Ho AM, Jain J, Rao A, et al. Expression of the transcription factor NFATp in a neuronal cell line and in the murine nervous system. J Biol Chem 1994;269:28181–28186.

16. Venkataraman L, Francis DA, Wang Z, et al. Cyclosporin-A sensitive induction of NF-AT in murine B cells. Immunity 1994;1:189–196.

17. Choi MS, Brines RD, Holman MJ, et al. Induction of NF-AT in normal B lymphocytes by anti-immunoglobulin or CD40 ligand in conjunction with IL-4. Immunity 1994;1:179–187.

18. Burd PR, Rogers HW, Gordon JR, et al. Interleukin 3-dependent and -independent mast cells stimulated with IgE and antigen express multiple cytokines. J Exp Med 1989;170:245–257.

19. Bradding P, Feather IH, Wilson S, et al. Immunolocalization of cytokines in the nasal mucosa of normal and perennial rhinitic subjects. The mast cell as a source of IL-4, IL-5, and IL-6 in human allergic mucosal inflammation. J Immunol 1993;151:3853–3865.

20. Gordon JR, Burd PR, Galli SJ. Mast cells as a source of multifunctional cytokines. Immunol Today 1990;11:458–464.

21. Galli SJ, Gordon JR, Wershil BK. Cytokine production by mast cells and basophils. Curr Opin Immunol 1991;3:865–872.

22. Gordon JR, Burd PR, Galli SJ. Mast cells as a source of multifunctional cytokines. Immunol Today 1990;11:458–464.

23. Emmel EA, Verweij CL, Durand DB, et al. Cyclosporin A specifically inhibits function of nuclear proteins involved in T cell activation. Science 1989;246:1617–1620.

24. Schreiber SL, Crabtree GR. The mechanism of action of cyclosporin A and FK506. Immunol Today 1992;13:136–142.

25. Hutchinson LE, McCloskey MA. Fc epsilon RI-mediated induction of nuclear factor of activated T-cells. J Biol Chem 1995;270:16333–16338.

26. Lewin I, Jacob-Hirsch J, Zang ZC, et al. Aggregation of the Fc epsilon RI in mast cells induces the synthesis of Fos-interacting protein and increases its DNA binding-activity: the dependence on protein kinase C-beta. J Biol Chem 1996;271:1514–1519.

27. Hughes MJ, Lingrel JB, Krakowsky JM, et al. A helix-loop-helix transcription factor-like gene is located at the *mi* locus. J Biol Chem 1993; 268:20687–20690.

28. Morii E, Takebayashi K, Motohashi H, et al. Loss of DNA binding ability of the transcription factor encoded by the mutant mi locus. Biochem Biophys Res Commun 1994;205:1299–1304.

29. Bockamp EO, McLaughlin F, Murrell AM, et al. Lineage-restricted regulation of the murine SCL/TAL-1 promoter. Blood 1995;86:1502–1514.

30. Murrell AM, Bockamp EO, Gottgens B, et al. Discordant regulation of SCL/TL-1 mRNA and protein during erythroid differentiation. Oncogene 1995;11:131–139.

31. Miller IJ, Bieker JJ. A novel, erythroid cell-specific murine transcription factor that binds to the CACCC element and is related to the Kruppel family of nuclear proteins. Mol Cell Biol 1993;13:2776–2786.

32. Martin DIK, Zon LI, Mutter G, et al. Expression of an erythroid transcription factor in mega-karyocytic and mast cell lineages. Nature (Lond) 1989;344:444–449.

33. Jippo T, Mizuno H, Xu ZD, et al. Abundant expression of transcription factor GATA-2 in proliferating but not in differentiated mast cells in tissues of mice: demonstration by in situ hybridization. Blood 1996;87:993–998.

34. Galson DL, Hensold JO, Bishop TR, et al. Mouse beta-globulin DNA-binding protein B1 is identical to a proto-oncogene, the transcription factor Spi-1/PU.1, and is restricted in expression to hematopoietic cells and the testis. Mol Cell Biol 1993;13:2929–2941.

35. Weiss DL, Hural J, Tara D, et al. Nuclear factor of activated T-cells is associated with a mast cell interleukin 4 transcription complex. Mol Cell Biol 1996;16:228–235.

36. Prieschl EE, Pendl GG, Elbe A, et al. Induction of the TNFα promoter in the murine dendritic cell line DC18 and the murine mast cell line CPII is differently regulated. J Immunol 1996; 157:2645–2653.

37. Shaw KT, Ho AM, Raghavan A, et al. Immuno-suppressive drugs prevent a rapid dephosphory-lation of transcription factor NFAT1 in stimulated immune cells. Proc Natl Acad Sci USA 1995;92:11205–11209.

38. Baumruker T, Pendl GG, Prieschl EE. Gene regulation after FcεRI stimulation in the murine mast cell line CPII. Int Arch Allergy Immunol 1997;113:39–41.

39. Jain J, Burgeon E, Badalian TM, et al. A similar DNA-binding motif in NFAT family proteins and the Rel homology region. J Biol Chem 1995;270:4138–4145.

40. Rao A. NF-ATp: A Transcription Factor Required for the co-ordinate induction of several cytokine genes. Immunol Today 1994;15:274–281.

41. Skerka C, Decker EL, Zipfel PF. A regulatory element in the human interleukin-2 gene promoter is a binding site for the zinc finger proteins Sp1 and EGR-1. J Biol Chem 1996;270:22500–22506.

42. Aramburu J, Azzoni L, Rao A, et al. Activation and expression of the nuclear factor of activated T cells NFATp and NFATc in human natural killer cells: regulation upon CD16 ligand binding. J Exp Med 1995;182:801–810.

43. Cockerill GW, Bert AG, Ryan GR, et al. Regulation of granulocyte-macrophage colony-stimulating factor and E-selectin expression in endothelial cells by cyclosporin A and the T-cell transcription factor NFAT. Blood 1995;86:2689–2698.

44. Nolan PG. NF-AT-AP-1 and Rel-bZIP: hybrid vigor and binding under the influence. Cell 1994;77:795–798.

45. Henkel G, Brown MA. PU.1 and GATA: components of a mast cell-specific interleukin 4 intronic enhancer. Proc Natl Acad Sci USA 1994;91:7737–7741.

46. Jarmin DI, Kulmburg PA, Huber NE, et al. A transcription factor with AP3-like binding specificity mediates gene regulation after an allergic triggering with IgE and antigen in mouse mast cells. J Immunol 1994;5720–5729.

47. McCaffrey PG, Jain J, Jamieson C, et al. A T cell nuclear factor resembling NF-AT binds to an NFκB site and to the conserved lymphokine promoter sequence "cytokine-1." J Biol Chem 1992; 267:1864–1871.

48. Woodrow M, Clipstone NA, Cantrell D. p21^{ras} and calcineurin synersize to regulate the nuclear factor of activated T cells. J Exp Med 1993; 178:1517–1522.

49. Izquierdo Pastor M, Reif K, Cantrell D. The regulation and function of p21^{ras} during T-cell activation and growth. Immunol Today 1995;16:159–164.

50. Tsai M, Chen RH, Tam SY, et al. Activation of MAP kinases, pp90rsk and pp70-S6 kinases in mouse mast cells by signaling through the c-*kit* receptor tyrosine kinase or Fc epsilon RI: rapamycin inhibits activation of pp70-S6 kinase and proliferation in mouse mast cells. Eur J Immunol 1993;23:3286–3291.

51. Offermanns S, Jones SV, Bombien E, et al. Stimulation of mitogen-activated protein kinase activity by different secretory stimuli in rat basophilic leukemia cells. J Immunol 1994;152:250–261.

52. Feig LA, Cooper GM. Inhibition of NIH 3T3 cell proliferation by a mutant ras protein with preferential affinity for GDP. Mol Cell Biol 1988;8:3235–3243.

53. Szeberenyi J, Cai H, Cooper GM. Effect of a dominant inhibitory Ha-ras mutation on neuronal differentiation of PC12 cells. Mol Cell Biol 1990;10:5324–5332.

54. McMahon SB, Monroe JG. Activation of the p21^{ras} pathway couples antigen receptor stimulation to induction of the primary response gene *egr*-1 in B lymphocytes. J Exp Med 1995;181:417–422.

55. Turner H, Reif K, Rivera J, et al. Regulation of the adapter molecule Grb2 by the Fc epsilon R1 in the mast cell line RBL2H3. J Biol Chem 1995;270:9500–9506.

56. Wodnar-Filipowicz A, Heusser CH, Moroni C. Production of the haemopoietic growth factor GM-CSF and interleukin-3 by mast cells in response to IgE receptor-mediated activation. Nature (Lond) 1989;339:150–152.

57. Wodnar-Filipowicz A, Moroni C. Regulation of interleukin 3 mRNA expression in mast cells occurs at the posttranscriptional level and is mediated by calcium ions. Proc Natl Acad Sci USA 1990;87:777–781.

58. Fruman DA, Wood MA, Gjertson CK, et al. FK506 binding protein 12 mediates sensitivity to both FK506 and rapamycin in murine mast cells. Eur J Immunol 1995;25:563–571.

59. Buccione R, Di Tullio G, Caretta M, et al. Analysis of protein kinase C requirement for exocytosis in permeabilized rat basophilic leukaemia RBL-2H3 cells: a GTP-binding protein(s) as a potential target for protein kinase C. Biochem J 1994;298:149–156.

60. Razin E, Szallasi Z, Kazanietz MG, et al. Protein kinase C-β and C-ε link the mast cell high-affinity receptor for IgE to the expression of c-fos and c-jun. Proc Natl Acad Sci USA 1994;91:7722–7726.

61. Haleem-Smith H, Chang E, Szallasi Z, et al. Tyrosine phosphorylation of protein kinase C-delta in response to the activation of the high affinity receptor for immuglobulin E modifies its substrate recognition. Proc Natl Acad Sci USA 1995;92:9112–9116.

62. Ozawa K, Szallasi Z, Kazanietz MG, et al. Ca(2+)-dependent and Ca(2+)-independent isozymes of protein kinase C mediate exocytosis in antigen-stimulated rat basophilic RBL-2H3 cells. Reconstitution of secretory responses with Ca^{2+} and purified isozymes in washed permeabilized cells. J Biol Chem 1993;268:1749–1756.

63. Lewin I, Nechushtan H, Ke Q, et al. Regulation of AP-1 expression and activity in antigen-stimulated mast cells: the role played by protein kinase C and the possible involvement of Fos interacting protein. Blood 1993;82:3745–3751.

64. Baranes D, Razin E. Protein kinase C regulates proliferation of mast cells and the expression of the mRNAs of fos and jun proto-oncogenes during activation by IgE-Ag or calcium ionophore A23187. Blood 1991;78:2354–2364.

65. Ludowyke RI, Peleg I, Beaven MA, et al. Antigen-induced secretion of histamine and the phosphorylation of myosin by protein kinase C in rat basophilic leukemia cells. J Biol Chem 1989;264:12492–12501.

66. Kawakami Y, Yao L, Tashiro M, et al. Activation and interaction with protein kinase C of a cytoplasmic tyrosine kinase, Itk/Tsk/Emt, on Fc epsilon RI cross-linking on mast cells. J Immunol 1995;155:3556–3562.

67. Woscholski R, Kodaki T, McKinnon M, Waterfield MD, Parker PJ. A comparison of demethoxyviridine and wortmannin as inhibitors of phosphatidylinositol 3-kinase. FEBS Lett 1994;342:109–114.

68. Serve H, Yee NS, Stella G, et al. Differential roles of PI3-kinase and Kit tyrosine 821 in Kit receptor-mediated proliferation, survival and cell adhesion in mast cells. EMBO J 1995;14:473–483.

69. Pendl GG, Prieschl EE, Harrer NE, et al. Effects of phosphatidylinositol-3-kinase inhibitors on

degranulation and gene induction of allergically triggered mouse mast cells. Int Arch Allergy Immunol 1996;112:392–399.

70. Marquardt DL, Alongi JL, Walker LL. The phosphatidylinositol 3-kinase inhibitor wortmannin blocks mast cell exocytosis but not IL-6 production. J Immunol 1996;156:1942–1945.

25
Regulation of IL-4 Expression in Mast Cells

Melanie A. Sherman and Melissa A. Brown

The Role of Interleukin-4 in Immunity

Cytokines play a critical role in determining the nature of an organism's response to a great variety of immunological challenges. A protective immune response requires not only that the appropriate cytokines are produced, but that the magnitude and timing of the production be carefully regulated as well. The dysregulation of cytokine expression at any level can have profound pathophysiological consequences.

Interleukin-4 (IL-4) is among the best studied of these proteins. First described as a T cell-derived factor, it is now recognized that mast cells, basophils, and eosinophils can produce IL-4.[1] This cytokine regulates immune responses in a number of ways. It plays a role in recruitment of inflammatory cells to sites of infection or injury either via direct chemotactic effects (on eosinophils, for example) or through the regulation of adhesion molecules on endothelial cells.[2,3] A great deal of attention has focused on the ability of IL-4 to direct T helper cell differentiation and influence the character of the immune response. Naive CD4+ T cells require the presence of IL-4 during priming to differentiate into a subset that expresses IL-4 on subsequent antigenic exposure.[4,5] This subset, termed Th2 cells, also produces IL-5, IL-10, and IL-13. These cytokines act to potentiate noncomplement-fixing antibody production (such as IgE and IgG 1) and eosinophil activa-

tion. It is the Th2 cytokines that play an important role in the defense against many extracellular pathogens. Th1 cytokines, including IL-2 and interferon (IFN-γ), are proinflammatory mediators that activate cell-mediated immune effector cells, including macrophages, and enhance production of neutralizing and complement-fixing antibodies.

There is a delicate balance in the local concentration of Th1 and Th2 cytokines that must be maintained for optimal immune protection because of the cross-regulatory effects exerted by several members of each cytokine subset. For example, although IL-4 is required to initiate and maintain a Th2 response, the exposure of naive T lymphocytes to IL-4 represses the development of Th1 cells and inhibits the activities of some of its associated cytokines.[4,5] Thus, IL-4 and other Th2 cytokines are postulated to have an important role in downregulating the immune responses, limiting potentially destructive chronic inflammation attributable to Th1 cytokines.[6] This view is supported by studies showing that IL-4 therapy protects mice from developing the Th1-mediated autoimmune diseases such as diabetes[7] and EAE (the murine model for multiple sclerosis).[8-10]

Mast Cells: Our Unsung Protectors?

It is widely acknowledged that mast cells and basophils are key cellular components in acute allergic responses. Cross-linking of the high-affinity IgE receptor (FcεRI) on these cells causes the immediate release of a number of mediators of hypersensitive reactions. However, recent studies have shown that mast cells clearly play a broader role in regulating inflammation through the release of inducible and preformed mediators. Their strategic location in areas that are most vulnerable to pathogenic invasion (such as the skin, lungs, and GI tract) suggest the mast cell's potential usefulness as a first line of defense in fighting infection. The appreciation of their expanded role in immune responses results, in part, from the discovery that mast cells have the ability to release a wide variety of cytokines, such as IL-1, IL-2, IL-3, IL-4, IL-5, IL-6, IFN-γ, GM-CSF, and tumor necrosis factor-α (TNF-α).[11] Like T cells, it appears that mast cells can selectively express distinct groups of cytokines depending on the factors present in the local environment. Preliminary studies with bone marrow-derived mast cells cultured in vitro show that the presence of IL-3 in the culture media results in the production of IL-4 mRNA upon stimulation. Substitution of IL-3 with stem cell factor or IL-12 allow the cells to adopt a Th1-like profile.[12,13]

The importance of mast cells in host defense has been illuminated by recent studies using mice congenitally deficient in c-kit receptor expression.[14,15] These mice consequently lack mast cells in their peripheral tissues.[16,17] A bacterial insult such as septic peritonitis[18] or *Klebsiella pneumoniae*[19] is deadly to such mice, but reconstitution with mast cells allows them to clear the bacteria normally. Antibody-blocking experiments show that the critical component is mast cell-derived TNF-α, which recruits neutrophils and facilitates the inflammation cascade.[19]

The immune response to *Schistosoma mansoni* eggs was shown to be dependent on mast cell-derived IL-5.[20,21] This cytokine causes local eosinophila and induces the production of eosinophil-derived IL-4, which in turn produces a strong T-cell response specific for the *Shistosoma* antigen. Such studies underscore the significance of mast cell-derived cytokines.

The Significance of Mast Cell-Derived IL-4

IL-4 is intimately associated with mast cell biology. Not only is IL-4 a mast cell growth factor,[22] but it is also required for IgE production by B cells.[23–25] Thus, IL-4 can indirectly influence the activation state of mast cells by promoting IgE synthesis. Most isotype switching is thought to occur in lymph node germinal centers, where it is triggered by T cell IL-4 secretion and by the cognate interaction of CD40 ligand (CD40L) on T lymphocytes and CD40 on B cells.[26,27] Mast cells also express CD40L and, in concert with their ability to produce IL-4, are able to induce B cell Ig switching in vitro.[28] These data suggest that Ig switching can also occur in peripheral tissues where mast cells reside.

As stated, IL-4 is required for IL-4 expression by Th2 cells. If Th2 cells need IL-4 to produce IL-4, which cells provide the initiating source of IL-4? Both mast cells and CD4+ NK1.1+ T cells have been implicated as a source of the early IL-4 required for Th2 development. NK1.1+ T cells release large amounts of IL-4 upon first antigen exposure and are required for the Th2-dominated response observed after intravenous injection of anti-IgD antibodies.[29–31] Mouse strains such as SJL and NOD are naturally deficient in these cells and show increased susceptibility to Th1-mediated autoimmune diseases.[31,32] However, recent studies argue against a role for NK1.1+ T cell-derived IL-4. β2-microglobulin knockout mice, which are deficient in NK1.1+ T cells, have normal Th2 responses to protein antigen, *Leishmania* infection, *Schistosoma* eggs, and allergens.[33–35]

Mast cells appear to be ideal candidates for providing a source of IL-4 that initiates and/or amplifies local Th2 responses for several reasons (1) they have widespread tissue distribution and can influence immune responses at many sites in the body, (2) unlike antigen-specific T cells, mast cell cytokine production can be induced in an antigen-independent manner by a wide variety of bacterial products and inflammatory mediators,[11] (3) the lag time between encountering antigen and releasing cytokine is negligible in mast cells, since some

IL-4 is preformed and can be stored and released upon degranulation,[36] and (4) in contrast to naive T cells, IL-4 and its associated STAT6 signaling pathway is not necessary for the development of an IL-4 producing phenotype in mast cells (M. Sherman et al., submitted for publication). Taken together, these data suggest the signaling requirements for IL-4 production in mast cells are much less stringent than those in conventional T cells. The central role of IL-4, and the likely importance of mast cell-derived IL-4, in regulating the character of local immune/inflammatory responses provides good rationale for studying the events that control IL-4 gene expression in mast cells.

Signaling Events Resulting in Cytokine Transcription

FcεRI cross-linking in mast cells results in an immediate cascade of intracellular signals[37–46] that ultimately lead to degranulation and the activation of genes involved in inflammation. The early events include tyrosine phosphorylation of both the IgE receptor itself[43,45] and phospholipase C-γ (PLC-γ).[44,45] MAP kinases, part of the ras/raf cascade,[47] are also activated by signaling through the FcεRI or c-kit receptor.[48,49] PLC induces the hydrolysis of inositol phospholipids, resulting in the activation of the second messengers inositol triphosphate (which mobilizes calcium) and diacylglycerol (which activates protein kinase C).

This second wave of intracellular events activate transcription factors involved in inducible gene expression in the mast cell. The pathways leading from cell surface to induction or activation of transcription factors are not completely understood. The use of agents that bypass the need for surface IgE receptor cross-linking has provided some clues. Mast cells can be activated to degranulate and initiate cytokine transcription in vitro by treatment with calcium ionophore.[50] In T cells, this calcium signal activates the phosphatase calcineurin, which subsequently induces members of the NF-AT family of proteins (nuclear factor of activated T cells) to upregulate transcription of cytokine genes.[51] As discussed next, NF-AT has a role in mast cell gene transcription as well and is likely to be the target of similar calcium-mediated signals. NF-AT can also be activated through the ras pathway in mast cells.[49,52]

Stimulation with phorbol esters, which activate protein kinase C, will synergize with the calcium signal, but phorbol esters will not stimulate mast cells without calcium mobilization.[42,53] Protein kinase C activation appears to upregulate the production of AP-1 family members such as Jun and Fos,[54,55] which activate transcription through AP-1 promoter binding sites.[56] AP-1 plays an important role in mitosis and can associate with other transcription factors to induce cytokine expression.[57,58] The observation that calcium ionophores alone stimulate mast cell IL-4 indicates that the protein kinase C (PKC) signal, and AP-1 transcription factors, are not required for IL-4 gene expression in mast cells.

Mast cells can also be activated by inflammatory mediators including C5a or bacterial products such as lipopolysaccharide (LPS) and f-met-leu-phe (FMLP).[59,60] The signaling pathways associated with these agents are not understood. LPS can induce Th2 cytokine expression independently of histamine release, suggesting that degranulation and gene expression are not necessarily linked.[60]

Transcriptional Regulation of Interleukin-4

The molecular explanation for tissue-specific gene expression lies with the transcription factors and their DNA-binding sites within the promoter region. Because the promoter elements are found in the chromosomal DNA of all cells, it is the presence or absence of specific activated transcription factors in the nucleus of the cell that determines whether a particular gene is expressed. The association of transcription factors with their specific promoter elements can enhance or inhibit the actions of the basal transcriptional machinery. The search for IL-4-specific transcription factors has been intensely competitive because these proteins are the key to the switch between Th1 and Th2 immune responses.

Mast Cells Utilize Cell-Specific IL-4 Transcriptional Regulatory Factors

Most studies of IL-4 gene transcriptional regulatory elements have been performed in T cells. These studies have defined multiple *cis-* and *trans*-acting elements that contribute to inducible gene expression in activated cells.[61-69] Our studies using long-term, IL-3-dependent mast cell lines such as CFTL15[70] or bone-marrow derived mast cells suggest that although mast cells utilize some similar components, there are unique elements that reflect the differences in the activating signals transmitted by the T cell receptor (TCR) and FcεRI as well as the distinct array of lineage-specific transcription factors expressed in these two cell types. Two regions in the IL-4 gene that associate with apparently mast cell-specific transcription factors have been the subject of intense investigation in our laboratory. A 5'-element termed ARE (an activation responsive element) is the target of mast cell activation signals. In addition, a sequence in the second intron of the mouse IL-4 gene was originally defined as an enhancer. Recent data suggest it can act as a promoter to drive expression of alternative IL-4 gene transcripts.

The ARE, which encompasses sequences between −88 and −60 relative to the transcription initiation site (TIS, designated as +1), is required for cell-specific, inducible promoter activity[71] in mast cells. This element contains P1, one of five sites with the consensus sequence ATTTTCCNNT, designated P0

through P4, that are present in the IL-4 promoter (Fig. 25.1).[63,65] Mutation of P1 within the context of a large segment of the IL-4 promoter severely reduces inducible reporter gene activity. Both P1 and P0 are essential for inducible T-cell IL-4 promoter activity[62,64] and have been implicated as targets for the selective IL-4 expression associated with a Th2 phenotype.[64,69,72] The contribution of P0 and other P sites has not yet been determined in mast cells.

NF-AT Associates with the P1 Element in Mast Cells

P1 contains a sequence that shares identity with the NF-AT binding site first defined in the IL-2 gene promoter. Our studies using in vitro DNA-binding assays in concert with specific antibodies have established that mast cells express an NF-AT-like protein that can associate with the P1 site of the IL-4 promoter.[71] The precise identity of this factor is still unknown.

NF-AT was originally described as a single, T cell-specific transcription factor responsible for activation-induced IL-2 gene expression. However, several reports now demonstrate that NF-AT is expressed in a wide variety of tissues and cell types, and is involved in the transcriptional regulation of cytokines such as IL-4, IL-5, IL-3, GM-CSF, and TNF-α. There are at least four genes encoding NF-AT family members (NF-AT1/NF-ATp, NF-AT2/NF-ATc, NF-AT3, and NF-AT4/c3/x) and that at least three of these have multiple isoforms.[51,73-75] Similar

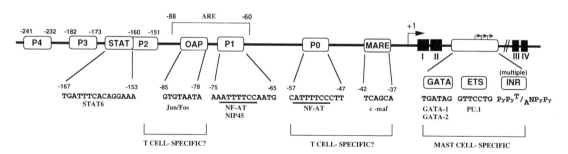

FIGURE 25.1. Multiple elements regulate the transcription of IL-4 in T and mast cells. Exons I–IV are *numbered*. The DNA elements and their coordinates are denoted by the *boxes* (all numbers are expressed relative to the TIS +1). Transcription factors that have been shown to associate in vitro with these elements are listed *below* the sequence of each element. *Py*, pyrimidine. INR sequences are denoted by forward arrows.

DNA-binding domains, homologous to the rel domain of the NF-κB family,[76,77] and a common region of SP repeats define these diverse proteins as family members.[51] The distinct sequences outside these regions imply that each factor can associate with unique accessory factors. The existence of multiple forms raises the question: do the various NF-AT family members have gene- and/or cell-specific activities or are they functionally redundant? This issue is difficult to address in vitro, because the Rel domain present in each NF-AT species can associate with the consensus sites in the promoters of many NF-AT-driven genes. There have been reports showing some discrete binding behavior of different NF-AT proteins,[73,78–80] but the immunocompetency of mice deficient in NF-AT1 suggests that other NF-AT members can at least compensate for loss of NF-AT1 in vivo.[81,82]

In studies of T-cell IL-2 gene expression, NF-AT is regulated by cellular location and phosphorylation state. Stimulation-induced increases in intracellular Ca^{2+} levels result in the dephosphorylation of NF-AT by the calcium-dependent phosphatase, calcineurin,[83,84] allowing both NF-AT and calcineurin to translocate to the nucleus as a complex.[85] This dephosphorylation event is also required for NF-AT binding to DNA.[86] Cyclosporin A (CSA) acts through this pathway to repress inducible IL-2 gene expression by inhibiting calcineurin activity.[83]

Most of the NF-AT sites in cytokine gene promoters are adjacent to nonconcensus AP-1 elements.[51] Electrophoretic mobility shift assays (EMSAs) using purified proteins and the IL-2 NF-AT binding site indicate that AP-1 factors are needed to stabilize the NF-AT–DNA complex.[87,88] Both AP-1 and NF-AT are required for IL-2 gene expression. AP-1 is part of the stimulation-dependent IL-4 NF-AT–P1 complex in T cells, suggesting it plays a similar role.[62,89]

Mast Cell NF-AT Is Unique

Our studies support the idea that an alternate form of NF-AT, distinct from that involved in T-cell IL-2 and IL-4 gene transcription, is associated with the mast cell IL-4 gene P1 site.[71] This is based on three key observations. (1) NF-AT binding activity is present in the nucleus of unstimulated mast cells, suggesting this protein is not regulated by activation-dependent cytoplasmic to nuclear translocation. Importantly, CsA inhibits binding of NF-AT in unstimulated cells, suggesting a calcium-independent role for this immunosuppressive agent. (2) DNA affinity chromatography in concert with western blot analysis using anti-NF–AT antisera indicates that a 41 kDa protein, which is present constitutively in the nucleus of mast cells but not T cells, preferentially associates with P1. This is significantly different from the size of "conventional" T cell-derived NF-AT, which is between 97 and 130 kDa.[51] (3) AP-1 factors are not associated with the NF-AT/P1 complex in mast cells. EMSA experiments using excess unlabeled AP-1 consensus sequence as a cold competitor show that the assembly of the mast cell ARE–protein complex is unaffected,[72] although AP-1 is a part of both the IL-2 and IL-4 NF-AT complexes in T cells.[62,89–91] Furthermore, Jun- and Fos-reactive antibodies block formation of the T cell P1–protein complex, yet have no effect on the mast cell complex. These data indicate that AP-1 transcription factors may not be important in IL-4 gene regulation in mast cells and provide a molecular explanation for why protein kinase C activation (which upregulates AP-1) is required for inducible T cell, but not mast cell, IL-4 expression. Similar results have been reported for the regulation of the IL-5 gene in mast cells. In this situation, NF-AT appears to cooperate with GATA instead of AP-1 to activate IL-5 transcription.[92] Additionally, the NF-AT elements present in the TNF-α promoter apparently work in the absence of AP-1 in T cells.[93]

The NF-AT species associated with the mast cell IL-4 promoter is either an alternate form of a previously characterized family member or the product of a gene that has not yet been described. To distinguish between these possibilities, a murine mast cell cDNA library was screened, under both high and low stringency conditions, with a probe corresponding to the NF-AT1 Rel domain. The majority of clones identified correspond to two NF-ATc (NF-AT2) isoforms that differ only in their amino terminal sequence (Sherman et al. J. Immunol., in press, 1999). Despite minimal discrepancies in the coding region, there are striking tissue and cell-specific differences in isoform expression patterns. Detection of NF-ATc. α mRNA is strictly dependent on cell activation signal in both T and mast cell line. In contrast the β isoform is expressed at very low constitutive levels in both cell types but is only upregulated in response to mast cell activation signals delivered through the FcεRI or via calcium ionophores. These results demonstrate another level of regulation with the NF-AT family that can contribute to cell-type specific gene expression. The contribution of these isoforms to the in vivo regulation of IL-4 in mast cells is currently under investigation.

Other Factors Synergize with NF-AT to Drive IL-4 Expression in T Cells: Role in Mast Cell IL-4 Transcription?

Th2 cells are distinguished from the Th1 subset by their cytokine production profile. The nuclear signaling events that lead to subset differentiation and selective expression of these cytokines are still not completely understood. The realization that NF-AT constitutes a family of transcription factors that have at least some redundant activities in regulating cytokines produced by both subsets raises the possibility that it is not NF-AT but the NF-AT-associated factors that are the key to regulating these differential responses. Two such candidates were recently identified using a yeast two-hybrid trap to identify proteins that interact with NF-AT1 (NF-ATp). The first, c-maf, is a 41-kDa proto-oncogene that is a member of the AP-1 family. This transcription factor, expressed in Th2 but not Th1 cells, is induced during the differentiation of Th2 cells from naive precursors.[72] Like AP-1, c-maf binds to a sequence adjacent to P0 termed the MARE element (located between −42 and −37). Mutation of this site inhibits promoter activity in T cells. In addition, this element confers Th2-specific reporter gene activity, consistent with the expression pattern.[64] Importantly, overexpression of NF-AT and c-maf in transfected B cells can induce expression of IL-4 protein.

The contribution of P0 or MARE to mast cell IL-4 gene transcription has not been formally examined. Preliminary studies in our laboratory indicate that c-maf is not expressed in mast cells and therefore, like AP-1, is not a component of NF-AT-mediated transcription in this cell lineage.

A second interacting protein, NIP45, is expressed in immune organs, although the precise T-helper cell expression pattern of NIP45 is not known.[94] NIP45, a 45-kDa factor, is constitutively present in the nucleus but its ability to directly bind DNA has not been established. Overexpression of NIP45 with NF-AT and c-maf in transfected B cells has a synergistic effect on endogenous IL-4 promoter activity compared with the effect of NF-AT and c-maf alone, resulting in very high levels of IL-4 protein production. It is not known how NIP45 works to enhance NF-AT activity. Possible mechanisms include a chaperone function (to recruit NF-AT to the nucleus) or a strong phosphatase activity (to maintain NF-AT in its activated state). The ability of c-maf and NIP45 to modulate the expression of other genes regulated by NF-AT in both T and mast cells remains to be examined.

Activated STAT6 Is Associated with Mast Cell IL-4 Expression

IL-4 signaling pathways in T and mast cells utilize other cell-type specific mediators that differentially affect IL-4 transcription in these cell types. As previously discussed, IL-4 and its associated signaling pathways are essential for the development of IL-4 producing T cells. These data suggest that the IL-4 gene itself, like Cε, FcεRI and the IL-4 receptor α chain, is an IL-4 responsive gene.[95-98] CD4+ T cells from STAT6-deficent mice are unable to produce IL-4 and other Th2-associated cytokines.[99-101] It was originally hypothesized that STAT6 acted directly on the IL-4 promoter to drive transcription. A consensus STAT6 binding site is present in the murine IL-4 gene between −153 and −167. In vitro DNA binding studies demonstrate that STAT6 is constitutively active in Th2, but not Th1 cells.[102] However, recent evidence does not support this contention. Mutation or deletion of the STAT6 site has no effect on the transcriptional activity of IL-4 promoter/reporter constructs in T cells.[103] Furthermore, the appearance of Th2-specific hypersensitive sites within the IL-4 chromosomal gene are dependent on the expression of STAT6.[104] These data support the idea that STAT6 does not act at the level of direct transcriptional activation in T cells. Rather, it may play a role in influencing chromatin accessibility of the IL-4 gene during early CD4⁺ T cell development.

Mast cells can develop into IL-4 producing cells in the absence of functional STAT6 (Sherman et al., submitted for publication). In fact, mast cells express a unique isoform of STAT6. This "mast cell STAT" is a product of the STAT6 gene but is only 65 kD in size and appears to lack the defined C-terminal transactivation domain. Despite the presence of the conventional 94 kD STAT6 molecule in extracts from IL-4 treated mast cells, it is the smaller isoform that preferentially associates with the IL-4 gene STAT6 binding site. (Sherman et al. J. Immunol. in press, 1999). Mutation of this site within the context of the 5′ IL-4 promoter/reporter construct results in a reproducible increase in promoter activity consistent with the idea that this truncated STAT6 isoform is a repressor of IL-4 gene transcription in mast cells (M. Sherman et al., submitted for publication).

A Mast Cell-Specific Regulatory Element Is Located in the Second Intron of the IL-4 Gene

Original studies of IL-4 gene regulation utilized transformed mast cells.[105] This unique group of cells includes ABFTL3, an Abelson MuLV transformed cell line, and the common T cytotoxic cell target, P815, a methyl cholanthrene-induced tumor mast cell line. These cells express high levels of IL-4 mRNA constitutively. Stimulation results in an increase in IL-4 mRNA and protein. Only marginal amounts of CAT gene activity were detected in transformed mast cells transfected with the 5′-IL-4 reporter gene constructs containing up to 6.3 kb of sequence, suggesting other regions were also important. To identify these, DNAse I hyper-

sensitivity site analysis was utilized. Such sites correlate with transcriptional regulatory elements in many genes.[106] A hypersensitive site was observed within the second intron of the murine IL-4 gene in both ABFTL 3 and P815 cell nuclei.[107] Functional assays established that this element acts as an enhancer. Two lines of evidence indicated this element is mast cell specific. (1) The enhancer is also active in several mast cell lines, including nontransformed mast cells, but not in T-cell lines examined. (2) It is composed of several functionally independent subregions defined by mutational analyses. The regions critical for activity include sequences that are similar to ets and GATA binding sites.[108] The ets family member, PU.1, and GATA-1 and GATA-2 are expressed constitutively in both transformed and nontransformed mast cells and can bind to the respective IL-4 intron sites. These factors are not expressed in T cells. Recently, Pit1a and STAT binding sites were also found to contribute to activity. Like PU.I and GATA-1 and -2, Pit1a is expressed in mast cells but not T cells (J. Hural, manuscript submitted). An in vivo function for these proteins remains to be demonstrated.

The contribution of this intronic regulatory element to mast cell-specific IL-4 gene transcription is unclear. There is complete sequence identity with the equivalent region in the human gene; this conservation within an intron suggests there is a physiological role for this element. In addition, using assays that assess open chromatin, a correlate of active gene transcription, two recent reports indicate that second intron is also a site of regulation of IL-4 in T cells. Bird et al. demonstrate this region is a site of hypomethylation in Th2 but not Th1 cells.[109] Agarwal et al. map two DNAse I hypersensitive sites to the second intron that are present only in developing Th2 cells.[104] Importantly, the position of these sites flank the DNAse I site that defines the enhancer in mast cells. Together with the results from our laboratory showing mast cell-specific transcriptional activity of this enhancer as well as mast cell-restricted expression of relevant transcription factors, we propose the sites of regulation in the second intron are distinct in T and mast cells.

Another potential role for this regulatory element was suggested by a recent report.[110] Using RNAse protection assays, it was observed that a number of embryonic mast cell lines and factor-dependent mast cell cultures express a truncated form of the IL-4 mRNA that initiates just downstream of the enhancer. RNAse protection assays demonstrated that this species is protected by probes containing IL-4 sequences encoding exons 3 and 4 but not exons 1 or 2. These observations, together with the fact that the intronic element does not efficiently interact with 5'-regulatory regions, has led us to hypothesize that the intronic regulatory element is actually a promoter that drives the transcription of an alternative transcript. In support of this idea, we have demonstrated that the element can act as a promoter in transient transfection assays (J. Hural, unpublished results). In addition we have identified several consensus initiator elements (Inr) within this sequence (see Fig. 25.1). Inr elements were first described in "TATA-less" promoters but are also found to act in concert with TATA boxes.[111–113] They bind basal transcription factors such as TFIID and are critical in the positioning of RNA polymerase II for correct transcription initiation. Inr-containing promoters have been shown to direct tissue specific expression of a variety of genes, including the immunologically relevant T- and B-cell restricted genes, RAG-1 and TdT.[112,114,115] These observations have lead to the reassessment of the physiological role of this regulatory region and its putative lineage-specific activity. It is intriguing to speculate that this alternative transcript has a role in regulating the production or activity of biologically active IL-4.

Conclusions

Because of their prevalence in immunologically strategic areas of the body and their ability to immediately release preformed IL-4 after activation, mast cells are likely the crucial players in initiating local Th2 responses. They may also contribute significantly to other IL-4-dependent processes previously thought to be the exclusive domain of T cells, such as IgE isotype switching and activation and recruit-

ment of inflammatory cells to sites of infection and injury. Several findings underscore the differences in signaling pathways leading to IL-4 gene transcription between mast cells and T cells and indicate that the study of IL-4 transcription is an excellent model system to study fundamental questions of tissue-specific gene expression. These include the apparently unique form of IL-4 gene-associated NF-AT, the absence of c-*maf* in mast cells, and the presence of an IL-4 gene intronic regulatory element. Future studies of mast cell-specific IL-4 expression will be aimed at (1) identifying additional cell-specific *cis*- and *trans*-acting factors, (2) determining the in vivo role of these elements, and (3) understanding how these nuclear targets are related to membrane proximal signaling events delivered through the FcεRI or other mast cell-activating agents. Insight regarding these issues will impact our ability to selectively augment (in infection control and regulation of pathological Th1 responses) or inhibit (in allergic reactions) IL-4 expression in immunotherapy in these cells.

References

1. Brown MA, Hural J. Functions of IL-4 and control of its expression. Crit Rev Immunol 1997;17:1–32.
2. Issekutz TB. Effects of six different cytokines on lymphocyte adherence to microvascular endothelium and in vivo lymphocyte migration in the rat. J Immunol 1990;144:2140–2146.
3. Schleimer RP, Sterbinsky SA, Kaiser J, et al. IL-4 induces adherence of human eosinophils but not neutrophils to endothelium: association with VCAM-1. J Immunol 1992;148:1086–1092.
4. Swain S. IL4 dictates T-cell differentiation. Res Immunol 1993;144:616–620.
5. Seder RA, Paul WE, Davis MM, et al. The presence of interleukin-4 during *in vitro* priming determines the lymphokine-producing potential of CD4+ T cells from T cell receptor transgenic mice. J Exp Med 1992;176:1091–1098.
6. Abbas AK, Murphy KM, Sher A. Functional diversity of helper T lymphocytes. Nature (Lond) 1996;383:787–793.
7. Mueller R, Krahl T, Sarvetnick N. Pancreatic expression of interleukin-4 abrogates insulitis and autoimmune diabetes in nonobese diabetic

(NOD) mice. J Exp Med 1996;184:1093–1099.

8. Racke MK, Bonomo A, Scott DE, et al. Cytokine-induced immune deviation as a therapy for inflammatory autoimmune disease. J Exp Med 1994;180:1961–1916.

9. Racke MK, Burnett D, Pak SH, et al. Retinoid treatment of experimental allergic encephalomyelitis. IL-4 production correlates with improved disease course. J Immunol 1995; 154:450–458.

10. Kuchroo VK, Das MP, Brown JA, et al. B7–1 and B7–2 costimulatory molecules activate differentially the Th1/Th2 developmental pathways: application to autoimmune disease therapy. Cell 1995;80:707–718.

11. Gordon J, Burd P. Mast cells as a source of multifunctional cytokines. Immunol Today 1990;11:458–464.

12. Smith TJ, Ducharme LA, Weis JH. Preferential expression of interleukin-12 or interleukin-4 by bone marrow mast cells derived in mast cell growth factor or interleukin-3. Eur J Immunol 1994;24:822–826.

13. Gupta AA, Leal-Berumen I, Croitoru K, et al. Rat peritoneal mast cells produce IFN-γ following IL-12 treatment but not in response to IgE-mediated activation. J Immunol 1996; 157:2123–2128.

14. Alexander WS, Lyman SD, Wagner EF. Expression of functional c-kit receptors rescues the genetic defect of W mutant mast cells. EMBO J 1991;10:3683–3691.

15. Nocka K, Tan JC, Chiu E, et al. Molecular basis of dominant negative and loss of function mutations at the murine c-kit/white spotting locus: W^{37}, W^v, W^{41} and W. EMBO J 1990;9:1805–1813.

16. Zhang Y, Ramos F, Jakschik B. Neutrophil recruitment by tumor necrosis factor from mast cells in immune complex peritonitis. Science 1992;258:1957–1959.

17. Kitamura Y, Go S, Hatanaka K. Decrease of mast cells in W/W^v mice and their increase by bone marrow transplantation. Blood 1978; 52:447–452.

18. Echtenacher B, Mannel D, Hultner L. Critical protective role of mast cells in a model of acute septic peritonitis. Nature (Lond) 1996;381:75–76.

19. Malavija R, Ikeda T, Ross E, et al. Mast cell modulation of neutrophil influx and bacterial clearance at sites of infection through TNF-α. Nature (Lond) 1996;381:77–80.

20. Sabin EA, Pearce EJ. Early IL-4 production by non-CD4+ cells at the site of antigen deposition predicts the development of a T helper 2 cell response to *Schistosoma mansoni* eggs. J Immunol 1995;155:4844–4853.

21. Sabin EA, Kopf MA, Pearce EJ. *Schistosoma mansoni* egg-induced early IL-4 production is dependent upon IL-5 and eosinophils. J Exp Med 1996;184:1871–1878.

22. Mosmann TR, Bond MW, Coffman RL, et al. T-cell and mast cell lines respond to B-cell stimulatory factor 1. Proc Natl Acad Sci USA 1986;83:5654–5658.

23. Snapper CM, Finkelman FD, Paul WE. Regulation of IgG1 and IgE production by interleukin 4. Immunol Rev 1988;102:51–75.

24. Snapper CM, Finkelman FD, Paul WE. Differential regulation of IgG1 and IgE synthesis by interleukin 4. J Exp Med 1988;167:183–196.

25. Schultz CL, Coffman RL. Control of isotype switching by T cells and cytokines. Curr Opin Immunol 1991;3:350–354.

26. Fuleihan R, Ahern D, Geha RS. Expression of the CD40 ligand in T lymphocytes and induction of IgE isotype switching. Int Arch Allergy Immunol 1995;107:43–44.

27. Klaus S, Berberich I, Shu G, et al. CD40 and its ligand in the regulation of humoral immunity. Semin Immunol 1994;6:279–286.

28. Gauchat J-F, Henchoz S, Mazzei G, et al. Induction of human IgE synthesis in B cells by mast cells and basophils. Nature (Lond) 1993;365:340–343.

29. Yoshimoto T, Paul WE. CD4+, NK1.1+ T cells promptly produce interleukin-4 in response to in vivo challenge with anti-CD3. J Exp Med 1994;179:1285–1295.

30. Yoshimoto T, Bendelac A, Watson C, et al. Role of NK1.1+ T cells in a Th2 response and in immunoglobulin E production. Science 1995;270:1845–1847.

31. Yoshimoto T, Bendelac A, Hu-Li J, et al. Defective IgE production by SJL mice is linked to the absence of CD4+, NK1.1+ T cells that promptly produce interleukin 4. Proc Natl Acad Sci USA 1995;92:11931–11934.

32. Gombert JM, Herbelin A, Tancrede-Bohin E, et al. Early quantitative and functional deficiency of NK1+-like thymocytes in the NOD mouse. Eur J Immunol 1996;26:2989–2998.

33. Brown D, Fowell D, Corry D, et al. Beta 2-microglobulin-dependent NK1.1+ T cells

are not essential for T helper cell 2 immune responses. J Exp Med 1996;184:1295–1304.

34. Weid T, Beebe A, Roopenian D, et al. Early production of IL-4 and induction of Th2 responses in the lymph node originate from an MHC class I-independent CD4+NK1.1− T cell population. J Immunol 1996;157:4421–4427.

35. Zhang Y, Rogers K, Lewis D. Beta 2-microglobulin-dependent T cells are dispensable for allergen-induced T helper 2 responses. J Exp Med 1996;184:1507–1512.

36. Bradding P, Feather IH, Howarth PH, et al. Interleukin-4 is localized to and released by human mast cells. J Exp Med 1992;176:1381–1386.

37. Beaven MA, Roger J, Moore JP, et al. The mechanism of the calcium signal and correlation with histamine release in 2H3 cells. J Biol Chem 1984;259:7129–7136.

38. Beaven MA, Moore JP, Smith GA, et al. The calcium signal and phosphatidylinositol breakdown in 2H3 cells. J Biol Chem 1984;259:7137–7142.

39. Beaven MA, Metzger H. Signal transduction by Fc receptors: the FcεRI case. Immunol Today 1993;14:222–226.

40. Stump RF, Oliver JM, Cragoe EJ, et al. The control of mediator release from RBL-2H3 cells: roles for Ca^{2+}, Na^+, and protein kinase C. J Immunol 1987;139:881–886.

41. Ali H, Collado-Escobar DM, Beaven MA. The rise in concentration of free Ca^{2+} and of pH provides sequential, synergistic signals for secretion in antigen-stimulated rat basophilic leukemia (RBL-2H3) cells. J Immunol 1989;143:2626–2633.

42. Cunha-Melo JR, Gonzaga HMS, Ali H, et al. Studies of protein kinase C in the rat basophilic leukemia (RBL-2H3) cell reveal that antigen-induced signals are not mimicked by the actions of phorbol myristate acetate and Ca^{2+} ionophore. J Immunol 1989;143:2617–2625.

43. Benhamou M, Gutking JS, Robbins KC, et al. Tyrosine phosphorylation coupled to IgE receptor-mediated signal transduction and histamine release. Proc Natl Acad Sci USA 1990; 87:5327–5332.

44. Park DJ, Min HK, Rhee SG. IgE-induced tyrosine phosphorylation of phospholipase C-γ1 in rat basophilic leukemia cells. J Biol Chem 1991;266:24237–24240.

45. Li W, Deanin GG, Margolis B, et al. Fcε-RI-mediated tyrosine phosphorylation of multiple proteins including phospholipase Cγ1 and the receptor β/γ 2 complex in RBL-2H3 cells. Mol Cell Biol 1992;12:3176–3182.

46. Hamawy MM, Minoguchi K, Swaim WD, et al. A 77-kDa protein associates with pp125FAK in mast cells and becomes tyrosine-phosphorylated by high affinity IgE receptor aggregation. J Biol Chem 1995;270:12305–12309.

47. Marshall CJ. MAP kinase kinase kinase, MAP kinase kinase, and MAP kinase. Curr Opin Genet Dev 1994;4:82–89.

48. Tsai M, Chen RH, Tam SY, et al. Activation of MAP kinases, pp90rsk and pp70-S6 kinases in mouse mast cells by signaling through the c-kit receptor tyrosine kinase or FcεRI: rapamycin inhibits activation of pp70-S6 kinase and proliferation in mouse mast cells. Eur J Immunol 1993;23:3286–3291.

49. Turner H, Cantrell DA. Distinct ras effector pathways are involved in FcεRI regulation of the transcriptional activity of elk-1 and NFAT in mast cells. J Exp Med 1997;185:43–53.

50. Sahara N, Siraganian RP, Oliver C. Morphological changes induced by the calcium ionophore A23187 in rat basophilic leukemia (2H3) cells. J Histochem Cytochem 1990;38:975–983.

51. Rao A, Luo C, Hogan P. Transcription factors of the NFAT family: regulation and function. Ann Rev Immunol 1997;15:707–747.

52. Prieschl EE, Pendl GG, Harrer NE, et al. p21ras links FcεRI to NF-AT family members in mast cells. J Immunol 1995;155:4963–4970.

53. Beaven MA, Guthrie DF, Moore JP, et al. Synergistic signals in the mechanism of antigen-induced exocytosis in 2H3 cells: evidence for an unidentified signal required for histamine release. J Cell Biol 1987;105:1129–1136.

54. Lewin I, Nechushtan H, Ke Q, et al. Regulation of AP-1 expression and activity in antigen-stimulated mast cells: the role played by protein kinase C and the possible involvement of fos interacting protein. Blood 1993;12:3745–3751.

55. Razin E, Szallasi Z, Kazanietz MG, et al. Protein kinases C-β and C-ε link the mast cell high-affinity receptor for IgE to the expression of c-fos and c-jun. Proc Natl Acad Sci USA 1994; 91:7722–7726.

56. Chiu R, Boyle WJ, Meek J, et al. The c-fos protein interacts with c-jun/AP-1 to stimulate transcription of AP-1 responsive genes. Cell 1988;54:541–552.

57. Gentz R, Rauscher FJ, Abate C, et al. Parallel association of fos and jun leucine zippers juxta-

poses DNA binding domains. Science 1989;243:1695–1699.

58. Landschulz WH, Johnson PF, McKnight SL. The leucine zipper: a hypothetical structure common to a new class of DNA binding proteins. Science 1988;240:1759–1764.

59. Bochner BS, Lichtenstein LM. Anaphylaxis. N Engl J Med 1991;324:1785–1790.

60. Leal-Berumen I, Conlon P, Marshall JS. IL-6 production by rat peritoneal mast cells is not necessarily preceded by histamine release and can be induced by bacterial lipopolysaccharide. J Immunol 1994;152:5468–5476.

61. Tara D, Weiss DL, Brown MA. An activation responsive element in the murine interleukin-4 gene is the site of an inducible DNA-protein interaction. J Immunol 1993;151:3617–3626.

62. Tara D, Weiss DL, Brown MA. Characterization of the constitutive and inducible components of a T cell activation responsive element. J Immunol 1995;154:4592–4602.

63. Chuvpilo S, Schomberg C, Gerwig R, et al. Multiple closely-linked NF-AT/octamer and HMG I(Y) binding sites are part of the interleukin-4 promoter. Nucleic Acids Res 1993;21:5694–5704.

64. Hodge MR, Rooney JW, Glimcher LH. The proximal promoter of the IL-4 gene is composed of multiple essential regulatory sites that bind at least two distinct factors. J Immunol 1995;154:6397–6405.

65. Szabo SJ, Gold JS, Murphy TL, et al. Identification of cis-acting regulatory elements controlling interleukin-4 gene expression in T cells: roles for NF-Y and NF-ATc. Mol Cell Biol 1993;13:4793–4805.

66. Kubo M, Kincaid RL, Ransom JT. Activation of the interleukin-4 gene is controlled by the unique calcineurin-dependent transcriptional factor NF(P). J Biol Chem 1994;269:19441–19446.

67. Rooney JW, Hodge MR, McCaffrey PG, et al. A common factor regulates both Th1- and Th2-specific cytokine gene expression. EMBO J 1994;13:625–633.

68. Wenner CA, Szabo SJ, Murphy KM. Identification of IL-4 promoter elements conferring Th2-restricted expression during T helper cell subset development. J Immunol 1997;158:765–773.

69. Bruhn KW, Nelms K, Boulay J-L, et al. Molecular dissection of the mouse interleukin-4 promoter. Proc Natl Acad Sci USA 1993;90:9707–9711.

70. Pierce JH, DiFiore PP, Aaronson SA, et al. Neoplastic transformation of mast cells by Abelson-MuLV: abrogation of IL-3 dependence by a nonautocrine mechanism. Cell 1985;41:685–693.

71. Weiss D, Hural J, Tara D, et al. Nuclear factor of activated T cells is associated with a mast cell interleukin 4 transcription complex. Mol Cell Biol 1995;16:228–235.

72. Ho I-C, Hodge M, Rooney JW, et al. The proto-oncogene c-maf is responsible for tissue-specific expression of interleukin-4. Cell 1996;85:973–983.

73. Hoey T, Sun Y-L, Williamson K, et al. Isolation of two new members of the NF-AT gene family and functional characterization of the NF-AT proteins. Immunity 1995;2:461–472.

74. Luo C, Burgeon E, Carew J, et al. Recombinant NFAT1 (NFATp) is regulated by calcineurin in T cells and mediates transcription of several cytokine genes. Mol Cell Biol 1996;16:3955–3966.

75. Park J, Takeuchi A, Sharma S. Characterization of a new isoform of the NFAT (nuclear factor of activated T cell) gene family member NFATc. J Biol Chem 1996;271:20914–20921.

76. Jain J, Burgeon E, Badaliam TM, et al. A similar DNA-binding motif in NFAT family proteins and the rel homology region. J Biol Chem 1995;270:4138–4145.

77. Wolfe SA, Zhou P, Dotsch V, et al. Unusual Rel-like architecture in the DNA-binding domain of the transcription factor NFATc. Nature (Lond) 1997;385:172–176.

78. Northrop JP, Ho SN, Timmerman LA, et al. NF-AT components define a family of transcription factors targeted in T cell activation. Nature (Lond) 1994;369:497–502.

79. Masuda ES, Naito Y, Tokumitsu H, et al. NFATx, a novel member of the nuclear factor of activated T cells family that is expressed predominantly in the thymus. Mol Cell Biol 1995;15:2697–2706.

80. Ho SN, Thomas DJ, Timmerman LA, et al. NFATc3, a lymphoid-specific NFATc family member that is calcium-regulated and exhibits distinct DNA binding specificity. J Biol Chem 1995;270:19898–19907.

81. Xanthoudakis S, Viola JPB, Shaw KTY, et al. An enhanced immune response in mice lacking the transcription factor NFAT1. Science 1996;272:892–895.

82. Hodge M, Ranger AM, Hoey T, et al. Hyperproliferation and dysregulation of IL-4 expression in NF-ATp-deficient mice. Immunity 1996;4:397–405.

83. McCaffrey PG, Perrino BA, Soderling TR, et al. NF-ATp, a T lymphocyte DNA binding protein that is a target for calcineurin and immunosuppressive drugs. J Biol Chem 1993; 268:3747–3752.

84. Jain J, McCaffrey P, Miner Z, et al. The T-cell transcription factor NFATp is a substrate for calcineurin and interacts with Fos and Jun. Nature (Lond) 1993;365:352–355.

85. Shibasaki F, Price E, Milan D, et al. Role of kinases and the phosphatase calcineurin in the nuclear shuttling of transcription factor NF-AT4. Nature (Lond) 1996;382:370–373.

86. Park J, Yaseen NR, Hogan P, et al. Phosphorylation of the transcription factor NFATp inhibits its DNA binding activity in cyclosporin A-treated human B and T cells. J Biol Chem 1995;270:20653–20659.

87. Northrop JP, Ullman KS, Crabtree GR. Characterization of the nuclear and cytoplasmic components of the lymphoid-specific nuclear factor of activated T cells (NF-AT) complex. J Biol Chem 1993;268:2917–2923.

88. Yaseen N, Park J, Kerppola T, et al. A central role for fos in human B- and T-cell NFAT (nuclear factor of activated T cells): an acidic region is required for in vitro assembly. Mol Cell Biol 1994;14:6886–6895.

89. Rooney JW, Hoey T, Glimcher LH. Coordinate and cooperative roles for NF-AT and AP-1 in the regulation of the murine IL-4 gene. Immunity 1995;2:473–483.

90. Boise LH, Petryniak B, Mao X, et al. The NF-AT-1 DNA binding complex in activated T cells contains Fra-1 and jun B. Mol. Cell Biol 1993;13:1911–1919.

91. Jain J, McCaffrey PG, Valge-Archer VE, et al. Nuclear factor of activated T cells contains Fos and Jun. Nature (Lond) 1992; 356:801–804.

92. Prieschl EE, Gouilleux GV, Walker C, et al. A nuclear factor of activated T cell-like transcription factor in mast cells is involved in IL-5 gene regulation after IgE plus antigen stimulation. J Immunol 1995;154:6112–6119.

93. McCaffrey PG, Goldfeld AE, Rao A. The role of NFATp in cyclosporin A-sensitive tumor necrosis factor-α gene transcription. J Biol Chem 1994;269:30445–30450.

94. Hodge M, Chun H, Rengarajan J, et al. NF-AT-driven interleukin-4 transcription potentiated by NIP45. Science 1996;274:1903–1905.

95. Kotanides H, Reich NC. Requirement of tyrosine phosphorylation for rapid activation of a DNA binding factor by IL-4. Science 1993;262:1265–1267.

96. Rothman P, Lutzker S, Cook W. Mitogen plus interleukin 4 induction of Ce transcriptis in B cells. J Exp Med 1988;168:2385–2389.

97. Conrad DH, Waldschmidt TJ, Lee WT, et al. Effect of B cell stimulatory factor-1 (interleukin 4) on Fcε and Fcγ receptor. J Immunol 1987;139:2290–2296.

98. Ohara J, Paul W. Up-regulation of interleukin 4/B-cell stimulatory factor 1 receptor expression. Proc Natl Acad Sci 1988;85:8221–8225.

99. Takeda K, Tanaka T, Shi W, et al. Essential role of Stat6 in IL-4 signalling. Nature (Lond) 1996;380:627–630.

100. Shimoda K, Deursen J, Sangster M, et al. Lack of IL-4-induced Th2 response and IgE class switching in mice with disrupted Stat6 gene. Nature (Lond) 1996;380:630–633.

101. Kaplan M, Schindler U, Smiley S, et al. Stat6 is required for mediating responses to IL-4 and for the development of Th2 cells. Immunity 1996;4:313–319.

102. Lederer JA, Perez VL, DesRoches L, et al. Cytokine transcriptional events during helper T cell subset differentiation. J Exp Med 1996;380:397–406.

103. Huang H, Hu-Li J, Chen H, et al. IL-4 and IL-13 production in differentiated T helper type 2 cells is not IL-4-dependent. J Immunol 1997;159:3731–3738.

104. Agarwal S, Rao A. Modulation of chromatin structure regulates cytokine gene expression during T cell differentiation. Immunity 1998;9:765–775.

105. Brown MA, Pierce JH, Watson CJ, et al. B cell stimulatory factor-1/interleukin-4 mRNA is expressed by normal and transformed mast cell lines. Cell 1987;50:809–818.

106. Gross DS, Garrard WT. Nuclease hypersensitive sites in chromatin. Annu Rev Biochem 1988;57:159–197.

107. Henkel G, Weiss DL, McCoy R, et al. A DNAse I hypersensitive site defines a mast cell enhancer. J Immunol 1992;149:323–330.

108. Henkel G, Brown MA. PU.1 and GATA: components of a mast cell-specific interleukin 4

intronic enhancer. Proc Natl Acad Sci USA 1994;91:7737–7741.

109. Bird JJ, Brown DR, Mullen AC, et al. Helper T cell differentiation is controlled by the cell cycle. Immunity 1998;9:229–237.

110. Siden E. Regulated expression of germline antigen receptor genes in mast cell lines from the murine embryo. J Immunol 1993;150:4427–4437.

111. Smale S, Baltimore D. The "initiator" as a transcriptional control element. Cell 1989;57:103–113.

112. Weis L, Reinberg D. Transcription by RNA polymerase II: initiator-directed formation of transcription-competent complexes. FASEB J 1992;6:3300–3309.

113. Ayoubi TAY, Van de Ven WJM. Regulation of gene expression by alternative promoters. FASEB J 1996;10:453–460.

114. Ernst P, Smale ST. Combinatorial regulation of transcription I: general aspects of transcriptional control. Immunity 1995;2:3 11–319.

115. Riley LK, Morrow JK, Danton MJ, et al. Human terminal deoxyribonucleotidyltransferase: molecular cloning and structural analysis of the gene and 5′ flanking region. Proc Natl Acad Sci USA 1988;85:2489–2493.

26
Arachidonic Acid Metabolism in Mast Cells

Clifton O. Bingham III, Jonathan P. Arm, and K. Frank Austen

Eicosanoids are potent inflammatory mediators derived from the 20-carbon unsaturated fatty acid, arachidonic acid (C 20:4), esterified at the sn-2 position of the glycerol backbone of membrane phospholipids. The liberation of free arachidonic acid from phospholipids requires the initial activity of phospholipase A_2 (PLA_2) enzymes. In mast cells, this free arachidonic acid is processed to unstable intermediates by 5-lipoxygenase (5-LO) in conjunction with 5-lipoxygenase activating protein (FLAP) and by the constitutive and inducible isoforms of prostaglandin endoperoxide synthase (cyclooxygenase; PGHS) for the subsequent activities of the terminal enzymes, leukotriene (LT) C_4 synthase to generate LTC_4 and prostaglandin (PG) D_2 synthase to generate PGD_2. Both human and mouse mast cells generate leukotrienes and prostaglandins in response to activation, although the relative amounts of these mediators are determined by the phenotype of the cells and their localization in situ. Studies with the mouse bone marrow-derived mast cell (mBMMC) have provided valuable insight into the regulation of the biosynthesis of these mediators in vitro.

Phospholipase A_2 Enzymes

Arachidonic acid is initially liberated from the sn-2 fatty acyl ester bond of membrane-bound glycerophospholipids by the activity of PLA_2 enzymes. This reaction is one of the rate-limiting steps in eicosanoid biosynthesis. The recognition of different classes of PLA_2 enzymes based on structure has led to a nomenclature for their further classification.[1,2]

The low molecular weight (approximately 14 kDa) mammalian PLA_2 enzymes are represented by groups I, II, and V.[3,4] These enzymes cleave not only arachidonoyl chains but also oleoyl, linoleoyl, and palmitoyl chains from the sn-2 position of phospholipids. Internal cysteinyl linkages, critical for tertiary structure, render their enzymatic activity sensitive to reducing agents such as dithiothreitol. Additionally, the low molecular weight enzymes share a highly conserved calcium-binding domain that is reflected in their common requirement for millimolar calcium concentrations for optimal activity as well as a conserved catalytic triad. Group I PLA_2 enzymes have 14 disulfide-linked cysteine residues and share a characteristic short "elapid" or "pancreatic" loop.[4] The mammalian group I PLA_2 is expressed most abundantly in pancreatic secretions, where it functions in digestion, and has also been identified in lung and spleen. Group I enzymes are synthesized as zymogens; the propeptide is suceptible to cleavage by trypsin. The genes for groups IIA, IIC, and group V PLA_2 enzymes are closely linked on the human chromosome 1p35–36 and on the syntenic distal region of the mouse chromosome 4.[5] Group II PLA_2 enzymes lack an elapid loop and are characterized by a 6-amino-acid carboxyl-terminal extension not present in the group I enzymes.[3] All are encoded as prepeptides with a hydrophobic leader sequence. Whereas the group IIA PLA_2

SEQUENCE ALIGNMENT OF RAT LOW MOLECULAR WEIGHT PLA₂ ENZYMES

```
                                        Mature Protein                    Ca²⁺-binding domain
                                                                                        **
rat group I          MKLLLLAALLTAGVTAHSISTR AVWQFRNMIKCTIPGSDPLREYNNYGCYCGLGGSGTPVDDLDRCCQTHDHCYNQ
rat group IIA                  --V---L-VVIMAFGSIQVQG. SLLE-GQ--LF.KT-KRADVS-GF--CHC-V--R-S-K-AT-WCCV---CC--R
rat group IIC        MDLLVSSG--GIAVFLVFIFCWTTSTLS.. SF---QR-V-H.-T-RSAFFS-YG--C-C----R-I---AT--CCWA--CC-HK
rat group V                    --R--TL-WFLACSVPAVPG.. GLLELKS--EK.VT-KNAVKN-GF--C-C-W--H---K-GT-WCCRM--RC-GL


                                                                *
rat group I          AKKLESCKFLIDNPYTNTYSYKCSGNVITCSDKNND...CESFICNCDRQAAICFSKVPYNKEY..KDLDTKKHC
rat group IIA        LE-RG.C.....GTKFL--KFSYR-GQ-SC-TNQDS...CRKQLCQC-KA--EC-ARNKKSYSL..-YQFYPNKFCKGKTPSC
rat group IIC        L-EYG.C.....QPIL-A-QFAIVNGTV-CGCTMGGGCLCGQKACEC-KLSVYC-KENLATY-KTF-Q-FPTRPQCGRDKLHC
rat group V          LEEKH.C.....AIR-QS-D-RFTQDLVICEHDSF....CPVRLCAC--KLVYCLRRNLWSYNR..LYQYYPNFLC
```

FIGURE 26.1. Sequence alignment of rat low molecular weight PLA₂ enzymes. The junction of the signal peptide with the mature protein is indicated by a *gap*; the activation peptide (AHSISTR) of the group I PLA₂ proenzyme is in *bold*; the conserved *Ca²⁺-binding domain* is indicated; the catalytic residues are indicated by an *asterisk*; cysteine residues are *shaded*.

enzyme contains 14 cysteines, the group IIC PLA₂ enzyme contains 16 cysteines.[6] The single group V PLA₂ enzyme contains 12 cysteines and lacks both the carboxyl-terminal extension found in the group II enzymes and the elapid loop characteristic of the group I enzymes[7] (Fig. 26.1).

The group IV PLA₂ enzyme was initially isolated from the cytosolic fraction of monocytes and platelets and has been termed cytosolic PLA₂ (cPLA₂).[8,9] The gene for cPLA₂ is distinct from those of the low molecular weight species and encodes a mature 85-kDa protein. This enzyme preferentially cleaves arachidonic acid from the *sn*-2 position of glycerophospholipids.[10] It does not contain the intramolecular cysteine linkages characteristic of the low molecular weight species and is thus resistant to reducing agents, and is active at submicromolar calcium concentrations characteristic of activated cells. Group VI calcium-independent PLA₂ (iPLA₂) has been isolated from macrophages and CHO cells.[11-13] The mammalian platelet-activating factor (PAF) acetyl hydrolase has been classified as group VII PLA₂.[1] The group IIA, IIC, IV, and V PLA₂ enzymes have been identified in mast cells and are described in detail here.

Group IIA PLA₂

The mammalian group IIA PLA₂ enzyme, a 14-kDa mature protein, was originally isolated from inflammatory synovial fluid.[14,15] PLA₂ activity attributed to group IIA has been recognized at inflammatory foci, in synovial fluid, and in the circulation of patients with septic shock and pancreatitis.[16] Transcripts are induced in mesangial cells, endothelial cells, synoviocytes, and astroglial cells by interleukin (IL-1α), tumor necrosis factor (TNF-α), IL-6, and substances that increase intracellular levels of cyclic 3′,5′-adenosine monophosphate (cAMP).[16] Transcripts are inhibited in mesangial cells by transforming growth factor (TGF-β).[17] Several strains of mice lack functional group IIA PLA₂ because of a single thymidine insertion in the third exon of the gene that causes a shift in reading frame, a premature stop codon, and transcripts that encode truncated and nonfunctional proteins.[18,19]

Initially the activity attributed to this enzyme was recognized to be secreted from cells and was termed secretory PLA₂ (sPLA₂). However, the enzyme is also localized to cellular membranes, mitochondria, and within granules of mast cells and platelets.[20-22] In resting platelets, group IIA PLA₂ is located on the inner leaflet of the plasma membrane but moves to the outer leaflet immediately after cell activation.[22] After being released from the cell, it may bind to cell-surface proteoglycans.[23] Site-directed mutagenesis studies of the mouse and rat enzymes have shown that rodent group IIA PLA₂ binds to heparin through critical lysine residues.[24] The fact that mutation of the heparin-binding domain does not affect the catalytic activity of the enzyme supports earlier studies demonstrating that heparin affinity can be used for purification without affecting enzymatic activity.[24] Although group IIA PLA$_2$ activity can be detected in stimulated cell supernatants, the measured activity can be augmented by the addition of heparin, which presumably competes with proteoglycans in the membrane to remove group IIA PLA$_2$ or another low molecular weight PLA$_2$ from the cell surface.[24,25]

Group IIA PLA$_2$ has activity distinct from its ability to cleave arachidonic acid from membrane phospholipids for prostaglandin and leukotriene generation. This enzyme also generates reactive lyso-phospholipids, which may facilitate fusion of the perigranular and cell membranes in exocytosis by rat serosal mast cells.[26] Group IIA PLA$_2$ may also bind to specific cell-surface receptors related to the macrophage mannose receptor.[27] A soluble form of the receptor that binds recombinant group IIA PLA$_2$ has been recognized. Transcripts for this receptor have been demonstrated in lung, liver, pancreas, kidney, and spleen, but are distributed differently among different species.[28] However, the signal transduction events mediated through the PLA$_2$ receptor have not been determined. Nonspecific pharmacological inhibitors of this PLA$_2$, which may affect other low molecular weight PLA$_2$ enzymes, inhibit exocytosis of both neutrophils and mast cells.[26,29,30]

Group IIC PLA$_2$

The gene for rat and mouse group IIC PLA$_2$ encodes a 14.7-kDa mature protein.[6] Both 2.3- and 5-kb alternatively spliced transcripts are expressed predominantly in the testis of rodents. The human group IIC PLA$_2$ gene has an incomplete third exon, and transcripts have not been found in the human tissues thus far examined; thus it is probably a pseudogene in humans.[6] The rat enzyme demonstrates enzymatic activity at neutral to basic pH, a dependence on millimolar concentrations of calcium, and preference for arachidonoyl phosphatidylinositol (PI) relative to arachidonoyl phosphatidylcholine (PC) and arachidonoyl phosphatidylethanolamine (PE) in an in vitro assay presenting phospholipid substrate in liposomes. Tran-

scripts for group IIC PLA$_2$ increase in developing rodent testis and may be important in spermatogenesis, production of seminal fluid, sperm capacitation, and the acrosomal reaction.[31]

Group V PLA$_2$

The human group V PLA$_2$, a predicted 13.6-kDa mature protein, is encoded by a predominant 1.2-kb transcript demonstrated principally in heart, but also in placenta and lung.[7] The expressed human and rat enzymes have structures and enzymatic activities that are similar to those of group IIA and group IIC PLA$_2$ with a requirement for millimolar concentrations of calcium, a pH optimum of 6.5 to 9, and a preference for oleoyl PC over arachidonoyl PC and arachidonoyl PE in vitro.[7]

Antisense constructs specific for a unique region of group V PLA$_2$ lead to a diminution of lipopolysaccharide (LPS)/PAF-induced PGE$_2$ generation in the P388D$_1$ cell line; thus, group V PLA$_2$ enzyme plays a role in mobilizing arachidonic acid in this cell.[32] A polyclonal antibody generated against human group IIA PLA$_2$, BQY-113A, identifies a cell-surface protein on 30% of P388D$_1$ cells at rest and on 50% of the cells within 10 min after LPS/PAF activation.[32] Because the P388D$_1$ cell line does not have transcripts for either the group IIA or the group IIC enzymes by reverse transcriptase-polymerase chain reaction, the antibody probably recognizes group V PLA$_2$ and not just the group IIA PLA$_2$, as was initally reported.

Group IV PLA$_2$

Group IV PLA$_2$, originally isolated from the cytosolic fraction of U937 monocytes, was termed cytosolic PLA$_2$ (cPLA$_2$).[33,34] The 3.4-kb transcript encodes an 85-kDa protein that migrates at 100 to 110 kDa on sodium dodecyl sulfate-polyacrylamide gel electrophoresis.[9] In contrast to the low molecular weight enzymes, cPLA$_2$ is selective for arachidonic acid in the sn-2 position of phospholipids, suggesting that it has a major role in eicosanoid generation. The activity of this enzyme increases dramatically with an increase in calcium concentration from 10^{-7}M to 10^{-6}M, suggesting that it is functional after stimulus-coupled activation and intracellular calcium flux.[8,9] Group IV PLA$_2$ is expressed constitutively in many cell types, including mast cells, and can also be induced by cell exposure to IL-1α, TNF-α, LPS, macrophage colony-stimulating factor, epidermal growth factor, platelet-derived growth factor, thrombin, and adenosine triphosphate.[8,9] The promoter, like that of other housekeeping genes, lacks a TATA box but also contains transcriptional regulatory elements including three sites for activating protein (AP-1) binding, two proto-oncogene Ets activator (PEA-3) motifs, and an asymmetric nuclear factor (NF-κB) site.[35]

The amino-terminus of cPLA$_2$ contains a calcium-dependent lipid-binding domain, similar to that found in protein kinase C, which is necessary for the translocation of the protein from the cytosol to the nuclear membrane and endoplasmic reticulum.[8,9] Studies in rat basophilic leukemia cells and macrophages suggest that cPLA$_2$ translocates preferentially to the nuclear envelope and endoplasmic reticulum, the site of leukotriene and prostaglandin synthesis within the cell.[36,37] cPLA$_2$ contains several serine residues as potential phosphorylation targets for mitogen-activated protein (MAP) kinases, of which Ser-505 is best characterized.[8,9] The activity of the enzyme is increased by phosphorylation at Ser-505 by p42/p44 MAP kinases.[38] Thus, a picture emerges of cPLA$_2$ being rapidly phosphorylated and translocated, leading to its acute activation in response to receptor/agonist cell stimulation. However, cPLA$_2$ has also been implicated in the slow supply of arachidonic acid over several hours to PGHS-2.[39,40]

In the P388D$_1$ cell line, both methyl arachidonoylfluorophosphonate (MAFP), an inhibitor of group IV and group VI PLA$_2$ enzymes, and LY311727, an inhibitor of low molecular weight PLA$_2$ enzymes, inhibit PGE$_2$ generation.[41] Antisense constructs specific for group V PLA$_2$ reveal that both group IV PLA$_2$ and the low molecular weight group V PLA$_2$ act sequentially to release intracellular arachidonic acid for PGE$_2$ generation.[32]

Prostaglandin Generation

PGHS-1 and PGHS-2

Arachidonic acid, made available by the activity of PLA$_2$ enzymes, is initially metabolized by the cyclooxygenase activity of prostaglandin endoperoxide synthase enzymes, PGHS-1 and PGHS-2, to form the unstable intermediate cyclic PGG$_2$, with further hydroperoxidation by the same enzymes to generate the unstable intermediate PGH$_2$.[42] PGHS-1 and PGHS-2 are encoded by distinct genes, and the proteins are 60% identical at the amino acid level and are 75% similar within species.[43] PGHS-1 has a longer signal peptide, while PGHS-2 has an 18-residue insertion at the carboxyl-terminus. Both proteins are approximately 70 kDa in size with similarities in the catalytic sites, but the transcript encoding mouse PGHS-1 is 2.8 kb and that for PGHS-2 is 4.4 kb. The gene for PGHS-1 is 22 kb in size and contains 11 exons; that for PGHS-2 is 8 kb and consists of 10 exons. The first 2 exons of PGHS-1 are encoded as a single exon in PGHS-2, and the remaining exon structure is similar; the differences in size are reflected in the intron size. The smaller size of the gene for PGHS-2 is consistent with other immediate-early genes. PGHS-1 has a TATA-less promoter, which is characteristic of other housekeeping genes, and reporter constructs have not shown significant regulatory elements[43]; however, our work indicates that in mast cells the expression of this enzyme is increased by KL with IL-3, IL-9, or IL-10.[44] In contrast, PGHS-2 has a highly inducible promoter with binding sites for NF-κB, NF-IL6, CAAT enhancer binding protein (C/EBP-β), Sp-1, and a cAMP-responsive element (CRE).[43,45] The 3'-untranslated region of PGHS-2 contains numerous AUUUA motifs, which may contribute to mRNA instability.[46]

PGHS-1 is constitutively expressed in many cell types and is traditionally associated with "housekeeping functions" such as platelet aggregation, regulation of renal blood flow, and maintenance of gastric cytoprotection.[42,43] Numerous cells including mast cells and platelets constitutively produce PGHS-1, which permits rapid generation of prostaglandins and thromboxanes (TX) in the immediate events following cell activation without a requirement for the induction of new genes.[43,47] Valeryl salicylate is a selective inhibitor of PGHS-1,[48] whereas aspirin and the traditional nonsteroidal antiinflammatory drugs (NSAIDs) inhibit both PGHS-1 and PGHS-2.[49] Aspirin irreversibly acetylates a serine residue in the catalytic site of both enzymes. PGHS-2 is inducible de novo in many cell types in response to proinflammatory cytokines, immunological stimuli, growth factors, LPS, and phorbol esters.[43,45] Dexamethasone inhibits the induction of PGHS-2 transcription with no effect on PGHS-1, and NS-398 selectively inhibits the enzymatic activity of PGHS-2 but not PGHS-1.[43,47] Acetylation by aspirin of the active site serine in PGHS-2 confers 15-lipoxygenase activity to the enzyme.[49]

PGD$_2$ Synthase

The unstable intermediate PGH$_2$, generated by the PGHS enzymes, is converted by cell-specific terminal synthases to PGD$_2$, PGE$_2$, PGI$_2$, and TXA$_2$.[42,43] In the mast cell, hematopoietic PGD$_2$ synthase, first described in spleen, converts PGH$_2$ into PGD$_2$.[50] This 28-kDa protein also exists in macrophages and Langerhans cells, although mast cells appear to be the major source of PGD$_2$. Hematopoietic PGD$_2$ synthase is distinguished from the brain type of PGD$_2$ synthase immunochemically and by its glutathione dependence. PGD$_2$ is a potent vasodilator and bronchoconstrictor and is chemotactic for neutrophils.[51] The measurement of the levels of PGD$_2$ or its receptor-active metabolite 9α, 11β-PGF$_2$ serves as a marker of mast cell activation. Elevated levels of either PGD$_2$ or its metabolite have been described in bronchoalveolar lavage fluid from patients with asthma at rest and after antigen challenge,[52] in nasal lavage fluid from patients with allergic rhinitis,[53] and in patients with systemic mastocytosis.[54] Transcripts for the PGD$_2$ receptor (DP), which binds PGD$_2$ with high affinity, have been localized to platelets, basophils, and neutrophils, and are found in lung, gastrointestinal tract, and uterus.[55] This receptor, like the other prostanoid receptors, has seven transmembrane domains and is linked to cAMP.[56] When activated, this receptor inhibits platelet aggregation.[56] PGD$_2$ and its metabolite may elicit bronchoconstriction through the thromboxane receptor (TP).[57]

PGD$_2$ may be converted intracellularly or extracellularly into PGJ$_2$ and its metabolites. These metabolites activate peroxisome proliferator-activated receptors (PPAR-γ) located in the nucleus to promote gene transcription.[58,59] Thus, in addition to serving as a transcellular mediator acting through plasma membrane receptors, PGD$_2$ and its metabolites may have direct intracellular functions.

Leukotriene Generation

5-Lipoxygenase and FLAP

The initial step in the generation of leukotrienes from arachidonic acid depends on two intermediate proteins acting in concert, 5-LO and FLAP. Arachidonic acid mobilized from phospholipids is initially presented to FLAP, an 18-kDa integral membrane protein localized to the perinuclear membrane.[60] FLAP binds arachidonic acid for the subsequent action by 5-LO. 5-LO is an 80-kDa enzyme that resides in the cytosol or euchromatin of the nucleus in resting cells, but rapidly translocates to the perinuclear membrane after the cell is activated.[61,62] A tyrosine phosphorylation of 5-LO is required for the translocation of this enzyme.[63] 5-LO sequentially converts arachidonic acid to the cyclic endoperoxide intermediate 5-hydroperoxy-eicosatetraenoic acid and then to the unstable epoxide intermediate LTA$_4$.

LTC$_4$ Synthase

LTC$_4$ synthase is an 18-kDa integral perinuclear membrane protein.[64] The human gene resides on chromosome 5q35, near the locus for the Th-2 type cytokines and receptors considered important in allergic inflammation and generation of airway hyperresponsiveness.[65] LTC$_4$ synthase displays 31% amino acid sequence homology to FLAP and is inhibited by the FLAP antagonist MK 886.[64] Whereas FLAP binds arachidonic acid for the subsequent activity of 5-LO, LTC_4 synthase opens the epoxide ring of LTA_4 by acid catalysis with Arg-51 and activates glutathione by base catalysis with Tyr-93, thereby conjugating these moieties to form LTC_4.[66] The microsomal glutathione-S-transferase (GST) II is bifunctional, has a homologous motif to LTC_4 synthase, and may allow cysteinyl leukotriene production by cells lacking 5-LO and normally engaged in xenobiotic function.[67] LTC_4 undergoes a probenecid-dependent export step from the cell, potentially through the multidrug resistance transporter.[68,69] Extracellular γ-glutamyl transpeptidase and dipeptidases convert LTC_4 to its receptor-active metabolites, LTD_4 and LTE_4, respectively.[70]

The cysteinyl leukotrienes LTC_4, LTD_4, and LTE_4 are recognized as the components of the slow-reacting substance of anaphylaxis (SRS-A) and are potent mediators of hypersensitivity reactions.[71] LTC_4, via conversion to LTD_4, is at least 1000 times more potent than histamine in the induction of wheal and flare reactions in the skin, bronchoconstriction, and vasodilation. There is now unequivocal evidence that cysteinyl leukotrienes are important mediators in bronchial asthma, especially in aspirin-sensitive asthma.[72]

Arachidonic Acid Metabolism in Mast Cells

Immediate Eicosanoid Generation

Mast cells activated through the high-affinity receptor for IgE (FcεRI) release both PGD_2 and LTC_4 within 10 min. The relative amounts of these mediators is influenced by the phenotype and tissue localization of the mast cells. Human lung mast cells containing tryptase (MC_T phenotype) preferentially generate LTC_4 with smaller amounts of PGD_2, whereas human mast cells containing both chymase and tryptase (MC_{TC} phenotype) in the uterus and skin produce PGD_2 in preference to LTC_4.[73–75] Mucosal mast cells from the intestine of helminth-infected rats generate LTC_4 in prefer-

ence to PGD_2, while rodent serosal mast cells preferentially generate PGD_2.[76] The more immature mBMMC generate LTC_4 in greater amounts than the prostanoid product.[77] Further heterogeneity is demonstrated in reponse to different agonists. Serosal mast cells but not mBMMC are activated by polybasic compounds and neuropeptides acting through G proteins. However, when mBMMC are cultured with fibroblasts in the presence of the ligand for the receptor tyrosine kinase c-*kit* (KL; stem cell factor), they respond to activation by polybasic compounds with the generation of both PGD_2 and LTC_4, although the amounts of these mediators are 10 fold less than that seen in response to FcεRI-dependent activation.[78] Both KL and fibroblasts are required to elicit this change in phenotype, but the fibroblast-derived factor has yet to be identified.

Our studies have demonstrated that mBMMC may be activated by KL for secretory granule exocytosis and lipid mediator generation.[77] The maximal response to 50 ng/ml KL is approximately 50% of the maximal response to activation through FcεRI. Activation by either stimulus is associated with phosphorylation of $cPLA_2$, maximal at 2 min, and translocation of 5-LO from a soluble fraction to a cell membrane fraction. Both events are substantially reversed 10 min after activation. The immediate production of PGD_2 depends on constitutively expressed PGHS-1 inasmuch as no protein or transcript is seen for PGHS-2 at this time, specific inhibitors of PGHS-2 such as NS-398 have no effect, and the specific inhibitor of PGHS-1, valeryl salicylate, inhibits this immediate phase of prostaglandin generation. The dependence of this immmediate phase on PGHS-1 has been confirmed by others.[79]

Both group IV $cPLA_2$ and group V PLA_2 have been implicated in the immediate phase of eicosanoid release. On the basis of the presence of a transient Ca^{2+} flux and the phosphorylation of $cPLA_2$ in the immediate phase of mast cell activation, we have concluded that this enzyme supplies arachidonic acid to PGHS-1 and 5-LO.[77] Furthermore, preincubation of mast cells with MAFP inhibits the immediate phase of

PGD$_2$ generation.[80] Preincubation of cells with either 12-epi-scalaradial or with heparin, both inhibitors of low molecular weight PLA$_2$ enzymes, does not affect the immediate phase of eicosanoid generation, and mice deficient in group IIA PLA$_2$ do not demonstrate impairment of the immediate phase, indicating that this PLA$_2$ does not play a significant role in the immediate phase.[25] Nevertheless, Fonteh et al. have demonstrated the capacity of low molecular weight PLA$_2$ enzymes to elicit or augment archidonic acid release from mast cells.[81,82] Reddy et al. have reported that the major transcript for a low molecular weight PLA$_2$ in mBMMC is that for the group V enzyme and that specific antisense constructs for the group V PLA$_2$ inhibit immediate PGD$_2$ generation in the MMC-34 cell line.[80] Thus, the group IIA PLA$_2$ is not required for immediate eicosanoid generation by mBMMC, but both group IV cPLA$_2$ and group V PLA$_2$ have been implicated. If there is a sequential action of group IV cPLA$_2$ and group V PLA$_2$ in a manner similar to that described for arachidonic acid release and PGE$_2$ generation in P388D$_1$ cells,[32] the requirement is not uniformly revealed during the rapid, immediate, and transient phase of mast cell arachidonic acid metabolism in all protocols.

Delayed Phase PGHS-2-Dependent PGD$_2$ Generation

When mBMMC are incubated with KL in combination with IL-10 and IL-1β, the immediate phase of exocytosis and eicosanoid generation is followed by the induction of transcripts for PGHS-2 from 2 to 5h later and the generation of PGHS-2 protein and a delayed phase of PGD$_2$ generation over 2 to 10h when both transcript and protein for PGHS-1 remain constant.[47] Both NS398, a PGHS-2 selective inhibitor, and dexamethasone, which inhibits the induction of PGHS-2 but not PGHS-1, suppress the delayed phase of PGD$_2$ generation. IL-4 inhibits the minimal induction of PGHS-2 mediated by KL + IL-10, but has no effect on the induction of PGHS-2 induced by the cytokine triad KL + IL-10 + IL-1β.[83]

mBMMC primed for 2h in KL plus IL-10 and then activated through FcεRI demonstrate PGHS-2 induction and delayed-phase PGD$_2$ generation.[84] The cytokine-independent mouse mast cell line MMC-34 also exhibits immediate and delayed PGD$_2$ generation dependent on PGHS-1 and PGHS-2, respectively, in response to FcεRI activation without a requirement for cytokine priming.[79] Biphasic generation of PGD$_2$ in response to FcεRI activation without a requirement for cytokine priming has also been reported in mBMMC derived from BALBc, C57BL/6, and AKJ mice.[79,85] However, in these protocols, mBMMC were withdrawn from cytokines before activation, suggesting that the absence of IL-3 alters the requirement for cytokine priming. Delayed-phase PGHS-2-dependent PGD$_2$ generation in response to activation with polybasic compounds, substance P, and 48/80 has also been reported from mBMMC cocultured with 3T3 fibroblasts and KL.[78]

The segregation of arachidonic acid metabolism between PGHS-1 in the immediate phase of PGD$_2$ generation and PGHS-2 in the delayed phase of PGD$_2$ generation has prompted an evaluation of the species of PLA$_2$ supplying substrate to each phase. Arachidonic acid is released from cells in two phases in reponse to KL + IL-10 + IL-1, which parallel the two phases of PGD$_2$ generation.[25] That the addition of 12-epi-scalaradial, an inhibitor of low molecular weight PLA$_2$, inhibits the second phase of PGD$_2$ generation suggests the participation of a low molecular weight species of PLA$_2$ in the delayed phase. The addition of heparin removes a PLA$_2$ activity from the surface of cells with the attendant inhibition of the delayed phase of PGD$_2$ generation.[25] Although transcripts for group IIA PLA$_2$ are induced in BMMC derived from BALBc mice, BMMC derived from C57Bl/6 mice, which are deficient in this enzyme, exhibit intact immediate and delayed phases of PGD$_2$ generation. Thus an alternative, heparin-sensitive low molecular weight PLA$_2$, notably group V PLA$_2$, functions in the delayed phase of PGD$_2$ generation. However, Reddy et al. reported that MAFP, an inhibitor of group IV cPLA$_2$ and group VI PLA$_2$, inhibited the delayed phase of PGD_2 generation, and that SB203347, an inhibitor of low molecular weight PLA_2 enzymes, did not.[85] Thus, in both immediate- and delayed-phase PGD_2 generation, there appears to be sequential, supporting, or alternative roles for the low molecular weight PLA_2 enzymes and group IV $cPLA_2$. Our more recent data indicate that both 12-*epi*-scalaradial and MAFP will inhibit the delayed phase of PGD_2 generation, invoking both classes of PLA_2, whereas only MAFP inhibits immediate generation.

Cytokine Upregulation of Immediate Eicosanoid Generation

In addition to its ability to elicit immediate and delayed activation of mBMMC, KL is a mast cell proliferation and differentiation factor, and drives mBMMC toward a cutaneous or serosal phenotype with expression of mouse mast cell protease-4 (mMCP-4) and synthesis of heparin proteoglycan[86] (also see Chapter 4). KL also primes human mast cells for exocytosis in response to cross-linking of FcεRI.[87] After 2 days of culture in KL alone, mBMMC developed with IL-3 respond to IgE/antigen activation with a 3-fold increase in PGD_2 and a 2-fold increase in LTC_4 compared to cells maintained in IL-3.[44] The addition of IL-3, IL-9, or IL-10 with KL results in a 6- to 10-fold increase in PGD_2 generation with no further increases in LTC_4 generation. Priming with KL alone is associated with increased expression of each of the enzymes of prostanoid biosynthesis, namely, group IV $cPLA_2$, PGHS-1, and PGD_2S but with no change in expression of 5LO, FLAP, or LTC_4S and no induction of PGHS-2.[44] Accessory cytokines IL-3, IL-9, and IL-10 selectively further increase the expression of PGHS-1. IL-4 inhibits priming through a selective inhibition of the increase in group IV $cPLA_2$ expression with no effect on intermediate or terminal enzymes, consistent with a role for group IV $cPLA_2$ in immediate eicosanoid biosynthesis.[83]

To uncover a role for IL-3 in expression of the eicosanoid pathways, mBMMC were derived from bone marrow with KL + IL-10 instead of IL-3.[88] When activated with IgE and antigen, these mBMMC generate small but equal quantities of PGD_2 and LTC_4. When the cells are maintained in KL + IL-10 but stimulated with IL-3 over a 2-week period, there is a marked increase in their IgE-dependent generation of LTC_4 but not PGD_2; this is accompanied by an upregulation in the expression of 5-LO and FLAP, which precedes the increased expression of group IV $cPLA_2$ and LTC_4 synthase. Thus, IL-3 upregulates the expression of the enzymes of the 5-LO/LTC_4 synthase pathway and KL regulates the PGHS/PGD_2 synthase pathway.

Conclusions

Three concepts regarding arachidonic acid metabolism in mast cells emerge from the studies with mBMMC. First, the activation of mast cells leads to the rapid, immediate generation of both LTC_4 and PGD_2 and is followed by an inducible, delayed pathway for the generation of PGD_2 but not LTC_4. The immediate generation of PGD_2 is mediated by constitutive PGHS-1, whereas the delayed phase of PGD_2 generation is mediated by induced PGHS-2. The functions of PGHS-1 and PGHS-2 in the supply of substrate to PGD_2S are segregated and are not interchangeable. Second, it seems likely that different combinations of PLA_2 enzymes are used by PGHS-1 and PGHS-2 for the source of arachidonic acid substrate, a finding initially described in the mast cell. Finally, during the development of immature mast cells from mouse bone marrow, IL-3 regulates the expression of 5-LO, FLAP, and LTC_4 synthase, the biosynthetic enzymes for the generation of the cysteinyl leukotriene parent, LTC_4; and stem cell factor (KL) regulates the expression of PGHS-1, PGHS-2, and PGD_2 synthase, which lead to the generation of the prostanoid product, PGD_2.

References

1. Dennis EA. The growing phospholipase A_2 superfamily of signal transduction enzymes. Trends Biochem Sci 1997;2:1–2.
2. Roberts MF. Phospholipases: structural and functional motifs working at an interface. FASEB J 1996;10:1159–1172.

3. Tischfield JA. A reassessment of the low molecular weight phospholipase A_2 gene family in mammals. J Biol Chem 1997;272:17247–17250.

4. Dennis EA. Diversity of group types, regulation, and function of phospholipase A_2. J Biol Chem 1994;269:13057–13060.

5. Spirio LN, Kutchera W, Winstead MV, et al. Three secretory phospholipase A_2 genes that map to human chromosome 1p35–36 are not mutated in individuals with attenuated adenomatous polyposis coli. Cancer Res 1996;56:955–958.

6. Chen J, Engle SJ, Seilhamer JJ, et al. Cloning and characterization of novel rat and mouse low molecular weight Ca^{2+}-dependent phospholipase A_2s containing 16 cysteines. J Biol Chem 1994;269:23018–23024.

7. Chen J, Engle SJ, Seilhamer JJ, et al. Cloning and recombinant expression of a novel human low molecular weight Ca^{2+}-dependent phospholipase A_2. J Biol Chem 1994;269:2365–2368.

8. Leslie CC. Properties and regulation of cytosolic phospholipase A_2. J Biol Chem 1997;272:16709–16712.

9. Clark JD, Schievella AR, Nalefski EA, et al. Cytosolic phospholipase A_2. J Lipid Mediat Cell Signal 1995;12:83–117.

10. Dietz E, Louis-Flamberg P, Hall RH, et al. Substrate specificities and properties of human phospholipases A_2 in a mixed vesicle model. J Biol Chem 1992;267:18342–18348.

11. Ackermann EJ, Kempner ES, Dennis EA. Ca^{2+}-independent cytosolic phospholipase A_2 from macrophage-like $P388D_1$ cells, isolation and characterization. J Biol Chem 1994;269:9227–9233.

12. Tang J, Kriz R, Wolfman N, et al. A novel Ca^{2+}-independent phospholipase A_2 contains eight ankyrin motifs. J Biol Chem 1997;272:8567–8575.

13. Wolf MJ, Gross RW. Expression, purification, and kinetic characterization of a recombinant 80 kDa intracellular calcium-independent phospholipase A_2. J Biol Chem 1996;271:30879–30855.

14. Seilhamer JJ, Pruzanski W, Vadas P, et al. Cloning and expression of phospholipase A_2 present in rheumatoid arthritic synovial fluid. J Biol Chem 1989;264:5335–5338.

15. Kramer RM, Hession C, Johansen B, et al. Structure and properties of a human nonpancreatic phospholipase A_2. J Biol Chem 1989;264:5768–5775.

16. Murakami M, Kudi I, Inoue K. Secretory phospholipases A_2. J Lipid Mediat Cell Signal 1995;12:119–130.

17. Schalkwijk C, Pfeilschifter J, Marki F, et al. Interleukin-1 beta and forskolin-induced synthesis and secretion of group II phospholipase A_2 and prostaglandin E_2 in rat mesangial cells is prevented by transforming growth factor-beta 2. J Biol Chem 1992;267:8846–8851.

18. MacPhee M, Chepenik KP, Liddell RA, et al. The secretory phospholipase A_2 gene is a candidate for the *Mom1* locus, a major modifier of Apc^{Min}-induced intestinal neoplasia. Cell 1995;81:957–966.

19. Kennedy BP, Payette P, Mudgett J, et al. A natural disruption of the secretory group II phospholipase A_2 gene in inbred mouse strains. J Biol Chem 1995;270:22378–22385.

20. Chock SP, Schumauder-Chock EA, Cordella E, et al. The localization of phospholipase A_2 in the secretory granule. Biochem J 1994;300:619–622.

21. Aarsman AJ, de Jong JGN, Arnoldussen E, et al. Immunoaffinity purification, partial sequence, and subcellular localization of rat liver phospholipase A_2. J Biol Chem 1989;364:10008–10014.

22. Bevers EM, Comfurius P, Zwaal RF. Changes in membrane phospholipid distribution during platelet activation. Biochim Biophys Acta 1983;736:57–66.

23. Sartipy P, Johansen B, Camejo G, et al. Binding of human phospholipase A_2 to proteoglycans; differential effect of glysoaminoglycans on enzyme activity. J Biol Chem 1996;271:26307–26314.

24. Murakami M, Nakatani Y, Kudo I. Type II secretory phospholipase A_2 associated with cell surfaces via C-terminal heparin-binding lysine residues augments stimulus-initiated delayed prostaglandin generation. J Biol Chem 1996;271:30041–30051.

25. Bingham CO III, Murakami M, Fujishima H, et al. A heparin-sensitive phospholipase A_2 and prostaglandin endoperoxide synthase-2 are functionally linked in the delayed phase of prostaglandin D_2 generation in mouse bone marrow-derived mast cells. J Biol Chem 1996;271:25936–25944.

26. Murakami M, Kudo I, Fujimori Y, et al. Group II phospholipase A_2 inhibitors suppress lysophosphatidylserine-dependent degranulation of rat peritoneal mast cells. Biochem Biophys Res Commun 1991;181:714–721.

27. Ancian P, Lambeau G, Mattei M-G, et al. The human 180-kDa receptor for secretory phospholipases A$_2$, molecular cloning, identification of a secreted soluble form, expression, and chromosomal localization. J Biol Chem 1995;270:8963–8970.

28. Higashino K, Ishizaki J, Kishino J, et al. Structural comparison of phospholipase A$_2$-binding regions in phospholipase A$_2$ receptors from various mammals. Eur J Biochem 1994;225:375–382.

29. Barnette MS, Rush J, Marshall LA, et al. Effects of scalaradial, a novel inhibitor of 14-kDa phospholipase A$_2$, on human neutrophil function. Biochem Pharmacol 1994;47:1661–1668.

30. Murakami M, Kudo I, Suwa Y, et al. Release of 14-kDa group II phospholipase A$_2$ from activated mast cells and its possible role in the regulation of the degranulation process. Eur J Biochem 1992;209:257–265.

31. Chen J, Saho C, Lazar V, et al. Localization of group IIC low molecular weight phospholipase A$_2$ mRNA to meiotic cells in the mouse. J Cell Biochem 1997;64:369–375.

32. Balboa MA, Balsinde J, Winstead MV, et al. Novel group V phospholipase A$_2$ involved in arachidonic acid mobilization in murine P388D$_1$ macrophages. J Biol Chem 1996;271:32381–32384.

33. Clark JD, Lin LL, Kriz RW, et al. A novel arachidonic acid selective cytosolic PLA$_2$ contains a Ca^{2+}-dependent translocation domain with homology to PKC and GAP. Cell 1991;65:1043–1051.

34. Sharp JD, White DL, Chiou XG, et al. Molecular cloning and expression of human Ca^{2+}-sensitive cytosolic phospholipase A$_2$. J Biol Chem 1991;266:14850–14853.

35. Tay A, Maxwell P, Li Z, et al. Isolation of promoter for cytosolic phospholipase A$_2$ (cPLA$_2$). Biochim Biophys Acta 1994;1217:345–347.

36. Glover S, deCarvalho MS, Bayburt T, et al. Translocation of the 85-kDa phospholipase A$_2$ from the cytosol to the nuclear envelope in rat basophilic leukemia cells stimulated with calcium ionophore or IgE/antigen. J Biol Chem 1995;270:30749–30754.

37. Peters-Golden M, McNish RW. Redistribution of 5-lipoxygenase and cytosolic phospholipase A$_2$ to the nuclear fraction upon macrophage activation. Biochem Biophys Res Commun 1993;196:147–153.

38. Lin LL, Wartman M, Lin A, et al. cPLA$_2$ is phosphorylated and activated by MAP kinase. Cell 1993;72:269–278.

39. Reddy S, Herschman HR. Prostaglandin synthase-1 and prostaglandin synthase-2 are coupled to distinct phospholipases for ther generation of prostaglandin D$_2$ in activated mast cells. J Biol Chem 1997;272:3231–3237.

40. Roshak A, Sathe G, Marshall LA. Suppression of monocyte 85-kDa phospholipase A$_2$ by antisense and effects on endotoxin-induced prostaglandin biosynthesis. J Biol Chem 1994;269:25999–26005.

41. Balsinde J, Denis EA. Distinct roles in signal transduction for each of the phospholipase A$_2$ enzymes present in P388D$_1$ macrophages. J Biol Chem 1996;271:6758–6765.

42. Smith WL, Garavito RM, DeWitt DL. Prostaglandin endoperoxide H synthases (cyclooxygenases)-1 and -2. J Biol Chem 1996;271:33157–33160.

43. Smith WL, DeWitt DL. Prostaglandin endoperoxide H synthases-1 and -2. Adv Immunol 1996;62:167–215.

44. Murakami M, Matsumoto R, Urade Y, et al. c-*kit* ligand mediates increased expression of cytosolic phospholipase A$_2$, prostaglandin endoperoxide synthase-1, and hematopoietic prostaglandin D$_2$ synthase and increased IgE-dependent prostaglandin D$_2$ generation in immature mouse mast cells. J Biol Chem 1995;270:3239–3246.

45. Fletcher BS, Kujubu DA, Perrin DM, et al. Structure of the mitogen-inducible TIS10 gene and demonstration that the TIS10-encoded protein is a functional prostaglandin G/H synthase. J Biol Chem 1992;267:4338–4344.

46. Ristimaki A, Garfinkel S, Wessendorf J, et al. Induction of cyclooxygenase-2 by interleukin 1α: evidence for post-transcriptional regulation. J Biol Chem 1994;269:11769–11777.

47. Murakami M, Matsumoto R, Austen KF, et al. Prostaglandin endoperoxide synthase-1 and -2 couple to different transmembrane stimuli to generate prostaglandin D$_2$ in mouse bone marrow-derived mast cells. J Biol Chem 1994;269:22269–22275.

48. Bhattacharyya DK, Lecomte M, Dunn J, et al. Selective inhibition of prostaglandin endoperoxide synthase-1 (cyclooxygenase-1) by valerylsalicylic acid. Arch Biochem Biophys 1995;316:19–24.

49. Meade EA, Smith WL, DeWitt DL. Differential inhibition of prostaglandin endoperoxide synthase (cyclooxygenase) isozymes by aspirin and other non-steroidal anti-inflammatory drugs. J Biol Chem 1993;268:6610–6614.

50. Urade Y, Ujihara M, Horiguchi Y, et al. Mast cells contain spleen-type prostaglandin D synthetase. J Biol Chem 1990;265:371–375.

51. Goetzl EJ. Oxygenation products of arachidonic acid as mediators of hypersensitivity and inflammation. Med Clin North Am 1981;65:809–828.

52. Murray JJ, Tinnel AB, Brash AR, et al. Release of prostaglandin D_2 into human airways during acute antigen challenge. N Engl J Med 1986;315:800–889.

53. Knani J, Campbell A, Enander I, et al. Indirect evidence of nasal inflammation assessed by titration of inflammatory mediators and enumeration of cells in nasal secretions of patients with allergic rhinitis. J Allergy Clin Immunol 1992;90:880–889.

54. Roberts LJ, Sweetman BJ, Lewis RA, et al. Increased production of prostaglandin D_2 in patients with systemic mastocytosis. N Engl J Med 1980;180:1400–1404.

55. Hirata M, Kakizuka A, Aizawa M, et al. Molecular characterization of a mouse prostaglandin D receptor and functional expression of the cloned gene. Proc Natl Acad Sci U S A 1994;91:11192–11196.

56. Hirata M, Ushikubi F, Narumiya S. Prostaglandin I receptor and prostaglandin D receptor. J Lipid Mediat Cell Signal 1995;12:393–404.

57. Beasley RC, Featherstone RL, Church MK, et al. Effect of a thromboxane receptor antagonist on PGD_2 and allergen-induced bronchoconstriction. J Appl Physiol 1989;66:1685–1693.

58. Kliewer SA, Lenhard JM, Willson TM, et al. A prostaglandin J_2 metabolite binds peroxisome proliferator-activated receptor γ and promotes adipocyte differentiation. Cell 1995;83:813–819.

59. Forman BM, Tontonoz PT, Chen J, et al. 15-deoxy-12,14-prostaglandin J_2 is a ligand for the adipocyte determination factor PPARγ. Cell 1995;83:803–812.

60. Vickers PJ. 5-Lipoxygenase activating protein (FLAP). J Lipid Mediat Cell Signal 1995;12:185–194.

61. Maliviya R, Maliviya R, Jakshik BA. Reversible translocation of 5-Lipoxygenase in mast cells upon IgE/antigen stimulation. J Biol Chem 1993;268:4939–4944.

62. Woods JW, Evans JF, Ethier D, et al. 5-Lipoxygenase and 5-lipoxygenase activating protein are localized in the nuclear envelope of activated human leukocytes. J Exp Med 1993;178:1935–1946.

63. Lepley R, Muskardin DT, Fitzpatrick FA. Tyrosine kinase activity modulates catalysis and translocation of cellular 5-lipoxygenase. J Biol Chem 1996;271:6179–6184.

64. Lam BK, Penrose JF, Freeman GJ, et al. Expression cloning of a cDNA for human leukotriene C_4 synthase, an integral membrane protein conjugating reduced glutathione to leukotriene A_4. Proc Natl Acad Sci U S A 1992;91:7663–7667.

65. Penrose JF, Spector J, Baldasaro M, et al. Molecular cloning of the gene for human leukotriene C_4 synthase: organization, nucleotide sequence, and chromosomal localization to 5q35. J Biol Chem 1996;271:11356–11361.

66. Lam BK, Penrose JF, Xu K, et al. Site-directed mutagenesis of human leukotriene C_4 synthase. J Biol Chem 1997;272:13923–13928.

67. Penrose JF, Baldasaro MH, Webster MH, et al. Molecular cloning of the gene for mouse leukotriene C_4 synthase. Eur J Biochem 1997;248:807–813.

68. Lam BK, Xu K, Atkins MB, et al. Leukotriene C_4 uses a probenecid-sensitive export carrier that does not recognize leukotriene B_4. Proc Natl Acad U S A 1992;89:11598–11602.

69. Leier I, Kedlitschky G, Buccholz U, et al. The MRP gene encodes an ATP-dependent export pump for leukotriene C_4 and structurally related conjugates. J Biol Chem 1994;269:27807–27810.

70. Orning L, Kaijsen L, Hammarstrom S. In vivo metabolism of leukotriene C_4 in man: urinary excretion of leukotriene E_4. Biochem Biophys Res Commun 1985;130:214–220.

71. Austen KF. From slow reacting substance of anaphylaxis to leukotriene C_4 synthase. Int Arch Allergy Immunol 1995;107:19–24.

72. Arm JP, Lee TH. Sulphidopeptide leukotrienes in asthma. Clin Sci 1993;84:501–510.

73. Benyon RC, Robinson C, Church MK. Differential release of histamine and eicosanoids from human mast cells activated by IgE-dependent and non-immunologic stimuli. Br J Pharmacol 1989;97:898–904.

74. Massey WA, Guo CB, Dvorak AM, et al. Human uterine mast cells. Isolation, purification, characterization, ultrastructure, and pharmacology. J Immunol 1991;147:1621–1627.

75. Lawrence ID, Warner JA, Cohan VL, et al. Purification and characterization of human skin mast cells. Evidence for human mast cell heterogeneity. J Immunol 1987;139:3062–3069.

76. Stevens RL, Austen KF. Recent advances in the cellular and molecular biology of mast cells. Immunol Today 1989;10:381–386.

77. Murakami M, Austen KF, Arm JP. The immediate phase of c-*kit* ligand stimulation of mouse bone marrow-derived mast cells elicits rapid leukotriene C_4 generation through post-translational activation of cytosolic PLA_2 and 5-lipoxygenase. J Exp Med 1995;268:4939–4944.

78. Ogasawara T, Murakami M, Suzuki-Nishimura T, et al. Mouse bone marrow derived mast cells undergo exocytosis, prostanoid generation, and cytokine expression in response to G-protein-activating polybasic compounds after co-culture with fibroblasts in the presence of c-*kit* ligand. J Immunol 1997;158:393–404.

79. Kawata R, Reddy ST, Wolner B, et al. Prostaglandin synthase-1 and prostaglandin synthase-2 both participate in activation-induced prostaglandin D_2 production in mast cells. J Immunol 1995;155:818–825.

80. Reddy ST, Winstead MV, Tischfield JA et al. Analysis of the secretory phospholipase A_2 that mediates prostaglandin production in mast cells. J Biol Chem 1997;272:13591–13596.

81. Fonteh AN, Samet JM, Chilton FH. Regulation of arachidonic acid, eicosanoid, and phospholipase A_2 levels in murine mast cells by recombinant stem cell factor. J Clin Invest 1995;96:1432–1439.

82. Fonteh AN, Bass DA, Marshall LA, et al. Evidence that secretory phospholipase A_2 plays a role in arachidonic acid release and eicosanoid biosynthesis by mast cells. J Immunol 1994;152:5438–5446.

83. Murakami M, Penrose JF, Urade Y, et al. Interleukin 4 suppresses c-*kit* ligand-induced expression of cytosolic phospholipase A_2 and prostaglandin endoperoxide synthase-2 and their roles in separate pathways of eicosanoid synthesis in mouse bone marrow-derived mast cells. Proc Natl Acad Sci USA 1995;92:6107–6111.

84. Murakami M, Bingham CO III, Matsumoto R, et al. IgE-dependent activation of cytokine primed mouse cultured mast cells induces a delayed phase of prostaglandin D_2 generation via prostaglandin endoperoxide synthase-2. J Immunol 1995;155:4445–4453.

85. Reddy ST, Herschman HR. Prostaglandin synthase-1 and prostaglandin synthase-2 are coupled to distinct phospholipases for the generation of prostaglandin D_2 in activated mast cells. J Biol Chem 1997;272:3231–3237.

86. Gurish MF, Ghildyal N, McNeil HP, et al. Differential expression of secretory granule proteases in mouse mast cells exposed to interleukin 3 and c-*kit* ligand. J Exp Med 1992;175:1003–1012.

87. Columbo M, Horowitz EM, Botana LM, et al. The human recombinant c-*kit* receptor ligand, rhSCF, induces mediator release from human cutaneous mast cells and enhances IgE-dependent mediator release from both skin mast cells and peripheral basophils. J Immunol 1992; 149:599–608.

88. Murakami M, Austen KF, Bingham CO III, et al. Interleukin-3 regulates development of the 5-lipoxygenase/leukotriene C_4 synthase pathway in mouse mast cells. J Biol Chem 1995;270:22653–22656.

27

Role of I_{CRAC} in the Regulation of Secretion

Chris Mathes, Francisco Mendez, Andrea Fleig, and Reinhold Penner

Signal transduction events mediate and regulate secretion in specialized cells, including mast cells. Secretion occurs when vesicles or granules containing molecular transmitters fuse with a specific region of the plasma membrane and release their contents into the cleft near an adjacent cell or into the extracellular milieu (i.e., bloodstream). This process is mediated by various messengers, which transfer a signal from the plasma membrane to these intracellular vesicles that they should move or fuse. Because secretion is so important for cellular communication and function, this process is also highly regulated by signal transduction intermediates and pathways. This chapter focuses on one aspect of this regulation, namely, the role of calcium influx through calcium-release-activated calcium current (I_{CRAC}) channels in regulating the secretory event in mast cells.

The activation and regulation of secretion has been studied extensively in mast cells. There is growing evidence for mast-cell heterogeneity based on biochemical differences in granule contents as well as expression of different marker proteins (see Chapter 4). For the purpose of this review, however, we will concentrate on two general classes of mast cells which are central to this topic of secretion (1) mucosal mast cells and (2) mast cells from connective tissue (also known as serosal mast cells). Information obtained from both types of mast cells is discussed, so it is important to define the similarities and differences between these cell types. The best-described example of

mucosal mast cells is the rat basophilic leukemia (RBL) cell, of which the RBL-1 and RBL-2H3 immortalized cell lines are the most popular representatives. These cells are readily grown in appropriate tissue culture conditions and have been studied both biochemically and physiologically.[1] Antigen stimulates secretion in RBL cells; however, they are not responsive to compound 48/80, which is a potent secretagogue for serosal mast cells. Rat peritoneal mast cells (RPMCs) are the best-studied example of serosal mast cells. Like RBL cells, RPMCs are responsive to antigen, but also to a variety of nonantigenic stimuli. In this chapter, we use RBL cells to represent mucosal mast cells and likewise RPMCs to represent serosal mast cells.

Antigen stimulates the signal transduction cascade leading to secretion by cross-linking IgE receptors on the surface membrane of mast cells. IgE is provided by B cells from the immune system when they are stimulated by T cells to become antibody producers. The Fc region of the IgE antibody binds specifically to Fc receptor proteins on the mast cell surface. This receptor activation activates several signal transduction mechanisms,[2] including the phospholipase C (PLC) and sphingomyelin (SM) pathways.[3] Activation of the PLC pathway leads to inositol 1,4,5-trisphosphate ($InsP_3$) and diacylglycerol (DAG) formation. These important second messengers activate calcium release from internal stores and protein kinase C (PKC), respectively. Activation of the SM pathway in mast cells leads to elevations of sphingosine and sphingosine-1-phosphate.[4] These sphingolipids initiate cell growth, proliferation, or survival.[5]

The calcium dependence of secretion in mast cells differs from that of neuronal transmitter release. Secretion in neuronal cells is strictly Ca^{2+} dependent and is mediated by voltage-gated Ca^{2+} channels.[6] The contribution of voltage-independent calcium influx is possible, in principle, as this type of calcium current has been identified in neuronal cells.[7] In nonneuronal cells, however, the Ca^{2+} dependence of secretion varies among cell types. For example, in RBL cells, removing external Ca^{2+} inhibits antigen-activated secretion,[8] implying that Ca^{2+} influx is critical for this process. Calcium-dependent secretion in mast cells is regulated only by voltage-independent Ca^{2+} influx. Removing external Ca^{2+} also inhibits secretion stimulated by antigen in RPMCs.[9] However, these cells can also secrete in response to various nonantigenic stimuli when intracellular Ca^{2+} is buffered to resting levels, which is an example of Ca^{2+}-independent secretion in mast cells.

Another term for vesicle secretion in mast cells is degranulation. Granules contain primarily histamine, but also serotonin, β-N-acetylglycosaminidase, proteases, and proteoglycans.[1] These agents cause vasodilation, permeability of small blood vessels, and smooth muscle contractions, leading to sneezing, runny nose, or asthma. The symptoms caused by degranulation can range in severity from minimal to pathological. The pathophysiology of degranulation indicates the importance of understanding its regulation. Hence, adverse improper mast cell function can lead to conditions such as mastocytosis (leading to high serum histamine levels),[10] severe asthma,[11] and uncontrolled cell growth,[12] all of which can have serious medical consequences.

There are two primary technical approaches for measuring secretion in mast cells: biochemical and physiological. An example of the biochemical approach involves loading mast cells with radioactively labeled serotonin and measuring the levels of secreted serotonin following stimulation. This is limited to population averages and has been used extensively with RBL cells. These cells do not possess an electrophysiologically measurable signal indicative for secretory events because secretion occurs so slowly. On the other hand, such an electrophysiological approach for measuring degranulation is feasible in RPMCs employing simultaneous whole-cell patch-clamp recording and measurements of intracellular free Ca^{2+} concentration (Fig. 27.1; see color plate). Patch-clamp recording allows voltage control and measurement of ionic currents in single cells. With the appropriate equipment, it can also be used to simultaneously record whole-cell capacitance, which reflects cell-surface area. In RPMCs, many granules (~1000) fuse with the plasma membrane within a few seconds, causing such an increase in membrane surface area that a measurable change in whole-cell capacitance can be observed (~threefold). Simultaneously measuring intracellular Ca^{2+} levels with Ca^{2+}-sensitive dyes, such as fura-2, allows for determination of the relationship between secretion and intracellular Ca^{2+} in single mast cells.

Figure 27.1 shows an example from a single RPMC in which intracellular Ca^{2+} and membrane capacitance were measured simultaneously during the secretory event. Pseudocolor images from a representative patch-clamped mast cell are illustrated on the left (read from left to right and from top to bottom). Line graphs of membrane capacitance and intracellular Ca^{2+} concentration are shown on the right. In the first couple of images, only the patch pipette is visible. Then, after breakin, the cell becomes dialyzed and $[Ca^{2+}]_i$ is low (~100 nM). In this particular cell, there is a biphasic increase in $[Ca^{2+}]_i$ before degranulation occurs. Notice how the cell increases in diameter by the end of the experiment. This size increase can also be readily seen on the capacitance graph and in the inset on the right, which shows a normal image of a mast cell before and after the dramatic degranulation event. This type of experiment demonstrates the most direct method for measuring the calcium dependence of secretion in single mast cells.

Calcium influx provided by I_{CRAC} regulates secretion in mast cells, and this is the specific focus of this chapter. I_{CRAC} is a highly Ca^{2+}-selective, voltage-independent Ca^{2+} current present in both RBL and RPMCs. This special current is activated during PLC pathway stimulation indirectly by internal Ca^{2+} store depletion.[13,14] Thus, when the stores are empty, I_{RAC} turns on and refills them, although the exact events leading to refilling remain unknown. The regulation of this current is discussed because of its impact on secretion and because it involves cross-talk between the PLC and SM pathways. The perspective of this chapter is from physiological studies on secretion and I_{CRAC} performed in our laboratory with RBL and RPMCs. We describe stimulus–secretion coupling and give an overview of the Ca^{2+} dependence of secretion in mast cells. Finally, we discuss the regulation of I_{CRAC} in relationship to its role in secretion.

Stimulus–Secretion Coupling in Mast Cells

Following the addition of secretagogue to an RPMC, there is a short delay (seconds) before degranulation begins (see Fig. 27.1; see color plate). Once it begins, this dramatic process of granule fusion lasts several seconds and can be readily observed under normal light microscope magnification. The kinetics of this reaction can be measured as an increase in whole-cell capacitance during patch-clamp experiments.[15,16] The capacitance change directly measures the time course and magnitude of the secretory response in single mast cells. Figure 27.2 shows examples of useful agents for activating capacitance changes in RPMCs, externally applied compound 48/80 or internally applied GTP-γS. Other known physiological secretagogues are antigen, substance P, polymyxin B, somatostatin, and mast cell degranulating peptide.[17]

Degranulation by these secretagogues involves guanosine nucleotide binding proteins (G proteins).[18,19] Evidence for the necessity of G proteins came from biochemical studies employing permeabilized mast cells[20] and whole-cell experiments in which GTP was excluded from the pipette (intracellular) solution.[21] Washout of endogenous GTP over time prevents secretion activated by 48/80 in RPMCs. GTP regulates G proteins by directly binding to them and their influence is regulated by hydrolysis.

There are two main functional types of G proteins, both of which appear to be important for secretion in mast cells: heterotrimeric G proteins that connect receptor activation to the PLC pathway (functionally termed G_p) and heterotrimeric G proteins that seem to be involved directly in the exocytotic fusion event (functionally termed G_e). Biochemical evidence suggests that G_e is critical for secretion in RBL cells. Apparently, this G protein is capable of regulating secretion directly and independently from receptor and PLC pathway activation.[19] There is good evidence to suggest that the molecular correlates to G_p and G_e may be the G proteins G_z[22] and G_{i3},[23] respectively. It seems that these G proteins can activate secretion directly and regulate it indirectly by stimulation of PKC and the elevation of intracellular Ca^{2+}, as is discussed here. However, it is likely that PLC or SM pathway intermediates regulate the G_e protein.[24] It has also been suggested that receptor-mediated mast cell secretion involves the small proteins Rac and Rho, which mediate actin redistribution and may be important for granule movements.[25] It is unknown, however, what the relative importance of Rac and Rho and G_e are in regard to stimulus–secretion coupling in mast cells.

Second messengers can also modulate secretion in mast cells. For example, in RPMCs cyclic adenosine monophosphate (cAMP) in the pipette solution prevents secretion in response to 48/80.[18] Intracellular cAMP levels are increased following receptor stimulation of the G protein G_s, which is in turn activates adenylate cyclase. This represents another signal transduction pathway that can potentially regulate secretion in mast cells. Messengers from the SM pathway also could potentially modulate secretion because antigen stimulates this pathway.[4] Nerve growth factor also stimulates both the SM pathway and tyrosine kinase activity in mast cells.[26,27] This growth factor promotes survival in mast cells, although it is not known

Color Plate

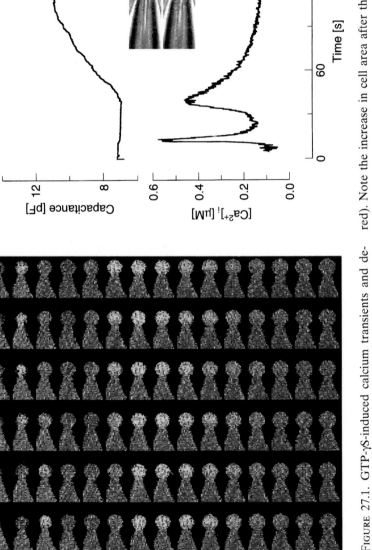

FIGURE 27.1. GTP-γS-induced calcium transients and degranulation in mast cells. A single mast cell was measured with combined whole-cell patch-clamp and calcium-imaging techniques. The pipette solution contained 10 μM GTP-γS to induce secretion. Calcium indicator dye fura-2 (200 μM) was monitored by dual-wavelength excitation (360 and 380 nm) at a frequency of 3 Hz. The *left panel* shows every third image acquired from the moment of breakin at a time interval of 1 s. Two transient increases in cytoplasmic free calcium concentration can be observed (changes in pseudocolor from blue over green and yellow to red). Note the increase in cell area after the second transient. Membrane capacitance and averaged cytoplasmic calcium concentration versus time after breakin are shown in the *right panel*. The standard pipette filling solution contained (mM): K-glutamate, 145; NaCl, 8; MgCl₂, 1; fura-2, 0.2; GTP-γS, 0.1. The extracellular solution consisted of (mM): NaCl, 140; KCl, 2.8; CaCl₂, 1; MgCl₂, 2; glucose, 11; HEPES-NaOH, 10. pH was adjusted with NaOH to 7.2. The pipette holding potential was 0 mV, and capacitance was measured using the fast automatic capacitance compensation of the EPC-9 patch-clamp amplifier.

FIGURE 27.2. Secretory and Ca^{2+} responses to compound 48/80 and GTP-γS stimulation in rat peritoneal mast cells. (A) A standard pipette solution containing GTP (300 µM) and 100 µM fura-2 was used, essentially corresponding to an "unbuffered" condition. Compound 48/80 (5 µg/ml) was applied at the time indicated.[42] (B) GTP-γS (100 µM) in the pipette solution stimulates Ca^{2+} release and an elevation of membrane capacitance (see also Fig. 27.1).[18] (C) Activation of PKC shifts the Ca^{2+} dependence of secretion. The cell was pretreated with PMA for 30 min. Intracellular Ca^{2+} was clamped to ~1 µM with a pipette solution containing Ca^{2+}-EGTA/EGTA at a ratio of 7:1 mM. With PMA treatment (PKC activation), degranulation occurs and the capacitance increases. Without PMA treatment this does not occur.[18]

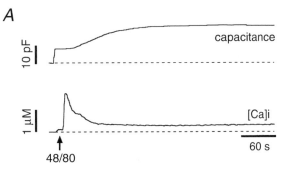

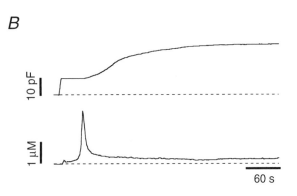

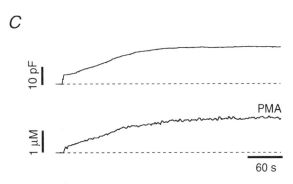

whether it regulates secretion in the process.[28,29] It will be of interest to see to what extent the SM pathway is involved in mast cell secretion.

Calcium Dependence of Secretion

Intracellular calcium is neither necessary nor sufficient for secretion in mast cells. When intracellular calcium is buffered to zero with EGTA in the patch-pipette solution in RPMCs, secretion in response to 48/80 is prevented,[18] but GTP-γS is still effective although degranulation is delayed and slowed.[30] This is presumably true in RBL cells as well, although it cannot be tested directly because these cells do not show an increase in capacitance to secretagogues in whole-cell experiments. Physiological intracellular Ca^{2+} levels (~1 µM) alone are insufficient to stimulate secretion in RPMCs.[18] When intracellular Ca^{2+} is buffered to about 1 µM, secretion does not occur spontaneously, although this amount of Ca^{2+} enhances or increases the rate of degranulation in response to GTP-γS or 48/80. Unphysiologically high levels of $[Ca^{2+}]_i$ (i.e., $10\mu M$) can induce secretion in RPMCs. This can be achieved with long applications of ionophore, such as ionomycin. Interestingly, too much $[Ca^{2+}]_i$ (i.e., $30\mu M$) is actually inhibitory.

Activation of PKC during stimulation of the PLC pathway can shift the Ca^{2+} dependence of secretion in RPMCs. This has been determined experimentally using the phorbol ester, phorbol 12-myristate 13-acetate (PMA), which activates PKC by binding directly to the DAG site on the regulatory subunit of PKC. PMA treatment (30 min at 37°C) allows secretion to occur at physiological $[Ca^{2+}]_i$ levels (i.e., ~1 μM), where it normally does not occur without the addition of a secretagogue. An experiment showing that PMA shifts the Ca^{2+} dependence of secretion is shown in Fig. 27.2c. PMA treatment, however, does not afford secretion by intracellular Ca^{2+} at resting levels (i.e., 200 nM). Therefore, agonists that activate PKC and elevate intracellular Ca^{2+} via the PLC pathway can more readily induce secretion in mast cells than secretagogues that do not stimulate both processes at the same time. The mechanism of PKC regulation of secretion remains at the moment unknown, although it has been suggested that PKC regulates the G protein G_e, which is critical for secretion.[24]

It is interesting that secretion can occur without Ca^{2+} release from internal stores in RPMCs. Evidence for this statement is that compound 48/80 elevates capacitance (reflecting secretion) in RPMCs buffered to resting $[Ca^{2+}]_i$ levels.[18] Moreover, PMA-treated cells can degranulate in response to 48/80 without a change in $[Ca^{2+}]_i$ (i.e., no Ca^{2+} spike). Therefore, despite the fact that most agonists or secretagogues, including antigen, stimulate both Ca^{2+} release and degranulation, the Ca^{2+} released into the cytoplasm from stores does not appear to play an important role in mast cell secretion.[6,18]

On the other hand, calcium entering from outside the cell is important for secretion. With intact RBL and RPMCs, removing external Ca^{2+} prevents secretion in response to antigen, demonstrating that calcium influx is essential.[8,9] Because degranulation can proceed in voltage-clamped RPMCs without a change in $[Ca^{2+}]_i$

levels, the fact that external Ca^{2+} is required for secretion could simply reflect that external elements (i.e., proteins, lipids) require Ca^{2+} and that Ca^{2+} influx is not necessary. Alternatively, there might exist a difference between the Ca^{2+} dependence of intact mast cells and voltage-clamped mast cells, which have been dialyzed by pipette solution.

Nevertheless, calcium influx through I_{CRAC} channels regulates the rate of secretion in RPMCs. Figure 27.3 shows representative experiments in which calcium influx was manipulated by changing the holding potential, instead of changing the external Ca^{2+} concentration. Secretion stimulated by GTP-γS is relatively slow when $[Ca^{2+}]_i$ is low, following the initial Ca^{2+} spike caused by Ca^{2+} release from internal stores (Fig. 27.3A). I_{CRAC} was largely reduced in this experiment by holding the membrane potential at $+20$ mV, where the driving force for Ca^{2+} is low and very little Ca^{2+} enters the cell. GTP-γS-stimulated secretion is considerably faster when Ca^{2+} is sustained at about $1\mu M$ (Fig. 27.3B). Here, the sustained Ca^{2+} elevation was achieved by holding the membrane potential at -40 mV, which is very permissive for I_{CRAC}. It has been demonstrated previously that degranulation in RPMCs proceeds faster when $[Ca^{2+}]_i$ is clamped to about $1\mu M$, as opposed to when it is clamped to resting levels.[30] These results demonstrate that Ca^{2+} influx plays a major role in regulating the rate of secretion in mast cells. Therefore, stimuli that modulate calcium influx in vivo would also modulate the kinetics of mast cell secretion, which may have important physiological consequences.

Mast cells express several other ion channel types that could potentially modulate secretion.[31] For example, RPMCs, but not RBL cells, express nonselective cation channels that are open and conductive during receptor stimulation. These cation channels are much less permeable to Ca^{2+} than I_{CRAC}. Moreover, in normal external solution (i.e., with 2 mM Ca^{2+}) currents through these cation channels are small and contribute very little to the plateau phase of the calcium signal. Therefore, while the exact physiological role for these cation channels is uncertain, it is clear that they play only a minor

A

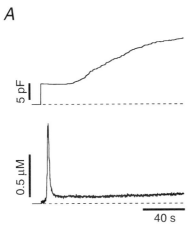

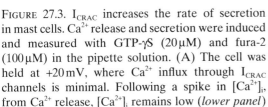

40 s

B

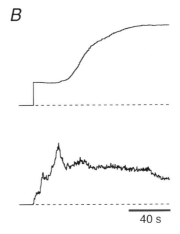

40 s

FIGURE 27.3. I$_{CRAC}$ increases the rate of secretion in mast cells. Ca^{2+} release and secretion were induced and measured with GTP-γS (20 μM) and fura-2 (100 μM) in the pipette solution. (A) The cell was held at +20 mV, where Ca^{2+} influx through I$_{CRAC}$ channels is minimal. Following a spike in [Ca^{2+}]$_i$, from Ca^{2+} release, [Ca^{2+}]$_i$ remains low (*lower panel*) and degranulation, shown as an increase in capacitance, proceeds slowly (*upper panel*). (B) When the holding potential is –40 mV, where Ca^{2+} influx through I$_{CRAC}$ channels is abundant, the Ca^{2+} signal is sustained (*lower panel*) and secretion occurs with a much faster time course (*upper panel*).

role for Ca^{2+} influx and their small contribution to Ca^{2+} influx might even be counterbalanced by the depolarization they effect when activated.

These experiments illustrate that Ca^{2+} influx pathways (most significantly, I$_{CRAC}$) are important for the regulation of the rate of exocytosis. In turn, the speed of degranulation determines the concentration of released mediators in the immediate vicinity of the secreting cell. A further important conclusion from these experiments is that Ca^{2+} influx itself is modulated by membrane voltage by setting the driving force for Ca^{2+} ions into the cell, where negative membrane potentials promote influx and positive voltages reduce entry. Therefore, other ion channels can enter the stage in regulating exocytosis indirectly by virtue of their impact on membrane potential (particularly, different types of Cl$^-$ and K$^+$ channels).

Both RPMCs and RBL cells express chloride channels, which may contribute indirectly to modulation of the Ca^{2+} dependence of secretion. When these channels are activated by secretagogues, Cl$^-$ ions enter mast cells and hyperpolarize the membrane potential. Hyperpolarization increases the driving force for Ca^{2+} entry through I$_{CRAC}$ channels and would therefore help to increase the rate of secretion, as described. The detailed mechanism for agonist activation of these chloride channels is unknown, although it has been shown that they can be activated by cAMP and high calcium in the absence of stimulus. Therefore, regulation of Cl$^-$ channels represents another potential mechanism by which mast cells can modulate the rate of secretion.

Another ionic current that could potentially regulate Ca^{2+} influx through I$_{CRAC}$ channels is the inward rectifier K$^+$ current in RBL cells. These channels are not found in RPMCs. At negative potentials (i.e., less than the K$^+$ equilibrium potential), these voltage-independent channels allow K$^+$ ions to flow from the outside to the inside of the cell. These positively charged ions depolarize the cell and force the potential back to its resting value. Therefore, this channel acts to clamp the membrane potential near the K$^+$ equilibrium value. This channel is constitutively active, but can be inhibited experimentally by the activation of pertussis toxin-sensitive G proteins. So far, agonists that modulate these channels are not known.

It is noteworthy to mention that on the basis of their different sets of expressed ion channels, RPMCs and RBL cells differ in their resting membrane potential: RPMCs appear to have no major resting conductance and are fairly depolarized at rest (close to 0 mV). On the other hand, RBL cells have a bistable membrane potential (either −80 mV or −40 mV) because of the peculiar current-voltage dependence of the the the inward rectifier K^+ current, which is the predominant resting conductance in this cell.[32] Other ion channels exist in mast cells,[31,33] but their contribution to secretion, if any, is difficult to determine. RPMCs have a small outward rectifying and voltage-gated K^+ current, which appears to be regulated by G proteins. Activation of this conductance would hyperpolarize, whereas a GTP-γS-activated Na^+ current that is also expressed in RBL cells[34] would depolarize the cells.

Regulation of I_{CRAC} and Consequences for Secretion

Regulation of I_{CRAC} represents a physiologically strategic point to regulate the rate of secretion because this current is the primary source of sustained calcium elevations in mast cells. The remainder of this chapter focuses on what is known about how I_{CRAC} is regulated and the consequences for this regulation on mast cell secretion.

I_{CRAC} is activated by an unknown mechanism following release of Ca^{2+} from internal stores. However, there is no linear correlation between store contents and degree of activation of I_{CRAC}. In fact, relatively low levels of $InsP_3$ can release a substantial fraction of stored Ca^{2+} without triggering activation of I_{CRAC}. Only after reaching a certain threshold level of $InsP_3$ (>1 μM) is I_{CRAC} activated and always complete.[34] Thus, I_{CRAC} activates in an all-or-none fashion and may therefore be considered the "action potential" of mast cells. This behavior of I_{CRAC} opens up several potential ways to regulate exocytosis by other signaling mechanisms that will affect the amount of $InsP_3$ production or the sensitivity of $InsP_3$ receptors to $InsP_3$. At the same time, such nonlinear activa-

tion requires a great deal of feedback control because Ca^{2+} influx must be fine-tuned to serve various cellular functions that have different Ca^{2+} requirements. Indeed, several modulatory mechanisms for I_{CRAC} exist and more are likely to be discovered in the future.

Calcium-dependent inactivation of I_{CRAC} has been described in mast cells and Jurkat-T cells.[14,35,36] Calcium entering through I_{CRAC} channels binds to the inner surface of the channel and somehow impedes the flow of calcium entry over time. Evidence for Ca^{2+}-dependent inactivation of I_{CRAC} came from experiments with BAPTA and EGTA, which are fast and slow Ca^{2+} buffers, respectively. Inactivation is hindered by the fast buffer BAPTA, whereas EGTA has less of an effect, because the fast buffer grabs Ca^{2+} ions as they pass through the channels and before they can bind and inactivate them. This process of inactivation helps to prevent overload of Ca^{2+} during long exposure to agonists that activate I_{CRAC}. With regard to secretion in mast cells, Ca^{2+}-dependent inactivation of I_{CRAC} would tend to decrease Ca^{2+} influx and hence reduce the ability of I_{CRAC} to speed up secretion during long time periods.

Small G proteins have been implicated in the regulation of I_{CRAC}, because GTP-γS in the pipette solution (100 μM; 2 min) inhibits the activation of ionomycin-induced I_{CRAC} in RBL cells.[37] Involvement of heterotrimeric G proteins was ruled out for the following reasons. (1) AlF_4^- ions, which activate trimeric G proteins, acted differently from GTP-γS; namely, AlF_4^- activated I_{CRAC}, presumably by activating G_p, stimulating IP_3 formation and depletion of internal Ca^{2+} stores. Moreover, unlike in mast cells (e.g., Fig. 27.3), GTP-γS does not seem to activate sufficient Ca^{2+} release, leading to store depletion and activation of I_{CRAC} in RBL cells.[37] (2) GDP-β-S, which prevents dissociation of trimeric G-protein subunits, has no effect on I_{CRAC}. (3) GTP-γS has no effect on I_{CRAC} once the current is activated, which would be expected for a heterotrimeric G-protein mechanism. (4) Finally, GTP antagonizes the GTP-γS effect, which is a hallmark feature for small G-protein function.

PKC regulates both I_{CRAC} and secretion. As mentioned, PKC shifts the Ca^{2+} dependence of secretion, allowing physiological levels of Ca^{2+} to trigger secretion in mast cells. PKC activation inactivates I_{CRAC},[38] as shown in Fig. 27.4A. Including ATP or ATP-γS in the pipette solution speeds inactivation of I_{CRAC}. This effect can be mimicked by PMA and prevented by the PKC inhibitors staurosporine and bisindolylmaleimide. Because agonists or secretagogues activate PKC via G_p in mast cells, inactivation of I_{CRAC} may have important consequences for secretion. For example, inactivation of CRAC might slow secretion rate, because calcium influx speeds up the rate of secretion. At the same time, less Ca^{2+} would be required to secrete when PKC is activated. Therefore, PKC activation appears to be an important regulator of secretion in mast cells, and part of its mechanism may be imparted by its inactivation of I_{CRAC}.

The energy status of mast cells also modulates I_{CRAC}. It has been shown in RBL cells that ADP inhibits I_{CRAC}, demonstrating that the ATP/ADP ratio, or cellular energy status, is critical for regulation of calcium influx.[39] This process is temperature- and calcium dependent. The possible consequences for secretion are likely to relate to how I_{CRAC} regulates secretion, because ATP itself is not necessary for the

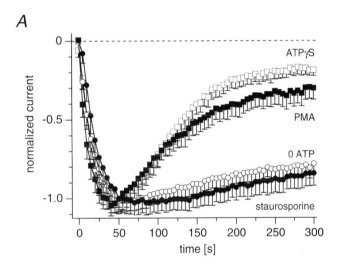

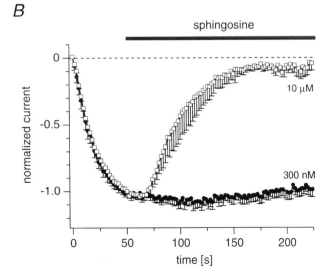

FIGURE 27.4. Examples of intermediates from the phospholipase C (PLC) and sphingomyelin (SM) pathways that regulate I_{CRAC} in RBL cells. (A) PKC mediates inhibition of I_{CRAC}. Activation of PKC with the phorbol ester PMA (100 nM) inactivates I_{CRAC} (with 2 mM ATP in the piptte solution). This inhibition can be mimicked by whole-cell breakin with 2 mM ATP-γS and prevented with the kinase inhibitors staurosporine (2 μM) and bisindolylmaleimide (500 nM; not shown).[38] (B) The SM pathway intermediate sphingosine (10 μM) inhibits I_{CRAC}. A lower concentration of 300 nM is not significantly different from controls (i.e., part A; labeled 0 ATP).[41]

final events of secretion.[40] By phosphorylation, ATP affects the affinity of proteins that bind Ca^{2+} and GTP, both of which are important for regulation of I_{CRAC} and/or secretion. Of course, ATP levels will affect PKC activity, which as mentioned, regulates I_{CRAC} and secretion. Finally, changes in the ADP/ATP ratio could modulate I_{CRAC} and therefore the secretion rate as well.

Activation of the SM pathway produces sphingosine and this sphingolipid also inhibits I_{CRAC} (Fig. 27.4B). Sphingosine inhibits CRAC directly and independently of S1P and ceramide.[41] Therefore, sphingosine could potentially reduce the rate of secretion by its block of I_{CRAC}. Antigen stimulates the SM pathway, but little is known about the regulation of secretion by SM pathway intermediates. As mentioned earlier, nerve growth factor (NGF) stimulates the SM pathway and inhibits I_{CRAC} via sphingosine elevation.[42] NGF is also involved in secretion, although it cannot induce secretion by itself. These data provide a preliminary link between the SM pathway and secretion that is worthy of further testing.

In conclusion, it appears that calcium influx through I_{CRAC} channels is important in mast cells. Indeed, Ca^{2+} influx is required for antigen-induced secretion in RBL and RPMCs. Even Ca^{2+}-independent secretion in RPMCs is influenced by I_{CRAC}, which increases the rate. The many levels of I_{CRAC} regulation afford "fine-tuned" regulation of secretion.

Acknowledgments. The authors thank Ann Hoffman for helpful discussions and comments. We acknowledge support by the Alexander von Humboldt Foundation (Bonn, Germany) to C. Mathes.

References

1. Beaven MA, Ludowyke R, Gonzaga HMS, et al. The orchestration of stimulatory signals for secretion in rat basophilic leukemia (RBL-2H3) cells. In: Vanderhoek JY, ed. Biology of Cellular Transducing Signals. New York: Plenum Press, 1990:313–322.
2. Beaven MA, Baumgartner RA. Downstream signals initiated in mast cells by Fc epsilon RI and other receptors. Curr Opin Immunol 1996; 8:766–772.
3. Beaven MA. Calcium signalling: sphingosine kinase versus phospholipase C? Curr Biol 1996; 6:798–801.
4. Choi OH, Kim JH, Kinet JP. Calcium mobilization via sphingosine kinase in signalling by the Fc epsilon RI antigen receptor. Nature (Lond) 1996;380:634–636.
5. Spiegel S, Milstien S. Sphingoid bases and phospholipase D activation. Chem Phys Lipids 1996; 80:27–36.
6. Penner R, Neher E. The role of calcium in stimulus-secretion coupling in excitable and non-excitable cells. J Exp Biol 1988;139:329–345.
7. Mathes C, Thompson SH. Calcium current activated by muscarinic receptors and thapsigargin in neuronal cells. J Gen Physiol 1994;104:107–121.
8. Fewtrell C, Kessler A, Metzger H. Comparative aspects of histamine secretion from tumor and normal mast cells. In: Weissman G, Samuelsson B, Paoletti R, eds. Advances in Inflammation Research. New York: Raven Press, 1979:205–221.
9. Pearce FL. Calcium and histamine secretion from mast cells. Prog Med Chem 1982;19:59–109.
10. Granerus G, Olafsson JH, Roupe G. Treatment of two mastocytosis patients with a histidine decarboxylase inhibitor. Agents Actions 1985; 16:244–248.
11. Assem ES. Inhibition of histamine release from basophil leucocytes of asthmatic patients treated with corticosteroids. Agents Actions 1985; 16:256–259.
12. Norrby K. Evidence of mast-cell histamine being mitogenic in intact tissue. Agents Actions 1985; 16:287–290.
13. Hoth M, Penner R. Depletion of intracellular calcium stores activates a calcium current in mast cells. Nature (Lond) 1992;355:353–356.
14. Hoth M, Penner R. Calcium release-activated calcium current in rat mast cells. J Physiol (Camb) 1993;465:359–386.
15. Fernandez JM, Neher E, Gomperts BD. Capacitance measurements reveal stepwise fusion events in degranulating mast cells. Nature (Lond) 1984;312:453–455.
16. Penner R, Neher E. The patch-clamp technique in the study of secretion. Trends Neurosci 1989; 12:159–163.
17. Lagunoff D, Martin TW, Read G. Agents that release histamine from mast cells. Annu Rev Pharmacol Toxicol 1983;23:331–351.

18. Penner R, Neher E. Stimulus-secretion coupling in mast cells. Soc Gen Physiol Ser 1989;44:295–310.

19. Aridor M, Traub LM, Sagi-Eisenberg R. Exocytosis in mast cells by basic secretagogues: evidence for direct activation of GTP-binding proteins. J Cell Biol 1990;111:909–917.

20. Lillie TH, Gomperts BD. Guanine nucleotide is essential and Ca^{2+} is a modulator in the exocytotic reaction of permeabilized rat mast cells. Biochem J 1992;288:181–187.

21. Penner R, Pusch M, Neher E. Washout phenomena in dialyzed mast cells allow discrimination of different steps in stimulus-secretion coupling. Biosci Rep 1987;7:313–321.

22. Hide M, Ali H, Price SR, et al. GTP-binding protein G alpha z: its down-regulation by dexamethasone and its credentials as a mediator of antigen-induced responses in RBL-2H3 cells. Mol Pharmacol 1991;40:473–479.

23. Aridor M, Rajmilevich G, Beaven MA, et al. Activation of exocytosis by the heterotrimeric G protein G$_{i3}$. Science 1993;262:1569–1572.

24. Buccione R, Di Tullio G, Caretta M, et al. Analysis of protein kinase C requirement for exocytosis in permeabilized rat basophilic leukemia RBL-2H3 cells: a GTP-binding protein(s) as a potential target for protein kinase C. Biochem J 1994;298:149–156.

25. Price LS, Norman JC, Ridley AJ, et al. The small GTPases Rac and Rho as regulators of secretion in mast cells. Curr Biol 1995;5:68–73.

26. Dobrowsky RT, Jenkins GM, Hannun YA. Neurotrophins induce sphingomyelin hydrolysis. Modulation by co-expression of p75NTR with Trk receptors. J Biol Chem 1995;270:22135–22142.

27. Blöchl A, Sirrenberg C. Neurotrophins stimulate the release of dopamine from rat mesencephalic neurons via Trk and p75Lntr receptors. J Biol Chem 1996;271:21100–21107.

28. Horigome K, Pryor JC, Bullock ED, et al. Mediator release from mast cells by nerve growth factor: neurotrophin specificity and receptor mediation. J Biol Chem 1993;268:14881–14887.

29. Horigome K, Bullock ED, Johnson EM Jr. Effects of nerve growth factor on rat peritoneal mast cells: survival promotion and immediate-

early gene induction. J Biol Chem 1994;269:2695–2702.

30. Neher E. The influence of intracellular calcium concentration on degranulation of dialysed mast cells from rat peritoneum. J Physiol (Camb) 1988;395:193–214.

31. Hoth M, Fasolato C, Penner R. Ion channels and calcium signaling in mast cells. Ann NY Acad Sci 1993;707:198–209.

32. Lindau M, Fernandez JM. A patch-clamp study of histamine-secreting cells. J Gen Physiol 1986;88:349–368.

33. Janiszewski J, Huizinga JD, Blennerhassett MG. Mast cell ionic channels: significance for stimulus-secretion coupling. Can J Physiol Pharmacol 1992;70:1–7.

34. Parekh AB, Fleig A, Penner R. The store-operated calcium current ICRAC: nonlinear activation by InsP3 and dissociation from calcium release. Cell 1997;89:973–980.

35. Zweifach A, Lewis RS. Rapid inactivation of depletion-activated calcium current (I$_{CRAC}$) due to local calcium feedback. J Gen Physiol 1995;105:209–226.

36. Zweifach A, Lewis RS. Slow calcium-dependent inactivation of depletion-activated calcium current. Store-dependent and -independent mechanisms. J Biol Chem 1995;270:14445–14451.

37. Fasolato C, Hoth M, Penner R. A GTP-dependent step in the activation mechanism of capacitative calcium influx. J Biol Chem 1993;268:20737–20740.

38. Parekh AB, Penner R. Depletion-activated calcium current is inhibited by protein kinase in RBL-2H3 cells. Proc Natl Acad Sci USA 1995;92:7907–7911.

39. Innocenti B, Pozzan T, Fasolato C. Intracellular ADP modulates the Ca^{2+} release-activated Ca^{2+} current in a temperature- and Ca^{2+}-dependent way. J Biol Chem 1996;271:8582–8587.

40. Gomperts BD, Churcher Y, Koffer A, et al. Intracellular mechanisms regulating exocytotic secretion in mast cells. Int Arch Allergy Appl Immunol 1991;94:38–46.

41. Mathes C, Fleig A, Penner R. Calcium-release-activated-calcium current (I$_{crac}$) is a direct target for sphingosine. J Biol Chem 1998;273:25020–25030.

42. Penner R, Neher E. Secretory responses of rat peritoneal mast cells to high intracellular calcium. FEBS Lett 1988;226:307–313.

28
Regulation of Secretion in Human Basophils

Donald MacGlashan, Jr.

In the long history of research on allergic disease, it has been appreciated for some time that the mast cell is involved in its genesis. Only more recently has the role of the human basophil been appreciated. The basophil has many similarities to the mast cell, possessing histamine-containing granules that are secreted when surface-bound IgE, bound to the same high-affinity receptor found on mast cells, is aggregated by antigens. However, while having these characteristics and sometimes serving as a surrogate to the mast cell because of these similarities, the basophil is a mobile leukocyte and its actual role in the progression of an allergic response has only recently been examined in some detail.[1-4] Now that mast cell activation is known to result in a transient inflammatory response that recruits eosinophils and basophils to the site of antigen exposure, the basophil response assumes a more direct relevance to understanding this disease. The basophil may have some characteristics that relate to mast cells, but as a leukocyte it has numerous differences. Indeed, it may be through the regulatory aspects of basophil secretion that we find the greatest differences between these two cell types. For convenience, regulation of secretion can be divided into those aspects of the biology that appear to be intrinsic to the cells themselves and those aspects which involve factors extrinsic to the basophil, such as cytokines. Naturally there can be an interrelationship between these two categories, such as the number of cytokine receptors on basophils determining cell responsiveness to this other-

wise extrinsic factor, but the distinction remains useful.

Over the past two decades, but primarily in the last decade, it has become apparent that the functional response of basophils is not restricted to the secretion of preformed mediators stored in the granules of the cell. In the early 1980s, when it became clear that many cells secrete metabolites of arachidonic acid, it was demonstrated that basophils secrete leukotriene C_4.[5] Limited searches for other arachidonic metabolites have not turned up any meaningful results.[6,7] Unlike mast cells,[8,9] there appear to be no identifiable prostaglandins secreted.[7] In the early 1990s, after it was discovered that murine mast cells could secrete a variety of cytokines, basophils were found to secrete IL-4.[10-12] Thus far, the list of cytokines is fairly restricted; IL-13[13] and macrophage inflammatory protein (MIP-1α)[14] appear to be secreted by basophils, but there is negative evidence for tumor necrosis factor-alpha (TNF-α), IL-8, interferon-gamma (IFN-γ), and IL-5[11] (and unpublished results).

Histamine secretion is generally expressed in terms of the percentage that is released of the basophil starting complement of stored hisatmine. This method is convenient because it relieves the investigator from matching basophil numbers between experiments or between donors; however, it does change the interpretation of the data. There is no similar method for expressing the data for the other mediators because they are not prestored. Therefore, the data are often expressed as the amount of mediator released per million basophils. It may be possible to find a loose upper bound for the secretion of newly synthesized mediators by using a stimulus shown to maximally activate essential second-messenger pathways, but this is not commonly done. For example, a combination of ionomycin and phorbol myristate acetake (PMA) will generally induce 100% histamine release and has been used to replace cell lysis.[15] It also induces significantly greater leukotriene release and may place some bounds on the amount of arachidonic acid generation possible.[7]

When considering regulation of basophil secretion, one must consider each of these classes of mediators. These mediators are distinguishable in two important ways. First, their secretion depends on very distinctive processes. Histamine release results from the degranulation of the basophil and as such is dependent on the biochemistry of granule fusion and extrusion. In recent studies, even this process appears to have subtleties. There are now two forms of histamine release; one involves the traditional fusion of granule membranes to the cell plasma membrane while the other appears to resemble the traditional constitutive process of small vesicle shuttling.[16] For example, in the early stages of f-met-leu-phe- (FMLP-) induced histamine release, the most obvious change is a large increase in the number of small vesicles that can be seen fusing with the plasma membrane.[17] These vesicles contain histamine and other higher molecular weight granule contents.[18–20] Thus regulation of histamine release may be dependent on the type of degranulation involved. FMLP-induced histamine release has characteristics of both piecemeal degranulation and anaphylactic degranulation while IgE-mediated degranulation has very few, if any, piecemeal characteristics.[21] Thus, the differences in pharmacological regulation of FMLP- and IgE-mediated histamine release may result from differences in the way control is exerted over the two types of degranulation.

Leukotriene (LTC) release requires the generation of arachidonic acid and the participation of a complex of enzymes that can metabolize the free arachidonic acid to LTC_4.

The details of where these reactions occur and which enzymes participate in the generation of the free arachidonic acid are currently under considerable debate and are explored in greater detail later. Whether there is any serious mixing of the biochemistries of degranulation and LTC_4 generation (following the initial steps in FcεRI aggregation and early tyrosine kinase activation) is not entirely clear. Several studies over the years have suggested some interdependency,[22,23] but often these studies are difficult to interpret clearly because of their use of nonselective pharmacological agents.

Cytokine generation, at least of the type exemplified by IL-4, involves another distinct group of steps. IgE-mediated IL-4 generation appears to require the transcription of new mRNA for IL-4,[11] although a formal proof of this assertion has not been attempted in human basophils (no nuclear run-on studies or mRNA stability studies have been done). Therefore, cytokine generation requires biochemistries that involve the regulation of mRNA transcription, protein synthesis, and protein export. This offers a level of control that is also distinct from that regulating histamine or leukotriene release. Indeed, it seems likely that weak correlation coefficients between the release of the various types of mediators[5,11,12] reflects the role of factors beyond the initial signaling events associated with FcεRI aggregation. The second characteristic that distinguishes the release of these three classes of mediators is the timing of their appearance. Although histamine and leukotriene release are temporally similar, occurring within minutes of stimulation, leukotriene release often lags behind histamine release and in some instances does not occur until histamine release is complete. The generation of cytokines occurs in a very different time frame; IL-4 reaches half-maximal secretion on average 2 h after IgE-mediated stimulation. IL-13[13] and MIP-α[14] secretion appear to take even longer, 12 to 24 h being typical. For the cytokines, the longer time frame probably reflects the need for mRNA transcription and protein synthesis, but the underlying causes for differences among the cytokines have not been studied.

On the basis of studies using other secretagogues, it is easy to see that the pathways leading to secretion of each of the mediator classes are distinct. While stimulation of basophils can be accomplished by a variety of secretagogues, the only constant among the secretagogues thus far is that they induce histamine release. The release of leukotrienes or cytokines is not a given. For example, C5a induces marked histamine release while inducing little or no leukotriene release[7,24,25] or IL-4 secretion.[10–12,26] FMLP, which has many similarities to C5a in terms of pharmacological regulation or signaling pathways, is one of the best secretagogues for leukotriene release[7,25,27] but is also a poor stimulus for IL-4 secretion.[10–12,26] IgE-mediated and histamine-releasing factor- (HRF-) mediated (which may require FcεRI aggregation at some level) secretion are the only secretagogues known to induce the secretion of all three classes of mediators. No stimulus has yet been found that induces leukotriene or IL-4 release in the absence of histamine release, although it is clear that IL-4 secretion can occur with concentrations of anti-IgE antibody that are suboptimal for histamine release. As noted, these differentiating features for the different classes of mediators ultimately require an examination of how the regulatory mechanisms interact with each distinct process, significantly complicating basophil studies.

Intrinsic Regulation of Secretion

Releasability

It is difficult to state with assurance that there are aspects to basophil secretion that could be considered totally intrinsic to the cell itself. Too much of basophil maturation is tied up in the interplay of various cytokines during development to know whether the basophils that are studied from peripheral blood are expressing a phenotype which is only connected to the individual genotype for basophils or to hidden extrinsic factors. However, for the time being, some aspects of the basophil response can be considered intrinsic and one of the most frequently studied, but still unexplained characteristics, has been termed releasability.[28] The appellation comes from studies indicating that the maximum histamine release which can be obtained from a donor's basophils seems to have little relationship to the density of FcεRI and endogenously bound IgE antibody (beyond some minimal level of expression). For example, although two donors may share the same density of cell-surface IgE, an examination of a full dose–response curve using anti-IgE antibody to stimulate the cells reveals that one donor's maximal response may be 100% histamine release while the second donor's basophil response shows only 10% histamine release.[29] The cells can be shown to release histamine equally if other stimuli are used so there seems to be no defect in the ability to degranulate. Whether or not the genesis for these differences is intrinsic or extrinsic remains to be elucidated.

This distinction between the degree of histamine release appears to extend to the other classes of mediators, although as noted the correlation between secretion of the various mediators is sufficiently poor that large differences in histamine release between two donors might not be as obvious when leukotriene release is examined. Differences between levels of IL-4 secretion appear to track more closely to histamine release than leukotriene release. These distinctions between the absolute amount of response for each of the mediators might be expected because the mechanisms for their generation differ so greatly. Therefore, it may be more useful to ask whether there are differences in sensitivity. The distinction being made is similar to that of potency versus efficacy for a drug and its receptor; potency could differ significantly while efficacy remained the same. For IgE-mediated stimulation, potency translates to the number of antigen-specific IgE molecules required to obtain half the basophil's maximum response.[5,30] Or, it could be placed in a more absolute domain by defining it as the number of antigen-specific IgE molecules required for a fixed amount of mediator release. Recent studies have shown that the two measures of response, maximal release and sensitivity, are independent parameters of the basophil response.[30] Like maximal release, donors differ in their sensitivity parameter.[30] However, this is a more difficult parameter to assess and it has not been applied to the release of IL-4. Preliminary studies of leukotriene release in this context indicate that the range of sensitivity among donors is even greater than for histamine release.[5]

Among donors studied for differences in maximal IgE-mediated histamine release, some do not release at all. This type of donor, termed nonreleasers, represents 10% to 20% of the population even if a very strict definition of nonrelease is used: no IgE-mediated histamine release above a signal-to-noise ratio of two for spontaneous release.[31] Whether these donors represent one end of the release continuum or a distinctive phenotype is not known. The basophils of these donors possess normal densities of FcεRI and cell-bound IgE, and the anti-IgE antibody used to probe release is clearly bound to the cell-surface IgE. Furthermore, there are lines of evidence that some signaling occurs; treating the nonreleasing basophil with anti-IgE, while not directly inducing release, alters the response of the basophils to two other non-IgE-dependent stimuli, FMLP and PMA.[31,32] As for the studies of releasability, other non-IgE-dependent stimuli function well on this type of donor as do nonphysiological stimuli such as ionomycin or PMA. It has been difficult to study the basophils of these donors for signal transduction differences because of the general difficulties of obtaining enough cells at high enough purities for this kind of study. However, the data to date suggest that aggregation of FcεRI results in weak cytosolic calcium ($[Ca^{2+}]_i$) signals.[32,33] This result implies that the mechanisms leading to the $[Ca^{2+}]_i$ response are deficient. Obvious candidates are the β-subunit of FcεRI or the associated tyrosine kinases, such as lyn and syk. In humans, FcεRI can be expressed in the absence of the β-subunit,[34] and recent studies in transfected cells have noted that the β-subunit acts as an amplifier of signaling: the αγ-subunits alone may signal but quite poorly.[35] Thus, in the absence of the β-subunit, an individual's basophils could express normal densities of cell-surface IgE and be cross-linked to similar levels of aggregates, and yet the signal generated would be quite

weak. The amount of β-subunit protein in nonreleasers has not been compared to releasers, but the amount of mRNA for the β-subunit has been compared and not found to be deficient.

Most recently, studies have begun to study the participation of lyn and syk kinases; thus, it should be possible to determine if there are deficiencies in these components in nonreleasing basophils. Certainly, the current paradigm of signal transduction for FcεRI would predict that, if these components were not expressed well or were deficient in some way, all IgE-mediated functions would be subnormal. This is true for all three classes of mediators in nonreleasers, so it seems likely that when differences are found, they will include differences in early signal transduction events. However, as noted, some IgE-mediated signal appears to be generated in these nonreleasers. Anti-IgE antibody can induce a sensitivity shift to PMA stimulation.[36,37] This sensitivity shift occurs equally in releaser and nonreleaser basophils[31] suggesting that at least one aspect of the signal transduction cascade is active. The components considered here included only those involved in activation biochemistries. It is plausible that components involved in downregulation, so-called desensitization, could account for less mediator release if they were overactive. This mechanism might explain the apparent similarity in the induced sensitivity shift to PMA between the two types of donor basophils while also resulting in poorer mediator release. This topic is considered next.

Desensitization

It has been noted since the early 1970s that basophils exhibited a process termed desensitization,[38] as revealed in basophil experiments by treating the cells with antigen in the absence of extracellular calcium. Because the extracellular calcium is required for all known mediator release, the cells do not secrete during this period of the experiment. When calcium is returned to the buffers, the cells begin secretion. However, the magnitude of the response is diminished and the extent of its inhibition is a function of the time the cells were treated in the absence of extracellular calcium. To completely desensitize the basophil response typically requires 45 to 90 min. During desensitization, there is no loss of cell surface IgE or its ability to bind antigen; that is, this is not downregulation by receptor shedding or endocytosis.[39] This process of desensitization is believed to occur during active secretion, being responsible for limiting the magnitude of the response. Indeed, it can be shown that desensitization is inhibited by a very narrow range of diisopropyl fluorophosphate (DFP) concentrations, and the same concentrations enhance histamine release in a manner consistent with desensitization processes regulating ongoing release.[40] These DFP studies also suggested that desensitization is an active process because it could be inhibited with a pharmacological agent. As the release of new mediators has been discovered, the characteristics of desensitization for these mediators has required examination. Furthermore, as each new signal transduction event is explored, the effects of desensitization also need exploration.

For signal transduction studies, there is the additional question of whether the event occurs in the absence of extracellular calcium, the conditions used to desensitize the basophil. The results of these newer studies have not yet shed light on the mechanism of desensitization, although the "behavioral characteristics" of this phenomenon are well described and place constraints on the types of mechanisms that can be postulated to explain its characteristics. Some of the motivation for its study comes from data showing an inverse correlation between the rate of desensitization and the maximal level of histamine release that can be obtained in a given donor. In other words, the faster the desensitization process, the lower the maximum histamine release observed in a normal challenge. Thus, desensitization processes could serve to create the releasability phenomenon described earlier, with differences in maximum release among donors resulting from differences in desensitization rates.

Most studies of desensitization have been carried out in the context of histamine release, but there are some differences if other mediators are used as the endpoint for these studies.

In a desensitization-style experiment, using anti-IgE antibody at the stimulus, histamine release decays in a exponential manner. However, the time course for the desensitization of LTC_4 release follows a curve in which the first minutes of "desensitization" actually result in enhanced LTC_4 release. Only after 10 to 15 min is the amount of LTC_4 secretion reduced by prior treatment of the cells with the stimulus, and by 20 to 30 min both histamine and LTC_4 release are similarly inhibited by the pretreatment with anti-IgE antibody. Thus, at the 5-min time point (5 min with anti-IgE antibody in the absence of extracellular calcium followed by 45 min of release in the presence of calcium), histamine release was inhibited while leukotriene release was enhanced. More recently, we have found that the desensitization of the $[Ca^{2+}]_i$ response follows a time course similar to leukotriene release. The average elevation in $[Ca^{2+}]_i$ was higher than nondesensitized controls when the basophils were first treated with anti-IgE in the absence of extracellular calcium for 5 min. Following 30 min of desensitization, however, there was nearly complete inhibition of the $[Ca^{2+}]_i$ response. Once again, at the point that the $[Ca^{2+}]_i$ response was enhanced (sometimes twofold), histamine release was modestly inhibited. These studies indicate the $[Ca^{2+}]_i$ response is not the sole determinant of histamine release but that it may be a strong determinant of leukotriene.

This same conclusion has been reached from the results of other experimental protocols unrelated to desensitization. The cause of the enhanced $[Ca^{2+}]_i$ response following the first minutes of desensitization is not known. Stimulation with anti-IgE antibody in the absence of extracellular calcium leads to a transient increase in cytosolic calcium that is likely to result from the release of internal stores. Without extracellular calcium to replenish these stores, the response rapidly decays to resting levels. These data indicate that the mechanisms leading to release of internal stores of calcium are intact in the absence of extracellular calcium, and one prediction is that activation of lyn and syk would be unabated by the absence of extracellular calcium, a conclusion reached in RBL cell studies.[41-45] The desensitization data also indicate that there is a process which ultimately serves to inhibit the generation of a $[Ca^{2+}]_i$ response. IL-4 secretion is markedly inhibited by 30 min of treatment in the absence of extracellular calcium but the desensitization rate is generally slower than for histamine release. Furthermore, there are differences in the relative desensitization rates for histamine and IL-4 secretion among donors; that is, there is significant discordance in the data for the two rates. For one donor, the desensitization rate for histamine release may be much faster than for IL-4 secretion and for a second donor the rates may be similar. These results suggest that there may be independent regulation of the two types of mediators. From one perspective, such a result would not be surprising because the process for secretion of the two mediators differs, but as desensitization data indicate modification of early signal transduction events, one might expect tighter concordance in the desensitization rates for all three mediators.

Although there is some difference in the rates of IL-4 and histamine release desensitization, in general the rates for desensitization are more similar than would be expected from a comparison of the secretion rates. If desensitization is a process that is ongoing during secretion and it serves to downregulate early signal transduction events, then all secretory events should end in approximately the same time frame. The fact that they do not suggests the presence of calcium or a sustained $[Ca^{2+}]_i$ elevation changes the behavior of the system at large. However, because cross-links are required throughout the process of secretion for each of the mediators,[46] including IL-4, the fact that histamine and leukotriene release end long before IL-4 secretion begins indicates that signaling is still active and yet some process has resulted in the selective cessation of histamine and leukotriene release. An additional hint that this selective process exists can be found in the desensitization experiments described here in which the calcium signal was enhanced while histamine release was significantly inhibited. These data suggest that it is this noncalcium-dependent event that is shut off

during desensitization for histamine release. To speculate further, the same event may be regulating leukotriene release but because this mediator is especially sensitive to calcium elevations (the 5-LO being directly sensitive to calcium), the enhanced cytosolic calcium response in the first 5 min of the desensitization experiment overcomes any other downregulatory process, temporarily enhancing leukotriene release.

Because desensitization ultimately results in inhibition of the $[Ca^{2+}]_i$ response, it is useful to look for signal transduction events that have been shown to downregulate $[Ca^{2+}]_i$ responses. One such possibility was suggested about 10 years ago.[47] Phorbol esters are known to activate protein kinase C (PKC), and phorbol esters were also shown to inhibit IgE-mediated elevations in cytosolic calcium.[33,47] Thus was born the idea that PKC might participate in human basophil desensitization. It has been shown to be involved in deactivation in RBL cells,[48] so it seems likely that a similar process might occur in human basophils. Because PKC may also be involved in activation biochemistries,[48-54] it is instructive to examine the state of knowledge concerning this ubiquitous enzyme.

Protein Kinase C

In studies of RBL cells, PKC has been shown to have both activation and deactivation roles.[48,49,55] There are many critical details to work out, and the problem is made more difficult by the multiplicity of isozymes present in most cell types. In RBL cells, PKC-α and PKC-ε are thought to play a downregulatory role while PKC-β and PKC-δ appear to participate in the activation cascade. A full characterization of the PKC isozymes present in human basophils is not yet complete, but recent studies indicate that they express PKC-βI, -βII, -δ, -ε, and -τ. There is no detectable PKC-γ, -η, or -ζ, and interestingly, PKC-α is either not present or present at very low levels (compared to lymphocytes or monocytes).[56] This apparent absence of PKC-α seems to occur in eosinophils and neutrophils,[57-59] but its significance to the cell biology is not clear. In this context, however, it is interesting to speculate that it might explain the relative sensitivity of basophils to direct activation of PKC with PMA. PMA alone does not readily induce secretion in RBL cells, while nearly 100% histamine release can be obtained in human basophils.[60] In RBL cells, the ratio of PKC-α to PKC-δ mass is approximately 1.0[49] while in the human basophil it is less than 0.02. If PKC-α is a downregulator, its absence relative to PKC-δ might lead to a situation in which a general activator of PKC such as PMA would result in an imbalance that leads to secretion. Until the relative roles of the various isozymes of PKC in basophils are delineated, the meaning of the PKC-α deficiency will remain unclear.

In early PKC studies, PKC was identified by its activity and the requirements for this activity.[61,62] The isozymes now known as the Ca-dependent PKCs required Ca^{2+}, phosphatidylserine, and diacylglycerol for activation, and so an ex situ phosphorylating activity with these requirements could be described as a PKC. The substrate for these measurements was histone III protein, not known to be an in situ physiological substrate for PKC. One hallmark of a canonical PKC study was the translocation of PKC-like activity from the cytosol to the membrane. In the first PKC studies in human basophils, a PKC-like activity was detected in the cytosol. When the cells were stimulated with FMLP, the activity translocated to the membrane; when stimulated with anti-IgE antibody, however, translocation was not apparent although a PKC-like activity increased in the membrane.[33,63] These studies suggested that PKC was activated during stimulation through FcϵRI but had different behavioral characteristics. Among different basophil preparations, there was a positive correlation for the peak of this activity during stimulation and subsequent histamine release, suggesting a proactivation role for the measured activity.

In more recent years, many investigations of PKC have focused on the location of the mass of a specific PKC isozyme before and after stimulation. Thus translocation has been cautiously equated with activation although this need not be true. Indeed, some PKC isozymes

have a high degree of association with the plasma membrane before stimulation, and receptor-mediated activation does not necessarily induce significant additional translocation (e.g., PKC-δ in RBL cells[54]). Because many factors regulate PKC activity, it is certainly possible that an isozyme already present in the membrane could become active when additional cofactors become available during activation. In human basophils, PKC-βI and -βII are not measurable in the plasma membrane in resting cells while both PKC-δ and PKC-ϵ are divided evenly between the two compartments in resting cells. Stimulation with PMA leads to marked translocation of PKC-βI and -βII, statistically significant translocation of PKC-ϵ, and little change in the location of PKC-δ. Stimulation with FMLP leads to translocation of PKC-βI and -βII and no statistically meaningful changes in PKC-δ or -ϵ. The translocation of βI and βII are not statistically different from controls until 30 s after stimulation, at a time when histamine release is nearly complete. Stimulation with anti-IgE antibody has not resulted in any observable changes of PKC-βI, -βII, -δ, or -ϵ. The results with anti-IgE antibody are surprising, but it is important to place these results in context. Stimulation with suboptimal concentrations of PMA and ionomycin lead to moderate histamine release. Under these conditions, where PMA alone induces little release, the translocation of PKC isozymes is not discernible. Therefore, the PKC translocation assay is relatively insensitive. Assuming a continuum of translocation, small changes that are adequate for histamine release are not measurable.

To further examine the role of PKC, some studies have examined the effects of various PKC inhibitors on receptor-mediated stimulation. As with all pharmacological studies, there are concerns about the selectivity of the agents. This is a particular concern for PKC inhibitors because they often mimic ATP and therefore can inhibit a wide variety of kinases, particularly tyrosine kinases. For example, staurosporin was once used as a PKC inhibitor but once it was determined that tyrosine kinases were required for early signal transduction and that staurosporin was a potent inhibitor of these kinases, its usefulness in this context was lost. Recent studies have indicated that the only compounds which seem to selectively inhibit PKC isozymes are the bisindolylmaleimides (Bis I, Bis II, or Ro 31 8220; Go 6976 and calphostin C appear to have other actions). When Ro 31-8820, Bis I or Bis II are used to inhibit PMA-induced histamine release, their potency agrees with expected values. However, there is no inhibition of IgE-mediated histamine release. This unexpected result suggests that the PKC activation, if it occurs, has no meaningful impact on degranulation in the context of an IgE-mediated response. Additional studies demonstrated that these inhibitors also did not alter the rate of desensitization, again a surprising result given the considerations noted earlier.

Most studies are done using histamine release as a marker of function. In the context of LTC_4 or IL-4 release, the results differ. PMA alone is not a stimulus for either of these two mediators, and in combination with a physiological secretagogue the results are complicated by the fact that PMA markedly suppresses the $[Ca^{2+}]_i$ response. This is particularly true for IgE-mediated release. Because $[Ca^{2+}]_i$ elevations appear mandatory for LTC_4 release and are probably also mandatory for IL-4 release, other approaches are necessary. Ionomycin alone is a good stimulus for these two mediators, and PMA has no effect on the $[Ca^{2+}]_i$ response that occurs as a consequence of transporting extracellular calcium through a membrane ionophore. However, PMA does alter the secretion of LTC_4 or IL-4. For LTC_4, PMA is modestly synergistic (unpublished results) while for IL-4 secretion, PMA causes at least 50% inhibition.[103] Recent studies on both the pathways used for secretion of these two mediators shed some light on how these results may occur.

LTC_4 secretion occurs when arachidonic acid (AA) is liberated by phospholipases. There is considerable disagreement over which phospholipases are responsible for its generation, and the specifics may depend on the particular cell type. In human basophils, there is strong evidence that supports AA generation resulting from the action of a secretory phospholipase A (PLA_2), originally thought to be type II $sPLA_2$, but now more likely to be type V $sPLA_2$, a very close relative.[104] Despite pharmacological evidence for the involvement of $sPLA_2$, there is other circumstantial evidence for the participation of $cPLA_2$. Part of this circumstantial evidence relates to temporal relationship between $cPLA_2$ phosphorylation, free AA generation, and LTC_4 release. The initial phase of LTC_4 release is slow relative to degranulation, and this is especially obvious when a fast stimulus such as FMLP is used. The LTC_4 release may be slow because the appearance of free AA is also slow. $cPLA_2$ appears to require two steps for proper activation, phosphorylation and translocation induced by a cytosolic calcium response.[64–66] For basophils stimulated with C5a, there is little or no LTC_4 release despite substantial histamine release, and phosphorylation of $cPLA_2$ is also slow. Interestingly, the $[Ca^{2+}]_i$ response following C5a is transient, returning to resting levels by 30 s, just about the time that histamine release stops and $cPLA_2$ phosphorylation begins to become apparent.[67] Although the $cPLA_2$ is eventually phosphorylated by this stimulus, the activation is not consummated by a coincident elevation in the $[Ca^{2+}]_i$ level; thus, no AA generation follows C5a stimulation. For FMLP, the $[Ca^{2+}]_i$ response is sustained, so the late phosphorylation of $cPLA_2$ is not a constraint to free AA generation by this enzyme. $cPLA_2$ can be phosphorylated by PMA, presumably through the activation of a PKC-dependent pathway, which may explain its ability to enhance ionomycin-induced LTC_4 secretion. This makes a very tidy explanation, but it implicates $cPLA_2$ in the AA generation responsible for LTC_4 release, and there is sufficient evidence for a role for $sPLA_2$ to make a strong conclusion impossible at this time.

IL-4 secretion from lymphocytes is minimal when the stimulus is ionomycin alone but is markedly enhanced when ionomycin is combined with PMA. Thus, it was surprising to find that PMA inhibited IL-4 secretion induced by ionomycin from human basophils.[103] In these experiments, PMA synergistically induces more histamine release while at the same time inhibiting IL-4 release. The addition of a selective PKC inhibitor reverses the enhancement of histamine release and the inhibition of IL-4 secretion. The mechanism underlying the generation of IL-4 is not yet known for basophils, so the reason that PKC activation downregulates this response is also unknown. Studies of the steady-state level of mRNA for IL-4 have been contradictory. Northern blot analysis indicates that PMA inhibits the increase in mRNA that follows ionomycin alone, while semiquantitative RT-PCR indicates no change in the increase of IL-4 mRNA when PMA is included in the reaction. More recent studies using a competitive RT-PCR assay support the dilutional PCR data. Recent studies of the IL-4 promoter in Jurkat cells noted that NF-κB acted as a competitive inhibitor of NFAT.[68] The general scheme for IL-4 promoter activation appears to include a requirement for NFAT so that competitive binding by NF-κB to the same promoter site used by NFAT would serve to downregulate transcription and possibly in only a modest way. Whatever the mechanism, the data suggested that inhibition of PKC activity during IgE-mediated secretion of IL-4 might enhance its release. This was found to be the case; the bisindoylmaleimides enhance IL-4 release induced by anti-IgE antibody. This is the only known effect of PKC inhibitors during IgE-mediated stimulation of basophils, and it indicates that there is indeed activation of PKC during this form of stimulation. Taken together with the studies noted previously, but excluding the possibility that an unmeasured PKC isozyme is active, it appears that aggregation of FcεRI results in a subtle level of PKC activation not detected by translocation studies, possibly detected by PKC activity studies (if the early studies of PKC activity by the artificial ex situ histone phosphorylation assay were actually detecting PKC activity), but not involved in either activation or desensitization events related to histamine release. Instead, the activity may affect the level of IL-4 secretion and it acts in a downregulatory capacity.

Regulation by Extrinsic Factors

Cytokines

Basophils are responsive to a variety of related or unrelated cytokines. Thus far, most cytokines serve to upregulate the basophil response. Cytokines known to alter basophil function and that share a common β-subunit include IL-3, GM-CSF, and IL-5.[69-73] The effects of IL-5 and GM-CSF are quite variable, while IL-3 is generally a very effective upregulator. IFN-γ,[73] IL-1,[74] and HRF also enhance histamine release a modest amount. There are some indications that IL-8 may inhibit IL-4-induced release with no effect on histamine release. As noted, the most active of these cytokines is IL-3. There is a large body of literature concerning the mechanisms by which IL-3 alters cell growth, and recent studies have implicated the JAK/STAT pathway.[75,76] The JAK tyrosine kinases associate with the β-subunit of these receptors, which induces a pathway that ultimately results in activation of STAT transcriptional factors. The enhancement induced by IL-3 in basophils has some interesting properties not at first appreciated from work on cell proliferation. IL-3 can enhance the response to all known secretagogues and results in the enhancement of all known mediators,[69-73] as well as other functions, such as adhesion.[77] This suggests that the mechanism of enhancement is generally required for function (or that many independent functions are altered).

In addition, there are three phases to the upregulation. The first phase occurs within minutes, the second phase requires hours (generally, overnight incubations are sufficient), and the third phase occurs in days, usually more than 2 days. The difference in the phases is both quantitative and qualitative. The second phase is marked by a slightly greater enhancement of histamine release (than occurs after short incubations) but marked additional enhancement of LTC_4 or IL-4 release.[11,12] Generally, LTC_4 release is enhanced approximately 2 fold after 5 min of treatment (and then holds constant for 1–4 hours) and 10 fold after 18 h. The results for IL-4 are similar although the enhancement of

IL-4 release is somewhat less after short treatments (with no enhancement being common). Longer treatments alter the basic phenotype of the basophil. As noted, some donor basophils do not respond to stimulation through FcεRI. Incubation with IL-3 for less than 48 h does not change this state, but by 3 to 4 days, these cells begin responding well to IgE-mediated stimulation.[78]

The changes induced by IL-3 are only beginning to be defined. Given the ability of this cytokine to alter the response to all secretagogues and enhance all mediator release, it seems likely that an element in the signal transduction steps common to all secretagogues is upregulated. A simple hypothesis was that ATP levels were increased by IL-3 treatment because ATP is required for many elements of active secretion. However, short preincubations with IL-3 had no effect on ATP levels in basophils.[27] Another common element for secretagogues is the elevation of $[Ca^{2+}]_i$. Here too, short pretreatments with IL-3 had no effect on the $[Ca^{2+}]_i$ response.[27] Indeed, PMA induces histamine release in the absence of a $[Ca^{2+}]_i$ response and IL-3 enhances PMA-induced release. However, overnight incubations with IL-3 do dramatically enhance the $[Ca^{2+}]_i$ response for all physiological stimuli thus far tested. This may explain the additional marked enhancement of LTC_4 and IL-4 release seen in these longer cultures (as noted earlier). However, the short-term enhancement of LTC_4 release following stimulation with C5a (where LTC_4 release shifts from none, without IL-3 treatment, to levels 50%–70% of those that occur naturally following FMLP) cannot be explained by a change in the calcium response. It was noted, however, that free AA release was also enhanced following IL-3 treatment and, in particular, the appearance of AA was accelerated to occur within seconds of stimulation. This enhanced rate of AA generation applies to all stimuli. For C5a, the kinetics of LTC_4 release, in IL-3-treated cells, precisely matches the kinetics of the transient $[Ca^{2+}]_i$ response; that is, it all occurs within 30 s of stimulation. Thus, the need for the $[Ca^{2+}]_i$ signal does not change, simply the timing of AA production. As noted, $cPLA_2$ phosphorylation is slow.

However, like PMA, IL-3 induces phosphorylation of $cPLA_2$, which suggests that it is this prephosphorylation step which allows IL-3 to accelerate AA production into the calcium phase of the response, resulting in the generation of LTC_4.

Once again, these results need to be interpreted cautiously because other studies suggest that the AA is derived from the activity of a type II(V) $sPLA_2$. Unfortunately, the steps to $sPLA_2$ activation are unknown, and this temporal relationship may be upheld for $sPLA_2$ activation as well. Although IL-3 induces phosphorylation of $cPLA_2$, it appears the pathway differs from PMA-induced changes. Bisindolylmaleimides inhibited the PMA-induced phosphorylation while having no effect on IL-3-induced phosphorylation, indicating that IL-3 causes these changes through a PKC-independent pathway.[67] These data may explain the effects of IL-3 on LTC_4 release but their relevance to histamine or IL-4 release is not clear, although short treatments with IL-3 have very modest effects on IL-4 release. It is interesting to note however that drugs that were designed to inhibit PLA_2 generally also inhibit histamine release, including early nonselective inhibitors as well as newer moderately selective inhibitors. With the exception of the calcium studies noted previously, there have been no other studies of signal transduction events in human basophils treated for longer periods. On the basis of this one qualitative shift in the calcium response and the multiplicity of studies on the changes induced by IL-3 (and similar cytokines) in other cell types, however, it is reasonable to speculate that the longer-term effects reflect a more elaborate induction of genes that modestly alter the apparent phenotype of the basophil. The observation that treatments lasting days can turn nonreleaser basophils into releasers further highlights this perspective. These later data also suggest that the deficiency in activating the nonreleaser cells may be caused by an absence of circulating IL-3. Control by other cytokines has not been investigated in detail beyond initial observations describing their respective effects on function.

Steroids

For human basophils, glucocorticosteroids are potent inhibitors of IgE-mediated secretion.[79] This is unlike human mast cells, where steroids have no apparent effect,[80] which can be seen in clinical studies of the acute-phase response to allergen challenge in steroid-treated patients.[1] In basophils, the steroid-induced inhibition is specific to certain secretagogues; IgE-mediated release is downregulated about 90%, FMLP-induced release is not affected,[81] and recent studies indicate that C5a induced is modestly inhibited.[82] Although the basic observation of glucocorticosteroid inhibition was made 15 years ago, no current studies have clarified the precise mechanism for the effect. It is interesting to note, however, that IL-3 can negate the inhibition by glucocorticosteroids; that is, IL-3-treated cells cannot be inhibited by subsequent treatment with glucocorticosteroids.[73] This is an interesting result, which bears some similarity to studies with cyclosporin A. Cyclosporin A (CsA) is an efficacious inhibitor of IgE-mediated histamine release but if basophils are first treated with IL-3, CsA has no effect.[83] IL-3 appears to have the ability to "re-wire" the signal transduction pathways leading to mediator release. For steroids, IL-3 must be present for many hours before the reversal of glucocorticosteroid inhibition is apparent, but for cyclosporin a short treatment with IL-3 is sufficient.

There are many studies of glucocorticosteroid action in other cell types, including RBL cells, but the results are so varied as to suggest that a clear understanding will require direct studies of human basophils. Like histamine release, both LTC_4 and IL-4[84] secretion are inhibited by glucocorticosteroids. However, unlike histamine and LTC_4 release, for which long incubations (>8h) with glucocorticosteroids are required to observe significant inhibition, IL-4 secretion is inhibited very rapidly and with markedly higher potencies. Indeed, treatments with glucocorticosteroids lasting only minutes (as fast as the steroid can be washed from the cells) at concentrations of 100nM are sufficient to inhibit IL-4 secretion by 40%. Given the appar-

ent requirement for IL-4 mRNA transcription, it seems likely that some early events induced by glucocorticosteroids are capable of inhibiting transcription.[84] Thus, there may be a multiplicity of changes induced by glucocorticosteroids of which histamine release modulation is a relatively late event.

cAMP

A mainstay of allergic disease therapy has been the use of agents that elevate cAMP levels in cells. Whether these compounds operate directly on mast cells and basophils to obtain the clinical effects or operate through other means, it is clear that elevations in cAMP have marked effects on basophil function.[85,86] There are several receptors on basophils that induce changes in cAMP; histamine H2[87] and β-receptors[88] are two that have been studied extensively. Elevations in cAMP induced by these agents are transient[89] so that the relative timing between the presentation of agonist and the stimulus is important if inhibition is to be optimal. Agents that elevate cAMP by inhibition of phosphodiesterases, which in human basophils appears to be predominately PDE-IV,[90] cause more prolonged elevations and are generally more efficacious inhibitors of secretion.[89] The precise significance of these behavioral characteristics to in vivo biology is not entirely clear. Indeed, the presence of an H2 receptor on basophils suggests a role in feedback inhibition, but the extent to which this plays a role in secretion is not clear. Many years ago it was shown that at low levels of release, antagonists of this receptor would indeed enhance histamine release modestly, if the release was far below the maximum possible.[91]

More recent studies have examined the effect of cAMP elevations on the $[Ca^{2+}]_i$ response in basophils.[92] For IgE-mediated changes in calcium, cAMP elevations had no effect on the first phase of the $[Ca^{2+}]_i$ response, which derived from internal stores, while markedly inhibiting the second phase of the response, which derived from the influx of extracellular calcium. The first phase of the $[Ca^{2+}]_i$ response is generally attributed to the generation of IP_3, which activates the IP_3 receptor located on ER-like internal calcium storage structures to release these internal stores. Because this phase of the response is not inhibited by cAMP elevations, one can expect that there is no inhibition of earlier steps in the signal transduction cascade, although this requires testing. However, the second phase of the calcium response is presumably also dependent on IP_3, although calcium must first enter the cell through pathway(s) that are presently poorly defined. If there is no effect on the IP_3 aspect of signal transduction, one would conclude that cAMP elevations inhibit the calcium influx pathway itself. This pathway is thought to be activated whenever internal stores are sufficiently depleted, the so-called capacitance model for calcium signaling.[93] If true, then any receptor-mediated $[Ca^{2+}]_i$ response that depends on this capacitance-driven influx should be inhibited by cAMP elevations. However, cAMP elevations do not inhibit the calcium response that follows stimulation with FMLP. Indeed, at lower levels of FMLP stimulation, where the second phase of the $[Ca^{2+}]_i$ response is more modest, elevations in cAMP often enhance the second-phase $[Ca^{2+}]_i$ response. Such data suggest that cAMP elevations act on influx or efflux pathways that differ for the two types of stimulation. Alternatively, the pattern of the $[Ca^{2+}]_i$ response following IgE-mediated stimulation may not reflect the canonical IP_3-driven system. It has recently been noted that IgE-mediated $[Ca^{2+}]_i$ responses in RBL cells use the generation of sphingosine-1-phosphate and its receptor on the internal calcium stores as a second pathway to calcium release.[94] When stimulated through FcεRI, the S1P pathway appeared to dominate the control of $[Ca^{2+}]_i$ elevations, but when stimulated through a GTP protein-linked receptor, IP_3 appeared to dominate control of the $[Ca^{2+}]_i$ response. Although these studies might offer an alternative explanation for the differential effects of cAMP elevations, follow-up studies have suggested that S1P may act on an external receptor, complicating interpretation of these results.

If the effect of cAMP is to regulate the second phase of the $[Ca^{2+}]_i$ response, it might be expected that mediator release that is particularly dependent on calcium signals would be more affected by cAMP elevations. With respect to LTC_4 release, agents that elevate cAMP are not more potent but they are more efficacious. With respect to IL-4 release, studies are not yet conclusive because these studies have thus far only been done in IL-3-containing media and IL-3 appears to reverse the effects of cAMP agonists. In the same context, the inclusion of cimetidine to antagonize the feedback inhibition of released histamine on basophil H2 receptors in the long-term cultures needed to study IL-4 secretion did not enhance IL-4 release. Forthcoming studies should clarify these issues.

IgE Antibody

Quite obviously, the presence of IgE antibody is necessary if the basophil is to respond to IgE-mediated stimulation. Until recently, it has not been appreciated that IgE has an influence beyond its passive binding to FcεRI and its subsequent binding to antigen to initiate secretion. Studies in the late 1970s noted a correlation between the circulating IgE levels and basophil FcεRI densities.[95] It is now appreciated that IgE antibody induces an increase in the level of FcεRI expression; this has been demonstrated in vitro and in vivo in mice[96–98] and in humans.[99,100] This expression probably has a significant impact on the function of the basophil.

A good illustration of this impact can be obtained from a recent trial of monoclonal antiIgE antibody therapy.[99] The antibody used in treating patients is designed to bind to IgE antibody at the site that interacts with FcεRIα. Thus, while IgE is in its solution phase, unbound to cells, it can bind to the therapeutic anti-IgE antibody and become part of a small aggregate, which should eventually be removed from the circulation. In addition, while still in circulation, it is unable to bind to FcεRI (or FcεRII as well). Importantly, it does not bind to IgE already bound to cells through either FcεRI or FcεRII, which prevents induction of anaphylaxis during its administration. As is now apparent after several hundred tested patients and the one long-term trial recently reported,[99] the antibody is well tolerated with no evidence of anaphylactic reactions. It is worth considering the logic underlying this approach to treating allergic diseases and whether is it reasonable to expect success based on what is known about the sensitivity of basophils or mast cells to IgE-mediated stimulation. Basophil sensitivity studies, that is, a determination of the number of IgE molecules per basophil required to obtain 50% of maximal histamine release, indicate an average value of about 3000 molecules per basophil.[30] For an antigen like the dust mite, 10% of the total circulating IgE titer may be specific for this antigen. The average starting IgE density of the basophils from dust mite-sensitive donors in this study was approximately 250,000 per basophil. To reduce the cell-surface IgE density to values that result in approximately 10% of the before-treatment basophil response to dust mite antigen, and factoring the affinity of IgE for FcεRIα, it can be calculated that a 2000- to 10,000-fold reduction in serum free IgE is required. This is a very demanding requirement and, as it turns out, it was not achieved in the published phase I toxicology trial. Only a 100-fold decrease was observed, and yet the median basophil response to dust mite antigen decreased to about 10% of the starting response.

The likely explanation for this successful effect on function despite the less-than-needed change in IgE titer was the downregulation of the FcεRI density on circulating basophils. This study demonstrated a decrease of about 26 fold in FcεRI density (from ≈250,000 to ≈8000) such that endogenous cell-bound IgE densities were below detectable limits of the assay, probably in the range of 500 to 1500 per cell, which would reduce antigen-specific levels to a range of 50 to 150 molecules per cell, precisely as would be expected for a 90% reduction in the basophil response. More recent studies of this same patient group indicate that if the FcεRI density rises to only somewhat higher levels, full antigen sensitivity returns to the basophil. Studies of mouse mast cells also demonstrate a marked increase in function following upregulation of FcεRI by IgE, indicating the upregulation of a fully functional receptor.[96,98] Therefore, the regulation of FcεRI expression by IgE is probably a critical compo-nent of regulating the function of basophils (and mast cells).

The mechanisms underlying regulation by IgE are currently unknown. Both in vitro studies using human basophils or mice mast cells indicate that upregulation is a slow process and linear within the time frame of study. For human basophils, 400 FcεRI molecules per hour appears typical, a rate that persists for up to 2 weeks. Thus the increase in upregulation observed is a function of the starting density; that is, at a constant rate of expression limited to 400 per hour, the amount of upregulation starting at 5000 per basophil appears more impressive than starting at 150,000 for typical periods of incubation (7 days). The dose–response curve for the upregulation lies in a range not entirely appropriate for the equilibrium constant of IgE for FcεRIα or FcεRII, being too high a concentration for FcεRI and too low for FcεRII. However, it is possible to show that if one considers a model similar to that found for IgE regulating the expression of FcεRII (IgE stabilizes its presence on the cell surface by protecting it from cleavage by a specific class of proteases[101,102]), a model that is therefore very sensitive to the rate that IgE binds to FcεRI, the dose–response curve for IgE versus upregulation is very appropriate for IgE interacting with FcεRI to cause its upregulation. Other experiments indicate that IgE is not interacting with other candidate receptors (FcεRII, eBP, FcγRII) to induce this upregulation. A similar conclusion can be drawn from studies in mice. In the near future, further studies should clarify which receptor is involved and how upregulation occurs.

Summary

During the past three decades, the number of identified extrinsic influences on inflammatory cells has skyrocketed. Likewise, the complexity of the identified signal transduction network within the cells has increased enormously. Many of the newly identified extrinsic factors such as cytokines, chemokines, and smaller molecules have been tested on human basophils because the basic assay of basophil function is relatively straightforward. Unfortunately, the same can not be said for intrinsic factors because these studies require both reasonably high numbers and purities for study. Thus, while a great deal can be said about the ability of a variety of extrinsic factors to alter basophil function, most of our understanding of how they may alter function must come from assumptions of similarities to other cell types. Of the extrinsic factors, only IL-3 and cAMP agonists have received deeper attention to their mechanisms of action. For intrinsic pathways of regulation, only elevations in $[Ca^{2+}]_i$ have been broadly studied. In the near future, studies of adhesion molecules and chemokines should receive further attention. The regulation of FcεRI expression in general and specifically by IgE antibody itself will probably receive enormous attention in the short term. The secretion of cytokines will undergo closer study by a wide variety of laboratories, and the regulation of IgE-mediated signaling will benefit from recent efforts to delineate the behavior of early tyrosine kinases, diacylglycerol generation, AA generation, and further studies to clarify the role of PKC.

References

1. Bascom R, Wachs M, Naclerio RM, Pipkorn U, Galli SJ, Lichtenstein LM. Basophil influx occurs after nasal antigen challenge: effects of topical corticosteroid pretreatment. J Allergy Clin Immunol 1988;81:580–589.

2. Iliopoulos O, Baroody F, Naclerio RM, Bochner BS, Kagey-Sobotka A, Lichtenstein LM. Histamine containing cells obtained from the nose hours after antigen challenge have functions and phenotypic characteristics of basophils. J Immunol 1992;148:2223–2228.

3. Guo CB, Liu MC, Galli SJ, Bochner BS, Kagey-Sobotka A, Lichtenstein LM. Identification of IgE bearing cells in the late response to antigen in the lung as basophils. Am J Respir Cell Mol Biol 1993;10:384–390.

4. Irani AA, Huang C, Zweiman B, Schwartz LB. Immunohistochemical detection of basophil infiltration in the skin during the IgE-mediated late phase reaction. J Allergy Clin Immunol 1997;99:S92.

5. MacGlashan DW Jr, Peters SP, Warner J, Lichtenstein LM. Characteristics of human basophil sulfidopeptide leukotriene release: releasability defined as the ability of the basophil to respond to dimeric cross-links. J Immunol 1986;136:2231–2239.

6. Warner JA, Freeland HS, MacGlashan DW Jr, Lichtenstein LM, Peters SP. Purified human basophils do not generate LTB4. Biochem Pharmacol 1987;36:3195–3199.

7. Warner JA, Peters SP, Lichtenstein LM, et al. Differential release of mediators from human basophils: differences in arachidonic acid metabolism following activation by unrelated stimuli. J Leukocyte Biol 1989;45:558–571.

8. Lewis RA, Holgate ST, Roberts LJ, Oates JA, Austen KF. Preferential generation of prostaglandin D2 by rat and human mast cells. Kroc Found Symp 1981;14:239–254.

9. Peters SP, MacGlashan DW Jr, Schulman ES, et al. Arachidonic acid metabolism in purified human lung mast cells. J Immunol 1984;132:1972–1979.

10. Brunner T, Heusser CH, Dahinden CA. Human peripheral blood basophils primed by interleukin-3 (IL-3) produce IL-4 in response to immunoglobulin E receptor stimulation. J Exp Med 1993;177:605–611.

11. MacGlashan DW Jr, White JM, Huang SK, Ono SJ, Schroeder J, Lichtenstein LM. Secretion of interleukin-4 from human basophils: the relationship between IL-4 mRNA and protein in resting and stimulated basophils. J Immunol 1994;152:3006–3016.

12. Schroeder JT, MacGlashan DW Jr, Kagey-Sobotka A, White JM, Lichtenstein LM. The IgE-dependent IL-4 secretion by human basophils: the relationship between cytokine production and histamine release in mixed leukocyte cultures. J Immunol 1994;153:1808–1817.

13. Li H, Sim TC, Alam R. IL-13 released by and localized in human basophils. J Immunol 1996;156:4833–4838.

14. Li H, Sim TC, Grant JA, Alam R. The production of macrophage inflammatory protein-1 alpha by human basophils. J Immunol 1996;157:1207–1212.

15. MacGlashan DW Jr. Graded changes in the response of individual human basophils to stimulation: distributional behavior of events temporally coincident with degranulation. J Leukocyte Biol 1995;58:177–188.

16. Dvorak AM. Basophils and mast cells: piecemeal degranulation in situ and ex vivo: a pos-

sible mechanism for cytokine-induced function in disease. In: Coffey RG, ed. Granulocyte Responses to Cytokines. New York: Marcel Dekker, 1992.

17. Dvorak AMJ, Warner JA, Kissell S, Lichtenstein LM, MacGlashan DW Jr. F-met peptide-induced degranulation of human basophils. Lab Invest 1991;64:234–253.

18. Dvorak AM, Morgan ES, Lichtenstein LM, MacGlashan DW Jr. Activated human basophils contain histamine in cytoplasmic vesicles. Int Arch Allergy Immunol 1994;105:8–11.

19. Dvorak AM, MacGlashan DW Jr, Morgan ES, Lichtenstein LM. Histamine distribution in human basophil secretory granules undergoing FMLP-stimulated secretion and recovery. Blood 1995;86:3560–3566.

20. Dvorak AM, Ackerman SJ, Letourneau L, Morgan ES, Lichtenstein LM, MacGlashan DW Jr. Vesicular transport of Charcot-Leydon crystal protein in tumor-promoting phorbol-ester stimulated human basophils. Lab Invest 1996;74:967–974.

21. Dvorak AM, Newball HH, Dvorak HF, Lichtenstein LM. Antigen-induced IgE-mediated degranulation of human basophils. Lab Invest 1980;43:126–139.

22. Marone G, Kagey SA, Lichtenstein LM. Effects of arachidonic acid and its metabolites on antigen-induced histamine release from human basophils in vitro. J Immunol 1979;123:1669–1677.

23. Sobotka AK, Marone G, Lichtenstein LM. Indomethacin, arachidonic acid metabolism and basophil histamine release. Monogr Allergy 1979;14:285–287.

24. Kurimoto Y, de Weck LA, Dahinden CA. Interleukin 3-dependent mediator release in basophils triggered by C5a. J Exp Med 1989;170:467–479.

25. MacGlashan DW Jr, Warner JA. Stimulus-dependent leukotriene release from human basophils: a comparative study of C5a and Fmet-leu-phe. J Leukocyte Biol 1991;49:29–40.

26. Ochensberger BS, Rihs S, Brunner T, Dahinden CA. IgE-independent interleukin-4 expression and induction of a late phase of leukotriene C4 formation in human blood basophils. Blood 1995;86:4039–4049.

27. MacGlashan DW Jr, Hubbard WC. Interleukin-3 alters free arachidonic acid generation in C5a-stimulated human basophils. J Immunol 1993;151:6358–6369.

28. Lichtenstein LM, MacGlashan DW Jr. The concept of basophil releasability. J Allergy Clin Immunol 1986;77:291–294.

29. Conroy MC, Adkinson NFJ, Lichtenstein LM. Measurement of IgE on human basophils: relation to serum IgE and anti-IgE-induced histamine release. J Immunol 1977;118:1317–1321.

30. MacGlashan DW Jr. Releasability of human basophils: cellular sensitivity and maximal histamine release are independent variables. J Allergy Clin Immunol 1993;91:605–615.

31. Nguyen KL, Gillis S, MacGlashan DW Jr. A comparative study of releasing and nonreleasing human basophils: nonreleasing basophils lack an early component of the signal transduction pathway that follows IgE cross-linking. J Allergy Clin Immunol 1990;85:1020–1029.

32. Knol EF, Mul FP, Kuijpers TW, Verhoeven AJ, Roos D. Intracellular events in anti-IgE nonreleasing human basophils. J Allergy Clin Immunol 1992;90:92–103.

33. Warner JA, MacGlashan DW Jr. Signal transduction events in human basophils—a comparative study of the role of protein kinase-C in basophils activated by anti-IgE antibody and formyl-methionyl-leucyl-phenylalanine. J Immunol 1990;145:1897–1905.

34. Miller L, Blank U, Metzger H, Kinet JP. Expression of high-affinity binding of human immunoglobulin E by transfected cells. Science 1989;244:334–337.

35. Lin S, Cicaia C, Scharenberg AM, Kinet JP. The FcεRIb subunit functions as an amplifier of FcεRIg-mediated cell activation signals. Cell 1996;85:985–995.

36. Schleimer RP, Gillespie E, Daiuta R, Lichtenstein LM. Release of histamine from human leukocytes stimulated with the tumor-promoting phorbol diesters. II. Interaction with other stimuli. J Immunol 1982;128:136–140.

37. Warner J, MacGlashan DW Jr. Persistence of early crosslink-dependent signal transduction events in human basophils after desensitization. Immunol Lett 1988;18:129–137.

38. Lichtenstein LM, De Bernardo R. IgE mediated histamine release: in vitro separation into two phases. Int Arch Allergy Appl Immunol 1971;41:56–71.

39. MacGlashan DW Jr, Mogowski M, Lichtenstein LM. Studies of antigen binding on human basophils. II. Continued expression of

antigen-specific IgE during antigen-induced desensitization. J Immunol 1983;130:2337–2342.

40. Kazimierczak W, Meier HL, MacGlashan DW Jr, Lichtenstein LM. An antigen-activated DFP-inhibitable enzyme controls basophil desensitization. J Immunol 1984;132:399–405.

41. Jouvin MH, Adamczewski M, Numerof R, Letourneur O, Valle A, Kinet JP. Differential control of the tyrosine kinase lyn and syk by the two signaling chains of the high affinity immunoglobulin E receptor. J Biol Chem 1994;269:5918–5925.

42. Benhamou M, Gutkind JS, Robbins KC, Siraganian RP. Tyrosine phosphorylation coupled to IgE receptor-mediated signal transduction and histamine release. Proc Natl Acad Sci U S A 1990;87:5327–5330.

43. Benhamou M, Stephan V, Gutkind SJ, Robbins KC, Siraganian RP. Protein tyrosine phosphorylation in the degranulation step of RBL-2H3 cells. FASEB J 1991;5:A1007.

44. Eiseman E, Bolen JB. Engagement of the high-affinity IgE receptor activates src protein-related tyrosine kinases. Nature (Lond) 1992;355:78–80.

45. Yamashita T, Mao SY, Metzger H. Aggregation of the high-affinity IgE receptor and enhanced activity of p53/56lyn protein-tyrosine kinase. Proc Natl Acad Sci USA 1994;91:11251–11255.

46. Dembo M, Goldstein B, Sobotka AK, Lichtenstein LM. Degranulation of human basophils: quantitative analysis of histamine release and desensitization, due to a bivalent penicilloyl hapten. J Immunol 1979;123:1864–1872.

47. Sagi-Eisenberg R, Lieman H, Pecht I. Protein kinase C regulation of the receptor-coupled calcium signal in histamine-secreting rat basophilic leukaemia cells. Nature (Lond) 1985;313:59–60.

48. Ozawa K, Yamada K, Kanzanietz MG, Blumberg PM, Beaven MA. Different isozymes of protein kinase C mediate feedback inhibition of phospholipase C and stimulatory signals for exocytosis in rat RBL-2H3 cells. J Biol Chem 1993;268:2280–2283.

49. Ozawa K, Szallasi Z, Kanzanietz MG, et al. Ca^{2+}-dependent and Ca^{2+}-independent isozymes of protein kinase C mediate exocytosis in antigen-stimulated rat basophilic RBL-2H3 cells: Reconstitution of secretory responses with Ca^{2+} and purified isozymes in washed

permeabilized cells. J Biol Chem 1993;268:1749–1756.

50. Lewin I, Jacob-Hirsch J, Zang ZC, et al. Aggregation of the Fc epsilon RI in mast cells induces the synthesis of Fos-interacting protein and increases its DNA binding-activity: the dependence on protein kinase C-beta. J Biol Chem 1996;271:1514–1519.

51. Razin E, Szallasi Z, Kazanietz MG, Blumberg PM, Rivera J. Protein kinases C-beta and C-epsilon link the mast cell high-affinity receptor for IgE to the expression of c-fos and c-jun. Proc Natl Acad Sci U S A 1994;91:7722–7726.

52. Lewin I, Nechushtan H, Ke Q, Razin E. Regulation of AP-1 expression and activity in antigen-stimulated mast cells: the role played by protein kinase C and the possible involvement of Fos interacting protein. Blood 1993;82:3745–3751.

53. Wolfe PC, Chang EY, Rivera J, Fewtrell C. Differential effects of the protein kinase C activator phorbol 12-myristate 13-acetate on calcium responses and secretion in adherent and suspended RBL-2H3 mucosal mast cells. J Biol Chem 1996;271:6658–6665.

54. Szallasi Z, Denning MF, Chang EY, et al. Development of a rapid approach to identification of tyrosine phosphorylation sites: application to PKC delta phosphorylated upon activation of the high affinity receptor for IgE in rat basophilic leukemia cells. Biochem Biophys Res Commun 1995;214:888–894.

55. Ozawa K, Szallasi Z, Blumberg PM, Beaven MA. The role of Ca^{2+}-independent isozymes of protein kinase C (PKC) in exocytosis in antigen-stimulated rat basophil RBL-2H3 cells. FASEB J 1992;6:A1633.

56. Miura K, MacGlashan DW Jr. Expression of Protein Kinase C Isozymes in Human Basophils: Regulation by Physiological and Nonphysiological Stimuli. Blood 1998;92:1206–1218.

57. Devalia V, Thomas SB, Roberts PJ, Jones M, Linch DC. Down-regulation of human protein kinase C alpha is associated with terminal neutrophil differentiation. Blood 1992;80:68–76.

58. Dang PM, Rais S, Hakim J, Perianin A. Redistribution of protein kinase C isoforms in human neutrophils stimulated by formyl peptides and phorbol myristate acetate. Biochem Biophys Res Commun 1995;212:664–672.

59. Bates ME, Bertics PJ, Calhoun WJ, Busse WW. Increased protein kinase C activity in low

density eosinophils. J Immunol 1993;150:4486–4493.

60. Schleimer RP, Gillespie E, Lichtenstein LM. Release of histamine from human leukocytes stimulated with the tumor promoting phorbol esters. I. Characterization of the response. J Immunol 1981;126:570–574.

61. Kishimoto A, Takei Y, Mori T, Kikkawa U, Nishizuka Y. Activation of calcium and phospholipid dependent protein kinase by diacylglycerol, its possible relation to phosphatidylinositol turnover. J Biol Chem 1983;255:2273.

62. Castagna M, Takai Y, Kaibuchi K, Sano K, Kikkawa U, Nishizuka Y. Direct activation of calcium-activated, phospholipid-dependent protein kinase by tumor-promoting phorbol esters. J Biol Chem 1982;257:7847–7853.

63. Warner JA, MacGlashan DW Jr. Protein kinase C (PKC) changes in human basophils. IgE-mediated activation is accompanied by an increase in total PKC activity. J Immunol 1989;142:1669–1677.

64. Lih-Ling L, Wartmann M, Lin AY, Knopf JL, Seth A, Davis RJ. cPLA2 is phosphorylated and activated by MAP kinase. Cell 1993;72:269–278.

65. Nalefski EA, Sultzman LA, Martin DM. Delineation of two functionally distinct domains of cytosolic phospholipase A2, a regulatory Ca-dependent lipid binding domain and a Ca-independent catalytic domain. J Biol Chem 1994;269:18239–18249.

66. Channon JY, Leslie CC. A calcium-dependent mechanism for associating a soluble arachidonoyl-hydrolyzing phospholipase A with membrane in the macrophage cell line RAW 264.7. J Biol Chem 1990;265:5409–5413.

67. Miura K, Hubbard WC, MacGlashan DW, Jr. Phosphorylation of cytosolic PLA2 (cPLA2) by Interleukin-3 (IL-3) is associated with increased free arachidonic acid and LTC4 release in human basophils. J Allergy Clin Immunol 1998;102:512–520.

68. Casolaro V, Georas S, Song Z, et al. Inhibition of NFATp-dependent transcription by NFκB. Implications for differential gene expression in T cell subsets. Proc Natl Acad Sci USA 1995;92:11623–11626.

69. Hirai K, Morita Y, Misaki Y, et al. Modulation of human basophil histamine release by hematopoetic growth factors. J Immunol 1988;141:3957–3961.

70. Hirai K, Yamaguchi M, Misaki Y, et al. Enhancement of human basophil histamine release by interleukin 5. J Exp Med 1990;172:1525–1528.

71. Bischoff SC, de Weck AL, Dahinden CA. Interleukin 3 and granulocyte/macrophage-colony-stimulating factor render human basophils responsive to low concentrations of complement component C3a. Proc Natl Acad Sci U S A 1990;87:6813–6817.

72. Bischoff SC, Brunner T, De WAL, Dahinden CA. Interleukin 5 modifies histamine release and leukotriene generation by human basophils in response to diverse agonists. J Exp Med 1990;172:1577–1582.

73. Schleimer RP, Derse CP, Friedman B, et al. Regulation of human basophil mediator release by cytokines. I. Interaction with antiinflammatory steroids. J Immunol 1989;143:1310–1317.

74. Massey WA, Randall TC, Kagey SA, et al. Recombinant human IL-1 alpha and -1 beta potentiate IgE-mediated histamine release from human basophils. J Immunol 1989;143:1875–1880.

75. Pazdrak K, Stafford S, Alam R. The activation of the Jak-STAT 1 signaling pathway by IL-5 in eosinophils. J Immunol 1995;155:397–402.

76. Ihle JN, Kerr IM. Jaks and Stats in signaling by the cytokine receptor superfamily. [Review]. Trends Genet 1995;11:69–74.

77. Bochner BS, Mckelvey AA, Sterbinsky SA, et al. IL-3 augments adhesiveness for endothelium and Cd11B expression in human basophils but not neutrophils. J Immunol 1990;145:1832–1837.

78. Yamaguchi M, Hirai K, Ohta K, et al. Culturing in the presence of IL-3 converts anti-IgE nonresponding basophils into responding basophils. J Allergy Clin Immunol 1996;97:1279–1287.

79. Schleimer RP, Lichtenstein LM, Gillespie E. Inhibition of basophil histamine release by anti-inflammatory steroids. Nature (Lond) 1981;292:454–455.

80. Schleimer RP, Schulman ES, MacGlashan DW Jr, et al. Effects of dexamethasone on mediator release from human lung fragments and purified human lung mast cells. J Clin Invest 1983;71:1830–1835.

81. Schleimer RP, MacGlashan DW Jr, Gillespie E, Lichtenstein LM. Inhibition of basophil histamine release by anti-inflammatory steroids.

II. Studies on the mechanism of action. J Immunol 1982;129:1632–1636.

82. Stellato C, Lichtenstein LM, MacGlashan DW Jr, Schleimer RP. Dexamethasone selectively inhibits IL-3 priming of basophil C5a responses. J Allergy Clin Immunol 1991;87:72a.

83. Cirillo R, Triggiani M, Siri L, et al. Cyclosporin A rapidly inhibits mediator release from human basophils presumably by interacting with cyclophilin. J Immunol 1990;144:3891–3897.

84. Schroeder JT, MacGlashan DW Jr, MacDonald SM, Kagey-Sobotka A, Lichtenstein LM. Regulation of IgE-dependent IL-4 generation by human basophils treated with glucocorticosteroids. J Immunol 1997; (in press).

85. Lichtenstein LM, Margolis S. Histamine release in vitro: inhibition by catecholamines and methylxanthines. Science 1968;161:902–903.

86. Bourne HR, Lichtenstein LM, Melmon KL, Henney CS, Weinstein Y, Shearer GM. Modulation of inflammation and immunity by cyclic AMP. Science 1974;184:19.

87. Lichtenstein LM, Gillespie E. Inhibition of histamine release by histamine controlled by H2 receptor. Nature (Lond) 1973;244:287–288.

88. Gillespie E, Lichtenstein LM. Histamine release from human leukocytes: relationships between cyclic nucleotide, calcium, and antigen concentrations. J Immunol 1975;115:1572–1576.

89. Peachell PT, MacGlashan DW Jr, Lichtenstein LM, Schleimer RP. Regulation of human basophil and lung mast cell function by cyclic adenosine monophosphate. J Immunol 1988;140:571–579.

90. Peachell PT, Undem BJ, Schleimer RP, et al. Preliminary identification and role of phosphodiesterase isozymes in human basophils. J Immunol 1992;148:2503–2510.

91. Tung R, Kagey SA, Plaut M, Lichtenstein LM. H2 antihistamines augment antigen-induced histamine release from human basophils in vitro. J Immunol 1982;129:2113–2115.

92. Botana LM, MacGlashan DW Jr. Differential effects of cAMP on mediator release in human basophils. FASEB J 1991;5:A1007.

93. Putney JW. A model for receptor-regulated calcium entry. Cell Calcium 1986;7:1–12.

94. Choi OH, Kim JH, Kinet JP. Calcium mobilization via sphingosine kinase in signaling by the FcεRI antigen receptor. Nature (Lond) 1996;380:634–636.

95. Malveaux FJ, Conroy MC, Adkinson NFJ, Lichtenstein LM. IgE receptors on human basophils. Relationship to serum IgE concentration. J Clin Invest 1978;62:176–181.

96. Hsu C, MacGlashan DW Jr. IgE Antibody upregulates high affinity IgE binding on murine bone marrow derived mast cells. Immunol Lett 1996;52:129–134.

97. Lantz CS, Yamaguchi M, Oettgen HC, et al. IgE regulates mouse basophil FcεRI expression in vivo. J Immunol 1997;158:2517–2521.

98. Yamaguchi M, Lantz CS, Oettgen HC, et al. IgE enhances mouse mast cell FcεRI expression in vitro and in vivo: evidence for a novel amplification mechanism in IgE-dependent reactions. J Exp Med 1997;185:663–672.

99. MacGlashan DW Jr, Bochner BS, Adelman DC, et al. Down-regulation of FcεRI expression on human basophils during in vivo treatment of atopic patients with anti-IgE antibody. J Immunol 1997;158:1438–1445.

100. MacGlashan DW Jr, White-Mckenzie J, Chichester K, et al. In vitro regulation of FceRIa expression on human basophils by IgE antibody. Blood 1998;91:1633–1643.

101. Lee WT, Conrad DH. Murine B cell hybridomas bearing ligand-inducible Fc receptors for IgE. J Immunol 1986;136:4573–4580.

102. Lee WT, Rao M, Conrad DH. The murine lymphocyte receptor for IgE. IV. The mechanism of ligand-specific receptor upregulation on B cells. J Immunol 1987;139:1191–1198.

103. Schroeder JT, Howard BP, Jenkens MK, Kagey-Sobotka A, Lichtenstein LM, MacGlashan DW Jr. IL-4 secretion and histamine release by human basophils are differentially regulated by protein kinase C activation. J Leukoc Biol 1998;63:692–698.

104. Hundley TR, Marshall L, Hubbard WC, MacGlashan DW Jr. Characteristics of arachidonic acid generation in human basophils: Relationship between the effects of inhibitors of secretory phospholipase A2 activity and leukotriene C4 release. J Pharm Exp Ther 1998;284:847–857.

29
Histamine Releasing Factors

Susan M. MacDonald

Asthma affects more than 15 million people in the United States.[1] The economic impact of the resulting morbidity, including days lost from work or school and the psychosocial problems of living with a chronic disease, as well as the mortality, is quite significant. Asthma accounted for more than 5100 deaths in the United States in 1991.[2] When asthma was viewed as a respiratory disease of airway obstruction caused by acute bronchospasm, the mast cell, basophil, and their products, including histamine, sulfidopeptide leukotrienes, and prostaglandins, became the primary foci for pharmacological intervention.[2] More recently, asthma has been defined as a disease of the airways characterized by "airway inflammation, airway hyperresponsiveness and reversible airway obstruction."[1] Because of this more recent definition of asthma, many investigators are examining models of inflammation. Antigen exposure in an allergic, asthmatic individual results in an acute reaction with a decrease in FEV_1 (forced expiratory volume in 1 s) that reverses over a few hours. If the initial antigenic stimulus is sufficiently high, there is, in many individuals, a return of symptoms in the absence of any renewed antigen exposure, termed the late-phase reaction (LPR). This occurs some 6 to 24 h later and is characterized by bronchial inflammation and enhanced bronchial responsiveness. Because these LPR are thought to more accurately reflect chronic, allergic, asthmatic reactions, and inflammation than do acute-phase responses, the LPR is a useful tool to explore the pathogenesis of

asthma.[2] Studies in our laboratory and others have shown that proinflammatory mediators are present in the LPR and that cytokines, able to cause eosinophil activation and basophil histamine release, are also found.[3-5]

There are several reasons why the LPR may be a more appropriate model of allergic inflammation than is the acute response to antigen, including its duration, susceptibility to oral steroids,[6] and cellular influx characterized by eosinophils, basophils, and monocytes.[7,8] However, the pathogenesis of this late reaction is complex and not well understood. An initial event in the human response to antigen might well be mast cell degranulation and release of mediators that, in turn, attract inflammatory cells and upregulate adhesion molecules. The basophils and eosinophils that infiltrate in the LPR are known to be activated or primed to be hyperresponsive to stimuli. There is growing evidence to suggest that the basophil, and not the mast cell, plays the more significant role in the LPR. First, the mast cell mediators that are found associated with the symptoms of the acute reaction, that is, histamine, prostaglandin D_2 (PGD_2), tryptase, and leukotriene C_4 (LTC_4) are clearly different than those found in the LPR, that is, histamine and LTC_4, but no PGD_2 or tryptase.[3] Basophils do not contain PGD_2 and have very little, if any, tryptase. Second, based on morphological and flow cytometric evidence, basophils clearly appear during antigen-induced LPR in skin chambers[9] and in the nose.[10] Furthermore, the IgE-bearing cells in late-phase nasal lavages express CD-18 leukocyte adhesion molecules.[11] This adhesion molecule is found primarily on basophils and other leukocytes, not on the lung mast cell.

Basophils also have been identified as the histamine-containing cells in bronchoalveolar lavage (BAL) fluids obtained during the LPR. Of Alcian blue-positive cells that increased 15 fold in the antigen versus the saline sites in late-phase BAL fluids, 96% were morphologically and functionally defined as basophils.[12] The numbers of BAL basophils were found to correlate with bronchial hyperreactivity as measured by methacholine challenge, and their histamine release to anti-IgE antibody correlated with baseline FEV_1 and the severity of the LPR to antigen.[13] In vivo studies by Maruyama et al. have shown an accumulation of basophils during airway narrowing in asthmatic patients.[14] Koshino et al. have shown an increase in basophils in biopsies of patients dying of asthma compared to other diseases. In addition, basophil numbers significantly correlated with bronchial hyperresponsiveness in those patients experiencing an asthmatic episode.[15] Additional studies have demonstrated abnormal basophil function in disease states such as asthma and urticaria.[16–18] These data suggest that the infiltration of proinflammatory cells, including basophils, is of critical importance in the LPR of allergic, asthmatic diseases. Because basophils are present in vivo in states of chronic inflammation, the stimuli for basophil activation, the so-called histamine-releasing factors (HRF) have come under intense investigation.

There is a large body of literature implicating the eosinophil as a cell central to the pathogenesis of asthma. The eosinophil has been documented in bronchial biopsies of asthmatics, in BAL fluids, and in peripheral blood of patients with mild to moderate asthma.[19] Bousquet et al. have shown a correlation between the degree of blood eosinophilia and the severity of asthma.[20] Both bronchial hyperresponsiveness and the clinical severity of asthma have been shown to be correlated with numbers of eosinophils expressing mRNA for IL-4, IL-5, and granulocyte-macrophage colony-stimulating factor (GM-CSF).[21] Also, eosinophils in BAL after segmental antigen challenge showed evidence of activation as demonstrated by the expression of markers such as FcγRII (CD 32), CD45, and CD11a.[22] Additionally, levels of eosinophil granule proteins including major basic protein (MBP) have been correlated with symptoms of asthma.[23] It is interesting to note that MBP also stimulates human basophils to release histamine.[24] In fact, elevated numbers of circulating eosinophil and basophil progenitors are found in the blood and sputum of asthmatic patients during disease exacerbations, and, after treatment with glucocorticoids, the numbers of these cells decrease in conjunction with resolution of symptoms.[20]

Mononuclear cells have also been shown to be significantly increased in LPR of antigen-challenged skin biopsy sites.[4] Peripheral blood monocytes and BAL macrophages are activated in symptomatic asthmatics as compared with allergic rhinitic and normal subjects.[25] Furthermore, asthmatic subjects have been shown to have activated monocytes and BAL macrophages 19 h after whole-lung antigen challenge, as detected by flow cytometry using phenotypic markers.[26] The monocyte has received increased attention because of the discovery of the differential expression of the low-affinity IgE receptor, FcεRII, on its cell surface. This receptor was found to be elevated on monocytes and alveolar macrophages of asthmatic as compared to normal subjects.[27] More recently, the expression of the high-affinity IgE receptor, FcεRI, has also been shown to be increased on monocytes of atopic individuals as compared to those from normal subjects.[28] Thus, allergic and, in particular, allergic asthmatic patients have monocytes that differ from those of normal individuals.

Even though there is a substantial body of literature documenting the presence of eosinophils, basophils, and monocytes in late asthmatic reactions, a single stimulus that would activate all these cell types has been sought intensely in many laboratories.

History

Studies of HRF date back to 1979 when Thueson et al. first described a histamine-releasing activity from cultured peripheral blood mononuclear cells stimulated with mitogens or antigens.[29] This HRF was further characterized and found to be very heterogeneous, with molecular weight species ranging from 15,000 to 50,000. Our laboratory reported that a similar factor that caused basophil histamine release was present in late-phase skin blister fluids.[30] This factor was not present in the early blister fluids when antigen was present. Subsequently, we found a similar activity in nasal lavages during the LPR and in culture supernatants of the human macrophage monocyte-like cell line, U937.[31,32] These initial reports of a HRF that is produced by cultured lymphocytes were confirmed by numerous investigators including Sedgwick et al.,[33] Alam et al.,[34] Ezeamuzie and Assem,[35] Kaplan et al.,[36] and Strickland et al.[37] Subsequently, HRFs were found to be produced in vitro by a variety of cell types such as T and B lymphocytes, mononuclear cells, alveolar macrophages, platelets, vascular endothelial cells, and various cell lines, including the RPMI 8866 B cell line and the U937 cell line mentioned earlier. This result initiated intense efforts to purify these factors by several laboratories, including our own.

In the context of purifying HRF, it became apparent that there were two distinct mechanisms of HRF-induced histamine release, one that is dependent on cell-surface IgE and another that is IgE independent. Our laboratory aggressively pursued the purification of the IgE-dependent HRF (see following). Two large groups of cytokines, the interleukins and the chemokines, contain the HRFs that cause basophil histamine release which is independent of IgE. These cytokines are discussed first.

Interleukins

Our previous work has shown that IL-1α and II-1β are not stimuli for human basophils but rather potentiate histamine release.[38] This result is different from the work of Subramanian and Bray, who observed direct release by IL-1.[39] The reason for this difference is unclear. IL-3, on the other hand, was shown to cause histamine release in a subset of donors at concentrations of 300 pg/ml to 300 ng/ml.[40] Kaplan and colleagues have confirmed this result.[41] However, the kinetics of release differentiate IL-3 from the IgE-dependent HRF. In our hands and based on the work from other laboratories, IL-2, IL-4, IL-5, IL-6, and G-CSF are not basophil-releasing agents. It should be noted that although they are not direct agonists of basophil histamine release, several cytokines are known to prime basophils for histamine release. These include IL-1,[38] as mentioned previously, IL-3 at low concentrations,[42] IL-5,[43] IL-6, and IL-7.[44]

Studies by Sarmiento et al. have shown that IL-3, IL-5, and GM-CSF, in a dose-dependent manner, enhance basophil histamine and leukotriene release induced by the eosinophil product MBP.[45] Furthermore, their work demonstrates that basophils and eosinophils stain brightly for the IL-3 receptor using an antibody to the α-chain of the receptor. These data suggest a link between the basophil and eosinophil in perpetuation of the inflammation seen in chronic allergic diseases. Tedeschi et al. have extended the observations of the effects of IL-3 on human basophils.[46] They showed that extracellular sodium acts as an inhibitor of IL-3-induced basophil histamine release and that gramicidin D, the sodium ionophore, markedly inhibits both IL-3- and anti-IgE-induced release. In another in vitro study, Columbo et al. demonstrated the importance of the kinetics of exposure to extracellular calcium and the secretion of histamine by platelet-activating factor (PAF).[47] Their data indicate that exposure of human basophils to extracellular calcium for 2 to 5 min leads to secretion by PAF and that longer exposure actually decreases release by this secretagogue. These data suggest that redistribution of extracellular calcium affects signal transduction in basophils in a complex manner. These authors also showed, as others have, that IL-3 augmented PAF-induced histamine release.

The importance of IL-3 regulation of basophil releasability was definitively shown by Yamaguchi et al.[48] Using basophils of enriched purity, they demonstrated that basophils from donors that did not release histamine to anti-IgE antibody converted to basophils which released histamine after 3 days in culture with 300 pmol/l of IL-3. They further showed that the defect in these basophils was not at the level of the high-affinity IgE receptor, FcεRI. The mechanism of this conversion to releasing basophils is presently unknown, but their data suggest that early signaling molecules, such as the tyrosine kinases Lyn and Syk, may be altered.

Finally, the importance of IL-3 on basophil function was shown in vivo by van Gils et al.[49] They studied IL-3 receptor regulation and cellular characterization of bone marrow and peripheral blood cells in rhesus monkeys undergoing IL-3 treatment. Their data demonstrated that in vivo therapy with IL-3 upregulated the number of high- and low-affinity receptors for IL-3 on basophils but not on mast cells or monocytes. This finding could provide a possible explanation for the increased incidence of allergic reactions documented with IL-3 administration.

Another group of molecules, neurotropic factors, have been shown to potentiate basophil mediator release. Nerve growth factor (NGF), the best-known neurotrophin, augments basophil histamine release from a variety of stimuli and induces LTC_4 production after C5a stimulation.[50] More recently, Bürgi et al. have demonstrated that NGF was by far more potent in priming basophils than other members of this family.[51] They also demonstrated that the action of NGF on human basophils is dependent on the tyrosine kinase receptor, trk, and not the low-affinity nerve growth factor receptor.

Chemokines

Another group of IgE-independent HRF include several of the chemokines, the so-called chemotactic cytokines. There is an ever-expanding literature on the chemokines.[52] Briefly, they consist of a superfamily of proteins with a molecular weight range of 6000 to 14,000 that are classified according to the position of cysteine residues. Currently, there are four subgroups. The C-X-C subgroup has two cysteines with an intervening amino acid, also called the alpha chemokines, and is found on chromosome 4. Platelet factor-4 (PF-4), IL-8, neutrophil-activating peptide-II (NAP-II), and connective tissue-activating peptide-III (CTP-III) are C-X-C chemokines. The C-C (two adjacent cysteines) or beta chemokines found on chromosome 17 include monocyte chemotactic protein-1 (MCP-1), MCP-2, MCP-3, MCP-4, regulated on activation, normal T-cell expressed, and secreted (RANTES), macrophage inflammatory protein-1α (MIP-1α), MIP-1β, eotaxin, and eotaxin-2. A recently defined chemokine group is the C-subfamily, of which lymphotactin is the only member to date.[53] Finally, the most recently discovered chemokine subgroup is a membrane-bound chemokine belonging to the C-X₃-C subgroup. Fractalkine, the only member of this group, was described as an endothelial cell surface-expressed chemokine.[54] The total number of chemokines presently exceeds 40, and the list is increasing. Table 29.1 describes the ability of chemokines to act as HRFs and cause histamine release from human basophils.

As is evident, the C-C chemokines are more active as stimuli for basophil histamine release but their potency varies. It has been known since the early part of this decade that MCP-1 is a potent agonist for basophil histamine release. MCP-3, which shares a high degree of homology with MCP-1, is both a chemotactic and a

TABLE 29.1. Chemokines as histamine-releasing factors (HRFs).

Chemokine subgroup		Reference
C-X-C		
IL-8	+/−	87
CTAP-III	+/−	87
NAP-2	+/−	87
C-C		
Eotaxin-2	+/−	62
MCP-4	+	57, 58
MCP-3	+++	55
MCP-2	++	59
MCP-1	+++	55
RANTES	+/−	87
MIP-1α	+/−	87
C		
Lymphotactin	?	53
C-X₃-C		
Fractaline	?	54

histamine-releasing agent.[55] A new member of the C-C chemokine family, MCP-4 (originally termed CKβ10), has been cloned, expressed, and some of its biological activities determined.[56] MCP-4 shares 56% to 61% sequence identity with MCP-1, MCP-2, and MCP-3 as well as 60% identity to eotaxin. It has biological properties similar to those of MCP-3 and eotaxin. Because MCP-1 and MCP-3 have previously been shown to be potent HRFs for human basophils, one might predict that MCP-4 would also stimulate basophil histamine release. Indeed, Stellato et al. have recently demonstrated that CKβ10 (MCP-4) was a potent chemotactic stimulus for human eosinophils and induced histamine release from IL-3-primed human basophils with a potency equal to MCP-3.[57] In another recent study, MCP-4 was shown to cause basophil histamine release from IL-3-primed basophils of two of four donors, as well as chemotaxis of human monocytes and eosinophils.[58] Another member of this family, MCP-2, has also been shown to induce histamine release from IL-3-primed basophils, as well as causing a calcium signal in eosinophils and basophils, but was less potent than MCP-1.[59] Thus, all members of the monocyte chemotactic protein family affect human basophils, but with varying degrees of efficacy. The reason for this is unknown but one explanation could be receptor occupancy by the various MCP homologs. Clearly, there are multiple chemokine receptors (see following), and they differ in their affinity for the various MCP members. Another possibility is structural differences in the amino-terminus of the molecule. Weber et al. have recently demonstrated that making amino terminally truncated analogs of MCP-1 dramatically altered the histamine-releasing capacity of this molecule.[60] Thus, removing the NH_2-terminal amino acids 2–76 led to a 50-fold reduction in both histamine and LTC_4 release from human basophils. Further truncation of amino acids 6–76 resulted in total loss of activity. Interestingly, while removal of amino acids 2–76 reduced the potency of MCP-1 on basophils, this analog maintained activity on eosinophils, as evidenced by chemotaxis and cytosolic free calcium changes.

RANTES, another C-C chemokine, causes chemotaxis of basophils and eosinophils, but is a weak stimulus for histamine release. The complexity of the biological functions of the chemokines and the multiple receptors with varying specificity may well explain the data of Kuna et al.[61] These investigators examined multiple C-C and C-X-C chemokines and showed that RANTES, MIP-1α, MIP-1β, CTAP-III, PF-4, IL-8, and inflammatory protein-10 (IP-10) could inhibit MCP-1-induced histamine release.

The newest member of the C-C chemokine family, eotaxin-2 (also termed CKβ6), induces chemotaxis of basophils and histamine and leukotriene C_4 release from IL-3-primed basophils.[62] As additional chemokines and receptors are discovered, their actions on human basophils will undoubtedly be explored, perhaps adding to the stimuli for these very responsive cells.

As mentioned earlier, in addition to the identification of new members of the chemokine family, the number of chemokine receptors found is also growing and presently surpasses 12. A review of the various receptors is beyond the scope of this chapter. To date, several laboratories have demonstrated the existence of several different chemokine receptors with overlapping specificities in varying cell types. At this point it is not clear whether activation of different receptors determines the diverse cell biology of the chemokines or whether different ligands binding to the same receptor alter signal transduction events.

As a final comment with respect to the HRF activity of chemokines, it is worthwhile to note that while murine MIP-1α was shown to cause histamine release from mouse peritoneal mast cells,[63] three different groups have confirmed the actions of various chemokines on basophils and have demonstrated no effect of MCP1-3, MIP-1α and MIP-1β, RANTES, and PF-4 on human mast cells. Hartmann et al.[64] looked specifically at skin mast cells, Okayama et al.[65] investigated lung and skin mast cells, and Fureder et al.[66] showed no effect of chemokines on mast cells obtained from lung, uterus, skin, or tonsils, either directly or when these mast cells were preincubated with stem cell factor.

Thus, in humans the chemokines act as IgE-independent HRFs solely through their actions on human basophils.

The IgE-Dependent HRF

As mentioned, the IgE-dependent HRF was found in vivo in LPR fluids from nasal lavages, skin blister fluids, and BAL. This HRF caused histamine release from basophils of a subpopulation of allergic donors. Originally, we thought that this HRF interacted with certain IgE molecules, which we termed IgE+. By definition, those basophils with IgE− on their cell surface were not responsive to this HRF. While HRF is a complete secretagogue for basophils possessing IgE+, we now have data indicating that HRF can act on all basophils, as well as on eosinophils from allergic donors and monocytes from some donors (see following).

For purification of the IgE-dependent HRF, 50 l of supernatant from cultures of the human monocyte/macrophage cell line, U937, were sequentially chromatographed and the resulting protein purified. Amino-terminal sequencing revealed 94% homology to p21, a protein from mouse tumor cells, as well as identity to p23, the human homolog. Both p21 and p23 were previously cloned on the basis of their abundant expression in tumor cells, and no function had been ascribed to either molecule. Both p21 and p23 were subcloned from the U937 cell line. Both recombinant proteins caused histamine release from the basophils of a subpopulation of donors, those who possessed IgE+.[67]

In addition to releasing histamine, recent studies have shown that basophils also generate and secrete high levels of IL-4,[68] a proinflammatory cytokine that has an immunoregulatory role in allergic inflammation. Unlike histamine release, which can be caused by many physiological stimuli, the secretion of IL-4 protein seems primarily dependent on IgE-mediated stimulation.[69–71] Not surprisingly, stimulation with human recombinant HRF (HrHRF) caused the basophils of IgE+ donors to produce IL-4 protein, which correlated with histamine release in 14 donors ($r =$ 0.9, $p = 0.001$).[72] As expected, basophils from IgE− donors neither release histamine or secrete IL-4 in response to HrHRF.

Despite the seeming dependence of the HrHRF on IgE to activate basophils, direct binding of HrHRF to IgE+ could not be detected. Both affinity chromatography and ELISAs failed to show direct binding of HrHRF to IgE+. Furthermore, a RBL cell line transfected with the alpha chain of the high-affinity IgE receptor[73] did not release histamine in response to HrHRF after passive sensitization with IgE+, while it did respond to polyclonal anti-IgE antibody. These findings suggest that HrHRF interacts with its own receptor, while IgE+ modifies the basophil response, making it more sensitive to some stimuli. In fact, similar to HrHRF, D_2O and IL-3 previously have been shown to act as complete stimuli for histamine release in basophils passively sensitized with IgE+.[74,75]

In addition to acting as a direct secretagogue on basophils bearing IgE+ in a manner similar to HrHRF, IL-3 is also known to prime basophils for histamine release and to have its own receptor.[76] Therefore, HrHRF ability to prime basophils for histamine release was examined and compared to that of IL-3. Human basophils bearing IgE− were primed with HrHRF or IL-3 for 15 min and then stimulated with polyclonal anti-IgE. Enhancement by HrHRF was dose dependent, averaging 38% ± 12% at a concentration of 0.03 μg/ml and 108% ± 26% at 30 μg/ml, with control anti-IgE-induced release being 23% ± 3% ($n = 11$).[77] HrHRF ability to enhance anti-IgE-induced histamine release paralleled that of IL-3. Interestingly, using basophils bearing IgE+ on which IL-3 and HrHRF act as complete secretagogues, no priming could be detected by either stimulus, even at low concentrations.

Not surprisingly, additional studies indicate that HrHRF also primes IgE− basophils to secrete IL-4.[78] Basophils ranging in purity from 5% to 25% were preincubated with HrHRF and then cultured for 4 h with polyclonal anti-IgE antibody. IL-4 protein secretion was enhanced by 50% ± 14% at 80 ng/ml of HrHRF with dose-dependent increases up to 10 μg/ml, for an optimal enhancement of 124%

± 10% above control IL-4 secretion (*n* = 5). This priming of IL-4 secretion is seen over a wide range of anti-IgE concentrations from 1 to 100 ng/ml. HrHRF-induced priming was also seen in response to another IgE-dependent stimulus, antigen E, but was not seen with f-met-leu-phe (FMLP), an IgE-independent stimulus.

In addition to IL-4, human basophils also produce IL-13 that, in humans, can substitute for IL-4 to induce B-cell switching for IgE production.[79,80] HrHRF also is a complete stimulus for IL-13 production from IgE+ donor basophils. Additionally, preincubation of IgE– donor basophils with HrHRF followed by stimulation with anti-IgE caused enhancement of IL-13 production, ranging from 17% to 52% above control levels (*n* = 5).[78] The aforementioned studies clearly show that HrHRF is an activator of human basophils, causing histamine release, as well as IL-4 and IL-13 secretion (Fig. 29.1).

In addition to basophils, eosinophils and monocytes are also recruited in the LPR, and HrHRF ability to interact with these cell types was also of interest to us. Thus, human eosinophils from allergic donors were purified using Percoll gradients and negative selection to greater than 95% purity, and studies of HrHRF ability to cause chemotaxis were undertaken. HrHRF induced chemotaxis of eosinophils equal to that of the positive control, the chemokine RANTES.[81] Additionally, human monocytes were obtained by lymphophoresis from anonymous donors and purified by elutriation. Subsequently, the monocytes were incubated with HrHRF, rabbit polyclonal anti-HRF, and a phycoerthyrin-labeled anti-rabbit antibody using a modification of our standard flow cytometry protocol. A shift in mean fluorescence intensity on flow cytometry, indicating binding of HrHRF to the monocytes, was observed when the cells were incubated with anti-HrHRF as compared to the irrelevant, negative control antibody (unpublished results). The body of data accumulated indicates that HrHRF is a novel cytokine that interacts with basophils, eosinophils, and monocytes, cells critical to the allergic inflammatory response. Thus, identification of the receptor for HrHRF will be an important step in understanding the role of HrHRF in chronic inflammation.

Initial investigation of the HrHRF receptor included comparative studies of various inhibitors of known signaling pathways on HrHRF activity on human basophils. Specifically, the effects of pharmacological agents were compared on HrHRF-mediated histamine release and that caused by other IgE-dependent stimuli, such as anti-IgE and antigen, as well as on the IgE-independent stimulus, FMLP. Human basophils were isolated from IgE+ donors by double Percoll gradients to a purity of 5% to

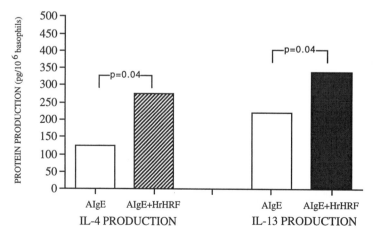

Enhancement of IL-4 and IL-13 production by HrHRF

FIGURE 29.1. HrHRF enhances IL-4 and IL-13 protein production from anti-IgE antibody- (AIgE-) stimulated basophils possessing IgE–. Basophils were isolated by double Percoll to purities of 5%–25%. Cells were preincubated with HrHRF or buffer for 15 min and then stimulated with AIgE. The cells were cultured for 4h or 18h, and the supernatants were harvested for IL-4 and IL-13 protein production, respectively. Results are shown as % enhancement above AIgE control. This graph is representative of five experiments.

40%. Pertussis toxin (2 μg/ml), an inhibitor of G proteins bearing an alpha-1 subunit, completely abolished FMLP-mediated histamine release ($n = 4$; $p < 0.05$), whereas it did not alter that by anti-IgE or HrHRF. In contrast, the general kinase inhibitor, staurosporine, blocked both anti-IgE- and HrHRF-mediated histamine release in a dose-dependent manner at concentrations of 50 to 500 nM ($n = 5$; $p < 0.05$). The staurosporine-derived PKC inhibitor, Go 6976 (300 nM), suppressed histamine release by both anti-IgE and HrHRF ($n = 8$; $p < 0.004$), while another staurosporine-derived kinase inhibitor, bisindolylmaleimide II (400 nM), had no effect. Interestingly, rottlerin (500 nM), a nonstaurosporine-derived kinase inhibitor, significantly enhanced histamine release by HrHRF ($n = 10$; $p < 0.009$), but did not affect histamine release mediated by the IgE-dependent stimuli, anti-IgE, or antigen.[82] The differential modulation of histamine release by rottlerin suggests a unique signal transduction pathway for HRF, providing further evidence for a specific receptor for HRF. Studies to directly identify and further characterize the HrHRF receptor are ongoing using human eosinophils.

Clinical Relevance of HRF

Chemokines with IgE-independent histamine-releasing activity have been found in vivo. Alam et al.[83] have shown increased levels of MCP-1, RANTES, and MIP-1α protein in BAL fluids of asthmatic patients when compared to normal subjects. Their work also shows that mRNAs for MCP-1, MCP-3, RANTES, MIP-1α, and IL-8 were detectable in BAL cells from both asthmatic and normal subjects. Sim et al. also have found that IL-8, MIP-1α, RANTES, IL-1β, and GM-CSF were elevated in nasal secretions during the LPR in 10 allergic patients. The increases in these chemokines correlated with symptom scores during the LPR; topical steroid treatment reduced both the symptoms and levels of these cytokines.[84] In other work, MIP-1α and IL-16 were detected in BAL fluids from asthmatic subjects 6 h after antigen challenge.[85]

Early studies linked basophil reactivity to the IgE-dependent HRF with atopy and LPR symptom intensity.[86] Additional data from Sampson et al. correlated spontaneous basophil histamine release and HRF production from mononuclear cells with the clinical status of food allergic patients with atopic dermatitis. When placed on a diet avoiding the offending food allergen, these patients showed clinical improvement. Furthermore, their basophils no longer spontaneously secreted histamine, nor did their cells produce HRF.[87] Thus, not only activation of cells by HRF, but also production of HRF, may play roles in the pathogenesis of allergic disease. In fact, investigation of HrHRF production indicates that IgE+ donor cells make more HRF than do those of nonallergic donors. The peripheral blood mononuclear cells from 12 donors (4 allergic IgE+, 4 allergic IgE–, and 4 nonallergic) were cultured for HrHRF protein production and processed for mRNA extraction and subsequent reverse transcription-polymerase chain reaction for HrHRF message. The quantity of mRNA for HrHRF did not differ among the groups when compared to beta-actin controls. However, using our histamine release bioassay, IgE+ patients were found to have more HRF protein in the cell culture supernatants than did nonallergic subjects, suggesting that all individuals make mRNA for HrHRF but that cells from atopic patients are more effective at translating it to protein.[88] Future studies will utilize ELISA technology to measure specific levels of HRF in similar fluids produced both in vitro and in vivo.

Conclusions

In summary, the IgE-dependent HRF is a unique molecule with no homology to any known interleukin, chemokine, or antigen. This molecule uncovered a functional heterogeneity of the IgE molecule, but the molecular basis of this heterogeneity is not understood at present.

The stimulus for basophils, eosinophils, and monocytes during the LPR has been an enigma. Interleukins and chemokines act as HRFs and also stimulate eosinophils, but there are few reports linking them to the LPR. As additional chemokines are discovered, elucidation of their biological actions will be pursued. We believe that the recently cloned IgE-dependent HRF, which activates all basophils, eosinophils from allergic donors, and monocytes from certain donors, and which is found in LPR fluids, plays an important role in perpetuating the LPR and the chronic inflammation that is the hallmark of allergic disease, in particular allergic asthma. Therefore, further characterization of HrHRF and identification of its receptor will increase our understanding of chronic allergic disease and may help to explain the varying severity of allergic responses.

References

1. Guidelines for the Diagnosis and Management of Asthma. Publication 91-3042. Bethesda: National Institutes of Health, 1991.
2. Busse WW, Calhoun WF, Sedgwick JD. Mechanism of airway inflammation in asthma. Am Rev Respir Dis 1993;147:520–524.
3. Naclerio RM, Proud D, Togias AG, et al. Inflammatory mediators in late antigen-induced rhinitis. N Engl J Med 1985;313:65–70.
4. Charlesworth EN, Hood AF, Soter NA, et al. Cutaneous late-phase response to allergen. Mediator release and inflammatory cell infiltration. J Clin Invest 1989;83:1519–1526.
5. Sim TC, Grant JA, Hilsmeier KA, et al. Proinflammatory cytokines in nasal secretions of allergic subjects after challenge. Am J Respir Crit Care Med 1994;149:339–344.
6. Behrens BL, Clark RAF, Marsh W, et al. Modulation of the late asthmatic response by antigen-specific immunoglobulin E in an animal model. Am Rev Respir Dis 1984;130:1134–1139.
7. McLean J, Garkowski A, Solomon W, et al. Changes in airway resistance on antigenic challenge in allergic rhinitis. Clin Allergy 1971;1:63.
8. Pipkorn U, Proud D, Lichtenstein LM, et al. Effect of short-term systemic glucocorticoid treatment on human nasal mediator release after antigen challenge. J Clin Invest 1987; 80:957–961.
9. Charlesworth EN, Kagey-Sobotka A, Schleimer RP, et al. Prednisone inhibits the appearance of inflammatory mediators and the influx of eosinophils and basophils associated with the cutaneous late-phase response to allergen. J Immunol 1991;146:671–676.
10. Bascom R, Wachs M, Naclerio RM, et al. Basophil influx occurs after nasal antigen challenge: effects of topical corticosteroid pretreatment. J Allergy Clin Immunol 1988;81:580–589.
11. Iliopoulos O, Baroody F, Naclerio RM, et al. Histamine-containing cells obtained from the nose hours after antigen challenge have functional and phenotypic characteristics of basophils. J Immunol 1992;148:2223–2228.
12. Guo CB, Liu MC, Galli SJ, et al. Identification of IgE-bearing cells in the late-phase response to antigen in the lung as basophils. Am J Respir Cell Mol Biol 1994;10(4):384–389.
13. Sparrow D, O'Connor GT, Rosner B, et al. Predictors of longitudinal change in methacholine airway responsiveness among middle-aged and older men: the normative aging study. Am J Respir Crit Care Med 1994;149:376–381.
14. Maruyama N, Tamura G, Aizawa T, et al. Accumulation of basophils and their chemotactic activity in the airways during natural airway narrowing in asthmatic individuals. Am J Respir Crit Care Med 1994;150:1086–1093.
15. Koshino T, Arai Y, Miyamoto Y, et al. Mast cell and basophil number in the airway correlate with the bronchial responsiveness of asthmatics. Int Arch Allergy Immunol 1995;107:378–379.
16. Findlay SR, Lichtenstein LM. Basophil "releasability" in patients with asthma. Am Rev Respir Dis 1980;122:53–59.
17. Kern F, Lichtenstein LM. Defective histamine release in chronic urticaria. J Clin Invest 1976; 57:1369–1377.
18. Gaddy JN, Busse WW. Enhanced IgE-dependent basophil histamine release and airway reactivity in asthma. Am Rev Respir Dis 1986;134:969–974.
19. Seminario M-C, Gleich GJ. The role of eosinophils in the pathogenesis of asthma. Curr Opin Immunol 1994;6:860–864.
20. Bousquet JB, Chanez P, Lacoste JY, et al. Eosinophilic inflammation in asthma. N Engl J Med 1990;323:1033–1039.
21. Robinson DS, Ying S, Bentley AM, et al. Relationships among numbers of bronchoalveolar lavage cells expressing messenger ribonucleic acid for cytokines, asthma symptoms, and airway methacholine responsiveness in atopic asthmatics. J Allergy Clin Immunol 1993;92:397–403.
22. Kroegel C, Liu MC, Hubbard WC, et al. Blood and bronchoalveolar eosinophils in allergic sub-

jects after segmental antigen challenge: surface phenotype, density heterogeneity, and prostanoid production. J Allergy Clin Immunol 1994;93:725–734.

23. Pin I, Radford S, Kolendowicz R, et al. Airway inflammation in symptomatic and asymptomatic children with methacholine hyperresponsiveness. Eur Respir 1993;6:1249–1256.

24. Gleich GJ, Adolphson CR, Leiferman KL. The biology of the eosinophilic leukocyte. Annu Rev Med 1993;44:85–101.

25. Viksman MY, Liu MC, Schleimer RP, et al. Alveolar macrophages (AM) are activated in patients with allergic respiratory diseases. J Allergy Clin Immunol 1995;95:170 (abstract).

26. Viksman MY, Liu MC, Schleimer RP, et al. Application of flow cytometric method using autofluorescence and a tandem fluorescent dye to analyze human alveolar macrophage surface makers. J Immunol Methods 1994;172:17–24.

27. Williams J, Johnson S, Mascali JJ, et al. Regulation of low affinity IgE receptor (CD23) expression on mononuclear phagocytes in normal and asthmatic subjects. J Immunol 1992;149:2823–2829.

28. Maurer D, Fiebiger E, Reininger B, et al. Expression of functional high affinity immunoglobulin E receptors (FcεRI) on monocytes of atopic individuals. J Exp Med 1994;179:745–750.

29. Thueson DO, Speck LS, Lett-Brown MA, et al. Histamine releasing activity (HRA). I. Production by mitogen- or antigen-stimulated human mononuclear cells. J Immunol 1979;123:626–631.

30. Warner JA, Pienkowski MM, Plaut M, et al. Identification of histamine releasing factor(s) in the late phase of cutaneous IgE-mediated reactions. J Immunol 1986;136:2583–2587.

31. Plaut M, MacDonald SM, Naclerio RM, et al. Characterization of human IgE-dependent histamine releasing factors. Fed Proc 1986;45:243 (abstract).

32. Plaut M, Liu MC, Conrad DH, et al. Histamine release from human basophils is induced by IgE-dependent factor(s) derived from human lung macrophages and an Fcε receptor-positive human B cell line. Trans Assoc Am Phys 1985;98:305–312.

33. Sedgwick JD, Holt PG, Turner KI. Production of a histamine-releasing lymphokine by antigen or mitogen stimulated human peripheral T cells. Clin Exp Immunol 1981;45:409–418.

34. Alam R, Rozniecki J, Salmaj K. A mononuclear cell derived histamine releasing factor (HRF) in asthmatic patients. Histamine release from basophils in vitro. Ann Allergy 1984;53:66–69.

35. Ezeamuzie IC, Assem ESK. A study of histamine release from human basophils and lung mast cells by products of lymphocyte stimulation. Agents Actions 1983;13:222–230.

36. Kaplan AP, Haak-Frendscho M, Fauci A, et al. A histamine-releasing factor from activated human mononuclear cells. J Immunol 1985;135:2027–2032.

37. Strickland DB, Turner KJ, Holt BJ, et al. Histamine-releasing factor—cellular source and kinetics of production are influenced by characteristics of antigen. Allergy Proc 1988; 9:317.

38. Massey WA, Randall TC, Kagey-Sobotka A, et al. Recombinant human interleukin 1α and 1β potentiate IgE-mediated histamine release from human basophils. J Immunol 1989;143:1875–1880.

39. Subramanian N, Bray MA. Interleukin 1 releases histamine from human basophils and mast cells in vitro. J Immunol 1987;138:271–275.

40. MacDonald SM, Schleimer RP, Kagey-Sobotka A, et al. Recombinant IL-3 induces histamine release from human basophils. J Immunol 1989;142:3527–3532.

41. Kuna P, Reddigari SR, Kornfeld D, et al. IL-8 inhibits histamine release from human basophils induced by histamine-releasing factors, connective tissue activating peptide III, and IL-3. J Immunol 1991;147:1920–1924.

42. Kurimoto Y, de Weck AL, Dahinden CA. Interleukin 3-dependent mediator release in basophils triggered by C5a. J Exp Med 1989;170:467–479.

43. Kirai K, Yamaguchi M, Misaki Y, et al. Enhancement of human basophil histamine release by interleukin 5. J Exp Med 1990;172:1525–1528.

44. Liao T-N, Hsieh K-H. Characterization of histamine-releasing activity: role of cytokines and IgE heterogeneity. J Clin Immunol 1992;12:248–258.

45. Sarmiento EU, Espiritu BR, Gleich GJ, et al. IL-3, IL-5 and granulocyte-macrophage colony-stimulating factor potentiate basophil mediator release stimulated by eosinophil granule major basic protein. J Immunol 1995;155:2211–2221.

46. Tedeschi A, Palella M, Milazzo N, et al. IL-3-induced histamine release from human basophils. Dissociation from cationic dye binding and

down-regulation by Na^+ and K^+. J Immunol 1995;155:2652–2660.

47. Columbo M, Horowitz EM, Kagey-Sobotka A, et al. Histamine release from human basophils induced by platelet activating factor: the role of extracellular calcium, interleukin-3, and granulocyte-macrophage colony-stimulating factor. J Allergy Clin Immunol 1995;95:565–573.

48. Yamaguchi M, Hirai K, Ohta K, et al. Nonreleasing basophils convert to releasing basophils by culturing with IL-3. J Allergy Clin Immunol 1996;97:1279–1287.

49. van Gils FC, van Teeffelen ME, Neelis KJ, et al. Interleukin-3 treatment of rhesus monkeys leads to increased production of histamine-releasing cells that express interleukin-3 receptors at high levels. Blood 1995;86(2):592–597.

50. Bishoff SC, Dahinden CA. Effect of nerve growth factor on the release of inflammatory mediators by mature human basophils. Blood 1992;79:2662–2669.

51. Bürgi B, Otten UH, Ochensberger B, et al. Basophil priming by neurotrophic factors. Activation through the trk receptor. J Immunol 1996;157:5582–5588.

52. Hedrick JA, Zlotnik A. Chemokines and lymphocyte biology. Curr Opin Immunol 1996;8:343–347.

53. Kennedy J, Kelner GS, Kleyensteuber S, et al. Molecular cloning and functional characterization of human lymphotactin. J Immunol 1995;155:203–209.

54. Brazen JF, Bacon KB, Hardleman G, et al. A new class of membrane-bound chemokine with a CX_3C motif. Nature (Lond) 1997;385:640–644.

55. Weber M, Dahinden CA. Basophil and eosinophil activation by C-C chemokines. Int Arch Allergy Immunol 1995;107:148–150.

56. Uguccioni M, Loetscher P, Forssmann U, et al. Monocyte chemotactic protein 4 (MCP-4), a novel structure and functional analogue of MCP-3 and eotaxin. J Exp Med 1996;183:2379–2384.

57. Stellato C, Collins P, Li H, et al. Production of the novel C-C chemokine MCP-4 by airway cells and comparison of its biological activity to other C-C chemokines. J Clin Invest 1997; 99:926–936.

58. Garcia-Zepeda EA, Combadiere C, Rothenberg ME, et al. Human monocyte chemoattractant protein (MCP)-4 is a novel CC chemokine with activities on monocytes, eosinophils, and basophils induced in allergic and nonallergic inflam-

mation that signals through the CC chemokine receptors (CCR)-2 and -3. J Immunol 1996; 157:5613–5626.

59. Weber M, Uguccioni M, Ochensberger B, et al. Monocyte chemotactic protein MCP-2 activates human basophil and eosinophil leukocytes similar to MCP-3. J Immunol 1995;154:4166–4172.

60. Weber M, Uguccioni M, Baggiolini M, et al. Deletion of the NH_2-terminal residue converts monocyte chemotactic protein 1 from an activator of basophil mediator release to an eosinophil chemoattractant. J Exp Med 1996; 183:681–685.

61. Kuna P, Reddigari SR, Rucinski D, et al. Chemokines of the α,β-subclass inhibit human basophils responsiveness to monocyte chemotactic and activating factor/monocyte chemoattractant protein-1. J Allergy Clin Immunol 1995;95:574–586.

62. Forssmann U, Uguccioni M, Loetscher P, Dahinden CA, Langen H, Thelen M, Baggiolini M. Eotaxin-2, a novel CC chemokine that is selective for the chemokine receptor CCR3, and acts like eotaxin on human eosinophil and basophil leukocytes. J Exp Med 1997;185(12):2171–2176.

63. Alam R, Forsythe PA, Stafford S, et al. Macrophage inflammatory protein-1 alpha activates basophils and mast cells. J Exp Med 1992; 176(3):781–786.

64. Hartmann K, Beiglbock F, Czarnetzki BM, et al. Effect of CC chemokines on mediator release from human skin mast cells and basophils. Int Arch Allergy Immunol 1995;108(3):224–230.

65. Okayama Y, Brzezinska-Blaszczyk E, Kuna P, et al. Effects of PBMC-derived histamine releasing factors on histamine release from human skin and lung mast cells. Clin Exp Allergy 1995;25:890–895.

66. Fureder W, Agis H, Semper H, et al. Differential response of human basophils and mast cells to recombinant chemokines. Ann Hematol 1995;70(5):251–258.

67. MacDonald SM, Rafnar T, Langdon J, et al. Molecular identification of an IgE-dependent histamine releasing factor. Science 1995; 269:688–690.

68. Brunner T, Heusser CH, Dahinden CA. Human peripheral blood basophils primed by interleukin-3 (IL-3) produce IL-4 in response to immunoglobulin E receptor stimulation. J Exp Med 1993;177:605–610.

69. Arock M, Merle-Béral H, Dugas B, et al. IL-4 release by human leukemic and activated normal basophils. J Immunol 1993;151:1441–1447.

70. MacGlashan DW Jr, White JM, Huang S-K, et al. Secretion of interleukin-4 from human basophils: the relationship between IL-4 mRNA and protein in resting and stimulated basophils. J Immunol 1994;152:3006–3016.

71. Schroeder JT, MacGlashan DW Jr, Kagey-Sobotka A, et al. IgE-dependent IL-4 secretion by human basophils. The relationship between cytokine production and histamine release in mixed leukocyte cultures. J Immunol 1994;153:1808–1817.

72. Schroeder JT, Lichtenstein LM, MacDonald SM. An immunoglobulin E-dependent recombinant histamine releasing factor induces IL-4 secretion from human basophils. J Exp Med 1996;183:1265–1270.

73. Gilfillan AM, Kado-Fong H, Wiggan GA, et al. Conservation of signal transduction mechanisms via the human FcεRIα after transfection into a rat mast cell lilne, RBL 2H3. J Immunol 1992;149:2445–2451.

74. MacDonald SM, White JM, Kagey-Sobotka A, et al. The heterogeneity of human IgE exemplified by the passive transfer of D_2O sensitivity. Clin Exp Allergy 1991;21:133–138.

75. MacDonald SM, Schleimer RP, Kagey-Sobotka A, et al. Recombinant IL-3 induces histamine release from human basophils. J Immunol 1989;142:3527–3532.

76. Kurimoto Y, DeWeck AL, Dahinden CA. The effect of interleukin 3 upon IgE-dependent and IgE-independent basophil degranulation and leukotriene generation. Eur J Immunol 1991;21:361–368.

77. MacDonald SM, Schroeder JT, Langdon JM, et al. Human recombinant IgE-dependent histamine releasing factor (HrHRF) primes basophils for histamine release. J Allergy Clin Immunol 1996;97:267.

78. Schroeder JT, Lichtenstein LM, MacDonald SM. Recombinant histamine-releasing factor enhances IgE-dependent IL-4 and IL-13 secretion by human basophils. J Immunol 1997;159:447–452.

79. Li H, Sim TC, Alam R. IL-13 released by and localized in human basophils. J Immunol 1996;156:4833–4838.

80. Zurawski G, de Vries JE. Interleukin 13, an interleukin 4-like cytokine that acts on monocytes and B cells, but not on T cells. Immunol Today 1994;15:19–26.

81. MacDonald SM, Schroeder JT, MacGlashan DW, et al. Human recombinant histamine releasing factor (HrHRF): a unique cytokine. J Invest Med 1996;44:222A.

82. Bheekha-Escura R, Chance SR, Langdon JM, et al. Pharmacologic regulation of histamine release by the IgE-dependent histamine releasing factor (HRF). J Allergy Clin Immunol 1997;99:S271.

83. Alam R, York J, Boyars M, et al. Increased MCP-1, RANTES, and MIP-α in bronchoalveolar lavage fluid of allergic asthmatic patients. Am J Respir Crit Care Med 1996;153:1398–1404.

84. Sim TC, Reece LM, Hilsmeier KA, et al. Secretion of chemokines and other cytokines in allergen-induced nasal responses: inhibition by topical steroid treatment. Am J Respir Crit Care Med 1995;152:927–933.

85. Cruikshank WW, Long A, Tarpy RE, et al. Early identification of interleukin-16 (lymphocyte chemoattractant factor) and macrophage inflammatory protein 1 alpha (MIP1 alpha) in bronchoalveolar lavage fluid of antigen-challenged asthmatics. Am J Respir Cell Mol Biol 1995;13(6):738–747.

86. MacDonald SM. Histamine releasing factors and IgE heterogeneity. In: Middleton E, Reed CE, Ellis EF, Adkinson NF, Yunginger JW, Busse WW, eds. Allergy: Principles and Practice. 4th ed. St. Louis: Mosby-Year Book, 1993:1–11.

87. Sampson HA, Broadbent KR, Bernhisel-Broadbent J. Spontaneous release of histamine from basophils and histamine-releasing factor in patients with atopic dermatitis and food hypersensitivity. N Engl J Med 1989;321:228–232.

88. Langdon J, Anders K, Lichtenstein LM, et al. Atopics translate mRNA for the IgE-dependent histamine releasing factor (HRF) more effectively than normals. J Allergy Clin Immunol 1995;95:336.

Index